Historical & Cultural Astronomy

The Historical & Cultural Astronomy series includes high-level monographs and edited volumes covering a broad range of subjects in the history of astronomy, including interdisciplinary contributions from historians, sociologists, horologists, archaeologists, and other humanities fields. The authors are distinguished specialists in their fields of expertise. Each title is carefully supervised and aims to provide an in-depth understanding by offering detailed research. Rather than focusing on the scientific findings alone, these volumes explain the context of astronomical and space science progress from the pre-modern world to the future. The interdisciplinary Historical & Cultural Astronomy series offers a home for books addressing astronomical progress from a humanities perspective, encompassing the influence of religion, politics, social movements, and more on the growth of astronomical knowledge over the centuries.

More information about this series at http://www.springer.com/series/15156

Kenneth I. Kellermann • Ellen N. Bouton • Sierra S. Brandt

Open Skies

The National Radio Astronomy Observatory
and Its Impact on US Radio Astronomy

 Springer

OPEN

Kenneth I. Kellermann
National Radio Astronomy Observatory
Charlottesville, VA, USA

Ellen N. Bouton
National Radio Astronomy Observatory
Charlottesville, VA, USA

Sierra S. Brandt
Consultant to the National Radio
Astronomy Observatory Archives
Providence, RI, USA

ISSN 2509-310X   ISSN 2509-3118   (electronic)
Historical & Cultural Astronomy
ISBN 978-3-030-32347-9   ISBN 978-3-030-32345-5   (eBook)
https://doi.org/10.1007/978-3-030-32345-5

This book is an open access publication.

Cover Caption: "NRAO's Very Large Array, located on the Plains of San Agustin, 50 miles west
of Socorro, New Mexico. The NRAO is a facility of the NSF operated under cooperative
agreement by AUI." Credit: NRAO/AUI/NSF.

This Springer imprint is published by the registered company Springer Nature Switzerland AG.
The registered company address is: Gewerbestrasse 11, 6330 Cham, Switzerland

We dedicate the book to the memory of Dave Heeschen, whose wise and forceful leadership and commitment to Open Skies led NRAO to become the world's premier radio astronomy observatory.

An optical telescope is visual and interesting. A radio telescope is an electronic instrument and you can't really see a lot of what makes it work. When it's done all you get is a computer printout.
Comments from an anonymous NSF referee.

FOREWORD

I know Ken Kellermann, well. We both did our PhDs in radio astronomy in the 1960s, supervised by the famous radio astronomer John Bolton. Ken got his lessons on how to build a telescope and do research when John was building the Owens Valley Observatory at Caltech while I started building the interferometer at Parkes when John and Ken moved from Caltech to Australia. Many years later I spent a challenging but rewarding 7 years working for NRAO as the first director of the newly completed VLA radio telescope in New Mexico. This was sandwiched between my time at Westerbork and the Australia Telescope. To Ellen Bouton, we owe a great debt for the legacy of NRAO's extensive archives of historical material which underpin the impressively detailed source material used in this book. Sierra Brandt's background as a historian of twentieth-century science nicely complements the contributions of the other two authors.

While this book is very clearly focused on the development of the US National Radio Astronomy Observatory (NRAO), it touches on many more broader issues, including the birth of a national facility, the open access policy for scientific research, the wider societal implications of searching for extraterrestrial intelligent life, and lessons learned from major construction projects. It is far more than just the history of NRAO. By discussing the development of NRAO in an international context the authors have also written a history of the development of radio astronomy as seen from a US perspective. They start from the well-covered ground when Karl Jansky of the Bell Telephone Laboratory discovered radio emission from the Milky Way in 1933, through the somewhat idiosyncratic but innovative experiments over the next decade by one individual, Grote Reber, to the major technology developments during World War II.

Radio astronomy had started in the USA, but in the immediate postwar period, other countries, notably the UK and Australia, embarked on vigorous programs of exploration of the radio sky, taking advantage of the influx into astronomy of the high caliber scientists and engineers who had developed the radar technology. Quoting the authors: "Early American radio astronomy did not have the same big impact as the programs in the UK and Australia." This external pressure was a major factor galvanizing the US scientific community into activity. In this time period, the obvious way forward was to build an even bigger dish than the British (250 foot) or the Australian (210 foot) radio telescopes. With the vision of building a very large antenna of perhaps 600 foot diameter this was "big science."

The establishment of a national radio astronomy facility in the USA is a fascinating story with many obstacles and detractors. The authors provide excellent context for the formation of NRAO by including detailed archival research on the steps that were involved. This is a well-informed and deep analysis of how decisions were being made. The National Science Board had already made a declaration for government support of large-scale basic scientific facilities (i.e., support for big science) and had given the construction of a major radio astronomy facility as an example. This impacted NSF policy and was the beginning of the national facility concept. A concept which was pioneered by the USA and later adopted in many other countries. However, establishing a national facility was not supported by all parties and throughout the book we can read lively accounts of the ongoing debate for and against the "big science" national facility concept instead of smaller groups of young innovative scientists based in the universities. This reached an extreme in the ongoing disagreements and confrontations between Merle Tuve (DTM) and Lloyd Berkner who was the president of AUI, the organization that ran the Brookhaven National Laboratory, a "big science" national facility for particle physics research.

Lloyd Berkner had a huge impact on the development of the National Radio Astronomy Observatory. The authors note that Berkner may not have the name recognition as some of the other American postwar science policy leaders such as Vannevar Bush, Robert Oppenheimer, or I.I. Rabi, but perhaps no one had a broader impact on mid-twentieth-century science policy. They include one extraordinary example: Berkner became the first Chair of the National Academy of Science Space Studies Board and he sent a strongly worded memorandum to NASA Administrator James Webb stating that "Scientific exploration of the Moon and planets should be clearly stated as the ultimate objective of the U.S. space program. ... Scientific exploration of the Moon and planets must at once be developed on the premise that man will be included. Failure to adopt and develop our national program upon this premise will inevitably prevent man's inclusion, and every effort should be made to establish the feasibility of manned space flight at the earliest opportunity." Less than 2 months later, in a special address to joint session of Congress, President John F. Kennedy conveyed Berkner's message stating that "this nation should set as a goal before this decade is out, of landing of a man on the moon and returning him safely to the Earth."

When David Heeschen became Director of NRAO in late 1962 he presented a clear view of the role of a National Facility. NRAO's multiple responsibilities included providing equipment and aid for visiting scientists, anticipating the need for future developments in radio astronomy, and playing a leading role in developing new instrumentation. As Director of one of NRAO's facilities (VLA) in the 1980s, I came to appreciate the value of Heeschen's vision, and I later implemented similar policies when establishing the CSIRO radio telescopes in Australia as a National Facility.

The book is most appropriately titled *Open Skies*, a concept which values open access to research facilities as the most effective way to make scientific progress. The NRAO has been the leading advocate for this concept in astronomy and has set the path which has been followed by almost the entire radio astronomy community worldwide. One of the authors, Ken Kellermann, has been a visionary advocate for *Open Skies* so it is no surprise that it features both in the title and throughout this book.

At the inception of the National Facility concept, Berkner and his deputy Richard Emberson proposed the open access model: "all qualified scientists without regard to institutional affiliation would have access to the facility," thus "insuring maximum scientific progress." This was the policy adopted by NRAO and is referred to as "open skies," following the nomenclature adopted by the international airlines governing reciprocal landing rights. In October 1959, Heeschen famously wrote to the editors of the three main astronomy journals in the USA requesting that they publish the following statement: "The facilities of the Observatory are open to any competent individual with a program in radio astronomy, regardless of institutional affiliation." NRAO then took this a step further by not requiring previous radio experience and by including international as well as national institutes.

For over 50 years almost all radio observatories in the world have obtained mutual advantage from this policy but, sadly, as discussed in the last chapter, the SKA participating countries are now questioning whether to continue this tradition.

When reading the account of the construction of the first big dish in Chap. 4, I was struck by the various comments about the need to finish construction before the telescope was obsolete. Similar comments had been made during the construction of the Parkes 210 foot telescope, and there seemed to be a feeling that there were only a few projects that the single dish would do well, so after a short burst of activity the big dishes would have no lasting impact. At the time, there seemed to be almost no realization that in reality it was these flexible instruments that would go on to study many of the new unanticipated discoveries.

Many of us in the broader radio astronomy community had heard about the problems encountered in the construction of the NRAO 140 Foot Telescope in the USA and the practical limitations that were imposed by some poor design decisions, such as the use of an equatorial mount rather than an alt-az mount. However, this is the first time we can read a frank and detailed account of how it overran its budget by a factor of 3, was 5 years late on a 2-year construction timescale, and still did not reach the specifications of the larger and cheaper 210 foot Parkes telescope completed in Australia a few years earlier. Bernard Burke described the 140 Foot radio telescope as having "served well, but its equatorial geometry is antique, its structural flexure is dreadful, its surface quality is inferior, its maintenance is expensive and man-power intensive, and its pointing is substandard." An interesting quote, and measure of the times, was related to the use of computers for coordinate conversion if using an alt-az mount: "... the operation of a precision scientific instrument should not be trusted to a computer." This is almost identical to the view that Bernard Mills expressed in Australia at that time when he suggested that aperture synthesis would never be practical if it depended on the use of an electronic computer.

But this book goes further; the authors have analyzed, like a scientific research paper, the reasons behind the decisions and the implications of the management structure that had been established. This concludes with an excellent summary of lessons learned; some written by Dave Heeschen, who was the senior member of the NRAO Scientific Staff during most of the 140 foot construction, and others summarized by the authors. Since it is clear that we have still not learned from these mistakes in our subsequent and even current management of big projects, I repeat them here in this Foreword for maximum impact.

In 1992, Dave Heeschen summarized the 140 Foot project as follows:

> The 140 Foot is a classic example of how not to design and build a telescope. The design specs were set by a committee of outside consultants who had no responsibility or accountability for the final result, and who gave liberally of poor advice. The 140-foot project leader, a very nice gentleman who was [assistant] to the president of AUI and responsible for the entire feasibility study that led to the establishment of NRAO, uncritically accepted all this advice. The telescope was originally going to have an alt-az mount because the consulting engineers thought that was the most feasible.... But the steering committee membership changed from time to time and finally had on it a prominent and outspoken scientist who insisted the mount should be equatorial.... Then the solar astronomer on the steering committee decided that the telescope should be able to observe the Sun from sunrise to sunset on June 22 each year. The errors made in bidding, contracting, and construction were even worse.... AUI wound up with a fixed price contract, for $4 million, with a company—E W Bliss—that really didn't want the job, except for one enthusiastic vice [president] who apparently bullied them first into accepting the final contract. He quit shortly afterward and AUI was left with a semi-hostile contractor.

Some important lessons were learned, or should have been learned, from the 140 Foot experience:

1. Beware of the lowest bidder
2. Be sure the contract is clear about who is responsible for what
3. Finish the design before starting construction

4. Establish clear points of contact, authority, and responsibility on both sides
5. Have a firm understanding of when the antennas will be delivered, with penalties for late delivery
6. Do not take committee advice too seriously
7. Have good in-house expertise

In October 1961, Joe Pawsey, from the CSIRO group in Australia, was offered the directorship of NRAO. He visited in March 1962 and started making plans for the future of NRAO, including the use of interferometers to increase angular resolution. Pawsey was diagnosed with a brain tumor in May 1962, returned to Australia, and died in November 1962. Dave Heeschen, from the NRAO staff, had been acting director and became director on October 19, 1962.

The concept of a radio telescope with angular resolution comparable to that obtained in the optical emerged in the early 1960s. While it was clear that a single dish could never achieve this, the radio interferometers certainly could, and this was already being demonstrated in Australia and in the UK. NRAO started down this path in competition with Caltech and its proposed Owens Valley Array based on John Bolton's highly successful interferometer. In 1970, NRAO finally won this competition (described as the period of the NRAO-OVRO wars) after they had hired some of the best young radio astronomers from Caltech.

Dave Heeschen certainly learnt from his own list of lessons and under his leadership NRAO brought the VLA project to completion in 1980, on schedule and close to the planned $78M budget appropriation. The VLA exceeded almost all its design specifications and has been by far the most powerful and most successful radio telescope ever built and, arguably, the most successful ground-based astronomical telescope ever built.

"The Bar is Open" was a well-known Heeschen alternative to an after-dinner talk which is used as the title for a chapter on one of the most productive periods of radio astronomy research at NRAO. When Dave Heeschen retired, he gave a talk: "advice to future directors and managers":

1. Hire good people, then leave them alone.
2. Do as little managing as possible.
3. Use common sense.
4. Do not take yourself too seriously
5. Have fun

I also recall another item of wise advice I received from Dave as I embarked on my first serious management role as VLA Director: "It's sometimes more important to make a clear decision than to get the decision right—but do try to get it right more than half the time!"

The chapter on millimeter wavelength (mm) astronomy is not just a summary of NRAO's involvement in mm astronomy, a field which was opened up and led by the USA, but an overview of all international mm astronomy efforts. The radio astronomy millimeter field developed in a very different way compared to the meter and centimeter wavelength astronomy. In the beginning millimeter wave receivers were bolometers with no spectroscopic capability and with poor sensitivity. As a result, there was a very limited scientific case with few observable sources, so the NRAO 36 foot at Tucson was a high-risk exploratory development. But with the unexpected discoveries of a plethora of spectral lines, millimeter radio astronomy became one of the hottest topics in radio astronomy, leading eventually to the billion-dollar ALMA project.

The book includes a great sequence of stories about building big telescopes for radio astronomy. These are all linked to NRAO, but reach well beyond NRAO as a result of NRAO's strong influence on developments in radio astronomy throughout the world.

The authors analyze in some detail the actual US funding process for a number of major proposals: 140 foot, VLA, VLBA, GBT, and ALMA. In particular, one author (KIK) was directly involved in the VLBA proposal and associated funding process. These actual situations illustrate the difference between the simplistic notions of proposal, review, and funding decision, with the real-life process. This chapter on the VLBA is not just a part of NRAO history but a great historical overview of all VLBI developments by an expert who lived through this era. VLBI experiments, especially those involving space missions, are among the most complex international projects that ever succeeded, involving different institutes, countries, and science agencies. These firsthand stories provide exceptional examples of successful scientific collaboration. Kellermann recounts his involvement in a VLBI collaboration with Russia in the peak of the Cold War.

The Sugar Grove 600 foot radio telescope, referred to by Harvard radio astronomer Edward Lilley as "a radio telescope fiasco," is also included. It is a story about a classified defense telescope being built near the Green Bank Observatory that has not been told before.

To maintain its viability through the period of traumatic delays in the construction of the 140 Foot dish, NRAO built a simpler, inexpensive 300 Foot transit antenna. At this time this 300 Foot Radio Telescope was one of the most powerful radio telescopes in the world and became an immediate success. For the first time, NRAO had a world-class instrument that was attractive to both visitors and NRAO staff. The successful completion of the 300 Foot transit radio telescope probably saved Green Bank from a premature closing resulting from the continued debacle with the 140 Foot antenna project. From the start of 300 Foot observations, the observatory operated as the first true visitor facility for radio astronomy.

But the 300 Foot story does not stop there. In 1988, NSF planned that the 27-year-old NRAO 300 Foot transit telescope be closed in order to provide funds for operating other new astronomical facilities. But when the 300 Foot

Telescope unexpectedly collapsed in November 1988, it was reported in the media as a national disaster for US astronomy. West Virginia's Senator Byrd demanded that the telescope be replaced. The NSF had other plans, but Byrd was able to include $75M in the 1989 Emergency Supplemental Appropriations Bill, which funded the world's largest (100-m) fully steerable dish for the Green Bank Telescope, and its chequered story makes another fascinating chapter.

As we approach the end of this book about the genesis of a national big science facility with all the fascinating stories about the many projects it engendered, we again return to the theme of collaboration at an international level with the construction of ALMA and the US involvement in the beginning of the SKA. The book concludes with discussions of closures and divestments, but this is offset by a vision for the Next Generation VLA (ngVLA) which is the US-proposed incarnation of the originally conceived SKA-high.

After 60+ years of progress we still find many lessons that still have not been learned, but this book may go some way to redressing this with its eloquent discussion of what happened in one field, coauthored by one of the experts who has participated directly in many of these developments.

Sydney, Australia   Ron Ekers

The original version of this book was revised. The correction to this book is available at https://doi.org/10.1007/978-3-030-32345-5_12

PREFACE

On April 27, 1933, at the annual meeting of the US National Committee for the International Union of Radio Science (URSI) in Washington DC, less than 50 years after Heinrich Hertz had demonstrated the propagation and detection of radio waves, Karl Guthe Jansky reported that he had detected radio signals from the center of the Milky Way. Jansky had no background in astronomy and was not searching for extraterrestrial radio signals, but was working for the AT&T Bell Laboratories to locate the source of interference to transatlantic telephone circuits. Only about 30 people were present to hear Jansky's dramatic announcement, one that would change the course of twentieth-century astronomy and lead to at least eight future Nobel Prizes. Although Jansky's discovery aroused great public interest, nearly two decades would pass before the American scientific community became sufficiently interested to invest in this emerging new field of astronomy. By that time, scientists trained in wartime radio and radar technology, primarily in Britain and Australia, had made a series of spectacular astronomical discoveries, and the USA was in danger of falling behind in this rapidly growing field, with obvious implications for technology, for military use, and for national prestige.

Prior to Jansky's discovery, astronomical research was confined to the narrow optical window between 4000 and 7000 Angstroms (400–700 nm), only about a factor of two in wavelength. With the spectacular postwar development in electronic instrumentation and the resulting march toward shorter and shorter wavelengths, radio observations today cover the broad spectrum between less than 1 mm to more than 10 m, a range of about 10^5 in wavelength. The subsequent rapid growth of space programs extended astronomers access to the entire electromagnetic spectrum from the infrared to the ultraviolet, X-ray, and gamma-ray, all of which are obscured by the earth's atmosphere. However, radio astronomy was the first outside the traditional optical window and resulted in the discovery of a wide range of previously unrecognized cosmic phenomena and many new previously unrecognized constituents of the universe. These included solar radio bursts, electrical storms on Jupiter, the so-called greenhouse effect on Venus leading to its extraordinarily hot surface temperature, precise tests of general relativity, the rotation of Mercury, the first exoplanets, radio galaxies, quasars, pulsars, cosmic masers, and cosmic evolution. The discovery of the microwave background by Arno Penzias and Bob Wilson in 1965 revolutionized cosmology and, interestingly, was made at the same AT&T Bell Laboratories as the initial discovery by Jansky of cosmic radio emission some three decades earlier.

The early pioneering radio astronomy observations were largely carried out by a single individual or small group who conceived of an experiment, designed and built the equipment, carried out the observations, and analyzed the resulting data. The 1956 establishment in the USA by the National Science Foundation (NSF) of the National Radio Astronomy Observatory (NRAO), operated by Associated Universities, Inc. (AUI), changed the culture and ultimately impacted *all* astronomy, not just radio astronomy. Before NRAO, US astronomical observatories were primarily privately funded through generous gifts from wealthy individuals like Charles Yerkes, James Lick, Percival Lowell, Andrew Carnegie, and John D. Rockefeller, and the philanthropic foundations which they established. The European-based observatories were often state supported, but they, like their private American counterparts, were used nearly exclusively by their own staff members.

Following the lead of the physicists who had created the Brookhaven National Laboratory (BNL), the NRAO was established to provide scientific instruments that were too costly for individual universities to build and operate. Although originally intended to provide opportunities for American radio astronomers to compete in the rapidly developing new field, with the availability of its first instrument, the 85 Foot Howard E. Tatel Telescope, NRAO developed a more broadly welcoming policy and announced its visiting scientist program on page 1179 of the October 30, 1959, issue of *Science*:

> The National Radio Astronomy Observatory was established by the National Science Foundation to make available to scientists from any institution facilities for research in radio astronomy…. The facilities of the observatory are open to any competent scientist with a program of work in radio astronomy, regardless of institutional affiliation.

Under the wise leadership of its young Director, Dave Heeschen, as additional telescopes were completed—the 300 Foot, the 140 Foot, the Green Bank Interferometer, the 36 Foot millimeter telescope, and eventually the Very Large Array (VLA)—NRAO facilities continued to be available to any scientists with a good program, independent of their institutional or national affiliation. This concept, which has become known as "Open Skies," after the Open Sky agreements which regulate international airline traffic, ultimately was adopted by nearly all major ground- and space-based American as well as international astronomical facilities. When two of us (Kellermann and Bouton) interviewed Dave Heeschen on July 13, 2011 (see https://science.nrao.edu/about/publications/open-skies), we asked, "What was the best thing you did during your years as Director, the thing that had the biggest impact?" Heeschen responded, "I think it's the establishment of the concept of the national observatory and the free use of the telescopes by people.... [T]hat I think was a really good thing that came out of it, and it's something which is persisting, you know, till this day. Everybody uses everybody else's telescopes in one way or another."

In his definitive book, *Cosmic Noise* (Cambridge: Cambridge University Press, 2009), Woodruff T. Sullivan III documented the explosive growth of radio astronomy from Jansky's unexpected discovery to the exciting postwar programs in Australia, in the UK, and, to a lesser extent, in the USA. Sullivan's book ends in 1953, 20 years after Jansky's discovery. The first discussions leading to the establishment of a national radio astronomy facility started in the early 1950s, and began the evolution of astronomy to a user-based, hands-off, big-science culture. In *Open Skies*, we have tried to pick up where Sullivan left off, describing the tumultuous circumstances leading to the creation of the NRAO, the difficult years which almost led to the closing of the observatory before it really got started, the later construction of the VLA, the Very Long Baseline Array (VLBA), and the Green Bank Telescope (GBT), along with the pioneering explorations into millimeter wavelength astronomy. It was a period which saw an unprecedented series of astronomical discoveries, mostly made possible by the explosive growth in radio astronomy techniques during the latter half of the twentieth century.

In planning the organization of *Open Skies*, we opted against a strictly chronological story, but instead deal with each major area of NRAO's contributions, arranged in separate chapters in approximate time sequence with each chapter organized in roughly—but not completely—chronological order. As background for those readers not familiar with the extensive literature on the early development of radio astronomy, we have included two introductory chapters about events which led to the start of discussions about establishing a national radio astronomy facility. A more detailed account of the people and activities during the early years in Green Bank is given in *But It Was Fun, ed. J. Lockman et al. (Green Bank: NRAO)*. We have included a listing of Abbreviations and Acronyms, which includes abbreviations used in the text and in endnote citations (Appendix A) as well as a Timeline (Appendix B) at the end of the book.

Two of the authors of *Open Skies*, Ellen Bouton and Ken Kellermann, have had a long association with NRAO. Bouton began work in the NRAO library in 1975; in 1983, she became the NRAO Librarian and since 2003 has been the NRAO Archivist. Kellermann joined the NRAO scientific staff in 1965. He was involved in most of the activities described in *Open Skies* since that time, especially the development of VLBI and planning for the VLBA, and from 1995 to 2003 was the NRAO Chief Scientist. Sierra Brandt brought her background in the history of astronomy to the NRAO/AUI Archives between 2011 and 2013 and has been a consultant since.

In researching the growth of NRAO and American radio astronomy, we have been aided by the collections in the NRAO/AUI Archives. Holdings include the formal records of NRAO, papers of many early NRAO staff members, as well as the personal papers of some of the pioneers of US radio astronomy, especially Grote Reber, as well as Ronald Bracewell, Bernard Burke, Marshall Cohen, John Kraus, and Gart Westerhout.

We are grateful to the AUI Board, and especially Patrick Donahoe and Robert Hughes, for making the records of AUI Board meetings available. We must also acknowledge the resources of many other institutional archives and of the Library of Congress. Shelly Erwin and Loma Karklins gave their generous help and support during many visits to the Caltech Archives to examine the papers of Jesse Greenstein, Lee DuBridge, Alan Moffet, and Gordon Stanley. Gordon's children, Teressa and Luise, kindly made available some of their father's personal papers. Janice Goldblum at the National Academy of Sciences Archives provided us with access to the records of the 1964 Whitford and the 1972 and 1973 Greenstein reviews of astronomy and astrophysics, as well as records of the early meetings of the USNC-URSI commission on radio astronomy. Elise Lipkowitz facilitated access to the records of the NSF's National Science Board (NSB) during the critical period surrounding the establishment and early years of NRAO. Shaun Hardy helped us research the papers of Merle Tuve at the Carnegie Institution Department of Terrestrial Magnetism. Robin McElheny at the Harvard University Archives facilitated our access to the papers of Harvey Brooks, Ed Purcell, Bart Bok, Donald Menzel, and Leo Goldberg. Brian Andreen at the Research Corporation kindly gave us access to their files related to their long-term support of Grote Reber. Bernard Schermetzler at the University of Wisconsin Archives made available Karl Jansky's letters written to his father in 1932 and 1933, which contain regular reports on the work leading to his remarkable discovery, along with snippets of life during those difficult depression years.

Tony Tyson and Bob Wilson provided much valuable information on Karl Jansky from the Bell Laboratories Archives. Karl Jansky's widow, Alice, and their children, Ann Moreau and David, shared with us memories of Karl Jansky and the environment at Bell Labs during the 1930s. Miller Goss kindly made available his extensive research on Joe Pawsey, including privately held Pawsey family papers provided by Joe Pawsey's son, Hastings Pawsey. Nora Murphy, MIT Archivist for Reference, Outreach, and Instruction, and Bonny Kellermann helped provide the recording of Otto Struve's 1959 Karl Taylor Compton Lecture. Steven Dick kindly gave us copies of Frank Drake's 1961 presentation at the Washington Philosophical Society, which led to the first conference on the search for extraterrestrial intelligence. We are also indebted to Richard Wielebinski for sharing with us his inside view of the activities leading to the construction of the German 100 m radio telescope.

Many of our colleagues read early versions of chapters, provided additional information, and made valuable suggestions for improvement. We thank Jaap Baars, Barry Clark, Marshall Cohen, Steven Dick, Bob Dickman, Pat Donahoe,

Phil Edwards, Miller Goss, Dave Hogg, Tony Kerr, Jay Lockman, Jeff Mangum, Jim Moran, Vern Pankonin, Marian Pospieszalski, Jon Romney, George Seielstad, Richard Schilizzi, Jill Tarter, Paul Vanden Bout, Cam Wade, Craig Walker, and Sandy Weinreb, each of whom reviewed one or more chapters. We are especially grateful to Tony Beasley, Ron Ekers, and Peter Robertson, who reviewed the entire manuscript and made many helpful suggestions, and whose careful eyes caught numerous errors others had missed. Rebecca Charbonneau, Heather Cole, Michele Kellermann, Sheila Marks, and Nicole Thisdell gave us valuable editorial help and advice, and Jeff Hellerman helped prepare images for both content and cover. Ramon Khanna, our Springer editor, provided valuable advice and support throughout our project. Of course, any remaining errors are the responsibility of the authors.

We are especially grateful to Woody Sullivan, who generously donated to the NRAO/AUI Archives the audio recordings and available written transcripts of the interviews he conducted between 1971 and 1988 with 225 different radio astronomers, as well as the working papers he compiled on the interviewees and on other scientists. Author interviews of Mike Balister, Bernie Burke, Fred Crews, Frank Drake, Cyril Hazard, Dave Heeschen, Dave Hogg, Bill Howard, Harvey Liszt, Tom Matthews, Kochu Menon, Mark Price, Grote Reber, Ted Riffe, and Cam Wade gave us additional valuable insights which were important in writing *Open Skies*.

We wish to thank Tony Beasley and AUI for their long support of the NRAO/AUI Archives and, in particular, for funding to make this work available as an open access publication. And finally, but certainly not least, we also want to thank Michele, Ron, and Dana for their long-suffering tolerance and support of this project.

Charlottesville, VA, USA   Kenneth I. Kellermann
Charlottesville, VA, USA   Ellen N. Bouton
Providence, RI, USA   Sierra S. Brandt

Contents

Correction to: Open Skies: The National Radio Astronomy Observatory and Its Impact on US Radio Astronomy   C1

Appendix A: Abbreviations and Acronyms   615

Appendix B: NRAO Timeline   625

Index   631

CHAPTER 1

A New Window on the Universe

In April 1933, at a small gathering at a meeting of the US National Committee of the International Scientific Radio Union (URSI), Bell Labs scientist Karl Guthe Jansky announced that he had detected 20.5 MHz (14.6 m) radio emission from the Milky Way. Jansky used a novel directional antenna based on an invention by AT&T Bell Labs colleague, Edmond Bruce, that rotated every 20 minutes to determine the direction and source of the interfering noise that was plaguing the telephone company. Jansky's remarkable discovery of what he called "star noise" was widely publicized in the media, but had little immediate impact in the astronomical community, as astronomers, who typically had little background in electronics or radio, saw no relation to their own work.

For more than a decade, the only significant progress was made by one individual, Grote Reber, who had just graduated from college with a degree in electrical engineering. His 32 foot parabolic dish, which he built in the yard next to his mother's house using his own funds, was the forerunner of the much larger radio telescopes later built in the UK, in Australia, and later the United States, as well as the millions of smaller dishes which have proliferated throughout the world for the reception of satellite-based TV broadcasting. With his home-built radio telescope, Reber detected galactic radio noise first at 160 MHz (1.9 m) then 480 MHz (62 cm), which he called "cosmic static." Reber recognized the nonthermal nature of the galactic radio emission, made the first radio maps of the Milky Way, discovered the intense radio emission from the Sun, and brought radio astronomy to the attention of the astronomical community.

1.1 STAR NOISE AT THE TELEPHONE COMPANY[1]

The first transatlantic telephone circuits were established by AT&T in 1927 between New York and London using very long wavelength 5 km (60 kHz) radio transmissions (Bown 1927). The following year, the AT&T Bell System

K. I. Kellermann et al., *Open Skies*, Historical & Cultural Astronomy, https://doi.org/10.1007/978-3-030-32345-5_1

inaugurated a short wavelength circuit to provide a greater capacity. Although the Bell System transatlantic telephone calls were very expensive, they were subject to interference and fading and were not very reliable. Little was known at the time about short wave radio propagation or the limits to weak signal reception. Some noise originated in the receiver systems, but some was external. Some of the external static clearly came from passing automobiles and airplanes or from local thunderstorms, but some came from an unknown origin (Oswald 1930). According to Al Beck (1984), members of the Bell Labs radio research staff were aware that when connected to an antenna, the receiver noise was greater than when connected to a load, and that the level of noise depended on the antenna and the time of the day. So, it was understood that at least some of the noise was apparently external to the receiver system.

Still only 22 years old, on 20 July 1928, Karl Jansky reported for his first day of work at the AT&T Bell Telephone Laboratories, joining the tightly knit members of the Radio Research Division (Fig. 1.1). By August, after a two-week orientation class, Jansky was working at the Cliffwood Laboratory in New Jersey, little realizing that he was about to embark on an engineering

Fig. 1.1 Members of the Bell Labs Radio Research Division. Shown in the first row from left to right: Art Crawford, Carl Feldman, Sam Reed, Joe Johlfs, Lewis Lowery, Russell Ohl, Bill Mumford, Karl Jansky, Merlin Sharpless, Archie King, Edmund Bruce, and Al Beck. In the second row are Carl Englund, Harald Friis, Douglas Ring, Otto Larsen, Carl Clauson, Morris Morrell, Carl Peterson, Maurice Collins, Dan Schenk, and Jim Morrell. Credit: Courtesy of J.A. Tyson

study leading to a series of discoveries that would fundamentally change our understanding of the Universe and its constituents. Radio galaxies, quasars, pulsars, the Cosmic Microwave Background (CMB), cosmic evolution, cosmic masers, gravitational lensing, electrical storms on Jupiter, and other now commonly known cosmic phenomena were all unknown until Karl Jansky opened the new radio window to the Universe.

Karl Guthe Janksy was born on 22 October 1905 in the territory of Oklahoma, where his father Cyril was Dean of the University of Oklahoma College of Engineering. Karl was named after Karl Guthe who was a former professor of his father at the University of Michigan. Karl grew up in Madison, Wisconsin, where his father became a Professor of Electrical Engineering at the University of Wisconsin.

Karl's older brother, Cyril Moreau Jansky, Jr., known as C.M. Jansky, Jr., and to his family and to his friends as "Moreau," received a BA in Physics in 1917 and MS in 1919, both from the University of Wisconsin. Following his graduation, he taught at the University of Minnesota, where one of his first students was Lloyd Berkner, who would go on to become the driving force behind the formation of the National Radio Astronomy Observatory as the first President of Associated Universities (Sect. 3.1). C.M. Jansky, Jr. worked briefly for Bell Labs, and played a leading role in the development of radio technology and regulation in the US. In 1930, with his former student, Stuart Bailey, he established the consulting firm of Jansky and Bailey, where he remained active until his death in 1975. In 1934, Jansky became President of the Institute of Radio Engineers (IRE), the predecessor to the current Institute of Electrical and Electronic Engineers (IEEE).

Like his brother, Karl also studied physics at the University of Wisconsin and received his BS in Physics in 1927, graduating Phi Beta Kappa with a thesis titled, "Conditions for Oscillations in a Vacuum Tube Circuit." While at Wisconsin, Karl was the fastest skater and a prolific scorer on the university ice hockey team, and later while working at Bell Labs, he was the table tennis champion of Monmouth County, New Jersey. After a year in graduate school,[2] Karl sought a job at Bell Labs. Although Bell Labs was initially unenthusiastic about hiring Karl, who had a chronic kidney disease, Moreau intervened on behalf of his brother and urged the Bell Labs president to hire Karl (Jansky 1957).[3] In recognition of his illness, instead of locating him at their main laboratory in industrial New York City, Karl was sent to their then small rural laboratory in Cliffwood Beach, New Jersey, where he began work to study the propagation of short wave radio transmissions and the noise limits to transatlantic telephone communications. He married Alice Larue Knapp the following year, and along with most of the other Bell Labs Cliffwood employees, Karl and his growing family lived in nearby Red Bank. Karl enjoyed the informal social life shared with his fellow engineers. In spite of his illness and against medical advice, he remained active. He played chess, tennis, and golf, enjoyed bowling and skiing, had the highest batting average on the softball team, and was a passionate bridge player who claimed to know what cards each of the players held. Karl was an avid Brooklyn Dodgers baseball fan and had very strong political opinions. He was critical of the Roosevelt administration and speculated that if the president's term was ten years instead of four that, "We might have another civil war."[4]

Throughout his career at Bell Labs, Karl's immediate boss was Harald Friis, a Danish-American radio engineer who had immigrated to the United States from Denmark and had established a reputation in radio antennas and propagation. Later Friis and his wife Inge became close personal friends of the Jansky family, and were godparents to Jansky's daughter Anne Moreau. With time, tensions developed between Jansky and Friis over Karl's work assignments, but apparently their personal relationship remained intact.

Jansky's discovery of cosmic radio emission is a classic example of the scientific method, complete with false leads, that George Southworth (1956) later compared to a Sherlock Holmes detective story. Jansky's story is described in his series of papers in the *Proceedings of the IRE* (Jansky 1932, 1933b, 1935), his laboratory notebook entries,[5] and regular weekly work reports, as well as the running account of his work documented in his detailed letters to his father back in Madison.[6] These letters provide a glimpse into the development of Karl's thinking as he acquired and interpreted new data, and reflected on the difficult economic challenges he and his young family faced during those trying depression years.

In order to determine the direction of interfering signals, Jansky needed a directional antenna whose orientation could be varied. His notebook entries for 22 to 29 June 1929 indicate that he devoted this period to the design of a rotating antenna. On 24 August, he noted, "Mr. Sykes was interviewed and will start work on the 'merrygoround' next Monday." The next months were spent designing and building the instrumentation with special attention to reducing receiver noise and obtaining good gain stability (Beck 1984). During this period, Jansky also planned the rotating Bruce Array[7] at Cliffwood Beach which was constructed by Carl Clausen, a member of the Bell Labs staff. Jansky's rotating array used a parasitic reflector to enhance the forward gain and directivity, and was mounted on the wheels and axles taken from an old Ford Model T car.[8] Motor driven, the array made a complete rotation in azimuth every 20 minutes. Jansky's work was interrupted by a decision to move the Laboratory to a new location at Holmdel, New Jersey, which would provide more room for the growing laboratory staff and less noise and local interference. A new circular track was constructed away from the laboratory building, and the rotating Bruce array was relocated to the new site (Fig. 1.2).

One of Karl's first tasks at the new site was to find a frequency free of interfering signals. On 10 May 1930, he wrote, "It was decided to operate upon a frequency of 20,689.7 [kc] or 14.5 meters."[9] He then calculated the size of the quarter wave antenna elements as "142.72 inches = 11 feet 10.72 inches or 11 ft 10¾ approx." A week later, on 17 May he wrote, "I designed the supporting framework for the array proper. The diagrams have been turned over to the shop office."

Fig. 1.2 Karl Jansky and his rotating Bruce Array known as the "merry-go-round," which he used in 1932 to detect radio emission from the Milky Way Galaxy. Credit: NRAO/AUI/NSF

During the summer and autumn of 1930, Karl used his rotating array to determine the direction of arrival of signals from transmitting stations in England and South America and easily detected static from nearby thunderstorms.[10] In November he became aware of static coming from a direction where there was no obvious weather disturbance, but apparently much of Jansky's time during his period was spent on other activities (Southworth 1956). Following an overhaul of his receiving system, he began in the summer of 1931 to keep more systematic records of displayed static on a running paper chart recorder (Fig. 1.3). During the 1931/1932 winter, after the summer thunderstorm activity had subsided, he noted that the anomalous noise was highly peaked in a direction that appeared to move with the time of the day, being strongest in the morning toward the east, toward the south at noon, and toward the west late in the afternoon. Jansky naturally concluded that it had something to do with the Sun, and wrote to his parents, "That would be interesting wouldn't it?"[11] Coincidentally, in December the Sun lies in the direction of the Milky Way, and over the following months the peak noise came earlier each day, and Jansky noted that it was well removed from the Sun. However, he apparently did not yet realize that the source was extraterrestrial.

Fig. 1.3   Karl Jansky examining the output of his paper chart recorder. Credit: NRAO/AUI/ NSF

At the April 1932 Washington meeting of the US National Committee for the International Union of Radio Science (URSI), Jansky presented a paper on "Directional Studies of Static on Short Waves,"[12] which he later published in the *Proceedings of the Institute of Radio Engineers* (*Proc. IRE*). In this, the first of his three classical papers, Jansky (1932) described in some detail his antenna and receiving system and reported that he had found three distinct groups of static.

> The first group is composed of the static received from local thunderstorms and storm centers. Static in this group is almost always of the crash type. It is very intermittent... The second group is composed of very steady weak static coming probably from [ionospheric] refractions from thunderstorms some distance away. The third group is composed of a very steady hiss type static the origin of which is not yet known.

Jansky then goes on to discuss the crash type static in some detail, but adds,

> The static of the third group is also very weak. It is, however, very steady, causing a hiss in the phones that can hardly be distinguished from the hiss caused by [receiver] noise.

He remarks that he did not recognize this third type of static until January 1932, but he was able to go back to reexamine his earlier data and recognized that the direction of arrival of this static coincided with … the direction of the sun. However, during January and February, the direction has gradually shifted so that now [March 1] it precedes in time the direction of the sun by as much as an hour.

Still assuming that the source of static was terrestrial, and not having any background in astronomy, Jansky continued to speculate that the hiss type static might be related to the Sun, and that the change in apparent direction might somehow be due to the Sun's changing declination after the winter solstice. He concluded, however, that, "the data as yet cover only observations taken over a few months and more observations are necessary before any hard and fast deductions can be drawn."

Throughout 1932, Jansky meticulously continued his observations. Indeed, after the summer solstice when the Sun reached its northern declination limit, instead of changing direction the shift in apparent position of the star noise continued. According to Southworth (1956) Jansky discussed his results with many of his associates, including his supervisor, Friis, Southworth, Edmond Bruce, Al Beck, Art Crawford, and probably most importantly, Melvin [Mel] Skellett. Skellett, who was a close friend of Jansky's, was studying for his PhD in astronomy at Princeton. He apparently recognized the sidereal nature of Jansky's data and advised Jansky to look at elementary astronomy text books. Jansky studied these text books and mastered the trigonometric transformations between terrestrial and celestial coordinates. He went back and reexamined his data, and on 21 December 1932 he wrote to his father,

I have taken more data which indicates definitely that the stuff, whatever it is, comes from something not only extraterrestrial but from outside the solar system. It comes from a direction that is fixed in space.

Karl apparently fully recognized the implications of his findings. Showing the same competitive spirit that his colleagues associated with his sports and bridge activities, he continued, "I've got to get busy and write another paper right away before someone else interprets the results in my other paper in the same way and steals my thunder from my own data." In an 18 January 1933 letter, he backed off somewhat, writing, "I have data which shows conclusively that the hiss type static comes from a direction which I know at least lies in a plane fixed in space and I think the direction is fixed in that plane but I am not sure of that as yet."[13]

Around this time, Karl's work was disturbed by a move to a new home in Little Silver with his wife and baby daughter, and an announced reorganization of Bell Labs with a threatened 30 percent cut in the engineering staff. He noted that a close friend and former roommate who had behaved "independently," but who had recently become a father and purchased a home was let "out."[14] This may have unnerved Jansky, who commented on the low morale at the lab and queried his father about possible teaching jobs at the University of Wisconsin or elsewhere. However, Karl escaped the layoffs and was able to present his remarkable results at the annual meeting of the US National Committee for URSI, which was held in Washington DC on 27 April 1933. Karl described URSI to his family as "an almost defunct organization … attended by a mere handful of old college professors and a few Bureau of Standards engineers."[15] "Beside this," Karl continued, "Friis would not let me give the paper a title that would attract attention but made me give it one ["A note on hiss type atmospheric noise"] that meant nothing to anybody but a few who were familiar with my work."

Karl's brother, Moreau, who himself was an influential leader of the URSI National Committee, was clearly impressed by his younger brother's paper and apparently convinced the AT&T publicity department to issue a press release describing Jansky's star noise.[16] As Karl wrote to his father, "the science editors of the N.Y. papers were alert enough to realize the importance of the subject and yesterday afternoon pestered the life out of the publicity department." The 5 May 1933 edition of the *New York Times* featured an "above the fold" article titled, "New Radio Waves Traced to the Centre of the Milky Way." Other headlines on the same front page ominously referred to the anticipated invasion of China by Japan and Nazi threats at the French border. The following day, the *Times* "Week in Science" section noted that Jansky's star noise was at the extreme end of the same electromagnetic spectrum that included the familiar visible spectrum that was the basis of all previous astronomical knowledge. On 15 May, the NBC Blue Network, which later became the ABC, interviewed Jansky and played three 10-second segments of Jansky's star noise received at Holmdel and sent over the AT&T Long Lines. In describing his star noise during his interview, Jansky explained,

> The observations show definitely that the maximum of hiss comes from somewhere on the celestial meridian designated by astronomer as "18 hours right ascension." … But my measurements further show that the radio hiss comes from a point on that 18-hour meridian somewhat south of the equator, that is at about minus ten degrees in declination … that seems to confirm Dr. Shapley's calculation that the radio waves seem to come from the center of gravity of our galaxy.[17]

Jansky's extraordinary discovery was reported in national and international newspapers as well as in the 15 May edition of *Time Magazine*. The publicity generated by the media exposure resulted in the usual crank letters. As he wrote to his father, "I received a letter today from spiritualist [who] thinks I am receiving messages from the 'other' world [and] from some crank mathematician [who] advises me to watch for numerical messages based on the factor '2, 4, 8 etc' indicating a 'superior' intelligence."[18] There is little doubt that Karl was aware of the impact of his discovery and that he relished the publicity. In October, he gave an invited lecture at the American Museum of Natural History in New York with the provocative title, "Hearing Radio from the Stars."

Perhaps sparked by the attention resulting from the *New York Times* article and realizing that his published *Proc. IRE* paper incorrectly suggested that the hiss type noise originated in the Sun, on 8 May, Karl sent a short note to *Nature*, titled, "Radio Waves from Outside the Solar System," (Jansky 1933a). In this paper, published on 8 July, Jansky states, "the direction of arrival of this disturbance remains fixed in space, that is to say the source of this noise is located in some region that is stationary with respect to the stars." He goes on to give the direction of the radio noise as "right ascension of 18 hours and declination of -10 degrees." In a paper meant for a more popular audience, Jansky (1933c) confidently used the more specific and provocative title, "Electrical Phenomena that Apparently Are of Interstellar Origin."

Jansky originally had wanted to announce his discovery at the Chicago IRE meeting which was held in June 1933, but Friis had rejected his request and "insisted" that he give the talk instead at the April URSI meeting. Following the attention resulting from the *New York Times* article and the NBC broadcast, at Karl's request, his brother Moreau again stepped in to use his influence to get Karl invited to the June IRE meeting.[19] At this point, ignoring Friis' reservations, but with the encouragement of more senior Bell Labs management, Karl decided on his own to change his title "to suit myself."[20] His IRE talk was published in the *Proc. IRE* (Jansky 1933b) as his now classic paper on "Electrical Disturbances Apparently of Extraterrestrial Origin." For the benefit of the IRE engineering readers, he first reviewed the relationship between terrestrial and astronomical coordinate systems and the difference between solar and sidereal time. In his introductory summary, he concludes "that the direction of arrival of these waves is fixed in space, i.e., that the waves come from some source outside the solar system," and here he gives this direction as the "center of the huge galaxy of stars and nebulae of which the sun is a member." Following his talk, Karl sent a copy of his paper to the well-known Princeton astronomer, Henry Norris Russell, and arranged to meet with Russell to discuss the meaning of his star noise.

For the next two years, Jansky was apparently preoccupied with other research activities, but found the time to analyze his data more carefully. In July 1935, he again gave a talk at the Annual IRE Convention in Detroit, and was able to report that the radio emission came from the entire galactic plane with the strongest radiation coming from the Galactic Center.[21] In his third *Proc. IRE* paper, following his Detroit talk, Jansky (1935) explained that the noise peaks correspond to those times when the antenna beam is oriented along the plane of the Milky Way, and second, that the largest peak comes from the "that section of the Milky Way nearest the center." Although he concluded that the "most obvious explanation of these phenomena … is that the stars themselves are sending out these radiations," he did not exclude the possibility "that the waves that reach the antenna are secondary radiations caused by some form of bombardment of the atmosphere by high speed particles which are shot off by the stars." In this paper, Jansky also made the first attempt to understand the physics behind his star noise noting that "one is immediately struck by the similarity between the sounds they produce in the receiver headset and that produced by the thermal agitation of electrical charge. In fact the similarity is so exact," he explained with some prescience, "that it leads one to speculate as to whether or not the radiations might be caused by the thermal agitation of charged particles."[22]

Jansky also commented on the important contrast between the optical and radio sky, pointing out that while visually the Sun appears brighter than the radiation from all the stars combined, the reverse was true at radio wavelengths. He realized that if all the stars in the Milky Way are like our Sun, that could not explain his observed noise from the Milky Way. He speculated that "a possible explanation ... [is] that the temperature of the sun is such that the ratio of energy radiated by it on the wavelengths studied to that radiated in the form of heat and light is much less than for some other classes of heavenly bodies found in the Milky Way." It is clear from this third *Proc. IRE* paper and from his short notes in *Nature* and *Popular Astronomy* (Jansky 1933a, c) that Jansky was not satisfied just to have solved the problem of short wave radio noise, but he wanted to understand and disseminate the implications to the astronomy and astrophysics communities as well as to the broader public.

More than a decade later, URSI President Sir Edward Appleton commented in his 1948 Presidential address, "Jansky's work seems to me to have all the characteristics of a fundamental discovery. In the first place he recognized something that was unexpected. In other words, he discovered something when he was actually looking for something else. But he then went further, for he recognized his unexpected result as being significant. And pursued it with zeal until much of its true meaning emerged."[23] As John Kraus (1981) later noted, Jansky's system contained all the elements of future radio telescopes: (1) a directional antenna, (2) a broad band low noise receiver, and (3) a radio quiet site. Somewhat later, Grote Reber (1988) added, "Reading Karl Jansky's articles is an enlightening example of how a first class human mind works, and how one hypothesis is discarded for another as more evidence rolls in."

Woodruff Sullivan (1978) later reanalyzed Jansky's data from 16 September 1932, and presented it in a modern form as a contour map in galactic coordinates (Fig. 1.4). Sullivan's map, which displays the concentration along the galactic plane and the maximum toward the Galactic Center, also shows the maximum later recognized as the Cassiopeia A supernova remnant as well as evidence for the Cygnus A/Cygnus X complex.

The Later Years In order to enhance Bell Labs' transatlantic radio communications capability, Friis developed the highly directional Multiple Unit Steerable Array (MUSA) rhombic antenna (Friis and Feldman 1937).[24] In his continuing study of short wave noise and radio wave propagation, Jansky (1937, 1939) used MUSA to show that in the absence of manmade interference, the sensitivity of short wave radio systems was limited by interstellar noise, and not by receiver circuit noise. During this period, he also tried to use the MUSA system with a goal of measuring the frequency dependence of his "star noise." The

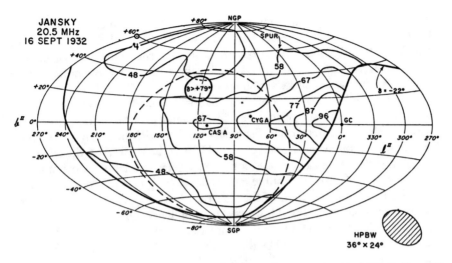

Fig. 1.4 Contour map of the 14.6 meter cosmic radiation derived by W.T. Sullivan III from observations made by Karl Jansky on 16 September 1932. The contours are normalized to a peak value of 100 corresponding to brightness temperature of 100,000 K. The dashed line shows the track of a typical 20 min rotation of the antenna. Credit: Courtesy of W.T. Sullivan III

published results were inconclusive, most likely due to the increasing solar activity and the corresponding D-layer ionospheric attenuation, but also to broadband interference generated by nearby diathermy machines. Al Beck, who worked with Jansky at the time, later noted that the cosmic signal was strongest at the longer wavelengths[25] but there is no contemporaneous record or documentation that Jansky or Beck had understood the nonthermal nature of the star noise. Interestingly, 1932 was near the bottom of the sunspot cycle. Had Jansky done his pioneering work five years earlier or five years later, the ionosphere probably would not have allowed him to detect extraterrestrial radio noise at 21 MHz (15 meters), although he would then probably have detected radio emission from the active Sun.

Although AT&T was quick to exploit the popular interest in Jansky's discovery, further investigation of "star-noise," other than as a noise floor, did not rank high in AT&T's priorities. Even during the critical period of 1932 and 1933, Jansky did not work full time on star noise, as his laboratory notebook shows considerable effort spent on understanding the direction of arrival of short wave radio transmissions. As noted by George Southworth (1956), "Somewhat later he was assigned to other duties and his work on radio astronomy came to an end. His interest nevertheless continued."

For some years after 1935, Jansky's "merrygoround" was used by Beck (1984) and others for testing antennas, but it ultimately fell into disrepair and the remnants were destroyed. Following a suggestion by Grote Reber[26] to Southworth, and with the encouragement and support of Al Beck, in 1964, a replica of Karl Jansky's antenna was erected at the entrance of the NRAO Green Bank Observatory. Years later, former Bell Labs scientists J. Anthony (Tony) Tyson and Robert (Bob) Wilson located the site of the original antenna at Holmdel, and on 8 June 1998, Bell Labs dedicated a memorial model antenna on the site of Jansky's original array.

With the rapidly deteriorating international situation in the late 1930s, Jansky became increasingly involved in classified defense work, particularly related to electronic detection of submarines, for which he received an Army-Navy citation. Following the end of World War II, he worked on the emerging AT&T microwave repeater network for long distance telephone communication and followed with interest the rapidly developing field of radio astronomy. After the Bell Labs invention of the transistor in 1947, Jansky was one of the first to use transistors to build low noise preamplifiers and received several patents on a radio direction finder or sextant based on the radio emission from the Sun which was later developed by the Collins Radio Co. for the Naval Research Laboratory. His deteriorating health limited his activities and, after a series of strokes, on 14 February 1950, Jansky succumbed to his long illness and died at the young age of 44. This was a year before the discovery by Ewen and Purcell (1951) of the 21 cm hydrogen line at Harvard (Chap. 2) brought radio astronomy to the attention of the broad astronomical community. Just before his death, Karl was transferred to a small group under George Southworth that also included Russel Ohl, who played a prominent role in the Bell Labs development of the transistor.

Controversy John Pfeiffer was a free-lance author who wrote the first popular book about the new science of radio astronomy called *The Changing Universe: The Story of the New Astronomy*. Prior to writing his book, Pfeiffer traveled to the major observatories and laboratories working in radio astronomy. In 1954 he visited Bell Laboratories and spent a day talking with Harald Friis and members of his Radio Research Section about Karl Jansky and his discovery of interstellar radio emission. In his book, Pfeiffer (1956, p. 17) claimed that Jansky "did all he could to convince his associates and superiors that the work was worth pursuing for practical reasons. But his arguments failed to produce results." Pfeiffer further incited the issue stating, "Rarely in the history of science has a pioneer stopped his work completely, at the very point where it was beginning to get exciting. Yet Jansky did just that."

In his review of Pfeiffer's book in *Science*, Frank Edmondson (1956), a well-known optical astronomer, writing from the National Science Foundation, threw coals on the fire when he hinted that "Jansky's failure to secure support for continued pure research at Bell Laboratories" may have led the US to be "lagging far behind other countries in the development of radio astronomy." This generated a strong rebuttal from Karl's associate J.C. Schelleng who wrote to the editor of *Science*, Graham DuShane,[27] arguing that, "It is news to Mr. Jansky's associates at Holmdel to be told that he tried seriously to continue the work and failed .... If the astronomers showed any excitement at the time, we saw no sign of it." Schelleng went on to point out that after reading Grote Reber's paper, Jansky "cheerfully" noted that "he had skimmed the cream," and "showed no sign of disappointment."

Karl's brother, C.M. (Moreau) Jansky, Jr. (1958) further incited controversy when he wrote the introductory article in the *Proc. IRE* special issue on radio astronomy describing his brother's discovery of extraterrestrial radio emission. In his paper, Jansky wrote that, "his superiors transferred his activities to other fields. He would have preferred to work in radio astronomy." Picking up on Jansky's remarks, the popular magazine, *Readers Digest,* reported, "Astronomers were slow to recognize [Karl] Jansky's discovery, [and that] his boss told him to stick to ordinary static, and reluctantly he did so." (Kent 1958). In a letter to Karl's brother Moreau,[28] and in a later paper in *Science,* Friis (1965) denied that this was true, and said Karl "was free to continue work on star noise if he had wanted to," and that Karl had never indicated to him "a desire to continue his star noise work." Moreover, noted Friis, during this period there had been no interest or encouragement from astronomers and it was not clear in what directions such research should go and what kind of equipment was needed. Essentially, he argued that Karl felt that having found the source of noise, he had completed that particular project. Much later, on several occasions Friis (1965, 1971) commented that Jansky should have received a Nobel Prize for his discovery of cosmic radio emission.[29]

Moreau reacted to Friis by writing to Karl's widow, Alice, to seek her comments.[30] Alice responded,

> Harald says that Karl never expressed to him a desire to continue work on his star noise. How incredible, how preposterous, how positively unbelievable. Periodically, over the years that Karl worked under Friis, he would come home and say, "Well, Friis and I had a conference today to discuss what my next project should be, and, as usual, Friis asked what I'd like to do, and as usual, I said, 'You know I'd like to work on my star noise,' and as usual, Friis said, 'Yes, I know, and we must do that some day, but right now I think—and—is more important, don't you agree?'"[31]

Moreau decided not to respond to Friis, but thanked Alice for "substantiating the statements made in the last two paragraphs of my paper," and added, "I have been informed that quite a number of people have expressed the opinion that they are glad that I said what I did."[32] It is perhaps significant that while Friis portrays a close personal relationship with Jansky, in all of his letters to his parents, Jansky always referred to "Friis," or "Mr. Friis," never using his first name. To Karl and Alice's children, Anne Moreau and David, they were Uncle Harald and Auntie Inge. However, at least at work, Friis apparently just called Karl "Jansky."

Radio astronomy again came to the forefront at Bell Labs in 1965 following the detection of the cosmic microwave background by Arno Penzias and Bob Wilson (Penzias and Wilson 1965). Bell Labs wanted to clarify the growing and clearly embarrassing controversy over Jansky. Ray Kestenbaum of their Public Relations Department conducted a series of interviews with Jansky's boss Harald Friis, his colleagues Al Beck, John Schelleng, and Art Crawford, as well as Karl's brother Moreau.[33] Everyone from Bell Labs described Karl as friendly, modest, and easy to get along with. They also noted that he was very competitive in sports and bridge. All, especially Friis, denied ever hearing any expression from Karl that he felt constrained from pursing his star noise research, and they all noted that it was not clear to any of them, Karl included, what might be the next steps. Karl, they maintained, was never "stopped" from pursuing further work on star noise, that he was happy with his work, that his primary research interests were in antennas and receiver noise, and that he enjoyed working at Bell Labs.

Although George Southworth was not interviewed, Kestenbaum later wrote to Southworth for his views on this controversial issue. His initial response was equivocal,[34] but recalling Southworth's (1956) earlier remark that Jansky "was assigned to other duties," Lloyd Espenschied, then retired from a senior management position at Bell Labs, later wrote to Southworth commenting on a paper in the Bell Labs Record that claimed that Jansky did not continue his star noise research "due to a lack of theoretical understanding … and by the inability to detect radio noise from the Sun."[35] "I wonder who concocted this fabrication as the 'cover-up' it is?" wrote Espenschied, "I remember very well how Karl's foundling babe was left out in the cold, much to his distress, simply because it was regarded as not being pertinent to the Bell system."

Southworth responded with a brief factual summary of Jansky's career at Bell Labs, his own satisfaction with Jansky's productivity at Bell Labs, but concluded that, "Other matters like the handling of Jansky's case bespoke many unpleasant, if not indeed tragic experiences. Most of us, without knowing why, thought intuitively that he had not been dealt with fairly."[36]

Were Friis and the other Bell Labs staff who were interviewed by Kestenbaum just trying to protect the reputation of the Labs? Or were Southworth and Espenschied expressing broader pent up dissatisfaction with the Labs? And were Moreau and Alice carried away with family loyalties and a desire to preserve and enhance Karl's reputation?

Contemporaneous insight to what happened between Friis and Jansky is documented in the series of Karl's letters to his parents which were often critical of Friis and do not support Friis' position. Friis was described by another Bell Labs employee, Russel Ohl, as a "dictator [who] wanted things done exactly the way he said."[37] When Karl complained to his father about the title of his 1933 URSI paper, he went on to mention that his brother Moreau reported that the IRE Board of Directors considered that "my paper was the outstanding paper of the Washington meeting," … and "that they all agreed that the title to the paper was too commonplace [a direct slap at Friis]." Karl's father expressed sympathy, but cautioned Karl, "Do not antagonize him. Keep on consulting him as formerly. He is your boss and loyalty ultimately pays no matter whether it is deserved."[38]

A month later, Karl wrote "I have not the slightest doubt that the original source of these waves whatever it is or wherever it is, is fixed in space. My data proves that, conclusively as far as I am concerned. Yet Friis will not let me make a definite statement to that effect but says I must use the expression '*apparently* fixed in space' or '*seems* to come from a fixed direction' etc. etc. ... But I suppose it is safer to do what he says."[39] Following the announced layoffs in the spring of 1933, Karl wrote to his parents, "Nothing has happened yet at the labs in the matter of firing some of the engineers. The ax is still hanging over our heads.... What a h - l of a way to run an organization." [40]

In January 1934, he wrote[41]

I now have what I think is definite proof that the waves come from the Milky way. However, I am not working on the interstellar waves any more. Friis has seen fit to make me work on the problems and methods of measuring noise in general. A fundamental and necessary work, but not very interesting as the interstellar waves, nor will it bring me near as much publicity. I'm going to do a little bit of theoretical research of my own at home on the interstellar waves however.

Grote Reber (1982) later reported that toward the end of World War II, he had met Jansky during a Washington URSI meeting, and that over lunch, Jansky had mentioned that in 1936 he wrote a memorandum proposing construction of a 100 foot diameter transit dish operating at 5 m wavelength (60 MHz). Apparently, as Jansky explained to Reber, he had been informed, "that the proposal was outside the realm of company business."[42] However, no record of such a proposal has been found in the Bell Labs Archives or among any of Karl's existing personal papers and letters.

To better appreciate the situation surrounding Jansky's important discovery, it is important to understand the political, social, and economic revolutions that were ravaging the world in 1933 and that only became worse over the following decade. Only a few months earlier Adolph Hitler had become Chancellor of Germany, and two days later he dissolved the German Parliament, starting a series of edicts and decrees that within a few years would lead to global catastrophe. Just six weeks before Jansky's announcement of his detection of radio noise from the Milky Way, Franklin Delano Roosevelt became the 32nd President of the United States, and promptly closed the banks. Later that year the United States would abolish the gold standard and prohibition; Babe Ruth would hit a home run to win the first major league all-star baseball game; the world's first drive-in theater would open in New Jersey; and the infamous John Dillinger would rob his first bank. Erwin Schrödinger and Paul Dirac won the 1933 Nobel Prize in Physics for their development of quantum mechanics, and the astronomy community was still absorbing the implications of Hubble's expanding Universe (Hubble 1929) with its emphasis on building large new optical telescopes to detect ever more distant galaxies.

The US and the world were in the throes of a major depression. It is said that fully one third of the US population was out of work. Rather than lay off staff, Bell Labs had cut its work week to four days, although many staff members, including Jansky, continued to work five days a week while getting paid for four. Bell Labs employees worried about their livelihood and many explored other opportunities. Anticipating the potential prospects for more time to pursue his research, but conscious of his chronic illness and uncertain life expectancy, Karl asked his father about possible positions at the University of Wisconsin or even teaching at a high school. With encouragement from Alice he applied for a position at Iowa State University, but was unsuccessful, and there is no evidence that he again seriously entertained leaving Bell Labs. By 1944, to meet the demands of the defense effort, Jansky was working overtime and enjoying the extra pay, which he needed to meet his increasing medical expenses.

While it is likely that Harald Friis did not encourage further work on Karl's star noise to the extent that Karl might have wished, neither did he apparently discourage Karl, other than by assigning him new tasks. Nevertheless, it seems Karl was not unhappy with his work or with Bell Labs. Aside from the critical comments to his father, he apparently did not push Friis hard to continue his star noise research. He understood the corporate nature of his employer and the constraints imposed first by the Depression, then by the War, as well as considerations of his health and the well-being of his young family. So while he may have shared his frustrations about Friis with his family, his father, and his wife, he also respected Friis as his boss, as well as valued him as a personal friend. It is also important to recognize that Friis, himself, was under pressure from senior management and directly from defense contractors to deliver on a multitude of contracts with limited staff, and Jansky was surely aware of this.[43] As many of his colleagues emphasized, the important thing may not be why Karl stopped his pioneering work after receiving broad national and international recognition, but that he accomplished so much in such a short time.

Recognition Although Karl Jansky's discovery of galactic radio emission did not have an immediate direct impact on astronomy, the discovery of radio noise from the Milky Way was widely recognized and discussed among astronomers as well as by the general public. Karl's father reported being at a talk by the University of Illinois astronomer Joel Stebbins, who drew attention to Karl's discovery.[44] Harlow Shapley, Director of the Harvard College Observatory (HCO), even wrote to Jansky asking for copies of his IRE paper and Jansky reported that they had a vigorous discussion at the Harvard Physics Department about Jansky's work.[45] At a later meeting in New York, Shapley asked about the cost of repeating the experiment, but was discouraged by Jansky's initial response. Subsequently, however, Jansky realized that the equipment he had used was originally built for another purpose, and he felt that it would be possible to confirm his results at much lower cost by using a commercial short wave receiver. But, apparently, by this time, Shapley had either lost interest or was discouraged by Harvard's unfamiliarity with anything to do with radio or electronics.

Two Harvard astronomers who thought hard about Jansky's star noise were graduate student Jesse Greenstein, and Fred Whipple, then a young Harvard faculty member. They had read Jansky's papers, and tried to interpret the radio noise as thermal radiation from cold dust (Whipple and Greenstein 1937). However, Jansky's star noise exceeded their model predictions by a factor of 10,000 (Sullivan 2009, p. 41). Henyey and Keenan (1940) tried to interpret the observations in terms of free-free emission from ionized hydrogen, and remarked, "in the case of Jansky's data the discrepancy is serious." Although these papers were unsuccessful in trying to explain the nature of the radio signals as thermal emission from interstellar dust, they underscored the need for unconventional nonthermal interpretations of cosmic radio emission. It would be another 15 years before the Russian scientists, Vitaly Ginzburg (1951) and Iosef Shklovsky (1952) would explain that the nonthermal galactic radio emission is due to synchrotron radiation from ultra-relativistic electrons moving near the speed of light in a weak magnetic field.[46]

At Caltech, Professor R.M. Langer was inspired by Jansky's papers to consider possible mechanisms to explain the observed radio emission from the Milky Way, and gave a talk to the American Physical Society proposing that Jansky's star noise was the result of free electrons combining with ionized dust particles (Langer 1936). Langer's ideas did not make the *New York Times* but did appear on the front page of the 13 March 1936 edition of the *Los Angeles Times*.

So Jansky's discovery was well known to the scientific community, and was certainly not ignored by astronomers. They considered it interesting and even important. However, no one, Karl included, appreciated the extent to which other new discoveries would follow from further research using this new window on the Universe, or, with the exception of a few individuals like Grote Reber, what should be the next step.

In 1948, Jansky was nominated for the Nobel Prize in Physics by the German physicist Winfried Schumann, who was known for his research on lightning-generated extremely low frequency radiation. But this was before the explosive growth of radio astronomy in the 1950s and the importance of Jansky's work was not widely appreciated. Perhaps not coincidently, Appleton's presentation at the 1948 URSI General Assembly in Stockholm may have been intended to call the Nobel committee's attention to Jansky's achievements. After the meeting, Karl's widow, Alice wrote to Appleton thanking him for his recognition of Karl's work with a passing reference to the Nobel Prize.[47] Most likely, had he lived longer, Karl Jansky would have been recognized with the Nobel Prize, which eight other scientists have subsequently received for work in radio astronomy.[48]

In 1959, the new laboratory building at NRAO was named the "Jansky Laboratory" and, after a major upgrade, the NRAO Very Large Array, was rededicated in 2012 as the "Karl G. Jansky Very Large Array." (Sect. 7.8) At the 1973 IAU General Assembly held in Grenoble, France, the Commission on Radio Astronomy passed the following resolution. "RESOLVED, that the name 'Jansky,' abbreviated 'Jy' be adopted as the unit of flux density in radio astronomy and that this unit, equal to 10^{-26} Wm^{-2} Hz^{-1}, be incorporated into the international system of physical units" (Contopoulos and Jappel 1974). Increasingly, the unit of Jansky has been used not only in the radio part of the spectrum, but at IR, optical, and X-ray wavelengths as well.

1.2   EARLY FOLLOW-UP TO JANSKY'S DISCOVERY

Since Jansky did not follow up his historic discovery, it was left to others to exploit the new window on the Universe that Jansky had opened. At the University of Michigan, John Kraus (1984) and his colleague Arthur Adel tried to detect the Sun at 1.5 cm wavelength just a few months after Karl's *New York Times* announcement. About the same time, Caltech Physics Professor Gennady Potapenko read about Jansky's work, and in October 1933, just six months after Jansky's *New York Times* announcement, Potapenko gave a talk at the Caltech Astronomy and Physics Club on "The Work of the Bell Laboratories on the Reception of Shortwave Signals from Interstellar Space."

In the spring of 1936, Potapenko and his student Donald Folland tried to reproduce Jansky's work, using first a pair of small loop antennas on the roof of the Caltech Physics Lab. Their receiver was tuned to 20.55 MHz (14.6 meters), close to the frequency used by Jansky, but due to ignition noise from passing automobiles they did not obtain any useful data. To get away from the noise of Pasadena, they moved their experiment out to the nearby Mojave Desert, where they fastened one end of a 35 foot wire to a 25 foot mast. One person walked the slanted wire around the pole to exploit the directivity of the arrangement while the other took data. Later Folland returned to his home in Utah, and in the summer of 1936 repeated the experiment. According to Jesse Greenstein,[49] Potapenko and Folland were able to detect a maximum in the Sagittarius region and later a second maximum in Cygnus.

Based on the success of their simple experiment, Potapenko, along with Fritz Zwicky, proposed constructing a rhombic antenna on a rotating mount designed by Russell Porter. Zwicky (1969, p. 90 and 91) estimated the cost to be about $200, although Greenstein later told Reber that it was more like $1000.[50] Zwicky and Potapenko tried to get funding from Caltech for their venture but were apparently turned down by Caltech President Robert Millikan, who, moreover, discouraged Potapenko from publishing his results.[51] There is no evidence that either Potapenko, Russell, or Folland ever returned to radio astronomy, although Zwicky, who remained at Caltech as a Professor of Astrophysics until his retirement in 1968, later maintained informal contact with the young Caltech radio astronomers and students. Considering that Caltech, with its strong basic research background and significant technical, academic, and financial resources, either did not chose to build on Jansky's discovery, or did not have a clear vision of what to do, it is perhaps not surprising that the telephone company also did not seize the initiative.

Around the same time, according to Grote Reber (1982), in 1936, Fred Whipple, at Harvard, "considered doing a test to confirm Jansky's discovery." His innovative plan was to put outriggers on the dome of the Harvard 60 inch telescope, then string wires around the ends of the outriggers as a rhombic antenna. When the dome was rotated, the antenna would scan around the horizon. However, the HCO Director, Harlow Shapley, was reluctant to support Whipple's proposed initiative and Whipple never pursued his plan.

Sullivan (2009, p. 113) discusses 1940 observations by John DeWitt, who had previously worked at Bell Labs and who, working alone, was able to detect galactic radio emission at 111 MHz using a simple rhombic antenna.[52] About the same time, Kurt Franz, working at the German Telefunkenen laboratories noticed an increase in the noise of his directional navigational system whose intensity shifted by four minutes a day. Franz (1942) knew about Jansky's work, and correctly realized that he had detected radio noise from the Galaxy.

1.3   Grote Reber and Cosmic Static[53]

Aside from the short experiments by Friis, by Potapenko and Folland, DeWitt, and Franz, the only known attempt to continue or expand on Jansky's discovery was by Grote Reber (Fig. 1.5), working by himself in Wheaton, Illinois. Reber had graduated in 1933 from the Armour Institute of Technology (now the Illinois Institute of Technology) with a degree in electrical engineering, specializing in the fledgling fields of electronics and communications. After graduation, he held a series of jobs with various Chicago companies, including

Fig. 1.5 Grote Reber in 1975 during one of his visits to Green Bank. Credit: NRAO/AUI/ NSF

General Household Utilities (1933–1934), the Stewart-Warner Corporation (1935–1937), the Research Foundation of the Armour Institute of Technology (1939), and finally the Belmont Radio Corporation. Initially, he worked on developing broadcast receivers, but later worked on military electronics. His starting salary after graduation was $25 per week. Reber enjoyed telling how his parents forgot to name him, so his birth certificate merely gives his name as "Baby Reber." Although he was called Grote by his parents, it wasn't until he was 20 years old that he officially had his name verified on a revised birth certificate by the authority of the Cook County Clerk, Richard E. Daley, who later became the infamous major of Chicago.

Reber (1958) later related that he had read Jansky's papers in the *Proceedings of the Institute of Radio Engineers* and had listened intently when Jansky's "star noise" was rebroadcast by the NBC Blue network. When he was only 16 years old, he obtained his amateur radio license, W9GFZ, signed by then Secretary of the Interior, Herbert Hoover. Reber (1958) recalled that in the late 1920s and 1930s, he noticed that if he connected an antenna to his receiver, the noise level would increase when the various receiver stages were tuned to the same frequency, but not when the antenna was disconnected. Probably he, as well as other radio amateurs, had detected Karl Jansky's galactic radio noise at 10 m wavelength, but did not realize this until many years later.

After contacting more than 60 countries with his amateur radio station, Reber was looking for new challenges. He was intrigued by the concept of cosmic radio emission, and in 1933, he wrote to Jansky to get more information about his work and to see if he could come to Bell Labs to work with Jansky,[54] but he was surprised and disappointed to learn that Bell Labs did not plan any further work in this area. Reber then contacted various observatories and university departments to see what they were doing, but like Jansky, he found little interest among the astronomers of the time who were busy with their own projects. He tried to interest Otto Struve and other astronomers at Yerkes Observatory, but they also showed little enthusiasm. As Reber later described it, "The astronomers were afraid, because they didn't know anything about radio, and the radio people were not interested, because it was so faint it didn't even constitute an interference—and so nobody was going to do anything. So I thought, well if nobody is going to do anything, maybe I should do something." [55]

Working by himself for nearly a decade, Grote Reber relentlessly pursued his own investigations of Jansky's fundamental discovery and set the stage for the extraordinary developments in radio astronomy which occurred over the next half a century. Reber did not wish to merely confirm Jansky's work but wanted to address two fundamental questions that still motivate radio astronomers today: "How does the intensity at any wavelength change with position in the sky," and "How does the intensity at any position change with wavelength?" (Reber 1958, 1982). He recognized that the then conventional wire arrays were effectively monochromatic, so borrowing techniques used in optical astronomy, as he later related, "I consulted with myself and decided to build a dish."[56] To supplement his background in engineering and to enhance his understanding of optics and astronomy, he took classes at the University of Chicago, including a course in astrophysics from Philip Keenan. As part of the requirements for Keenan's class, Reber prepared a survey paper titled "Long Wave Radiation of Extraterrestrial Origin," in which he discussed the results of Jansky, Friis and Feldman, and Potepenko.[57]

Cosmic Static Reber first experimented with a paraffin lens, but found that it was too heavy and too unwieldy.[58] During the summer of 1937, he took leave from his Chicago job, and using his own funds, designed and built a 32 foot parabolic transit dish in a vacant lot next to his mother's house. Except for the galvanized iron reflecting surface and fasteners, Reber constructed his antenna entirely out of wood. Like Jansky, Reber made use of scrapped parts from an old Model T truck as part of the elevation drive system. Curious neighbors could only speculate about the purpose of the unfamiliar structure rising in the small town of Wheaton, but Reber's mother found it a convenient place to hang her wash. Before her marriage, his mother, Harriet Grote, was an elementary school teacher. Among her seventh and eighth grade students at Longfellow School in Wheaton were Edwin Hubble and Red Grange, later to become a legendary football hero. Reber recalled that, as a teenager, Grange delivered ice to their home, and later he corresponded with Hubble to question the interpretation of redshifts.[59]

In April 1937, Reber wrote to Jansky asking about the gain of Jansky's equipment so that he could better estimate the sensitivity he would need to detect interstellar radio emission.[60] Using his experience and skills as an electrical engineer and radio amateur, he designed, built, and tested a series of radio receivers which he placed at the focal point of his antenna, with the connecting wires running through a coal chute to his observing room in the basement of his mother's house. Although Jansky's work was carried out at a wavelength of 15 m (20.6 MHz) in the short wavelength band, Reber initially decided to observe at a much shorter wavelength of 9 cm (3300 MHz) which, at the time, was the shortest feasible wavelength for existing technology. At 9 cm, he would get better angular resolution than Jansky had,[61] and he also expected that, following the Rayleigh-Jeans radiation law, the celestial radio noise would be very much stronger at the shorter wavelength.[62] His receiver, which he installed at the focal point of his antenna, used a homemade crystal detector followed by an amplifier.

By the spring of 1938 his antenna and 9 cm receiver were completed, but Reber was unable to detect any radio noise from the Galaxy, from several bright stars, nor from the Sun, the Moon, or the nearby planets. Although his observations gave negative results, Reber (1958, 1982) was able to draw the important conclusion that "the celestial radiation did not conform to the Rayleigh-Jeans Law." He rebuilt his receiver operating at the longer wavelength of 33 cm (910 MHz) where more sensitive and more stable instrumentation was available, but still he had no success. Undaunted, he built a new

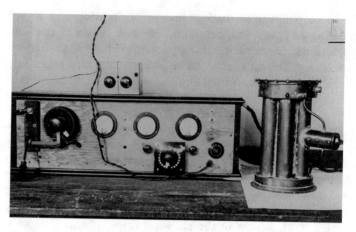

Fig. 1.6 Some of Reber's Wheaton equipment used to study cosmic radio emission. On the left is his receiver monitor and control system. On the right is his 160 MHz amplifier. Credit: NRAO/AUI/NSF

receiver to operate at 1.9 m (160 MHz), where he used a newly developed RCA tube as a radio frequency amplifier to give better sensitivity (Fig. 1.6). Finally, in the spring of 1939, just a few months before the start of World War II, on his first night of observing with his newest receiver, Reber succeeded in detecting Jansky's galactic radio noise at 1.9 m wavelength, which he called *cosmic static*.

While working to develop military electronics in Chicago, Reber continued to make 1.9 meter observations at home in Wheaton. Attempts to detect radio emission from a few bright stars such as Vega, Sirius, and Antares, or Mars as well as the Sun were unsuccessful, and he concluded that there was little correspondence between the brightness of the sky at radio and optical wavelengths. Grote's younger brother, Schuyler, who was a student at the Harvard Business School, put Grote in contact with Fred Whipple and Harlow Shapley at the Harvard College Observatory.[63] Although Whipple expressed interest in Reber's accomplishments,[64] Shapley, who had already been in contact with Jansky several years earlier, remained reluctant to get involved with something that no one at Harvard knew anything about, and claimed that they could not start any new activities as they were already over-committed to other programs.

Reber submitted the results of his findings in a short paper titled "Cosmic Static" to the *Proc. IRE* where Jansky had published most of his pioneering papers. However, also wanting to reach out to astronomers, Reber sent a similar paper with the same title to the *Astrophysical Journal (ApJ)*. The editors of both journals questioned the validity of his interpretation of Jansky's results, although the *Proc. IRE* promptly accepted the paper for publication in the full form as submitted (Reber 1940a). The second paper was received with skepticism by *Astrophysical Journal* editor, Otto Struve, who asked Bart Bok from Harvard to act as the referee. According to Jesse Greenstein, by then a young astronomer at the Yerkes Observatory, since Reber had no academic connection and unclear credentials, his paper produced a flurry of excitement at the *Astrophysical Journal* editorial offices at the Yerkes Observatory located north of Chicago.[65] Reber (1982) later commented that since the astrophysicists didn't understand how the radio waves could be generated, they felt that "the whole affair was at best a mistake and at worse a hoax."[66] At various times, Bart Bok, Otto Struve, Chandrasekhar, Philip Keenan, Jesse Greenstein, and Gerard Kuiper traveled to Wheaton to evaluate Reber's radio observations and equipment and also to evaluate Reber. As later claimed in a footnote by Reber (1982),

> Otto Struve didn't reject my 160 MHz paper. He merely sat on it until it got moldy. I got tired of waiting, so I sent some other material to the Proceedings of the IRE. It was published promptly in the February, 1940 issue [Reber 1940a]. From a much slower start, this beat the ApJ by four months. During the early days of radio astronomy, the astronomy community had a poor track record. The engineering fraternity did much better!

Bart Bok cautioned Struve that he could not afford to turn down the paper because it might "be a great success" (Levy 1993, p. 45). Following an exchange of correspondence with Philip Keenan, Reber's paper was finally published as a short note in the *ApJ* (Reber 1940b) along with the companion paper by Henyey and Keenan (1940) that discussed Jansky and Reber's data in terms of free-free emission from interstellar ionized hydrogen. In a later paper, sent to *Proc. IRE*, Reber (1942) published a more detailed report of his observations along with an extensive technical description of his instrumentation and a discussion of the impact of automobile ignition noise. As in his two previous papers, as well as those to follow, Reber again used the title, "Cosmic Static."

Among the few traditional astronomers who paid serious attention to Reber were Bengt Strömgren, who was then visiting Yerkes from Denmark, Otto Struve, Bart Bok, and Jesse Greenstein. Greenstein, also had been fascinated by Jansky's discovery of cosmic radio noise, and following his visit to inspect Reber's equipment, Greenstein and Reber became "moderately good friends."[67] Grote Reber had forged the first lasting links between radio scientists and astronomers.

In 1940 with the encouragement of Otto Struve and Jesse Greenstein, Reber tried to negotiate with the University of Chicago and the Office of Naval Research (ONR) to move his antenna to a quieter site at the McDonald Observatory in Texas. But they could not agree on how to recover the cost of moving the antenna and operating a radio observatory in Texas. Greenstein and Struve suggested that Grote receive an appointment at the University of Chicago, so the university could administer the program and collect overhead costs from ONR. Reber insisted on preserving his independence and was not interested in working for the university. He explored the possibility of continuing his astronomy research while remaining an employee of his company, which he proposed would administer ONR funding, but this never came to fruition.

Following the onset of World War II, Reber worked for a limited time at the Naval Ordnance Laboratory in Washington on the electronic protection of naval vessels. Due to a chronic hearing impairment, he was exempt from military service, and in 1943 Reber returned to Wheaton, to his job in Chicago, and to continue his radio astronomy investigations. Encouraged by his earlier success, he purchased a chart recorder to relieve him of the task of writing down the receiver output every minute. After making further improvements to his receiver and feed system, Reber went on to systematically map the 160 MHz cosmic static. In order to keep the cost down, Reber's homebuilt antenna was limited to motion in elevation only. He laboriously observed the entire sky visible from Wheaton by changing the elevation of the antenna each day and letting the rotation of the earth scan the sky. Automobile ignition noise interfered with Reber's measurements, so he observed only at night. In the daytime Reber returned to his job designing broadcast radios at Stewart Warner in Chicago, where he commuted by train. The train journey to Chicago took one hour each way.[68] Upon returning home, Reber would catch a few hours' sleep each evening before returning to his night's observing. On weekends, he analyzed his data, and converted his fixed elevation scans to a two dimensional map of the sky which he published in the *Astrophysical Journal* (Reber 1944).

Reber's maps (Fig. 1.7) clearly showed the pronounced maxima at the galactic center and what were later recognized as the Cygnus A/Cygnus X complex of sources, as well as the Cas A radio source (Reber 1944). As a result of his job as a radio engineer, Reber had access to state-of-the-art test equipment and the latest microwave vacuum tubes. In order to improve his angular resolution, he built new equipment to work to a shorter wavelength of 62 cm (480 MHz). Over a 200 day period in 1946 he repeated his observations at the

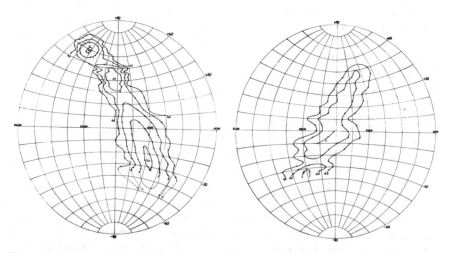

Fig. 1.7 Reber's (1944) contour maps of the 160 MHz radio emission from the Milky Way Galaxy. Contours are shown as a function of right ascension and declination. Credit: NRAO/AUI/NSF

shorter wavelength producing a more detailed map of the galactic radio emission which now also showed what was later recognized as the strong radio galaxy Cygnus A (Reber 1948a).

In his classic paper in the *ApJ* (Reber 1944), and later in the *Proc. IRE* (Reber 1948a), he noted that his radio maps trace the distribution of material in the Milky Way, commented on the evidence for spiral arms, and confirmed that the strongest radio noise was coming from the center of the Galaxy. He noted, with surprise and disappointment, that the 62 cm radiation was weaker than at 1.9 meters, indicating that the celestial radiation was nonthermal in origin. Later he remarked, "If the data doesn't fit the theory, change the theory not the data."[69]

In 1947, Reber and Greenstein (1947) published what became the first review of radio astronomy. It included not only summaries of the pioneering investigations by Jansky and by Reber, but also covered the emerging postwar results now coming from radio scientists in England and Australia (Sect. 2.1). Trying to reach a broader audience, the following year Reber published an account of "Cosmic Radio Noise" in the semi-popular magazine *Radio-Electronic Engineering* (Reber 1948b), followed by papers in *Sky and Telescope* (Reber 1949a), *Scientific American* (Reber 1949b), and in the popular *Leaflets of the Astronomical Society of the Pacific* (Reber 1950).

1.4   IMPACT OF KARL JANSKY AND GROTE REBER

Karl Jansky was the first person to look at the Universe outside the traditional visual wavelength band. Following the initial flurry of public interest and activity resulting from the 1933 *New York Times* article and *NBC* broadcast, there was little reaction or encouragement from the astronomical community. As evidenced by Potapenko, Folland, Zwicky, Greenstein, and Whipple, and perhaps Jansky himself, it was apparently not so much a lack of interest, but no one seemed to know what to do next. The gap between the radio scientists and the astronomers was just too great. Moreover, at the time the astronomical community was preoccupied with Edwin Hubble's announced expansion of the Universe and the need for larger optical telescopes to study more distant galaxies. Jansky's reports of radio emission from the Galaxy were received with interest, but did not really didn't fit into the then current mainstream astronomy. No one knew what to do next until Reber demonstrated radio observations could reveal new information about the Galaxy.

As John Pierce wrote on the occasion of the 1998 commemoration of the Jansky Memorial at Holmdel, "Jansky's work shows that very important phenomena can be disregarded when they don't find a niche in the science of their times."[70] As a result of the Depression, the threat of war, a chronic and ultimately fatal disease, and the commitment of AT&T to its commercial and defense related obligations, Jansky spent the rest of his short career working in other areas. He became a recognized expert on noise, and was decorated for his defense work. Jansky's seminal papers in *Proc. IRE* were widely read throughout the world. In his third IRE paper (Jansky 1935) he commented, "this star static ... puts a definite limit upon the signal strength that can be received from a given direction at a given time and when a receiver is good enough to receive that minimum signal it is a waste of money to spend any more on improving the receiver." Later, scientists working on the development of radar facilities in the UK, as well as in Germany, recognized that their inability to lower the noise temperature of their meter wave radar, navigational, and communication systems was due to the Galactic radio emission which they referred to as "Jansky noise" (Lovell 1984).

Although he operated outside the mainstream of the astronomical community and often ridiculed conventional research, Grote Reber, unlike Karl Jansky, was recognized by the astronomy profession with most of its major prizes, including the Catherine Bruce Medal of the Astronomical Society of the Pacific, the prestigious Elliot Cresson Medal of the Franklin Institute, and the 1962 Russell Prize of the American Astronomical Society. Reber received an honorary Dr. Sc. Degree from Ohio State University, and in 1999 he was named by the Illinois Institute of Technology as a Man of the Millennium. In 1987, he was inducted into the DuPage County (Illinois) Heritage Gallery Hall of Fame, and was inducted posthumously to the National Inventors Hall of Fame in 2013.

Grote Reber was the world's first radio astronomer, and for nearly a decade the only person in the world devoting significant effort to this new field of astronomy. He went to great effort to demonstrate the importance of his work to the astronomical community. His maps of the radio emission from the Milky Way and his report of intense nonthermal radio emission from the Sun provided much of the incentive for the dramatic growth in radio astronomy following the end of World War II and also stimulated the theoretical research which led to Henk van de Hulst's prediction of the 21 cm hydrogen line. His 32 foot home-built radio telescope was the largest parabolic dish ever built at that time, and his development of focal plane feeds and receivers set the stage for later generations of radio telescopes and the ubiquitous home satellite receivers.[71] His innovative receiver designs became known to British radar workers and were implemented in World War II radar systems.[72]

Reber also had another, perhaps less well recognized, impact on radio astronomy. Of the many astronomers who were first exposed to radio astronomy through early visits to Wheaton, three went on to play major roles in the future development of the field. Otto Struve would become the first director of NRAO; Bart Bok, perhaps influenced by his student, Jesse Greenstein, began the radio astronomy program at Harvard where many of the early NRAO staff got their start in radio astronomy; and Greenstein himself went on to start the radio astronomy group at Caltech. Both Greenstein and Bok were also to play major roles in the creation of NRAO (Chap. 3). Later, Reber's contacts with science policy leaders such as Lloyd Berkner, Vannevar Bush, and Merle Tuve may have helped to stimulate their interests in radio astronomy and their contributions to the establishment of and large investment in radio astronomy in Australia, at Caltech, and at NRAO.

Although Karl Jansky was the first to detect cosmic radio emission, it was Grote Reber, who through his innovative experiments, forceful personality, and stubborn persistence finally convinced astronomers that radio astronomy might be important, thus opening a new window on the Universe. He worked alone in a previously unexplored part of the electromagnetic spectrum, designing and building his own equipment, and he was surely the only astronomer, perhaps the only scientist, in the modern era to accomplish so much while working alone as an amateur.

Over a period of less than a decade, working alone and only part time, Grote Reber established the nonthermal nature of the galactic radio emission, recognized that the radio sky was very different from the visual sky, based on his radio maps speculated on the spiral arm nature of the Galaxy, and published the first observations of radio emission from the Sun and later the remarkably intense solar radio emission associated with the active Sun. He showed, for the first time, that radio observations were more than a curiosity. He was the first to appreciate the potential of shorter wavelengths for radio astronomy and was the first to introduce the parabolic dish for radio astronomy. Throughout his life, he argued in support of long wavelength radio astronomy and the development of phased arrays over expensive steerable dishes. Microwave technology had not yet caught up with his ambitions. It would take the wartime development of microwave electronics to open up the microwave, then later millimeter-wave, bands for radio astronomy, and it would take the later development of high speed digital electronics before radio astronomers again seriously considered observing the meter and decameter wavelength sky.

Both Jansky and Reber made the profound observation that the radio sky is very different from the visual sky, foreshadowing the remarkable radio astronomy discoveries of the next half century. Ironically, after the end of World War II, it was in the UK and Australia that former radar scientists followed up on Jansky and Reber's observations of galactic and solar radio emission. In the US the few radio astronomy programs were largely driven by Cold War defense interests and, until the NSF initiative to develop a national radio astronomy facility, the embryo US radio astronomy programs were largely funded by the military – the Office of Naval Research (ONR) or the Air Force Office of Scientific Research (AFOSR).

NOTES

1. This section is based in part on correspondence between Karl and his parents held at the University of Wisconsin Archives, from Jansky's notes and reports held in the Bell Labs Archives, from other Jansky family correspondence, and from discussions and correspondence with former Bell Labs Scientist, J.A. (Tony) Tyson, as well as with Karl's wife, children, and sister. We are grateful to David Jansky who has kindly made family correspondence and documents available to us.

2. Jansky did not receive his Master's degree until 1936, after submitting a thesis based on his work at Bell Labs.

3. H. Friis confirmed the influence of C.M. Jansky, Jr. in hiring Karl at Bell Labs during his 1965 interview with R. Kestenbaum, NAA-KGJ, Bell Labs Oral Interviews, and also in a 22 November 1955 letter to J.R. Howland, NAA-KGJ.

4. KGJ to parents, 5 May 1933, NAA-KIK, Open Skies, Chap. 1. Karl and his wife Alice regularly corresponded with Karl's parents. As a Professor of Electrical Engineering at the University of Wisconsin, Karl's father was interested in Karl's work, which Karl carefully described in his letters. Although addressed to Karl's mother and father as well as his younger siblings still living at home, Karl's technical remarks were clearly directed to his father. When Karl's parents sold their home in Madison, the new owners found these letters in the attic, and, aware of their historical value, donated them to the University of Wisconsin Archives. We are indebted to the University Archivist, Bernard Schermetzler, for making these letters available.

5. Jansky's notebook and his work reports were located in the Bell Labs Archives by J.A. Tyson, who kindly made copies available to the authors. NAA-KIK, Open Skies, Chap. 1.

6. KGJ to parents, NAA-KIK, Open Skies, Chap. 1.

7. Edmond Bruce was a Member of the Bell Labs Technical Staff who had invented the antenna array which bears his name. There is some evidence that he had independently detected excess noise which appeared to be extraterrestrial, but he never followed this up.

8. The classical Bruce Array was bidirectional. By adding a parasitic reflector element, Jansky's array was omni-directional. The array had an azimuth beamwidth of 24 degrees and elevation beamwidth of 36 degrees (Sullivan 1978).

9. Jansky's notebook entry, pg. 81. Bell Laboratory Archives, Photocopy: NAA-KIK, Open Skies, Chap. 1. Jansky later found interfering signals at 14.5 meters and changed his observing wavelength to 14.6 meters.

10. KGJ to his parents, 5 December 1930, NAA-KIK, Open Skies, Chap. 1.

11. KGJ to his parents, 27 November 1932, NAA-KIK, Open Skies, Chap. 1.

12. Karl's older brother Moreau later recalled that he was probably the chair of the session. R. Kestenbaum interview with C.M. Jansky, Jr., 11 May 1965, NAA-KGJ, Bell Labs Oral Interviews. See also note 29.

13. KGJ to parents, 18 January 1933, NAA-KIK, Open Skies, Chap. 1.

14. KGJ to parents, 31 July 1932, NAA-KIK, Open Skies, Chap. 1.

15. KGJ to parents, 5 May 1933, NAA-KIK, Open Skies, Chap. 1. We are indebted to J. Goldbaum, NASEM Archivist, for providing us with a copy of the 27 April 1933 URSI agenda.

16. Russell Ohl, oral history conducted in 1965 by F. Polkinghorn, IEEE History Center, Hoboken, NJ; R. Ohl interview by L. Hoddeson, 20 August 1976, AIP.

17. Transcript of the 15 May 1933 NBC WJZ interview with Karl Jansky, NAA-KIK, Open Skies, Chap. 1.

18. KGJ to parents, 10 May 1933, NAA-KIK, Open Skies, Chap. 1.

19. C.M. Jansky Jr. to C.M. Jansky, 16 May 1933; KGJ to parents, 25 May 1933, NAA-KIK, Open Skies, Chap. 1.

20. KGJ to his parents, 10 May 1933, NAA-KIK, Open Skies, Chap. 1.

21. John Kraus (1984) later reported that less than two dozen people heard Jansky's historic talk.
22. KGJ to his parents, 22 January 1934, NAA-KIK, Open Skies, Chap. 1.
23. Copies of Sir Edward Appleton's 1948 URSI Presidential Address were found in the papers of C.M. Jansky, Jr., donated to the NRAO/AUI Archives by Karl's son David. NAA-KGJ, Correspondence, Family Correspondence.
24. The MUSA array near Manahawkin, New Jersey was used together with an identical array in the UK by President Roosevelt to communicate with British Prime Minister Winston Churchill during WWII.
25. R. Kestenbaum interview with A. Beck, 10 February 1965, NAA-KGJ, Bell Labs Oral Interviews.
26. GR to G. Southworth, 15 April 1960, NAA-GR, General Correspondence. https://science.nrao.edu/about/publications/open-skies#section-1
27. J.C. Schelleng to G. Du Shane, 16 October 1956, and 8 November 1956, NAA-KIK, Open Skies, Chap. 1.
28. H. Friis to C.M. Jansky, Jr., 23 January 1958, NAA-KIK, Open Skies, Chap. 1.
29. In his 1965 interview with Ray Kestenbaum and in the referenced publications, Friis firmly and repeatedly held the position that Jansky never asked to continue his star noise work, but at the same time he noted that had Karl lived longer, he surely would have received the Nobel Prize for his discovery of galactic radio emission. NAA-KGJ, Bell Labs Oral Interviews.
30. C. M. Jansky, Jr. to Alice Jansky Knopp, 7 February 1958, NAA-KIK, Open Skies, Chap. 1 (Alice had remarried Otto Knopp).
31. Alice Jansky Knopp to C.M. Jansky, Jr., 11 February 1958, NAA-KIK, Open Skies, Chap. 1.
32. C.M. Jansky, Jr. to Alice Jansky Knopp, 26 March 1958, NAA-KIK, Open Skies, Chap. 1.
33. Audio recordings of these interviews were retained at the Bell Labs Archives. In 1997, J.A. Tyson kindly made copies available to NRAO. With Tyson's permission, the recordings have been transcribed and posted on the Web, See NAA-KGJ, Bell Labs Oral Interviews.
34. G. Southworth to R. Kestenbaum, 10 March 1965, NAA-KIK, Open Skies, Chap. 1.
35. L. Espenschied to G. Southworth, 21 February 1965, NAA-KIK, Open Skies, Chap. 1.
36. G. Southworth to L. Espenschied, 23 February 1965, NAA-KIK, Open Skies, Chap. 1.
37. R. Ohl, 1976, op. cit.
38. C.M. Jansky, Jr. to Karl, Alice and Anne Moreau, 9 May 1933, NAA-WJS, Cosmic Noise, Working Papers, K.G. Jansky.
39. KGJ to parents, 10 June 1933, NAA-KIK, Open Skies, Chap. 1.
40. KGC to parents, 5 May 1933, NAA-KIK, Open Skies, Chap. 1.
41. KGJ to his parents, 22 January 1934, NAA-WJS, Cosmic Noise, Working Papers, K.G. Jansky.
42. G. Reber to J. Pfeiffer, 31 January 1955, NAA-GR, General Correspondence I.
43. R. Ohl, 1975 op. cit.
44. Cyril Jansky to Karl, Alice and Anne Moreau, 9 May 1933, NAA-WJS, Cosmic Noise, Working Papers, K.G. Jansky.
45. KGJ to his parents, 22 January 1934, op. cit.

46. The first suggestions that the nonthermal cosmic radio emission was due to synchrotron radiation was by Alfvén and Herlofson (1950) and Kippenheuer (1950), but it was the quantitative papers by Ginzburg (1951) and Shklovsky (1952) that convinced the astronomical community of its relevance.

47. Alice Jansky Knopp to Sir Edward Appleton, 17 April 1950, NAA-KGJ, Correspondence, Family Correspondence. The letter indicates that Alice had previously written to Appleton while Karl was still alive to promote his case for the Nobel Prize.

48. Martin Ryle and Anthony Hewish in 1974, for aperture synthesis and pulsars respectively; Arno Penzias and Robert Wilson in 1978, for their discovery of the cosmic microwave background; Russell Hulse and Joseph Taylor in 1993, for their discovery of a new type of pulsar that opened up new possibilities for the study of gravitation; and John Mather and George Smoot in 2006, for their discovery of the blackbody form and anisotropy of the cosmic microwave background.

49. JLG to GR, 13 March 1962, NAA-GR, General Correspondence I. https:// science.nrao.edu/about/publications/open-skies#section-1

50. Ibid.

51. Oral interviews WTS with Gennady Potapenko, 26 December 1974, 21 August 1975, NAA-WTS, Working Files, Interviewees, Potapenko. https://science. nrao.edu/about/publications/open-skies#section-1

52. Additional material on DeWitt's measurements may be found in Sullivan's papers in the NRAO Archives, NAA-WTS, Working Papers, DeWitt.

53. This section on Grote Reber is adapted from Kellermann (2004), with permission from the Astronomical Society of the Pacific, and is based in part on Reber's (1958) paper describing his early experiments in Wheaton. In his paper, which was published in a special edition on radio astronomy in the *Proc. IRE*, Reber describes the details of his antenna, feeds, and the receivers that he constructed and used to study celestial radio emission during the period 1937–1947.

54. GR oral interview with KIK, 13 June 1994, NAA-KIK, Oral Interviews.

55. Alberta Adamson interview with GR at Wheaton History Center, 19 October 1985, transcribed at NRAO, NAA-GR, Oral History Interviews.

56. Ibid.

57. NAA-GR, Notes and Papers, 1938. https://science.nrao.edu/about/publica-tions/open-skies#section-1

58. GR to R. Langer, 7 July 1937, NAA-GR, General Correspondence. https:// science.nrao.edu/about/publications/open-skies#section-1

59. GR to Edwin Hubble, 5 January 1939, NAA-GR, General Correspondence. https://science.nrao.edu/about/publications/open-skies#section-1

60. GR to KGJ, 26 April 1937, NAA-GR, General Correspondence. https://sci-ence.nrao.edu/about/publications/open-skies#section-1

61. The angular resolution depends on the size of the telescope aperture measured in wavelengths, so for a fixed antenna size the resolution is better at shorter wavelengths.

62. According the Rayleigh-Jeans radiation law the intensity of the thermal radia-tion from a hot body is inversely proportional to the square of the wavelength. So by going to the shortest wavelengths technically feasible, Reber hoped to achieve the best possible angular resolution and also anticipated that for any given temperature that the radio radiation would be about 25,000 times stron-

ger than that detected by Jansky. Since the galactic radiation is weaker at the shorter wavelength, it is considered "non-thermal."

63. S. Reber to GR, 16 December 1938, NAA-GR, General Correspondence I. https://science.nrao.edu/about/publications/open-skies#section-1
64. F. Whipple to GR, 22 December 1938, NAA-GR, General Correspondence I. https://science.nrao.edu/about/publications/open-skies#section-1
65. JLG, *I Was there in the Early Years of Radio Astronomy*, unpublished and undated manuscript, NAA-KIK, Open Skies, Chap. 1.
66. GR oral interview with Alberta Adamson, Wheaton History Center, 19 October 1985, NAA-GR, Oral Interviews.
67. JLG unpublished manuscript, op. cit.
68. In 2019, it still takes 51 minutes.
69. KIK Oral interview with GR, 25 May 1991, NAA-KIK, Oral Interviews.
70. J. Pierce to J.A. Tyson, 23 April 1998, *Dedication of the Jansky Monument* (Holmdel: Bell Labs).
71. Development of the German Wurzburg-Riese 7.5 meter (25 feet) radar antennas used during WWII probably started before 1940, but of course this was not known to Reber.
72. Private communication, Sir Bernard Lovell to KIK, May 2007.

BIBLIOGRAPHY

REFERENCES

Alfvén, H. and Herlofson, N. 1950, Cosmic Radiation and Radio Stars, *Phys. Rev.*, **78**, 616

Beck, A. 1984, Personal Recollections of Karl Jansky. In *Serendipitous Discoveries in Radio Astronomy*, ed. K.I. Kellermann and B. Sheets (Green Bank: NRAO/AUI), 32

Bown, R. 1927, Transatlantic Radio Telephony, *Bell Laboratory Technical Journal*, **6**, 248

Contopoulos, G. and Jappel, A. eds. 1974, *Transactions of the IAU XV – B* (Dordrecht: Reidel)

Edmondson, F. 1956, Review of The Changing Universe. The Story of the New Astronomy, *Science*, **124**, 541

Ewen, H.I. and Purcell, E.M. 1951, Observation of a Line in the Galactic Spectrum, *Nature*, **168**, 356

Franz, K. 1942, Messung der Empfangempfindlichkeit, *Hochfrequenztechnik. und. Elektroakustik*, **59**, 143

Friis, H.T. 1965, Karl Jansky: His Career at Bell Telephone Laboratories, *Science*, **149**, 841

Friis, H.T. 1971, *Seventy-five Years in an Exciting World* (San Francisco: San Francisco Press)

Friis, H.T. and Feldman, C.B. 1937, A Multiple Unit Steerable Antenna for Short-Wave Radio Reception, *Bell Tech J.*, **16**, 337

Ginzburg, V.I. 1951, Cosmic Rays as a Source of Galactic Radio Emission, *Dokl. Akad. Nauk SSSR*, **76**, 377 (in Russian, English translation in W. Sullivan, *Classics in Radio Astronomy* (Dordrecht: Reidel), 93

Henyey, L.G. and Keenan, P.C. 1940, Interstellar Radiation from Free Electrons and Hydrogen Atoms, *ApJ*, **91**, 625

Hubble, E. 1929, A Relation between Distance and Radial Velocity among Extra-Galactic Nebulae, *PNAS*, **15**, 168

Jansky, C.M. Jr. 1957, The Beginnings of Radio Astronomy, *American Scientist*, **45** (1), 5

Jansky, C.M. Jr. 1958, The Discovery and Identification by Karl Guthe Jansky of Electromagnetic Radiation of Extraterrestrial Origin in the Radio Spectrum, *Proc. IRE*, **46**, 13

Jansky, K.G. 1932, Directional Studies of Atmospherics at High Frequencies, *Proc. IRE*, **20**, 1920

Jansky, K.G. 1933a, Radio Waves from Outside the Solar System, *Nature*, **132**, 66

Jansky, K.G. 1933b, Electrical Disturbances Apparently of Extraterrestrial Origin, *Proc. IRE*, **21**, 1387

Jansky, K.G. 1933c, Electrical Phenomena that Apparently Are of Interstellar Origin, *Popular Astronomy*, **41**, 548

Jansky, K.G. 1935, A Note on the Source of Interstellar Interference, *Proc. IRE*, **23**, 1158

Jansky, K.G. 1937, Minimum Noise Levels Obtained on Short Wave Receiving Systems, *Proc. IRE*, **25**, 1517

Jansky, K.G. 1939, An Experimental Investigation of the Characteristics of Certain Types of Noise, *Proc. IRE*, **27**, 763

Kellermann, K.I. 2004, Grote Reber (1911-2002), *PASP*, **116**, 703 (https://doi.org/10.1086/423436)

Kent, G. 1958, A New Window on the Sky, *Readers Digest*, **72**, 91

Kippenheuer, K.O. 1950, Cosmic Rays as the Source of General Galactic Radio Emission, *Phys. Rev.*, **79**, 738

Kraus, J. 1981, The First 50 Years of Radio Astronomy, Part 1: Karl Jansky and His Discovery of Radio Waves from Our Galaxy, *Cosmic Search*, **3** (4), 8

Kraus, J. 1984, Karl Guthe Jansky's Serendipity, Its Impact on Astronomy and its Lessons for the Future. In *Serendipitous Discoveries in Radio Astronomy*, ed. K.I. Kellermann and B. Sheets (Green Bank: NRAO/AUI), 57

Langer, R.M. 1936, Radio Noise from the Galaxy, *Phys. Rev.*, **49**, 209

Levy, D.H. 1993, *The Man Who Sold the Milky Way: a Biography of Bart Bok* (Tucson: University of Arizona Press)

Lovell, A.C.B. 1984, Impact of World War II on Radio Astronomy. In *Serendipitous Discoveries in Radio Astronomy*, ed. K. Kellermann and B. Sheets (Green Bank: NRAO/AUI), 89

Oswald, A.A. 1930, Transoceanic Telephone Service-Short Wave Equipment, *Trans. AIEE*, **49**, 629

Pfeiffer, J. 1956, *The Changing Universe: The Story of the New Astronomy* (New York: Random House)

Penzias, A. and Wilson, R.W. 1965, A Measurement of Excess Antenna Temperature at 4080 Mc/s, *ApJ*, **142**, 1149

Reber, G. 1940a, Cosmic Static, *Proc. IRE*, **28**, 68

Reber, G. 1940b, Cosmic Static, *ApJ*, **91**, 621

Reber, G. 1942, Cosmic Static, *Proc. IRE*, **30**, 367

Reber, G. 1944, Cosmic Static, *ApJ*, **100**, 279

Reber, G. 1948a, Cosmic Static, *Proc. IRE*, **36**, 1215

Reber, G. 1948b, Cosmic Radio Noise, *Radio-Electronic Engineering*, July 1948

Reber, G. 1949a, Galactic Radio Waves, *S&T*, **8**, 139

Reber, G. 1949b, Radio Astronomy, *Scientific American*, **181**, 34

Reber, G. 1950, Galactic Radio Waves, *ASP Leaflet No. 259*

Reber, G. 1958, Early Radio Astronomy in Wheaton, Illinois, *Proc. IRE*, **46**, 15

Reber, G. 1982, A Timeless, Boundless, Equilibrium Universe, *PASAu*, **4**, 482

Reber, G. 1988, A Play Entitled the Beginning of Radio Astronomy, *JRASC*, **82,** 93

Reber, G. and Greenstein, J.L. 1947, Radio Frequency Investigations of Astronomical Interest, *Observatory*, **67**, 15

Shklovsky, I.S. 1952, On the Nature of Radio Emission from the Galaxy, *Astr. Zhur. SSSR*, **29**, 418

Southworth, G.C. 1956, Early History of Radio Astronomy, *Scientific Monthly*, **82** (2), 55

Sullivan, W.T. III 1978, A New Look at Karl Jansky's Original Data, *S&T*, **56**, 101

Sullivan, W.T. III 2009, *Cosmic Noise* (Cambridge: CUP)

Whipple, F.L. and Greenstein, J.L. 1937, On the Origin of Interstellar Radio Disturbances, *PNAS.*, **23**, 177

Zwicky, F. 1969, *Discovery, Invention, Research* (New York: Macmillan)

FURTHER READING

Jansky, C.M. Jr. 1979, My Brother Karl Jansky and his Discovery of Radio Waves from Beyond the Earth, *Cosmic Search*, **1** (4), 12

Jansky, D. 1984, My Father and His Work. In *Serendipitous Discoveries in Radio Astronomy*, ed. K. Kellermann and B. Sheets (Green Bank: NRAO/AUI), 4

Kellermann, K.I. 1999, Grote Reber's Observations on Cosmic Static, *ApJ*, **525**, 37

Kellermann, K.I. 2003, Grote Reber (1911-2002), *Nature*, **421**, 596

Kellermann, K.I. 2005, Grote Reber (1911-2002): A Radio Pioneer. In *The New Astronomy: Opening the Electromagnetic Window and Expanding our View of Planet Earth*, ed. W. Orchiston (Dordrecht: Springer), 43

Kraus, J. 1982, The First 50 Years of Radio Astronomy, Part 2: Grote Reber and the First Radio Maps of the Sky, *Cosmic Search*, **4** (1), 14

Kraus, J. 1988, Grote Reber, Founder of Radio Astronomy, *JRASC*, **82,** 107

Parsons, A.M.J. 1984, Personal Recollections for the Green Bank Symposium. In *Serendipitous Discoveries in Radio Astronomy*, ed. K. Kellermann and B. Sheets (Green Bank: NRAO/AUI), 22

Reber, G. 1948, Solar Intensity at 480 Mc, *Proc. IRE*, **36**, 88

Reber, G. 1955, Fine Structure of Solar Transients, *Nature*, **175**, 78

Sullivan, W.T. III 1984a, Karl Jansky and the Beginnings of Radio Astronomy. In *Serendipitous Discoveries in Radio Astronomy*, ed. K. Kellermann and B. Sheets (Green Bank: NRAO/AUI), 39

Sullivan, W.T. III 1984b, *The Early Years of Radio Astronomy* (Cambridge: CUP)

CHAPTER 2

The Postwar Explosion in Radio Astronomy: The US Falls Behind

During the Second World War, a number of radar scientists independently discovered powerful radio emission from the Sun. Following the cessation of hostilities, and making use of their wartime experience, scientists, mostly at Jodrell Bank and Cambridge in the UK and in Sydney, Australia, used discarded radar systems to further investigate the complex solar radio emission, discovered powerful radio emission from old supernova explosions, and even more powerful radio sources from what later became known as radio galaxies. Encouraged by their early successes with relatively primitive equipment and the potential for new discoveries, scientists in the UK, Australia, the USSR, and the Netherlands developed plans to build more powerful radio telescopes and sophisticated new instrumentation.

In the US, support for radio astronomy was largely driven by Cold War defense concerns, with funding mostly from the Office of Naval Research (ONR) and the Air Force Office of Scientific Research (AFOSR). Radio astronomy projects were begun at several universities, research institutes, and government and military laboratories. Despite the considerable investment in resources under the direction of skilled scientists and engineers, the early American radio astronomy research did not have the same impact as the programs in the UK and Australia.

In this chapter we discuss the postwar explosion in radio astronomy leading to concerns that the United States had fallen behind other countries, namely the UK and Australia, in this rapidly developing field of astronomy with obvious commercial and military implications to the growing Cold War environment. These concerns, whether real, imagined, or invented, would lead to efforts, begun in the mid-1950s, to establish a national radio astronomy facility to compete with the rapidly developing programs in Europe and Australia. Edge (1984), Edge and Mulkay (1976), Elbers (2017), Frater et al. (2017), Kellermann (2012), Orchiston et al. (2007), Robertson (1992), and especially Sullivan (2009) have described these developments in more detail, while more

© The Author(s) 2020, corrected publication 2021
K. I. Kellermann et al., *Open Skies*, Historical & Cultural Astronomy,
https://doi.org/10.1007/978-3-030-32345-5_2

personal accounts are given by Bolton (1982), Bowen (1984, 1987), Bracewell (2005), Christiansen (1984), Denisse (1984), Ginzburg (1984), Graham-Smith (2005), Haddock (1984), Hanbury Brown (1991), Hey (1973), Kraus (1984, 1995), Lovell (1984a, b, 1990), Mills (1984, 2006), Salomonovich (1984), Tanaka (1984), Wang (2009), and in the compilations by Kellermann and Sheets (1984) and Sullivan (1984).

2.1   POSTWAR RADIO ASTRONOMY

By the end of WWII, it was increasingly apparent that the discoveries by Jansky and Reber, as well as the wartime serendipitous discoveries of powerful radio emission from the Sun, presented a new way to study the Universe. Scientists and engineers in a number of countries, particularly in Australia, the UK, The Netherlands, Russia, and in the US turned their attention to following up these opportunities. With a few exceptions, namely in the Netherlands and at Harvard University and Berkeley, these programs were initiated by scientists and engineers with backgrounds in physics and radio science, rather than by astronomers.

The MIT Radiation Laboratory The MIT Radiation Laboratory, known as the "Rad Lab," was a remarkable example of wartime success that employed over 3500 skilled scientists and spent nearly $4 million a month developing and producing radar systems. As the end of WWII approached, Robert (Bob) Dicke, then a young PhD from the University of Rochester, together with other Rad Lab colleagues, conducted a series of groundbreaking experiments that would later have a major impact on the development of radio astronomy. Using a small 1.5 foot (46 cm) dish, Dicke and Berringer (1946) measured the 1.25 cm (24 GHz) radio emission from the Sun and the Moon, and reported black body temperatures of 11,000 K and of 292 K respectively. During a partial solar eclipse on 9 July 1945, they demonstrated that the 1.25 cm emission from the Sun came from the same region as the visual solar disk. These measurements were at by far the shortest wavelength that had been used for radio astronomy. During the same period, Dicke et al. (1946) observed, for the first time, the thermal radio emission from the atmosphere at 1.0 cm, 1.25 cm, and 1.5 cm, confirming that the suspected opacity that had been indicated by erratic 1.3 cm radar operation was due to atmospheric water vapor. They were unable to detect any radio emission from any stars, and also put an upper limit of 20 degrees on any "radiation from cosmic matter," although Dicke did not appreciate its importance until two decades later.

In order to suppress the effect of receiver gain instabilities which could mimic the response of a cosmic source, in all of these studies, Dicke (1946) rapidly switched his receiver between the antenna and a reference source 30 times per second, and measured only the difference signal which was nearly independent of receiver gain fluctuations. This synchronous detection scheme which has come to be known a "Dicke switch" became the basis of all non-interferometric radio astronomy continuum measurements and, in a modified form, spectroscopic observations as well.

This series of observations, which took place during the summer of 1945, led to seven key results: the first radio astronomy observations at short centimeter wavelengths, the first detection of radio emission from the Moon, the first measurements of microwave attenuation by atmospheric water vapor, the development of the Dicke switch, the quantitative development of the fundamental radiometer equation, calibration using a variable temperature resistive load, and the first upper limit on any isotropic cosmic background radiation. While these contributions of Dicke and his colleagues marked the beginning of the postwar era of radio astronomy, at the time they went largely unnoticed by the astronomical community.

Radio Astronomy in the UK Following the end of WWII, radio astronomy in Britain was pursued in three places: at Cambridge University under Martin Ryle, at the University of Manchester's Jodrell Bank Observatory under Bernard Lovell, and at the UK Army Operational Research Group (AORG) under J. Stanley Hey. Ryle, Lovell, and Hey had each been involved in wartime radar research. Only Ryle and Lovell returned to university life after the War, while Hey began a radio astronomy program within the AORG.

Hey's background was in physics, having received a Masters' degree in X-ray crystallography at Manchester University (Hey 1973). He had no previous training in radio engineering prior to a six week military training program, and no background in astronomy. After seven years leading the AORG radio astronomy program at Richmond Park, just 10 miles from downtown London, Hey built a research group at Malvern where he later constructed a two-element variable spacing radio interferometer with baselines up to 1 km.

Lovell returned to the University of Manchester to set up a radar system first to study cosmic rays and then to study the ionization trails left by meteors. To enhance their sensitivity, Lovell and J.A. Clegg built a fixed 218 foot (66 m) reflector at Jodrell Bank near Manchester. Although Lovell planned to use the telescope for a meteor scatter program, he was joined by Robert Hanbury Brown and Cyril Hazard, who used the dish for a variety of radio astronomy programs at 1.89 meters (159 MHz), including observations of M31 (Hanbury Brown and Hazard 1951) and a survey that disclosed 23 discrete radio sources (Hanbury Brown and Hazard 1953). Inspired by the success of the 218 foot dish, Lovell went on to build the iconic fully steerable 250 foot (76 meter) dish, which went into operation in 1957 (Sect. 6.6) following an agonizing engineering and financial near-fiasco that almost sent Lovell to prison (Lovell 1984b, 1990). Later, under the leadership of Henry Palmer, Jodrell Bank radio astronomers also built a series of radio-linked interferometers of ever increasing angular resolution (Sect. 8.1).

Radio astronomy at Cambridge was established under the forceful leadership of Martin Ryle, who, over a period of several decades, guided the development of interferometer systems of ever increasing sophistication and angular resolution, coupled with the formulation of the concepts of aperture synthesis (Sect. 7.1). With these innovative radio telescopes, Ryle and a group of talented students made a series of radio source surveys, including the 1C (Ryle et al. 1950), 2C (Shakeshaft et al. 1955), 3C (Edge et al. 1959), and its revision 3CR (Bennett 1962). The 3CR survey, in particular, became for many years the basis of detailed radio source studies, at least in the Northern Hemisphere.

The Australian CSIRO Radiophysics Laboratory As the War was winding down, E.G. (Taffy) Bowen left the MIT Radiation Lab to take up an appointment at the Australian Commonwealth Scientific and Industrial Research Organization (CSIRO, then called CSIR or Council for Scientific and Industrial Research). Two years later he became Chief of the CSIRO Radiophysics Laboratory, where he began programs in cloud physics and radio astronomy. Joseph L. Pawsey, who had joined the Radiophysics Lab in 1940, led the new radio astronomy program. John Bolton (Kellermann 1996; Robertson 2017), Paul Wild (Frater and Ekers 2012), Bernard (Bernie) Mills (Frater et al. 2017, p. 23), and Wilbur Norman (Chris) Christiansen (Frater et al. 2017, p. 59) all joined the group, and along with Pawsey, began their own independent research programs.

Pawsey (1946), with support from Bolton and Gordon Stanley, observed the Sun from field stations at Collaroy and Dover Heights in the Sydney suburbs, confirming that the intense radio emission from the Sun was associated with sunspot activity. Although their radar antenna did not have sufficient angular resolution to determine the location of the radio emission on the Sun, Pawsey used a trick to enhance the resolution of his simple antenna.

During WWII, shipboard radar operators would sometimes note that when an aircraft was approaching at low elevation, two signals were received, one directly from the aircraft and one reflected off the ocean. The two signals interfered with each other, and as the angle of approach changed due to the changing differential path length, there would be a series of interference maxima and minima.[1] Pawsey exploited this "sea interferometer" effect to study the structure of solar radio emission using a small antenna located on a Dover Heights cliff overlooking the Pacific Ocean (McCready et al. 1947).[2]

Both Christiansen and Wild continued to study solar radio emission. Initially, Wild concentrated on studying the solar dynamic spectra, and established the now standard nomenclature used to classify solar radio bursts. He later built a 96-element circular array to obtain dynamic images of the Sun with a resolution of up to about 2 arcmin (Wild 1967), and went on to become first Chief of the CSIRO Division of Radiophysics and then Chairman of CSIRO. Christiansen built a series of interferometer arrays of increasing sophistication, and developed novel synthesis imaging techniques complementing the development of aperture synthesis by the Cambridge group (Sect. 7.1). Bernie Mills developed an alternate approach to obtain high resolution by building large cross arrays (known as a Mills Cross) with each arm built of many simple dipoles (Mills and Little 1953). John Bolton, assisted by Gordon

Stanley and Bruce Slee, identified the first radio sources with optical counterparts (Sect. 2.3). Together with Stanley, Bolton left Australia in 1955 to start the radio astronomy program at Caltech, only to return six years later to direct the operation of the Parkes 210 foot radio telescope (Kellermann 1996; Robertson 2017). However, when Bowen decided to concentrate the Radiophysics resources on the new Parkes 210 foot radio telescope, Mills and Christiansen left CSIRO to join the schools of physics and electrical engineering at the University of Sydney (Robertson 1992; Frater et al. 2017).

Radio Astronomy in Europe Although recovery from the devastating effects of WWII was slow, modest radio astronomy programs began in the Netherlands and in France using captured German radar systems, particularly the 7.5 meter Würzburg parabolic dishes.

The French radio astronomy program began under the leadership of Jean-François Denisse and Jean-Louis Steinberg, who were later joined by Émile Blum and others. As France was occupied from the early days of WWII, French scientists were somewhat isolated from the technical advances that led to the rapid development of radio astronomy in the "Anglo-Saxon world" (Denisse 1984). As elsewhere, the first programs were in solar radio astronomy, largely because the Sun was the brightest source in the sky and did not require sophisticated equipment. Other observations included a 900 MHz survey of the galactic plane using a 7.5 meter Würzburg antenna (Denisse et al. 1955). Later Lequeux et al. (1959) built a variable spacing interferometer at the Nançay Observatory using two Würzburg antennas with baselines up to 1.5 km, which was used by Lequeux (1962) at 1.4 GHz to observe the structure of 40 discrete sources, many for the first time. The Nançay interferometer observations were among the first to demonstrate the double nature of many extragalactic radio sources. However, with limited sensitivity and lack of phase stability, the Nançay interferometer could not compete with the almost contemporaneous Caltech Owens Valley Interferometer.

Unlike those in other countries, the Dutch radio astronomy program began not with radio physicists, but under the leadership of the well-known astronomer, Jan Oort. Indeed, due to the lack of scientists experienced in radio instrumentation, the Dutch program initially suffered until the electronics engineer C. Alexander (Lex) Muller was recruited from Dutch industry by Oort. As in other countries, the Dutch radio astronomy program was initially focused on the Sun. But for the decade after the 1951 detection of the hydrogen 21 cm line (Sect. 2.4) it was nearly exclusively devoted to 21 cm research, with the goal of using Doppler shifts to determine the kinematics of the Galaxy.

Radio astronomy in Sweden started in 1949 when Olof Rydbeck brought five Würzburgs from the remote Norwegian coast to his Onsala Space Observatory (OSO) which was part of the Chalmers University of Technology. Early research was concentrated on the Sun and 21 cm observations of galactic structure. Following the construction of 25 and 20 meter telescopes, the OSO became one of the European leaders in very long baseline interferometry

(VLBI) (Sect. 8.1) and molecular spectroscopy. Some of the first successful maser amplifiers used for radio astronomy were developed at the OSO. Hein Hvatum, later to become head of NRAO technical programs, began his career at Chalmers under the strong supervision of Rydbeck. Like many of the other pioneers of radio astronomy, Rydbeck's background was in electronics and radiophysics (Radhakrishnan 2006). Although Sweden was neutral during WWII, Rydbeck was one of the leaders in developing the wartime radar defenses needed to maintain Sweden's neutrality.

In Germany, research in radio physics was initially forbidden by the occupying Allied powers but developed starting in the 1950s, especially after 1962 when Otto Hachenberg became director of the University of Bonn's 25 meter Stockert radio telescope (Sect. 9.2).

Radio Astronomy in Japan As related by Tanaka (1984), just as in Germany, the start of radio astronomy in Japan was delayed by their defeat in WWII. Separate projects using relatively simple equipment to study the Sun were started by Minoru Oda, T. Hatanaka, A. Kimpara, Koichi Shimoda, and Haruo Tanaka. Later Tanaka led the development of a five-element interferometer for high resolution solar imaging which was expanded to eight elements in 1954. As pointed out by Ishiguro et al. (2012), perhaps one of the most important Japanese contributions to early radio astronomy was the development by Shintaro Uda and his professor Hidetsugu Yagi of the Yagi-Uda antenna, more commonly known as the Yagi antenna, which was used world-wide in many of the early radio astronomy systems, as well as in a multitude of other short wave communication systems. With the creation of the Nobeyama Radio Observatory in 1970 following the construction of the 84-element heliograph, the six-element millimeter array of precision 10 meter dishes, the 45 meter radio telescope (Sect. 10.5), and the establishment of a vigorous space VLBI (Sect. 8.9) program, Japan became a major player in cosmic as well as in solar radio astronomy, and a pioneer in the development of millimeter wave interferometry (Sect. 10.5).

Radio Astronomy Behind the Iron Curtain As in Western countries, early radio astronomy research in the USSR was primarily the domain of scientists proficient in radio physics and electronics who had developed their skills while working on wartime radar programs. Although postwar support for radio astronomy in essentially all countries had its origin in military radar, in most countries it was implemented in universities and in civilian research laboratories. However, in the USSR radio astronomy remained within military-oriented and tightly controlled laboratories that limited the distribution of research findings. Even when results of radio astronomy investigations were published in the open literature, critical data on the instrumentation and techniques used in the observations were often restricted, leading to skepticism or disregard of Soviet papers by Western scientists. Within the Soviet Union itself, there was little communication among the different radio astronomy groups who were competing for limited resources, and even less contact with the broader international astronomical community.

Starting in 1958, many of the most important Soviet journals were translated into English, but Soviet observational results had little impact outside of the USSR. In contrast, however, the theoretical work of people like Iosef Shklovsky, Solomon Pikel'ner, Vitaly Ginzburg, and Yakov Zel'dovich in Moscow, as well as Viktor Ambartsumian in Armenia, and later their students, Nikolai Kardashev, Igor Novikov, Vyacheslav Slysh, and Rashid Sunyaev, was widely recognized and greatly influenced both theoretical thinking as well as motivating new observational programs in the US, Europe, and Australia. Indeed, the English language translation of Shklovsky's book, *Cosmic Radio Waves* (Shklovsky 1960) was used by generations of students around the world.

Postwar Soviet observational radio astronomy programs were led by S.E. Khaikin (Lebedev Physical Institute), N.D. Papalesky (Lebedev), V.S. Troitsky (Gorky), and V.V. Vitkevich (Lebedev). One of the earliest programs involved a 1947 eclipse expedition to Brazil. Using a 1.5 meter (200 MHz) phased dipole array mounted on the deck of their ship, Khaikin had the ship's captain maneuver the vessel to track the Sun during the solar eclipse and was able to demonstrate that the radio emission came from the much larger coronal region rather than the totally eclipsed Sun (Salomonovich 1984). Starting in 1948, under the leadership of Vitkevitch, Lebedev radio astronomers set up a series of observing stations on the south coast of Crimea, including a refurbished captured German Würzburg-Riese antenna. In 1959, Lebedev constructed a 22 meter (72 foot) precision dish at one of the sites on the shore of the Black Sea that operated at wavelengths as short as 8 mm, and for a long time was the largest radio telescope in the world operating at such short wavelengths. Twenty years later, it became the focal point of the first US-USSR VLBI observations (Sect. 8.2).

Radio Astronomy in China As in other countries, early post-war radio astronomy in China concentrated on solar research at meter wavelengths, but with instrumentation imported from the USSR. Even more than in the USSR, radio astronomy activities in China were cloaked in secrecy and hidden from Western scientists. Likewise Chinese scientists had little contact with the West (Wang 2009). But in remarkable contrast to pervading policies, starting in 1963, Chris Christiansen from Australia made more than a dozen visits to China, where he helped implement a series of advanced radio astronomy arrays. As a result of his frequent trips to China, Christiansen for many years was denied visas to visit the United States. Following the end in 1976 of the Cultural Revolution in China and the resumption of diplomatic relations with Western countries, exchanges between Chinese and Western scientists flourished; a new generation of young radio astronomers were trained, and China went on to develop strong programs in radio astronomy.

2.2   RADIO WAVES FROM THE SUN[3]

Although the Sun is the strongest radio source in the sky other than the Milky Way Galaxy, especially during periods of solar activity, neither Jansky nor Reber, until 1944, were able to detect radio emission from the Sun. However, during the 1920s and 1930s US, British, and Japanese amateur radio operators, mostly operating in the 10 m (28 MHz) band, reported noise or hissing that was probably due to radio emission from the Sun during periods of intense solar activity. Probably the first documented report of solar radio emission was by the British radio amateur D.W. Heightman who operated the amateur radio station G6DH. Heightman (1936) reported a "smooth hissing sound when listened to on a receiver," which he suggested, "apparently originates on the Sun, since it has only been heard during daylight."

Solar radio emission was independently detected and documented on multiple occasions during World War II by radar operators in the UK, in Germany (Schott 1947), in Japan (Tanaka 1984), and in the South Pacific by Elizabeth Alexander.[4] Hey (1973, p. 15) and Lovell (1984a) have described one of the best known and perhaps the most dramatic serendipitous detection of solar radio emission which occurred in February 1942 when Britain was on high alert for an impending German invasion. On 12 February two German warships, the Scharnhorst and the Gneisenau, passed through the English Channel undetected by the jammed 4.2 m (71.4 MHz) British coastal radar. Concerned about the effectiveness of their coastal defense radar chain, the UK War Office assigned the AORG's J.S. Hey the job of understanding and eliminating the jamming. On 27 and 28 February, the coastal radar systems again became inoperable due to apparent German anti-radar activity, but Hey noticed that the apparent jamming occurred only in the daytime when the radar antennas were pointed toward the Sun, and that it was a time of exceptional sunspot activity. He concluded that the Sun was a source of powerful 4–8 m (37–75 MHz) radio emission about 100,000 times greater than expected from a 6000 K black body. Hey described his discovery in a secret report[5] which became known to scientists in other Allied countries. However, wartime secrecy precluded any publication of his discovery in the scientific literature until the close of hostilities (Hey 1946).

Just a few months later, on 29 June 1942, Bell Labs scientists, A.P. King and George Southworth succeeded in detecting the thermal radio emission from the quiet Sun, first at 9.4 GHz (3.2 cm) and later at 3.06 GHz (9.8 cm). As with Hey's discovery, Southworth's study of the radio emission from the quiet Sun wasn't published until after the War (Southworth 1945, 1956). However, Southworth's 1942 classified report was distributed to various groups, including the Harvard College Observatory and the British War Office for further distribution within the British Commonwealth.[6] During this period, many Bell Labs visitors, including Bowen and Pawsey from Australia, became aware of Southworth's detection of solar radio emission.

In his 160 MHz (1.9 m) survey paper, Reber (1944) commented, only in passing, that, in spite of daytime interference, he was able to detect radio emission from the Sun, and realized that the radio emission was more than 100 times more intense than expected from a 6000 K black body.[7] Since the earlier observations of solar radio emission by Hey and Southworth remained under military classification until after the end of WWII, Reber's (1944) paper was the first published report of solar radio emission. With great perception, he noted that even if all the stars in the Galaxy radiated with the same intensity as the Sun, that would fail to account for the observed radio emission from the Milky Way. Again, in his later paper on his 480 MHz (62 cm) survey, Reber (1944) comments only in the last sentence of the paper that he also obtained data on solar radio emission. Although there had been numerous classified observations of solar radio emission in the UK, in the South Pacific, and in the US, Grote Reber (1944), unrestricted by wartime security constraints, was the first to report radio emission from the Sun in the scientific literature and the first to recognize and report the observation of the near 1 million degree solar corona.

While demonstrating his equipment to National Bureau of Standards (NBS) visitors on 21 November 1946, Reber was surprised to observe that intense radio bursts from the Sun drove his chart recorder off scale.[8] Re-inspection of his earlier records indicated that a similar, but much weaker phenomenon had also occurred on 17 October. In surprising contrast to his earlier slow meticulous work and his 1944 indifference to solar radio emission, Reber dashed off a hastily written letter to *Nature* (Reber 1946).[9] In this report, Reber also noted that "the apparent solar temperature [of the quiet Sun] was about a million degrees," in good agreement with the Australian work (Pawsey 1946).

2.3   RADIO STARS AND RADIO GALAXIES

Grote Reber's 160 MHz (1.9 m) and especially his 480 MHz (62 cm) maps indicated several peaks in the radio emission in addition to the one near the Galactic Center, but it would be another serendipitous discovery by J. Stanley Hey, that suggested the first evidence for discrete sources of cosmic radio emission. Hey first learned about Jansky's work from his supervisor when he reported his 1942 discovery of solar jamming of the British coastal radar systems, and had since read Jansky's and Reber's papers in *Proc. IRE*. Just a few months after the end of the War in Europe, Hey, Phillips, and Parsons (1946b) used a 64 MHz (5 m) radar antenna to map out the galactic radio emission between declinations −3 and +60 degrees. With a beamwidth of 6 by 15 degrees, they observed the peaks previously noted by Reber at the position of the Galactic Center and also in the Cygnus region near right ascension $20^{hr}30^m$ and declination +35 degrees. Continued observations with improved equipment led to a remarkable discovery. Hey, Parsons, and Phillips (1946a) found that the observed emission from the Cygnus region appeared to fluctuate by about 15% on time scales considerably less than one minute and concluded that, "such marked variations could only originate from a small number of discrete sources."

Halfway around the world, in Sydney, Australia, Hey's discovery excited John Bolton. Assisted by Gordon Stanley, Bolton constructed an antenna at Dover Heights overlooking the Pacific Ocean which they used as a 100 MHz sea interferometer to show that the Cygnus source was less than 8 arcmin in extent (Bolton and Stanley 1948). Using this same sea interferometer, Bolton (1948) went on to discover six new discrete sources. Simultaneous observations made from Australia and New Zealand, as well as from sites at Cambridge and Jodrell Bank showed that the intensity variations were independent at the two sites, confirming that they were not intrinsic to the source, but were the result of the signal propagation through the Earth's ionosphere.[10] After months of painstaking observations, Bolton, Stanley, and Slee (1949) succeeded in measuring the positions of three strong radio sources with accuracy better than half a degree. For the first time it was possible to associate radio sources with known optical objects. They identified the strong radio sources, Taurus A, Centaurus A, and Virgo A with the Crab Nebula, and the galaxies NGC 5128 and M87 respectively. NGC 5128, with its conspicuous dark lane, and M87, with its prominent jet, were well known to astronomers as peculiar galaxies. In their *Nature* paper, Bolton et al. mostly discussed the nature of the Crab Nebula, but in a few paragraphs near the end of their paper, they commented,

NGC 5128 and NGC 4486 (M87) have not been resolved into stars, so there is little direct evidence that they are true galaxies. If the identification of the radio sources are [*sic*] accepted, it would indicate that they are [within our own Galaxy].

As implied by the title, of their paper, "Positions of Three Discrete Sources of Galactic Radio-Frequency Radiation," Bolton et al. incorrectly dismissed the extragalactic nature of both Centaurus A and M87. When asked many years later by one of the authors (KIK) why he did not recognize that he had discovered the first radio galaxies, Bolton responded that he knew they were extragalactic, but that he also realized that the corresponding radio luminosity would be orders of magnitude greater than that of our Galaxy and that he was concerned that, in view of their apparent extraordinary luminosity, a conservative *Nature* referee might hold up publication of the paper.[11] Nevertheless, in spite of their stated reservation, their 1949 paper is generally regarded as the beginning of extragalactic radio astronomy (Bolton 1982). Yet, for the next few years the nature of discrete radio sources remained controversial within the radio astronomy community. While Ryle and his group at the Cavendish Laboratory continued to refer to *radio stars*, the extragalactic nature of NGC 5128 and M87 and their powerful radio luminosity was broadly recognized and caught the attention not only of astronomers, but of the physics community as well. How could there be such strong radio sources? Where did the energy come from? How was this energy converted into radio emission?

The nature of the broader population of radio stars remained uncertain. Among the four brightest sources, there was one galactic supernova remnant, two nearby galaxies, and one unidentified source with no apparent optical counterpart. Mills and Thomas (1951), using a two-element interferometer, succeeded in measuring a more accurate position of the Cygnus A radio source. Mills communicated an apparent identification with a faint galaxy to Rudolph Minkowski, a well-known astronomer at Caltech's Mt. Wilson and Palomar Observatories. But Minkowski could not accept that such a faint, and presumably distant galaxy could be such a strong radio source, and dismissed Mills' identification. Only after F. Graham-Smith (1951) used his two-element interferometer in Cambridge to obtain an even more accurate radio position did Minkowski accept the identification with what appeared to be two galaxies in collision. However, the classic paper by Baade and Minkowski (1954) which reported the identification of Cygnus A gave only footnote recognition to Mills' previous identification.

The identification of the second strongest radio source in the sky with a faint distant galaxy was remarkable. Already both Cambridge and Australian radio astronomers were cataloguing radio sources as much as a 100 times fainter than Cygnus A. The galaxy identified with Cygnus A was near the limit of what was then the largest optical telescope in the word, the 200 inch Palomar telescope. Radio telescopes were apparently poised to open a new approach to exploring the distant and correspondingly early Universe.

The Source Count Controversy After years of considering that the radio stars were part of a galactic population, Martin Ryle was finally convinced, by the observed isotropy of the source distribution, that they were powerful extragalactic objects that could be used to address broad cosmological problems. During the 1950s, there were two fundamentally different approaches to cosmology. On the one hand, Herman Bondi, Thomas Gold, and Fred Hoyle were promoting their non-evolving Steady-State cosmology in which the Universe is, and always was, everywhere the same. A unique characteristic of the Steady-State model was that to maintain the requirement of an unchanging but expanding Universe, their model required the controversial continuous creation of new matter by an unknown process. This stood in contrast to what Hoyle sarcastically called the "big bang" cosmology which has a characteristic time scale beginning with a mysterious singularity and a Universe that evolves with time.

Although even the simplest radio telescopes were able to record data from very great distances, there was a serious problem in using radio measurements for cosmological investigations, since radio observations alone give no indication of distance. Any one radio source might be a nearby galactic supernova, a relatively close low luminosity radio galaxy, or a much more powerful distant radio galaxy. Without optical identifications of radio sources and measurement of their redshift (distance), radio astronomers were unable to determine the distance or even the nature of their radio sources, and the positions of only a few of the strongest radio sources were known with sufficient accuracy to identify an optical counterpart. Although most of the fainter radio sources had no optical identification, Martin Ryle and Peter Scheuer at Cambridge reached a profound conclusion from the count of the observed number of radio sources as a function of limiting flux density.

In a static, uniformly filled universe with Euclidian geometry, the number of radio sources, N, above a given flux density, S, should increase as $S^{-1.5}$ or log $N = -1.5$ log S.[12] Based on the Cambridge 81 MHz 2C survey of 1936 sources, of which 1906 sources were small diameter and isotropically distributed, Ryle and Scheuer (1955) concluded that since the observed exponent was closer to -3 than -1.5, "Attempts to explain the observations in terms of a steady-state theory have little hope of success." In other words, the Cambridge survey contained many more faint, presumably more distant, radio sources than would be expected in a universe with a uniformly distributed population. This meant that in the past, there were either more radio sources or that they were more luminous than at present, concepts referred to as density evolution or luminosity evolution respectively.

However, contemporaneous observations, from the Sydney group led by Bernie Mills using his Mills Cross at essentially the same frequency, disagreed. The Sydney source count had a slope of only -1.7 which they argued, in view of experimental errors, did not differ significantly from -1.5 (Mills and Slee 1957). Moreover a one-to-one comparison of the overlapping survey region, indicated very little agreement, and Mills and Slee concluded that the "discrepancies ... reflect errors in the Cambridge catalogue, and accordingly deductions of cosmological interest derived from its analysis are without foundation."[13] In the ensuing years, Martin Ryle had to deal with challenges from two directions: first from the theoretical side from his Cambridge colleague, Fred Hoyle, who questioned his interpretation of the data, and second, probably more seriously, from the Australian radio astronomers who questioned his data. In several well-publicized lectures Ryle (1955, 1958) vigorously defended his position that the source-count was inconsistent with Steady-State cosmology. Further observations indeed showed that probably three-fourths of the Cambridge 2C sources were not real, but were blends of weaker sources, known as confusion.

A new more reliable Cambridge 3C survey (Edge et al. 1959) of 242 sources away from the galactic plane, made with higher angular resolution, reduced the claimed source count slope to -2.0. However, the complete Sydney survey of 1658 sources indicated a slope of -1.5 after correction for the effects of noise and confusion, and concluded that "the source counts indicate no divergence from uniformity and no obvious cosmological effects" (Mills et al. 1960). However, Ryle and Clarke (1961) still maintained that "the results appear to provide conclusive evidence against the Steady-State model." The new and more reliable Cambridge 4C and 5C surveys, as well as the Parkes 408 MHz survey, each indicated a slope near -1.8, consistent with the Sydney value, but still apparently in excess of the steady-state theory (Hewish 1961; Ryle 1968).

Mills and Hoyle continued to argue that the source count could equally well be interpreted as a local deficiency of only a few dozen strong sources rather than a cosmic excess of weak sources. Ryle responded that even these strong sources are so distant that their distribution is cosmologically significant.[14] The debate, which went on for more than a decade, was intense, bitter, and personal (Kragh 1996, p. 305), and was only decided in 1965, not from radio source counts, but from the discovery by Penzias and Wilson (1965) of the cosmic microwave background, which was almost universally accepted as convincing evidence for an evolving or Big Bang universe. Ryle was right; we live in an evolutionary universe. However, his arguments were wrong and based on unreliable data. Mills had much better data, consistent with contemporary source counts, but he got the wrong answer.

It is interesting to look back on the source count controversy in light of current data and understanding. Both the Sydney and, especially, the Cambridge data contained serious instrumental errors as well as errors of interpretation. Because there are more weak radio sources than strong sources, even random errors due to noise or confusion make more weak sources appear stronger and strong sources appear weaker at any given flux density level, thus making the observed source count appear steeper than the real value. Secondly, both Ryle and Mills continued to use cumulative counts, where each point was the sum of all the stronger sources. So the data points were not independent and the estimated errors were unrealistically too small. Moreover, features in the source count at some flux density, S, would propagate to lower flux densities.[15] Finally, in a real expanding universe, whether Steady-State or Big Bang, the effect of the redshift is to make sources weaker than would be the case if their flux density fell off as an inverse square of the distance, so the expected slope, in a non-evolving universe should be smaller than -1.5. Modern radio source counts go about a million times fainter than the Cambridge or Sydney surveys. Except for the strongest one hundred or so sources, the slope of the radio source count is -1.5 or flatter, so it is not clear that the steeper slope reported by the Cambridge observers has any relevance.

The long controversy over the radio source counts did not help the image of radio astronomy within the broader astronomical community. Among radio astronomers, the 2C fiasco gave interferometers a bad name, which took years to overcome. Although the evidence for an evolving universe is now widely accepted, it is interesting to remember that the Steady-State model made the specific predication that the rate of expansion must increase with time, as confirmed only in 1998 with the Nobel Prize winning discovery that the expansion of the Universe is indeed accelerating and the recognition that 70% of the Universe is composed of so-called dark energy (Riess et al. 1998; Perlmutter et al. 1999).[16] It is interesting, however, to speculate how the history of cosmology might have been different if the acceleration of the Universe had been discovered before the discovery of the cosmic microwave background by Penzias and Wilson (1965).

2.4   The 21 cm Hydrogen Line: The Beginning of Radio Spectroscopy

When Jan Oort in the Netherlands learned about Grote Reber's (1944) paper, he suggested to his student, Henk van de Hulst, that he might consider the possibility of spectral line radiation from interstellar hydrogen, the most abundant element in the Universe.[17] Van de Hulst's (1945) classic paper, written in 1944 and published after the end of the war, primarily dealt with thermal emission from stars and interstellar dust (called "smoke" by van de Hulst), free-free (Bremsstrahlung) emission from ionized hydrogen, neutral hydrogen recombination lines from $\Delta n = 1$ atomic transitions, and in only two paragraphs, he discussed the 21 cm (1420 MHz) hyperfine line from the spin flip transition in neutral hydrogen atoms.[18] Due to a calculation error (Sullivan 1982, p. 299), Van de Hulst (1945) pessimistically concluded that the recombination lines "are unobservable" due to the Stark effect, and that the hyperfine line "does not appear hopeless," but that "the existence of this line remains speculative."

Shortly after the end of the war, Oort wrote to Reber asking his advice on the construction of an antenna to study interstellar gas.[19] Keeping his cards close to his chest, Oort made no mention of their plan to search for the 21 cm hydrogen line. However, about the same time, van de Hulst was visiting the Yerkes Observatory, and went to Wheaton to ask Reber's opinion on whether the 21 cm line of neutral hydrogen could be detected (Reber 1958). Van de Hulst's inquiry apparently ignited Reber's interest in searching for the hydrogen line himself. He consulted Greenstein[20] about the probable strength of the 21 cm line and started to build a 21 cm receiver and horn feed. He also had ambitious plans to return to 3 cm and even to work at 1 cm, but unfortunately, he never completed any of these projects, perhaps discouraged by van de Hulst's pessimism about detecting the hydrogen line.

Later, Iosef Shklovsky (1949) published a paper in Russian discussing the possibility of detecting monochromatic radio emission. Shklovsky had not seen the Dutch language paper by van de Hulst, but was aware only of the brief mention by Reber and Greenstein (1947) in their review of radio astronomy. Working independently of van de Hulst, Shklovsky calculated the transition frequency, but reached a more optimistic conclusion, writing,

> In summary, we must say that the resources of contemporary radio techniques fully allow one to detect and measure the monochromatic radio emission of the galaxy…. It is of fundamental importance that the interstellar hydrogen can in this case be directly studied in its ground state. In direct contrast, all methods now existing in astrophysics allow one to determine the population of only the excited states of the hydrogen atom…. Study of the intensity distribution of monochromatic radio emission from the Milky Way would give the distribution of interstellar gas in various regions of the galaxy…. Soviet radiophysicists and astronomers should endeavor to solve this absorbing and important problem.

Shklovsky's paper did not go unrecognized in the US. Following a year teaching astronomy at Amherst and three years as a Second Lieutenant in the US Naval Air Corp, where he took courses in electronics at Princeton and MIT, Harold I. (Doc) Ewen, while maintaining an appointment as a Naval Reserve officer, entered Harvard graduate school to pursue a PhD in physics.[21] His commanding officer was Donald Menzel, then a Captain in the Naval Reserves. Menzel sent Ewen for a two week assignment at the National Bureau of Standards, where he worked under the supervision of Grote Reber. Returning to Harvard, Ewen contacted Edward Purcell, who suggested that Ewen investigate the literature on the neutral hydrogen hyperfine line. Using his naval connections, Ewen was able to obtain a translation of van de Hulst's (1945) paper from the Naval Research Laboratory (NRL). Although discouraged by van de Hulst's conclusion, Ewen and Purcell agreed that Ewen's thesis topic would be to assemble a microwave spectrometer with the objective of setting the "level of non-detectability" of galactic hydrogen. Meanwhile, to earn money, Ewen worked nearly full time building a proton accelerator at the Harvard Cyclotron Lab.

Resigned to a negative thesis, Ewen later claimed that he had little interest in the project until he saw Shklovsky's 1949 paper.[22] Although he had assumed that the Dutch were not going to try to detect the hydrogen line, he then worried that the Soviets, encouraged by Shklovsky's remarks, were probably already hard at work in pursuit of a detection. Full of enthusiasm and excitement, Shklovsky had tried to convince Victor Vitkevich, the leading Soviet radio astronomer, to attempt to detect the 21 cm line. But, after showing some initial interest, according to Shklovsky (1982), Vitkevich suggested that while it was easy for theoreticians to throw out ideas, experimental radio astronomy is real work and that he had enough ideas of his own. Years later, Shklovsky noted, Vitkevich admitted that he had been intimidated by Lev Landau, the famous Russian physicist, who claimed that the whole idea was "pathological."

Purcell was able to obtain a $500 grant from the American Academy of Arts and Science that enabled Ewen to purchase components to build a 21 centimeter receiver and to construct a pyramidal horn antenna (Figs. 2.1 and 2.2). For the next year, while continuing to work trying to produce a proton beam at the Cyclotron Lab in the daytime, on weekends Ewen worked on his hydrogen line system. He built a horn out of plywood and copper foil, which he pointed outside the window of the Lyman Physics Building at a declination of −5 degrees. The instrumentation needed to measure the 21 cm hydrogen line was developed using test equipment "borrowed" from the Cyclotron Lab on Friday afternoon and returned by Monday morning. At Purcell's suggestion, Ewen designed and built a frequency switched system patterned after Dicke's load switching radiometer.[23] His search for the hydrogen line was aided by the recent laboratory measurement of the line frequency of 1420.405 MHz (Kusch and Prodell 1950), so he knew precisely where to tune his receiver. Due to Doppler broadening, the galactic hydrogen emission is smoothed over a broad frequency range. With comparable signal strength in both the line and reference

Fig. 2.1 Doc Ewen and his 21 cm receiver at Harvard University, March 1951. Credit: NRAO/AUI/NSF

channels separated by only 10 kHz, Ewen initially missed seeing the 21 cm hydrogen signal. Supported by another $300 from Purcell's personal funds, Ewen purchased a commercial shortwave receiver to use as the final stage of his receiver. This permitted him to use a sufficiently wide 75 kHz separation between the signal and reference frequency to avoid any galactic hydrogen radiation in the reference channel.

On 25 March 1951, Ewen detected the galactic 21 cm radiation peaking near 18 hours right ascension, although the observed frequency was about 150 kHz above the laboratory value. A quick consultation with the Harvard astronomers confirmed Ewen's suspicion that the discrepancy was due to the Doppler shift caused by the Earth's orbital motion and the motion of the Solar System. At the time, van de Hulst happened to be visiting Harvard, where he was teaching a course in astronomy. There had been no previous contact between Ewen and van de Hulst until, at Purcell's suggestion, Ewen informed him of their successful detection. Ewen recalled that van de Hulst seemed "shocked" and did not appear particularly "delighted to learn that his 'long

Fig. 2.2 Doc Ewen and his original horn antenna at NRAO in Green Bank, West Virginia, May 2001. Credit: NRAO/AUI/NSF

shot speculation' proved to be correct."[24] Only then did Ewen learn that the Dutch had been unsuccessfully trying for years to detect the 21 cm line. Although van de Hulst had not expected that the hydrogen line could be detected, he explained that Oort appreciated its importance, and continued to encourage the attempt at Leiden. Van de Hulst peppered Ewen with questions about where he found hydrogen emission in the Galaxy and how to build the receiver. Lex Muller adopted the Harvard frequency switching concept, and on 11 May 1951, using a 7.5 meter Würzburg antenna at Kootwijk in central Holland, Muller and Oort (1951) were able to confirm the Harvard result. Characteristically, not satisfied with a simple confirmation, the Dutch paper included observations made at different positions in the Galaxy and commented on the distribution of hydrogen gas and the kinematics of the Galaxy.

Another Harvard visitor at this time was Frank Kerr from the CSIRO Radiophysics Laboratory. Following a visit to Ewen's lab in early April 1951, Kerr informed Joe Pawsey in Australia about Ewen's successful detection of galactic hydrogen. Kerr explained Purcell's frequency comparison technique and requested possible confirmation from Australia. Pawsey wrote back to Purcell that they had not previously been working on the hydrogen line detection, but upon hearing from Kerr of Harvard's success, he informed Purcell they had initiated two separate programs to confirm Ewen's result.

Ewen claimed that he wrote his 47 page thesis in three days and defended it in May, just two months after his detection of the hydrogen line. After managing to finagle his way through his German language exam,[25] Ewen received his PhD in Physics from Harvard. Ewen and Purcell (1951a) presented their discovery at the Schenectady meeting of the American Physical Society on 16 June 1951. The Harvard (Ewen and Purcell 1951b) and Dutch (Muller and Oort 1951) papers were sent to *Nature* on June 14 and 20 respectively and published in successive papers in the 1 September 1951 issue. The Dutch paper was followed by a brief note from Joe Pawsey, dated 12 July, announcing that Christiansen and Hindman had also confirmed Ewen's discovery but with no details. Christiansen and Hindman (1952a, b) published a brief report in *Observatory* followed by a more detailed paper in the *Australian Journal of Scientific Research*. Using a tracking antenna, they showed that the distribution of galactic hydrogen was concentrated along the galactic plane and based on observed frequency splitting of the line profiles, they suggested evidence for the galactic spiral arms.

All three programs were enabled by using frequency switching, and all three programs used a variable capacitance to scan in frequency. Ewen later commented that there was never a real competition to make the first detection. But rather "we gave them the Heath kit, the parameter values of the signal, and instructions on where to look. The Australians freely admit this, the Dutch never have."[26]

A few months after these papers appeared, Paul Wild from Australia submitted a paper to the *Astrophysical Journal* (Wild 1952) discussing expected frequencies for various possible hydrogen atom transitions, including high order recombination lines, fine structure lines, as well as the spin-flip hyperfine structure line and the effect of Zeeman splitting in a magnetic field. Although not submitted until after Wild became aware of the actual detections at Harvard, Leiden, and Sydney, his paper was based on two pre-discovery papers which he wrote in 1949 for private distribution.[27]

When Grote Reber learned about Harvard's successful detection of hydrogen, he wrote to congratulate Ewen and urged him to try for the 92 cm (327 MHz) deuterium line.[28] But Ewen had other ideas. He enrolled in the Harvard Business School to pursue an MBA to complement his PhD in Physics. He was called to active duty during the Korean War, but served for only a month before negotiating his return to Cambridge as a civilian. Although he later participated in several radio astronomy conferences, and took part in the search for a site for NRAO, Ewen never continued his pioneering research on the 21 cm hydrogen line, but helped Bart Bok start the Harvard radio astronomy project and served as co-director with Bok. Together with his friend, Geoff Knight, Ewen started his own business, the Ewen-Knight Company to construct a new hydrogen line receiver for Harvard. Ewen Knight went on to be a major supplier of equipment to many radio observatories including NRAO. Building on Karl Jansky's patent, Ewen later developed a form of radio sextant which was used on Polaris submarines when they needed to know their location, independent of weather conditions, before firing their missiles.

2.5   EARLY US UNIVERSITY RADIO ASTRONOMY PROGRAMS

Unlike in other countries, US postwar radio astronomy was largely funded through contracts between universities and the military research programs at the Office of Naval Research (ONR) and the Air Force Office of Scientific Research (AFOSR). Still buoyed by the scientific successes of WWII and further stimulated by the Korean War, in the 1950s US universities, including Caltech, Cornell, Stanford, the University of California, and the University Illinois, received generous support from the ONR and the AFOSR to begin modest radio astronomy programs. Other programs in the US began at the Carnegie Institution's Department of Terrestrial Magnetism (DTM) and at the Naval Research Laboratory (NRL). Starting in 1948, Grote Reber pursued radio astronomy programs at the National Bureau of Standards (NBS) until 1951 and then, with modest support from the New York based Research Corporation, in Hawaii. Except at Harvard and Ohio State, the National Science Foundation played little role in these early years of US radio astronomy.

Harvard University As Ewen later remarked, "The discovery of the 21 cm interstellar hydrogen line radiation in 1951 initially generated little interest at Harvard."[29] However, after several years of military service, Campbell Wade had just returned to Harvard to complete his undergraduate degree in astronomy. While in the military, Wade had acquired some skills in radio technology. He heard about Ewen's discovery and was excited about the prospects for further work in what he foresaw as an exciting new field of astronomy. He naturally went to Bart Bok, who was widely recognized among the student body as probably the friendliest Harvard astronomy professor, and who had just returned from an extended period in South Africa. Bok contacted his friend Ed Purcell to discuss the prospects for establishing a radio astronomy program at Harvard, but he had no encouragement from his peers. Walter Baade suggested that at age 51, Bok was too old to get involved in this new field with uncertain prospects. After all, Jansky and Reber had measured the continuous radiation from the Galaxy and Ewen and Purcell had discovered the 21 cm radiation from galactic hydrogen. Nothing was left! Nevertheless, with Purcell's encouragement, Bok sent Wade, along with two senior graduate students, David Heeschen and Edward Lilley, to see Ewen and learn how to operate his equipment. However, the three of them found the room locked and learned that all of the equipment had been sent to DTM in Washington, where Merle Tuve wanted to start a new radio astronomy program.[30]

A few months later Bok obtained the first NSF grant for radio astronomy, for $32,000 to purchase a 24 foot (7.3 meter) antenna, but he needed additional funds for the instrumentation. He convinced Mabel Agassiz, a wealthy Harvard benefactor, to contribute another $40,000 toward the construction and operation of the telescope that was located in the town of Harvard, Massachusetts, about 30 miles distant from the Harvard University campus in Cambridge. Bok was able to get additional support from the University, as well as from the Research Corporation, to operate the telescope and to provide additional instrumentation. Ewen built a new hydrogen line receiver and Heeschen, Lilley, and Wade helped to put the telescope into operation and conducted the first research programs. Heeschen and Lilly went on to be among the first Americans to receive their PhD degree for work in radio astronomy, when they both passed their Harvard oral exams on 26 November 1954.[31] They were followed by T.K. (Kochu) Menon in 1956, Thomas Matthews in 1956, W.E. (Bill) Howard III in 1958, May A. Kaftan-Kassim in 1958, Nannielou (Nan) Hepburn Dieter Conklin in 1958, Frank Drake in 1958, and Campbell (Cam) Wade in 1958.

In 1954, Bok and Dave Heeschen convinced Mrs. Agassiz to help fund the design of a 60 foot (18 meter) radio telescope. Only in this way, they explained, could American astronomers keep up with radio astronomers in England, and the Netherlands, who were using their captured German Würzburg antennas for a variety of research programs. Bok received another NSF grant of $132,000 for the construction and initial operation of the telescope. When completed in the spring of 1956 by D. S. Kennedy and Co., the Harvard 60 foot dish became one of the largest fully steerable radio telescopes in the world and the largest equatorially mounted telescope.[32] The Ewen-Knight Corporation built new receiving equipment for 21 cm research, including a 20 channel spectrometer. NSF Director Alan Waterman spoke to more than 200 people at the dedication of Harvard's new radio telescope, referring to radio astronomy as "a new window on the universe," and acknowledging the responsibility of the Federal Government to support basic research "when such support is necessary and in the interest of science."[33]

Unlike the radio astronomy programs in Australia and the UK, the Harvard project was managed by astronomers, not radio scientists. Even at the other American universities involved in radio astronomy the programs were led by people with backgrounds in radiophysics such as William Gordon, Charlie Seeger, and Marshall Cohen at Cornell, John Kraus at Ohio State, and Ronald Bracewell at Stanford. Bok insisted that Harvard students had to learn astronomy as well as radio technology. The research programs at the 24 and 60 foot antennas were heavily concentrated on studying galactic H I emission, particularly in the direction of various nebulosities, star clusters, and the Galactic Center along with the delineation of the Galaxy's spiral structure. However, they found time to search, unsuccessfully, for the galactic hydroxyl (OH) molecule based on a calculated frequency provided by Charlie Townes (1957), and Heeschen observed H I in the Coma cluster (Fig. 2.3).

Dave Heeschen, Frank Drake, Kochu Menon, Cam Wade, and Bill Howard all went on to become prominent members of the NRAO Scientific Staff, with Heeschen later becoming the NRAO director from 1962 to 1978. Drake became director of the Cornell Arecibo Observatory. Nan Dieter went to the

Fig. 2.3 Kochu Menon, Dave Heeschen, Russell Anderson, and Bart Bok at the base of Harvard's 60 foot Agassiz telescope in early 1956. Credit: NRAO/AUI/NSF

nearby Air Force Cambridge Research Lab, then to Berkeley where she played a key role in the discovery of cosmic hydroxyl masers (Dieter et al. 1966). Tom Matthews went to Caltech where he contributed to the discovery of quasars. John (Jack) Campbell became one of Ewen's first employees before going on to become Systems Engineer for the VLA. Cam Wade, while still a graduate student, worked for Ewen-Knight before taking up a postdoctoral position in Australia, followed by a long career on the NRAO Scientific Staff where he led the VLA site search and procurement. After a postdoc at NRAO followed by a faculty position at the State University of New York, May Kaftan-Kassim returned to her native Iraq where she became Director of the Iraqi National Astronomy Observatory Project,[34] and later the Iraqi representative to the 1968 UN Vienna Conference on the Exploration and Peaceful Uses of Outer Space. After a period at NRL and Yale, Ed Lilley returned to join the Harvard faculty where he spent the rest of his career.

Cornell University The Cornell solar radio astronomy program began in the Electrical Engineering Department in the late 1940s with support from the AFOSR and ONR. Under the leadership of Charles Burrows, William (Bill) Gordon, and Charles (Charlie) Seeger, older brother of the well-known folk singer, Pete Seeger, this was probably the first US university based program in radio astronomy. Martha Stahr, who was a Professor of Astronomy and the first female faculty member in the Cornell School of Arts and Sciences, conducted daily observations of the Sun at 205 MHz which she related to sunspot activity (Stahr 1949). Later the group extended their research program to include mapping the galactic radiation (Seeger and Williamson 1951). From 1948 to 1960, Stahr (later Stahr Carpenter) also issued a valuable series of bibliographic reports listing publications in radio astronomy that brought radio astronomy work, which at the time was largely published in engineering journals, to the attention of the broader astronomical community.[35]

After Seeger left Cornell for Sweden in 1950, then the Netherlands, Marshall Cohen, who had received his PhD in physics from Ohio State, joined the Cornell group in 1954 and began a program to study the polarized solar radio emission. Cohen later recalled, "What we were doing at Cornell when I got there—I found out very quickly was really quite primitive compared to what was going on in Australia." And in fact, part of the Cornell program was terminated not long afterward.[36]

Starting in 1958, under the leadership of Bill Gordon, Cornell scientists began discussing the construction of a 1000 foot (305 meter) diameter back scatter radar system to study the ionosphere. Cohen realized that with the proposed antenna it would also be possible to get radar echoes from Mars and Venus and pointed out other passive radio astronomy applications of the Arecibo Ionospheric Observatory (See Sect. 6.6) which was later funded by the AFOSR.

The University of California at Berkeley The radio astronomy program at the University of California began somewhat later than the other university programs, and, like the program at Harvard, was initiated and initially managed by classical astronomers with backgrounds in optics rather than radio engineering. The Berkeley Astronomy Department was rejuvenated after WWII by Otto Struve, who came to Berkley as department chair in 1950. The following year, he brought Harold Weaver to Berkeley from the Lick Observatory, and in 1954, at Weaver's invitation, Ron Bracewell came to Berkeley from the CSIRO Radiophysics Laboratory and taught a course in radio astronomy. With the urging of Otto Struve, who was perhaps reflecting on his unsuccessful effort a decade earlier to help Grote Reber get established, the Dean established a faculty committee to investigate how the university could enter the new field of radio astronomy.[37] A favorable committee report provided the incentive for Otto Struve and Harold Weaver to propose a new observatory, but considerably larger than the one envisioned in the committee report. In 1958, following Weaver's sabbatical at Harvard and the Carnegie Department of Terrestrial Magnetism, Weaver and Struve found the funding from ONR needed to establish the Berkeley Radio Astronomy Laboratory. They selected a site at Hat Creek in northern California to build first a 33 foot (10 meter) radio telescope, and in 1962 an 85 foot (26 meter) telescope constructed by the Philco Corporation. Doc Ewen built the 1420 MHz and 8 GHz receivers.

Over the next three decades Weaver led a vigorous observational program concentrating on 21 cm galactic research, and in 1965, Hat Creek made the first observations of interstellar OH masers, opening a whole new field of research. In January 1991, the Hat Creek 85 foot radio telescope collapsed in a storm with 100 mile per hour winds. But by that time, under the leadership of Jack Welch, the emphasis at Hat Creek had moved to millimeter wavelengths (Sect. 10.4) and the telescope was not replaced.

Stanford University Following a one year visiting appointment at Berkeley, Ron Bracewell went to Stanford, and, with funding from AFOSR, built the Stanford Microwave Spectroheliograph. The Spectroheliograph, based on the design by Christiansen, consisted of two orthogonal arrays of sixteen 10 foot (3 m) dishes each 375 feet (114 m) long (Bracewell and Swarup 1961). Operating at 3.3 GHz (9.1 cm) the Stanford array was used until 1980 to make daily maps of the microwave emission from the solar corona with an angular resolution of 3.1 arcmin. During this period, more than 200 astronomers visited the site and engraved their names on the telescope piers. In 2012, ten of the piers were moved to the VLA site in New Mexico where they form part of an innovative sundial designed by Woody Sullivan (Frater et al. 2017, p. 1; Sullivan et al. 2019) (Fig. 2.4).

Ohio State University John Kraus was a radio engineer and an expert on antenna design who was known for his invention of the helical antenna. Also, as a prominent radio amateur, he designed the popular W8JK antenna. Kraus was present at Jansky's 1935 Detroit IRE presentation, where he recalled that there were less than two dozen people in the audience (Kraus 1984). In 1941, Kraus and Grote Reber worked together at the Naval Ordnance Laboratory degaussing naval ships, and they lived in the same rooming house in Washington. Kraus (1995, p. 114) later recalled Reber's "contagious enthusiasm," about his equipment and observations. Apparently, his exposure to these two pioneers of radio astronomy ignited Kraus' interest and excitement about this new field of research, and a decade later he initiated a radio astronomy program at Ohio State University.

Kraus was captivated by the ongoing source count controversy and the desire to push radio catalogues to weaker sources. This meant building the largest possible collecting area for a given cost, which Kraus translated into a transit radio telescope. His initial radio telescope consisted of an array of 96 helical antennas mounted on a tiltable ground plane. Over a period of three years, Ko and Kraus (1957) used this early radio telescope to map the sky north of −40 degrees declination at 1.2 meters wavelength (250 MHz) with a resolution of 1 by 8 degrees. In addition to tracing out the radio emission from the galactic plane, the Ohio State Survey isolated a number of discrete sources, which they noted are concentrated along the galactic plan.

Fig. 2.4 Ron Bracewell touches up paint on a pier at the Stanford Spectroheliograph with the signatures of visiting astronomers, including one of the present authors (KK). Ten of the original piers are now part of a sundial at the VLA site. Credit: NRAO/AUI/ NSF

2.6   US GOVERNMENT AND MILITARY RADIO ASTRONOMY PROGRAMS

National Bureau of Standards Grote Reber's mother died in 1945 and there was little reason for him to remain in Wheaton. He sought support from the Carnegie Institution's Department of Terrestrial Magnetism, Harvard, MIT, and various government agencies. In a 1946 letter to Harlow Shapley, Reber estimated the cost, including labor, of reproducing his telescope to be $15,278.[38] At Harvard, Shapley discussed Reber's proposal with Donald Menzel, who had a background in radiophysics and had many contacts in the military industrial complex (Needell 2000, p. 60). As with Reber's earlier approach to Harvard, Shapley decided they were too busy and had insufficient resources to take on any new projects. Reber's letter stirred up interest for both Merle Tuve and Lloyd Berkner at DTM as well as their Carnegie boss, Vannever Bush. Probably inspired by Reber, Tuve started the DTM radio astronomy program led by Lloyd Berkner. A few years later, Berkner would become President of Associated Universities, Inc. (AUI), where he forcefully led the effort to establish the National Radio Astronomy Observatory, an initiative first started by Bush and Lee DuBridge (Chap. 3).

Reber realized that he was no longer able to compete with these larger government and university activities. He accepted an offer from E.U. Condon to set up a radio astronomy program at the National Bureau of Standards (NBS) Central Radio Propagation Laboratory (CRPL) in Washington, DC, with the prospect of building a 75–100 foot diameter dish. He sold his Wheaton dish and all instrumentation, including a 1400 MHz (21 cm) amplifier and feed, to the US government for $18,570. Everything was moved from Wheaton to Sterling, Virginia, near the location of the current Dulles Airport. His dish was reassembled on a turntable so it could be moved in azimuth as well as elevation, but there is no record that it was ever used again by Reber or by anyone else for radio astronomy. Reber's dish was later disassembled and transported to Boulder, where it remained unassembled until it was moved to Green Bank to be erected under Reber's supervision at the entrance to NRAO in 1959–1960 (Fig. 2.5).

Arguing that a 100 foot dish would only be a small improvement over his Wheaton dish, one of Reber's first actions in Washington was to develop detailed plans for building a large, fully steerable dish 220 feet in diameter operating to wavelengths as short as 10 cm (3 GHz), which he later estimated might cost $650,000 to build.[39] Realizing that an equatorial mount for an antenna of this size would be prohibitively expensive, he proposed using an alt-azimuth mount with an innovative "combination of cams and levers" to control the telescope (Fig. 2.6).

Reber approached CRPL management, the Pentagon, and Merle Tuve at Carnegie, but his ambitious initiative was either rejected or ignored. Later, Lloyd Berkner at AUI expressed interest, but on the advice of his steering committee, AUI pursued a different approach (Sect. 4.4), and Reber was never able to obtain the funds to build his 220 foot antenna. Indeed, it would be nearly another half a century before a fully steerable antenna of this size would be built in the US (Chap. 9).

While working at NBS, Reber participated in a NRL expedition to observe a total solar eclipse on the island of Attu, at the westernmost end of the Aleutian Islands. Multi-wavelength observations on 12 September 1950 showed that the radio Sun was larger than the optical disk, and that the apparent size increased with increasing wavelength (Hagen et al. 1951). As Reber later reported, this "was probably the first total solar eclipse of the Sun which was successfully observed in a pouring rain during a hurricane."[40]

Naval Research Laboratory 50 Foot Radio Telescope[41] Perhaps the biggest organized postwar radio astronomy program in the United States, or at least the most expensive, was at NRL, which began as an outgrowth of wartime radar and communications research. The operation of NRL is led by a Naval Captain, but the research program is led by a civilian. Early NRL postwar research programs concentrated on the Sun, partly because the military was interested in the Sun and how it affected communications, but perhaps, like elsewhere, more because that was all they could see with their limited

Fig. 2.5 Grote Reber in 1960 beside his reconstructed 32 foot Wheaton radio telescope at the entrance to NRAO in Green Bank. Credit: NRAO/AUI/NSF

equipment. In 1951, under the leadership of John Hagen, who had led a wartime NRL centimeter wavelength radar development program, NRL obtained a 50 foot (15 meter) antenna from the Collins Radio Company. The antenna was designed by University of Iowa engineering professor Ned Ashton, whose previous experience was primarily in designing bridges. Ashton would go on to play a major role in the troubled NRAO 140 Foot Telescope project (Sect. 4.4). With a background in radio engineering rather than astronomy, NRL chose an alt-az mount for its radio telescope, using a surplus five inch gun mount for the azimuth bearing and an analogue coordinate converter to provide pointing instructions. But the 1.5 arcminute pointing accuracy made operation difficult at 8 mm, where the beamwidth was only 3 arcminutes, and

Fig. 2.6 Grote Reber's model for a 220 foot telescope design. Credit: Feld, J. Radio Telescope Structures. 1962. *Ann. NY Acad. Sci.* **93**: 353–456. Used with permission of the NYAS

gave alt-az mounts a bad name during later debates over the nature of NRAO's 140 Foot Telescope. The machined surface, which was composed of 30 separate solid aluminum panels, had a maximum deviation from a true paraboloid of 0.8 mm in the inner 20 foot diameter and 1.2 mm in the outer part of the dish (Holzschuh 1958).

When completed in 1951, the NRL antenna (Fig. 2.7) was the largest filled aperture radio telescope in the world, and would remain so until the Harvard 60 foot and the 25 meter Dwingeloo dish went into operation four years later (Fig. 2.8). However, working primarily at short centimeter wavelengths, the sensitivity and research opportunities were limited. As Graham-Smith remarked to Joe Pawsey "it was the world's most expensive radio telescope and all it can see is the Sun and the Moon" (Haddock 1984). Led by Fred Haddock and later Ed McClain and Cornell Mayer, the 50 foot telescope was used by NRL scientists at 9.4 cm (3.1 GHz) to study the thermal radio emission from galactic H II regions and the planets Venus (Mayer et al. 1958a), Mars, and Jupiter (Mayer et al. 1958b), to make the first observations below 21 cm of a few strong radio galaxies (Haddock et al. 1954), and the first measurement of radio polarization (Mayer et al. 1957). NRL also pioneered 21 cm absorption line studies (Hagen and McClain 1954). However, in what proved to be an embarrassment, Lilley and McClain (1956) reported the detection of the redshifted 21 cm absorption line from Cygnus A which not only was never confirmed by others, but it turned out Lilley and McClain had incorrectly used the optical redshift to calculate the radio frequency of the expected absorption and were observing at the wrong frequency.

As it was at the time the most important and the most influential American radio astronomy observatory, NRL scientists, particularly John Hagen, had considerable influence on the later formation of NRAO (Chap. 3). The NRL telescope still sits on its original location on the roof of a laboratory building overlooking the Potomac River. For more than half a century it served as an NRL icon, visible from commercial aircraft landing over the Potomac River at nearby Reagan National Airport.

The location of the NRL dish in Washington and near National Airport meant it was subject to considerable RFI; NRL needed to find a new site for radio astronomy. In 1955, NRL obtained an 84 foot (26 meter) antenna from the D.S. Kennedy Company, patterned after Harvard's Kennedy-built 60 foot radio telescope (McClain 1958). The site for the 84 foot dish was chosen following a survey using the same equipment that had been used in 1955 to

Fig. 2.7 NRL's iconic 50 foot radio telescope on the roof of the main laboratory building, visible from commercial flights into and out of nearby Reagan National Airport. Credit: NRL

Fig. 2.8 Prof. Jan Oort and Queen Juliana leaving the 1956 dedication of the Dwingeloo 25 meter radio telescope. Credit: NFRA

locate the Green Bank site (Sect. 3.4). The chosen site at Maryland Point, Maryland, about 45 miles from the NRL Washington Lab, was found to be the quietest site within 50 miles of Washington. The Kennedy dish had a perforated aluminum surface and was effective at wavelengths as short as 10 cm.

2.7   PRIVATE INITIATIVES

Carnegie Institution Department of Terrestrial Magnetism (DTM) Shortly after the end of WWII, Lloyd Berkner and Merle Tuve initiated a modest program in radio astronomy at DTM. According to Burke (2003), Tuve rejected

Berkner's ambitious plan to build a large dipole array, probably contributing to the long-running animosity between Tuve and Berkner that would soon impact Berkner's plans to develop a national radio astronomy facility (Chap. 3). As explained in Sect. 2.4, before Bok and his students got involved in 21 cm research, Tuve had talked Purcell into sending him Ewen's 21 cm receiver, which he initially used on a German Würzburg. DTM's Howard E. Tatel designed an equatorial mount for the 7.5 meter telescope, making it unlike the Würzburgs used in European radio observatories.

In 1953, Tuve was joined by Bernard Burke, a recent graduate from MIT, and by F. Graham-Smith, on a one-year leave from Cambridge. Together, Graham-Smith and Burke built a 22 MHz Mills Cross at a site in Maryland. After Graham-Smith returned to Cambridge, Burke, working with Ken Franklin, who was on leave from New York's Hayden Planetarium, accidently discovered the strong 22 MHz bursts from Jupiter (Burke and Franklin 1955). In 1959 DTM purchased a 60 foot (18 meter) dish from Blaw-Knox, which was used by Tuve and others for mapping the distribution of galactic hydrogen.

Grote Reber Goes to Hawaii[42] After only a few years working at the NBS, Grote Reber became discouraged by the lack of support for his planned large radio telescope. He had also become increasingly frustrated working under government bureaucracy, as well as by the deteriorating atmosphere in Washington reflected by the growing impact of McCarthyism. He was intrigued by the discoveries reported by Australian radio astronomers using the sea interferometer technique (Sect. 2.1) which could determine the positions of radio sources from the precise timing as they rose or set over the sea. He realized that in order to obtain the accurate two-dimensional coordinates needed for identification with optical counterparts, he would need to make measurements at both rising and setting, and that the best place to do this was from a mountaintop in Hawaii. In a letter to Joe Pawsey, Reber wrote "I got tired of working for Uncle Harry [Truman] and his Boys so I took a vacation in Hawaii. Things looked so good I decided to stay."[43] He went on vacation to Hawaii, and never came back to NBS, apparently abandoning his last paycheck.

From 1951 to 1954, Reber worked on top of Haleakala on the Island of Maui, Hawaii, where he built a rotating antenna to observe between 20 and 100 MHz. By replacing the fixed antenna used by the Australians with one that rotated in azimuth, and by working on the top of a 3000 meter high island mountain, he could observe sources both rising in the east and setting in the west with an interferometer having an effective baseline of 6000 meters. In principle his sea interferometer had a resolution of about one arcminute, but he was plagued by ionospheric refraction and terrestrial interference and was only able to obtain useful results on a few of the strongest radio sources (Reber 1955, 1959). As a result of his experience, he finally concluded that mountain tops were not suitable for radio telescopes. But Reber's sea interferometer was the first mountain-top telescope in the Hawaiian Islands and set the stage for the proliferation of optical and radio telescopes later constructed on both Haleakela and Mauna Kea. The site of Reber's radio telescope was preserved until 2014, when it was destroyed to make room for a parking lot for the Daniel K. Inouye Solar Telescope.

2.8 Why Did the US Fall Behind the UK and Australia? Or Did It?

Looking back after more than half a century, it is natural to ask—did the United States really fall behind the radio astronomy programs being pursued elsewhere even in other, much smaller countries such as the Netherlands and Australia? Or, did American radio astronomers just move in a different direction, or just imagine that they were behind, or perhaps even exaggerate the situation in an attempt to acquire more resources? In the UK, Australia, and the Netherlands, radio astronomy programs were generally concentrated at one or two major laboratories. In both Australia and the UK, early postwar radio astronomy research was conducted primarily at meter wavelengths, using relatively simple and relatively inexpensive technology with equipment derived from wartime radar research. In the United States, reflecting the broad enthusiasm to get into this exciting and promising new field of research, modest radio astronomy programs sprang up at a few universities, but primarily at government and military laboratories. The early postwar US radio astronomy programs were mostly funded by the military, and so were driven, at least in part, by Cold War military needs. This translated into facilities that operated at centimeter wavelengths where there was considerable expertise and experience carried over from the War, but where receiver sensitivities were much poorer than at the meter wavelengths used in the UK and Australia, and where the cost and the complexity of antennas and instrumentation were greater.

At least for the first decade following the end of WWII, there was no focused effort in the US to develop any major radio astronomy research programs of the type being pursued at Cambridge and Jodrell Bank or in Sydney or Leiden. The US entered the postwar era with relatively great wealth, especially when compared with Australia or war-torn Europe. So why did this new field that had been pioneered by the Americans Karl Jansky and Grote Reber appear to thrive in the UK and Australia, but not in the US?

In part, the answer may lie in the enormous prestige of postwar American science and scientists, brought about by the very successful atomic energy program which had led to an abrupt and unanticipated, but greatly welcomed, end to the war in the Pacific. The related areas of atomic and nuclear physics overshadowed the equally important—some will say more important—developments in radiophysics and electronics resulting from the wartime radar research at the MIT Radiation Laboratory, Bell Labs, and elsewhere. Although, as Lee DuBridge remarked, "the bomb ended the war, but radar won the war,"[44] atomic and nuclear physics became the golden areas of research for American scientists, with a seemingly endless flow of money leading to powerful particle accelerators being built at Harvard, the University of California, Caltech, Cornell, Rochester, and other universities, as well at the new Brookhaven National Laboratory.

At the end of 1945, the MIT Rad Lab was shut down. Robert Dicke (Princeton), Lee DuBridge (Rochester), E.O. Lawrence (Berkeley), Edwin McMillan (Berkeley), Ed Purcell (Harvard), and I.I. Rabi (Columbia), all left for universities where they applied their skills in microwave electronics to atomic spectroscopy and particle accelerator physics. Charlie Townes, who was one of the first to recognize the possibility of observing cosmic molecular lines in the radio spectrum (Townes 1957), moved to Columbia University, where he worked on quantum electronics and the development of the maser and later the laser. Taffy Bowen, the Rad Lab liaison to the British radar program, left for Australia, where he started the very successful CSIRO radio astronomy program.

For nearly a decade, almost no one in the US paid much attention to radio astronomy after Jansky and Reber and the casual experiments of Potapenko and Folland. Reber, in particular, was unable to obtain funding or even recognition of his ambitious plan to build a 220 foot dish for radio astronomy, and he ultimately left the United States for Tasmania, where he carried out a program in long wavelength radio astronomy as well as publishing papers in botany and archaeology. NRL did build the then largest steerable radio telescope in the world, and one that operated at much shorter wavelengths than any other contemporary radio telescope, but the motivation was more to address military needs than astronomy. The NRL 50 foot dish also pioneered the use of the alt-az mount in astronomy but, working at short centimeter and millimeter wavelengths, the scientific returns were perceived to be limited.

Much good research was done by US radio astronomers. Berkeley, Cornell, Harvard, Ohio State, and Stanford started important university programs in solar radio astronomy. NRL made the first detailed studies of the thermal radio emission from the Moon and planets. Franklin and Burke at DTM discovered intense radio bursts from Jupiter. Ewen and Purcell at Harvard were the first to detect the 21 cm hydrogen line, beating the Australian and Dutch radio astronomers and starting the new field of cosmic radio spectroscopy. Ewen and Purcell also invented the important frequency switched spectral line radiometer, following Robert Dicke's invention of the switched radiometer, since used by generations of radio astronomers for continuum observations. However, none of these programs captured the attention of the US astronomy community as did the Australian and British discoveries of supernova remnants and distant powerful radio galaxies. As it developed, the cosmological Cambridge-Sydney $\log N - \log S$ debate was a red herring, full of observational errors and naive interpretation on both sides, but for over a decade, the apparent impact to cosmology captivated the astronomical community.

The discovery of the 21 cm hydrogen line by Ewen and Purcell at Harvard did create a unique opportunity, but one that was exploited first by the Dutch and Australian radio astronomers in mapping the dynamics of the Milky Way. While Purcell continued his research on nuclear magnetic resonance, Ewen went into business, manufacturing and selling radio astronomy receivers. Bart Bok built the Harvard radio astronomy project around 21 cm research, but the Australian and Dutch radio astronomers skimmed the cream, and the Harvard program suffered from lack of leadership after Bok went to Australia. However, Bok did develop what was probably the first graduate program in radio astronomy, a program from which many of the first generation of students went on to lead the establishment of the National Radio Astronomy Observatory in Green Bank.

NOTES

1. This is equivalent to the 1834 classical Lloyd's mirror optical effect.
2. Because the signals from different parts of the radio source have slightly different path lengths, the interferometer response is sensitive to source dimensions. The paper by McCready et al. was apparently submitted with Pawsey as the first author but the order was changed by the editor. We thank Ron Ekers and Miller Goss for bringing this historical note to our attention.
3. A more extensive discussion of the various discoveries of solar radio emission is given by Sullivan (2009) pp. 79–99, and for later observations, pp. 284–314.
4. In 1945, Elizabeth Alexander detected radio emission from the Sun while operating a 200 MHz (150 m) radar on Norfolk Island in the South Pacific, as described in her "Report on the Investigation of the Norfolk Island Effect," NAA-WTS, Working Papers, E. Alexander. See also Orchiston (2005).
5. A copy of Hey's 1942 secret report is located in NAA-WTS, Sullivan Publications, Classics in Radio Astronomy.
6. Even the published results in 1945 were devoid of any technical details including the observing frequencies and antenna.
7. GR to JLG, 3 November 1946, NAA-GR, General Correspondence. https://science.nrao.edu/about/publications/open-skies#section-2
8. GR to JLG, 21 November 1946, NAA-GR, General Correspondence I. https://science.nrao.edu/about/publications/open-skies#section-2
9. In this *Nature* letter, which carries a submission date of 24 November, Reber reports on observations made on 23 and 24 November. No year is given anywhere, but presumably all dates refer to 1946. The paper was published in the 28 December 1946 edition of *Nature*.
10. The ionosphere introduces apparent fluctuations in the intensity of small radio sources in much the same way as the atmosphere causes the twinkling of stars. Since telescopes more than a hundred or so miles apart see a different ionosphere the fluctuations appear uncorrelated. If the variations were intrinsic, they would appear the same at different sites.
11. Interview with KIK, August 1989.

12. In a static universe with normal Euclidean geometry, since the number of sources, N, within a given distance, D, is proportional to the volume, D^3, while the flux density, S, is related to $1/D^2$, it follows that N is proportional to $S^{-3/2}$ or $\log N = -1.5 \log S$. Thus when plotted on a log–log plot of number vs. flux density, the expected slope is -1.5.

13. To address what he saw as a legitimate criticism of the 2C catalog, Scheuer (1957) developed a statistical analysis of the interferometer data which was not subject to the errors of noise and confusion that contributed to cataloging individual sources. Hewish (1961) used Scheuer's method to calculate a slope of -1.8 for the new 4C survey, in good agreement with the value obtained from counting individual sources.

14. We now know that indeed there are structures in the distribution of galaxies on scales of hundreds of Megaparsecs.

15. Crawford, Jauncey, and Murdoch (1970) showed how to derive the differential source count from ungrouped data.

16. See also the Nobel Prize lectures by Adam Riess, https://www.nobelprize.org/prizes/physics/2011/riess/lecture/, Brian Schmidt, www.nobelprize.org/prizes/physics/2011/schmidt/lecture/, and Saul Perlmutter, https://www.nobelprize.org/prizes/physics/2011/perlmutter/lecture/

17. Modern radio spectroscopy typically employs narrow band filters to separate the narrow band radiation from atomic and molecular transitions, which are referred to as "spectral lines," by analogy with optical Fraunhofer lines observed in solar and stellar visual spectra.

18. Radio recombination lines (RRL) are an extension of the familiar Lyman and Balmer series in the optical spectrum, but with quantum numbers much larger than 1. The detection of RRLs is discussed further in Chap. 6. The 1420 MHz (21 cm) hyperfine transition occurs when the electron and proton spins flip from being in the same direction to opposite directions with a release of energy, corresponding to $E/h = 1420.404$ MHz. where h is the Planck constant.

19. J. Oort to GR, 30 August 1945. NAA-GR, General Correspondence I. https://science.nrao.edu/about/publications/open-skies#section-2

20. GR to JLG, 19 November 1950, NAA-GR, General Correspondence I. https://science.nrao.edu/about/publications/open-skies#section-2

21. A personal account of Ewen's career and his detection of the 21 cm line can be found in https://science.nrao.edu/about/publications/open-skies#section-2, his interview with the AIP at https://www.aip.org/history-programs/niels-bohr-library/oral-histories/6659, his interview with Woody Sullivan at, https://science.nrao.edu/about/publications/open-skies#section-2,   and Ewen's 19 February 1978 handwritten notes to Purcell held at the Harvard University Archives in the papers of Edward Purcell.

22. Ewen to Purcell, 19 February 1978, op. cit.

23. In order to mitigate the effect of receiver instabilities, Ewen's radiometer recorded the difference between two nearby frequencies. However, his initial choice of only 10 kHz between the signal and reference frequencies proved to be too close, and so most of the galactic 21 cm signal was canceled out.

24. H. Ewen to E. Purcell, 2 February 1978, op. cit.

25. https://science.nrao.edu/about/publications/open-skies#section-2

26. H. Ewen to E. Purcell, 2 February 1978, op. cit.

27. These two reports, *The Radio-Frequency Line-Spectrum of Atomic Hydrogen, I. The Calculation of Frequencies of Possible Transitions*, and *2. The Calculation of Transition Probabilities* were distributed as RPL 33 (February 1949) and RPL 34 (May 1949) respectively. It is not clear if these papers were known to Ewen and Purcell or to Oort and van de Hulst, but were surely known to Wild's Radiophysics colleagues Bowen and Pawsey.

28. GR to H. Ewen and E. Purcell, 8 October 1951, NAA-GR, General Correspondence I. https://science.nrao.edu/about/publications/open-skies#section-2. The 327 MHz line of deuterium escaped detection by radio astronomers for more than another half century (Rogers 2007).

29. https://science.nrao.edu/about/publications/open-skies#section-2

30. The initiative by Wade to involve Bok and Ewen in pursuing further radio astronomy work at Harvard was described by Wade, T.K. Menon and David Heeschen in their interviews with the authors 21 March 2015, 27 July 2012, and 31 May 2011 respectively, NAA-KIK, Oral Interviews. https://science. nrao.edu/about/publications/open-skies#section-2

31. Probably the first PhD based on work done in radio astronomy was John Hagen, who received his PhD in 1949 based on a observations of solar radio emission.

32. Only the 25 meter (82 foot) Dwingeloo Telescope completed in 1955 was larger.

33. A. Waterman, 28 April 1956, "Windows on the Future," Remarks at the Dedication of the George A. Agassiz Telescope. NAA-NRAO, Founding and Organization, Planning Documents. https://science.nrao.edu/about/publications/open-skies#section-2

34. The partially built 30 meter Iraqi radio telescope was destroyed in a bombing attack by the Iranian Air Force.

35. See *Bibliography of Radio Astronomy, Bibliography of Extraterrestrial Radio Noise*, and *Bibliography of Natural Radio Emission from Astronomical Sources*.

36. Cohen, Marshall H. Interview by Shelley Erwin. 1996, 1997, 1999, Oral History Project, Caltech Archives. http://resolver.caltech.edu/CaltechOH: OH_Cohen_M

37. Committee members included the physicist Luis Alvarez, Sam Silver from Electrical Engineering, and Harold Weaver from Astronomy.

38. GR to H. Shapley, 27 July 1946, NAA-GR, General Correspondence I. https:// science.nrao.edu/about/publications/open-skies#section-2

39. GR to O. Struve, 16 July 1946; GR to H. Shapley, 27 July 1946; GR to C. Schauer, 12 October 1954, NAA-GR, General Correspondence I. https:// science.nrao.edu/about/publications/open-skies#section-2

40. GR to O. Struve, 3 October 1950, NAA-GR, General Correspondence I. https://science.nrao.edu/about/publications/open-skies#section-2

41. A detailed description of the NRL 50 foot dish is given by Holzschuh (1958).

42. This section on Grote Reber is adapted from Kellermann (2004), with permission from the Astronomical Society of the Pacific.

43. GR to J. Pawsey, 19 December 1951, National Archives of Australia. We are indebted to Miller Goss for bringing this letter to our attention.

44. Quoted in *Rabi, Citizen and Scientist* (Rigden 1987), p. 164.

BIBLIOGRAPHY

REFERENCES

Baade, W. and Minkowski, R. 1954, Identification of the Radio Sources in Cassiopeia, Cygnus A, and Puppis A, *ApJ*, **119**, 206

Bennett, A.S. 1962, The Revised 3C Catalogue of Radio Sources, *Mem. Roy. Astr. Soc.*, **58**, 163

Bolton, J.G. 1948, Discrete Sources of Galactic Radio Frequency Noise, *Nature*, **162**, 141

Bolton, J.G. 1982, Radio Astronomy at Dover Heights, *Proc. Astron. Soc. Austr.*, **4**, 349

Bolton, J.G. and Stanley, G.J. 1948, Variable Source of Radio Frequency Radiation in the Constellation of Cygnus, *Nature*, **161**, 312

Bolton, J.G., Stanley, G.S., and Slee, O.B. 1949, Positions of Three Sources of Galactic Radio-Frequency Radiation, *Nature*, **164**, 101

Bowen, E.G. 1984, The Origins of Radio Astronomy in Australia. In *The Early Years of Radio Astronomy*, ed. W.T. Sullivan III (Cambridge: CUP), 85

Bowen, E.G. 1987, *Radar Days* (Bristol: A. Hilger)

Bracewell, R.N. 2005, Radio Astronomy at Stanford, *JAHH*, **8**, 75

Bracewell, R. and Swarup, G. 1961, The Stanford Microwave Spectroheliograph Antenna, a Microsteradian Pencil Beam Antenna, *IRE Trans. Ant. Prop.*, **AP-9**, 22

Burke, B.F. 2003, Early Years of Radio Astronomy in the U.S. In *Radio Astronomy from Karl Jansky to Microjansky*, ed. L.I. Gurvits, S. Frey, and S. Rawlings (Les Ulis: EDP Sciences), 27

Burke, B.F. and Franklin, K.L. 1955, Observations of a Variable Radio Source Associated with the Planet Jupiter, *JGR*, **60**, 213

Christiansen, W.N. 1984, The First Decade of Solar Radio Astronomy in Australia. In *The Early Years of Radio Astronomy*, ed. W.T. Sullivan III (Cambridge: CUP), 113

Christiansen, W.N. and Hindman, J.V. 1952a, 21 cm Line Radiation from Galactic Hydrogen, *Observatory*, **72**, 149

Christiansen, W.N. and Hindman, J.V. 1952b, A Preliminary Survey of 1420 Mc/s. Line Emission from Galactic Hydrogen, *Austr. J. Sci. Res.*, **A5**, 437

Crawford, D.F., Jauncey, D.L., and Murdoch, H.S. 1970, Maximum-Likelihood Estimation of the Slope from Number-Flux Density Counts of Radio Sources, *ApJ*, **162**, 405

Denisse, J.F. 1984, The Early Years of Radio Astronomy in France. In *The Early Years of Radio Astronomy*, ed. W.T. Sullivan III (Cambridge: CUP), 303

Denisse, J.-F., Leroux, E. and Steinberg, J.-L. 1955, Observations du Rayonnement Galactique sur la Longueur d'onde de 33 cm, *Comptes Rendus*, **240**, 278

Dicke, R.H. 1946, The Measurement of Thermal Radiation at Microwave Frequencies, *Rev. Sci. Instrum.*, **17**, 268

Dicke, R.H. and Beringer, R. 1946, Microwave Radiation from the Sun and the Moon, *ApJ*, **103**, 375

Dicke, R.H. et al. 1946, Atmospheric Absorption Measurements with a Microwave Radiometer, *Phys. Rev.* **70**, 340

Dieter, N.H., Weaver, H., and Williams, D.R.W. 1966, Secular Variations in the Radio-Frequency Emission of OH, *AJ*, **71**, 160

Edge, D.O. 1984, Styles of Research in Three Radio Astronomy Groups. In *The Early Years of Radio Astronomy*, ed. W.T. Sullivan III (Cambridge: CUP), 351

Edge, D.O. et al. 1959, A Survey of Radio Sources at a Frequency of 157 Mc/s, *Mem. Roy. Astr. Soc.*, **68**, 37

Edge, D.O. and Mulkay, M.J. 1976, *Astronomy Transformed: The Emergence of Radio Astronomy in Britain* (New York: Wiley)

Elbers, A. 2017, *The Rise of Radio Astronomy in the Netherlands* (Cham: Springer Nature)

Ewen, H.I. and Purcell, E.M. 1951a, Radiation from Hyperfine Levels of Interstellar Hydrogen, *Phys. Rev.*, **83**, 881

Ewen, H.I. and Purcell, E.M. 1951b, Observation of a Line in the Galactic Radio Spectrum, *Nature*, **168**, 356

Frater, R.H. and Ekers, R.D. 2012, John Paul Wild AC CBE FAA FTSE. 17 May 1923—10 May 2008, *Biog. Mem. R. Soc. London*, **10**, 229 https://royalsocietypublishing.org/doi/pdf/10.1098/rsbm.2012.0034

Frater, R.H., Goss, W.M., and Wendt, H.W. 2017, *Four Pillars of Radio Astronomy, Mills, Christiansen, Wild, Bracewell* (Cham: Springer Nature)

Ginzburg, V.L. 1984, Remarks on My Work in Radio Astronomy. In *The Early Years of Radio Astronomy*, ed. W.T. Sullivan III (Cambridge: CUP), 289

Graham-Smith, F. 1951, An Accurate Determination of the Positions of Four Radio Stars, *Nature*, **168**, 555

Graham-Smith, F. 2005, The Early History of Radio Astronomy in Europe. In *Radio Astronomy from Karl Jansky to Microjansky*, ed. L.I. Gurvits, S. Frey, and S. Rawlings (Les Ulis: EDP Sciences), 1

Haddock, F.T. 1984, US Radio Astronomy Following World War II. In *Serendipitous Discoveries in Radio Astronomy*, ed. K.I. Kellermann and B. Sheets (Green Bank: NRAO/AUI), 115

Haddock, F.T., Mayer, C.H., and Sloanaker, R.M. 1954, Radio Emission from the Orion Nebula and Other Sources at λ 9.4 cm, *ApJ*, **119**, 456

Hagen, J.P., Haddock, F.T., and Reber, G. 1951, NRL Aleutian Radio Eclipse Expedition, *S&T*, **10** (5), 111

Hagen, J.P. and McClain, E.F. 1954, Galactic Absorption of Radio Waves, *ApJ*, **120**, 368

Hanbury Brown, R. 1991, *Boffin* (Bristol: A. Hilger)

Hanbury Brown, R. and Hazard, C. 1951, Radio Emission from the Andromeda Nebula, *MNRAS*, **111**, 357

Hanbury Brown, R. and Hazard, C. 1953, A Survey of 23 Localized Sources in the Northern Hemisphere, *MNRAS*, **113**, 123

Heightman, D.W. 1936, Observations on the Ultra High Frequencies, *T&R Bulletin*, **May 1937**, 496

Hewish, A. 1961, Extrapolation of the Number-Flux Density Relation of Radio Stars by Scheuer's Statistical Method, *MNRAS*, **123**, 167

Hey, J.S. 1946, Solar Radiations in the 4-6 Metre Radio Wave-length Band, *Nature*, **157**, 47

Hey, J.S. 1973, *The Evolution of Radio Astronomy* (New York: Science History Publications)

Hey, J.S., Parsons, S.J., and Phillips, J.W. 1946a, Fluctuations in Cosmic Radiation at Radio Frequencies, *Nature*, **158**, 234

Hey, J.S., Phillips, J.W., and Parsons, S.J. 1946b, Cosmic Radiations at 5 Metre Wavelength, *Nature*, **157**, 296

Holzschuh, D.L. 1958, The NRL Precision "Big Dish" Antenna, *IRE International Convention Record*, **32**

Ishiguro, M. et al. 2012, Highlighting the History of Japanese Radio Astronomy. 1: An Introduction, *JAHH*, **15**, 213

Kellermann, K.I. 1996, John Gatenby Bolton (1922-1993), *PASP*, **108**, 729

Kellermann, K.I. 2004, Grote Reber (1911-2002), *PASP*, **116**, 703 (https://doi.org/10.1086/423436)

Kellermann, K.I. ed., 2012, *A Brief History of Radio Astronomy* (Dordrecht: Springer), (translation of Russian edition, ed. S.Y. Braude et al.)

Kellermann, K. I. and Sheets, B. eds. 1984, *Serendipitous Discoveries in Radio Astronomy* (Green Bank: NRAO/AUI)

Ko, H.C. and Kraus, J.D., 1957, A Radio Map of the Sky at 1.2 Meters, *S&T*, **16** (4), 160

Kragh, H. 1996, *Cosmology and Controversy: the Historical Development of Two Theories* (Princeton: Princeton Univ. Press)

Kraus, J.D. 1984, Karl Guthe Jansky's Serendipity, Its Impact on Astronomy and Its Lessons for the Future. In *Serendipitous Discoveries in Radio Astronomy*, ed. K. I. Kellermann and B. Sheets (Green Bank: NRAO/AUI), 57

Kraus, J.D. 1995, *Big Ear Two* (Powell, Ohio: Cygnus-Quasar Books)

Kusch, P. and Prodell, A.G. 1950, On the Hyperfine Structure of Hydrogen and Deuterium, *Phys. Rev.*, **79**, 1009

Lequeux, J. 1962, Mesures Interférométriques a Haute Résolution du Diamétre et de la Structure des Principales Radiosources, *Annales D'Astrophysique*, **25**, 221

Lequeux, J. et al. 1959, L'intérférométriques à Deux Antenna à Espacement Variable de la Station de Radioastronomie de Nançay, *Comptes Rendus*, **249**, 634

Lilley, A.E. and McClain, E.F. 1956, The Hydrogen-Line Red Shift of Radio Source Cygnus A, *ApJ*, **123**, 172

Lovell, A.C.B. 1984a, Impact of World War II on Radio Astronomy. In *Serendipitous Discoveries in Radio Astronomy*, ed. K.I. Kellermann and B. Sheets (Green Bank: NRAO/AUI), 89

Lovell, A.C.B. 1984b, The Origins and Early History of Jodrell Bank. In *The Early Years of Radio Astronomy*, ed. W.T. Sullivan III (Cambridge: CUP)

Lovell, A.C.B. 1990, *Astronomer by Chance* (New York: Basic Books Inc.)

Mayer, C.H., McCullough, T.P., and Sloanaker, R.M. 1957, Evidence for Polarized Radio Radiation from the Crab Nebula, *ApJ*, **126**, 468

Mayer, C.H., McCullough, T.P., and Sloanaker, R.M. 1958a, Observations of Venus at 3.15-CM Wave Length, *ApJ*, **127**, 1

Mayer, C.H., McCullough, T.P., and Sloanaker, R.M. 1958b, Observation of Mars and Jupiter at a Wave Length of 3.15 cm, *ApJ*, **127**, 11

McClain, E.F. 1958, The Naval Research Laboratory's 84-foot Radio Telescope, *S&T*, **17** (12), 608

McCready, L.L. et al. 1947, Solar Radiation at Radio Frequencies and its Relation to Sunspots, *Proc. Roy. Soc. A*, **190**, 357

Mills, B.Y. 1984, Radio Sources and the Log N – Log S Controversy. In *The Early Years of Radio Astronomy*, ed. W.T. Sullivan III (Cambridge: CUP)

Mills, B.Y. 2006, An Engineer Becomes an Astronomer, *ARAA*, **44**, 1

Mills, B.Y. and Little, A.G. 1953, A High-Resolution Aerial System of a New Type, *Austr. J. Phys.*, **6**, 272

Mills, B.Y. and Slee, O.B. 1957, A Preliminary Survey of Radio Sources in a Limited Region of the Sky at a Wavelength of 3.5 m, *Austr. J. Phys.*, **10**, 162

Mills, B.Y., Slee, O.B., and Hill, E.R. 1960, A Catalogue of Radio Sources between Declinations −20° and −50°, *Austr. J. Phys.*, **13**, 676

Mills, B.Y. and Thomas, A.B. 1951, Observations of the Source of Radio-Frequency Radiation in the Constellation of Cygnus, *Austr. J. Sci. Res. A*, **4**, 158

Muller, C.A. and Oort, J.H. 1951, The Interstellar Hydrogen Line at 1,420 Mc./sec and an Estimate of Galactic Rotation, *Nature*, **168**, 356

Needell, A.A. 2000, *Science, Cold War and the American State, Lloyd V. Berkner and the Balance of Professional Ideals* (Amsterdam: Harwood Academic)

Orchiston, W. 2005, Dr. Elizabeth Alexander: First Female Radio Astronomer. In *The New Astronomy*, ed. W. Orchiston (Dordrecht: Springer), 71

Orchiston, W. et al. 2007, Highlighting the History of French Radio Astronomy, *JAHH*, **10**, 221

Pawsey, J.L. 1946, Observations of Million Degree Thermal Radiation from the Sun at a Wavelength of 1.5 Metres, *Nature*, **158**, 633

Penzias, A. and Wilson, R.W. 1965, A Measurement of Excess Antenna Temperature at 4080 Mc/s, *ApJ*, **142**, 419

Perlmutter, S. et al. 1999, Measurements of Omega and Lambda from 42 High-Redshift Supernovae, *ApJ*, **517**, 565

Radhakrishnan, V. 2006, Olof Rydbeck and Early Swedish Radio Astronomy, *JAHH*, **9**, 139

Reber, G. 1944, Cosmic Static, *ApJ*, **100**, 279

Reber, G. 1946, Solar Radiation in 489 Mc/sec, *Nature*, **158**, 945

Reber, G. 1955, Tropospheric Refraction near Hawaii, *IRE Trans. Ant. Prop.*, **AP-3**, 143

Reber, G. 1958, Early Radio Astronomy at Wheaton, Illinois, *Proc. IRE*, **46**, 15

Reber, G. 1959, Radio Interferometry at Three Kilometers Altitude Above the Pacific Ocean, *JGR*, **64**, 287

Reber, G. and Greenstein, J.L. 1947, Radio Frequency Investigations of Astronomical Interest, *Observatory*, **67**, 15

Riess, A.G. et al. 1998, Observational Evidence from Supernovae for an Accelerating Universe and a Cosmological Constant, *AJ*, **116**, 1009

Rigden, J.S. 1987, *Rabi Scientist and Citizen* (New York: Basic Books)

Robertson, P. 1992, *Beyond Southern Skies – Radio Astronomy and the Parkes Telescope* (Cambridge: CUP)

Robertson, P. 2017, *Radio Astronomer – John Bolton and a New Window on the Universe* (Sydney: NewSouth Publishing)

Rogers, A.E.E. 2007, Observations of the 327 MHz Deuterium Hyperfine Transition, *AJ*, **133**, 1625

Ryle, M. 1955, Radio Stars and Their Cosmological Significance, *Observatory*, **75**, 137

Ryle, M. 1958, Bakerian Lecture. The Nature of the Cosmic Radio Sources, *Proc. Roy. Soc.*, **248**, 289

Ryle, M. 1968, Counts of Radio Sources, *ARAA*, **6**, 249

Ryle, M. and Clarke, R.W. 1961, An Examination of the Steady-State Model in the Light of Some Recent Observations of Radio Sources, *MNRAS*, **122**, 349

Ryle, M. and Scheuer, P.A.G. 1955, The Spatial Distribution and Nature of Radio Stars, *Proc. Roy. Soc.*, **230**, 448

Ryle, M., Smith, F.G., and Elsmore, B. 1950, A Preliminary Survey of the Radio Stars in the Northern Hemisphere, *MNRAS*, **110**, 508

Salomonovich, A.E. 1984, The First Steps of Soviet Radio Astronomy. In *The Early Years of Radio Astronomy*, ed. W.T. Sullivan III (Cambridge: CUP), 269

Scheuer, P.A.G. 1957, A Statistical Method for Analyzing Observations of Faint Radio Stars, *Proc. Camb. Phil. Soc.*, **53**, 764

Schott, E. 1947, 175 MHz Solar Radiation, *Phys. Blatter*, **3** (5), 159

Seeger, C.L. and Williamson, R.E. 1951, The Pole of the Galaxy as Determined from Measurements at 205 Mc/sec, *ApJ*, **113**, 21

Shakeshaft, J.R. et al. 1955, A Survey of Radio Sources Between Declinations −38° and +83°, *MNRAS*, **67**, 106

Shklovsky, I.S. 1949, Monochromatic Radio Emission from the Galaxy and the Possibility of Its Observation, *Astr. Zh.*, **26**, 10. English translation in Sullivan, W.T. III 1982, *Classics in Radio Astronomy* (Cambridge: CUP), 318

Shklovsky, I.S. 1960, *Cosmic Radio Waves* (Cambridge: Harvard Univ. Press)

Shklovsky, I.S., 1982, On the History of the Development of Radio Astronomy in the USSR, *News on Life, Science, and Technology*. No. 11 (Moscow: Izd. Znanie), 82 (in Russian)

Southworth, G. 1945, Microwave Radiation from the Sun, *J. Franklin Inst.* **239**, 285

Southworth, G.C. 1956, Early History of Radio Astronomy, *Sci. Monthly*, **82** (2), 55

Stahr, M. 1949, The Variation of Solar Radiation at 205 Megacycles, *AJ*, **54**, 195

Sullivan, W.T. III 1982, *Classics in Radio Astronomy* (Cambridge: CUP)

Sullivan, W.T. III ed. 1984, *The Early Years of Radio Astronomy* (Cambridge: CUP)

Sullivan, W.T. III 2009, *Cosmic Noise: A History of Early Radio Astronomy* (Cambridge: CUP)

Sullivan, W., Goss, W.M., and Raj, A. 2019, A "Radio Sundial" for the Jansky Very Large Array in New Mexico, *Compendium*, **23** (3) https://science.nrao.edu/about/publications/sundial

Tanaka, H. 1984, Development of Solar Radio Astronomy in Japan up until 1960. In *The Early Years of Radio Astronomy*, ed. W.T. Sullivan III (Cambridge: CUP), 335

Townes, C.H. 1957, Microwave and Radio-Frequency Resonance Lines of Interest to Radio Astronomy. In *IAU Symposium No.4, Radio Astronomy*, ed. H.C. van de Hulst (Cambridge: CUP)

van de Hulst, H.C. 1945, Origin of Radio Waves from Space, *Nederlandsch Tijdschrift voor Natuurkunde*, **11**, 210. English translation in Sullivan, W.T. III 1982, *Classics in Radio Astronomy* (Cambridge: CUP), 302

Wang, S. 2009, Personal Recollections of W.N. Christiansen and the Early Days of Chinese Radio Astronomy, *JAHH*, **12**, 33

Wild, J.P. 1952, The Radio-Frequency Line Spectrum of Atomic Hydrogen and Its Applications in Astronomy, *ApJ*, **115**, 206

Wild, J.P. 1967, The Radioheliograph and the Radio Astronomy Programme of the Culgoora Observatory, *Proc. Astron. Soc. Austr.*, **1**, 38

FURTHER READING

Allison, D.K. 1981, *New Eye for the Navy: The Origin of Radar at the Naval Research Laboratory* (Washington, NRL) https://apps.dtic.mil/dtic/tr/fulltext/u2/a110586.pdf

Bolton, J.G. 1960, Radio Telescopes. In *Telescopes, Stars and Stellar Systems*, **1**, ed. G.P. Kuiper and B.M. Middlehurst (Chicago: University of Chicago Press), 176

Bok, B.J. 1956, The George R. Agassiz Radio Telescope of the Harvard Observatory, *Nature*, **178**, 232

Bok, B.J., Ewen, H.I., and Heeschen, D.S. 1957, The George R. Agassiz Radio Telescope of Harvard Observatory, *AJ*, **62**, 8

Conklin, N.D. 2006, *Two Paths to Heaven's Gate* (Charlottesville: NRAO)

Heeschen, D.S. 1956, Harvard's New Radio Telescope, *S&T*, **15**, 388

Kellermann, K.I. 1972, Radio Galaxies, Quasars, and Cosmology, *AJ*, 77, 531

Kraus, J.D. 1988, Grote Reber, Founder of Radio Astronomy, *JRASC*, **82**, 107

Leverington, D. 2017, *Observatories and Telescopes of Modern Times* (Cambridge: CUP)

Levy, D.H. 1993, *The Man Who Sold the Milky Way* (Tucson: University of Arizona Press)

Orchiston, W. and Slee, O.B. 2005, The Radiophysics Field Stations, and the Early Development of Radio Astronomy. In *The New Astronomy*, ed. W. Orchiston (Dordrecht: Springer), 119

Scheuer, P.A.G. 1968, Radio Astronomy and Cosmology. In *Stars and Stellar Systems*, **9**, ed. A.R. Sandage et al. (Chicago: University of Chicago Press), 725

Scheuer, P.A.G. 1990, Radio Source Counts. In *Modern Cosmology in Retrospect*, ed. B. Bertotti et al. (Cambridge: CUP), 331

Strom, R. 2005, Radio Astronomy in Holland before 1960. In *The New Astronomy*, ed. W. Orchiston (Dordrecht: Springer), 93

Sullivan, W.T. III 1990, The Entry of Radio Astronomy into Cosmology: Radio Stars and Martin Ryle's 2C Survey. In *Modern Cosmology in Retrospect*, ed. B. Bertotti et al. (Cambridge: CUP), 309

Van Worden, H. and Strom, R. 2006, The Beginnings of Radio Astronomy in the Netherlands, *JAHH*, **9**, 3

Weaver, H.F. 2006, Early Days of the Radio Astronomy Laboratory and the Hat Creek Radio Observatory. In ASPC **356**, *Revealing the Molecular Universe: One Antenna is not Enough*, ed. D.C. Backer et al. (San Francisco: ASP), 75

CHAPTER 3

A New Era in Radio Astronomy

By the early 1950s the US Department of Defense, especially the Navy, and the newly created National Science Foundation (NSF) began to play a major role in American science, especially in astronomy. Meanwhile, Associated Universities, Inc. (AUI), which founded and operated the Brookhaven National Laboratory, was looking for new business. During these Cold War times, the United States could not afford to fall behind in the exciting and rapidly developing new area of radio astronomy. Caltech, MIT, Harvard, and Naval Research Laboratory scientists discussed how to get the United States more involved in this emerging field that had clear commercial and military applications as well as extraordinary opportunities for basic research. Two key conferences held at the end of 1953 and the start of 1954 provided the catalyst for an NSF-funded feasibility study aimed toward the goal of establishing a national radio astronomy facility.

After an exhaustive search, a site was chosen in a remote part of West Virginia between the small hamlets of Arbovale and Green Bank. Following more than two years of confrontational discussions about the nature of the proposed national radio astronomy facility and how it would be managed, the NSF awarded a contract in November 1956 to AUI to manage the construction and operation of the National Radio Astronomy Observatory.

3.1   The Business of Science

As a result of their widely recognized contribution to the development of nuclear weapons and radar, American scientists emerged from WWII with a prestige that afforded them great influence during the post-war decades of the Cold War. Their influence increased after the 1957 Soviet launch of Sputnik as more money started flowing to US science and technology. Prior to the war, the center of physics was in Europe, but as a result of the physical devastation in Europe brought about by six years of conflict and the migration of many

© The Author(s) 2020, corrected publication 2021
K. I. Kellermann et al., *Open Skies*, Historical & Cultural Astronomy,
https://doi.org/10.1007/978-3-030-32345-5_3

eminent scientists from Europe to the United States, the US emerged after 1945 as the dominant scientific power in the world, fueled by unprecedented government spending on science. Although research was mostly concentrated at the universities, the generous government financial support, which had been prompted by the urgencies of the war, continued during the post-war period. Initially much of the federal support for science came from the military for defense of the country, driven, at least in part, by competition among the services. At this time, federal leadership in science fell to the Atomic Energy Commission (AEC), the National Bureau of Standards (NBS), and the military services. The Office of Naval Research (ONR), in particular, developed close ties with American universities to support a variety of basic and applied research programs, even those that had no direct bearing on defense programs.

The National Science Foundation[1] As early as 1942, Senator Harley Kilgore from West Virginia introduced a bill to create a National Science Foundation to distribute grants and contracts supporting both basic and applied research. Kilgore's bill paid particular attention to a broad geographic distribution of the funds. By contrast, the Vannevar Bush (1945) classic report, *Science—The Endless Frontier*, which argued for continued federal support for research, was more elitist than the Kilgore bill, and argued that the most public good would come from supporting only the best scientists and the best universities and laboratories.

It would be eight years before Congress and the Truman administration could agree on language addressing issues such a geographic diversity, the inclusion of the social sciences, applied research, patent rights, and administrative control. A 1947 bill, giving control to a board of scientists that appointed a director who reported to the board, was vetoed by President Truman, who wanted to appoint the director himself, with the board acting only in an advisory capacity.

The National Science Foundation was finally established in 1950. Alan Waterman, former Chief Scientist at ONR, was appointed as the first NSF director and recruited other senior ONR personnel to fill key positions at the new organization. In particular, Admiral Tom Owen, who later became Assistant Director for the Division of Astronomy, Atmospheres, Earth, and Oceans (AAEO), colloquially referred to as "Earth, Air, Fire, and Water," later played a major role in funding the Very Large Array (Chap. 7). Dan Hunt, who was a former Navy Captain, headed the NSF National Centers and then later the Astronomy Division. Although the 1950 NSF enabling act authorized an annual budget of $15,000,000, the initial budget for fiscal year 1951 was only $224,000. By 1951, the Korean War had begun and national priorities were turning elsewhere. The NSF budget was increased to $3.5 million for fiscal year 1952, still well below the authorization level, but sufficient to begin a modest grant program to support individual investigators. But with the limited NSF funds, many highly qualified proposals were left unfunded.

Waterman was looking for ways to enable the new NSF to make an impact on American science beyond just supporting research grants for individual investigators. Coming from ONR, he was aware of the emerging opportunities in radio astronomy, with its obvious cutting edge commercial and military applications and its potential impact on basic research. He was also aware of the ambitious plans in the UK and Australia to construct large new facilities for radio astronomy research and the potential for important new discoveries. Under Waterman's leadership, the NSF gave early support to the construction of the Harvard 24 foot and then 60 foot radio telescopes, as well as to Ohio State University for John Kraus's helical array. But the bulk of federal support for radio astronomy still came from the defense related programs at ONR and the Air Force Office for Scientific Research (AFOSR), as well as the privately funded program at the Carnegie Institution's Department of Terrestrial Magnetism (DTM).

An important NSF policy change occurred in May 1955 when the National Science Board (NSB) issued a statement declaring that[2]:

1. The NSF should recommend as a national policy the desirability of government support of large-scale basic scientific facilities when the need is clear, and it is the national interest, when the merit is endorsed by panels of experts, and when the funds are not readily available from other sources,
2. A national astronomical observatory, a major radio astronomy facility, and university research installations of computers, accelerators and reactors are examples of such desirable activities for the NSF.

Although there was no specific Congressional authorization for the NSF to use its funds to pay for scientific equipment and facilities, apparently Congress did not object (Lomask 1976, p. 139). More importantly, the NSF was explicitly forbidden by law to operate any facilities or laboratories,[3] so any funds used for this purpose had to be routed through a university or other research institution. What was not specified, however, was the degree of control that the NSF would have over the construction, operation and maintenance of any facility that they might fund.

Recognizing that there might be opposition to the diversion of funds from the NSF mandate to support individual investigators to the funding of construction and operation of large scale facilities, the NSB further stipulated that "Funds for such large-scale projects should be handled under special budgets."[4] In August, NSF Director Alan Waterman informed the White House that the "1957 NSF budget request will include an item for support of certain research facilities now urgently needed," and that "this item is in addition to the established program of the foundation in support of research by grants for basic research."[5] Unfortunately, the separation of research grants and facility funding was never firmly implemented at the NSF and remained a matter of contention for many decades.

The establishment of the 1995 Major Research Equipment or MRE (later Major Research Equipment and Facilities Construction or MREFC) funding line (Sects. 9.6 and 10.7) somewhat addressed the problem of funding the construction of new facilities, but left unresolved the source of annual funding to operate these large and expensive facilities, leading to increased tensions and competition for funds between facilities and investigators.

The Association of Universities for Research in Astronomy (AURA) and the National Astronomy Observatory[6] The broad ranging 1955 NSF policy decision was motivated largely by the growing demands of astronomy. Even before radio astronomers had begun to rally around the construction of one or more large (and expensive) radio telescopes to be located at a national radio astronomy facility, optical astronomers began what would be a parallel effort leading toward a National [Optical] Astronomy Observatory. At the time, the major US optical astronomy facilities, which were constructed largely through philanthropic support, were available primarily to scientists located only at those institutions that owned and operated the telescope, such as the Caltech Mt. Wilson and Palomar Observatories (MWPO), the University of California (Lick Observatory) and the Universities of Texas and Chicago (McDonald and Yerkes Observatories). Astronomers from other universities, particularly from the East Coast or Midwest, had at best access to smaller less competitive facilities and unfavorable skies. Indeed, with only limited opportunities for research in observational astronomy, universities were not able to attract the most talented students. Even at Harvard University, which once had the largest telescope in the world and probably the largest astronomy faculty in the US, the instrumentation had deteriorated, and the site 25 miles from Boston suffered from light pollution and poor weather.

Concerned about the lack of opportunity for the disenfranchised astronomers from the East and Midwest, John Irwin (1952) suggested that a clear site should be found in Arizona for photoelectric photometry that could be used by all astronomers. In August 1952, NSF convened an "Ad Hoc Meeting of Astronomical Consultants," which was later constituted as the *NSF Advisory Panel for Astronomy* with Jesse Greenstein from Caltech as the Chair.[7] However, apparently concerned that a search for a site would not yield any astronomical research results, a joint $21,200 proposal from Ohio State University and the Universities of Arizona and Indiana to search for a suitable location was not approved by Greenstein's Committee (Edmondson 1991; Goldberg 1983). The following year, during a meeting at Lowell Observatory on photoelectric photometry, Leo Goldberg stressed the need for a co-operative center for astronomy located at a good site with a 100-inch-class telescope to provide better opportunities for all optical observing, not just photometry, particularly for those without other access to competitive facilities and clear skies. In January 1954, following a recommendation from the NSF Advisory Panel for Astronomy, Waterman appointed an additional *Advisory Panel for a National Astronomical Observatory* with Robert McMath, Director of the University of Michigan's McMath-Hulbert Observatory as the Chair (England 1982,

p. 281).[8] As conceived, the national observatory would be owned by the US government but operated by a consortium of private universities.[9] With grants totaling $818,400 to the University of Michigan, Aden Meinel and Helmut Abt studied some 150 potential sites located throughout the Southwest leading to the choice of Kitt Peak, Arizona, some 60 miles west of Tucson (Goldberg 1983).

The McMath Panel proceeded with ambitious plans to construct a 30 inch telescope within 18 months, to be followed by an 84 inch telescope and also a large telescope dedicated to solar observations. The proposed 1957 NSF budget included funding for what was called the "American Astronomical Observatory," and the NSF wrote to McMath that "the Foundation would welcome a proposal from a group of universities organized for the purpose of managing and operating an astronomical observatory facility" (Goldberg 1983). Following an organizational meeting in March 1957, the Association of Universities for Research in Astronomy (AURA) was incorporated. Seven universities, California, Chicago, Harvard, Indiana, Michigan, Ohio State, and Wisconsin, became charter members of AURA, and the first contract with the NSF to operate the national observatory was signed in December 1957. Although the McMath Panel initially anticipated that the new observatory would be federally funded but privately owned, the NSF decided that the federal government would maintain ownership.

Princeton, citing their commitment to space astronomy, and Caltech, with their own 100 and 200 inch telescopes, did not initially join AURA.[10] In March 1958, Kitt Peak was selected as the site of the Kitt Peak National Observatory (KPNO) which was soon to house 36 inch, 84 inch, and 158 inch telescopes, and later the McMath Solar Telescope. Nicholas Mayall was appointed as the first Director of KPNO, and when he retired in 1971, Leo Goldberg became KPNO Director. In 1963, AURA formed the Cerro Tololo Inter-American Observatory (CTIO) in Chile and purchased the land surrounding the Observatory as well as land in the nearby town of La Serena to house the Observatory headquarters. In 1983, AURA reorganized their observatories to include the KPNO, CTIO, and the National Solar Observatory as parts of the National Optical Astronomy Observatory (NOAO). AURA moved their corporate offices from Tucson to Washington to be closer to the political action, and John Jefferies became the new director of NOAO.

Associated Universities Inc.[11] Following the end of the Second World War, American research in nuclear physics was concentrated at the University of California Radiation Laboratory at Berkeley, Los Alamos, the University of Chicago Argonne Laboratory, and at the Oak Ridge National Laboratory in Tennessee. Scientists from the Northeast felt excluded from pursuing the rapidly developing opportunities in nuclear and high energy particle physics, in spite of the earlier pioneering work at Harvard, MIT, and Columbia University. Following separate initiatives to develop two new regional centers for nuclear physics in the New York and Cambridge areas, it became clear that the newly created Atomic Energy Commission would not fund more than one new center, and if the two groups could not agree, there would be no new nuclear physics laboratory in the Northeast. On 23 March 1946, scientific and administrative representatives of nine Northeast universities[12] assembled at Columbia to promote the establishment of a government facility for nuclear physics. Lee DuBridge from Rochester, former Director of the MIT Radiation Laboratory, was chosen to lead the discussion which culminated in a resolution to join together to further their mutual interests.

The group wisely rejected a suggested name, PYJOHMITCH, based on the initials of the nine founding universities, and instead adopted the name of "Associated Universities Incorporated" (AUI). Except for Harvard, each AUI member university agreed to contribute $25,000 toward the new laboratory. When it was pointed out to George Kistiakowski, the Harvard representative, that Harvard was the richest of the nine institutions, Kistiakowski quickly retorted, "How do you think it got that way?"[13]

Initially AUI was incorporated in New Jersey, since it was thought that the new nuclear physics laboratory would be located in New Jersey, but when it was decided to locate the laboratory at Camp Upton on Long Island, Associated Universities Inc. was created as a New York corporation on 18 July 1946. The 18-member Board of Trustees included one administrator and one scientist from each member university (Fig. 3.1). Edward Reynolds, Vice President for

Fig. 3.1   First meeting of the AUI Board of Trustees in 1947. Credit: NRAO/AUI/NSF

Administration at Harvard and retired Brigadier General in the Army Medical Supply Services, was elected as the first Chairman of the Board and President of AUI, while MIT Professor Philip Morse was named as the first Director of the new Brookhaven National Laboratory. AUI Secretary Norman Ramsey later noted that "initially almost all the details and plans had to be made by the Trustees themselves. After there existed a Brookhaven staff, it was a somewhat difficult transition for the Trustees to acquire the confidence in the Brookhaven staff and to allocate to that staff its full responsibility."[14] Many of the same issues were to arise a decade later with the newly formed NRAO.

For the first five years, AUI did not have a separate President. Rather the position of President was assumed by the Chairman of the Board of Trustees. But, wanting to expand its activities beyond Brookhaven, AUI recruited Lloyd Berkner (Fig. 3.2), who started as its full time President in February 1951. His mandate was to bring in new business needed to smooth out the financial fluctuations in the AEC Brookhaven contract, new business which might also act as a deterrent to AUI micromanagement of Brookhaven.[15] Berkner was on the short list, and apparently the choice to be the first NSF Director, but withdrew from consideration to accept the AUI appointment (England 1982, p. 124).

Fig. 3.2 AUI President Lloyd V. Berkner led the effort to establish a national radio astronomy facility managed by AUI. Credit: NRAO/ AUI/NSF

Lloyd Viel Berkner[16] was born in 1905 in Milwaukee, Wisconsin, and grew up in Sleepy Eye, Minnesota and in North Dakota. Like many youths of the time, Berkner became fascinated with airplanes and radio, and held an amateur radio license, 9AWM. Following a year spent as a shipboard radio operator, he attended the University of Minnesota where he received his bachelor's degree in electrical engineering in 1927. One of his professors was Curtis M. Jansky, who had worked hard to persuade Bell Labs to hire his younger brother Karl in spite of Karl's chronic illness (Sect. 1.1). After graduation, Berkner worked on developing aircraft VHF radio navigation systems. Later, while working at the Bureau of Standards, he provided communications support for Admiral Richard E. Byrd and accompanied Byrd on his first Antarctic expedition as the expedition's radio engineer. His contributions were later recognized by the naming of what was then the most southerly known island in the world as Berkner Island. With perhaps somewhat less success, Berkner also became involved in assisting with radio support for Amelia Earhart on her airplane adventures. While at the Bureau of Standards, Berkner began his long research program in ionospheric physics, leading to the discovery of the F layer of the ionosphere responsible for long distance short wave radio propagation. Moving to the Carnegie Institution's Department of Terrestrial Magnetism (DTM), and building on techniques developed earlier by Merle Tuve, Berkner developed an advanced ionospheric sounding system which he deployed in Washington, DC, at Huancayo, Peru, at Watheroo in Western Australia, and near Fairbanks, Alaska to help predict optimum frequencies for short wave radio communication. During this period, he also worked with Tuve in the development of the proximity fuse, which was to play a critical role during WWII.

Tuve, who later started a radio astronomy program at DTM (Needell 2000, p. 265) is credited with having developed ionospheric sounding techniques, and, along with James Van Allen, for the WWII invention of the proximity fuse. But by this time, according to Needell (1987 p. 267), Tuve had become concerned about the growth of Federal investments in "big science" and the justification of basic research "as a basis for the defense of the free world," which he argued compromised individual creativity. Prior to the War, Tuve was Lloyd Berkner's boss at DTM, where their very different research styles led to conflicts and lasting tensions. Moreover, questions of priority and credit surrounding Berkner's contribution to both ionospheric sounding and the proximity fuse may have contributed to their subsequent confrontations around the establishment of NRAO a decade later.

During the War, Berkner had served as head of the Radar Section in the Navy Bureau of Aeronautics, where he developed aircraft radar systems and was instrumental in starting a program at the MIT Radiation Laboratory to protect American ships from Japanese attacks. It was in this capacity that he got to know I. I. Rabi and others who later became leaders of AUI.[17] As a result of his continuing work on naval electronics, Berkner rose to the rank of Rear Admiral within the Naval Reserve and was recognized by the British Government with the "Order of the British Empire."

Following the end of WWII, Berkner became head of the DTM Section for Exploratory Geophysics of the Atmosphere and also led DTM's fledging radio astronomy program. He was an active member of URSI Commission V on "Extraterrestrial Radio Noise" (later changed to "Radio Astronomy") as well as Chair of the US National Committee for URSI. As a regular advisor to the Department of Defense and to NATO, Berkner was instrumental in establishing the Distant Early Warning (DEW) line radar system and in organizing the NATO Military Assistance Program. From 1956 to 1959, he served as a member of the President's Scientific Advisory Committee (PSAC) where he led a study leading to the detection of underground nuclear tests and helped draft the 1959 Antarctic Treaty. He was a Fellow of the Arctic Institute, a Member of the National Academy of Sciences, and a Fellow of the American Philosophical Society. He was rumored to be part of a 1947 alleged secret group, known as the Majestic-12, appointed by President Truman to investigate the nature of UFOs. Later, he became Treasurer of the National Academy of Sciences, and was the founder and Chair of the 1957 International Geophysical Year, president of the International Council of Scientific Unions (1955) where he led the effort to found COSPAR, the joint IAU/URSI Committee on Space Science, president of the American Geophysical Union (1959), and president of the Institute of Radio Engineers (1961).

In 1958, Berkner became the first Chair of the National Academy of Science Space Studies Board (SSB). Following a three-year study, on 31 March 1961, on behalf of the SSB he sent a strongly worded memorandum to NASA Administrator James Webb stating that[18]:

> Scientific exploration of the Moon and planets should be clearly stated as the ultimate objective of the U.S. space program. … Scientific exploration of the Moon and planets must at once be developed on the premise that man will be included. Failure to adopt and develop our national program upon this premise will inevitably prevent man's inclusion, and every effort should be made to establish the feasibility of manned space flight at the earliest opportunity. … The Board strongly urges official adoption and public announcement of the foregoing policy and concepts by the U.S. government.

Less than two months later, in a special address to joint session of Congress, President John F. Kennedy conveyed Berkner's message stating that "this nation should set as a goal before this decade is out, of landing of a man on the moon and returning him safely to the Earth."[19]

Although Berkner may not have the name recognition as some of the other American post-war science policy leaders such as Vannevar Bush, Robert Oppenheimer, or I. I. Rabi, perhaps no one had a broader impact on mid-twentieth century science policy than Lloyd Berkner. He held numerous high level national and international positions which he was able to use to promote the case for increased Federal government funding of science. He died on 4

Fig. 3.3 AUI Assistant
to the President, Richard
M. Emberson, ca. 1962.
Credit: © 1962 IEEE;
used with permission

June 1967, following a heart attack suffered while attending a meeting of the
Council of the National Academy of Sciences.

Among Berkner's first moves as AUI President was to transfer the AUI
office from Brookhaven to the Empire State Building in New York City to
escape the day-to-day issues surrounding the operation of this complex operation and to hire Richard (Dick) Emberson (Fig. 3.3) as Assistant to the
President. A year later Emberson also became Assistant Secretary of the AUI
Corporation. After receiving his PhD in Physics from the University of Missouri
in 1936, Emberson spent three years building infrared detectors at the Harvard
College Observatory. He was at the MIT Radiation Laboratory during the
War, then at NRL. Immediately before joining AUI, Emberson spent five years
at the Department of Defense Research and Development Board.

3.2   First Steps Toward a National Radio Astronomy Facility[20]

Oddly, the first stimulus toward creating a national radio astronomy facility in
the US, came not from American scientists, but from Australia. By the late
1950s it was becoming clear that Australia could no longer remain competitive

Fig. 3.4 Chief of the CSIRO Division of Radiophysics, E.G. (Taffy) Bowen played a major role in alerting US scientific leaders to the growing opportunities in radio astronomy. Credit: CSIRO Radio Astronomy Image Archive

in the rapidly growing field of radio astronomy by simply continuing their modest individual investigator-initiated programs discussed in Chap. 2. While a small country like Australia could not compete in popular, but expensive scientific research areas such as nuclear or high energy physics, the modest CSIRO radio astronomy program was probably the most successful area of scientific research in Australia, and one that had brought considerable visibility and prestige to this country of only ten million people. CSIRO Radiophysics Chief Edward (Taffy) G. Bowen (Fig. 3.4) wanted to maintain the prominence of Australian radio astronomy, but was aware of the growing radio astronomy programs in the UK and in Europe. He wanted to make an impact, and in particular not be outdone by Bernard Lovell's planned 250 foot radio telescope under construction in the UK. But he knew that he would need to look beyond Australia for the kind of funding he needed to compete with Lovell at Jodrell Bank and Martin Ryle at Cambridge.

Bowen had been a key player in the British and later Allied radar effort during WWII. He was part of the secret 1940 Tizard Mission to the United States to discuss British radar developments, and carried plans and a prototype for a cavity magnetron, a key component to the development of microwave wavelength radar (Bowen 1987). During the following War years, Bowen spent considerable time in the US acting as liaison between the US and UK efforts to develop microwave radar systems. He helped set up the MIT Radiation Laboratory, which became the center for American radar development. During a three year stay at MIT he became close friends with influential American radar scientists, including Radiation Lab director Lee DuBridge, Robert Bacher, Alfred Loomis, Chair of the American Microwave Committee, and Vannevar Bush, later to become President of the Carnegie Institution and probably the most influential science policy maker in the post-war period. During this time,

Bowen visited a number of laboratories involved in radar development, including Bell Laboratories where he met Karl Jansky (Bowen 1987, p. 170). In 1944, Bowen joined the Australian Council for Scientific and Industrial Research (CSIR, reorganized in 1949 as Commonwealth Scientific and Research Organization, CSIRO) to help develop Australia's radar defenses. In 1946, he became Chief of the Division of Radiophysics where he led the transition from wartime radar toward a variety of peaceful programs, including radio astronomy in which Australia quickly became a world power (Hanbury Brown et al. 1992).

Following the War, Lee DuBridge became president of the California Institute of Technology (Caltech), and Robert Bacher head of the Caltech Division of Physics, Mathematics, and Astronomy. Caltech operated the Mt. Wilson and Palomar Observatories (MWPO) together with the Carnegie Institution of Washington (CIW), but Caltech had no formal astronomy program in spite of owning the largest and most powerful telescope in the world, the 200 inch Mt. Palomar telescope. Soon after he took over at Caltech, DuBridge hired Jesse Greenstein, then a young but already highly respected astronomer, away from the Yerkes Observatory to begin an astronomy program at Caltech and to train students in astronomy. Greenstein quickly recruited a strong faculty, and the department started to turn out a series of very successful PhD graduates. Their work was all based on research done at the MWPO's 60 and 100 inch telescopes on Mt. Wilson and the 200 inch telescope on Mt. Palomar. When Berkner left Carnegie's DTM for AUI, it was apparently a setback to Tuve's radio astronomy ambitions, so with the support of his boss, Vannevar Bush, Tuve exchanged ideas with his Carnegie counterpart, MWPO Director Ira Bowen, about initiating some sort of cooperative radio program in conjunction with MWPO optical astronomers (Needell 1991, p. 63). But Ira Bowen was concerned that any support for radio astronomy might come at the expense of the more traditional MWPO programs in optical astronomy (Needell 1991, p. 65).

The Pasadena radio astronomy ambitions began at Caltech, two miles distant from the MWPO offices. While he was still a graduate student at Harvard, Jesse Greenstein had become fascinated with Karl Jansky's 1933 discovery of galactic radio emission. Although he went on to be recognized for his pioneering optical and theoretical research on interstellar dust, the chemical composition of stars, and the physical properties of white dwarfs (Gunn 2003; Kraft 2005), perhaps more than any other single individual, Jesse Greenstein (Fig. 3.5) later became a powerful force in raising the US to world leadership in radio astronomy, even though he himself never did any observational or experimental work in radio astronomy. Later Greenstein recalled that as a youth he had gazed at the giant radio antennas on the New Jersey shore being used by AT&T for transatlantic telephone communications (Greenstein 1984a, b) and that he had illegally operated an amateur radio station without bothering to obtain the required FCC license. Together with Fred Whipple, while still a graduate student, Greenstein wrote the first theoretical paper in radio astron-

Fig. 3.5 Caltech Professor Jesse L. Greenstein started the radio astronomy program at Caltech and was instrumental in gaining support for a national radio astronomy facility. Credit: Greenstein-10.12-12, Courtesy of the Archives, California Institute of Technology

omy, trying to explain galactic radio emission as the result of thermal emission from interstellar grains, but their predictions were low by nearly a factor of 10,000 (Whipple and Greenstein 1937).

As discussed in Sect. 1.3 Greenstein had been part of the delegation that Otto Struve sent to Wheaton to evaluate Grote Reber's work. Following his several trips to Wheaton, Greenstein and Reber ultimately became good friends, and together with Reber, Greenstein wrote the first review paper in radio astronomy which served as an important trigger in bringing the work of radio astronomers to the attention of the broader astronomical community (Reber and Greenstein 1947). At Caltech, aware of the exciting discoveries in radio astronomy being made in the UK, in Australia, and elsewhere in the US, Greenstein argued that Caltech needed to get into this exciting new field (Gunn 2003; Greenstein 1994).[21] Although Greenstein had strong support from the MWPO astronomers Walter Baade and Rudolph Minkowski, President DuBridge wanted "assurance that radio astronomy would uncover enough in the way of new, fundamental knowledge."[22] Ultimately, DuBridge would be convinced, but the persuasive arguments were to come not from Greenstein, but from DuBridge's old wartime friends, Taffy Bowen and Vannevar Bush.

Faced with the need to raise funds for his planned large radio telescope, Taffy Bowen suggested that Caltech needed to get into radio astronomy, and pointed out the unique opportunity that would be created by combining a large radio telescope with the unique facilities of the MWPO. Bowen suggested that he could come to Caltech along with some of "his boys" to take charge of running the radio telescope. To press his case, during a 1951 visit to the US, Bowen met not only with DuBridge, but also with his other former Radiation Lab friend, Robert Bacher, and with MWPO Director, Ira Bowen (no relation). DuBridge and Bacher were impressed by Taffy Bowen's enthusiasm for radio astronomy, although Ira Bowen, an atomic spectroscopist still unimpressed by the potential of radio astronomy, commented that radio astronomy was too hard and too complicated for his observatory.[23] Earlier, Bowen had apparently discouraged Charles Townes (1995, p. 195) from working in radio astronomy with the remark, "I don't think radio waves are ever going to tell us anything about astronomy."

Following his Caltech visit, Taffy Bowen traveled to the East Coast where he met with Vannevar Bush, Alfred Loomis, a Trustee of both the Carnegie and Rockefeller Foundations, as well as other former Rad Lab colleagues Ed Purcell at Harvard, and Jerry Wiesner at MIT. Bowen was pursuing a two-pronged approach to support his ambitious radio telescope project. Either he would get American backing to finance the building of a radio telescope in Australia, or at least convince the Americans to build one in the US that he, along with "his boys" would come and help run.

The identification by Baade and Minkowski (1954) of the strong radio sources Cas A and Puppis A with Galactic nebulosities and of Cygnus A with a distant galaxy sparked interest among Caltech and MWPO astronomers including, finally, Ira Bowen. Encouraged by Vannevar Bush, Alfred Loomis, and Ira Bowen,[24] DuBridge wrote to [Taffy] Bowen asking him to "estimate where the subject stands at the present time and where it ought to go from here."[25] Although a large steerable dish of the order of 200 feet in diameter remained the centerpiece of Caltech's focus, DuBridge was already thinking more broadly of a "radio astronomy laboratory or radio astronomy observatory" and suggested to Bowen that they were "particularly enthusiastic about your being the director of such a laboratory."[26]

Bowen responded with a thoughtful "Draft Programme for a Radio Observatory."[27] In his 11 page document, Bowen pointed out that radio astronomy "has tended to grow up in radio laboratories which are not closely associated with astronomical observatories," but that "the time appears ripe, therefore, to bring the radio and visual observations into closer contact." Bowen added that "the biggest single advance in the technique of radio astronomy is likely to come from the use of a very large radio telescope ... 200 to 250 feet in diameter." Following a brief summary of the outstanding observational areas of radio astronomy, including the Sun, the so-called "radio stars," and the recently discovered 21 cm hydrogen line, Bowen outlined the design concepts of a 200 to 250 foot diameter dish which he estimated could be built for about $1,000,000 and operated by an initial staff of 13 scientists, engineers, and technicians, plus clerical and support staff, at an annual cost of about $100,000. Bowen's report was well received by Bacher and DuBridge, although, around this time, DuBridge became aware of Lovell's plans to build a 250 foot radio telescope at the University of Manchester, and wrote to Bowen asking for his reaction to Lovell's telescope.[28] Ever confident and enthusiastic, Bowen responded with more details of a possible antenna design, but noted that since "there is no point in making a telescope smaller than the one at Manchester, there may be some point in adopting a new unit of size namely 100 yards (or perhaps 100 meters)."[29]

But Bowen had not given up hope of building his antenna in Australia, and he also asked Bush "if there is any possibility" of getting support from the Carnegie Foundation.[30] In spite of Bush's initial uncertainty about Carnegie funding, Bowen ultimately received grants from the Carnegie and Rockefeller Foundations and secured further funds from Australia to build his Giant Radio Telescope near the farming community of Parkes, several hundred miles west of Sydney (Sect. 6.6), and he dropped any further discussions with Caltech (Robertson 1992, p. 115). But he had planted the seeds of ambition in Pasadena, as well as in Cambridge and Washington, to bring the United States back into the field pioneered by Janksy and Reber. However, it remained unclear how Caltech might actually organize and operate a radio astronomy program.

Meanwhile, Greenstein and Bacher continued to pressure DuBridge to begin a radio astronomy program at Caltech. Greenstein recognized the potential power of combining a radio astronomy program with the optical facilities of the MWPO, and tried to convince DuBridge that Caltech needed to get more involved in this new window on the Universe. Greenstein was also the Chair of the new NSF Advisory Committee for Astronomy, and wrote to DuBridge that at their meeting on 5 and 6 February 1953, the committee discussed the need for a new observatory "in the southwest for use by scientists from other institutions and specifically devoted to photoelectric research" and the "need for a larger national effort in the field of radio astronomy."[31] Greenstein conveyed to DuBridge that his committee had pointed out that "we lag far behind Australia, Great Britain, and the Netherlands" and the need for "closer collaboration among the radio engineers and physicists who have thus far led this field, and astronomers who must interpret and use the results; the very great talents in applied electronics in the United States; and that important technical advances in electronics may arise in the course of this work."[32]

Raymond Seeger was then the NSF Acting Assistant Director for Mathematical, Physical, and Engineering Sciences (MPE) and a close confident of Waterman. Looking for an entry into radio astronomy with its broad potential industrial and military implications, Seeger wrote to Greenstein, suggesting that Greenstein explore with DuBridge whether he "would sponsor a meeting of a group of active workers in the field, and of astronomers, to discuss our national situation in this field and to work out in outline a national program."[33] Seeger even suggested that the NSF would be willing to support such a meeting. Seeing the opportunity to establish Caltech's authority in the field, DuBridge responded favorably, and saying he felt it "desirable to have this conference in Pasadena."[34] But DuBridge could not ignore the fact that the active radio astronomy programs in the US were at Cornell, Harvard, NRL, the Bureau of Standards, Ohio State, and DTM, and more broadly, that the power center for American science was on the east coast, particularly in Washington and Cambridge.

As Mt. Wilson and Palomar Observatories were jointly operated by Caltech and the CIW there was naturally close contact between the Caltech and Carnegie managements. With the encouragement and support of Greenstein, Bacher and Ira Bowen, DuBridge wrote on 20 March 1953 to Merle Tuve, DTM Director, as well as to Ed Purcell and Jerry Wiesner, offering to hold an international conference in Pasadena on radio astronomy in the autumn of 1953 and inviting them to be members of the organizing committee with himself as Chair and Greenstein as Executive Officer. In his letter, DuBridge explained that the purpose of the conference would be to

1) obtain a broad picture of the current status of the experimental work in radio astronomy in the United Sates,
2) to attempt to evaluate the probable major goals of radioastronomical [sic] work for the near future and the probable contributions to fundamental knowledge of the universe,
3) In light of the above, to reach conclusions relative to undertaking further research in radioastronomy [sic] in the United States.[35]

Ed Purcell responded with enthusiastic support, but commented that he was aware "of one or two other projected US conferences in radio astronomy," including ones at the National Academy of Sciences and at URSI, both planned for April 1953 in Washington, DC, and he commented that "radio astronomy conferences are springing up all over the place." Purcell also noted that while the proposed Caltech conference appeared more comprehensive than the others, there might be a problem in that some of the potential international participants are unlikely to attend two meetings on radio astronomy held in the US around the same time period. Purcell also passed a copy of DuBridge's letter to Bart Bok (Fig. 3.6), head of the Harvard radio astronomy program.[36] But Bok wrote to Greenstein that he had already organized a radio astronomy symposium at the American Association for the Advancement of Science (AAAS) to be held in Boston starting on 26 December 1953, and he invited Greenstein to be a speaker.[37]

Merle Tuve (Fig. 3.7) responded to DuBridge confirming that, indeed, he was already organizing a radio astronomy meeting at the National Academy of Science (NAS) on 29 April 1953 which would include talks by Martin Ryle, Walter Baade, Ed Purcell, H. C. van de Hulst, and John Hagen (Fig. 3.8), and also that John Hagan had scheduled three sessions on radio astronomy at the Spring URSI meeting in Washington on 27–29 April 1953.[38] DuBridge and Greenstein were surprised to learn that three other radio astronomy meetings had already been planned without their knowledge, especially considering that their MWPO colleague Walter Baade was an invited speaker at the NAS confer-

Fig. 3.6 Harvard Professor Bart Bok started the radio astronomy program at Harvard where many of the future NRAO leaders were trained, and also was a vocal supporter of AUI's plan to manage NRAO. Credit: NRAO/AUI/NSF

Fig. 3.7 DTM Director Merle Tuve, 1946. Tuve, who chaired the NSF Radio Astronomy Advisory Committee, expressed reservations about the planned role of AUI as the manager of NRAO. Credit: Courtesy of Carnegie Institution, Department of Terrestrial Magnetism

Fig. 3.8 John Hagen, Head of radio astronomy at the Naval Research Laboratory and Chair of the AUI Radio Astronomy Steering Committee. Credit: NRL

ence, that DuBridge was a member of the NAS, and that Greenstein and Bok were long-time friends. Trying to salvage some role for Caltech, Greenstein proposed that perhaps they could combine Bok's scheduled short meeting with the more comprehensive one that he and DuBridge had planned and suggested that perhaps they "could join forces, and apply to the NSF for travel funds."[39] Although Bok was supportive of Greenstein's suggestion,[40] citing the Christmas holiday on one end and the AAS Nashville meeting on other end, DuBridge declined the opportunity for a joint meeting with Bok's AAAS conference. Instead, following discussions with Seeger at the NSF, DuBridge, at Vannevar Bush's invitation, reluctantly agreed to hold his proposed meeting during January 1954 at the Carnegie Institution in Washington and not in Pasadena as he had proposed.[41]

Greenstein and DuBridge discussed possible speakers and the need for travel funds, but, as Greenstein complained, they had their signals crossed with regard to publication.[42] With the endorsement of the NSF, Greenstein and Bok had discussed the possibility of publishing some of papers from the January 1954 Washington meeting in the proceedings of the December 1953 Boston meeting. But DuBridge claimed he wanted to keep things informal and had declined the opportunity.[43] An organizing committee for the Washington meeting was established, consisting of Greenstein as Chair, DuBridge, Bok, Hagen (NRL), Tuve, Wiesner (MIT), and Seeger (NSF).

The NSF agreed to provide financial support particularly for participant travel expenses. Although Merle Tuve volunteered to provide administrative support from the Carnegie Institution, including administering the NSF funds, he had very different views from Greenstein and DuBridge on the nature and purpose of the Washington meeting. As it was the last of four US meetings, Tuve echoed Graham Smith's comment that the Washington conference "can hardly be intended for their benefit; they surely feel no strong need for getting together to inform and stimulate each other."[44] While he agreed with the goal of bringing together "astronomers and electronic physicists into warm and stimulating contact with the current state of radio astronomy," Tuve, characteristically, was strongly opposed to "the idea that one of the purposes of the conferences might be to figure out what kind of a large-scale effort the United States might, or could, or should undertake in this field." Rather, he wrote Greenstein, "The only way to nurture a subject is by finding and encouraging young men who are interested in the subject." Tuve wrote that he was concerned that Greenstein and DuBridge "have not really moved away from the original idea of holding a conference in order to figure out what large-scale equipment and activity is appropriate for the United States, and presumably appropriate for NSF support," and he went on to say that "I do not wish to be a party of any such thing," and urged that "you and Lee should reexamine your own ideas before you write the invitation letters." In suggesting participants for the meeting, Tuve again stressed that, "the meeting is not for radio astronomy workers to inform each other of their latest activities, but rather to interest and stimulate toward active participation investigators from astronomy and research electronics, especially with the hope of inducing some young men to work in this interesting new area."[45] This was to be the start of Tuve's three year struggle against a major government-funded national facility for radio astronomy, first with Greenstein and DuBridge, and later, even more forcefully, with Lloyd Berkner and AUI.

The Washington Conference Sets the Ball Rolling Greenstein maintained that one of the objectives of the Washington conference was to explore opportunities for constructing a large antenna to allow American scientists to compete with those in England and Australia. However, in response to Tuve's concern about the goals of the meeting, Greenstein agreed that the invitations would state only that the meeting was intended to "encourage increased activity and participation of various American groups in this … area of research."[46] But Tuve's definition of increased activity was to "encourage young people who want to do something, rather than the deans and other officials who would like to start up a given activity, whether they have competent and enthusiastic personnel or not." Tuve went on to point out that "radio astronomy has been given quite a bit of financial help in this country." Indeed, as it would turn out, over the next half century, radio astronomers would receive a disproportional share of the NSF astronomy budget for new construction.

Bok's meeting was held in Boston on 26 and 27 December 1953, at the American Academy of Arts and Sciences, and was sponsored by the American Association for the Advancement of Science. Both organizations were known as the *AAAS*, contributing to some confusion. Bok, characteristically full of enthusiasm and undaunted by holidays, initially planned for an early morning start on 26 December, the day after Christmas. But facing objections from the participants, he moved the start to the afternoon so participants could travel overnight after celebrating Christmas with their families. Participants were also challenged by some of the coldest Boston temperatures ever recorded with overnight temperatures reaching −26F (−34C). But the meeting went off without incident and included reviews by John Hagan, Harold (Doc) Ewen, John Kraus, Merle Tuve, Graham-Smith and Bernard Mills. Grote Reber, although invited, decline to come, characteristically citing his preference to remain in Hawaii to take data, but interestingly, a week later he did participate in the Washington meeting. Graham Smith and Bernard Mills, who were both on long term visits to DTM, inspired the group with their reports of the exciting work going on at Cambridge and Sydney. In spite of Bok's announced good intentions, there were no publications from the Boston meeting, nor was there any attempt made in Boston to influence any policy decisions regarding future US activities in radio astronomy.

The program of the Washington meeting was arranged primarily by Greenstein, following consultation with Tuve, with only minor input from the other members of the organizing committee. Greenstein convinced DuBridge to act as the titular Chairman of the meeting and, as such, to chair at least the opening and closing sessions. The conference, which was more elaborate than Bok's Boston meeting, ran from 4 to 6 January 1954 and was attended by about 75 people from universities, government, and industry. With the support of Vannevar Bush, Greenstein was able to secure funding from the new National Science Foundation to help support the participation of internationally prominent scientists such as Fred Hoyle, Graham Smith and Hanbury Brown from the UK, Hannes Alfvén from Sweden, Taffy Bowen and Bernard Mills from Australia, and Henk van der Hulst from the Netherlands. In addition to the organizers, other American participants included John Kraus from Ohio State, as well as representatives from NRL, DTM, Harvard, Princeton, MWPO, and the US Naval Observatory. The conference ignited the enthusiasm, not only of the astronomers present, but, perhaps more importantly, the influential east coast science power brokers. It would be this conference that provided the impetus toward establishing a US national radio astronomy facility, but the path would be a tortuous one, fraught with turf battles and long-standing personality conflicts.

The published reports of the conference in Volume 59 of *Journal of Geophysical Research* (pp. 149–198) only included short abstracts of the presentations (See also Hagan 1954). There were no formal discussions or recommendations from the conference about planning for the future, but after the close of the conference, a small group got together at the NSF "to consider some questions of national policy in this field."[47] In a report authored by Greenstein, citing recent US work on discrete sources, the 21 cm hydrogen line, ionized gas regions, and the Sun, the group was unanimous in their belief that there were important new scientific results to be expected in the field of radio astronomy. Most, but notably not all, participants agreed that the existing US effort was inadequate, and argued that since existing radio astronomy programs at Harvard and Ohio State, as well as at several research centers would provide adequate training for young scientists, "consideration should be given to constructing at least one major research center with large equipment, such as a 250-foot steerable paraboloid." Noting the limitations of private, industrial, and Department of Defense funding, they recommended that the NSF appoint a committee to "be responsible for a more detailed estimate of the budgetary needs, suggesting opportunities for immediate expansion and planning for the ultimate large scale capital expansion."

Fig. 3.9 NSF Director Alan T. Waterman speaking on 28 April 1956 at the dedication of the Harvard 60 foot antenna; Donald H. Menzel is seated behind him. Credit: NAA-DSH, Photographs

Alan Waterman (Fig. 3.9) had become the first director of the NSF just two weeks after Ewen and Purcell's detection of the 21 cm hydrogen line, and saw radio astronomy as an opportunity for the young NSF to make an impact. In February 1954, following a visit to Caltech and the MWPO, Waterman lost no time in responding to the January recommendation and created an NSF Advisory Panel for Radio Astronomy with Merle Tuve as Chair.[48]

The Menzel Report Inspired by the Washington meeting, Donald Menzel (Fig. 3.10), Director of the Harvard College Observatory, together with Ed Purcell, Doc Ewen, Fred Whipple, Cecilia Gaposchkin, and Bart Bok, met with MIT Vice President Julius Stratton and MIT Professor Jerome Wiesner, to discuss the possibility of a joint Harvard-MIT effort to acquire and operate a large research tool for radio astronomy."[49] But they soon realized that their ambitious plans were probably too big, not only for a combined Harvard-MIT

Fig. 3.10 Harvard
College Observatory
Director Donald Menzel,
who authored the 1954
report that initiated
Berkner's plans to develop
a national facility for radio
astronomy. Credit:
Harvard-Smithsonian
Center for Astrophysics

effort, but even if NRL were brought into the picture. Following Stratton's suggestion to establish "a radio observatory to be operated on behalf of all United States scientists" (Emberson 1959), Harvard Vice President for Research and Harvard member of the AUI Board, Edward Reynolds, suggested to Menzel that Associated Universities, Inc. might undertake the job of creating and operating a research facility.

Donald Menzel was trained as a physicist and was known primarily for his work in stellar spectroscopy, but he was not unknown to the radio community. In his youth, he held an amateur radio license (W1JEX), and as early as 1937, he had speculated on the possibility of communication with Martians by short wave radio. During WWII, working for the US Navy, Menzel studied the relation between radio propagation and solar activity. In this capacity he got to know Lloyd Berkner, who was in charge of naval aviation electronics. Menzel also remembered Grote Reber's 1936 letter to Harlow Shapley as well as Reber's 1946 proposal to build a 200 foot diameter radio telescope, and after the detection of the 21 cm hydrogen line by Ewen and Purcell, Menzel supported Bart Bok's efforts to start the first radio astronomy program at an American university.

Menzel traveled to Washington to meet with Berkner and Emberson to plan a May organizing meeting. In preparation for the May meeting, Menzel drafted a report for Berkner titled, "Survey of the Potentialities of Cooperative Research in Radio Astronomy." He ended his covering letter to Berkner with, "I hope this is the beginning of a new and important era in radio astronomy."[50]

In his introduction to the report, Menzel stated that the field of radio astronomy

encompasses most of astronomy, stars, cosmic evolution, geophysics of the atmosphere, aerodynamics, astroballistics, electronics, radio communication, electromagnetic and hydrodynamic properties of gases, statistical mechanics, thermodynamics, interaction of atoms and radiation, cosmic rays, properties of the atomic nucleus, and the speed of chemical reactions.

Menzel went on to discuss options for solar, planetary, and galactic studies, including prescient remarks on the prospects for radio astronomy investigations of the hydroxyl (OH) molecule and deuterium as well the opportunities for further research based on the neutral hydrogen 21 cm line. Interestingly, OH would not be detected in the interstellar medium for nearly another decade (Weinreb et al. 1963) while deuterium remained elusive for more than half a century (Rogers et al. 2005). He also discussed the possibilities for research in the related areas of ionospheric physics, active lunar and planetary radar, as well as "allied laboratory and theoretical studies."

In his report, Menzel noted that "although several individual scientists in the United States were preeminent in the early development of radio astronomy, there has been no broad, coordinated attack in this country on the basic problems of this field," and he argued that "other nations (in particular Britain, Holland, and Australia) now lead in this important area." He speculated that part of the reason for "the lag" was the "enormous expense of the tools," and commented that the objective of his survey was to bring together "the various scientific groups who have an interest in the field, in order to pave the way for a coordinated attack on the problems, perhaps through the medium of a formal organization, similar to that of Associated Universities Incorporated."

As he had already discussed with Berkner and Emberson, Menzel's report suggested that "a small committee of interested scientists should be formed initially to discuss the details of the program, to formulate a working polity, and to make recommendations for future extension of the work." He added that the initial participants might include, "in addition to representatives from Harvard, MIT and other members of Associated Universities Incorporated, … scientists from the Franklin Institute, Penn State, NRL, the Carnegie Institute [DTM] and possibly others." Caltech was noticeably absent from Menzel's list.

In a carefully laid out plan, Menzel thoughtfully emphasized the need to find a suitable site for the new radio facility. Although he commented on the need for "freedom from local radio interference," he considered that the "primary consideration be given to accessibility." Perhaps somewhat gratuitously, he added that "the council of the Harvard College Observatory wishes to express the opinion that its facilities at the Agassiz Station might be expanded to meet the needs of the proposal." He also suggested that the radio facility include several optical telescopes, and stated that Harvard might have some spare mirrors to donate. He laid out the requirements for recruiting staff and a director as well as the need to train students in this new field, pointing out that "only at Harvard is there at present an academic program in radio astronomy."

Menzel concluded his report to Berkner by noting that with the establishment and operation of Brookhaven, AUI had "effectively solved a similar problem" and "the suggestion that the Associated Universities Inc. should itself support the proposal merits most serious consideration." Menzel proposed that AUI set up a Steering Committee to develop plans to set up a radio observatory and suggested some 20 possible members of the Steering Committee—all from the Northeast. But as an apparent afterthought, he added DuBridge, Greenstein, and Otto Struve from UC Berkeley.

Contentious AUI and NSF Committees Berkner lost no time in reacting to Menzel's report. Two weeks later he and Menzel met with the NSF Director, Alan Waterman, and other NSF staff as well as Jerrold Zacharias from MIT and Emanuel (Manny) Piore from ONR. They convinced Waterman to sponsor a small meeting to "discuss the possibility of building a large dish in the near future," and to entertain a proposal from AUI to set up a group to further develop the planning.[51] While the NSF anticipated that they or the Department of Defense might pay to construct such a facility, they expected that private funds would be sought "for the maintenance for the upkeep of the facility."

Following the 26 April 1954 meeting at the NSF, Berkner appointed an "Ad Hoc Group for Cooperative Research in Radio Astronomy" which he promptly convened on 20 May 1954 in AUI's New York Offices on the 69th floor of the Empire State Building. Berkner's meeting was attended by 37 scientists representing 28 different institutions, including Minkowski and DuBridge from Caltech, John Hagen from NRL, and Merle Tuve from DTM.[52] Alan Waterman and Ray Seeger representing the NSF and Manny Piori from ONR came as observers. Jesse Greenstein chose not to give up his 200 inch observing time to participate in the meeting, but he later confided to John Hagan[53] that while he supported anything that would bring new resources to radio astronomy, he was concerned about the role of AUI and particularly Harvard, and worried more about the lack of experienced people than the shortage of expensive telescopes.

Berkner conveyed AUI's interest in building a 250–300 foot dish, which he estimated might cost between $2,000,000 and $5,000,000, that AUI would make available for use by all universities, not just members of AUI. Although it was understood that construction funds would need to come from the federal government, probably the NSF, Berkner suggested that AUI might seek a $5,000,000 endowment whose income of about $200,000 per year could be used for operation of the observatory. Technical discussions were concentrated on the size of the dish and whether or not it should be on an equatorial or alt-az mount. Princeton physicist Robert (Bob) Dicke made some insightful comments and suggestions.[54] First he pointed out that to avoid unacceptably high sidelobes, it was necessary to taper the illumination and thus reduce the effective area that one had worked so hard to build. Second, he suggested that instead of trying to keep the feed support structure sufficiently rigid to avoid flexure, one might consider a servo system to stabilize the support structure. Finally, with great prescience, Dicke pointed out the advantages of using an interferometer system composed of a number of small dishes instead of a single large dish. Dicke specifically suggested an interferometer composed of three 87 foot dishes moving on a circular track which would have the equivalent collecting area as a single 150 foot antenna, but which, he argued, would only cost about 1/5 as much, although he conceded that savings would be at least partially offset by the cost of the instrumentation. This seems to be the first ever suggestion to build a radio interferometer composed of multiple steerable parabolic antennas, and predates by several years the Caltech two-element interferometer or the Cambridge One-Mile Radio Telescope. However, Dicke's memo was ahead of its time, and appears to have had little serious impact to the subsequent discussions, which continued to focus on large fully steerable dishes.

Lee DuBridge surprised the participants by announcing that Caltech planned to create their own radio astronomy observatory with probable funding from the Office of Naval Research (ONR) (Sect. 6.6). The proclamation by DuBridge perhaps further motivated the Northeast science establishment and the NSF to try to establish their own facility, now needed to compete not only with Australia and the UK, but also with Caltech and ONR. It also obviated any argument to locate the proposed radio astronomy facility in the western part of the US. The group, which included the members of the NSF Advisory Panel for Radio Astronomy, supported the concept of establishing a national facility for radio astronomy and agreed that AUI should propose to the NSF to do a feasibility study for establishing a cooperative radio observatory. Emberson (1959) later wrote that the 20 May conference concluded that a three step process was needed, "(i) a feasibility study on objectives and organization, sites, and facilities, (ii) final design of facilities and equipment, and (iii) construction of the observatory."

The following day, the AUI Executive Committee authorized Berkner to apply to the NSF for a grant to study the selection of a site and to prepare a preliminary design for a radio astronomy facility,[55] and at their next meeting on 18 June 1954, the Executive Committee approved Berkner's request to appoint Dick Emberson, as "Acting Director of a radio astronomy project," along with a supporting salaried staff. With his strong background in radiophysics, Emberson was a natural choice to guide the AUI radio astronomy program. The Executive Committee also approved the appointment of an ad hoc committee recommended by Menzel to "work in close liaison with the project staff" and with a committee of the AUI Board. Recognizing that their lack of expertise in radio astronomy might open them to criticism from Tuve and others, AUI considered the need to extend the composition of the Board of Trustees, but Berkner argued that the committee could keep the Board informed through a small committee of the Board.[56]

Berkner's choice of John Hagen to chair the AUI Committee was a logical move. Starting in 1935, Hagen had worked at the Naval Research Laboratory where, as described in Sect. 2.6, he had begun a program of research in radio astronomy. As the Chair of the US Radio Astronomy Commission V of URSI, Hagen was a natural choice to chair Berkner's study committee. The other 10 members of the AUI committee came primarily from the major East Coast universities and laboratories.[57] At the suggestion of Bart Bok, David (Dave) Heeschen, who was soon to become one of the first Americans to receive a PhD degree in radio astronomy, was appointed as a consultant to the committee. Both Emberson and Heeschen worked with the Steering Committee and Berkner to plan the proposed feasibility study. Notably missing were Lee DuBridge and Jesse Greenstein, who had initiated the January 1954 meeting which had started the sequence of events leading to AUI's involvement in radio astronomy.

The Boston and Washington meetings had discussed "small science" programs typically carried out by small close-knit teams, but in just a few short months the discussion had grown to anticipate big science with a national facility, government funding, and management by committee. After years of neglect and following a series of high level meetings held over a period of four months, the United States suddenly had two national advisory committees for radio astronomy—one reporting to AUI and one reporting to the NSF.[58] Sometime they worked together, but at other times they were in conflict. These were in addition to the two NSF committees for optical astronomy, the Advisory Panel for Astronomy and the Advisory Panel for a National Astronomical Observatory. Conflicts also arose between the East and West Coast establishments, between radio and traditional (optical) astronomers, between advocates for and opponents to big government spending for science, and even between long standing personal rivals. Four individuals, Bok, Hagen, Kraus, and Tuve, served on both the AUI and NSF radio astronomy committees, which led to further tensions and mistrusts. Leo Goldberg, a member of the NSF radio astronomy panel who also served on the Advisory Panel for a National Astronomical Observatory as well as the NSF Division Committee for Mathematical, Physical, and Engineering, expressed concern about radio astronomers taking funds from "real" astronomers.[59] Much of the deliberations of the two committees and interaction between them were by written letters, sometimes private between only two committee members, other times with multiple carbon copies sent to all members of one or both committees. Little formal documentation remains from these committee/panel deliberations, although the sense of the deliberation can be reconstructed from the private papers of the participants, particularly those of Greenstein and Tuve. The situation was confused by the loose and changing definition of the committee names. At various times each committee was referred to as a "panel" or "committee" and as "advisory committee/ panel" or "steering committee/panel," often leading to confusion over who was speaking for which committee or panel and what hat they were wearing.

The NSF and AUI committees each addressed similar questions: what instruments to build at the national radio facility, where to build the facility, who would manage the facility, and how to recruit staff, particularly a director. But over the next two years the discussion invariably returned to the size and cost of the proposed radio telescope, how it would be managed, and by whom.

To discuss the draft proposal that had been prepared by Berkner and Emberson, the AUI Steering Committee met together with the members of the NSF Advisory Panel on 26 July 1954 at the AUI offices in New York. Merle Tuve, Chair of the NSF Panel was unable, or perhaps, unwilling to attend. After reviewing the broad range of potential scientific research programs, the participants dismissed the need for co-locating the proposed radio facility with optical facilities, but in order to meet all of the scientific objectives, they suggested that a range of equipment, including a high gain antenna as well as a number of smaller antennas, a standing fan-beam antenna for surveying, and an interferometer, would be needed. They also discussed the criteria for siting the facility, recognizing the need to balance observing conditions (weather and freedom from RFI) with practical accessibility.[60]

The next day, Berkner hand-carried 15 copies of an AUI proposal, along with minutes of the January and 20 May meetings, to Waterman at the NSF. The proposal requested $105,000 for a one-year study to investigate the feasibility of establishing and operating a national radio astronomy facility. AUI intended that the Phase I proposal would produce (a) a consensus of research objectives, (b) a list of the instrumentation needed, (c) an examination of possible sites, (d) an examination of other costs that might be required, e.g., roads, power, buildings, etc., (e) a determination of the cost of a Phase II study for the detailed design and construction, and (f) consideration of operating costs.[61] Dick Emberson was specified as the Principal Investigator. Berkner and Emberson referred to the January and 20 May 1954 meetings as justification for the study, but interestingly, they did not mention the December 1953 meeting in Boston.

Waterman apparently quickly recognized the opportunities for the NSF and called Berkner back the next day to explain that if funds for Phase II, which Berkner estimated to be between $1 and $1.5 million, would be needed starting in 1956, Waterman would need that information now, with "as much justification as feasible," and that he needed to know by the spring of 1955 what funds Berkner would need for fiscal 1957. Over the following week, after discussions within the Foundation, it was agreed that "plans for radio astronomy should be worked out as a specific NSF program," and that the NSB and Bureau of the Budget be kept informed. Waterman was keen that "this schedule can move forward rapidly," and hoped that the NSB would give him the "authority to carry on with the Bureau of the Budget."[62] But first Waterman had to deal with Tuve and his NSF Advisory Panel for Radio Astronomy.

Berkner's proposal was sent to more than 12 referees, including members of Tuve's committee, as well as to the Chairs of other related NSF advisory committees. Although many of the reviewers noted that they had no experience in radio astronomy, there was a general consensus that the United States had fallen behind in this important new field of astronomy and that to compete with the new and planned international facilities such as the Jodrell Bank 250 foot antenna, there was an urgent need for a cooperative large facility for radio astronomy. But some members questioned the need for the proposed study, asking whether it was an appropriate use of NSF research funds, or saying that it was too expensive. Several respondents asked whether AUI, which had no experience in radio astronomy, was the appropriate organization to carry out the study, while others noted AUI's success in managing Brookhaven. A few reviewers recognized the likelihood that AUI's study would lead to requests for millions of dollars, much more than the NSF could likely afford.[63] As Rudolph Minkowski later commented, "The most severe criticism came from those referees who are least familiar with radio astronomy."[64]

Before consulting with his Panel, Tuve paid a quiet visit to the NSF to express his concerns to Waterman, Seeger, and Peter van de Kamp, the new program manager for astronomy, that the AUI proposal was overly ambitious, claiming that, with the exception of Harvard, there was little interest even among the AUI universities. He also noted that AUI's first responsibility was to Brookhaven, and he did not see how the proposal for radio astronomy was related to Brookhaven. Finally, Tuve claimed that since his Panel "had informed itself rather thoroughly as to the potentials of radio astronomy and the plans that might be undertaken," that "much of the feasibility study proposed by AUI was unnecessary."[65] Waterman explained to Tuve that he "was looking entirely to the Tuve Committee [i.e. Panel] on Radio Astronomy" for guidance on "how the general program of the country in radio astronomy might be developed." Seeger added that they urgently needed some guidance about the 1956 radio astronomy program if funds were to be needed in the next fiscal year. Somewhat contradictorily, Tuve responded that his committee could not "make such a recommendation without having had a meeting on the subject," but that "they would probably not recommend as a major program as indicated."[66]

In summarizing the reports of the reviewers to his Panel, Tuve argued that no one at AUI was active in radio astronomy, that the "proposal is being made by administrators not researchers," and that "there is no visible basis for continuity of any activity on radio astronomy in the AUI." Tuve also called attention to the construction problems with the Manchester 250 foot project, saying it would be prudent to await the outcome of that effort "before another project of the same kind is initiated."[67] Once again, having been alerted by Tuve, Taffy Bowen stepped forward with an offer to volunteer his services to lead a jointly financed feasibility study for a large steerable radio telescope, and Tuve suggested that Bowen, rather than AUI could be "in full charge of the design project."[68] Tuve, who opposed big government support of science, and was reacting to his long standing suspicion of Berkner's motives, argued that since the reviews were negative they should give AUI only $15,000. But Greenstein came to AUI's rescue, pointing out that the average rating of the proposal was between good and excellent and that Tuve's report reflected only his own views.[69]

Following his meetings with Waterman, Berkner spent the next 2½ months in Europe, primarily to attend scientific meetings in Belgium, the Netherlands, and in the UK. Although he was able to take advantage of the opportunity to talk with Ryle, Lovell, and Oort and to learn about their progress and plans, his lengthy absence delayed meeting with the NSF to respond to questions from the reviewers and the NSF Advisory Panel. However, during Berkner's absence, Emberson did meet with the NSF and provided an estimate of the funds that would be needed for construction, for land acquisition and the rate at which funds would need to be made available. This information was subsequently forwarded by the NSF to the Bureau of the Budget to aid in their planning for the FY56 and FY57 budget cycles. But Emberson noted that Tuve's Panel was a "real block to early action on the AUI proposal."[70] Emberson also shared these details with the members of the ad hoc AUI committee, who interestingly pointed out the additional costs that would be needed to acquire a suitably large parcel of land to minimize any local sources of interference as well as the cost of providing advanced electronic instrumentation.

Aside from Berkner's absence from the US, although the NSF had established a Radio Astronomy Panel, no funds had been allocated for the Panel to meet until Tuve proposed that the Carnegie Foundation should administer a grant to provide for it to hold three meetings over the next year.[71] It was not until mid-November 1954 that the NSF Panel was able to meet to discuss the AUI proposal. In preparation for their 18 November 1954 meeting, a subset of the group, together with Berkner and Emberson, met on the evening of 4 November at the Cosmos Club in Washington to discuss revisions to the AUI proposal. As described by Bok, "it was a long and difficult evening."[72] In response to the issues raised by the reviewers and the Advisory Committee, Berkner addressed the criticism that AUI had no active radio astronomers by pointing out that the AUI proposal was based on a "a number of meetings, conferences, and discussions in which every U.S. leader in radio astronomy has participated to some extent," and that as a collaborative effort he anticipated extensive participation from the university community as well as from Ryle, Lovell, Bowen, and Oort, all of whom, he said, had offered to supply their construction plans.[73]

Prior to the 18–19 November meeting of the NSF Advisory Panel, Tuve informed its members that he felt that the AUI proposal should be for "a" facility and "not necessary for 'the' national facility of the USA" and went on to emphasize the importance of the existing facilities and that the subject is "astronomy and astrophysics," and not "automatic gadget engineering."[74] Tuve's Panel was hardly unbiased or disinterested. Greenstein, who was involved in planning Caltech's own major radio astronomy project, responded in favor of the cooperative facility but expressed reservations about starting with a large antenna.[75] Six members of the Panel were also members of the AUI ad hoc committee that had met at AUI to help to develop the AUI proposal, while Bok and Kraus themselves had large grant proposals before the Foundation. At their 18–19 November meeting, which was held at the Carnegie Institution, the Advisory Committee (a) reviewed the previous relevant meetings including the evening "rump meeting" on 4 November, (b) reviewed the AUI proposal as modified on 8 November in response to the 4 November meeting, (c) considered proposals from Harvard for their 60 foot radio telescope along with informal proposals from Kraus at Ohio State, and possible needs for other radio astronomy projects at Cornell, NRL, DTM, Caltech, the National Bureau of Standards, Michigan, Princeton, Stanford, Penn State and the University of Alaska.[76] Rather surprisingly, just before the meeting, Greenstein stated that he was resigning from the NSF Panel citing both internal and external responsibilities, but Tuve asked him to reconsider.[77]

The Panel struggled with the realization that the AUI proposal could not be treated in the same way as a typical research grant, so there was considerable debate about to what extent the proposed AUI study should be "supervised" by the Panel, by the NSF itself, or whether AUI should be free to act independently.[78] While Tuve wanted AUI to concentrate on antenna design, Berkner responded that "If all the panel wants is a cost estimate, …. I could get Hughes Aircraft to submit one." But Tuve argued that "We can't turn this whole job over to AUI," to which Berkner replied, "You must have confidence in what we are trying to do," and appealed to the success of Brookhaven, and that "nothing should be done by the Panel to question AUI's judgment." It was becoming clear at this meeting that radio astronomy might be in a very privileged position for rapid growth. Although the NSF 1954 total budget for astronomy was less than $200,000, Peter van de Kamp pointed out that "radio astronomy was considered at present to be in a special class of subjects, and that it might not be impossible for the Foundation to give a hearing to recommendations which total more than $200,000 during the present fiscal year."[79]

Apparently encouraged by van de Kamp's remarks and not wanting to miss an opportunity, the NSF Panel agreed that the AUI study should concentrate on a steerable antenna and not a Mills Cross type of radio telescope, and that the cost of the antenna might be in the vicinity of $3 million, but that they should also consider antennas that might be built for $1 and $6 million, as well as consider the "the largest steerable antenna that might be built without special regard to cost, its limits being based on the strength of materials and similar considerations."[80] They also specified that the antenna should work to at least 21 cm over the whole surface and down to 3 or 10 cm over a limited area and be able to see the entire northern sky "at least as far south as 10 degrees below the Galactic Center." Moreover, the committee suggested that the "search area for the site be confined to an area within 300 miles of Washington, D.C." The reason given for the geographical restriction was to minimize travel time by scientists and students from northeast universities, although by this time jet travel was already making single day transcontinental travel feasible. But the panel members were also likely seeking balance between the national radio facility and big west coast optical astronomy facilities as well as the planned new radio astronomy program beginning at Caltech which they felt would give west coast radio astronomers adequate access. Also, with a possible location near Washington, the NSF could watch with a closer eye and exert some control over the radio facility.

The Panel then quibbled over the size of the grant. Following Tuve's suggestion that the AUI member organizations should demonstrate their interest by contributing to the study, they recommended that AUI be given only

$85,000 and not the requested $105,000, but only as an initial grant which could be supplemented by another $15,000 "if AUI found it difficult to obtain this portion from their participating members." However, this recommendation was subject to AUI more specifically defining the scope of their study to evaluating only the feasibility of a national radio facility and the criteria by which the site should be chosen. The Panel then went on to recommend that the NSF award grants to Bok for a total of $128,000 for the operation of the Harvard 24 foot radio telescope and toward the construction of a 60 foot radio telescope, to Kraus at Ohio State for $23,000 for further studies of standing parabolic antennas, and to Tuve to administer a grant to host visiting radio astronomers to the United States and to send US students to other countries. Following a request from van de Kamp, the committee assigned the following ranking to the proposals "in order of their estimated importance: 1) Harvard, 2) Ohio State, 3) AUI, 4) grants for visitors."

That same day, Berkner sent to the NSF a three page letter essentially identical to the draft that that he shared with the Tuve's group at the Cosmos Club meeting, repeating the justification for the feasibility study, the endorsement of the radio astronomy community, and saying that, if successful, he expected this Phase I study to "generate the basic plans for the construction, management, and operation of the required facility," to be followed by a more detailed Phase 2 for the actual antenna design and choice of a contractor.[81] While he was careful to reassure Tuve and the NSF that funding of this Phase 1 proposal was not a foot-in-the door toward AUI operation of the radio astronomy facility, Berkner went on to add, "I would be less than frank in saying that in undertaking Phase 1 under the present proposal, AUI would expect to make a further proposal under Phase 2 if Phase 1 proves that such a further proposal is feasible." Of course, as had already been discussed earlier in the day at the Advisory Panel meeting, the fact that Lovell was currently building a 250 foot radio telescope, that a slightly smaller one was being planned by Bowen in Australia, and that Caltech was developing plans, already demonstrated clearly that it would be feasible to construct a national radio facility in the United States with a "large" antenna as the centerpiece. And Lloyd Berkner was determined that AUI would have a major role in building and operating the national facility.

Without waiting to receive formal notification of the NSF grant, Berkner lost no time in planning for a national radio astronomy facility and organized a small meeting at AUI on 10 December 1954 with Charles Husband, who had designed the Jodrell Bank 250 foot antenna and was supervising its construction, and Michael Karelitz from Brookhaven as a consultant to AUI. The assumption was that the basic instrument would be a "big dish" as the most versatile way "to solve certain problems," but that "the need for interferometers ... must not be ignored."[82] Berkner pointed out that the AUI task would be to determine "what should be built and at what cost." Husband optimistically advised "not to be scared by engineering problems in the structural steel phase," and that the "big engineering difficulties are in the control gears and the instrumentation." Based on his design and experience with the Jodrell Bank alt-az antenna, Husband outlined design concepts and cost estimates for steerable antennas up to 500 feet in diameter. Karelitz argued that "Serious consideration, has to be given to advantages of an equatorial mount," but at the time, an equatorial option seems not to have been seriously discussed. At this stage both the NSF and AUI were thinking of only a "collaborative" observatory, where the construction funds would somehow be provided by the government (NSF or ONR), but ongoing operations would be the responsibility of the universities interested in radio astronomy who would provide people and resources.

3.3   CREATING THE NATIONAL OBSERVATORY

Sensitive to the accusations that AUI had no expertise in radio astronomy or, for that matter in antenna design, Berkner sought advice from Bell Laboratories, and in particular from Janksy's old boss at Bell Labs, Harald Friis, who advised that "425 to 450 feet is about the limit for a reflector or dish of conventional design which would probably cost close to ten million dollars."[83]

In January 1955, Berkner learned from Waterman that the NSF would grant $85,000 to AUI for their proposed feasibility study.[84] Assuming that this planning and feasibility study did not require National Science Board approval, the NSF had not consulted the NSB, apparently ruffling some feathers of those NSB members from southern universities who looked at AUI as an elite private institution (England 1982, p. 282). Upon learning of the grant award, Berkner relieved Emberson of his other tasks as Assistant to the President so that he could assume full time responsibility as the Project Director for radio astronomy. In February 1955, following approval of the NSF funding, the AUI ad hoc committee was formally reconstituted as the AUI Steering Committee for Radio Astronomy. With the support of the Steering Committee, following a meeting with the NSF staff on 9 April 1955, AUI put the project on a fast track with the intention of presenting to the Foundation, "a proposal looking to the immediate establishment of a radio astronomy facility to be operated by AUI," and with a goal of presenting it to the 20 May 1955 meeting of the National Science Board.[85] But the NSB presented hurdles that would first need to be overcome: "whether the Foundation, as a matter of policy, should embark on a program involving continuing support of a large scale project, and if this question is answered in the affirmative, the Board will need to … chose from a selection of several proposals which will then be before it for consideration."[86]

From the beginning, AUI's goal was to construct a facility that "should provide research opportunities not available elsewhere."[87] This may be contrasted with AURA, which was organized to create a national optical observatory to provide observing opportunities for astronomers who did not otherwise have access to their own facilities. AUI also recognized that some research groups might be interested in bringing their own instruments to the new observatory at their own expense, providing that the site was sufficiently radio quiet. However, the emphasis was on building a large antenna whose cost was beyond the means of an individual university and would be at least competitive with the large dishes being built in Manchester and Sydney. One of the first questions asked was "Are there any real technical limitations to the maximum size of steerable radio telescopes?" Following consultation with structural experts, AUI optimistically, even if not realistically, concluded that diameters up to several thousand feet would be feasible.

Knowing that Lovell was already building a 250 foot telescope at Jodrell Bank set an unspoken lower limit on Berkner's ambition. When the AUI Steering Committee met on 26 March 1955, they agreed that size needed to be balanced against cost. But, perhaps seduced by the prospects of generous federal funding, the Steering Committee initially focused on a 500 or 600 foot diameter radio telescope which they felt was justified, especially by the opportunities for both galactic and extragalactic 21 cm hydrogen line research. The Committee also noted the need for high resolution and sensitivity for continuum studies at 21 cm and shorter wavelengths that would be afforded by a very large steerable paraboloid.

Jacob Feld, an independent contractor, outlined the factors to be considered in the design of large radio telescopes and proposed to undertake a design study for a 600 foot antenna with full sky coverage, a surface tolerance of 1 inch,[88] and a pointing accuracy of 7 arcsec (5% of the 10 cm half power beam width). Some concession was made by limiting the sky coverage to 8 hours about the polar axis if an equatorial mount was adopted, although the declination range was rather unrealistically and unnecessarily specified to reach from 5 degrees below the North Pole to 5 degrees below the horizon.

But the Steering Committee also recognized that that it would be challenging to construct a 600 foot antenna and that to establish its viability, the new national radio astronomy facility needed to have observing facilities sooner than a 600 foot instrument could be erected. It is not clear to what extent the discussions about a 600 foot diameter radio telescope were based on knowledge of the NRL 600 foot antenna planned for Sugar Grove (Sect. 9.3). At least some members of the AUI Steering Committee, in particular John Hagen from NRL, were surely aware of this then-classified project, as apparently were Emberson and senior NSF staff (McClain 1960, 2007; Needell 2000, p. 284).

Since modest size antennas in the range of 50 to 85 feet in diameter were being built at Harvard, DTM, and NRL as well as for the military, both Merle Tuve[89] and AUI confidently assumed that their "design might be extrapolated to 100 to 150 feet and be essentially off-the-shelf."[90] They agreed to take immediate steps to procure a radio telescope with an aperture of about 150 feet and to simultaneously conduct feasibility studies for apertures in the range of 300 to 500 feet and greater.[91] The Steering Committee reviewed a detailed design submitted by Grote Reber for a 220 foot dish, as well as Reber's thoughts about 500 feet and larger dishes.

For both the 600 foot and 150 foot designs, the Steering Committee considered the pros and cons of placing the antenna in a radome to protect a much simpler and less expensive antenna structure from the weather. Two novel radome designs were discussed, one an air supported fiber-glass or nylon fabric, another based on a geodesic dome proposed by Buckminster Fuller. But they soon concluded that the cost of a radome would more than offset the decreased cost of the antenna structure. Moreover, they recognized that it would be difficult to build a radome that would not attenuate incoming signals over the broad frequency range of interest to radio astronomy.[92]

Both Berkner and Emberson were nervous about confining design considerations to extrapolations of antennas that had already been built and agreed that they needed to receive independent outside advice. Nevertheless, Emberson proposed an ambitious schedule which called for requesting an additional grant of $300,000 to continue the feasibility study in 1956, with a goal of starting a five-year construction and operating plan on 1 July 1957 "involving an expenditure of about twenty million dollars," although it was recognized by the AUI Board of Trustees that "20 million dollars is considerably in excess of the entire annual budget of the National Science Foundation," and that "the Corporation has never formally decided that it should undertake the management of an enterprise of this character." The Board also noted that the AUI charter would prevent its operating outside the state of New York, so they voted to amend the charter to allow AUI to operate in a state other than New York. At the same time they authorized Berkner to "enter into contractual arrangement for the construction and operation of a facility for research in radio astronomy," for which they anticipated AUI would receive a management fee "of between $35,000 and $50,000 a year."[93]

To support their ambitious plans, the Steering Committee asked Bart Bok to convene a sub-committee to "promptly" prepare "a report on a scientific justification for a variety of large steerable parabolic reflectors, with aperture of 150 feet, 300 feet, and 500 feet."[94] Bok's Panel noted that high gain might be more important than narrow beam width, but stressed the need to maintain sufficient precision to operate at least to 21 cm if not shorter wavelength. Not surprisingly, they emphasized the impact that 21 cm observations with a large radio telescope could make to the understanding of galactic structure, and commented on "the potentialities for research into the structure and dynamics of neighbor galaxies and the fainter galaxies," as well as speculating on the possibility of other spectral lines such as deuterium at 327 MHz and OH at 1668 MHz. They also pointed out the importance of measuring precise positions, flux density, and spectra of the discrete radio sources, but there was no consensus about the relative merits of interferometers and Mills Cross instruments for this type of work. Although they considered that "the primary justification for a paraboloid antenna with an aperture of 150 feet or more rests on the research potential of the instrument for galactic and extra-galactic studies," they also called attention to the "great possibilities for solar work," and especially the importance of a large paraboloid for planetary radar. The panel report concluded that "there are few problems to be expected in the construction of [a 150 foot paraboloid]" and that "the acquisition of new and superior equipment, designed to round out or fill in the picture obtained with present equipment, has led to new discoveries."

Emberson and Berkner met again on 8 April 1955 with Waterman and other NSF staff to discuss their planned proposal, but Berkner rejected the NSF suggestion that Tuve participate in the meeting.[95] Emberson explained that their proposed mode of operation would closely parallel the academic, corporate, and government partnership that had been so successful at Brookhaven, and that "all qualified scientists without regard to institutional affiliation would have access to the facility," thus "insuring maximum scientific progress." To review the effectiveness of both staff and visitor research, Emberson and Berkner proposed to institute a Visiting Committee patterned after the successful Brookhaven Visiting Committees.[96] Following their 8 April meeting, based on a preliminary draft of the Planning Document, Emberson submitted a seven page detailed statement of AUI's thoughts on "The Establishment and Operation of a National Radio Astronomy Facility."[97]

Tuve's skepticism became apparent to Bok at Tuve's 24 April 1955 presentation at the NSF, which was also attended by Bok and Hagen. Bok became concerned about how Merle Tuve and his NSF Advisory Panel would treat the AUI proposal. Without strong support from Tuve's Panel, it was not going to get very far with the NSB at their 19–20 May 1955 meeting. Bok wrote to Tuve urging that the NSF Panel report should, (a) urge continuing support for the existing university facilities (Ohio State, Harvard, and Cornell) as well as extending support to others such as Michigan, Illinois, Caltech, and Berkeley in order to "insure a steady research productivity," and a "steady flow of PhD's;" (b) support "the prompt establishment in the Eastern United States of a National Radio Observatory," to be operated by AUI, (c) that "a 120 to 150 foot paraboloid reflector be promptly built for the Observatory, ... and at a later date .... a very large at least partially steerable paraboloid." More specifically, Bok urged that AUI be given a grant in 1956 to design and construct "a 120 to 150 foot instrument" ... operating to 10 cm, "to purchase the land for a National Radio Observatory," and to "continue the inquiry into the construction of a large dish."[98] But aside from completing the design of the 120–150 foot antenna, Tuve wrote a bold "No!!!" beside most of Bok's ambitious proposals.[99]

In anticipation of the 19–20 May NSB meeting, Tuve summarized his understanding that the NSF Panel[100]

a) was enthusiastic about the prospects for radio astronomy including the need for "special budget support,"
b) agreed that "very high priority, probably ahead of anything else, must be given to the support of existing activities in radio astronomy at universities and research institutions, along with the encouragement of one or two new additional groups,"

   c) agreed to endorse "the construction of a reflector of intermediate dimensions, in the range 120 to 150 feet in diameter (probably 140 feet [43 meters])," and said that

   d) "the proposal for a very large dish (250 to 600 ft. in diameter) is a project of uncertain value."

As an afterthought, Tuve added, "We regard radio astronomy as part of astronomy, and do not believe that it should compete with the proposed National [optical] Observatory. ... Radio astronomy is a study of the heavens, not just glorified electronics." Tuve acknowledged that there was no firm agreement on whether AUI was the appropriate organization to develop and operate the intermediate size dish, and suggested that operation by two state universities "might be a better arrangement." He went on to suggest that funds be made available to purchase at least options on a suitable site "preferably within reasonable distance of Charlottesville, Virginia," and that planning for the very large dish should not impact the construction of the intermediate size antenna. But he added, "I refuse to be pressured into any detailed approval or disapproval of AUI proposals on such a schedule," to which Greenstein noted on his copy of Tuve's letter, "I absolutely agree."

Tuve's letter triggered responses from Hagen, Greenstein, and Minkowski. Hagen took strong issue with Tuve's opposition to a large antenna.[101] Elaborating on Bok's scientific justification, Hagen listed 11 research areas, ranging from the Sun and other solar system objects to galactic and extragalactic problems, that required a large dish. As he acknowledged, Hagen had initially opposed the intermediate size dish on the grounds that it "might prejudice our chances of arriving at our real goal, which is to obtain an antenna of at least 300 ft. aperture." But he did not oppose the majority decision to start with the 150 foot antenna, provided that it had sufficient precision. Minkowski's response[102] primarily addressed the Kraus antenna which he felt unjustified, since he argued there were enough surveys, and that the need was to identify optical counterparts and have frequency flexibility to enable measurements of radio spectra. Minkowski also contended that the AUI budget proposal was unrealistic and included too large a staff, particularly for administrative personnel. In a separate letter, Greenstein and Minkowski pointed out the "prime purpose" for the AUI initiative was to consider the feasibility and cost of large antennas, and urged that "AUI concentrate cost studies on the 300 and 600 foot dishes," but they recognized that "experience with the Manchester 250 foot and the proposed U.S. National 140 foot" might be needed before the costs were known. Acknowledging the equal demands of optical and radio astronomy, they claimed that the AUI plan to spend "$25 million for several large reflectors," was "inspiring in scope but is not realistically justified by the capabilities in the United States."[103] Moreover, they argued, because of the small number of radio astronomers in the United States, there was no justification for a Brookhaven-type facility. Rather, they claimed, the "intermediate size dish" could be "administered by one or two state universities, for example

Virginia and Michigan, with the site probably near Virginia." Like Tuve, they felt "it dangerous to leave the direction of the activities of a new cooperative institution to a group familiar with the problems of large physics laboratories." Of particular concern was the staffing needs of a large national facility, which they speculated would "destroy all going institutions in the field," and that it was "improbable that student training could be carried on away from the universities with the very large equipments proposed." Many of their concerns would be echoed over the following decades as NRAO indeed grew to dominate US radio astronomy.

Tuve's Panel finally suggested a total of $3.93 million for what they referred to as the "Eastern Radio Astronomy Facility."[104] Of this nearly four million dollars, they suggested that $1.7 million would be needed for the 140 foot "intermediate" sized reflector and its auxiliary equipment and $300,000 for continued engineering studies of a "super" reflector. Another $1.25 million was suggested for the construction, operation, and maintenance of the university facilities. Peter van de Kamp, then head of NSF Astronomy, was charged with preparing a written report for the NSB. Van de Kamp's report[105] closely followed Tuve's recommendations, but gave a somewhat more positive twist to the idea of a very large dish by saying that "the feasibility of a large dish (250 to 600 ft. in diameter) deserves careful study," rather than Tuve's "is a project of uncertain value." Based on Tuve's recommendations, van de Kamp went on to suggest a budget of $3.5 million over four years starting in FY 1957 for the construction and operation of an intermediate sized radio telescope, $200,000 for a 70 × 700 foot standing parabolic reflector at Ohio State, $700,000 for construction, maintenance and operation of a number of smaller facilities, and $300,000 for engineering studies for a large dish type radio telescope. However, van de Kamp added that the ultimate cost of the proposed AUI Radio Astronomy Observatory, including the construction and operation for 5 years of a large steerable dish, was estimated to be $24 million, and that Kraus envisioned building a 4,000 × 400 foot standing paraboloid at an estimated cost of $8 million.

Following the recommendations of the AUI Steering Committee, on 6 May 1955, on behalf of AUI, Emberson submitted to the NSF a detailed five-year plan centered about "at least one very large and very precise radio reflector," but which also included, "a precision surface 20–50 foot" dish, a "100–150 foot reflector," a "250-foot reflector that would be a scale model of the fourth reflector, which would be perhaps 600 feet."[106] Emberson included Bok's summary of the "research objectives," arguments supporting the "need for a radio astronomy program in the United States," and a detailed discussion of the proposed "organization of a national facility for radio astronomy research." He reiterated the importance of radio astronomy to understanding the Universe, as well as to electronic communication and military security, and that the United States was being challenged by new radio astronomy facilities under construction in Australia, England, and the Netherlands. He did not miss the opportunity to point out that "the British 250-foot antenna, built by a nation with a fraction of our resources is scheduled for operation in 1955." With great prescience, Emberson noted that because "among the radio sources already discovered there appear some that appear not to follow the classical laws of radiation by hot bodies, and further that radio frequencies have been identified through molecular beam experiments in the physics laboratory, one can sense that these two branches of science may join in solving some of the riddles of nature." Perhaps more important, and prophetic, he went on to state, "It would be a serious error to suppose that all possible discoveries have been made in this new and expanding field of science."

Although Emberson presented eloquent arguments for establishing a large radio facility patterned after the successful Brookhaven operation, he also recognized the need to maintain the smaller university facilities to train students, to conduct research appropriate to the smaller facilities, and to provide "a breeding ground" for "great researchers" to use the "great facilities." The AUI plan projected an "ultimate" staff of 106 individuals with an annual operating budget of about $700,000. Emberson suggested that they would need another $300,000 in 1956 for Phase 2 of the feasibility study, followed by $6,899,000 in 1957 which included site acquisition, buildings, roads, etc. as well as $800,000 for construction of a 150 foot reflector. Perhaps to mitigate "sticker shock," he offered, "It is our hope that non-government sources of support may be found, both for part of the initial construction of the facility and for its operation, [and] that the NSF be asked to underwrite the operations for a five-year period, with the implicit understanding that if and as other sources of support materialize during that period the Foundation's obligation would be proportionatly [*sic*] reduced." The AUI plan called for a scientific staff "a nucleus of permanent employees whose efforts would be supplemented by qualified visiting scientists." Finally, as a major policy directive, the plan proposed that, "All qualified scientists without regard to institutional affiliation would have access to the facility, thus permitting its efficient use and insuring maximum scientific progress." This "Open Skies" policy, as it became known, would characterize NRAO for the next half century, and to an extent would be adopted by other radio and later optical observatories, although often only after "encouragement" by the NSF in return for funding. To counter criticism about the lack of radio astronomy experience, the proposed AUI plan called for a periodic review by a Visiting Committee appointed by the AUI Board.

The 19 May 1955 meeting of the NSB Committee for MPE Sciences began with a closed session where the NSF MPE Acting Assistant Director Raymond Seeger summarized the recommendations for the proposed large scale radio and optical facilities. He was followed in an open session by separate presentations from R.R. McMath, Chairman of the NSF Advisory Panel for the National Optical Astronomy Observatory, and by Tuve, supported by Bok, Greenstein, Hagen, and Minkowski, for the radio astronomy Panel. Tuve's Panel recommended that the AUI concentrate on the 140 foot telescope and only do feasibility studies for the larger instrument. One issue faced by the NSB was not only the large dollar amounts involved in these proposed national facilities, but that, unlike research grants which may typically provide support for limited periods, the national facilities appeared to require a continuing commitment by the NSF for a large operating budget. This tension between national facilities and research grants would continue in the astronomical community over the next half century. The university scientists wanted access to the unique facilities provided by the national observatories, but not if it came at the expense of their research grants. However, the research grants were of limited value if there were no telescopes to use.

The AUI Steering Committee met on Saturday, 28 May 1955 to review the specifications for the intermediate size telescope, whose size seemed to oscillate from 140 feet to 150 feet depending on the committee, the audience, and the day. The AUI Committee also discussed a plan for the operation of the facility, the choice of a site, and how to reconcile the planned budgets with those recommended by the NSF Committee. Bok noted that the NSF Advisory Panel was concerned that AUI was putting too much emphasis on the very large radio telescope, apparently at the expense of the 140 foot dish, but he was reassured by Berkner and Hagen that they were serious about an early construction of the 140 foot radio telescope.

In reviewing the draft specifications for what was now considered the "small" 140 foot telescope, Bok reported that the NSF Panel did not consider the specifications for construction to be sufficiently precise. But the AUI Committee was concerned that if "detailed materials and construction specifications were prepared, a manufacturer would strive to give only what might conform to the specifications without regard as to whether the completed instrument would perform as desired." Wisely, the AUI Committee agreed that they would continue with the use of performance specifications,[107] but questions of this nature would continue to plague NRAO when faced with contracting for future radio telescopes. Charles Husband, who designed and was building the Jodrell Bank 250 foot radio telescope, had been contracted to advise AUI, and wrote,[108] "We have tended to carry out improvements to the design as actual construction work proceeded. For many reasons this is not a good thing to do." Perhaps, less obviously, he also commented, "I think you would save a great deal of money by preparing a design in considerable detail before inviting bids," and went on, "The client being responsible for producing the design is practically always the more economical in the long run." This latter philosophy was fundamental to all of NRAO's future antenna Requests for Proposals, except in the case of the Green Bank Telescope. The extenuating circumstances surrounding the GBT funding and the perceived need to begin construction before the design was finalized led to a huge cost increase (Chap. 9). The AUI Committee also struggled with the question of equatorial vs. alt-azimuth design. Based on his experience with the NRL 50 foot dish, Hagen expressed concerns that[109] "the alt-azimuth mounts require more complex computers and servo-mechanisms and that maintenance would therefore be more costly and time consuming." But Goldberg and Haddock argued that[110] "low altitudes would be highly desirable for some solar work, lunar occultations, and eclipses," that would be more difficult with an equatorial mount, and Bill Gordon[111] "stressed that atmospheric and ionospheric problems required access to very low altitudes, particularly to the north." Unable to decide between the two options, the Committee agreed to seek two bids, one for each type of mount.

Although the budget recommended by the NSF Advisory Panel did not differ significantly from the AUI plan, Emberson defended the slightly larger AUI figures by pointing out that AUI had included essentially[112] "everything that might be considered for the Facility." Nevertheless, there was concern that the AUI was planning for a much larger staff and more extensive operation than recommended by the NSF Panel. Following discussions among Seeger, van de Kamp, and Tuve, and in recognition of an expected increase in the FY1957 NSF budget, Seeger agreed to an increase in the budget for salaries and maintenance, but Tuve expressed concern that the proposed budgets for radio astronomy would[113] "equal or exceed the figure currently planned for astronomy," and that the proposed budget for radio astronomy "should not be allowed to reduce the sum being apportioned to optical astronomy, but should be added to it, thus serving to greatly increase the grand total for astronomy. Our panel still considers radio astronomy as a branch of astronomy and not a substitute or competitor."

By the time of the 17 June 1955 meeting of the AUI Executive Committee, AUI had not received any formal notification of the results of the 19–20 May meeting of the NSB, but, through the joint membership of the AUI and NSF committees, Emberson reported that he had learned that the NSB[114] "had a lively interest in radio astronomy and that the Foundation's radio astronomy panel had proposed that $3,300,000 be allocated for a national radio astronomy facility." Emberson also informed the Executive Committee that he had initiated a contract with Jacob Feld for design studies of a 600 foot reflector and had engaged two consultants from the University of Pennsylvania to evaluate Reber's design.

To continue the feasibility studies started under the first NSF grant, AUI applied for another grant for $234,500, which included funds for two independent designs for the 140 foot telescope along with a preliminary design for the 600 foot telescope, and for site evaluations including options to purchase land. In discussions with Waterman and other NSF staff, Emberson stressed that since the feasibility study for the 600 foot antenna had been successfully completed under Phase 1, omission from the Phase 2 grant would "constitute a time delay somewhat greater than six months."[115] The NSF was already beginning to appreciate that the cost of the new radio astronomy facility would be significantly more than Tuve's panel had indicated.[116] At the same time, the NSF was considering comparable levels of funding for the proposed American Astronomical Observatory to be located somewhere in the Southwest to provide clear skies and suitable facilities for photoelectric observing, and this would surely add to the NSF budget burden.[117] Nevertheless, at its August 1955 meeting, the NSB approved the AUI budget proposal for $3.5 million to be spent over a four year period starting 1 July 1956 and asked the Bureau of the Budget that this be included in the NSF's FY1957 budget.

Bart Bok, with his usual enthusiasm, wrote in support of full funding for the AUI Phase 2 proposal, citing the successful progress of the Phase 1 study and the urgency of moving ahead so that the United States did not lose momentum at what he called "this critical stage."[118] Jesse Greenstein indicated that since funds were limited he could not "go along with the wholehearted endorsement of a $234,000 grant,"[119] and that it was premature to invest too much money in the design of the 140 Foot Telescope until it was clear that construction funds would be approved by Congress. John Kraus's strongest statement in support of the AUI proposal was that "continued support for studies and planning on a multi-million dollar national facility is unavoidable," but he spent most of his two-page report arguing for the smaller university facilities, especially his standing parabolic reflector at Ohio State.[120] Minkowski pointed out that there was serious concern that "300 feet is the maximum possible size of a sufficiently rigid dish," and that no "further funds be diverted from the core of the study."[121] Merle Tuve, reporting on a series of telephone conversations with his Panel, expressed concern about the impact of the AUI grant on astronomy research grants and suggested that funds be specifically earmarked for Kraus's antenna at Ohio State. He reported that the Panel recommended that the NSF grant to AUI be cut to only $140,500 since some of proposed activities could be deferred, but that AUI should add "one or two high level members to their professional staff, preferably appointing the director who is to take charge of the evolution of this project." The challenge to find a director turned out to be more difficult than anticipated, and would end up taking another five years. However, Tuve's panel went on record "to use AUI as the vehicle for bringing the Eastern National Radio Astronomy Facility into being, but that the actual title to the property and decisions as to the future ownership and control be held in abeyance, probably for several years." Unwilling to give AUI a free hand, Tuve added, "It is understood, however, that this ownership and control would be in the hands of universities, whether through the AUI mechanism or directly with one or two selected institutions."[122,123]

At a meeting with Waterman on 6 December 1955, Berkner outlined the proposed scientific and administrative structure of the facility, the nature of the expected contract between the NSF and AUI, and said he hoped to have a director appointed by 1 July 1956.[124] They also began a discussion on how the land for new facility would be acquired, whether it should be public or private land, and who would hold title to the property. Although no decision had been reached, or even formally discussed about who would manage the radio astronomy facility, both Berkner and Waterman tacitly assumed that it would be AUI.

Meanwhile, Waterman was juggling inquires and pressures from both Berkner and Tuve, each promoting their own agendas. Even the name of the new facility was contentious. Tuve had started to use the name "Eastern Radio Astronomy Facility" to make it clear that it wasn't a government operated "national" facility, but Waterman didn't want it to appear as a facility only for the East Coast. "Green Bank Observatory" was suggested, but Tuve argued that would be inappropriate if the facilities were dispersed. They both agreed that it was "difficult to pick a proper name until the site was selected, property purchased etc." Waterman agreed not to call it a "National Observatory," and suggested to just call it the "Radio Astronomy Facility."[125] In his phone conversation with Waterman, Tuve also complained about "Berkner's attitude toward the whole thing," arguing that "Berkner is trying to run everything his way and that Hagen and Bok are going along for political reasons." Waterman cautioned Seeger, "that we should watch AUI carefully as they have a technique for pushing things through and it is hard to do anything about it. Berkner lines up a few members of Tuve's committee beforehand and then calls committee meeting and everyone usually agrees with him, but major points should be considered by the Foundation before giving him a 'yes.'" He advised Seeger, "In any conversation with Berkner in which he wants a quick OK, to tell him that he must check with me first." Waterman explained that he saw a need "to encourage something else than AUI," as "this will keep AUI in bounds." Tuve and Seeger also did not miss an opportunity to exchange a few criticisms, with Tuve expressing concern about "Seeger's willingness to agree to all of Berkner's suggestions," and Seeger's criticism of the way Tuve was running his Panel.[126]

The AUI Steering Committee met again on 11–13 December 1955, together with NSF and AUI staff as well as invited consultants from the US Geological Survey and elsewhere. Much of the meeting was devoted to reviewing the site studies and a planned 12 December trip to a candidate site at Massanutten, Virginia (see following section "Choosing the Site").[127] But the Committee also discussed the design of the 140 foot telescope and the difficult problem of guidance and control. Tuve argued for an equatorial mount since he contended "that precision positioning could be more easily achieved … on an equatorial mount than on an alt-az mount." Unfortunately, neither Tuve nor the rest of the participants appreciated that while the drive system might be simpler and more accurate, structural distortions on an equatorially mounted telescope could lead to much larger positioning errors than those introduced by the control system on an alt-az mount. The Committee did recognize, however, that an equatorial mount would lead to "something less than full hemispheric coverage," a reasonable constraint that somehow was to be forgotten when it came to defining the final design.

Emberson described the progress on the three ongoing commercial 140 foot design projects, all of which were for an alt-az mount. Numerous arguments were presented against an equatorial design, including the limited sky coverage, the need for counterweights, and most importantly that "problems of stiffness and rigidity would be more difficult." It was also "pointed out that since there was little prospect for a 600 foot equatorial mounted dish, … an equatorial 140 foot would teach little of the problems that would be met with larger instruments." Although it would "delay the 140 foot project by at least two months, … the Committee unanimously recommended the above equatorial program be initiated," although this would mean diverting funds from the large telescope design. Unable to agree on the size of the planned "smaller" telescope to be mounted on the Laboratory building, the "the Committee concluded that two small instruments were desirable, one in the 25 foot range, and the other 60 foot." The announced schedule optimistically called for completion of the 140 foot design and call for bids in March 1956 followed by the award of a construction contract in May or June. Emberson reported that he expected construction to start in September 1956 after review of the detailed design, and had set a target date of December 1957 for the completion of the 140 foot radio telescope. Sadly, it would be almost ten years before the 140 foot radio telescope would be completed and available for research. Perhaps recognizing that the Steering Committee had outlived its purpose of consolidating community support for Berkner's ambitions, a few days later, at their 14 December 1955 meeting, the AUI Executive Committee approved Berkner and Emberson's request to disband the informal Steering Committee and replace it with an "Advisory Committee to the Board of Trustees."[128] This step, perhaps, was facilitated by John Hagen's announcement that he would be resigning from the Committee, as he was about to assume charge of the ill-fated Project Vanguard.[129] Nevertheless, Tuve was "dumbfounded" when he learned from Seeger that the AUI Steering Committee would be dissolved.[130]

In reporting to his NSF Panel about the Steering Committee meeting, Tuve expressed his frustration about the decision to "build an extensive community in the deep woods," rather than have "at least an auxiliary laboratory and administration building adjacent to some nearby university, probably in Charlottesville, or in Northwest Washington." Tuve complained, "I was consistently and vigorously opposed in this by a variety of arguments," and went on to quote verbatim the two resolutions passed by the Steering Committee, but he neglected to mention that he had proposed the motions.[131]

The NSF Advisory Committee met again on 16–17 January 1956 to address the delicate question of the management and location of the radio astronomy facility. They asked if the facility should be managed by a single university, a group of universities, or other non-profit institutions, with or without experience in radio astronomy; to what extent was it necessary to choose a site that required building an extensive and expensive nonscientific infrastructure; and to what extent should the country's radio astronomy facilities be concentrated at one site.[132] The Panel and Waterman expressed concern that AUI, with its membership confined to the Northeast, did not properly represent all of the major institutions in the country with interests in radio astronomy,[133] and the Panel concluded that "there are serious problems with AUI as the corporate organization."[134] Waterman considered the possibility of letting AUI construct the facility and some other group or university handle the "detailed administration," but Berkner again argued that an appropriately chosen Visiting Committee would satisfy the need for national representation.[135] Following assurances of cooperation in securing the land from the Governor of West Virginia, Berkner was anxious "to proceed with plans for a contract for construction, operation, and maintenance of the observatory." But he was informed by the NSF that although they would consider the AUI plans, the NSF could not proceed with contracts until they received the House report on the

Foundation's budget request.[136] Meanwhile, Tuve's concerns about AUI were supported by the NSB, which unanimously recommended that "to the extent possible and practicable, the governing body for any major facility receiving substantial assistance from the Foundation should be as representative as possible of interested universities and institutions throughout the country with experience in the field."[137] Waterman also explained the Board's policy.

> In supporting large-scale facilities, the Federal Government should aim to limit its support to established construction costs, but might recognize that in individual instances continuing support may be required, preferably on a diminishing scale. Once a sound venture is started, the Government should stand behind it, but not necessarily be its sole support.[138]

Although AUI and the NSF agreed that while the discussions about a national radio astronomy facility were not secret or classified, "there should be no public announcement covering the construction of facilities"[139] until Congress appropriated the funds. However, someone leaked the news, and a long story appeared in the 26 January 1956 issue of *The New York Times* under the headline, "President Recommends Radio Telescope Funds for New Astronomy Research."[140] *The Times* article reported that, if approved, the 140 foot telescope would cost about $7 million, but that the total cost of the facility might cost up to $30 million. The article turned out to be somewhat of an embarrassment, as the House Appropriations Committee had approved only $4.5 million for the total cost of the radio astronomy facility.[141]

Meanwhile, no steps were taken toward disbanding or renaming the AUI Steering Committee, although Bart Bok replaced Hagen as the Committee Chair. The Steering Committee unanimously agreed with Berkner's suggestion that that the new facility be called "The Karl G. Jansky Radio Astronomy Observatory."[142] A week later, the NSF agreed to Berkner's suggestion to name the observatory after Jansky, but there is no evidence that name was ever considered further.[143]

On 10 January 1956, at the request of Waterman, AUI submitted their draft *Planning Document for the National Radio Astronomy Facility*.[144] With unsuppressed optimism and enthusiasm, Berkner suggested a five-year budget for the construction and operation of four radio telescopes with diameters of 25–50 feet, 140 feet, 250 feet, and 600 feet. Berkner's plan was reviewed by an equally enthusiastic small subcommittee of the Advisory Committee composed of Bok, Leo Goldberg, and Ed McClain. The final *Plan for a Radio Astronomy Observatory* submitted to the NSF in August 1956 included:[145]

1. A "standard size" 28-ft dish to develop and test electronic equipment as well as making "worthwhile observations."
2. A 60 foot dish which had now also reached the status of "standard size" by virtue of the 60 foot Kennedy dish recently completed at Harvard.
3. A 140 foot dish.

4. A 250 to 300 foot diameter antenna, possibly a scale model of the 600-ft antenna.
5. A 600-ft radio telescope following the specifications given for the Feld study.

AUI's "Plan" was apparently the first use of the word "Observatory" in any AUI or NSF document. While the NSF was openly discussing the establishment of a National [Optical] Astronomy Observatory, all reference to the parallel radio program had previously referred only to a "facility." In addition to specifying the planned radio telescopes, the AUI "Plan" carefully laid out a management plan which was received with some alarm by the NSF. Basically, the AUI plan called for government (e.g., NSF) funding and ownership of the facility, but all aspects of the operation would be controlled by AUI. Under the AUI plan the NSF would have no control over the selection of the director or advisory committees, operating policies, expenditure of funds or auditing. While the NSF recognized that "good administration requires a single line of authority," and that the AUI plan "does this with remarkable thoroughness. It also succeeds in divorcing the National Science Foundation from any practical measure of control over responsibility for use of the funds provided."[146] Also, of continuing concern to the Foundation, was AUI's repeated reference to itself as "scientists and institutions in the eastern part of the United States" and their repeated insistence against expanding the AUI Board which appeared to stand in contrast to the NSF goal of a "national" radio astronomy facility. Berkner reacted "very strongly" to these NSF criticisms, alluding to "interference of government with research in general and with the proposed facility in particular."[147]

At their 27–28 March 1956 meeting, the NSF Advisory Committee, led by Merle Tuve, was not enthusiastic about AUI's ambitious plan, and rejected the two larger radio telescopes proposed. They suggested instead that the initial construction include only four more modest sized instruments: one 28 foot antenna, two 60 foot antennas, and one 140 foot antenna. The two 60 foot antennas were included because the committee projected a heavy demand for observing time from university staff and students, and also anticipated their possible use as an interferometer. The NSF Advisory Committee also had their own ideas about how to build a radio telescope. Specifically, some members of the Committee had concerns that the coordinate conversion for an alt-azimuth mounted antenna would not meet the stringent pointing requirements, so AUI was instructed to develop at least one design based on an equatorial mount.[148] Berkner anticipated that AUI would first issue an RFP for an alt-azimuth mounted 140 foot dish, but defer a construction contract until an equatorial design was available. Following a later review of the Feld, Husband, and Kennedy designs, as well as the separate designs for the drive and control systems by a team of consultants, it was realized that none of the designs could meet all the requirements, so the proposal to the NSF acknowledged that the design "is still open for study."[149]

In addition to debating the number and size of the major instruments, the AUI proposal gave careful consideration to how the observatory would be run, what staff was needed, and of course how big an annual budget would be required. The AUI proposal delegated considerable authority to a Director who "must bear the overall responsibility for the Observatory," and went on to specify that "he must be not only be a research scientist of recognized ability; he must also have proved his ability to administer scientific projects; and that he possess special aptitude in selecting and supervising scientific personnel." Characteristic of the time, only the pronoun "he" was used; there was no recognition of the possibility that the Director, or any of the scientists, might be a woman, and indeed, none were considered in the subsequent lengthy Director search.

The proposal included a Chief Scientist who would also act as the Deputy Director, a Chief Engineer, and a Business Manager to round out the senior staff. AUI anticipated that the scientific staff would be approximately equally divided between resident and visiting scientists. Responding to repeated reminders from the NSF as well as from Merle Tuve and his Panel that neither the AUI staff nor any of the Trustees had any expertise in radio astronomy, AUI finally agreed to add two at-large Trustees with experience in radio astronomy, but Berkner refused to consider any changes in the Board's basic organization. Tuve expressed concern about AUI appointing the Visiting Committee, since, as he argued, "this mechanism can be self-biasing." Tuve also expressed concern about AUI's offer to "contribute recreational facilities costing hundreds of thousands of dollars," since he correctly appreciated that this "would be quite a hurdle if a new contractor had to buy out the facilities which belonged to AUI." Somewhat surprisingly, Tuve also suggested that some two to three million dollars be invested in placing a 24 inch Schmidt along with a larger 72 inch reflector in Green Bank to facilitate related optical studies.[150] While such a scheme had obvious scientific merit, the very different environmental requirements of radio and optical observatories rendered this impractical.

Not dissuaded by the concerns of their Advisory Committee, the NSF Deputy Director, Charles Sunderlin, leaked to Berkner and Emberson that Waterman intended to recommend to the Foundation's Committee on Physical Science and to the National Science Board that AUI be selected "as the operating agency for the National Radio Astronomy Facility,"[151] although this was not fully supported by all of the NSF staff, some of who preferred a "rather loose type of organization that would permit a greater degree of control by the Foundation."[152] In response to a question from Sunderlin, Berkner emphasized "the great importance of having the operating institution in full control during the construction phase," and that, "AUI would not be prepared to act simply as a construction contractor," although he was prepared to accept a contract for "three to five years with provision for extension."[153] Sunderlin also informed Berkner that because the House of Representatives had voted to reduce the NSF budget by a factor of two from $7 million to $3.5 million, and because it was uncertain whether or not the Senate would try to restore some or of all of the deleted funds, it was unclear how much money would be available for the radio facility in 1957. Berkner was therefore told to consider three 1957 budgets levels of $2.095 million, $3.895 million, and $5.170 million, respectively, as an absolute minimum cost for a facility that would compare with others, the minimum cost for a facility including the 140 foot telescope, and the cost of fully carrying out AUI's plans for the development of the facility.[154] Emberson encouragingly commented that in his opinion "the Foundation does not wish to omit the 140 foot radio telescope."[155]

Recognizing the magnitude of the effort on which they were about to embark and noting that AUI was constituted to manage the Brookhaven National Laboratory, not to do radio astronomy, Berkner asked the AUI Board if it was necessary to consult the founding universities, although he speculated that "in his opinion, the language of the agreement is sufficiently broad to cover almost any contractual obligation." The Board agreed, and told him only that "the universities should be kept advised."[156] Signals from the NSF were encouraging. In his address at the 28 April 1956 dedication of the new Harvard 60 foot radio telescope, (Sect. 2.5) the NSF Director, Alan Waterman, publically acknowledged the NSF plans to build a 140 foot diameter radio telescope to be operated by a "group of universities," and said that "substantial funds for this purpose have been included in our 1957 budget."[157] Waterman went on to refer to radio astronomy as "A new window on the universe," and acknowledged the responsibility of the Federal Government to support basic research "when such support is necessary and in the interest of science."[158]

Very shortly after the dedication of the Harvard radio telescope, Berkner gave a talk at the 94th meeting of the American Astronomical Society describing the need and plans for a national radio astronomy facility. In his short two-page published paper, Berkner (1956) referred to the 600 foot reflector as an "ultimate compromise between cost and operating characteristics," and expressed his believe that "it would be entirely feasible to construct a radio telescope of this size with adequate precision."

3.4   Choosing the Site

Discussions on the important question of site selection for the national radio astronomy facility began very early. Already, in November 1954, the NSF Advisory Panel on Radio Astronomy dictated that the site should be within 300 miles of Washington. Following discussions with Harold Alden, Director of the University of Virginia's McCormick Observatory, Carl Seyfert from Vanderbilt, John Hagen from NRL, and Peter van de Kamp from the NSF, Emberson and the AUI Steering Committee defined an additional set of criteria deemed important for the successful operation of a radio observatory. These included[159]:

(a) *Radio Noise*: To minimize the impact of radio frequency interference (RFI), AUI specified that the telescopes should be located in an area with a small local population, should not be in view of any high voltage lines, should be in a valley surrounded by mountains, be at least 50 miles distant from any city, and not be near any commercial air route. Furthermore, AUI noted that "the quietness of the site must be assured for the future; for example, by appropriate zoning regulations."

(b) *Weather:* Low humidity to minimize erosion of steel structures and impact to electrical insulation, low winds, and low occurrence of hurricanes and tornadoes along with little ice and snowfall to minimize loads on the telescopes were specified. "Reasonably mild" weather was desirable to facilitate maintenance. At the time, little was understood about the impact of tropospheric water vapor[160] on centimeter wavelength radio astronomy measurement, and so was not considered in evaluating the quality of each of the sites considered.

(c) *Latitude:* Lower latitudes were preferred to maximize the amount of available sky and in particular facilitate access to the important region near the center of the Milky Way Galaxy. However, AUI noted that northern latitudes would facilitate research on "aurorae, ionospheric scintillation, and polar blackouts."

(d) *Social and Professional Amenities:* To the extent possible, AUI sought "as many as possible of the attributes of a university campus, including laboratories, shops, libraries, conference rooms," and proximity to scientists working in broadly defined related areas. Easy access to "housing … stores, theaters, and recreational facilities" was also considered to be "desirable."

(e) *Access:* The chosen site should be "easy to reach by plane, rail, or automobile" and as previously specified by the NSF, the site was to be within 300 miles of Washington.

(f) *Availability and Size:* A total area of five to ten thousand acres was desired to allow for adequate separation of future telescopes and arrays. Such a parcel of land would need to be available either as existing government property or by purchase from private owners.

AUI recognized that it would not be possible to meet all of these criteria, and that indeed, some were "mutually contradictory and incompatible,"[161] so some compromises would be necessary. Consideration of known weather patterns, and population distributions along with the other geographical constraints suggested a valley somewhere in the Appalachian Mountains located within an area approximately 300 miles by 100–150 miles in extent oriented in roughly a northeast-southwest direction and west of Washington, DC (Fig. 3.11).

After receiving his second NSF grant in the spring of 1955, Berkner appointed an ad-hoc panel to evaluate potential sites for how well they met the criteria set out by the Steering Committee. Panel members included the state geologists from Virginia, West Virginia, and Tennessee, representatives of the University of Virginia and the NSF as well as radio astronomers from NRL and

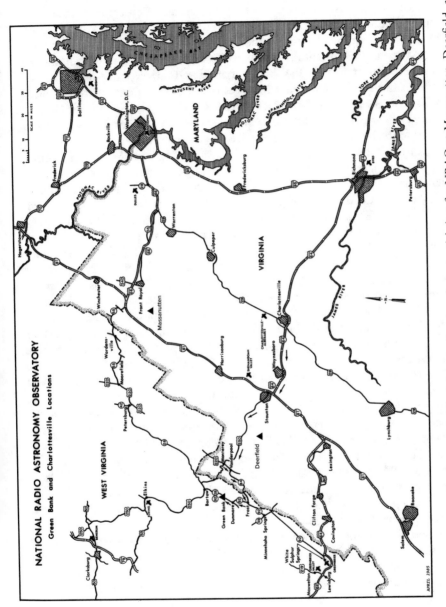

Fig. 3.11 Map of the Maryland-Virginia-West Virginia area showing the potential sites for NRAO at Massanutten, Deerfield, and Green Bank marked by triangles. Credit: NRAO/AUI/NSF

Harvard, along with two optical astronomers with experience in astronomical site testing. Richard Emberson represented AUI.[162] The panel evaluated 20 different potential sites in Georgia, Tennessee, Virginia, and West Virginia. Much of the groundwork was done by Hagen and Emberson who spent a couple of weeks in what Hagen later referred to as the "wilds of West Virginia, Tennessee and Virginia looking at the potential sites, many of which were easily rejected due to their proximity to urban or industrial centers, power lines, radio or radar transmitters."[163] One site was considered unacceptable due to a high southern horizon which would have restricted observations of the southern sky. The US Forest Service, the Park Service, the Geological Survey, the General Services Administration, and the Tennessee Valley Authority all provided valuable support, but the Army Map Service was less helpful.[164]

For the most part, the study proceeded objectively, with little political interference, except from a few Congressmen who gratuitously offered what they each claimed were ideal choices for the planned radio facility. West Virginia Congressman C. M. Bailey requested consideration of the marshy area known as Cranberry Glades which had previously been rejected due to inadequate shielding from mountains and inadequate terrain to support large radio telescopes, but Bailey insisted on meeting with the NSF along with "a small committee of business men," to present the advantages of their proposed site.[165] One of the attendees at their NSF meeting on 14 March 1956 was the warden of the local prison, who noted the availability of prisoner labor.[166] Following a site visit urged by the NSF to keep the Congressman happy, Emberson reported that "if one jumps up and down … with water oozing up ankle deep, persons standing twenty feet away can feel the waves passing by."[167] Long time Congressman Harley Staggers, also from West Virginia, called Waterman to point out the desirable features of Spruce Knob, the highest point in West Virginia which looks out over three counties. Rather than try to explain to Staggers that the technical requirements for a radio telescope were not the same as for an optical telescope, Waterman merely told him that "the determination of the site was in the hands of a committee which is highly qualified scientifically," and that he "would be glad to bring his invitation to their attention."[168] Once Green Bank was chosen as the future site for the radio facility, Staggers helped to dispel the local fears about the impact of a big federally supported program in their quiet valley, and emphasized the new opportunities for employment (Kenwolf 2010).

Based on their initial review, the panel narrowed their consideration first to five sites, four in Virginia and one in West Virginia, some of which were also being considered by the Naval Research Laboratory as potential locations for their planned facility which ultimately went to Sugar Grove, WV.[169] The consulting firm of Jansky & Bailey[170] was hired to evaluate the levels of RFI at each site, using equipment borrowed from NRL. During the latter half of 1955, Jansky & Bailey made measurements between 50 MHz and 10 GHz, but due to changing propagation effects, they found significant differences in RFI between measurements made in daytime or nighttime and between measurements made in the summer or winter months. Differing amounts of commercial activity in the areas surrounding each site also contributed to the day-night differences. Only one site, at Massanutten, VA was studied in both the summer and winter, and this was used to try to normalize the other studies to determine the differences in RFI among the five sites. The Green Bank, WV site was judged to be superior "to all others in regard to radio noise" and also had by far the smallest surrounding population, and compared with Deerfield, VA, less likely to have population or industrial growth.[171] As described by Bart Bok, "the rather remote valley near Greenbank [*sic*] and Arbovale in West Virginia (at an altitude of 2600 and 2700 feet) seems to be the answer to the radio astronomers' prayers."[172] As it turned out, the region chosen has one of the highest percentages of cloud cover in the United States, and this has greatly limited the effectiveness of the Green Bank radio astronomy program at centimeter wavelengths.

The isolation of Green Bank from population centers was recognized as both an advantage and disadvantage. The isolated site offered better observing conditions in terms of lower radio noise, but clearly more difficult access and less attractive living conditions. Tuve, claiming that "Berkner would like a large community in the woods,"[173] perhaps more than others appreciated the practical difficulties that would be faced by an isolated staff and urged that at least some of the facilities be placed in Charlottesville, Virginia. In fact, the NSF had some informal discussions with Professor Jesse Beams from the University of Virginia who conveyed the interest of both the physics and astronomy communities in seeing UVA as a possible manager of the proposed facility. To strengthen their position Beams offered that within a few years the University was likely to appoint a radio astronomer to the faculty.[174] Berkner, on the other hand, noted that placing some of the administrative facilities in Charlottesville would keep the Green Bank staff even more isolated, and instead envisioned what Tuve called a Los Alamos type of operation. But the Bureau of the Budget reportedly did not like the idea of providing housing at government expense,[175] so Berkner began to discuss seeking private funding for housing and other non-operational infrastructure. Concerns about the social impact of living in rural Appalachia would continue to plague the Observatory for the next decade, finally leading to migration of most of the scientific staff to Charlottesville, and effectively creating Tuve's model (Sect. 4.7).

In preparation for the planned AUI Steering Committee meeting scheduled for 13 December 1955, on 2 December six groups set out in different cars to personally inspect the five most promising sites. Due to recent rain and snow they were unable to reach the Massanutten site, which probably should have been a message about the suitability of the site for building a radio telescope. A lengthy debriefing session was held in Washington on Sunday evening 11 December, which included some other AUI Steering Committee members, some AUI Trustees, along with C. E. Curtis and Helen Sawyer Hogg from the NSF. The following morning a group of 15 people left by bus for one final inspection of the Green Bank site. During the trip they again reviewed all aspects of the site question, including radio quietness, geographic require-ments, and the nature of the operation of the facility. In view of the problems encountered in reaching the Massanutten site, the representatives from the US Geological Survey surprisingly reported that the five sites appeared to be about the same regarding geology and the effect of inclement weather on access.

The Steering Committee met the next day at the NSF for further discussion, after which Merle Tuve moved and Bart Bok seconded a motion which read[176]:

(1) It is the recommendation of the Committee that the site near Green Bank, West Virginia, subject to verification down to very low field intensities of the expected low radio interference level, be selected specifically for the proposed 140-foot parabolic reflector, and possibly for two or three antennae rays [sic] or other equipment of modest cost; and

(2) This recommendation is made without prejudice to the possible location or locations which may in the future be recommended if this National Radio Astronomy Facility grows to include other specialized equipment or labora-tory facilities.

The resolution was unanimously passed by the Committee, although it was later realized that the existing noise measurements were the best that could be made with existing equipment, so the requirement for additional noise mea-surements was rescinded. The second part of the resolution showed great pre-science in recognizing the need for the future extension of NRAO to Arizona for the 36 foot millimeter wave antenna (Chap. 10) and later to New Mexico for the VLA (Chap. 7). To ensure against protection from locally generated interference, the Steering Committee also unanimously agreed to another motion, again proposed and seconded by Tuve and Bok respectively, which read

It is the recommendation of the Committee that all or nearly all of the land in the Green Bank, West Virginia valley shown in the attached map be acquired by direct purchase or as an alternative, suitable controls e.g., by some agency of the State of West Virginia, be arranged to insure [sic] the future continued suitability of this valley for the National Radio Astronomy Facility; and the Committee is confident that the U.S. Forest Service will assist in maintaining radio quietness in the surrounding forest and overlooking mountain heights.[177]

Recognizing that it would be useful to consider contingency sites in case the Green Bank proved unsuitable, and, as Berkner pointed out, consideration of alternate sites would strengthen the bargaining position in acquiring the land, the Committee unanimously agreed to include Deerfield and Massanutten, in that order, as alternates. In parallel with the site search, AUI commissioned the New York City-based architectural firm of Eggers and Higgins to prepare a site development plan based on four radio telescopes of 25–50 feet, 140 foot, 250 foot, and 600 foot diameter, a staff of about 100, a central laboratory and administrative building, site maintenance and telescope maintenance buildings, telescope control buildings, along with a dormitory and apartment building in combination with a cafeteria, as well as several on site residences for key staff and visiting scientists. Based on a generic site taken from a selection of 14 potential sites being considered by AUI, Eggers and Higgins determined that the total cost of site development for the antenna foundations, roads, buildings, power, and water would be close to $10 million.[178]

A more intensive site inspection took place in January 1956 with a two-day visit by Emberson, Heeschen, Bok, and A. Doolittle from the consulting firm of Eggers & Higgins.[179] As a measure of the "human activity in the area" and potential sources of interference, they counted somewhat over one hundred houses in the valley, and laid out potential locations for the planned 25 foot, 60 foot, 140 foot, 200–300 foot, and 600 foot radio telescopes, along with laboratory, maintenance, housing, and cafeteria buildings. Characteristic of the area, the nearest hotel the group could find was in Elkins, WV, 50 miles distant over Cheat Mountain.

Earlier, Emberson had reported that the "Foundation fully recognizes the desirability of moving rapidly, and would not object to the acquisition of a site with private funds."[180] Still facing competition over who the NSF would chose as the managing organization, in March 1956 AUI acquired options to purchase 6,200 acres of land at a cost of $502,000, thus apparently squeezing out any potential competitors including those representing gas and oil interests. Three months later, Congress appropriated $3.5 million to acquire the land, erect buildings, and design and construct a 140 foot radio telescope. However, not willing to accept Berkner's preemptive strike in obtaining the land options, the NSF invoked the right of eminent domain, and authorized the Army Corps of Engineers to seize the land, and the AUI options became irrelevant. This not only had the effect of neutralizing Berkner's tactical maneuver, but also removed the prospects of having to deal with landowner holdouts for higher prices. But the seemingly government land grab also had negative consequences. The Corps of Engineers was instructed to start their acquisition at the center and work outward until the authorized sum of $550,000 was committed (Emberson 1959). Most of the privately owned land was bought at market value, but there were a few holdouts who only relinquished their land under pressure or perceived pressure. Although the affected land owners were all compensated, the non-negotiable seizure of private farms and homes that had been owned by generations of the same families would impact relations between NRAO and the local population for decades to come. Some years later, a small portion of land was returned to the local cemetery, accepting the local argument that the occupants of the cemetery "won't bother you none" (Lomask 1976, p. 142).

Even before being formally selected to manage NRAO, Berkner was lobbying with his long-time friend, William Marland, then the Governor of West Virginia, for the protection of the Green Bank site from RFI. Following the urging of a number of influential Harvard alumni in the state government, on 9 August 1956 Governor Marland convened a special session of the West Virginia state legislature to enact the "West Virginia Radio Astronomy Zoning

Act (WVRAZA)" which prohibited the use of any electrical equipment within two miles of a radio observatory "if such operation causes interference with reception by said radio astronomy facility, of radio waves emanating from any non-terrestrial source."[181] In addition, the Zoning Act set limits on the field strength of any electrical equipment operated within ten miles of the radio observatory. Although a fine of $50 a day was specified for any knowing violation, there is no evidence that anyone was ever prosecuted for violating the West Virginia Zoning Act. Nor is it clear that the Zoning Act was legal, since the 1934 Federal Communications Act placed the regulation of radio transmissions only in the hands of the federal government and the Federal Communications Commission. Nevertheless, the WVRAZA has remained valuable in restricting the use near NRAO of a growing list of potentially harmful devices, such as TV antenna mounted preamps, microwave ovens, and most recently cell phones. The initiative of the West Virginia legislature was the first legislation in the world specifically intended to protect basic research, and it set a precedent for the National Radio Quiet Zone (NRQZ) established in 1958 by the Federal Communications Commission to protect both NRAO and the Sugar Grove Naval Information Operations Center (Fig. 3.12). Within the NRQZ, all applications for licensed radio transmitters are referred to NRAO for comment. NRAO does not have veto power over radio transmissions within the NRQZ, but can provide comments to the FCC on applications for radio licenses, and often successfully negotiates with the applicant to modify their frequency or antenna beams to reduce the impact of their transmissions.

3.5   Confrontation and Decision

Continued friction between the AUI and NSF Radio Astronomy Committees, between Berkner and Tuve, and between the NSF radio and optical astronomy committees, led the NSF to call for another meeting to iron out the differences among the various individuals and the different advisory bodies. As later described by Heeschen (1996), "It really boiled down to a disagreement between Berkner and Tuve over just what the observatory should be. Berkner was the archetypal 'big science' scientist," and "wanted the observatory to be an institution like Brookhaven, which would provide extensive facilities and services for visiting scientists, and have its own scientific and technical staffs engaged in research and development." By contrast, "Tuve was the classical 'string and sealing wax' scientist, and had established a very successful and productive research group at DTM along these lines. Tuve wanted the observatory to consist of just a telescope, with minimal staff, facilities and services." According to Heeschen, Berkner and Tuve had "developed a mutual dislike and distrust" dating back to the time they had worked together before and during WWII.[182] But Tuve also resented the way that Berkner grabbed the momentum which he, DuBridge, and Greenstein had started when they organized the January 1954 Washington meeting, and the possible impact that this might have to the major radio astronomy initiative now being pursued by

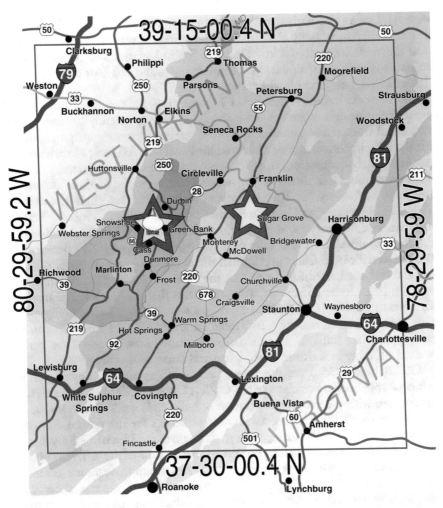

Fig. 3.12 Map showing the boundaries of the 13,000 sq. mile (34,000 sq. km) National Radio Quiet Zone. The NRAO Green Bank, West Virginia and the Sugar Grove, Virginia Naval Station are shown as stars. Credit: NRAO/AUI/NSF

DuBridge and Greenstein at Caltech together with DTM's sister institution, the MWPO. Ed McClain (2007) later speculated that Berkner put Tuve on the AUI Steering Committee fully expecting that he would get criticism from Tuve, but that the criticism would come from a member of an AUI committee, and not from an outsider.

Although the AUI "Plan for a Radio Astronomy Observatory" was not formally submitted to the NSF until August 1956, early drafts were circulated to the NSF and others, so the 11 July 1956 meeting participants, especially members of Tuve's Panel, all had access to the ambitious AUI plan. On the table were AUI vs. an alternative contractor; equatorial vs. alt-az mount for the 140 foot antenna; the viability of a 600 foot class antenna; national facilities vs. support for small university facilities; radio vs. optical astronomy; Green Bank vs. co-location with the national optical observatory; and not least of all the prestige of Merle Tuve and Lloyd Berkner.

In the weeks leading up to the 11 July meeting, the confrontations further expanded through vigorous lobbying.[183] Tuve, apparently unaware that Waterman was already committed to AUI, started a letter campaign to undercut the AUI initiative. Again, Tuve told his Panel that AUI was a northeastern, not a national organization, and as "a large-scale atomic energy activity … their Trustees cannot be expected to give major attention to the needs of astronomy and radio astronomy."[184] Tuve also wrote to Irvin Stewart, President of West Virginia University[185] and to Jesse Beams at the University of Virginia[186] claiming that "our Advisory Panel for Radio Astronomy has taken the position that the trustees of AUI are too fully occupied with the large-scale operations at Brookhaven for us to feel confident that the radio astronomy activities will have full and unbiased attention," and that the NSF was looking for an alternative to AUI. In his letter to Stewart, Tuve remarked that the "Oak Ridge Institute for Nuclear Studies [ORINS] has expressed some interest in becoming a possible sponsor for this activity." To Beams, he ended with, "I trust that you will take appropriate initiative in this matter." Beams, Stewart and Tuve had all worked together in wartime Washington and knew each other well.

On 28 June 1956 Tuve also wrote Leo Goldberg and Ira Bowen, "as the principal able bodied members of the Optical Observatories Panel," expressing his concern with the growing separation of radio and optical astronomers, and suggesting that they consider a "single trusteeship for the optical and radio astronomy ventures."[187] Bowen was sympathetic to the idea of a jointly administered radio and optical facility, but noted that in view of the different observing requirements (RFI vs. sky transparency) any common headquarters might be far removed from at least one of the actual observatories themselves, although he noted that with the expected introduction of commercial jet transport that "it will require little if any more time than the ground transportation from a nearby city to the point of operation regardless of whether the installation is in the East or West."[188] Bowen, however, underestimated the need not to separate the engineers and technicians who, of necessity must be located close to the facility, from the observers who may prefer a more academic environment. It was this issue which a decade later restricted the relocation of the NRAO headquarters to a location "close" to Green Bank (Sect. 4.7).

Bok responded to Tuve on 20 June 1956, with copies to the Panel and to the NSF staff stating that "I do not share these misgivings" and that "At least half dozen of AUI's Trustees … have in the past year shown a profound personal interest in the problems of the National Radio Observatory, and I have nowhere found any indication that anyone at AUI is inclined to favor the northeastern part of the United States over other sections."[189] The next day, Bok wrote again, this time in response to Tuve's and Stewart's letters, and somewhat more forcefully he stated, "I must say frankly that I do not remember our Panel actually taking this stand, and I for one feel that this is not a fair statement."[190] Bok was careful not to appear to offend Tuve, since he was also lobbying Tuve's committee for increased support for his Agassiz project.[191] On 25 June 1956 John Hagen responded more forcefully to Tuve's assertion about the Panel's position on AUI, by saying that he was "greatly disturbed by your two recent letters," and that "This is just not a fact, the Advisory Panel has never taken such a position. It is my feeling that the majority of the members of the Advisory Panel feel that AUI has done an excellent job to date in planning this activity and would look with favor upon continuation of the NSF relations with AUI." "Moreover," wrote Hagen, "I feel that when as radical a move as an attempt to generate a combination of two universities and the Oak Ridge Institute for Nuclear Studies with the implied support of the National Science Foundation is taken by the Chairman of the NSF Panel, he should first call a meeting of the Panel, discuss the matter and have agreements within the Panel."[192]

Jesse Greenstein was not able to attend the July meeting, but wrote expressing great concern over the performance and reliability of a computer-controlled telescope and strongly favoring an equatorial mount for the 140 foot telescope, even if it cost more. Greenstein did not express any great concerns about AUI management of the radio astronomy facility, but more broadly he worried about the impact to the funding of the university facilities, both radio and optical. Moreover, he remarked "I am absolutely aghast at the enormous expenditure for buildings," which he noted had more sleeping rooms (32) than the number of radio astronomers working in the United States."[193] Both Greenstein and Bok challenged Tuve's suggestion to build optical telescopes or to locate optical astronomers in Green Bank, pointing out the very poor sky condition in that part of the country. But, they both concurred with Tuve in the need to bring optical and radio astronomers together. Ed Purcell[194] and Bill Houston[195] had more tempered responses. Purcell conveyed confidence in having AUI act as the manager of the radio facility, but expressed concern about what he felt were the inappropriately tight AUI specifications for the 140 foot pointing accuracy and the need for a very large fully steerable antenna, suggesting that a very large antenna with limited sky coverage would be cheaper and more useful than a somewhat smaller fully steerable instrument, and that, if needed, a separate instrument could be built for solar astronomy. Houston expressed confidence in AUI, but recognized the possible advantages in having a broader group of 30–40 universities involved, and perhaps combining the management of both the national radio and optical facilities.

In a second letter on 29 June 1956, addressed to Bok but apparently as an afterthought also sent to his Radio Astronomy Panel and NSF staff, Tuve indicated that when two years earlier, the Panel "removed the onus from Berkner and the AUI of a political decision as to regional location by stating that the search for sites should be within 300 miles of Washington, D.C.," they had not realized that they were "going to set up something comparable to Palomar in cost and complexity." Now, he argued, "since we have moved into the range of 4.5 to 5.0 millions of dollars," he wanted to explore the possibility of a "joint installation with the national optical observatory, evidently in Arizona."[196] Tuve went on to point out the "relatively sad state of the 140 foot dish," and that neither Berkner nor Emberson had made the necessary "crucial engineering decisions," and accused Berkner of knowingly setting the specifications too high. He accused AUI of "using clerical procedures and a checkbook," to purchase "some exceedingly expensive studies ... so warped by the aim for a 600 foot dish that they have produced only a series of contradictory suggestions instead of a design." Moreover, Tuve again argued that there was no one at AUI who would use the new facility, and that the other radio qualified astronomers at Harvard, DTM, Ohio State, Caltech, and Stanford were all building their own facilities with funds from ONR, NSF, and the Air Force. "Where," he asked, "are the sound research men who will need this new five million dollar facility? ... Shall we now decide we must import them from Manchester or Sydney.... Can we name even one first-rate man who is prepared to accept personal responsibility to make this added 'National Facility' a wise and fruitful venture for the NSF." Tuve, of course, did not miss the opportunity to remind Bok that the two strongest advocates for the national radio facility and the two most qualified people to assume the position of director were Bok and Hagen, and that Bok was leaving the US to go to Australia and Hagen was about to leave radio astronomy for a "satellite adventure."

Three days later, claiming that "radio astronomy activities have largely been 'captured' during the past two years by Berkner, Hagen, and Bok" and their plans for a $25 to $30 million facility, Tuve laid out a more detailed plan for how the "NSF and its advisors in astronomy should firmly and flatly dissociate themselves from this kind of thinking," arguing that "the AUI proposal can wreck the entire NSF if it is adopted."[197] Instead, Tuve proposed forming a "University Corporation for Astronomy," and suggested the possible names of a number of "men with some immediate active connection with Radio Astronomy." But he specifically added, "Do not include, Bok, Hagen, or Berkner."

Meanwhile, Bok, feeling offended, did not delay, and responded on 3 July 1956 to Tuve's letter of 29 June 1956.[198] Reacting to the perceived allegation that he "was running away from the project," Bok felt obliged to point out that in January he had informed everyone of "his decision to leave Harvard" and that he had "in mind either going to Australia or joining the staff of the National Radio Observatory." When Bok received a formal offer for the Directorship of the Mt. Stromlo Observatory in Canberra, he let it be known that he would accept the Australian position after two weeks, "unless good reason could be given why I should not do so." John Hagan apparently wrote to Waterman that it was important to keep Bok in the United States, but as Bok described, "nothing developed, ... [and] I decided that I was free to accept the [Australian] Directorship."[199] More to the point, Bok laid out a seven-point plan of action that he wished might come out of the scheduled July meeting.

1. NSF should purchase the land in Green Bank,
2. Settle the administrative issues with preference given to AUI as the contractor,
3. Decide on the needed precision for the 140 foot telescope,
4. Obtain an engineering study for an equatorial mount making use of the experience with the Harvard 60 foot and Caltech 90 foot antennas,
5. Appoint a Director,
6. Build up a staff and instrumentation,
7. Increase NSF support for university facilities.

Bok also challenged Tuve's accusation that the radio astronomers were isolated from optical astronomers by pointing out that the projects at Harvard, Caltech, Ohio State, and Michigan were led or co-led by an optical astronomer, but that "the contact [between radio and optical astronomers] may not be a close one for the radio astronomers working in and near Washington, D.C." Bok also expressed that, in view of the heavy concentration of optical facilities in the west, he was "strongly opposed" to Tuve's suggestion of co-locating the national radio observatory together with the optical observatory near Phoenix, Arizona, and with a final dig, he pointed out that "some of you people in Washington, fairly remote from university contacts, do not apparently realize the terrific boon to the development of student interest in the Eastern United States in radio astronomy as a result of the [planned] Greenbank [*sic*] operation."

Greenstein took a more intermediate approach.[200] While not opposed to having AUI as the contractor for the construction and operation of the radio observatory, he agreed with Tuve that the NSF should establish a single Board for both observatories to apportion resources between the two fields, and that consideration be given to locating the 140 foot dish in the same area as the optical facility although travel from the East would be more expensive, "but hardly more time-consuming than road travel to Greenbank [*sic*]." Like Bok, Greenstein felt that there were a sufficient number of young men in the United States who would use the radio observatory, and that "if we lack the ideal leader for the group, I would, personally, not be embarrassed at importing one."

The "shootout" to decide on the management for the national facility was held on 11 July 1956 in Washington and was organized by the NSF Acting MPE Director, Raymond Seeger along with Frank Edmondson, at that time the NSF Program Director for Astronomy.[201] Prior to the meeting, AUI decided to take a "passive role," and not do any advance lobbying in response to Tuve's letters.[202] More than 40 people attended, including not only members of Tuve's Advisory Panel and the AUI Steering Committee, but also John Bolton from Caltech and many of the major players in optical astronomy, including Seyfert, Lyman Spitzer, and Albert Whitford. Institutional leaders attended as well, perhaps looking to grab a piece of the action: the Presidents of Rice and Ohio State Universities, and the Universities of Pennsylvania, Virginia, and West Virginia, along with the Directors of DTM (Tuve) and the Oak Ridge Institute for Nuclear Studies (Pollard), and a number of University Vice Presidents. AUI was represented by President Lloyd Berkner, Frank Emberson, and Charles Dunbar, the AUI Corporate Secretary. Jerry Wiesner and Julius Stratton, two future MIT presidents, represented MIT, while Harvard was represented by Menzel and Ed Reynolds, the Harvard Vice President for Administration and member of the AUI Board of Trustees. Alan Waterman and eight other staff attended from the National Science Foundation. NRL and ONR sent representatives. Detlev Bronk, President of the National Academy of Sciences and Chairman of the National Science Board, presided. Greenstein and Purcell were unable to attend.

The morning session opened with a brief review from Waterman explaining that with the approval of the Administration, the Foundation now accepted responsibility for the construction of new scientific facilities, although that was not part of its original charter to support research. He then added that, based on the recommendation of Tuve's Advisory Panel for Radio Astronomy, the NSF had financed a feasibility study for a radio astronomy facility, but since the Foundation was not allowed by law to operate research laboratories itself,[203] the NSB would consider the AUI study, and would make a final decision about how to operate the new facility at its August meeting.

Tuve, perhaps not to anyone's surprise, claimed that there already was good support for new radio astronomy facilities at universities around the country, that the need was to integrate radio and optical astronomy, and to keep control within the universities and away from "the hands of a professional management group." Tuve contrasted management by "a true university operation … in the hands of university research men" and a "special laboratory operation, nominally in university hands, but actually controlled and guided by small self-approving groups of experts." He went on to argue that "very high priority, probably ahead of anything else, must be given to the support of existing activities in radio astronomy at universities," and that NSF money should go to active research astronomers, not to physicists, engineers, or administrators. Departing from his prepared remarks, Tuve contended that "the AUI plan is a poisonous whitewash," adding that, "a very large dish (250-600-ft) is a project of uncertain value," and that "the AUI plan for a 600-ft radio telescope is a power bid by people who love to manage things." (England p. 284). Tuve concluded his speech by stating, "My frank advice against accepting AUI as the contracting agency for radio astronomy is not based on their list of trustees, but on their very large plans and on the pattern of self-generated approval and on their automatic initiative for larger and larger projects of the same technical type."

Bart Bok followed with brief history of the development of radio astronomy, both in the US and abroad, including a detailed description of the status of solar, planetary, 21 cm, continuum research, and various radio astronomy activities underway at each of ten different US facilities. Contrary to Tuve, Bok claimed that, "the majority of radio-astronomical centers in the Eastern and Central United States are not contemplating the construction in the near future of new large scale equipment, because they expect to make use of the Greenbank [*sic*] facilities."

Berkner again reviewed the history leading to the current meeting, reminding everyone that AUI was approached by Harvard, MIT, and NRL to undertake a study on how a radio observatory might be organized to provide large scale facilities that seemed beyond the capacity of any one university to undertake. Naturally, he described the success of AUI in operating Brookhaven, and noted the addition of Dave Heeschen[204] to the full time staff at AUI to supplement the radio astronomy expertise of the Steering Committee. But he added that the AUI Planning Document was a general report with respect to the character of the management, and it did not recommend that AUI operate the facility.

Emberson summarized the content of the Planning Document, including the process which led to the choice of the Green Bank site and the status of the 140 foot design. Following Rudolph Minkowski's brief outline of the interdependence of radio and optical astronomy, Leo Goldberg spoke of the need for balance between national and university facilities, for both radio as well as for optical astronomy.

Although the morning session went off without much acrimony, the seeds of discontent had clearly been sown. Detlev Bronk led off the afternoon deliberations with the premonition, "I have the unsatisfactory feeling that after the end of the afternoon conference you may have an unsatisfactory feeling of nothing have been accomplished or decided." More to the point, while acknowledging that neither he nor the NSB had expertise in radio astronomy, Bronk reminded those present that it would be the NSB that made the final decision on the development of radio astronomy through the national facility, so the NSB would of necessity depend on the advice received. Moreover, he pointed out that as the first large NSF research facility, the decisions made on how to operate the radio astronomy facility would have broader impact "than merely for radio astronomy itself." Noting that because the NSF is funded on a year to year basis, the construction of a research facility can provide no legal responsibility for its future operation, but Bronk nevertheless warned that "I can conceive a time when the National Science Foundation would be using all the money it could get from the National Treasury in the support of the activities which it had created in the past." Waterman added that "astronomy being pure research is in the position of requiring probably more support on the part of the Government than most facilities would have." Berkner spoke at great length about his interpretation of a national facility as one that responds to broad interests and pressures from individual scientists and not from universities, but is not an independent laboratory with its own agenda or programs. M.V. Houston, President of Rice University and NSB member, cynically commented that "Washington is a long way off from most of the country," and since you can't have more than one national facility for radio astronomy, "one has to make it clear that the facilities are available to all who wish to use them, ... not as a favor, but available ... as a matter or right."

Menzel and Wiesner reminded the group that AUI involvement came about only because the instrumental needs were too great even for the combined resources of MIT and Harvard, and that Edward Reynolds, Harvard's Financial Vice President, had suggested that they approach AUI, not the other way around as Tuve had contended. Wiesner recalled that their small MIT-Harvard group had discussed forming a new corporation, but realized that the management cost would be much less working with AUI, and that it would take much longer to set up a new organization. Menzel acknowledged that the 600 foot dish was a long range vision, but that the immediate need was for a 150 foot class antenna, and argued that the design of the intermediate size antenna should not be affected by any consideration that it be a prototype for the 600 foot telescope.

William Pollard, Director of the ORINS, suggested instead that they form "a new corporation of national scope" to manage the new radio astronomy facility. Pollard acknowledged that he was "not a radio astronomer or for that matter an astronomer of any kind," nor was he even aware of the proposed radio astronomy initiative until learning about it from Merle Tuve just a month earlier. Pollard and Tuve had worked together at Los Alamos during the war, and Tuve, who expressed concern about his "Christian responsibility" had contacted Pollard, a recently ordained Episcopal priest, as Tuve later reported, "to help me orient my thoughts constructively" (Wang 2012). Apparently the previous NSB Vice Chair, Paul Gross, had lobbied with Seeger in support of Pollard's initiative (Needell 1987, p. 279) and he received further support at the 11 July meeting from Houston and Morris, the other southern members of the NSB present at the meeting. At the meeting, Pollard passed around a descriptive brochure of his proposed new organization that included an "Agreement for Incorporation," along with draft by-laws, and suggested a procedure for the organization to be chartered under West Virginia law. Pollard argued that his new organization would be more open to new membership than AUI and suggested that since the new organization could be established in "three to five weeks," it would not introduce any significant delay in establishing the national radio astronomy facility. Irvin Stewart, President of West Virginia University, added that "if you are going to locate in our backyard," the University has "an interest in this," and spoke in support of the Pollard proposal. Tuve claimed that since the technical specifications for the 140 foot antenna were not yet decided and that "if it is built according to the present internal contradictions ... it will be a white elephant. ... Therefore, you need not think you are holding things up if it takes a while to get a sponsoring agency."

Tuve returned to his concerns that creating a national radio facility separate from the planned national optical facility would further split astronomy, but was challenged by Bok, who claimed that "many of us do not agree that there is even a limited divorce between optical and radio astronomy at the present time," and suggested a joint meeting of the two panels to discuss this question. Bok was supported by Hagen, who argued that "the connection between radio astronomy and astronomy is much, much better today than it has ever been and it is steadily improving." Tuve read from Greenstein's letter expressing concern about the competition between the universities and the national facility for funds, but under pressure from Wiesner, he also read from the very strong letter from Purcell, who wrote, "I don't know of any organization in the country that has had experience and success comparable to that of AUI in this special sort of enterprise." Bok added that the real issue was the concern about management by committee, and the solution was to find a good director, give him the power and responsibility to make the decisions and "fire him if you don't like him."

Asked about the position of the optical community, Leo Goldberg agreed that because of the obvious different technical requirements, it would be inappropriate to co-locate the two national facilities, but he could conceive of a single management structure possibly including a national solar facility as well.

Finally, following up on their agreed strategy, Theodore Wright, Vice President for Research at Cornell and long-time AUI Trustee, spoke passionately saying,

> The Trustees of AUI would not want to undertake the management of this new facility if they were not to be enthusiastically supported by the NSF, by its Advisory Committee and by the scientists and universities in the country who will do the research work in the laboratory itself. AUI is not out to get the job. It feels however, it can make a real contribution and if it is backed, it will go ahead enthusiastically.

"Furthermore," added Wright,

> I do not think that AUI would wish to undertake the construction phase and then bow out. So, in the interest of time saving and after all I have heard here, I am convinced that the National Science Foundation will not go astray if it looks favorably upon contracting with AUI to manage the contemplated National Radio Astronomy Facility.

Wright was supported by Harvard's Ed Reynolds, who, pointing his arm at Tuve, added (England 1982, p. 285)

> The atmosphere here has disturbed me very, very much indeed. ... My disturbance goes so far that I am going to enlarge on what Dr. Wright has just said. ... We don't want this contract under this atmosphere. You are going to have to want us enthusiastically if we are to take it on. Then we can do a good job for you.

In reflecting over the afternoon's tumultuous session, MIT's Julius Stratton was said to have remarked, "Never have so many thought so differently on so few matters" (England 1982, p. 284). Bronk summarized the meeting with the important conclusion that there was agreement that the managing organization would be responsible to both the NSF and to the astronomical community. But considering the lack of agreement about if or how the NSF-sponsored radio and optical astronomy initiatives should be coordinated, Bronk concluded the meeting with a call for the NSF Panel for Radio Astronomy and for the two Optical Astronomy panels to meet jointly and consider what they would like to recommend to the National Science Board as to the relationship between the three panels "and the desirability or undesirability of having some sort of more formal common association in this general undertaking."

Returning from the 11 July meeting, Menzel wrote a scathing letter to Tuve pointing out the many contradictions between his testimony and his prior written statements.[205] Menzel complained that "many of us are greatly disturbed that you chose to inject regional competition into the situation by writing Dr. Stewart and Dr. Beams in your letters of June 19 inviting opposition to the AUI plan." Menzel specifically chastised Tuve for his claim, "Our Advisory Panel for Radio Astronomy has taken the position that the trustees of AUI are too fully occupied with the large-scale operations at Brookhaven for us to feel confident that the radio astronomy activities will have the full and unbiased attention," whereas in fact wrote Menzel, "You did not even consult your panel, before taking this unilateral action." Later, Berkner shared his frustrations with Menzel, expressing his concerns about Tuve's attempts to undercut AUI's plans to construct and manage the national radio astronomy facility.[206]

Meanwhile, Pollard and Stewart, apparently undiscouraged by the lack of reaction to their bold suggestion, circulated a formal proposal to form a new corporation to be known as the Association of Universities for Radio Astronomy, or AURA, with its headquarters to be located in Green Bank. Pollard's invitation went to 12 prominent scientists and science administrators, including Menzel, Goldberg, Hagen, Struve, and Greenstein, as well as Colgate Darden, President of the University of Virginia, and Howard Bevis, President of Ohio State University, inviting them to become an "incorporator" of AURA. Pollard and Stewart included a proposed "Agreement for Incorporation" along with seven pages of proposed "By-Laws." But aside from his co-conspirators Irvin Stewart and Merle Tuve, there was no support from within the astronomy community for this initiative. Donald Menzel was particularly incensed, and accused the conspirators of "southern pork barrel interest" and "selfish motives on the part of the Carnegie-Cal Tech [sic] axis" for the attempt to discredit AUI (England 1982, p. 286).

Interestingly, the acronym AURA was to be soon reconstituted as the Association of Universities for Research in Astronomy, which would manage the Kitt Peak National [optical] Observatory, the sister organization of NRAO. However, Pollard's initiative was supported by the influential National Science Board Vice Chair, Paul Gross, who also happened to be a member of the ORINS Board. Earlier, encouraged by Tuve, Gross had told Waterman that he supported what he (incorrectly) claimed was the view of the NSF Radio Astronomy Advisory Committee that the national radio astronomy facility should be managed by a group with broader representation than AUI (Needell 1987, p. 227).

Sensitive to the anti-AUI sentiment, Waterman spoke with both West Virginia University President Stewart and University of Virginia President Darden to explore their interest in having a role in managing the radio facility. Both expressed possible interest in some joint management scheme, but neither Stewart nor Darden seemed interested in assuming responsibility for sole management of the facility (England 1982, p. 287).

Following the 11 July meeting, Berkner wrote to Waterman stating AUI "will make a proposal to construct and manage a National Radio Astronomy Facility," but only "if desired and requested to do so by the National Science Foundation."[207] However, Berkner emphasized, "it would not be appropriate to undertake the management of the Facility through an initial phase only, ... after which some other agency would manage the operation itself." In response to the criticism raised by Tuve and others that AUI did not have sufficient expertise in radio astronomy, Berkner repeated that AUI "would appoint an Advisory or Visiting Committee for the Radio Astronomy Facility, similar in composition and function to the visiting committees for the various departments at Brookhaven," and that "appropriate representation will be given to astronomy" on the AUI Board of Trustees.

Meanwhile, Leo Goldberg and others were still concerned about the developing divide between the radio and optical factions of the astronomy community, as well as the deepening rift emerging over who would manage the new national radio observatory. Goldberg, perhaps more than anyone else, appreciated that unless the two communities could come together, chances of funding for either the national radio or optical facilities would be dubious. Following the 11 July closing directive from Detlev Bronk, the NSF Radio Astronomy Panel, the NSF Advisory Panel for Astronomy, and the NSF Astronomical Observatory Panel agreed to meet in a joint session. The combined meeting of the three NSF astronomy panels was held on 23 July at the University of Michigan in Ann Arbor and was attended by Waterman along with other members of the NSF staff. Waterman opened with the announcement that unless there were objections, "the NSF was ready to proceed with acquisition of the Green Bank site," and made it clear to the assembled panel members that the advisory panels were purely advisory to the NSF.[208] Ever the optimistic statesman, Goldberg argued, "I think our positions should be that we need them all, and that the country is rich enough to afford them. I have been rather impressed by the timid attitude on the part of astronomers which seems to take it for granted that funds must always be limited." According to Greenstein,[209] the Ann Arbor meeting "went off very well in a mood of friendliness and compromise. There was no debate of any kind about the details of management, and this is now definitely in the hands of the National Science Foundation and its board for final decisions." Greenstein reported a consensus agreement on the choice of Green Bank for the radio astronomy facility, but that "beginning with the next fiscal year, applications for operating funds for university installations in radio astronomy will be handled by the regular advisory panel for astronomy." Two days later the NSF issued a press release announcing that Green Bank had been selected as the site for a new radio astronomy facility.[210]

After the 11 July meeting, and again after the joint meeting in Ann Arbor, Tuve's Panel met to recommend operating funds for various university radio astronomy programs as well as funds to help support international radio astronomers planning to attend the 1957 URSI General Assembly, to be held in Boulder Colorado, and for Greenstein to visit Australia. In accord with the discussions toward integrating all of the astronomy advisory functions, the Radio Panel agreed that future yearly research grants in radio astronomy should be handled by the reconstituted NSF Advisory Panel for Astronomy. However, Tuve made it clear that the Radio Panel would retain the responsibility for facility grants. The Panel also made special mention of "the great importance" that they attached to "the fact that Dr. David Heeschen has now joined this activity," and urged that he "retain his connection with the NSF activity whether or not the management sponsor is AUI or some other agency."[211]

However, the issue of the relationship between radio and optical facilities, the level of NSF control over these facilities, and the constitution of AUI continued to fester. With a combined management of the radio and optical facilities, the NSF might maintain a higher level of influence and control, which Berkner opposed. Greenstein, concerned that since Berkner would never give up control over the radio facility, worried that this would lead to AUI management of both the radio and optical facilities, and also opposed any joint operation of the two national facilities.[212]

After making revisions resulting from the 11 July meeting and the subsequent informal exchanges between AUI and NSF staff, AUI delivered 50 copies of the *Planning Document* to the NSF in time for consideration by the NSB at its August meeting. Among the modifications to the earlier version, AUI, realizing that the restricted nature of AUI could be a deal killer, finally broke down and agreed to add three "at-large members" to the AUI Board. Initially, these would be Otto Struve from the University of California, Leo Goldberg from the University of Michigan, and Carl Seyfert from Vanderbilt University.

On 24 August 1956 the NSF National Science Board met to discuss three options for the new radio astronomy facility[213]:

(a) Management by a university. (The NSF Director, Alan Waterman had already held discussions with both West Virginia University President Stewart and the University of Virginia President Colgate Darden and Physics Department Chair, Jesse Beams.)

(b) Management by a new organization such as the Oak Ridge Institute for Nuclear Studies [ORINS].

(c) Management by AUI.

Also on the table were the long debated issues of the nature and performance specifications for the first radio telescopes, and to what extent the management of the national radio and national optical facilities should be combined.
Waterman presented the AUI plan, reported that AUI had obtained options on
the land, and that the West Virginia Zoning Act had been passed (England
1982, p. 288). Merle Tuve testified on behalf of the NSF Advisory Panel claiming that there is "no value in a telescope capable of operation at less than
10 cm," that AUI was "too busy and has no experience," and went on to ask,
"where will the staff come from? UK? Australia?" There is "no need to rush,"
he argued, but Hagen responded that Tuve "had misrepresented the committee and was speaking only for himself."[214] The NSB realized that there was no
really viable option to AUI. This was the first big NSF project. Both Waterman
and the NSB needed a success, but they weren't going to give AUI a blank check.

On 4 September 1956 Waterman notified Berkner that the NSB had decided
that it would take too long to form a new organization and authorized the NSF
Director to enter into negotiations with AUI for the establishment and operation of the Radio Astronomy Observatory.[215] But there were strings attached.
Concerned about the lack of any AUI experience or involvement in astronomy,
and that AUI only represented a limited number of universities, all in the
Northeast, the NSB stipulated that AUI would need to agree to appoint three
at-large members to the Board of Trustees and that AUI would operate the
facility for the "use of the nation's scientists," independent of their institutional
affiliation. Waterman also conveyed that the NSB decision to contract with
AUI for the construction and initial management of the radio facility was

> subject to a clear understanding with AUI that the Foundation will give serious
> consideration to the possibility of establishing at the end of that time a common
> management for the Radio Astronomy Observatory and for the Optical
> Astronomy Observatory; it is to be further understood that the selection of the
> Director and of the AUI Advisory Committee for the Observatory will be made
> in consultation with the Director of the National Science Foundation.[216]

Berkner responded firmly, informing Waterman that "the detailed planning,
construction, and management of the Observatory will be handled ... by
AUI," although he added that "AUI will, of course welcome basic policy recommendations by the NSF concerning the general direction of the program."[217] Berkner went on to remind Waterman that while the proposed $4
million initial budget "will cover the creation of an effective radio astronomy
facility," $6.7 million "represents a realistic estimate of what is needed to pro-
vide the facilities that radio astronomers in the United States really need." Not
to be intimidated, Waterman stated that "authority for major decisions with
respect to the radio astronomy observatory are lodged at the National Science
Board and the Director," and that "the Foundation is not in any position at
this time to comment on the magnitude of future commitments."[218]

At its meeting on 21 September 1956, the AUI Executive Committee authorized Berkner to "negotiate and execute … a contract for the construction and operation of a Radio Astronomy Observatory."[219] On 17 November 1956, AUI and the NSF signed a contract for the "Construction and Operation of the Radio Astronomy Observatory (Fig. 3.13)."[220] The five-year contract initially committed the amount of $5.13 million, but noted that "The Foundation may increase or decrease this obligation at its discretion from time to time." Indeed, the signed contract obligated only $4 million, although a month before signing, the NSF Director had authorized an additional amount of $1.13 million. No indication was made in the contract that this was to be a "national observatory," although it was specified that the observatory would be made available to qualified personnel to the extent possible. The contract was written in very general terms and did not include any details of the instrumentation to be built other than, "The observatory shall contain facilities and equipment appropriate for the conduct of research in radio astronomy (including one or more radio telescopes, at least one of which shall have a diameter of at least 140 feet) and appropriate ancillary buildings and facilities as mutually agreed upon from time to time." There was no mention in the contract of any possible larger instrument such as a 600 foot telescope. The NSF also specified that AUI "consult with the Director of the Foundation" before appointing the

Fig. 3.13 17 November 1956 signing of the first AUI 5-year contract to operate NRAO. Seated on the left is AUI President Lloyd Berkner and on the right is the NSF Director, Alan Waterman. Watching are members of the NSF staff. Credit: NRAO/AUI/NSF

Director of the Observatory. After this initial five-year period, the NSF stipulated that the choice of the management would be reconsidered, possibly for a joint operation with a national optical observatory.

For more than half a century, until 2015, the NSF would renew the AUI contract or Cooperative Agreement every five years based on a non-competitive proposal from AUI. There was never any further serious discussion of combining the management of the national radio and optical astronomy observatories, although for a period of years in the mid-1960s, the NRAO and KPNO staffs met annually for exchange visits. However, due to budget pressures and increased emphasis on visitor use at both observatories, and less attention given to in-house research programs, these exchange visits ended after a few years. The national radio and optical observatories continued as they had begun, with independent management by AUI and AURA respectively, reflecting the differing technology and different scientific emphasis at the two national observatories.

Notes

1. This section is based in part on *The Science Foundation: A Brief History*, NSF88-16 by George T. Mazuzan (1994). https://www.nsf.gov/about/history/nsf50/nsf8816.jsp. Maruzan was a former NSF Historian at the Office of Legislative and Public Affairs. More detailed background on the NSF can be found in the books by J. Merton England (1982) and by Milton Lomask (1976).
2. Minutes of the NSB 34th meeting, 20 May 1955.
3. Public Law 507-81st Congress, Chapter 171-2D Session, S. 247.
4. Ibid.
5. Waterman to Gabriel Hauge, Administrative Assistant to the President, 26 August 1955, LoC-ATW, Box 26.
6. Events leading to the establishment of the Kitt Peak Astronomical Observatory (KPNO) operated by the Association of Universities for Research in Astronomy (AURA) are discussed in more detail by England (1982), Goldberg (1983), Edmondson (1991, 1997) and Leverington (2017).
7. Other panel members were Lawrence Aller (Michigan), Dirk Brower (Yale), Martin Schwarzschild (Princeton), Gerard Kuiper (Arizona), Fred Whipple (Harvard), and Gerald Kron (Lick). Leo Goldberg, a member of the NSF Mathematical, Physical, and Engineering Sciences (MPE) Divisional Committee, was effectively an ex officio member.
8. Other members of the Panel were Ira Bowen (MWPO), Bengt Stromgren (Yerkes), Otto Struve (UC Berkley Leuschner Observatory), and Albert Whitford (University of Wisconsin).
9. Waterman notes to diary, February 1954, LoC-ATW, Box 26.
10. Tensions between the well-equipped Caltech/MWPO and KPNO continued for many years, although Caltech later became a member of AURA in 1972. By 2017, there were 44 member institutions of AURA.

11. This section is adopted from *The Early History of Associated Universities and Brookhaven National Laboratory, BNL 992 (T-421)*, which is based on a talk given by Norman Ramsey at Brookhaven on 30 March 1966.
12. Columbia, Cornell, Harvard, Johns Hopkins, MIT, Pennsylvania, Princeton, Rochester, and Yale.
13. Ramsey op cit.
14. Ramsey op. cit.
15. AUI-BOT, 18 July 1952.
16. This section is based in part on Hales (1992) and Needell (2000).
17. See Needell (1987) p. 264.
18. Berkner to James Webb, 31 March 1961, The Space Studies Board, Compilation of Reports, National Academies of Science, Engineering, Medicine, Washington, DC.
19. President John F. Kennedy, speech to a Joint Session of Congress, 25 May 1961.
20. The circumstances leading up to the creation of NRAO are discussed by DeVorkin (2000), England (1982), Kenwolf (2010), Lockman et al. (2007), Lomask (1976), Malphrus (1996), Munns (2003, 2013) and Needell (1987, 1991, 2000). Numerous factual errors in the Malphrus book have been noted by Baars (NAA-KIK, Open Skies) and by Findlay (NAA-JWF, Miscellany, Theses).
21. See also Greenstein oral history at Caltech. CITA.
22. Greenstein, unpublished manuscript, "I Was There in the Early Years of Radio Astronomy", NAA-KIK, Open Skies.
23. Robert F. Bacher oral history, CITA. http://oralhistories.library.caltech.edu/93/
24. According to Needell (2000, p. 265), during an early 1952 visit to Pasadena, Bush met with DuBridge and Ira Bowen to discuss the possibility of building a large radio telescope close to the Caltech/Carnegie MWPO. Apparently similar discussions had taken place earlier between Merle Tuve and Ira Bowen.
25. Bush to DuBridge, 14 February 1952, CITA-LAD, Box 35, Folder 1.
26. DuBridge to Bowen, 21 February 1952, CITA-LAD, Box 35, Folder 1.
27. E. Bowen, Draft Programme for a Radio Observatory, 22 May 1952, CITA-LAD, Box 35, Folder 2.
28. DuBridge to Bowen, 11 June 1952, CITA-LAD, Box 35, Folder 2.
29. Bowen to DuBridge, 30 August 1952, CITA-LAD, Box 35, Folder 2.
30. Bowen to Bush, 22 August 1952, CITA-LAD, Box 35, Folder 2.
31. Greenstein to DuBridge, 19 February 1953, CITA-JLG, Box 113, Folder 1.
32. Ibid.
33. Ibid.
34. DuBridge to Greenstein, CITA-JLG, Box 113, Folder 1.
35. DuBridge to Tuve, Purcell and Wiesner, 19 and 20 March 1953, CITA-LAD, Box 34, Folder 15.
36. Purcell to DuBridge, 26 March 1953, CITA-LAD, Box 34, Folder 15.
37. Bok to Greenstein, 30 March 1953, CITA-JLG, Box 3, Folder 26.
38. Tuve to DuBridge, 2 April 1953, DTMA, Series 1, Box 4, Folder 3.
39. Greenstein to Bok, 9 April 1953, CITA-LAD, Box 34, Folder 15.
40. Bok to Greenstein, 13 April 1953, CITA-LAD, Box 34, Folder 15.
41. DuBridge to Bok, 5 May 1953, CITA-LAD, Box 34, Folder 15.

42. Greenstein to DuBridge, 10 June 1953, CITA-LAD, Box 34, Folder 15.
43. Ultimately, papers from the December 1953 Boston meeting were never published. The January 1954 Washington meeting papers were published in the *Journal of Geophysical Research*, Vol. **59**, p. 149 (1954).
44. Tuve to Greenstein, 23 September 1953, DTMA, Series 1, Box 4, Folder 8.
45. Ibid.
46. Tuve to Greenstein, 14 October 1953, DTMA, Series 1, Box 4, Folder 8.
47. Participants included: P. E. Klopsteg, J. H. McMillan, and R. J. Seeger from the NSF, R. Minkowski from the MWPO, Merle Tuve from DTM, Greenstein and DuBridge from Caltech, Bart Bok from Harvard, John Kraus from Ohio State, and Taffy Bowen from Australia. The group's report to the NSF was written by Greenstein (CITA-LAD, Box 34, Folder 15).
48. Other members of the NSF Advisory Panel were Bart Bok (Harvard), Jesse Greenstein (Caltech), John Hagan (NRL), John Kraus (Ohio State), Rudolph Minkowski (MWPO), and Ed Purcell (Harvard). Peter van de Kamp represented the NSF on the Panel.
49. Historical account presented by Richard Emberson at the 28 May 1955 meeting of the AUI Steering Committee, NAA-NRAO, Founding and Organization, Planning Documents, and by Menzel at the 11 July 1956 NSF meeting, NAA-NRAO, Founding and Organization, Meeting Minutes. Also see Emberson (1959).
50. Menzel to Berkner, 13 April 1954, "Survey of the Potentialities of Cooperative Research in Radio Astronomy," NAA-DSH, Radio Astronomy History, NRAO. https://science.nrao.edu/about/publications/open-skies#section-3
51. Diary notes of Raymond Seeger, 26 April 1954, LOC-ATW, Box 26.
52. The discussion of the 20 May 1954 meeting is based on diary notes of Raymond Seeger, LOC-ATW, Box 26.
53. Greenstein to Hagan, 1 June 1954, CITA-JLG, Box 14, Folder 1.
54. Dicke to AUI Committee on Radio Astronomy, 7 June 1954, CITA-JLG, Box 111, Folder 7.
55. AUI BOTXC, 21 May 1954.
56. AUI BOTXC, 18 June 1954.
57. Other members of the AUI Steering Committee were Bart Bok (Harvard), Armin Deutsch (MWPO), Harold (Doc) Ewen (Harvard), Leo Goldberg (Michigan), William Gordon (Cornell), Fred Haddock (NRL), John Kraus (Ohio State), Aden Meinel (Yerkes Observatory), Merle Tuve (DTM), Jerry Weisner (MIT), and H.E. Wells (DTM).
58. The NSF Advisory Panel was at times referred to as "Committee" rather than "Panel," but we have consistently used "Panel" to avoid confusion with the AUI Steering Committee appointed by Berkner.
59. Goldberg to van de Kamp, 4 August 1954.
60. Summary Notes AUI Steering Committee, 26 July 1954, DTMA, Series 1, Box 4, Folder 2.
61. Research Proposal to the National Science Foundation for a Grant in Support of a Feasibility Study for a National Radio Astronomy Facility, 26 July 1954, NAA-NRAO Founding and Organization, Planning Documents. https://science.nrao.edu/about/publications/open-skies#section-3
62. Waterman to diary, 3 August 1954, LOC-ATW, Box 26.
63. The total NSF 1954 budget was $8 million.

64. Minkowski to Tuve, 19 October 1954, DTMA Series 1, Box 4, Folder 1.
65. Waterman to diary, 7 September 1954, LOC-ATW, Box 26.
66. Ibid.
67. Tuve to NSF Radio Astronomy Panel, (Purcell, Bok, Greenstein, Minkowski, Hagen, and Kraus), 5 October 1954, DTMA, Series 1, Box 4, Folder 1.
68. Ibid.
69. Greenstein to Tuve, 14 October 1954, DTMA, Series 1, Box 4, Folder 1; CITA-JLG, Box 39, Folder 12.
70. AUI-BOTXC, 11 September 1954; AUI-BOT, 28 October 1954.
71. Tuve to van de Kamp, 29 October 1954, DTMA, Series 1, Box 4, Folder 1.
72. Bok to Greenstein, 8 November 1954, CITA-JLG, Box 3, Folder 26. Other participants included Bok, Purcell, Seeger, Hagen, and Tuve.
73. Berkner to Seeger, 18 November 1954, Appended to AUI-BOTXC minutes, 19 November 1954.
74. Tuve to NSF Radio Astronomy Committee [i.e. Panel], 17 November 1954, DTMA, Series 1, Box 4, Folder 2.
75. Greenstein to Tuve, 16 November 1954, DTMA, Series 1, Box 4, Folder 1.
76. Tuve to NSF Radio Astronomy Committee [i.e. Panel], 11 November 1954, DTMA, Series 1, Box 4, Folder 1; LOC-ATW, Box 26. Participants were Bok (Harvard), Hagen (NRL), Kraus (Ohio State), Minkowski (MWPO), Tuve (DTM), and van de Kamp (NSF).
77. Greenstein to Tuve, 11 November 1955 telegram, DTMA, Series 1, Box 4. Starting with his friendship and collaboration with Grote Reber, two early publications on radio astronomy, his organization of the Washington Conference, his role in starting radio astronomy at Caltech and in the discovery of quasars (Chap. 6), his early support of the AUI radio astronomy initiatives, and his influence at the 1970 NAS Decade Review of astronomy (Chap. 7) that recommended the VLA as its top priority, Greenstein, perhaps more than anyone else in the United States, was a central figure in the development of the American radio astronomy program.
78. Notes on NSF Advisory Committee [i.e. Panel] meeting dated 19 November 1954 and minutes of the meeting dated 20 November 1954, DTMA, Series 1, Box 4, Folder 1.
79. Seeger in fact had estimated that between $200,000 and $300,000 would be available from the NSF out of Fiscal Year 1955 funds. Seeger to diary, 16 November 1954, LOC-ATW, Box 26.
80. NSF Advisory committee minutes, 19 November 1954, op. cit.
81. Berkner to Seeger, 18 November 1954, op. cit.
82. Dunbar to Berkner, 28 December 1954, DTMA, Series 1, Box 4, Folder 1. Summary of meeting 28 December 1954. Also in attendance were Peter van de Kamp from the NSF, John Hagen from NRL, and other AUI officers.
83. AUI-BOT, 20 January 1955.
84. Ibid.
85. AUI-BOT, 14 April 1955.
86. Ibid.
87. *Plan for a Radio Astronomy Observatory*, p. 31, NAA-NRAO, Founding and Organization, Planning Documents. https://science.nrao.edu/about/publications/open-skies#section-3

88. It was not specified, but presumably this referred to the maximum, not rms, deviation. One inch maximum deviation would have meant approximately an 8 mm rms surface accuracy roughly consistent with a minimum operating wavelength of about 10 cm.

89. Tuve to DuBridge, 10 March 1955, CITA-LAD, Box 35, Folder 2.

90. *Plan for a Radio Astronomy Observatory*, op. cit., p. 32.

91. Research Objectives for Large Steerable Paraboloid Radio Reflectors (prepared by Bok and his Panel for the AUI Steering Committee), 22 April 1955, NAA-NRAO, Founding and Organization, Planning Documents. https://science.nrao.edu/about/publications/open-skies#section-3; AUI-BOT, 14 April 1955.

92. AUI-BOT, 14 April 1955.

93. Ibid.

94. Research Objectives for Large Steerable Paraboloid Radio Reflectors, op. cit.

95. Seeger to diary, 1 April 1955, LOC-ATW, Box 26.

96. Emberson to Seeger, 21 April 1955, CITA-JLG, Box 111, Folder 7.

97. Ibid.

98. Bok to Tuve, 28 April 1955, with copies to Hagen, Greenstein, Minkowski, Purcell, Kraus, Ewen, CITA-JLG, Box 111, Folder 6.

99. Bok to Tuve, 28 April 1955, DTMA, Series 1, Box 4, Folder 1.

100. Tuve to NSF Panel, 2 May 1955, CITA-JLG, Box 111, Folder 6; DTMA, Series 1, Box 4, Folder 1.

101. Hagen to NSF Panel, 10 May 1955, CITA-JLG, Box 111, Folder 6; DTMA, Series 1, Box 4, Folder 1.

102. Minkowski to Tuve, 4 May 1955, CITA-JLG, Box 111, Folder 6. Both Minkowski and Hagen incorrectly argued against interferometers on the grounds that, unlike a parabolic dish, interferometers were narrow band.

103. Greenstein and Minkowski to NSF Committee, 18 May 1955, CITA-JLG, Box 111, Folder 6; DTMA, Series 1, Box 4, Folder 1.

104. An initial report to the NSF Divisional Committee was presented by Tuve, Bok, and Hagen on Sunday 24 April (Tuve to NSF Advisory Committee, 2 May 1955, CIT-JLG, Box 111, Folder 6). On 21 June 1955, Tuve sent a more detailed report to the Committee, CITA-JLG, Box 111, Folder 7.

105. van de Kamp, "Inter-University Radio Astronomy Observatory," May 1955, CITA-JLG, Box 111, Folder 6.

106. Emberson to Seeger, 6 May 1955, LOC-ATW, Box 26.

107. Minutes of the AUI Steering Committee for 28 May 1955, NAA-NRAO, Founding and Organization, Meeting Minutes.

108. Ibid.

109. Ibid.

110. Ibid.

111. Ibid.

112. Ibid.

113. Tuve to van de Kamp, 23 June 1955, CITA-JLG, Box 111, Folder 6.

114. AUI-BOTXC, 17 June 1955.

115. Waterman to diary, 16 August 1955, LOC-ATW, Box 26.

116. van de Kamp to Waterman, 14 July 1955, LOC-ATW, Box 26.

117. Paul H. Kratz to diary, 24 August 1955, LOC-ATW, Box 26.

118. Bok to Charles Cutts, NSF Engineer, 23 September 1955, CITA-JLG, Box 3, Folder 26.

119. Greenstein to Bok, 29 September 1955, CITA-JLG, Box 111, Folder 6.

120. Kraus to Tuve, 24 September 1955, NAA-JDK, Notes and Papers, US Radio Astronomy; DTMA, Series 1, Box 4, Folder 1; CITA-JLG, Box 111, Folder 6.

121. Minkowski to Tuve, 8 October 1955, CITA-JLG, Box 111, Folder 7.

122. Seeger to diary, 27 September 1955, LOC-ATW, Box 26.

123. Tuve to Seeger, 29 September 1955, CITA-JLG, Box 111, Folder 6; DTMA, LOC-ATW, Box 26.

124. Sunderlin to diary, 9 December 1955, LOC-ATW, Box 26. Also in attendance were Bok, Tuve, Emberson, and from the NSF Seeger, Cutte, Hogg, and Sunderlin.

125. Waterman to diary, 21 and 22 December 1955, LOC-ATW, Box 26.

126. Ibid.

127. Minutes of the AUI Steering Committee, 30 December 1955, CIT-JLG, Box 113, Folder 2.

128. AUI-BOTXC, 14 December 1955. Members named to the New AUI Advisory Committee were: Bart Bok (Harvard), W. E. Gordon (Cornell), Merle Tuve (DTM), E. F. McClain (NRL), Leo Goldberg, (Michigan), John Kraus (Ohio State), A.B. Meinel (National Optical Observatory), A.J. Deutsch (Caltech, MWPO), J.B. Wiesner (MIT).

129. Under Hagen's leadership, Project Vanguard was an NRL program, designed as part of the International Geophysical Year, which intended to launch the first artificial satellite into Earth orbit. After the naval rocket collapsed in flames during a nationally televised launch failure, Explorer 1 was launched by Army Ballistic Missile Agency under Werner von Braun's leadership a few months after the successful launch of Sputnik 1and 2 by the USSR, and became the first US orbiting satellite.

130. Seeger to diary, 25 January 1956, LOC-ATW, Box 26.

131. Tuve to NSF Advisory Panel, 20 December 1955, CITA-JLG, Box 111, Folder 6. Tuve's reference to Green Bank as the "deep woods" was perhaps a sarcastic reference to the mythical home of the Phantom, a popular comic strip character of the 1950s.

132. Advisory Panel for Radio Astronomy, Agenda, 16–17 January 1956, CITA-JLG, Box 111, Folder 6; LOC-ATW, Box 26.

133. Waterman to diary, 18 January 1956, LOC-ATW, Box 26.

134. Hogg, 4 May 1956 Summary Notes on the Radio Astronomy Observatory, LOC-ATW, Box 26.

135. Waterman to diary, 19 January 1956, LOC-ATW, Box 26.

136. Sunderlin to diary, 23 January 1956, LOC-ATW, Box 26.

137. Hogg, op. cit.

138. Hogg, op. cit.

139. Minutes of the December 11–13 1955 AUI Steering Committee, CITA-JLG, Box 111, Folder 2.

140. *New York Times*, 22 January 1956, p. B15.

141. Seeger to diary, 30 March 1956, LOC-ATW, Box 26.

142. Minutes of AUI Steering Committee, 25–27 March 1956, NAA-NRAO, Founding and Organization, Meeting Minutes.
143. NSF-AUI meeting, 2 April 1956, NAA-NRAO, Founding and Organization, Meeting Minutes. https://science.nrao.edu/about/publications/open-skies#section-3
144. *Planning Document for the National Radio Astronomy Facility*, NAA-NRAO, Founding and Organization, *Planning Documents*. A second draft was submitted on 21 May 1956.
145. *Plan for a Radio Astronomy Observatory*, p. 39. https://science.nrao.edu/about/publications/open-skies#section-3
146. G.H. Hickox to diary, 15 February, 1956, LOC-ATW, Box 26.
147. Seeger to diary, 23 March 1956, LOC-ATW, Box 26.
148. Seeger to Berkner, 22 March 1956, NAA-NRAO, Founding and Organization, Meeting Minutes; LOC-ATW, Box 26.
149. *Plan for a Radio Astronomy Observatory*, op. cit. p. 39.
150. Tuve to Seeger, 14 June 1956, CITA-JLG, Box 111, Folder 6.
151. AUI-BOT, 19 April 1956.
152. AUI-BOTXC, 18 May 1956; Dunbar to Berkner, 14 May 1956, NAA-NRAO, Founding and Organization, Meeting Minutes.
153. Ibid.
154. AUI-BOTXC, 19 April 1956.
155. Ibid.
156. Ibid.
157. Waterman, Remarks at the Dedication of the George R. Agassiz Telescope, NAA-NRAO, Founding and Organization, Planning Documents. https://science.nrao.edu/about/publications/open-skies#section-3
158. Ibid.
159. The criteria for selecting the site were outlined in the *Plan for a Radio Astronomy Observatory* submitted to the NSF in August 1956.
160. Precipitable water vapor is a measure of the total water vapor content in path through the atmosphere and differs from the relative humidity (often just called "humidity") which is a measure of the fraction of water vapor in the local atmosphere compared with the maximum amount possible.
161. *Plan for a Radio Astronomy Observatory*, op. cit., p. 18.
162. Those who actively participated in the search included H.L. Alden, E.R. Dyer, and W. Nelson (University of Virginia), J.E. Campbell (Tennessee Valley Authority), C. Cutts and P. van de Kamp (NSF), R.M. Emberson (AUI), H.I. Ewen (Harvard), F.T. Hadock and J.P. Hagen (NRL), W. Hardiman (Tenn. State Geologist), R.A. Laurence (Knoxville Geological Survey), W. McGill (Virginia State Geologist), P.H. Price (West Virginia State Geologist), and C.K. Seyfert (Vanderbilt). August 1956, *Plan for a Radio Astronomy Observatory, Section III B.*
163. NAA-WTS, Sullivan interview with John Hagen, 27 August 1976. https://science.nrao.edu/about/publications/open-skies#section-3
164. *Plan for a Radio Astronomy Observatory*, op. cit., p. 16.
165. James Mitchell, Diary Note, 20 February 1956, LOC-ATW, Box 26.
166. Hogg, Diary Note, 14 March 1956, LOC-ATW, Box 26.

167. Emberson to Seeger, 9 April 1956, LOC-ATW, Box 26.
168. Waterman to diary, 7 February 1956, LOC-ATW, Box 26.
169. The other potential sites were Burkes Garden, VA, Little Meadows, VA, Massanutten Mountain, VA, Green Bank, WV, and Deerfield, VA.
170. Curtis M. Jansky, founder of Jansky and Baily, was Karl Jansky's older brother who had intervened to get Karl a job at Bell Laboratories.
171. Kenwolf (2010) gives a colorful, though sometimes misleading, description of the Green Bank area and the impact of NRAO on the local population.
172. Bok, 7 August 1956, Toward a National Radio Observatory, NAA-NRAO, Founding and Organization, Planning Documents. https://science.nrao.edu/about/publications/open-skies#section-3
173. Seeger to diary, 21 December 1955, LOC-ATW, Box 26.
174. Seeger to diary, 30 January 1956, LOC-ATW, Box 26.
175. Waterman to diary, 22 December 1955, LOC-ATW, Box 26.
176. Minutes of the AUI Steering Committee Meeting, 13 December 1955, CITA-JLG, Box 113, Folder 2.
177. Ibid.
178. Feasibility Report for the National Science Foundation on Construction of a National Radio-Astronomy Facility, 7 May 1955, NAA-NRAO, Founding and Organization, Planning Documents. https://science.nrao.edu/about/publications/open-skies#section-3
179. Emberson, 16 January 1956 memo to radio astronomy file, NAA-NRAO, Founding and Organization, Correspondence. https://science.nrao.edu/about/publications/open-skies#section-3
180. AUI-BOT, 14 April 1955.
181. West Virginia Legislature, Chapter 37A, Article 1, Radio Astronomy Zoning Act, 1956. http://www.wvlegislature.gov/wvcode/code.cfm?chap=37a&art=1
182. Needell (1987) gives a detailed account of the long-lasting conflict between Berkner and Tuve.
183. In spite of the apparent 2–3-day postal delivery time, some of the hastily written letters and responses overlapped in time.
184. Tuve to NSF Panel, 14 June 1956, CITA-JLG, Box 111, Folder 6.
185. Tuve to Stewart, 19 June 1956, CITA-JLG, Box 111, Folder 6.
186. Tuve to Beams, 19 June 1956, CITA-JLG, Box 111, Folder 6.
187. Tuve to Goldberg, 28 June 1956, CITA-JLG, Box 111, Folder 6.
188. Bowen to Tuve, 28 June 1956, CITA-JLG, Box 111, Folder 6.
189. Bok to Tuve, 20 June 1956, NAA-NRAO, Founding and Organization, Correspondence; CITA-JLG, Box 111, Folder 6.
190. Bok to Tuve, 22 June 1956, NAA-NRAO, Founding and Organization, Correspondence; CITA-JLG, Box 111, Folder 6.
191. Bok to Tuve, 5 July 1956, NAA-NRAO, Founding and Organization, Correspondence.
192. Hagen to Tuve, 25 June 1956, NAA-NRAO, Founding and Organization, Correspondence.
193. Greenstein to Tuve, 21 June 1956, CITA-JLG, Box 111, Folder 6.

194. Purcell to Tuve, 27 June 1956, NAA-NRAO, Founding and Organization, Correspondence; CITA-JLG, Box 111, Folder 6.
195. Houston to Tuve, 26 June 1956, CITA-JLG, Box 111, Folder 6. M. V. (Bill) Houston was a member of the National Science Board from Rice University.
196. Tuve to Bok, 29 June 1956, NAA-NRAO Founding and Organization, Correspondence; CITA-JLG, Box 111, Folder 5.
197. Tuve to NSF Panel, 1 July 1956, CITA-JLG, Box 111, Folder 5.
198. Bok to Tuve, 3 July 1956, NAA-NRAO Founding and Organization, Correspondence; CITA-JLG, Box 111, Folder 5.
199. Ibid.
200. Greenstein to NSF Panel, 5 July 1956, CITA-JLG, Box 111, Folder 5; Greenstein to Tuve, 6 July 1956, CITA-JLG, Box 111, Folder 5.
201. The following discussion of the 11 July meeting is taken from the Minutes of the Conference on Radio Astronomy Facilities, NAA-NRAO, Founding and Organization, Meeting Minutes.
202. Dunbar to Berkner, 2 July 1956, NAA-NRAO, Founding and Organization, Correspondence; AUI-BOTX, 20 July 1956.
203. The enabling legislation which established the NSF states "The Foundation shall not, itself, operate any laboratory or pilot plants." Section 14(c).
204. Heeschen had joined AUI only ten days earlier.
205. Menzel to Tuve, 19 July 1956, CITA-JLG, Box 111, Folder 2.
206. Berkner to Menzel, 24 July 1956, NAA-NRAO, Founding and Organization, Correspondence.
207. Berkner to Waterman, 20 July 1956, Appended to the minutes of BOTXC, 20 July 1956.
208. Minutes of the Joint Meeting of NSF Advisory Panel for Astronomical Observatory and NSF Panel on Radio Astronomy, 23 July 1956, NAA-NRAO, Founding and Organization, Meeting Minutes.
209. Greenstein to Menzel, 27 July 1956, CITA-JLG, Box 111, Folder 5.
210. NSF Press release, 26 July 1956, CITA-JLG, Box 113, Folder 2.
211. Tuve, 25 July 1956, Minutes of the 22–23 July 1956 NSF Advisory Panel on Radio Astronomy, NAA-NRAO, Founding and Organization, Meeting Minutes; CITA-JLG, Box 111, Folder 5.
212. Greenstein to Tuve, 5 November 1956, CITA-JLG Box 113, Folder 2.
213. Minutes of the NSB, 24 August 1956.
214. Hagen to Tuve, 25 July 1956, CITA-JLG, Box 111, Folder 5.
215. Referenced at the start of 8 October 1956 Berkner to Waterman letter appended to AUI-BOTX, 18 October 1956.
216. Appendix to the Minutes of the 16–17 October 1956 meeting of the AUI Advisory Committee, NAA-NRAO, Founding and Organization, Meeting Minutes.
217. Berkner to Waterman, 8 October 1956, Attached to AUI-BOTXC, 18 October 1956.
218. Waterman to Berkner, 12 October 1956, Attached to AUI-BOTXC, 18–19 October 1956.
219. AUI-BOTXC, 21 September 1956.
220. NAA-NRAO, Founding and Organization, Planning Documents. https://science.nrao.edu/about/publications/open-skies#section-3

Bibliography

References

Baade, W. and Minkowski, R. 1954, Identification of the Radio Sources in Cassiopeia, Cygnus A, and Puppis A, *ApJ*, **119**, 206

Berkner, L.V. 1956, Radio Telescopes, Present and Future - Plan for a National Radio Astronomy Facility, *AJ*, **61**, 165

Bowen, E.G. 1987, *Radar Days* (Bristol: IOP Publishing)

Bush, V. 1945, *Science, The Endless Frontier* (Washington: United States Office of Scientific Research and Development)

DeVorkin, D.H. 2000, Who Speaks for Astronomy? How Astronomers Responded to Government Funding after World War II, *Historical Studies in the Physical and Biological Sciences*, **31** (1), 55

Edmondson, F. 1991, AURA and KPNO: The Evolution of an Idea, 1952-1958, *JHA*, **22**, 68

Edmondson, F. 1997, *AURA and its US National Observatories* (Cambridge: Cambridge University Press)

Emberson, R.M. 1959, National Radio Astronomy Observatory, *Sci*, **130**, 1307

England, J.M. 1982, *A Patron for Pure Science: The National Science Foundation's Formative Years, 1945-57* (Washington: National Science Foundation)

Goldberg, L. 1983, Founding of Kitt Peak, *S&T*, **65** (March), 228

Greenstein, J.L. 1984a, Optical and Radio Astronomers in the Early Years. In *The Early Years of Radio Astronomy*, ed. W.T. Sullivan III (Cambridge: CUP), 67

Greenstein, J.L. 1984b, An Astronomical Life, *ARAA*, **22**, 1

Greenstein, J.L. 1994, The Early Years of Radio Astronomy at Caltech, *AuJPh*, **47**, 555

Gunn, J. 2003, Jesse Greenstein (1909-2002), *BAAS*, **35**, 1463

Hagan, J.P. 1954, Radio Astronomy Conference, *Sci*, **119**, 588

Hales, A.L. 1992, Lloyd Viel Berkner, February 1, 1905-June 4, 1967, *BMNAS*, **61**, 3

Hanbury Brown, R., Minnett, H.C., and White, F.W.G. 1992, Edward George Bowen 1911–1991, *Historical Records of Australian Science*, **9**, No. 2, (also appeared in the *Biographical Memoirs of Fellows of the Royal Society of London*, 1992)

Heeschen, D.S. 1996, The Establishment and Early Years of NRAO, *BAAS*, **28**, 863

Irwin, J.B. 1952, Optimum Location of a Photoelectric Observatory, *Sci*, **115**, 22

Kenwolf, L.G. 2010, *A Social and Political History of the National Radio Astronomy Observatory at Green Bank, WV* (Morgantown: West Virginia Libraries)

Kraft, R.P. 2005, Jesse Leonard Greenstein, 1909-2002, *BMNAS*, **86**, 1

Leverington, D. 2017, *Observatories and Telescopes of Modern Times* (Cambridge: Cambridge University Press)

Lockman, F.J., Ghigo, F.D., and Balser, D.S. eds. 2007, *But It Was Fun* (Green Bank: National Radio Astronomy Observatory)

Lomask, M. 1976, *A Minor Miracle: An Informal History of the National Science Foundation* (Washington: National Science Foundation)

Malphrus, B. 1996, *The History of Radio Astronomy and the National Radio Astronomy Observatory: Evolution toward Big Science* (Malabar, Fla.: Krieger Pub.)

McClain, E. 1960, The 600-Foot Radio Telescope, *SciAm*, **202** (1), 45

McClain, E. 2007, A View from the Outside. In *But It Was Fun*, ed. F.J. Lockman, F.D. Ghigo, and D.S. Balser (Green Bank: NRAO)

Munns, D.P.D. 2003, If We Build It, Who Will Come? Radio Astronomy and the Limitations of 'National' Laboratories in Cold War America, *Historical Studies in the Physical and Biological Sciences*, **34** (1), 95

Munns, D.P.D. 2013, *A Single Sky: How an International Community Forged the Science of Radio Astronomy* (Cambridge: MIT Press)

Needell, A.E. 1987, Lloyd Berkner, Merle Tuve, and the Federal Role in Radio Astronomy, *OSIRIS Ser 2*, **3**, 261

Needell, A.E. 1991, The Carnegie Institute of Washington and Radio Astronomy: Prelude to an American National Observatory, *JHA*, **22**, 55

Needell, A.E. 2000, *Science, Cold War and the American State: Lloyd V. Berkner and the Balance of Professional Ideals* (Amsterdam: Harwood Academic)

Reber, G. and Greenstein, J.L. 1947, Radio-frequency Investigations of Astronomical Interest, *Obs*, **67**, 15

Robertson, P. 1992, *Beyond Southern Skies – Radio Astronomy and the Parkes Telescope* (Cambridge: CUP)

Rogers, A.E.E. et al. 2005, Deuterium Abundance in the Interstellar Gas of the Galactic Anti-center from the 327 MHz Line, *ApJ*, **630**, L41

Townes, C. 1995, *Making Waves* (Washington: American Institute of Physics)

Wang, J. 2012, Physics, Emotion, and the Scientific Self: Merle Tuve's Cold War, *Historical Studies in the Natural Sciences*, **42**(5), 341

Weinreb, S. et al. 1963, Radio Observations of OH in the Interstellar Medium, *Nature*, **200**, 829

Whipple, F.L. and Greenstein, J.L. 1937, On the Origin of Interstellar Radio Disturbances, *PNAS*, **23**, 177-181

FURTHER READING

Bok, B.J. 1956, A National Radio Observatory, *SciAm*, **195** (4), 56

Bok, B.J. 1956, Toward a National Radio Observatory, 7 August 1956, NAA-NRAO Founding and Organization, Planning Documents https://science.nrao.edu/about/publications/open-skies#section-3

Dick, S.J. 1991, National Observatories - an Overview, *JHA*, **22**, 1

Emberson, R.M. and Ashton, N.L. 1958, The Telescope Program for the National Radio Astronomy Observatory at Green Bank, West Virginia, *Proc IRE*, **46** (1), 23

CHAPTER 4

Growing Pains

The search for the director of NRAO turned out to be unexpectedly difficult, as various astronomers turned down offers, citing the remoteness of the Green Bank site and the need to give up their own research programs. Finally, in 1959, Otto Struve, a distinguished optical astronomer and member of AUI's Search Committee, agreed to take on the job. But in 1961 both he and AUI President Lloyd Berkner resigned. I.I. Rabi, the new AUI President, appointed Australian Joe Pawsey as NRAO director, but due to a fatal illness he never served, and in 1962, Dave Heeschen became the new NRAO Director.

The initial intent of the founders was that NRAO would build a very large antenna of the order of 600 feet in diameter. But in order to gain operating experience while the large antenna was being designed and constructed, AUI decided to first build an intermediate size dish, about 150 feet in diameter, which they thought could be easily and quickly constructed. Contentious arguments about the mounting and operating wavelength finally led to a contract to build a 140 foot equatorially mounted telescope with what turned out to be unrealistically optimistic cost estimates and time schedules. An uncertain funding schedule, poor design, use of faulty materials, contractor strikes, and management by committee, led to numerous delays and cost overruns which threatened the continued existence of the new radio observatory. To maintain its viability, NRAO first purchased a commercial 85 foot radio telescope, then built a 300 foot transit antenna which was limited to observations at relatively long wavelengths. Under Heeschen's leadership, problems were painstakingly solved and the 140 Foot Telescope was finally completed in 1965 at a cost of $14 million, three times the initially budgeted cost. With both the 300 and 140 Foot Telescopes operational, NRAO was finally able to fulfill its role as a national radio astronomy center for visiting scientists.

However, living and working in rural Green Bank proved a challenge for the NRAO staff and their families. In 1965, much of the scientific and engineering staff moved to the new NRAO headquarters in Charlottesville, Virginia, which greatly changed the sociology of the Observatory.

4.1   FINDING A DIRECTOR

As early as 1955, Merle Tuve urged Lloyd Berkner to appoint a director for the new radio astronomy facility, but Berkner correctly pointed out that it would be difficult to recruit someone until AUI and NSF agreed on a contract and operating procedures. In early 1956, the AUI Steering Committee informally discussed the choice of director. Bart Bok was the overwhelming favorite, with John Hagen second. Berkner wanted someone with more radio astronomy experience, and approached Hagen to be Director of the radio astronomy facility and Bart Bok to be the Principal Scientist.[1] Hagen, however, had just accepted the Vanguard job (Sect. 3.3). Bok was considering an opportunity to go to Australia to head the Mt. Stromlo Observatory and did not consider Berkner's inquiry as a formal offer. In July 1956, hoping to defuse the issue of choosing a director for the radio astronomy facility, Waterman again approached Bok, asking if he might not change his mind about going to Australia and instead become the director. According to Bok, his reply was, "NUTS," and that he was very happy with his decision to go to Australia.[2]

After he learned that the National Science Board would select AUI as the manager of the new Radio Astronomy Observatory, Berkner realized the need to name a director who would "partake in the formation of policies and plans for the Observatory and play a major role in the technical decisions," and he wrote to Donald Menzel for his assistance. Berkner asked Menzel to form "an ad hoc nominating committee" to "nominate a panel of not less than three and preferably five individuals," and "take into account the ages and known commitments of possible Directors."[3] Berkner did not hesitate to express his own views on the qualities of the NRAO Director, stating that "in my personal opinion, the Director should be a US citizen." Somewhat optimistically, Berkner speculated that by the start of 1957, AUI "will already have bids for the construction of the 140-foot telescope and it is probable that a compromise between costs, altazimuth /equatorial mounts, surface and control tolerances, and similar factors may be necessary," so he saw the need to act quickly. He asked Menzel to provide nominations in time for the AUI Trustees Executive Committee meeting scheduled for 15 November 1956. Berkner appreciated that "in the temporary absence of a public announcement of AUI's future involvement, it is not proper that AUI openly take steps toward the selection of the Director" and cautioned Menzel that "it seems necessary that the matter be held closely."

Menzel wrote to Ira Bowen, C.D. Shane, W.W. Morgan, Otto Struve, and Jerry Wiesner to solicit their participation in a search committee to choose a director.[4] In his initial letter, Menzel stated, "It is for us to determine … whether this director should be primarily an astronomer or primarily a radio engineer." He then went on to express his view that, "Certainly, direction of the scientific program, in the long run, is more important than the ability to design electronics." The Committee debated the relative merits of previous research accomplishments, understanding of electronics, experience in radio astronomy, executive experience, good judgement, personality, and age. They noted that the most qualified individuals, such as Taffy Bowen, Joe Pawsey, and R. Hanbury Brown were not American citizens, but Menzel cautioned that, "the justification for support by the National Science Foundation was the need of <u>American</u> astronomers for the radio facility. It would look mighty queer, if among American astronomers, we could not find someone with proper qualifications."[5]

Bart Bok again clearly emerged as the top candidate and was formally approached by Menzel, who reported "His answer was definite and direct" that his commitment to Australia was firm.[6] Having a traditional background in astronomy, while initiating and leading what was then the pre-eminent US university radio astronomy program and training a new generation of radio astronomers, Bok was clearly uniquely qualified for the job. Had a formal offer from AUI been made earlier, most likely Bok would have accepted. As evidenced by his July 1956 letter to Tuve,[7] he was clearly interested, but again, Bok reiterated that he had made a commitment to go to Australia which he intended to honor.[8]

Just a week before the signing of the AUI-NSF contract, after several mail ballots and exchanges of letters commenting on the various candidates, the Search Committee met at AUI's New York offices on 12 November 1956 to consider 26 names. The Committee recognized that the top leadership should include both an astronomer and someone skilled in radio astronomy instrumentation. Their report stated that if the choice of director was an astronomer, then the second position should go to "someone in the electronics field" and "if the top post went to a physicist ... or engineer ... the appointment of an astronomer to the second position was imperative."[9] Leo Goldberg, director of the University of Michigan Observatory, emerged as the unanimous choice for NRAO Director.

In order to induce Goldberg to accept the appointment, AUI offered him a generous salary, a liberal allowance to spend at least one month every year at some other institution, and a semester leave every fourth year. But Goldberg declined, citing the loss of research opportunities and the freedom which he enjoyed in a university position.[10] Ironically, 15 years later, Goldberg would become Director of the National Optical Astronomy Observatory, which had its own broad range of administrative and operational problems. Albert Whitford from the University of Wisconsin was the Committee's second choice, but he delayed responding to Berkner's offer. Finally, after several months, following pressure to reply from Leeland Haworth, Brookhaven Director and an AUI Trustee, Whitford also declined. Jesse Greenstein and Fred Whipple were close alternates, and were followed by Charles Townes, Horace Babcock, and John Hagen in another close group. Greenstein declined the offer citing the remoteness of Green Bank and the loss of opportunities for personal research.[11] Charlie Townes received strong support from Greenstein himself as well as the East Coast scientists on the committee, but he was not well known by the West Coast astronomers, Bowen, Struve, and Shane. Berkner apparently gave serious consideration to offering Townes the directorship but there is no record that he ever actually discussed the possibility with Townes.

At the time the AUI contract was signed with the NSF in November 1956, AUI had not yet identified a director. Under pressure from the NSF, Berkner declared himself "Acting Director," although he probably assumed that this would be for only a short term. The AUI Advisory Committee and the Trustees repeatedly declared that "the appointment of a Director is the most important problem to be solved by the corporation."[12] By early 1959, with no solution in sight, the several Trustees conferred among themselves and suggested to Berkner that Otto Struve might consider an appointment as NRAO Director. Struve had been brought in as an at-large Trustee in response to the agreement to broaden the AUI Board to include expertise in astronomy and astrophysics. Following consultation with the NSF and NRAO senior staff, at their 17 April 1959 meeting, the AUI Board appointed Otto Struve (Fig. 4.1) as NRAO Director and Vice President of AUI, effective 1 July 1959.

Otto Struve came from a long line of famous astronomers and observatory directors, and was then a Professor of Astronomy and Director of the Leuschner Observatory at the University of California. Before that he had been the long-time Director of the Yerkes Observatory and editor of *The Astrophysical Journal*. He had an outstanding reputation as a scientist for his contributions to

Fig. 4.1 Otto Struve, NRAO Director from 1 July 1959 to 1 December 1961. Credit: NRAO/ AUI/NSF

astrophysics, particularly in stellar spectroscopy and stellar variability. He served as President of the American Astronomical Society from 1947 to 1950 and as President of the International Astronomical Union from 1952 to 1955. He was a strong supporter of the effort to establish a national optical astronomy observatory, and in 1954, he became a member of the NSF Advisory Panel for Astronomy.

Struve was also an experienced and accomplished science administrator. As Director of the Yerkes Observatory, he recruited an outstanding staff including Subrahmanyan Chandrasekhar, Jesse Greenstein, Gerard Kuiper, and Bengt Stromgren, and also attracted such distinguished visitors as Jan Oort, Pol Swings, and Henk van de Hulst. While still Yerkes Director, Struve established the McDonald Observatory in Texas, and served simultaneously as the Director of both Observatories. As a member of the AUI Board of Trustees he had taken an active interest in the affairs of the Observatory. Struve was not a radio astronomer, but he had a history of association with radio astronomy and radio astronomers. He had tried to help Grote Reber transfer his Wheaton dish to a more suitable site in Texas in the 1940s, and sought an appointment for Reber at the University of Chicago. As early as 1953 he had written to Greenstein about the need for better instrumentation for radio astronomy (DeVorkin 2000, p. 82). While at Berkeley, he helped Harold Weaver initiate a radio astronomy program and hired Ron Bracewell to teach a course in radio astronomy in the 1954–1955 academic year.

Even earlier, Struve (1940) wrote, "that the small astronomical observatory is compelled to search for something that it is able to do, instead of doing what is scientifically important and interesting." And he went on to advocate a cooperative approach to observational astronomy to provide telescope access to astronomers, especially young astronomers and students, who were at institutions without good instruments and with poor skies. Struve was known as a strong supporter of what later became to be known as "big science," and had the right background and philosophy to lead NRAO. Most importantly, he brought a lot of prestige to NRAO at a time when it was badly needed.

As had happened earlier at Brookhaven (Sect. 3.1), AUI management was unwilling to delegate authority to their new NRAO Director. Responsibility for the 140 Foot construction remained with AUI, which expected Struve to concentrate on operation of the NRAO radio telescopes, overseeing the scientific program at Green Bank, and recruiting the scientific and engineering staff needed to carry out the NRAO mission as a radio observatory for visiting scientists. Struve agreed to this split of responsibility between AUI and NRAO, feeling it would give him more opportunity to concentrate on scientific matters, but according to Heeschen,[13] he later regretted this decision.

Shortly after he became NRAO Director in the summer of 1959, Struve (1960) had already noted that NRAO was a national observatory in name only, and that there were more powerful facilities available elsewhere. Especially concerned about the isolation of Green Bank, he explained to the AUI Board that unless NRAO were to expand substantially, "Green Bank will become simply one of about a dozen observatories working in the field of radio astronomy," and added that both Caltech and Cornell looked more attractive to a scientist wishing to work in radio astronomy. He was supported by Board member Carl Seyfert who added that NRAO "should not be just one of a number of generally similar institutions."[14] Struve also cautioned Heeschen and other senior NRAO staff that, "NRAO is rapidly becoming one of many radio-observatories, some of which are better equipped than our own. I sense that even now the NRAO does not provide enough of an incentive for many visiting astronomers. If I were a young astronomer, I should be equally attracted by research opportunities at Stanford, Cal Tech [sic], Cornell, Illinois, and several others. Perhaps this is what the U.S.A. really needs; but if so, the original purpose of NRAO is not being met. My feeling is that an enormous amount of very clever thinking by many able physicists and astronomers has resulted in the invention and construction of a wide variety of excellent, but very expensive instruments."[15]

Unfortunately, Struve's short tenure as NRAO Director was marred by his conflict with AUI, and especially AUI President I.I. Rabi, over issues surrounding the 140 Foot Telescope construction, his continued attention to his research and publication on matters unrelated to NRAO, and by his declining health, due in part to wounds and multiple diseases acquired during WWI and then during the Russian Civil War when he served as an artillery and cavalry officer, first in the Russian then in the White Army.

4.2   Getting Started

Following the signing of the contract with the NSF in November 1956, the AUI Board established the new Advisory Committee for Radio Astronomy, which replaced the old Steering Committee.[16] To oversee the effort of building the Observatory in Green Bank, AUI established an office in Marlinton, a 45-minute drive and some 25 miles to the south of Green Bank. The Marlinton office was shared with the Army Corps of Engineers, who were charged with obtaining the land rights in Green Bank. The first office in Green Bank was opened on 14 May 1957. Initially, the NRAO staff included four people: Dick Emberson and his secretary Mary Beth Fennelly, John Carrol, an engineer working on the design of the 140 Foot Telescope, and Dave Heeschen (2008) (Fig. 4.2). Heeschen and Edward Lilley were the senior radio astronomy students in Bart Bok's radio astronomy group at Harvard, and the first to receive a PhD in radio astronomy at a US university. Following a year spent as an instructor at Wesleyan University in Middletown, Connecticut, Heeschen returned to Harvard as a Lecturer and Research Associate. While still at Harvard, encouraged by Bart Bok, he became a consultant to Dick Emberson at AUI, attended meetings of the AUI Steering Committee, and participated in the December 1955 NRAO site search. On 1 July 1956, four months before the NSF awarded the contract to AUI, Dave Heeschen accepted a full-time appointment with AUI in the office of the President, and then became the first member of the NRAO scientific staff.

Fig. 4.2 David
S. Heeschen, NRAO
Acting Director from 1
December 1961 to 19
October 1962; NRAO
Director from 19 October
1962 to 1 October 1978.
Credit: NRAO/AUI/
NSF

John Findlay was the next person to join the NRAO staff. During WWII, Findlay had served in the Royal Air Force, where he supervised the installation of radar stations throughout North Africa, the Middle East, and South East Asia. Following his undergraduate education at Cambridge, Findlay started his academic career as a PhD student working for Lord Rutherford, who unfortunately died a few months after Findlay arrived in Cambridge in 1937, although according to Findlay there was no connection between these two events.[17] After two years of research in ionospheric physics under Jack Ratcliffe, Findlay became involved in the development of radar, during which time he met and got to know Martin Ryle. Following the War, Findlay led a radar development group in Britain and did research in ionospheric physics at Cambridge University. In 1952, he spent time at the Carnegie Institution Department of Terrestrial Magnetism, where he first met Merle Tuve and Lloyd Berkner, who were at the forefront of American ionosphere research. While working at the Ministry of Supply in Great Britain, under pressure from his American wife to return to the United States, Findlay wrote to Berkner and others asking about employment possibilities in the United States. Having just obtained the NSF contract to study the feasibility of establishing a new radio astronomy facility in the United States, Berkner offered Findlay a job to help establish the observatory, and at the end of 1956 he joined Berkner, Dick Emberson, and Dave Heeschen at the AUI offices on the 72nd floor of the Empire State Building in New York City.

Heeschen and Findlay were soon followed by other Harvard graduates, first by Frank Drake, later by Jack Campbell,[18] William E. Howard III, May Kaftan-Kassim from Iraq, T. K. Menon from India, and Campbell Wade, and they

Fig. 4.3 Early NRAO staff. Left to right from the top: John Findlay, Frank Drake, Cam Wade, Hein Hvatum, Sebastian von Hoerner, Frank Low, Dave Hogg, Sandy Weinreb, Ken Kellermann, Barry Clark, Mort Roberts, Bill Howard. Credit: NRAO/AUI/NSF

formed the core of the early NRAO scientific staff (Fig. 4.3). When asked why the early staff was all from Harvard, Heeschen explained that he was unable to recruit anyone else, but acknowledged that he had not tried to attract any of the then-recent Caltech graduates such as Alan Moffet or Robert Wilson.[19]

One of AUI's first tasks was to activate some sort of instrument, as the West Virginia Zoning Act grandfathered any source of interference existing before the Observatory went into operation. Working out of a small house in Green Bank that had been taken over by the US Army Corps of Engineers, and using equipment on loan from NRL, Heeschen, together with NRL's Ed McClain and Ben Yaplee, built a simple 30 MHz interferometer. It consisted of two half wave dipoles and a Hallicrafter's communications receiver, and they used it to observe the Sun on 25 and 26 October 1956. Although these observations occurred a few weeks before the signing of the actual contract between NRAO and the NSF and a year before the official 17 October 1957 NRAO dedication, this became the first radio astronomy observation made from the

DEDICATION OF RADIO ASTRONOMY OBSERVATORY
Green Bank, W. Va. - October 17, 1957

L to R: Dr. R. M. Emberson, Dr. L. V. Berkner,
G. A. Nay, Dr. J. W. Findlay, Prof. N. L. Ashton, Dr. D. S. Heeschen, H. Hockenberry

Fig. 4.4 Dedication of NRAO in Green Bank, 17 October 1957. Left to right: Richard Emberson, Lloyd Berkner, G. Nay, John Findlay, Ned Ashton, Dave Heeschen, and H. Hockenberry, with a model of the planned 140 Foot Telescope. Credit: NRAO/AUI/NSF

Green Bank site.[20] A year later, John Findlay resurrected this simple interferometer to record the radio transmissions from the Russian Sputnik spacecraft.

In July 1957, three months before the official dedication of NRAO, Findlay and Heeschen moved to West Virginia to begin the task of building the Observatory. Six months later they were joined by Frank Drake. By this time, the Observatory had moved its offices from Marlinton to Green Bank. The formal groundbreaking and dedication of the Observatory, which took place on 17 October 1957, appears to be the first use of the name "National Radio Astronomy Observatory"[21] (Fig. 4.4). Prior to that time, AUI and NSF referred to the "radio astronomy project," "radio astronomy facility," or "radio astronomy observatory." Indeed, the concept of a truly "national" observatory remained somewhat contentious, at least until the actual dedication, when it became known as the "National Radio Astronomy Observatory." Although 17 October was billed as the "groundbreaking," actual construction work on the site had begun earlier with the building of roads, sewers, wells, and the preparation of the sites for the 85 foot and 140 Foot Radio Telescopes.

The day before the dedication, the AUI Advisory Committee met with a number of the visiting radio astronomers to discuss their observing plans for the soon to be completed 85 foot and 140 Foot Radio Telescopes. Emphasis was on solar and 21 cm work, but several participants also mentioned studies of H II regions, planets, and accurate source positions. Two participants indicated they were setting up new radio astronomy groups at the University of Virginia and the University of Pennsylvania that would be based on use of the Green Bank instruments, and they expected to hire radio astronomers to exploit these opportunities. Anticipating the future use of maser amplifiers, Charles Townes pointed out the need for accommodating receivers cooled by liquid helium or nitrogen at the antenna focus.

In November 1957, Heeschen laid out a detailed plan for the Observatory. Based on planned starting dates for scientific observing of 1 September 1958 for the 85 foot telescope and 1 January 1960 for the 140 Foot, Heeschen outlined the expected growth of the scientific, technical, business, and supporting staffs, which he projected would reach 55 to 62 persons by mid-1960.[22] In a separate document, he described his views on how the NRAO should be organized, pointing out NRAO's multiple responsibilities: to provide equipment and aid for visiting scientists, to anticipate the need for future developments in radio astronomy, to play a leading role in developing new instrumentation, to provide absolute flux density measurements, and to be a source of national standards for radio astronomy. He outlined a comprehensive research philosophy that guided the growth of NRAO for the next half century. There would be no "Observatory Program," but each staff member should expect to carry out their own independent research in their particular field of interest, and the staff would share the obligations of the Observatory in equipment development, long range planning, calibration, and assisting visitors.[23] At the same time, NRAO laid out a site development program that would accommodate the 85 foot and 140 foot antennas as well as the "very large antenna" that was expected "in the next four or five years." Buildings were planned to accommodate "an outstanding scientific staff, both permanent and visiting, ... an equally competent ... engineering and technical staff," along with "the minimum auxiliary and supporting staff and equipment necessary to manage and maintain the site." With enviable optimism as to the staff requirements, as well as construction schedules, the site plan including the need for three radio astronomers to support the 85 foot telescope by 1 July 1958, three more by 1 July 1960 to support the 140 Foot, and a total of nine by the time "the very large antenna is constructed in late 1961."[24]

4.3   THE 85 FOOT TATEL RADIO TELESCOPE (AKA 85-1)

Recognizing that the 140 Foot Telescope would take some time to construct, and needing to get established with an observational capability, the initial plan for NRAO called for the quick construction of a more modest sized instrument of 60 foot diameter, presumably to be modeled after the Harvard 60 foot radio telescope. However, Heeschen became aware of a commercially designed, equatorial mounted 85 foot diameter antenna that the Blaw-Knox Corporation was building for the University of Michigan. NRAO determined that a copy could be built in Green Bank for $255,730, or well within the budgeted $310,000. On 1 October 1957, two weeks before the Observatory ground-breaking, AUI signed a contract with Blaw-Knox to construct an 85 foot radio telescope.

The conceptual design of the Blaw-Knox 85 foot antennas was initially developed by DTM scientist Howard Tatel for the DTM 60 foot radio tele-scope. Unfortunately, Tatel died before the antenna could be constructed, and the detailed engineering design was completed by Blaw-Knox engineer Robert (Bob) Hall, who would later play major roles in the design and construction of future NRAO antennas, including the 300 Foot Green Bank transit telescope, the Tucson 36 foot mm telescope, and, finally, the 100 meter Green Bank Telescope. There was no specified delivery date in the AUI contract, although Blaw-Knox had initially claimed that 15 July 1958 "was easily in reach." As Bob Hall preferred to work by himself, the design took longer than planned, but he anticipated that the lost time could be made up during fabrication. However, in an effort to reduce their expenses, Blaw-Knox laid off half their factory workers and went from two daily shifts to one. Worried that they also might be laid off when the jobs were complete, the remaining crew slowed their work, while the unions forbad any overtime when the plant had just laid off a shift.[25] The delivery date slipped to 1 October 1958. Apparently confident of the October date, AUI scheduled a dedication timed to coincide with previ-ously scheduled Green Bank meetings of the AUI Board and the NRAO Advisory Committee.

The dedication of the 85-1 antenna, as it later became known at NRAO, took place as planned on 16 October 1958, just a year after the dedication and groundbreaking of the Observatory, and it was formally named "The Howard E. Tatel Radio Telescope" (Fig. 4.5). Although Michigan had ordered their 85 foot antenna before NRAO, for reasons the authors have been unable to deter-mine, Blaw-Knox delivered the first dish to Green Bank. Since the wiring and instrument installation by Blaw-Knox contractors and NRAO staff was not yet complete, the Green Bank antenna was not available for use until four months later, seven months behind the original schedule. This was "a source of consid-erable embarrassment" to NRAO and AUI, which had planned observing pro-grams by visitors and staff to begin in the summer, with the intention of announcing results at the forthcoming Paris Symposium on Radio Astronomy (30 July–6 August 1958) and at the triennial International Astronomical Union meeting in Moscow (12–20 August 1958).[26]

The initial observations with the Tatel Telescope, which marked the begin-ning of the first real research capability at NRAO, took place on 13 February 1959. Even then, the antenna failed to meet many of the performance specifi-cations, and there were still serious deficiencies in the structure which needed fixing, including loose bolts, leaking lubricant, gear misalignment, excessive backlash, and the need to replace the feed support structure. Nevertheless, in view of the already scheduled observing programs, it was agreed that the

Fig. 4.5 Dedication of the 85 Foot Howard E. Tatel Telescope, 16 October 1958. The telescope, still lacking surface panels at dedication, was completed in February 1959, with first observations on 13 February 1959. Credit: NRAO/AUI/NSF

scheduled observations could take place before AUI accepted the antenna. It was not until the end of 1959 that NRAO would agree to accept the antenna, although even by then, Blaw-Knox had not met all of the contract performance specifications. In hindsight, the schedule delay in the construction of the Tatel Telescope was a forerunner of the much longer and more serious delays later encountered by the 140 Foot, 36 Foot, and 100 meter GBT projects.

Since NRAO did not yet have an engineering staff, the 85 foot antenna was initially equipped with commercial receivers and a feed system consisting of a 1.4 GHz (21 cm) receiver built by Airborne Instruments Laboratory, an 8 GHz (3.75 cm) receiver built by Ewen-Knight, and a dual band 3/21 cm feed developed by Jasik Laboratories. According to Heeschen[27] this was the only time in its history that NRAO used completely commercially built receivers, although for many years following, NRAO would purchase low noise maser and parametric amplifiers.

Blaw-Knox had earned their reputation by building road-paving equipment and had no previous experience in radio astronomy. Indeed, the design was well underway when they realized that, due to the latitude difference, the Green Bank and Michigan equatorial mounted antennas could not be identical.

However, they did go on to produce a number of 85 foot antennas of the same design as the NRAO and Michigan antennas. The next one was for the Jet Propulsion Laboratory (JPL) Goldstone tracking station. One was for the Harvard College Observatory near Fort Davis, Texas, to support a program in solar radio astronomy by Alan Maxwell, a New Zealander, who had arrived at Harvard from Jodrell Bank. More than 10 other nearly identical antennas were located at the NASA Deep Space Network (DSN) tracking stations around the world. Comparably-sized radio telescopes were constructed by other manufacturers for the University of California at Hat Creek, California, and the Naval Research Laboratory at Maryland Point, Maryland. Some of these antennas remain in operation after more than 50 years. Years later the DSN antennas in South Africa and in Australia were decommissioned from their role in spacecraft tracking, and ownership was transferred to the host countries, where some were used for radio astronomy.

Initially, scientists wanting to use the 85 foot Tatel Telescope were unencumbered by the need for detailed scientific proposals, referees, time assignment committees, etc., so planning the Observatory's first scientific programs was a simple affair. Over a meal at the local diner, the NRAO scientific staff discussed the initial observing program that included the planets, the Galactic Center, H II regions, planetary nebulae, and the spectra of extragalactic sources. The observations were scheduled by Heeschen, based on handwritten requests from NRAO and visiting staff members for specific days and times, but unaccompanied by any scientific justification. The first controversial proposal came from Doc Ewen, who requested a large block of observing time for a Venus radar experiment. As Chair of the NRAO Astronomy Department, Heeschen had been assigning telescope time, but was reluctant to assign such a large block of time to one observer, and sought Berkner's advice. Due to the time-dependent nature of his planned radar experiments, Ewen ultimately withdrew his proposal, but, by the middle of 1960, the anticipated demands for nighttime observing on the Tatel Telescope caused Heeschen to inform the staff, "I think it would be well if each person with an observing program (or contemplated program) would write a short description of it—including present status, why it's being done, and with what receiver, and an estimate of the telescope time required to complete it."[28]

The spring of 1959 saw NRAO's first visiting observers, including George Field from Princeton, Hein Hvatum from Chalmers Technical Institute in Sweden, T. K. Menon from the University of Pennsylvania, Morton Roberts from the University of California, and Gart Westerhout from Leiden University. Although the Tatel telescope had only modest capabilities compared with NRAO/AUI's ambitions, as Heeschen (1996) pointed out, it served a variety of purposes. NRAO learned how to develop and manage a visiting observer program, and instrumentation being developed for the 300 Foot and 140 Foot Telescopes was tested on the 85 foot antenna. In particular, in fall 1961 Sandy Weinreb, then an MIT graduate student, brought his original 21 channel digital autocorrelation spectrometer to use on the 85 foot Tatel antenna in an attempt to detect the Zeeman splitting of the 21 cm hydrogen line[29] (Weinreb 1962a) and interstellar deuterium (Weinreb 1962b). Frank Drake's 1960 Project Ozma on the Tatel Telescope brought national recognition to NRAO and Green Bank, although it was not all favorable (Chap. 5). Later the Tatel Telescope became the first antenna of the three element Green Bank Interferometer (Chap. 8).

Characteristic of common practice at the time, NRAO started its own Observatory publication series, called "Publications of the National Radio Astronomy Observatory," which was privately distributed. Volume 1, covering the period from April 1961 to August 1963, included 17 publications. Reflecting what was then a general isolation of radio astronomy from mainstream astronomy, the absence of external peer review, and their limited distribution, the NRAO Publications were not widely recognized or cited, and from 1963 onwards NRAO staff and visitor publications appeared only in recognized peer-reviewed astronomy and technical journals.

4.4   THE 140 FOOT SAGA[30]

Planning for the New Radio Telescope During the years leading up to the formation of NRAO, a range of antenna sizes up to the ambitious dimension of 600 feet were discussed. Both the NSF and AUI committees provided advice on the flagship telescope for the new national observatory, but they were not always in agreement. Although the value and feasibility of the 600 foot telescope were hotly contested, there was a consensus to start with a so-called "intermediate" size reflector in the 150 foot range. However, the tradeoffs between size and precision and between an equatorial and alt-az mounting were debated for years. For smaller antennas, it was agreed that a traditional equatorial mount was optimum, while for larger antennas, it was clear that only an alt-az mount could work. The planned antenna was in the awkward class that Grote Reber said was neither big enough to be interesting nor small enough to easily build. With some foresight, he commented that being of intermediate size, much time and effort would be wasted fighting over whether it should be on an equatorial or alt-az mount.[31] The initial specification for a 150 foot diameter dish was redefined to be 40 meters, then rounded up to 140 feet (42.7 m) because, according to Struve et al. (1960), someone felt that the size of the national instrument should be expressed in feet, not in meters.

At this point, essentially everyone concerned was convinced, unrealistically as it turned out, that the construction of an intermediate-size dish would be a straightforward extrapolation of existing 60 to 82 foot designs. Initially, the main scientific driver for building the 140 Foot antenna was to do both galactic and extragalactic 21 cm research. As project planning developed, there was increasing interest in going to shorter wavelengths to investigate radio galaxy spectra, to study the thermal emission from the planets and H II regions, and to study polarization. Interestingly, in a 1957 letter to the NSF, with foresight Heeschen noted that the 140 Foot would be uniquely capable of detecting spectral lines at centimeter wavelengths.[32] At the same time, the exciting work being done on radio galaxies in the UK and Australia (Sect. 2.1) led to increased interest in the identification of discrete radio sources and greater emphasis on precise pointing of the antenna to measure accurate source positions. This gradual specification creep led to a telescope that would not be available off-the-shelf, but would be at the cutting edge of technology and both more difficult and more expensive to build than anyone at AUI anticipated.

At their 28 May 1955 meeting, the AUI Steering Committee prepared a Request for Proposals for a 140 foot diameter radio telescope which specified that the pointing should be better than 30 arcsec and the surface accuracy 1/4 inch over the inner half and 3/8 inch over the remaining part.[33] After considerable debate, the decision between an equatorial and altazimuth mount was left open, but full sky coverage remained a stated requirement. In June, Emberson reported that he had requested proposals from 21 commercial companies for the design and construction of an intermediate-sized dish.[34] However, the resultant cost estimates for the intermediate-sized dish varied widely, and none of the companies seemed prepared to produce an antenna meeting all of the challenging AUI specifications. Most of the manufacturers contacted indicated they would work on only a specific portion of the project, and could not deliver the full antenna. Significantly, several companies stated that AUI would need to supply a complete design before they could make a firm bid.[35] At this point, the Steering Committee agreed to defer the issue of construction and to concentrate just on the design. AUI anticipated that a 140 Foot Telescope would cost $2.2 million to design and construct, although the bids ranged from $1.3 million (D.S. Kennedy & Company) to $3.3 million (American Machine and Foundry). The full costs of developing the observatory, including other research equipment, roads, power, water, and buildings, was estimated to be about $4.6 million. Annual operating costs, which included salaries for the Director, two radio astronomers, one electrical engineer, and two technicians, as well as one mechanical engineer, were projected to reach $259,000 by 1959.[36]

Designing the Telescope After receiving the second NSF grant of $140,000 in July 1955, AUI decided to proceed in two phases: first, develop the design for a 140 Foot Radio Telescope, and second, contract to build the antenna. AUI issued three contracts for the design. At this time, everyone agreed that an equatorial mount would be too complex for a telescope as large as 140 feet, so the requested designs were all for an altazimuth mount using an appropriate coordinate converter. Jacob Feld from New York agreed to scale his 600 foot design to 140 feet. Charles Husband was asked to develop a design for a 140 foot telescope based on his experience with the Jodrell Bank telescope, which was soon to become the largest, fully steerable radio telescope in the world. D.S. Kennedy & Co. of Cohasset, Massachusetts, had already built several modest sized antennas for radio astronomy, including the Harvard 60 foot telescope, as well as other antennas for commercial and military applications, and agreed to develop a third design for a 140 foot radio telescope. In the Feld design, the dish was supported by two towers that rotate on a track, while the

Husband design followed the concept outlined nearly a decade earlier by Grote Reber (Chap. 9) for his proposed 220 foot antenna, in which the dish is supported at the two ends of a single supporting ring. In the Kennedy design, the dish was supported by two rings on a rotating structure mounted on a concrete pedestal (Fig. 4.6).

In order to operate at wavelengths shorter than 10 cm, even as short as 3 cm, the surface accuracy was now specified as ±1/4 inch, without defining whether that referred to peak-to-peak or rms deviations. The pointing accuracy was initially specified to be 10 arcsec, corresponding to 5% of the HPBW at 3 cm. The designers indicated that they could easily meet a 30 arcsec specification, could probably obtain 20 arcsec pointing accuracy, but that 10 arcsec pointing "may have to wait on the development of improved technology and components."[37]

By this time the NRL 50 foot alt-az dish had been in operation for four years, and it was known that the Dwingeloo, Jodrell Bank, and Parkes dishes were to be alt-az mounted. Moreover, as was certainly known by at least some members of the AUI Steering Committee, digital computer-controlled alt-az mounted antennas were already in common use in military radar systems.[38] Nevertheless, the AUI Committee, led by John Hagan and Dick Emberson, expressed concern that the required pointing precision of the Green Bank telescope could not be met with an alt-az design and mechanical coordinate converter.[39] Hagan's concern was no doubt based on the recognized pointing issues with the NRL 50 foot dish, so he argued, probably correctly, that an analogue computer of the type used at NRL and Dwingeloo could not meet the desired 140 foot pointing specifications. Although digital computers were already available, there was a widely held view that the operation of a precision scientific instrument should not be trusted to a computer.

The concerns about an alt-az design were shared by Merle Tuve and his NSF Panel, who questioned the potential precision of any servo system that would be needed in the alt-az design. According to Heeschen (1996), Tuve's lack of faith in servo drives was at least in part the result of his wartime experiences

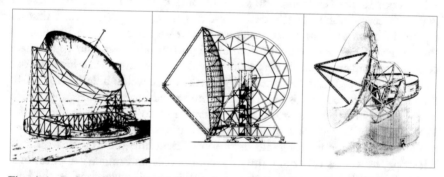

Fig. 4.6 Left to right: Feld, Husband, and Kennedy telescope designs for the Green Bank antenna. Credit: NRAO/AUI/NSF

with servo-driven gun mounts. Further, the optical astronomers on both the AUI and NSF committees, and the Harvard astronomers in particular, who had experience with their 24 and 60 foot equatorial dishes, viewed the equatorial mount more favorably. So Tuve and his NSF Advisory Panel told the NSF to specify that at least one design should be for an equatorial mount, although they acknowledged that the cost would be greater than for an alt-az mount. Moreover, Tuve claimed that the specified pointing precision was unrealistic and that AUI was "accepting from all advisers their desired performance characteristics and then asking the designers to meet these high specifications." He also made the valid point that the same specifications must be used in evaluating the two designs.[40] In view of the engineering challenges for both designs, Tuve proposed to relax the 140 foot pointing specifications and to eliminate the requirement that the dish be able to track the Sun, Moon, and planets. But Berkner was firmly opposed to relaxing the tolerances laid down in the design specifications, since he argued "this would diminish the range of experiments for which the instrument could be used."[41] Tuve was supported by Greenstein, who claimed that the "equatorial design might not be feasible," but that "the problems on the altazimuth mount are far from solved," and he informed Tuve that the Caltech 90 foot dishes would be on equatorial mounts.[42] Following pressure from Tuve's Panel, the NSF instructed Berkner to divert $10,000 originally planned for studies of the large reflector to "an immediate study of a polar type mount for the 140 foot reflector."[43]

At their March 1956 meeting,[44] the AUI Advisory Committee reviewed the Feld, Husband, and Kennedy designs. The Husband design adopted the same welded steel surface that later turned out to be so unsuccessful on his Jodrell Bank 250 foot antenna.[45] Tuve, claiming to speak for the NSF Advisory Panel, noting the difficulty in achieving the design specifications, urged that the design be relaxed to allow the telescope to work at 7 or 8 cm rather than 3 cm and stated that he saw "no particular value in having a telescope capable of operating at less than 10 cm wavelength." Haddock, who was using the NRL dish for 3 cm wavelength observations of planetary nebulae and H II regions,[46] was "greatly disturbed at the suggestion of relaxing the tolerances," and wanted to go to even shorter wavelengths where the thermal radiation would be stronger. Bok argued that "as accurate a reflector as possible is what is wanted." Tuve retaliated, advocating the need to determine what observations were desired, and then build with that objective in mind. Similar arguments and counterarguments relating to the relative importance of solving old problems and discovering new questions would be expressed about almost every proposed new radio telescope over the next half century.

All three designers agreed that the additional cost of an equatorial mount would outweigh the cost savings from the simpler drive and control mechanism on an alt-az antenna, and that further funds would be needed to explore the equatorial designs in more detail. Aside from the cost differences, Berkner reminded everyone that an alt-az design for the planned 140 Foot Telescope would give valuable experience in the design of bigger antennas. Pointing out that AUI had one or more workable alt-az designs, Berkner asked for authorization to proceed on the basis of one of those designs rather than delay further the establishment of the National Radio Astronomy Facility which was still in the proposal stage. Although it was agreed that if AUI followed the urging of the NSF Advisory Committee to obtain an equatorial design, it would take another four to six months. Bok pointed out that since construction funds were not available anyway, it would be "desirable to undertake a study of an equatorial mount." Berkner relented, and as instructed by the NSF, he agreed to "use funds in the amount of $10,000, previously allocated for studies of large dishes, to obtain a preliminary design of an equatorial mount." However, already at this time it was clear that at best, the FY57 NSF budget request of $3.5 million for the radio facility was more than a million dollars below the amount Emberson felt was needed for land acquisition, site development, equipment, and antenna construction.[47]

Berkner, Emberson, and much of the AUI Steering Committee favored the structurally simpler alt-az mounting, in part because it would also serve to evaluate the design of the planned and much larger 600 foot telescope. Tuve, on the other hand was concerned about the more complex variable speed drive and computer systems needed for the alt-az mount, and argued that "no comparison between equatorial and altazimuth suitability can be made unless at least one of them is a complete design." He was particularly concerned about the cost of a computer to do the coordinate conversion and the requirements for the servo system, which he claimed was "outside the contract," for the three designs provided by Feld, Kennedy, and Husband. "Unless a second independent computer is added," he argued, "there is no way of knowing at all times where the dish is actually pointing."[48] As Frank Drake[49] later pointed out, Tuve apparently did not appreciate that the direction of the forces on the declination bearings, and on the fork supporting the dish of a polar mounted telescope, changes as the telescope moves in hour angle, so that an equatorial mount is much more difficult to design than an alt-azimuth mount. The problem is compounded because adding more weight to give stiffness increases issues of the load on the bearing.

The AUI and NSF committees debated the minimum operating wavelength and the tradeoff with antenna diameter and cost. Fred Haddock argued that 1 cm would be ideal to study the thermal emission from planetary nebulae and H II regions, but 2 or 3 cm would be a reasonable compromise with cost. John Bolton claimed that even a factor of two improvement in surface accuracy would double the cost of the reflector, but Dave Heeschen and Bart Bok countered that in practice, previously built radio telescopes, such as the one at Harvard, had exceeded their design specifications, so that it should be possible to reach wavelengths as short as a few centimeters at no increase in cost.

As early as July 1956, Merle Tuve discussed the 140 Foot challenges with Blaw-Knox. Using the "Tatel-Carnegie design" for the DTM 60 foot antenna, he asked if Blaw-Knox would be able to give a "quotation or estimate" on the cost of a 140 foot equatorially mounted parabolic dish. Tuve reiterated his often stated view that the AUI specifications were too tight, and asked how the cost might be reduced by relaxing the specifications, or if a larger dish could be built with reduced specifications "without extreme increase in costs." Tuve also asked that the telescope be able to point down to the horizon over a range of 210 to 240 degrees of azimuth, a specification which would later turn out to be very expensive.[50] Around this time, Emberson was becoming concerned about the apparent difficulties inherent in the equatorial design, and predicted that "the alt-azimuth design will be the one chosen in the end."[51] But only a month later, based on optimistic input from Feld, Emberson expressed more confidence in an equatorial approach and proposed setting up a small group to further investigate an equatorial design for the 140 Foot Telescope.[52] As later described by Emberson and Ashton (1958), the choice between an alt-az and equatorial mount depended on whether one had more confidence in a complex mechanical structure or a complex drive and control system, and since the telescope would be used by astronomers who may have had "unhappy experiences with electronic and electrical equipment," AUI decided in favor of an equatorial mount.

Initially, some of the Advisory Committee members wanted to retain true full sky coverage, even for an equatorially mounted dish, in order that observations of transient celestial phenomena or geophysical activity could be studied at lower culmination. This would have meant tipping the dish structure over the zenith, but fortunately, at the October 1956 AUI Advisory Committee,[53] it was agreed to restrict the sky coverage appropriate to the equatorial mount. Measuring the surface accuracy was recognized as a possible problem, and a number of approaches were considered. By this time, there were sufficient uncertainties surrounding the 140 Foot project, which was about to become a contractual responsibility, that it did not seem feasible to simultaneously push the studies of the much larger (600 foot) telescope. Nevertheless, Berkner was confident that the construction of the 140 Foot Telescope would not take more than two years, and perhaps could be completed in as little as 18 months. However, some members of the AUI Board were already expressing concern over the lack of a firm cost estimate, so Berkner agreed that "commitments be limited to land acquisition, construction of essential roads, and installation of electric power."[54]

Under the guidance of the newly constituted AUI Advisory Committee on Radio Astronomy, in October 1956 AUI contracted with Ned Ashton for a "definitive" design of an equatorial mounted 140 foot radio telescope. Ashton was a structural engineer from the University of Iowa who had also designed the NRL 50 foot radio telescope, which was, when built, the largest in the world. A special ad hoc committee, including Howard Tatel and Merle Tuve from DTM, along with John Bolton and Bruce Rule from Caltech, was constituted in October 1956 to give "advice" to Ashton.[55] Bruce Rule was a Caltech structural engineer who had been involved in the construction of the Palomar 200 inch, and he was in charge of the design and construction of the new twin 90 foot equatorially mounted radio telescopes for Caltech's Owens Valley interferometer. Rule and Bolton offered to make the design of the Caltech 90 foot antennas available to Ashton and the Committee. The committee gave careful consideration to how both the polar and declination axes would be aligned and the methods by which the reflector surface would be fabricated, measured, and adjusted. But it was acknowledged that it would strain known optical or microwave procedures to adjust the surface to the needed accuracy.[56]

With an equatorial mount, there could be considerable cost savings by restricting the sky coverage, but "full sky coverage" was argued to be a fundamental requirement for a general purpose radio telescope. As a compromise between sky coverage and cost, the Advisory Committee initially agreed that the hour range would be restricted to six hours from the meridian. However, Donald Menzel, who carried a lot of weight with both AUI and the NSF, insisted that the telescope must be able to follow the Sun when it was above the horizon at any time of the year,[57] thus increasing the hour angle limit to seven hours. Tuve pointed out that this would seriously impact the performance and cost. Ashton denied that this was true (Heeschen 1996), although the increased range of hour-angle required longer yoke arms, which greatly increased the weight of the movable structure and the load on the polar bearing. Deciding to include the capability to observe the Sun at any time probably more than doubled the cost of the antenna and compromised both the surface and pointing accuracy, as well as being responsible for the long delay in its completion. According to Dave Heeschen, it was arguably the worst decision made in NRAO's long history.[58] Ironically, by time the 140 Foot Telescope was completed in 1965, solar studies with the limited resolution of filled aperture telescopes were no longer of great interest. As noted by Heeschen (1996), the 140 Foot Telescope had gradually "evolved from a quick-off-the-shelf instrument to one that pushed the state of the art of telescope construction and would become extremely costly and time consuming to build."

Without an NRAO director, management of the 140 Foot program remained in the hands of AUI. As discussed previously, when Otto Struve became NRAO director in July 1959, it was agreed that the "director's primary task will be to assemble a staff to operate the Observatory when it is completed,"[59] and the responsibility for the construction of the 140 Foot Telescope remained with Emberson and AUI. This unfortunate agreement, which removed the day-to-day management of the construction from Green Bank, was perhaps the most significant contribution to the series of miscalculations and mismanagements that resulted in multiple cost increases and schedule delays, and nearly led to the closing of NRAO before it even began any effective operation. According to Heeschen (2008), Struve apparently regretted the agreement and later unsuccessfully tried to transfer management of the telescope construction back to Green Bank. Only after both Struve and Berkner had resigned in 1961 did the project management shift from AUI in New York to NRAO in Green Bank. But by then millions of dollars had been wasted and five years had been lost.

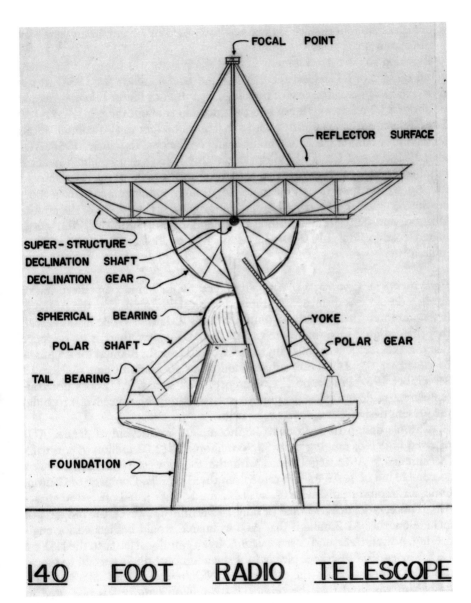

FOCAL POINT

REFLECTOR SURFACE

SUPER-STRUCTURE
DECLINATION SHAFT
DECLINATION GEAR

SPHERICAL BEARING

POLAR SHAFT

TAIL BEARING

YOKE

POLAR GEAR

FOUNDATION

140 FOOT RADIO TELESCOPE

Fig. 4.7 Ned Ashton design for the 140 Foot Telescope. Credit: NRAO/AUI/NSF

Ashton's design called for a large yoke supporting the dish structure attached to the polar axis, which was supported by a large spherical bearing sitting on nine (later reduced to four) oil supported pads (Fig. 4.7). This was much like the iconic 200 inch Palomar telescope, but, of course, the 140 Foot was much larger and heavier, and also had more precise pointing specifications. The yoke, polar axis, and spherical bearing were to be made of steel, and the dish backup structure of aluminum. The surface was composed of 72 aluminum panels set in three concentric rings.

Planning for the development of the Observatory was based on an antici-pated budget of $4 million for FY1957 and $1.13 million for 1958. At the time, AUI estimated the cost of building the 140 Foot Radio Telescope would be about $2.2 million.[60] But due to the uncertainty about the price of the 140 Foot Telescope, it was difficult with the limited budget to also commit funds for a control building, instrumentation, and operations. In January 1958, AUI issued "Estimated Costs of Construction" for NRAO that included detailed requirements for roads, water, electric power, buildings, housing, cafeteria, and staffing. By this time, due to the more ambitious requirements to work at short centimeter wavelengths, the increased appreciation of the impact of the remote location, and better understanding of the engineering problems, AUI's esti-mated cost for the 140 Foot Telescope had ominously increased to $6.8 million.[61]

Contracting for Construction An NSF suggestion for an independent review of Aston's design was dismissed by Emberson on the grounds that the AUI ad–hoc committee "had already accomplished this review."[62] While it was agreed that there should be a written record of the AUI review, there is no evidence that this ever happened. On 1 August 1957, AUI issued a Request for Proposals to construct the 140 Foot antenna following Ashton's design. At the 12 September 1957 Pre-Proposal Conference, AUI stated "AUI does not con-template a re-design phase to precede construction; the objective is to build the present design."[63]

Ashton's design turned out to be incomplete and devoid of details. AUI received nine bids ranging from $3.96 million to $12.02 million to construct the antenna.[64] After negotiations with the four lowest bidders, the lowest acceptable bid of $4.75 million came from the E.W. Bliss Company of Canton, Ohio, an amount nearly twice the available funds, but the bid offered an attrac-tive, if unrealistic, delivery time of only 14 months. However, the bid did not include another $1.2 million that AUI estimated would be needed for engi-neering, power, taxes, and cost escalation. In an emotional letter to the NSF on 12 February 1958, Berkner pleaded for the additional funds needed to accept the Bliss proposal.[65] At the subsequent NSF review on 13 March 1958, Emberson admitted that the original $2.2 million figure "was first used in almost casual conversation," and was for a "smaller less precise instrument." Struve kept up the pressure by pointing out to the NSF the ultimate need for "a very large antenna with a diameter well in excess of 1000 feet," and said that the "140 foot telescope should be regarded as merely a stopgap."[66] AUI was initially authorized to contract with Bliss for $145,000 for engineering (Phase I), with an option for fabrication (Phase II) which was contingent on upon an additional Congressional appropriation, and gave Bliss a letter of intent autho-rizing Phase I in December 1957. After Congress had passed a supplemental appropriations bill, AUI was able to let a fixed price contract to E.W. Bliss for

$4.75 million for fabrication of the 140 Foot Radio Telescope. Although not included in the contract, completion was promised in 24 months. On 9 June 1958, representatives from Bliss, AUI, and NSF gathered in Green Bank to sign the contract for the construction of the 140 Foot Telescope. Groundbreaking was two months later, on 14 August 1958 (Fig. 4.8).

The contract with Bliss was poorly formulated. There were no penalties for delay or incentives for early completion. AUI maintained that Bliss was responsible for completing the design, and would be free to depart from the Ashton design, provided that the stringent performance specifications were met. Nevertheless, Bliss repeatedly claimed that they were unwilling and unable to provide designs not covered by Ashton. As described by Heeschen (1996),

> The exact relation between Ashton's design and that of the contractor was never adequately spelled out, neither in the RP nor in the subsequent contract with the successful bidder, but it was clear that Ashton's concept was to be used. Ashton's detailed designs could be changed by the contractor, but only with the approval of AUI. The degree of responsibility for performance was also worded ambiguously.

Fig. 4.8 Groundbreaking for the 140 Foot Telescope, 14 August 1958. From left to right, Eugene Hallik (AUI), unidentified (Bliss Co.), Frank Callender (NRAO), Richard Emberson (AUI), and John Findlay (NRAO). Credit: NRAO/AUI/NSF

Signs of Trouble Apparently neither Ashton nor Bliss nor AUI had considered how the antenna was to be fabricated or erected, or how the large components would be shipped to Green Bank. From the beginning, Bliss had problems in completing the design and in fabricating the spherical bearing. As early as mid-1959, John Findlay reported that the completion date of the 140 Foot had slipped from early 1960 to the "fall of 1960," but that "construction on the site and in the contractor's shops is proceeding satisfactorily."[67] By the time the promised two year construction period had passed in mid-1960, fabrication of the spherical bearing, yoke, and polar shaft were proceeding in the Bliss plant, but already the project was nearly two years behind schedule (Heeschen 2007a). Due to design problems, work on the drive and control system, which had been subcontracted to the Electric Boat Division of General Dynamics, had not yet started, and Bliss also reported difficulty in developing a procedure for fabricating the surface panels.

On 5 May 1960, Bliss wrote to AUI that small cracks had developed in the spherical bearing, and that in cold temperatures, the type A-373 steel being used to fabricate the bearing and the polar shaft might be subject to what was called "brittle fracture." It was known that in North Atlantic winters during WWII, brittle fracture caused merchant ships constructed from A-373 steel to fracture and sink. Green Bank winters can be very cold, with temperatures reaching −30F (−34C). Since Bliss apparently had a supply of A-373 steel and experience in using it on other projects, they used it to fabricate the spherical bearing and the polar shaft. AUI was concerned that although the cracks in the bearing were minute, there was a slight danger that under cold weather the cracks might propagate, leading to failure of the bearing which supported the whole movable structure. An additional complication arose when neither Ashton nor Bliss were able to develop a procedure for welding the polar shaft to the yoke and sphere. Bliss wanted to bolt the pieces together, but this required modification of the components that already had been fabricated. Ashton argued that a bolted structure would not be sufficiently strong. Bliss maintained that the problem was due to the unsatisfactory Ashton design, and that AUI would be responsible for any additional costs and delay. AUI promptly responded that the problems were not due to design faults, but to unsatisfactory shop and welding procedures, and since, according to the contract, Bliss was responsible for the final design, they should be responsible for finding a solution and for any additional expenses involved.

Both AUI and Bliss hired consultants, but they disagreed on the optimum technical solution and on legal responsibility. On 24–25 May 1960, Lewis Burchill, the AUI Controller, and Emberson met with Bliss to discuss renegotiating the contract to provide a firm completion date, but they agreed that this would not be possible until the technical issues were resolved.[68]

Over the next months, AUI, the NSF, and Bliss discussed a number of possible solutions:

1) heating the enclosure surrounding the sphere, which could be risky if power were lost to the heating system during cold weather;

2) drilling small "mouse holes," or coating the sphere with epoxy to reduce stresses;

3) using a process known as "normalizing" to heat at least the spherical bearing, and perhaps all of the steel components, to a temperature of 1600 degrees F, which would change the structure of the steel and greatly decrease the possibility of brittle fracture;

4) scrap the sphere and start over with a new type of steel; or

5) do nothing and take a chance that brittle fracture would not occur. Meanwhile, the polar shaft, which had been fabricated with the same defective steel, had already been delivered to Green Bank.

On 30 August 1960, AUI and NRAO staff, along with consultants, met with the NSF to address the problems and discuss whether to terminate the contract with Bliss. Although Ned Ashton claimed that the chances of brittle fracture were negligible, and that the safety factor included in the design of the sphere excluded the possibility of catastrophic failure, Emberson maintained that this was potentially a very serious problem and he did not want AUI to take any chances, however small. The antenna was already way behind schedule and over budget. NRAO, and especially Struve, expressed concern that by time the 140 Foot Telescope was finished, it would no longer be cutting edge. They were well aware of the growing number of 85 foot radio telescopes around the world, as well as the Jodrell Bank 250 foot dish, and the Parkes 210 foot telescope then under construction. Struve contended that NRAO was not able to attract new staff members because the existing equipment was not sufficiently attractive or unique, and that there was a danger current staff would leave. Moreover, he argued that radio astronomy was progressing so fast that, after 1962, the 140 Foot would be "outdated." Perhaps somewhat irresponsibly, but no doubt with good intentions, he maintained that "the urgency of need" outweighed the "ideal technical or engineering solutions."[69]

Encouraged by Ashton's reassurance, recognizing that refabricating the major structural elements would be costly and introduce yet further delays, and concerned about minimizing the delay, Struve insisted that Bliss be allowed to continue with the erection of the antenna using the existing yoke, bearing, and polar shaft without normalizing. Although Emberson expressed dissatisfaction with the decision, the NSF agreed to proceed, and Emberson ordered that the spherical bearing and other fabricated components be shipped to Green Bank for integration and assembly. But the shipment was delayed by a railroad strike and by the continuing contention over who was responsible for the design faults and subsequent delays.[70] Lloyd Berkner, meanwhile, was on a two-month trip to Europe and the Middle East attending to some of his other national and international responsibilities. When NRAO and AUI met with the NSF on 30 August 1960, Berkner was in London as head of the US URSI delegation, and did not participate in this important discussion. After being updated by Emberson, three days later, Berkner telegrammed his agreement with Ashton and Struve to use the existing structures, and thus avoid further delay and cost increases which would "damage the health and effectiveness of the Observatory." But he encouraged Emberson to "investigate heating of the bearing house to prevent low temperatures that may induce brittle fracture."[71] By this time, although probably unknown to most of the principals, Berkner had already decided to resign as AUI President.[72]

After his return to the United States in late September 1960, Berkner, along with Emberson and Burchill, met with Struve and the Observatory staff in Green Bank to develop plans to either renegotiate the contract with Bliss or decide instead to give notice to terminate the contract.[73] Following a 22 September 1960 meeting between AUI and Bliss, with both sides backed by their respective legal teams, AUI asked Bliss to propose a contract revision that included a firm fixed price for completion and a firm date for final delivery, "preferably not substantially later than November 1, 1961."[74] As a prerequisite to any negotiation, AUI required that all finished parts be shipped to Green Bank, that work on the sub-contracted drive and control system be completed, and that assembly in Green Bank be resumed.[75] AUI was ready to assume responsibility for Ashton's design, provided that they had the opportunity to inspect progress and that Bliss agree to a price adjustment if they were not able to complete the telescope by the agreed date.[76] AUI appreciated that any renegotiation would require additional funds, which would need NSF approval, and that this would likely mean further delay, especially if the negotiations involved normalizing any of the steel components. By this time, Bliss was claiming over $1 million in excess costs they argued that they had already incurred, due, they claimed, to defects in Ashton's design. Burchill was confident that that they would settle for less, but the NSF agreed to "make additional funds available to you in such amount as may be necessary for a full settlement," and to provide up to an additional $100,000 for assembly of the telescope in Green Bank.[77] Meanwhile, at their 23 September 1960 meeting, the AUI Board authorized Berkner, at his "discretion," to terminate the contract with Bliss, although they recognized that this would be "fraught with difficulty."[78]

Finding a Solution Berkner's and Struve's apparent confidence was not shared by Emberson or the NSF. On 25 September 1960, Berkner and Emberson, along with Lewis Burchill and NRAO Business Manager Frank Callender, met with NSF Director, Alan Waterman, and other NSF staff, to review the deteriorating situation in Green Bank. After further review at the NSF, and probably influenced by Emberson's concerns, on 5 October, Waterman wrote to Berkner that while[79]

> we fully understand and sympathize with your desire, and that of the NRAO staff, to place the telescope in operation at the earliest possible date, ... the interests of the Federal Government must be protected by assuring that sound procedures are followed in fabricating any part of the telescope whose failure might result in

death or injury of personnel, severe financial loss and long delay in putting the telescope into operation.

After reviewing the options, Waterman went on to state

Because of the complexity, the serious consequences of major mistakes, and the possibility that our judgements may be overly influenced by the pressures we are all under to complete the job rapidly, the Foundation feels it is desirable at this time to have the technical problems reviewed by a highly qualified committee of experts in order that AUI may have the best possible basis for a decision on this problem.

Waterman added that he had appointed Dr. Augustus B. Kinzel, an engineer and Vice President of Union Carbide, to chair a committee charged with addressing the apparent problems with the spherical bearing and polar shaft.[80]

Discouraged by the deepening problems surrounding the 140 Foot project, on 23 September 1960 Berkner quietly informed AUI that he was resigning as AUI President.[81] Berkner's resignation was officially accepted at the 20–21 October 1960 Annual Meeting of the AUI Board of Trustees.[82] Struve was also upset by the delays in the project. Citing ill health, frustration over the NSF reversal of the 30 August 1960 decision to continue construction with the existing bearing, which, he noted, "is a severe disappointment to me personally," and feeling that important decisions were being made without consulting him, Struve stated that he was unwilling to assume the responsibility for the success or failure of the 140 Foot Telescope unless he could be sure of control. Noting that he had previously taken the position that he would not continue as Director unless the telescope could be finished by 1 July 1962, and that "nothing had happened which gives him any confidence that this condition will be met," Struve told AUI that he "cannot continue to serve as director of the Observatory."[83] Although Struve's effectiveness as NRAO Director was questioned, in view of the worsening 140 Foot situation, AUI apparently did not want to deal with the potentially embarrassing, nearly simultaneous resignation of both the AUI President and the NRAO Director.

At their 18 November 1960 meeting, the AUI Executive Committee reassured Struve that the "Director should be in complete charge at the Observatory and that Dr. Struve could count on the unwavering support of the Trustees in his administration of the Observatory."[84] The Trustees noted that in the earlier absence of a full time director, "practices had grown up at Green Bank and had simply been continued without any thought of acting contrary to the wishes of the director." The minutes of the 18 November AUI Executive Committee reported that, following the unanimous vote of their "complete confidence in Otto Struve, their satisfaction at the progress of the Observatory under his direction, their hope that he will find it possible to continue as Director," Struve "expressed his pleasure at the action taken and emphasized that he had never felt any lack of confidence on the part of the Trustees, but on the part of the contracting agency, the NSF." Later Struve (1961) would write, "the Observatory does not yet fulfill its intended function of serving as a 'national laboratory.' This is due to several causes, the most important of which is the delay in completing the 140 foot telescope." Heeschen recalled that Struve "remained very upset," over the reversal, without consulting him, of the 30 August decision, and that "his growing dissatisfaction with his role in the project would lead to his departure just a year later."[85]

These were certainly the darkest days in the history of NRAO. There was no clear route to completing the 140 Foot Telescope without major redesign and refabricating most of the structure, which would be both costly and lengthy. The AUI President who had been the driving force behind the creation of NRAO had resigned, and the NRAO Director was threatening to resign. The AUI contract with the NSF to operate NRAO was coming up for renewal. The cost overruns and delays already incurred by the Green Bank 140 Foot antenna, as well as those surrounding the Navy's 600 foot antenna at Sugar Grove, West Virginia (Sect. 9.3), came to the attention of the White House.[86] The NSF and the American radio astronomy community were becoming increasingly disillusioned with the lack of progress in Green Bank, and NRAO was becoming the subject of scorn and ridicule. At Caltech, the two-element interferometer was already in operation and was making a wide range of exciting solar system, galactic, and extragalactic discoveries (Sect. 6.6), while NRAO and the 140 Foot were being dismissed by the Caltech staff and students. The situation was so serious that the Observatory was threatened with closure.[87] Annoyed at the likely delays that would be introduced by the NSF decision to appoint an external committee to review the 140 Foot situation, Berkner took matters into his own hands, and did not hesitate to make Kinzel aware of his concerns about introducing further delays in the completion of the telescope. Seemingly ignoring the fact that he had just announced his wish to resign as AUI President, or maybe recognizing that he had nothing to lose, Berkner boldly wrote to Kinzel that, "you will learn of the technical aspects of the problem in your forthcoming meetings with members of the AUI staff, representatives of the E. W. Bliss Company, and consultants."[88] Suggesting that there were more important issues at stake than just the metallurgy, Berkner, hinting that the urgent need to use the telescope was more important than dealing with the metallurgical issues, wrote, probably inappropriately, to Kinzel,

> It is important, however, that you and members of your committee understand the unique position the 140-foot telescope has taken in the life of the Observatory and why we believe completion at the earliest possible date is necessary even if some measure of perfection may have to be sacrificed.

Continuing, he added,

> Therefore, a 140-foot telescope has become a symbol for the staff of the observatory in generating the opportunity at a national institution which would keep the

U.S. in the forefront of radio astronomy development. It is moreover, a symbol in the eyes of the scientific community generally including members of the NSF Board and to Congress and the public which is of course the ultimate source of support for such an institution as the NRAO.

But on 31 October 1960, in a letter to Leeland Haworth, who was to succeed him as AUI President, Berkner made it clear that he was relinquishing immediately all responsibility for the 140 Foot to Haworth and Struve, and informed Haworth that "AUI has employed long-time Trustee, Ted Reynolds, as an independent business consultant to study the AUI-Bliss relations."[89]

In November 1960, Bliss wrote to Emberson and Waterman requesting that the contract be revised to "to incorporate the changes which have incurred to date and to restate the consideration to be paid Bliss for the contract, as amended, on a cost plus fixed fee basis."[90] Following a meeting between AUI and Bliss on 23 November, Bliss estimated that their cost to complete the telescope, including the "additional cost to alter the present sphere and to accomplish a bolted field joint between the sphere and the shaft," would be just over $7 million, and proposed a total price, including their 7% fee, of $7.558 million.[91] Aware that "the expedient completion of this project is of great importance," Bliss claimed that "erection could be completed by at least by November 1962," but only with the additional "judicious expenditure of $600,000 for overtime premium, shift differentials, multiple work areas, additional facilities, and extra supervision." Another half a million was needed to complete the design work. Haworth noted that allowance for contingency further increased the likely contract price to at least $9 million, but that this was still less than some of the original bids.[92] This meant that AUI would need to get another $3.7 million from the NSF, but apparently, as Haworth explained, because West Virginia was considered a "distressed area," it was "desirable to speed up shipments from Canton so that work can go forward in the field using local labor."[93] AUI had no confidence in the engineering capabilities at Bliss, and began to explore alternate arrangements to provide satisfactory engineering supervision for the remainder of the work.[94]

Not surprisingly, the appointment of the Kinzel Committee had delayed any further work on the telescope. At one of the Committee meetings, Ashton "strongly urged that the original plan for welding the sphere to the shaft be followed, and, indeed, insisted that no other plan was feasible."[95] The Committee refused to accept Ashton's position, and their January 1961 report concluded that the spherical bearing should indeed be normalized and that it should be bolted rather than welded to the polar axis.[96] Due to the ensuing contract renegotiations, redesign efforts would not get underway until April 1961, and construction did not begin again until autumn 1961. At least another year had been lost.

Although the NSF had appointed the members of the Kinzel Committee, by agreement, their report was advisory to AUI. But the AUI consulting engineers were concerned about bolting the polar axis to the spherical bearing, which had been designed to be welded, not bolted. Not only was there a danger that the bolts might become loose, but it would be necessary to heat both the sphere and polar shaft. Instead, AUI concluded that it would be better to redesign the polar axis and the yoke, along with the spherical bearing, and fabricate all of the components using steel not subject to brittle fracture. After a meeting at the Bliss plant in Canton, still unable to reach a consensus, AUI, the NSF, Bliss, and Kinzel agreed to pursue both concepts. Following several meetings in early 1961 between AUI and Bliss, the contract with Bliss was terminated with a lump sum settlement, and it was agreed that "all future work performed by the E. W. Bliss Company in connection with the construction of the 140-foot telescope will be covered by a cost-plus-fixed fee contract."[97] But according to the 20 April 1961 minutes of the AUI Executive Committee, "serious disagreement developed over the degree of control which AUI will have to exercise over procedures followed by Bliss in fabrication and erection, as well as over procurement, subcontracting, and other matters." Bliss maintained that AUI's position "violated the understanding arrived at ... on March 1, 1961, ... [and] threatened to bring suit for breach of contract."[98]

New Management and a New Contractor In April 1961, AUI hired the Stone & Webster Engineering Company (S&W) to manage the whole project, and in particular to act on behalf of AUI in dealing with Bliss on all technical, financial, and management issues.[99] AUI finally acknowledged that it did not have the expertise to manage the project with its limited in-house staff. In particular, AUI was seeking engineering advice on the "problems raised by the polar shaft and main bearing," and noted that "no power of decision was vested in the Kinzel Committee."[100] After reviewing the work to date, S&W concluded that Ashton had prepared only a conceptual design. Since Bliss had refused to accept any design responsibility, Ashton's drawings became the de facto design.[101] Consequently, in yet another decision reversal, S&W decided that it would be necessary to fabricate a new polar shaft, yoke, and spherical bearing, which would all be bolted rather than welded together, and would use steel not subject to brittle fracture. The S&W design maintained only the broad character of Ashton's design. Meanwhile all construction work had again ceased. With the assistance of S&W, AUI negotiated a new agreement, withdrawing work from Bliss, as well as transferring the Electric Boat sub-contract for the drive and control system from Bliss to AUI.[102] AUI finally assumed responsibility for the design, and S&W became the agent for AUI in all future dealings with Bliss.

NRAO's Frank Drake was particularly frustrated with the delays, and did not see any solution to the problems connected with the polar axis and spherical bearing. He also noted that the 140 Foot was already becoming obsolete at wavelengths longer than about 10 cm, and that Green Bank was not a very good site for short wavelength work. In a memo to Struve and the entire Green Bank scientific and senior technical staff, Drake made two radical suggestions:

convert the design to an alt-azimuth mount, and move the telescope to a better site near Tucson, Arizona.[103]

Following AUI's acceptance of Berkner's resignation in October 1960, AUI Trustee Edward (Ted) Reynolds was put in direct charge of the project. Nevertheless, relations between the NRAO Green Bank staff and AUI remained troublesome. Basically AUI didn't trust Struve, but was unwilling and probably unable to replace him. At the same time AUI leadership was evolving. BNL Director Leeland Haworth succeeded Berkner as AUI President, and was followed first by Ted Reynolds and then I.I. Rabi. Decisions based on recommendations from consultants and committees were being made in New York by AUI without involving or even informing Struve, Heeschen, and Findlay in Green Bank. Heeschen finally led a revolt of the entire NRAO Scientific Staff, spearheading a 9 May 1961 letter to Struve.[104]

> The situation with regard to the 140-foot telescope is of great concern to us. There has been essentially no progress whatsoever for at least a year, and there appears to be no basis for expecting progress in the near future. There is, we feel, valid reason for questioning whether the AUI-Bliss contribution will succeed in completing the telescope before it is obsolete.

> We feel very strongly that the principle [*sic*] source of the 140-foot troubles lies in the way it is being managed by AUI. The lack of action in the past year and the apparent inability to make decisions is appalling and inexplicable to us. The decisions that are needed to get the telescope completed cannot be based solely on technical considerations. The scientific needs of the Observatory are not being given sufficient consideration largely because there is virtually no contact between the Observatory and the management of the telescope project.

> We do not understand why the 140-foot telescope has not been placed under your direction, and we have been questioned repeatedly by other scientists as to why the Observatory and the Director are not more immediately involved. Everyone seems to agree that in principle the job should be run from Green Bank. We reject the argument that the job cannot be managed by NRAO because NRAO has an inadequate engineering staff. Engineering advice is available from many sources. Stone and Webster for example can report as easily to you as to anyone else. The real problem is in deciding which of the conflicting engineering opinions should be acted on.

The next day, Struve forwarded to Rabi the letters from both Heeschen and Drake, along with a strong note criticizing AUI management of the 140 Foot Telescope project. Struve wrote[105]:

> AUI has ruled that the project should not be directed from Green Bank, and I have accepted this decision, despite the fact that I am not personally in favor of it.

> During the past few months I have not been fully informed concerning the negotiations with the Bliss Company, the Stone and Webster Engineering Management

firm and the NSF. This creates an intolerable situation that cannot continue much longer.

It is perhaps appropriate to mention that the NRAO suffers from having had virtually no AUI president during the past six months or so. Mr. Berkner, though absent for considerable intervals of time, had a good grasp of the whole project and managed to accomplish a great deal and provide a stabilizing influence on the whole organization. Matters of vital importance to the Observatory have from time to time centered around persons who lack understanding of the scientific competence to make major decisions but whose advice and influence have been obviously accepted by those who have the authority to act.

Struve had clearly crossed the line with his censure of AUI and, by inference, criticism of Rabi, which may have led to Struve effectively being summarily fired by Rabi six months later (Sect. 4.6). Frustrated with AUI, Struve sent Heeschen to represent NRAO at the 19 May 1961 meeting of the AUI Executive Committee, where Heeschen continued his criticism of AUI's management of the 140 Foot project. Apparently of particular concern was the relative authority and responsibility of Struve and Emberson. Responding to Heeschen's criticism, Reynolds acknowledged the need to "improve communication with the Observatory." Although the Board reaffirmed that Reynolds was in overall charge of the 140 Foot project, "Reynolds explained that Dr. Emberson will continue to serve as Project Manager, but that, *as in the past,* he is responsible to the Director of the Observatory."[italics added][106] At Struve's suggestion,[107] Dave Heeschen was appointed as the "scientific project director," which Heeschen later described as "meaningless."[108] Apparently, at least three individuals, Struve, Emberson, and Heeschen, now had some authority and responsibility for the 140 Foot antenna construction, but there was no clear division of responsibility.

Meanwhile, the NSF was becoming increasingly concerned with the repeated reversals of decisions on how to deal with engineering problems surrounding the 140 Foot project, as well as the growing cost and apparent management issues. Faced with the emerging S&W report recommending that all of the components already fabricated by Bliss be scrapped, the AUI Trustees circled the wagons, and agreed "that the only sensible choice is to follow the conservative Stone & Webster recommendations in all proposals to the NSF. The responsibility for taking risks should be placed on the Foundation and not be assumed by AUI."[109] Heeschen suggested that a "radical alternative" was to instead scrap the whole project, and build a copy of the Haystack 120 foot antenna, which he initially estimated might be done for $2 million. However, following a visit to Haystack, Heeschen and Findlay concluded in their 17 July 1961 report that it would be better not to make an entirely new start, but to salvage what little they could from the Ashton-Bliss enterprise. According to the minutes of the meeting, "Mr. Reynolds emphasized the importance of avoiding any discussion of scrapping the 140' Telescope. This possibility should not even be mentioned unless there is an alternative which is clearly more desirable."[110]

At the 20–21 July 1961 AUI Executive Board meeting, Reynolds reported that "Ashton still insists that his design is entirely feasible and can be completed in less time and for less money than the one now being prepared by Stone and Webster," and that Ashton had gone to the NSF "objecting to what he claimed to be waste of time and money involved in the present plans."[111] At the same time, AUI was concerned that the S&W recommendations involved rejecting the recommendations of the Kinzel Committee, and that following a meeting with Kinzel on 17 July 1961, Reynolds reported that "Kinzel insisted that the solution his committee recommended is still satisfactory," although he agreed that the S&W design "was in some respects superior."[112] Closing his report, Reynolds wrote that he had informed the NSF that the total cost of the 140 Foot Telescope would be close to $11 million.

By September 1961, AUI had taken over from Bliss the contract with Electric Boat for the drive and control system, but it was becoming clear that Bliss was also having difficulty fabricating the surface panels, while Aston continued to be critical of S&W. To complicate matters, the AUI contract to operate NRAO was due to expire on 16 November 1961. AUI was concerned, as they probably should have been, that the NSF was likely to impose stricter controls on the operation of the Observatory, and was unwilling to agree to a long term extension of the contract until the NSF agreed to AUI's plan for completing the 140 Foot Telescope.[113]

At Bliss, the President and other senior management all resigned. AUI decided to terminate the contract with Bliss since "there is no work in connection with the newly designed components which Bliss is capable of doing satisfactorily."[114] Ending the contract with Bliss was no easy matter, and had to be justified to the NSF, the Bureau of the Budget, and to Congress. S&W was unwilling to help document the case, as it did not want "to place itself in the position of apparently profiting by taking work away from E.W. Bliss."[115] S&W was now estimating the cost to finish would be over $12 million, or about $2.5 million more than allocated by the NSF based on estimates made only six months earlier. Even this price was predicated on a November 1963 completion, and Reynolds emphasized that "if AUI and NSF want this instrument at the estimated price, unusual efforts will have to be made to prevent the construction schedule from going into an additional year."[116] At this point, the Trustees seriously debated "whether completing the telescope ... was the best use of funds available for the support of the Observatory." They concluded, however, that, "the adverse effect, from the public relations point of view, of abandoning the project," and the instrument's promise "to be a very valuable research tool for many years to come," were sufficient motivation to complete the construction of the 140 Foot Telescope.

Following the resignation of Otto Struve as NRAO Director, and pending the arrival of the new Director, Joe Pawsey, Dave Heeschen became NRAO Acting Director on 1 December 1961 (Sect. 4.6). Maxwell Small, the BNL

Business Manager and former Construction Manager for Brookhaven's Alternating Gradient Synchrotron and High Flux Beam Reactor, was recruited as the 140 Foot Telescope Construction Manager. Small set up an office in his home in Boston near S&W, with an agreed goal of being the single point of contact between AUI and S&W. Ensuing discussions among AUI, NRAO, Bliss, S&W, and the NSF, led the new Bliss management to agree to terminate the contract with AUI.[117]

The NSF agreed to the AUI/S&W approach and on the level of financial support required to complete the telescope, in "an amount not to exceed $12,095,000."[118] However, the bids from Bethlehem Steel to re-fabricate and erect the polar axis and spherical bearing were significantly higher than the S&W estimates, as were the estimates from Electric Boat for the drive and control system. Meanwhile, the NSF was becoming increasingly concerned that they were not being fully informed about the negotiations with Bethlehem Steel, nor about the schedule and cost for completing the telescope, but AUI could only respond that these remained uncertain.[119] Unfavorable media coverage and confrontational correspondence with Ashton led to further anxieties at AUI and the NSF.[120] By time of the April 1962 meeting of the AUI Executive Committee, the cost estimated by S&W had risen to $13.3 million.

Completing the Job After S&W had completely redesigned the spherical bearing, the yoke, and the polar shaft, fabrication was subcontracted to various firms.[121] Ashton's 22 foot diameter spherical bearing design, which had been so controversial, had been reduced to 17.5 feet, the largest that could fit with three inch clearance through a rail tunnel near Droop Mountain, WV, on the way to Green Bank (Fig. 4.9). It was the largest nickel steel casting ever poured. A specially built railway car was used to transport the 167 ton bearing from Eddystone, Pennsylvania, where it was cast, to the Westinghouse foundry near Pittsburgh where it was machined to a precision of less than 0.003 inches (Heeschen 2007a, b). The spherical bearing, the polar axis, and the yoke were all shipped to Green Bank, first by rail to the nearby town of Durbin, and then by road over the last 13 miles to the telescope site. A small bridge over a creek near Green Bank needed to be rebuilt in order to bear the 55 foot long 90 ton load of the massive polar axis. Following the advice of Small, a contract for aluminum surface panels was let to the D.S. Kennedy Company, while Electric Boat continued as the subcontractor for the drive and control system. All that remained of the Ashton-designed, Bliss-fabricated telescope was the aluminum backup structure and the already completed 5800 ton concrete and steel foundation, which extended 30 feet below the ground level. The original Bliss polar shaft and yoke were discarded. Some pieces were sent to Brookhaven as shielding for BNL accelerators; the rest was buried in Green Bank where it remains as a memorial to the troubled 140 Foot Radio Telescope.

In planning for the assembly and erection in Green Bank, Heeschen and Small felt that since the ultimate responsibility for the project was with AUI, that AUI rather than S&W should assume more control. Furthermore, they

Fig. 4.9 Testing a model of the 17.5 foot spherical bearing for clearance in the rail tunnel through Droop Mountain, 1961. Credit: NRAO/AUI/NSF

argued, "much would be gained by having members of the Observatory staff actively working on the Telescope at the earliest possible date." Although the "Trustees doubted the desirability of diminishing in any way the responsibility of Stone & Webster,"[122] at the 15 February 1963 meeting, AUI decided to modify the contract with S&W to give AUI responsibility for all field operations.

Max Small moved to Green Bank in May 1963 to take charge of completing the fabrication of the backup structure on site, the assembly of the declination and polar shafts, and of the yoke and spherical bearing, all scheduled to arrive by rail from the various plants where they were being manufactured. An anticipated railway strike, which would further delay the construction schedule, threatened to interrupt shipments before the winter, but as it turned out, new problems in fabricating the major components ended up in delaying their shipment and erection in Green Bank.

In September 1963, authority for the 140 Foot project was transferred from the AUI President in New York to the NRAO Director in Green Bank, where the project was handled in the same way as other Observatory projects. Max Small hired two engineers and two clerks to facilitate interaction with S&W, as well as with the various manufacturers. At the 17 October 1963 meeting of the AUI Executive Committee, Small reported that the design was 95% complete, fabrication 70%, and construction 20%.[123] The polar shaft was finally delivered and on site, but due to delays in the receipt of the other major telescope pieces, Small was forced to renegotiate a new erection schedule with Pacific Crane and Rigging. The new contract, which assigned increased responsibility to Pacific Crane for "all work necessary to bring the telescope to completion after the components have been completed and delivered," included additional compensation of $1 million. But D.S. Kennedy reported "a wide variety of problems" in fabricating the surface panels, and stated that they were going out of business after delivering the panels.[124]

The spherical bearing was finally received in late April 1964. It was bolted to the polar axis and then lifted into place without incident. This was followed by separate lifts to hoist the two parts of the yoke into place. On 4 November 1964, a large crowd gathered to witness the final lifting of the telescope backup structure to fasten onto the yoke arms. Pacific Crane and Rigging company was in charge of the lift. However, the 266 ton structure proved too heavy, and when lifted just off the ground a cable snapped. There was no damage except to the cable. Five days later, after repairs and the revision of the lifting procedure (including cutting off a portion of one of the lifting cranes!), the backup structure was successfully lifted into place, and was bolted to the yoke the following day.[125] The 140 Foot major structural work was finally complete. All that remained was the installation of the 72 surface panels. Preliminary tests indicated that the panels distorted due to solar heating, but this was largely mitigated by the use of a special paint designed to scatter the incoming solar radiation and radiate strongly in the infrared, thus keeping the dish surface below ambient air temperature. By the end of 1964, all the surface panels were in place on the backup structure, and the two 167 foot tall cranes were dismantled. On 23 December 1964, the telescope was moved for the first time, and pointed to the zenith, with the entire 2700 ton weight of the rotating structure sitting on four oil pads, floating on a thin film of oil only 0.005 inches thick and under 3000 pounds per square inch pressure.

The 140 Foot Telescope (Fig. 4.10) construction was finally completed in the spring of 1965. The many delays and huge cost overrun of the project, which required unexpected additional funding for NRAO had challenged the credibility of the Observatory, and even the concept of a national federally funded facility. But this quickly changed with the introduction of a vigorous visitor program and the resultant flow of scientific results, particularly in the area of centimeter wavelength spectroscopy.

Fig. 4.10 Completed 140 Foot Radio Telescope. Credit: NRAO/AUI/NSF

The first astronomical observations with the 140 Foot Telescope were made on 22 May 1965 at 234 MHz (1.3 meters) and 405 MHz (74 cm) to study the Crab Nebula during a lunar eclipse. In July 1965, Bertil Hoglund, a visiting scientist from Sweden, and NRAO staff member Peter Mezger, detected a long sought hydrogen recombination line at 5 GHz (6 cm) (Sect. 6.2). Painting the telescope took place throughout the summer of 1965, while observations continued at 11, 6 and 2 cm. The aluminum panels were set during the summer nights to minimize the effect of solar heating and distortion.

The final cost of the 140 Foot Telescope was about $14 million. The dedication was held on 13 October 1965. Unlike the dedication of Green Bank eight years earlier, it was a beautiful sunny day. More than 150 visitors joined the NRAO staff by a podium erected under the telescope, surrounded by the splendid West Virginia fall foliage. One of the authors, (KIK) clearly recalls that in his speech at the dedication, Dave Heeschen remarked, "This isn't the largest radio telescope in the world, but it is the largest equatorial mounted radio telescope in Pocahontas County, West Virginia." Indeed, no radio telescope larger than 85 feet was ever again built with a polar mount, and since the late 1990s, starting with the Keck 10 meter telescope on Mauna Kea, all large optical telescopes now use the much simpler alt-az mountings which had been rejected by AUI for the 140 Foot Telescope.

By the end of 1965, the telescope had been used at wavelengths as short as 9 mm. However, due to large scale deformations when the telescope was tilted from the zenith, the aperture efficiency decreased substantially at wavelengths shorter than about 3 cm. Due to the nature of the equatorial mount, the efficiency changed differently with hour angle or declination. In 1976, as a test bed for the VLA antennas, the 140 Foot Telescope was modified so it could be used at the Cassegrain focus. Two years later, the fixed subreflector was replaced with a deformable subreflector whose surface was adjusted to partially compensate for the loss of gain at low elevations. The 140 Foot Telescope was originally designed to have a pointing accuracy better than 10 arcsec so that it could be used to determine accurate radio source positions (Struve 1960; Drake 1960), although by the time it was completed in 1965, it had become clear that radio source positions are best determined by interferometric means. In fact, 140 Foot pointing errors up to 30 arcsec, especially in the daytime, continued to plague observers, especially when observing at short centimeter wavelengths where the beamwidth was only a few arcminutes across.[126]

Until AUI relinquished control of the project to NRAO in September 1963, the 140 Foot Telescope was the first and only construction project managed directly by AUI, rather than by Brookhaven or NRAO. Throughout the project, maintaining that operation at short centimeter wavelength with a 140 foot diameter dish would provide unique opportunities, the NRAO staff had always argued against relaxing the performance specifications in order to limit the ever-increasing cost. The 140 Foot was smaller than the Jodrell Bank 250 foot, the Australian 210 foot, and Canadian 150 foot antennas, and had only a small fraction of the collecting area of the Arecibo 1000 foot dish (Sect. 6.6). Neither the pointing precision nor the surface accuracy were as good as either the Haystack 120 foot or the Canadian 150 foot telescopes, both of which were completed earlier and in routine operation well before the completion of the 140 Foot Telescope. By time the 140 Foot was completed in 1965, the Parkes 210 foot radio telescope was already in operation at wavelengths as short as 6 cm, and by 1972 at 1.3 cm. But as described in Sect. 6.2, as a result of the outstanding instrumentation and a competitive "open skies" access policy, until the completion of the VLA in 1980, the 140 Foot Telescope was arguably the most productive radio telescope in the world, with a growing oversubscription rate. Of particular interest was the use of the deformable subreflector and 1.3 cm maser radiometer for a wide variety of programs to study interstellar water vapor and ammonia. From 1967 until the completion of the VLBA in 1993, the 140 Foot Telescope was the backbone of the growing national and international VLBI effort.

In July 1999, for lack of operating funds, the 140 Foot Telescope was closed as an NSF-supported user facility for astronomical observations. For several years an independently funded MIT group used the antenna for ionospheric studies. Later, with financial support from the Russian Astro Space Center, it was resurrected in 2013 as a ground station for the Russian RadioAstron space VLBI mission (Sect. 8.9).

John Findlay later remarked, "no one with hindsight will deny" that "the choice of an equatorial mount was idiotic,"[127] and that the choice of a hydrostatic bearing which was copied from Palomar was foolish, while Bernard Burke described the 140 Foot Radio Telescope as having "served well, but its equatorial geometry is antique, its structural flexure is dreadful, its surface quality is inferior, its maintenance is expensive and man-power intensive, and its pointing is substandard."[128] Grote Reber was a bit more colorful in his appraisal of NRAO and the 140 Foot project. Writing to John Findlay, he remarked, "If such an affair had happened during the days of Elizabeth I, there would have been some public hangings."[129]

In 1992 Dave Heeschen summarized the 140 Foot project, by saying[130]

The 140 foot is a classic example of how not to design and build a telescope. The design specs were set by a committee of outside consultants who had no responsibility or accountability for the final result, and who gave liberally of poor advice. The 140 ft project leader [Dick Emberson], a very nice gentleman who was [assistant] to the president of AUI and responsible for the entire feasibility study that led to the establishment of NRAO, uncritically accepted all this advice. The telescope was originally going to have an az-el [*sic*] mount because the consulting engineers thought that was the most feasible.... But the steering committee membership changed from time to time and finally had on it a prominent and outspoken scientist [Tuve] who insisted the mount should be equatorial.... Then the solar astronomer [Menzel] on the steering committee decided that the telescope should observe the sun from sunrise to sunset on Jun 22 each year.

The errors made in bidding, contracting, and construction were even worse.... AUI wound up with a fixed price contract, for $4 million, with a company—E W Bliss—that really didn't want the job, except for one enthusiastic vice [president] who apparently bullied them first into accepting the final contract. He quit shortly afterward and AUI was left with a semi-hostile contractor.

Some important lessons were learned, or should have been learned, from the 140 Foot experience:

1) beware of the lowest bidder;
2) be sure the contract is clear about who is responsible for what;
3) finish the design before starting construction;
4) establish clear points of contact, authority, and responsibility on both sides;
5) have a firm understanding of when the antennas will be delivered, with penalties for late delivery;
6) don't take committee advice too seriously; and
7) have good in house expertise.

Regrettably, many of these same issues arose with the ill-fated 600 foot Sugar Grove antenna and resurfaced 25 years later with the Green Bank Telescope (Chap. 9).

4.5   THE 300 FOOT TRANSIT RADIO TELESCOPE

By early 1958, the 140 Foot Telescope was no closer to completion, and it was not even clear if it would be completed (Heeschen 1996). While NRAO scientists were able to do some interesting observations with the Tatel 85 foot telescope, it was by no means a state of the art facility that would attract visiting observers in the way that had been expected for the US national observatory. Not only did the University of Michigan operate a similar, and indeed, a somewhat better 85 foot antenna, but competing facilities were coming on line throughout the world, even within the United States. Both Germany and the Netherlands were operating 25 meter class radio telescopes; Jodrell Bank had their 250 foot dish; and planning for the Parkes 210 foot radio telescope was well along. At Caltech, with financial support from the Office of Naval Research, John Bolton was building a novel two-element interferometer in the Owens Valley, capable of operating at centimeter wavelengths. With a modest budget that was dwarfed by the generous NRAO NSF budget, Caltech scientists would begin an ambitious radio astronomy program that would make the Owens Valley Radio Observatory the dominant radio astronomy facility in the US.

With only a modest radio telescope, essentially no visiting observers, and facing increasing concerns about when, or even if, the 140 Foot Telescope would be completed, John Findlay and Dave Heeschen thought a fixed 300 foot miniature Arecibo type antenna could be built for about $300,000. But their 1958 proposal was not well received by the NRAO Visiting Committee or by the NSF.

A year later, Findlay and Heeschen developed a bold plan to build the best antenna that they could for not more than about $1 million, which they thought to be the largest amount of money the NSF would approve without long delays. Following his appointment as NRAO Director in July 1959, Otto Struve was able to sell the project to the Visiting Committee and then to the NSF (Heeschen 2007b, 2008). A 300 foot transit antenna, movable only in elevation, thus simplifying its design and limiting the construction cost, seemed to offer the best compromise between opportunity for scientific returns and price. Funding was approved in the 1961 NSF budget. John Findlay became the project manager, and recruited Bob Hall to design the telescope.

Hall had just left Blaw-Knox for a new position at the Rohr Corporation, which was anxious to get into the antenna business. Between his jobs at Blaw-Knox and Rohr, with the aid of five Blaw-Knox engineers, Hall spent six weeks at the end of 1960 working for NRAO from his home in Chula Vista, CA, to design the 300 Foot antenna. Later, Ed Faelten was retained to complete the engineering drawings necessary for construction bids. To meet the limited construction budget, the design was kept simple. Specifically, the height of the supporting towers was limited, which constrained elevation motion to 60 degrees from the zenith and the corresponding observable declination range from the north pole to minus 19 degrees. The antenna was driven in elevation by a 230 foot long quadruple chain which wrapped around the antenna elevation wheel. The construction cost was further reduced by using standard steel members, simple joints and bearings, and constructing the reflecting surface from chicken wire rather than solid panels. It was anticipated that the future would see a large, fully steerable, radio telescope at NRAO, so the useful scientific life of the 300 Foot was expected to be not more than about five years. Emphasis was on getting it completed and on the air quickly, rather than longevity. The 3/8 inch chicken-wire holes would restrict the operation to wavelengths longer than about 20 cm, but the 300 Foot antenna nevertheless became a powerful facility for both galactic and extragalactic 21 cm H I research, as well as for continuum source observations.

In April 1961, NRAO contracted with Bristol Steel and Iron Works to construct and erect the antenna. Groundbreaking in Green Bank was on 27 April 1961. Under Findlay's leadership, construction took less than 18 months at a cost of about $850,000, a record construction time for any large radio telescope project. On 21 October 1962, the 300 Foot Telescope was handed over to the Green Bank operations staff and began its first astronomical observations. The next day, President John F. Kennedy announced the US naval blockade of Cuba in response to the discovery of Soviet missile bases a week earlier. Two days later, the US military went on the highest military alert since 1945, and the start of observations with the world's largest parabolic dish went relatively unnoticed by the nation.

The 300 Foot Radio Telescope, now one of the most powerful radio telescopes in the world, became an immediate success. For the first time, NRAO had a world class instrument that was attractive to both visitors and NRAO staff. The successful completion of the 300 Foot transit radio telescope probably saved Green Bank from a premature closing resulting from the continued debacle with the 140 Foot antenna project. From the start of 300 Foot observations, the Observatory operated as the first true visitor facility for radio astronomy. One of the earliest visiting observers was Bernard Burke who, with his colleagues from DTM, brought a 100 channel receiver for 21 cm spectroscopy in November 1962. Gart Westerhout, who had recently arrived in the US from the Netherlands to start a radio astronomy program at the University of Maryland, became a regular user of the 300 Foot Telescope. At the start of each summer, Westerhout would arrive in Green Bank with his family and a cadre of students to help observe and reduce data, and to escape the heat and humidity of the eastern Maryland summers. According to Heeschen (2007b), the 300 Foot taught NRAO how to manage an oversubscribed telescope, train operators, provide calibration and documentation, and, in general, deal with visitors. Unlike the 85 foot and 140 Foot antennas, the 300 Foot was the first of a series of antennas and arrays that would be conceived by NRAO staff and built under the direction of NRAO.

Initially, the rim of the 300 Foot antenna would hit the ground at low elevations, so the antenna could not be moved over the full 60 degree range of zenith angle allowed by the drive system. In October 1962, at Frank Drake's urging, the Observatory started to dig a pit at the south side of the structure in order to lower the antenna elevation limit. But the excavation reached bedrock after digging only six feet below the surface. The Green Bank site manager, Bob Elliot, suggested deepening the pit further by using dynamite, much to the chagrin of the conservative scientists who were concerned that the blast might destroy their new telescope. Indeed, the explosion sent rock and other debris more than 100 feet into the air, scattering the crowd that had assembled to watch the big event. Fortunately, no one was hurt, but one of the rocks put a one foot hole in the dish surface.[131] However, the exercise was successful in extending the southern declination limit to minus 19 degrees.

By the summer of 1966 several structural deficiencies had become apparent, and the backup structure was strengthened with the addition of 20,000 pounds of steel, 120 sections of rib structure were replaced, and welding added another 7000 pounds to the structure. The wire mesh surface, which was irregular to start with, further deteriorated as a result of staff walking on the antenna surface, and in October 1966 the surface was removed and flattened by laying sections on the ground and running over them with a steam roller (Fig. 4.11).

Although the 140 Foot antenna later became the NRAO workhorse, it was primarily used for centimeter wavelength spectroscopy. For continuum observations at longer wavelengths, the 140 Foot was limited by confusion,[132] while for 21 cm spectroscopy, observers preferred the greater collecting area and better resolution of the 300 Foot. The 300 Foot was designed to be used in a "drift-scan" mode, where the antenna would be driven in elevation to the declination of interest, shortly ahead of meridian transit, and the rotation of the Earth would allow the source or area of interest to drift through the antenna beam, typically in about one minute. To increase the available integration time, in 1969 a so-called "traveling feed" was constructed, which would allow the antenna beam to track equatorial sources for up to an hour depending on the wavelength of observation. The travelling feed was later replaced in 1980 with a unit that supported heavier cryogenically cooled receivers.

By 1970 it was becoming clear that the basic antenna structure and pointing were sufficiently precise to allow operation at a shorter wavelength than 21 cm, but the chicken wire surface was too porous and too irregular for this purpose. In 1970, NRAO replaced the original chicken-wire surface with perforated aluminum panels. The contract for the new surface was placed with a new, relatively unknown company, Radiation Systems Inc. (RSI) from Sterling, VA. Richard (Dick) Thomas, the president and principal owner of RSI, was anxious for business, and apparently underbid for the contract. NRAO Associate Director Ted Riffe recalled that Thomas realized that he was about to lose a large amount of money, and appealed to the NRAO to renegotiate the contract. Riffe, who had come to NRAO from the West Virginia coal mining industry, was a hard-nosed business man who sat quietly while Thomas explained where he had made an error, how his error would hurt his employees and the economy of Northern Virginia, how his family would be deprived, his children not able to go to college, and so on. Finally, Riffe looked Thomas in the eye and replied, "Bull shit!"[133] NRAO did not adjust the price, but RSI

Fig. 4.11 Working in vain effort to smooth the surface of the 300 Foot Telescope, October 1966. Credit: NRAO/AUI/NSF

weathered the storm and went on to become a major player in the antenna industry, including constructing the surface panels for the VLA (Chap. 7), and contracting to build the ten VLBA antennas (Chap. 8). But 20 years later, again in his anxiety to win the contract, Thomas would underbid for the GBT (Chap. 9), leading to a huge cost overrun and eventually the end of RSI.

Over the next years there were many minor 300 Foot repairs, including additional weldings and more reinforcing structures, and the structure was repainted multiple times. An unanticipated use of the 300 Foot Telescope, one that would later prove fatal, was to take advantage of the great sensitivity by rapidly moving the antenna in elevation to cover a wider area of the sky than possible with only Earth rotation drift scans. When first proposed by the Green Bank scientists, Fred Crews, head of telescope operations, was reluctant to introduce the unplanned stresses on the telescope that would result from nearly continuously driving the antenna and from the rapid reversals of direction which would occur at the end of each scan.[134] Naturally, the scientists wanted to limit the time wasted in turning the telescope around at the end of each scan. Following review by the cautious telescope operations staff, operators began the practice of slowing the antenna to a gradual stop before reversing its direction, thus minimizing any sudden decelerations or accelerations.

The 300 Foot antenna was built to withstand snow or ice loads up to 10 pounds per square inch (Lockman et al. 2007, pp. 103–105). While dry snow might fall through the chicken wire surface, wet snow or ice could become a serious problem. During the first few years, the dish was tipped during snow-storms, and most of the snow would fall out. But on two occasions, a small army of site personnel had to use brooms to sweep up the snow, and it was difficult to avoid damaging the surface by walking only along the ribs. On one occasion small fires were built under the dish to melt the snow, but the melting snow dripped down and extinguished the fires. At Findlay's semi-serious sug-gestion, NRAO acquired a surplus jet engine, which was used to blow snow off the dish. Needless to say, for nearby households, the sound of a jet engine run-ning all through a snowy winter night was like trying to sleep next to an airport where the same plane was continuously taking off. The use of the jet engine to de-ice the dish was abandoned after a few years, in deference to residents of Green Bank and Arbovale, and because of the considerable maintenance required to keep the engine operational. Following the installation of the new more robust aluminum panel surface in 1970, snow accumulation was less of a concern.

4.6   JUMPING SHIP

Lloyd Berkner Resigns as AUI President Even while juggling the two simulta-neous jobs as President of AUI and Acting Director of NRAO, Berkner assumed many other national and international responsibilities. In 1955, he became president of the International Council of Scientific Unions (ICSU), then president of the International Union of Radio Science (URSI), and was the leader of the 1957–1958 International Geophysical Year. From 1958 to 1962, he was Chair of the newly created National Academy Space Studies Board, and from 1956 to 1959 a member of President Eisenhower's Science Advisory Committee. In 1957 he became a member of the Board of Texas Instruments, and in 1958 he returned to Antarctica to prepare a report for Eisenhower that became the basis for continuing the US Antarctic program. Following his resignation from AUI in late 1960, in the midst of the 140 Foot construction problems and corresponding unrest among the Green Bank staff, (Sect. 4.4) Berkner went to Dallas, TX, to become the first president of the new Graduate Research Center of the Southwest. On 22 November 1963, Berkner was waiting to have lunch with President Kennedy, whom he had wel-comed to Dallas earlier in the day before he began his fateful motorcade.

Although Berkner's resignation as AUI President did not become effective until 30 November 1960, as discussed in Sect. 4.4, AUI records show that the Board of Trustees had already accepted his resignation at their annual meeting on 21 October 1960, following his announced resignation on 23 September 1960 at a closed session of the Executive Committee.[135] Presumably, Berkner must have sometime earlier begun discussions with the group in Texas. At this same meeting on 21 October, the AUI Board appointed a selection committee chaired by Trustee Edward Reynolds, the Administrative Vice President at Harvard and a retired brigadier general, to "nominate one or more individuals" to succeed Berkner. At their 18 November meeting, probably recognizing the time required to recruit a new President, the Board appointed Brookhaven Director and AUI Vice President, Leeland Haworth as AUI President. Haworth served as President for only four months, until 30 March 1961 when he took a leave of absence from AUI to accept a position on the Atomic Energy Commission. During this entire period Haworth also continued in his demanding role as Brookhaven Director. He was succeeded as President by Edward Reynolds. Since Berkner's resignation, Reynolds had taken charge of the 140 Foot project, but, by previous agreement, he only served as AUI President for three weeks. AUI stayed inside in choosing their next president, and, on 21 April 1961, named I.I. Rabi, Trustee from Columbia, to be President.

Rabi was known as a no-nonsense individual who demanded and accepted nothing less than excellence from his students and colleagues. He was well known for his 1944 Nobel Prize-winning discovery of nuclear magnetic resonance. In 1947, Rabi and his students were the first to make a laboratory measurement of the 1420 MHz hyperfine line in hydrogen, which led to his development of the hydrogen maser atomic clock. During WWII, he worked on microwave radar at the MIT Radiation Laboratory where he became Deputy Director, and later worked as a consultant to the Manhattan Project, which brought him to the 1945 Trinity test at Alamogordo, NM. In 1946, along with MIT and Harvard scientists, Rabi founded the Brookhaven National Laboratory, and became a founding Trustee of Associated Universities, Inc. He was later influential in the formation of CERN, and as a statesman for national and international cooperation in science (Ramsey 1993). Rabi served as AUI President until 19 October 1962, at which time he became Chairman of the AUI Board, a position which he held for a year. Rabi was succeeded as AUI President by Gerald Tape, who had been the AUI Vice President and Deputy Director at Brookhaven where he had oversight of the large reactors and accelerators. Like Rabi, Tape had worked at the MIT Radiation Laboratory during the War where he was an important liaison between the Rad Lab scientists and Army and Navy officers responsible for implementing these new instruments of electronic navigation. Tape left AUI on 10 July 1963 to become the US Ambassador to the International Atomic Energy Commission (IEAC), at which time Edward Reynolds again chaired a search committee for a new AUI President. AUI Trustees Curry Street, Frank Long, and Norman Ramsey all declined, so Reynolds assumed the position on an interim basis starting 10 July 1963. Expressing frustration over the rapid turnover of Trustees and lack of active participation of the Trustees in the affairs of the corporation, Reynolds notified Rabi that he did not want to continue as President after the October 1963 Board meeting,[136] but he was convinced to remain until 1 December 1964, when he was succeeded by Theodore Wright, who served from 1

December 1964 to 1 October 1965. Reynolds was succeeded by Theodore Wright (1 December 1964 to 1 October 1965), T. Keith Glennan (1 October 1965 to 30 June 1968), and then Franklin Long as Acting President (1 July 1968 to 1 May 1969), After completing his term as IEAC Ambassador, Jerry Tape returned to AUI as President from 1 May 1969 finally bringing much needed stability to AUI. Tape retired as AUI President on 10 October 1980 and was succeeded by Robert Hughes, a chemist from Cornell, and former NSF Assistant Director for Mathematical and Physical Sciences.

Dick Emberson, who had been so instrumental in leading the feasibility studies leading up to the 1956 establishment of NRAO, and then became director of the 140 Foot construction project, remained as 140 Foot project director and as NRAO Acting Deputy Director, but according to Dave Heeschen (1996) his role was diminished after S&W became involved. After Heeschen became NRAO Director in October 1962, Emberson left AUI to become Director of Technical Services, and then Executive Director and General Manager, of the Institute of Radio Engineers (IRE) later to become the Institute of Electric and Electronic Engineers (IEEE).

Otto Struve Resigns as NRAO Director In his Biographical Memoir for the National Academy of Science, Kevin Krisciunas (1992) described Struve as "both dedicated and demanding," and said "his physical appearance and demeanor" were imposing and intimidating. Although he was greatly respected by Dave Heeschen and the NRAO scientific staff, Struve's European background and somewhat conservative life style did not mesh well with the much younger and enthusiastic NRAO scientific staff or with life in rural Appalachia. He became frustrated by his inability to recruit scientists to the NRAO staff, a problem he recognized was due in large part to the lack of world class observing facilities, an absence which was exacerbated by the delays in the 140 Foot Telescope construction project and the isolated location of Green Bank. In addition to his demanding responsibilities as the NRAO Director, throughout his tenure Struve tried to maintain an active program of personal research and publication in areas unrelated to the NRAO mission, and became increasingly frustrated by the need to attend to administrative matters and by seemingly endless meetings.

Following Berkner's resignation and Struve's 1960 attempt to resign over the escalating issues surrounding the 140 Foot Telescope construction (Sect. 4.4), the NRAO Scientific Staff recognized that the reputation of the Observatory, as well as their relations with Struve, were being questioned. Led by Heeschen, the staff sent a very carefully worded letter to Struve expressing their confidence and their wish that he "continue as director as long as possible."[137] But recognizing both Struve's declining health and endurance, and his limited understanding of radio astronomy instrumentation, they suggested that AUI appoint a Deputy Director as soon as possible to assist Struve.

In January 1961, Haworth and the Trustees Committee on NRAO, went to Green Bank to personally assess the situation at the Observatory. The

Committee report recommended that "all engineering activities ... be centered in Green Bank as rapidly as feasible," that a mechanical engineer be recruited to "participate in the completion of the 140' Telescope" and "become the leader of a permanent group which will have the responsibility for the design, construction, and maintenance of a continuing series of mechanical structure."[138] But they added that "for the present the negotiations with E.W. Bliss Company and the National Science Foundation regarding the 140' telescope must of necessity be carried out by the officers of AUI." Finally, fearful of the adverse publicity and reaction of the NSF if Struve were to resign, they stressed "that the steady guiding hand of Dr. Struve is essential to the success of the National Radio Astronomy Observatory." Noting that "if certain conditions are met, Dr. Struve might reconsider his earlier announced wish to retire," the Committee urged "the President to explore with Dr. Struve the measures that would make it possible for him to continue." Struve responded to the Trustees that he would reconsider his request to be relieved of the Directorship in the summer of 1961, and, health permitting, to continue for at least one to two years more.[139] Prompted by Heeschen's letter and a proposal from Struve, AUI appointed John Findlay as Deputy Director of NRAO, and Dave Heeschen as Assistant to the Director replacing Findlay. At this point Findlay, who was ten years older, clearly outranked Heeschen, but the Board instructed Struve to explain to Findlay "that the appointment as Deputy Director should not be regarded as a stepping stone to the Directorship, because the Trustees believed that the director of the Observatory should be an astronomer."[140]

Although there appear to be no further questions about Struve's position as NRAO Director by either AUI or by Struve himself, Rabi had apparently concluded that he didn't want Struve to remain as Director of NRAO, and he began to work behind the scenes to find a replacement. In June 1961, during a visit to Green Bank, probably at Rabi's suggestion, Bart Bok wrote to Joe Pawsey to informally ask if he might be interested in the NRAO directorship.[141] Pawsey responded unenthusiastically, claiming that Green Bank was too "committed to big paraboloids," and that his "interest is in techniques," but he left the door open for possible further discussion.[142] Pawsey came to the United States during the summer and early autumn of 1961 to attend the August meeting of the International Astronomical Union that was held in Berkeley, CA, and also to visit various radio observatories and his brother-in-law, Ted Nicoll, who lived in Princeton, NJ. During a visit with Pawsey at the home of Pawsey's brother-in-law, Rabi discussed the Green Bank situation with Pawsey and asked him for his impressions of the situation there.

Following a three-day visit to Green Bank, Pawsey wrote to Rabi on 5 October 1961 that he "found the situation disappointing," and that none of the staff are "really first rate." But he was impressed by two visitors, T.K. Menon and Sander [Sandy] Weinreb, who was using the Tatel Telescope for his PhD research.[143] Pawsey reported that "the director's experience has not been in the field in which the staff is weak,—the technical radio and experimental physics side." Interestingly, and perhaps with some envy, he went on to say that, "Both

Greenbank [*sic*] and U.S. radio astronomy suffer from the same basic difficulty: a dearth of good radio astronomers and the too ready availability of elaborate equipment." Consistent with his earlier remarks to Bok, Pawsey only gave lukewarm support to the 140 Foot and 300 Foot antennas then under construction. However, he noted that for prestige and staff morale, both the 140 Foot and 300 Foot Telescopes "should be completed as soon as practicable." But the best he would say about the 140 Foot was that it will be "a thoroughly useful instrument." Pawsey stopped short of suggesting that the 140 Foot equatorial design should be abandoned in favor of the "master equatorial" idea used by Freeman Fox in designing the Parkes 210 foot or the Algonquin Park 150 foot dishes, but he did suggest that the Green Bank "engineers should be informed of the existence of the Freeman Fox design," as "there could be snags in the original Greenbank [*sic*] design and this could be a replacement." Citing the report of the NSF Pierce Committee (Keller 1960) Pawsey declared that "future emphasis should be directed toward instruments having a resolution of 1′ or less," and went on to suggest that NRAO might try to hire W.C. [Bill] Erickson and Sander Weinreb.

It is not clear what transpired between Rabi and Pawsey during their Princeton conversation and to what extent Pawsey's Green Bank visit was considered by either of them as being only in an advisory capacity, or whether either or both of them thought of it as a response to Pawsey's consideration of the possibility of becoming NRAO Director. Pawsey's use of phrases such as, "I should also like to encourage," or "My view is," and "I envisage" as well as "During my initial preparatory phrase," certainly suggests that Pawsey was responding to Rabi's overtures about the NRAO Directorship, and that he was seriously considering the possibility.

Three weeks later, the minutes of the 26 October 1961 meeting of the AUI Board of Trustees Executive Committee, reported that "the Committee held an executive session, at which it was directed that no report be made."[144] During a discussion of the renewal of the AUI contract with the NSF, Rabi expressed concern that if the contract were extended for only two years, that this might "impede the selection of a new director to replace Dr. Struve."[145] This was the first indication that changes might be forthcoming, although there was apparently no further indication from either Rabi or Struve that either might be aware of any imminent change. During his presentation to the Executive Committee, Struve commented that "the entire staff of the Observatory supported the continuation of AUI management of the observatory. The consensus of the Trustees was "that every effort should be made to obtain an extension of at least three and preferably five years." Struve reported on the good progress being made with construction of the 300 Foot dish (Sect. 4.5) and his concern about turnover in the scientific staff, while AUI Vice President Edward Reynolds gave a report on the status of the 140 Foot project (Sect. 4.4). After the formal adjournment, the minutes report that Trustees moved to another building to hear scientific presentations from Frank Drake on Venus observations, from Dave Heeschen on extragalactic radio source spectra, from MIT graduate student Sandy Weinreb on his attempt to detect the 327 MHz deuterium line and from T.K. Menon on H II regions. Following the formal signature of the Corporate Secretary, Charles Dunbar, attesting that the meeting was adjourned a note was added that,

> At the conclusion of the scientific presentation the Trustees met again in executive session. The Secretary was instructed to make the following report: The President informed the Trustees that Otto Struve had asked to be relieved of his responsibilities as Director of the National Radio Astronomy Observatory on December 1, 1961, or as soon thereafter as could be arranged. The choice of a successor to Dr. Struve was extensively discussed as well as interim arrangements to be made pending the appointment of a new Director. At the conclusion of the discussion, on motion duly made and seconded, all Trustees present voting, it was unanimously VOTED: That the President be and hereby is authorized to offer to Dr. Joseph Pawsey of Sydney, Australia, an appointment as Director of the National Radio Astronomy Observatory on such terms as the president deems appropriate. Dr. Rabi said that he would write without delay to Dr. Pawsey and report the result as soon as possible.

Unlike the frustrating three-year long search that preceded Struve's appointment, there was no search committee, and there is no evidence that Rabi sought the advice of anyone else. At this point, there apparently was no longer any agonizing over whether or not the NRAO director need be an American citizen. Merle Tuve must have been gratified when he heard the news that his 1956 question to the NSB, "Where will the staff come from? UK? Australia?" (Sect. 3.5) had been answered.

Interestingly, this important action was taken by only the AUI Board Executive Committee and not the full Board of Trustees. There is no indication from the agenda that was distributed to the Board on 4 October 1961 that there would be any discussion of Struve's successor, although it seems likely that this was the purpose of the executive session called at the start of the day. There was no statement of who moved and who seconded the motion to appoint Pawsey as the Director of NRAO. The only evidence that Struve had again asked to be relieved of his responsibilities was the announcement by Rabi at the late afternoon unscheduled session.

The next day, 27 October 1961, the full Board of Trustees met for what appeared to be a routine meeting primarily addressing corporate matters and business related to the operation of Brookhaven. Again, there was no scheduled discussion of Struve's successor, nor, according to the minutes, any announcement of the resolution passed on the previous afternoon by the Executive Committee to appoint Pawsey (Fig. 4.12). The only reference to Struve's tenure as Director came up during Struve's presentation of the NRAO Visiting Committee report, where the minutes recorded, "Dr. Struve said that since he is retiring as Director in less than a year, he would like to remind the Trustees of some of the problems he has encountered, in the hope that they

Fig. 4.12 Joseph
L. Pawsey at IAU
Symposium 4, Jodrell
Bank, 1955. Credit:
NAA-WTS Working Files,
Interviewees

might be avoided in the case of his successor."[146] At the time Struve gave no indication that he expected to retire in only two months.

Apparently, Rabi informed Struve of the outcome of Executive Committee action and suggested that Struve resign on 1 December 1961, as, four days later, on 31 October 1961, perhaps in a face-saving move, Struve wrote to Rabi[147]

to ask you whether my services at the National Radio Astronomy Observatory could be dispensed with about this December 1, either temporarily or permanently, in order to give me an opportunity to engage more actively in research than I have found it possible to do during the past two and a quarter-years. I feel that I must try to catch up with recent developments in astrophysics and I am unable to do so while I am compelled to spend nearly all of my time in non-scientific meetings and investigations by single persons and groups. These leave me in a state of continuous fatigue which is the cause of other health problems. A leave of absence for several months would of course be the most desirable arrangement from my point of view, if you feel that I deserve it. If this would not be possible, then according to our records I believe that I shall have 45 days of vacation pay due me if I leave by December 1. In either case the Observatory would not have to fear that the scientific staff would be left without guidance and protection, since I gathered from your remarks last week that the problem of my successor has virtually been solved. If you should wish it, I could return to Green Bank for a short time either before Dr. Pawsey arrives or soon afterward.

On 31 October 1961, the same day that Struve wrote to Rabi offering his resignation, Rabi wrote to Pawsey that the AUI Board had approved his nomination to be Director of the National Radio Astronomy Observatory, and that he had already obtained the approval of the NSF Director, Alan Waterman."[148] From the content of Rabi's letter, it is clear that he and Pawsey had previously discussed the matter in London, if not also in Princeton, although Rabi's formal offer letter was written the same day that Struve wrote his letter of resignation. Rabi did not disclose to Pawsey that he expected Struve to retire on 1 December 1961, but rather mentioned that "Dr Struve has asked for retirement as of October 1, 1962."

Pawsey promptly responded that before making a decision, he first wanted to talk with E.G. Bowen and CSIRO Chairman Fred White.[149] Two weeks later, on 17 November 1961, at an executive session of the meeting of the AUI Executive Committee, Rabi reported on the letter he had received from Struve, and the Committee voted unanimously, "pursuant to his own request," to relieve Struve of his duties as NRAO Director.[150] At the same time, David Heeschen was appointed as Acting Director of NRAO, effective 1 December 1961, in anticipation of Pawsey's arrival as the Director in October 1962. Immediately after the AUI Board meeting, Rabi wrote to Struve that the Trustees had accepted his resignation,[151] and also informed the NRAO staff that "at his request," Struve had "been relieved of executive responsibility" as Director of NRAO effective 1 December.[152] Fred White reluctantly agreed to Pawsey accepting the NRAO position, but was concerned that it should appear that Pawsey was accepting an invitation to help the Americans, and not that he was leaving due to any discontent with CSIRO or Australia.[153] Pawsey then accepted a three-year appointment at NRAO, "with the possibility of returning to my employment in Australia at the end of that time," but he made it clear that he wanted to maintain close ties with Australia and would seek increased cooperation between the US and Australia.[154] Since he was unable to take up the position until October 1962, he agreed that, as suggested by Rabi, he make a visit to Green Bank in the spring of 1962. Rabi responded that "your decision leaves us all rejoicing, Waterman, Scherer, myself, the whole board of AUI, the Staff of the Observatory, etc."[155] Heeschen was apparently losing patience with the situation and wrote to Rabi, "I do not intend to take care of all the administrative dirty work at the expense of my own radio astronomy interests and then be simply a rubber stamp to either the NSF or a distant director-to-be on the interesting things."[156]

Expressing his frustration with the 140 Foot situation, Rabi had no patience for Struve's complaining, and after receiving Pawsey's acceptance, Rabi sent a handwritten letter to Taffy Bowen, congratulating him on the recent dedication of the Parkes 210 foot antenna and commenting that "Pawsey is saving my life by coming as director of NRAO next year. The present incumbent although a great optical astronomer has no administrative talent and no knowledge of radio astronomy. We hope now we are off to a better start."[157] Rabi's rather disparaging remarks were clearly unfounded and inappropriate. The problems with Green Bank and the 140 Foot Telescope project preceded Struve's appointment and, in any event, AUI had barred Struve from any role in the 140 Foot project. Moreover, aside from his distinguished career in astronomical research, Struve had been the very effective director of the University of Chicago Yerkes Observatory. According to Osterbrock (1997, p. 159) "Otto Struve resurrected the Yerkes Observatory." He founded the McDonald Observatory in Texas and served simultaneously as a strong Director of the McDonald and Yerkes Observatories (Evans and Mulholland 1986). As mentioned earlier, he was a strong advocate of large astronomical facilities. Dave Heeschen (2008) later described Struve as a "fine person and a great astronomer," who had an important impact on NRAO, but that "his final years were made so stressful by the problems with the 140 foot telescope that he had inherited."

As discussed in Sect. 4.5, after Heeschen and Findlay's proposal to build the 300 Foot dish had been rebuffed in 1958 by the AUI Visiting Committee and the NSF, it was Otto Struve who was able to convince both AUI and the NSF to fund the construction during the depths of despair about the lack of progress with the 140 Foot Telescope. It was also Struve who obtained the money and solicited donations of privately held books and journals for the NRAO library, which today arguably holds the largest collection of any radio astronomy library in the world, including complete runs of many journals such as *Nature*. Before the days of the Internet, the presence of a comprehensive scientific library had an immeasurable impact on scientific life in isolated Green Bank. In 1961, during a time of widespread global tensions, Struve also organized a US-USSR conference on radio astronomy that was held in Green Bank, opening a door which led a few years later to the decades-long NRAO-USSR collaboration in very long baseline interferometry (Sect. 8.2). After leaving NRAO, Struve held visiting positions at Princeton and Caltech, then returned to Berkeley, where he continued to work at the University of California until his death on 6 April 1963.

Joe Pawsey Appointed NRAO Director Although Joe Pawsey had been the founder and leader of the very productive radio astronomy group at the CSIRO Radiophysics Laboratory, he, Bernard Mills, and Wilbur (Chris) Christiansen were irritated with Bowen's focus on the large expensive programs to build the Parkes 210 foot telescope and Wild's solar heliograph at Culgoora (Robertson 1992, p. 198) at the expense of the more traditional Radiophysics innovative programs that had been pursued by the small groups. Two years earlier, fearing that Mills might leave Radiophysics, Pawsey had written to Struve to suggest a joint program to build a Mills Cross at Green Bank, noting that Sydney had the expertise and NRAO had the money.[158] However, in 1960, both Mills and Christiansen left Radiophysics to accept professorships, in physics and electrical engineering departments respectively, at the University of Sydney. Pawsey was particularly angered by the decision, which was made without consulting him, to appoint Bolton as head of the Parkes radio telescope. Pawsey had been made irrelevant by Bowen (Robertson 1992, p. 200; 2017, p. 227) and was open to Rabi's offer. Dave Heeschen (2008) and the NRAO staff were pleased with the selection of Pawsey and were looking forward to his coming to Green Bank as the NRAO Director.

As agreed, after stopping off in Pasadena to meet with Struve, Pawsey met with Rabi at the Princeton home of his brother-in-law, then attended the 16 March 1962 meeting of the AUI Executive Committee, after which he traveled to Green Bank with AUI Vice President Gerald Tape. One morning, about a week after he had arrived in Green Bank, Campbell Wade observed that Pawsey was dragging his foot,[159] and Frank Drake (Drake and Sobel 1992, p. 26) saw that Pawsey "was partially paralyzed on the left side of his body." Following medical evaluation in Washington, Rabi reported that Pawsey appeared to improve, and "there was every reason to suppose that his recovery will be sufficient to permit him to assume full-time duty at the Observatory on October 1 as planned."[160] However, at the 18 May 1962 meeting of the AUI Executive Committee, Rabi reported that Pawsey's illness "had taken a decided turn for the worse," and, "he is now at the Massachusetts General Hospital under the care of [AUI Trustee] Dr. William Sweet, and that on May 16 he was operated on for a brain tumor. The result of the operation is still uncertain."[161] In an executive session, Rabi expressed the opinion that "The chance that Dr. Pawsey's health will ever get to be sufficiently good to permit him to assume the duties of Director appears to be negligible," and "it will be necessary to find a new Director for the National Radio Astronomy Observatory." Rabi's sober appraisal of Pawsey's condition was confirmed the following month by Dr. Sweet.[162]

Pawsey remained optimistic. While still recovering from the surgery at Massachusetts General Hospital, he wrote a letter for Tape to read and deliver to Rabi following up on their earlier March discussion about his ideas on achieving high angular resolution.[163] Although he had hoped to return to Green Bank to discuss his plans for the Observatory, after his discharge from the hospital on 12 July 1962 Pawsey went to Princeton to recuperate with his sister and brother-in-law. Pawsey and Tape met in Princeton on 18 July, together with Pawsey's wife Lenore and his brother-in-law. Dave Heeschen was also present, and they all agreed that Pawsey should not undertake the NRAO Directorship, but while in Australia, Pawsey would "keep a hand in NRAO programs," and in particular, "the possibility of a cooperative program with CSIRO."[164] Before leaving for Australia, Pawsey prepared a short report outlining his 'general objectives in coming to Green Bank.[165] In his report, Pawsey outlined the following priorities for NRAO:

a) Develop a first-class scientific team
b) Provide extremely powerful radio astronomy equipment
c) The stimulation of US research in radio astronomy

Pawsey discussed two major Green Bank projects: completion of the 140 Foot Telescope, and a project which he said "is not well defined," but which he described as "the high resolution project" with the specific objective of "furthering the study of radio galaxies."[166] Interestingly, Pawsey gave higher priority to the high resolution project than to completing the 140 Foot antenna, but he also commented on a possible future solar program and, with great insight, he endorsed the plans by Frank Drake and Frank Low to develop a millimeter wavelength capability, as well as noting the opportunities for low-frequency radio astronomy.

Accompanied by his wife, Lenore, and Paul Wild from the Radiophysics staff, Pawsey returned to Australia on 27 July 1962.[167] Wild had been in the US for the previous two weeks and had visited Green Bank. Before their departure Rabi tried to recruit Wild to come to Green Bank as the NRAO Director but Wild declined, citing his opportunities and obligations at CSIRO.[168] AUI also considered Robert Hanbury Brown and Henk van de Hulst as possible NRAO directors. But it was recognized that Hanbury Brown was committed to building his intensity interferometer in Australia, and that van de Hulst would not be an effective director at an American national observatory.[169] In November 1962, both Heeschen and Tape made a long planned visit to Australia, where they reviewed the Australian radio astronomy and nuclear research programs, especially the work with the newly completed Parkes radio telescope, and they were able to visit with Pawsey before his death in Sydney on 30 November 1962.

Dave Heeschen Becomes NRAO Director There was little option left for Rabi but to appoint David Heeschen as the Director of NRAO, effective 19 October 1962. Again, there was no search committee or consultation within the community, although Rabi's selection was strongly supported by the AUI Board. Heeschen was only 36 years old, but had been involved with NRAO from the very beginning, first as a consultant, then as an employee of AUI, later as Head of the NRAO Astronomy Division, and finally, since 1 December 1961, as Acting Director of NRAO. In 1960, he received tenure from AUI, but was almost drawn away from NRAO by an attractive offer from the University of Virginia as a Full Professor, Chairman of the Astronomy Department, and Director of the University of Virginia Observatory.[170]

But Rabi had appointed Heeschen over the older and more senior John Findlay, who had originally been recruited by Berkner. Findlay was furious, and never forgave Rabi, who apparently disliked Findlay, and the relationship between Findlay and Heeschen became strained. Findlay developed a drinking problem, and although he continued to take a leadership position in the planning for the long-awaited, large fully steerable telescope (Sect. 9.4), and in the initial planning for the 36 foot millimeter telescope (Sect. 10.2), he was gradually relieved of his responsibilities as Deputy Director, although Heeschen later claimed that he never had the nerve to fire or replace him.[171]

4.7 EXODUS FROM GREEN BANK

The site chosen for NRAO is actually situated between Green Bank and the even smaller village of Arbovale. From the earliest years, NRAO has used a Green Bank postal address, but the telephone exchange was in Arbovale. For reasons which remain lost to history, the Observatory came to be associated with Green Bank rather than Arbovale. Perhaps "Green Bank" sounded more colorful than "Arbovale," or perhaps AUI was trying to draw an analogy with "Jodrell Bank." Located in Pocahontas County, only about five miles from the Virginia border, local loyalties during the American Civil War were mixed, and the sympathies of the two villages leaned in opposite directions. The local churches were the center of political debate and, as a result of the allegiances founded in the Civil War, there remain today separate United Methodist Churches in Green Bank and Arbovale, located less than two miles apart and sharing the same pastor.

According to the Pocahontas County Historical Society, there is no evidence that "Pocahontas ever set foot in present-day Pocahontas County," but when the then-Virginia county was established in 1821, the Governor of Virginia was Thomas Randolph, who was a direct descendant of Pocahontas. In the early part of the twentieth century, during the heyday of the lumber industry, the small town of Cass, located about ten miles from the Observatory, was a thriving town with a population of more than 2000, which, by the late 1950s had declined to less than 400. The local sawmill was once the largest double band sawmill in the world, producing 1.5 million board-feet of lumber per week. In 1911, West Virginia had more than 3000 miles of logging railroad line, more than any other state in the country. By 2010, only the 11 mile long Cass Scenic Railway line remained. Although remote by almost any measure, a local business proprietor once boasted that half of the people in the United States lived within a day's drive of Cass. Nearby were the colorfully named communities of Clover Lick and Stony Bottom. At the time NRAO arrived in 1956, there were no bars in West Virginia. Alcohol could not be served in public except in restaurants and private clubs. Each county was allowed a liquor store, but the hours were kept limited. Frank Drake later recalled asking a Cass local if he knew where the liquor store was and being told, "Yup, but I ain't gonna tell ya." (Drake and Sobel 1992). On another occasion, after a party at his home, an NRAO scientist had put the empty bottles out in the trash, only to find them removed the next morning. After some investigation, his local cleaning lady confessed that she was so embarrassed that his neighbors might find out that there had been drinking in his house that she hid the bottles. There were no secrets in Green Bank. All telephone calls had to go through an operator in Cass, who, if she wished, could, and probably did, listen in to phone calls going through her exchange; those who shared one's multi-party telephone line undoubtedly listened as well.

In 1996, Dave Heeschen (1996) gave the following colorful description of life in Green Bank when he arrived in 1956 with his wife, Eloise, and their young children.

In 1956, Green Bank and Arbovale were small villages relatively isolated from the outside world by the mountains and the poor roads. The local residents were almost all descended from a few original families, and spoke a form of English that had been frozen in time by the isolation of the area. That and their distinctive accent made them difficult to understand at times. Like all true Americans anywhere in the world, when they weren't understood they simply spoke faster and louder. Farming and hunting were the only activities, the latter carried out with much more enthusiasm than the former, and with almost total disregard for formal hunting seasons and the laws. In fact, about the only recognition of the existence of hunting laws was that the local schools closed on the opening day of deer season. Green Bank consisted of two small stores, a post office, an Oldsmobile dealership, and a few homes. Arbovale was smaller, having one store with a post office in the store. With a few exceptions the locals were very friendly and welcomed the coming of the Observatory. Many looked at it as a potential source of badly needed jobs.

As most of the NRAO scientific and technical staff came from more urban environments, they, and particularly their families, often did not easily fit into the Green Bank culture. Shopping, medical, and dental facilities were a 40- to 60-minute drive from the Observatory, although a local Arbovale general store carried a wide, yet limited, range of supplies, from food to hardware to equipment needed to castrate bulls. AUI provided funds to help subsidize a local doctor by contracting to provide annual physical exams for all of the resident staff. However, it was even more difficult to recruit physicians than scientists and engineers. At one point a married physician pair were recruited from Norway, but they only lasted a few years. There were long periods when the only local doctor was an osteopath (DO) rather than an MD. More than one child was born on the way to the nearest hospital in Elkins, 50 miles distant over a sometimes treacherous Cheat Mountain.

Winters in Green Bank were cold and the ground could be covered by snow for many months. The only guaranteed frost-free month was July. Opportunities for employment for observatory wives were limited, although some spouses found satisfying work teaching in the local schools.[172] Some parents felt that the local school system was inadequate for their ambitious children, although in fact many of the discipline problems common in larger city school systems were not found in Green Bank. Particularly in the primary grades, there was strong emphasis on the fundamentals. However, owing to the relatively small fraction of the local children that went on to universities, the opportunity to take the advanced level courses thought to be needed for acceptance to elite universities was limited. Many of the local children, when legally allowed, left school to raise families and work. Most of the local children were on reduced-cost lunches, reflecting the generally low income levels in the community.

When she was the Green Bank site director from 1981 to 1983, Martha Haynes frequently told colleagues that the nearest McDonald's was an hour away—over a mountain. The nearest TV stations were also very distant, and the Observatory site had been specifically chosen because it was shielded by the surrounding mountains. Even when using the largest Radio Shack antennas, picture quality from Roanoke and Pittsburgh TV stations was at best marginal.

If incoming staff were required to build or buy homes, it could be a dangerous investment, particularly for scientific staff on term appointments. Unlike most areas of the country where the population was growing, the population of Pocahontas County was decreasing due to the lack of employment opportunities. NRAO reluctantly built 24 houses on the site, mostly in an area that became known as "the rabbit patch"[173] and provided them for staff at a very modest rent. Although intended in principle for young scientists on term appointments, in practice they were also occupied by tenured scientists, engineers, and administrative staff on indefinite appointments. AUI also shared the financial risk with those who chose to build their own homes, and partly guaranteed the mortgages so that local banks could provide low interest loans. In an effort to compensate for the limited medical support in the area, AUI purchased an ambulance and trained staff to serve as EMTs. But even with this additional support, it remained difficult to recruit or retain scientific, and particularly engineering, staff.

In some respects, Green Bank could be an attractive place to live and work, especially if one enjoyed outdoor recreation activities. Opportunities for hunting and fishing were excellent. NRAO families often got together for weekend hikes in the surrounding mountains. A small group of scientists and engineers became active spelunkers and enjoyed investigating the local limestone caves. As in many small towns, local school sports were very popular and often the focus of community activity. One could walk, bicycle (and in the winter, ski) to work if you lived in one of the nearby Observatory houses. Observatory rents were attractive, and maintenance was readily available from the Observatory maintenance staff. Children were free to play and ride bicycles on Observatory grounds without any concerns by them (or their parents) for traffic or other urban dangers. Starting in 1962, NRAO operated a kindergarten on the site. Costs were shared by the NSF, AUI, and the parents.

Nevertheless, the lack of social, medical, and educational opportunities soon became apparent. Staff members, and especially their families, were not happy. Some, like John Findlay, chose to live elsewhere and make the long commute home on weekends. Others, such as Frank Drake, left NRAO to pursue opportunities elsewhere. Interestingly, although there were a few hardy Americans (including one of the present authors—KIK), it was particularly hard to hire American scientists and engineers, who wanted the ambiance of universities and cities. In the mid-1960s, the resident scientific staff in Green Bank had come largely from Germany, the Netherlands, Sweden, Poland, Norway, France, India, and Iraq. By 1962, NRAO was beginning to plan for the design and construction of the VLA and the 36 foot millimeter telescope, and it did not make sense to have either the engineering staff for these projects, or the growing administrative staff, isolated in remote Green Bank. Dave Heeschen made the difficult decision that if NRAO were to maintain its viability, it would be necessary to move the NRAO headquarters from Green Bank to a more desirable location. But who should move and where should they go?

One of the successes of the Green Bank operation was the close collaboration between the scientists and engineers, facilitated, at least in part, by virtue of the fact that they and their families lived, worked, and played together. In particular, the NRAO scientists and engineers worked closely together designing and commissioning new instrumentation. In principle, the scientists might be able to live away from the telescopes in Green Bank, but then they would lose contact with the engineers. There was no distinction at NRAO between design engineers and those who maintained the equipment. An engineer who designed and constructed a piece of equipment was responsible for its operation. If it failed in the middle of the night, the engineer was expected to show up at the telescope and fix it. So there was no obvious way to keep some engineers in Green Bank and not others. Moreover, it was the engineers more than the scientists that motivated moving from Green Bank. The scientific opportunities for research with the unique NRAO facilities in Green Bank might have been sufficient to attract members of the scientific staff, but the same was not true for engineers. Although there were attractive opportunities for engineering research and development at NRAO, salaries were not competitive with the rapidly expanding, attractive opportunities in the military-industrial complex characteristic of the Cold War period.

In May 1962, while still only the Acting Director, Heeschen raised with AUI the idea of moving the NRAO headquarters from Green Bank. Trying to minimize the impact of a split operation, he looked for nearby possibilities, and considered Charlottesville, VA, home of the University of Virginia (UVA) and Morgantown, WV, home of West Virginia University. Both cities were about a two and a half-hour drive from Green Bank. Charlottesville seemed to offer better living conditions and had the stronger university.[174] Moreover, the West Virginia University did not have an astronomy program. Although UVA had one of the oldest Astronomy Departments in the country and operated a 26 inch telescope that had been the gift of Leander McCormick, the department languished for many years while the university pursued excellence in the humanities and related areas. In 1960, Heeschen had seriously considered the offer from UVA to head the astronomy department, so he was familiar with UVA and with Charlottesville.

After a long search, UVA had finally hired Larry Fredrick in 1962 to rejuvenate the Astronomy Department and the McCormick Observatory. Shortly after Fredrick arrived at UVA, Heeschen called him on the phone to ask if "they could chat."[175] When he arrived in Fredrick's office, Heeschen closed the door and hesitatingly asked if Fredrick would object if he brought the NRAO headquarters and the scientific staff to UVA. Fredrick was enthusiastic and set up a meeting with UVA President Edgar Shannon, who indicted that UVA could provide the land and attractive financial conditions to construct a building for NRAO on the UVA Grounds. A local bank indicated they would finance mortgages for staff purchasing homes in Charlottesville. Everything was set, but Heeschen needed to first convince AUI, who five years earlier had gone out on a limb with the NSF and Merle Tuve to "build an extensive community in the deep woods." (Sect. 3.3).

Heeschen's recommendation to move the NRAO headquarters from Green Bank turned out to be very controversial. At its 17 May 1963 meeting, the AUI Executive Committee debated the arguments pro and con for moving the NRAO headquarters to Charlottesville. The Trustees were split on whether or not to move, and those who did support a move were split on the relative merits of remaining close to Green Bank or locating at a prestigious university. Some Trustees considered easy access to Green Bank as the strongest factor in determining the best location; others argued that "a congenial and stimulating environment for the scientific staff ... should be the primary consideration." AUI President Gerald Tape expressed concern that if the new headquarters became part of a university "there would be little reason for the AUI-NRAO mechanism," and that the Observatory might "easily deteriorate into simply one more radio astronomy laboratory," and lose its identity as a National Observatory. The Trustees concluded their meeting with a request to Heeschen "to prepare for consideration by the Trustees a detailed statement of operating plans under a split location arrangement," saying that "considerations should be given to several possible headquarters locations."[176]

Although his presentation to the AUI Executive Committee was scheduled for the afternoon of 18 July 1963, a small subgroup of Trustees asked Heeschen to be available in the morning for "a full discussion" and "following that distribute his memorandum with whatever revisions the morning discussion has produced."[177] The formal presentation and discussion stretched over two days. Instead of presenting several possible locations as instructed, Heeschen opened his four page report with, "I recommend that NRAO establish offices in Charlottesville, Virginia and that scientists and certain other members of the staff be transferred there."[178] After summarizing the need to move, his report weighed the advantages of a "stimulating and intellectual environment" against "closeness to Green Bank," and argued that it was essential to be close to Green Bank. He also noted that the planned VLA and 36 foot millimeter telescopes would mean that Green Bank would be less important as the administrative, technical, and scientific headquarters for a broadly geographically dispersed Observatory. After extensive discussion among the Executive Committee, Ed Reynolds, who had taken over the AUI Presidency from Tape, appointed a committee to guide a detailed study of the proposed move. The committee was chaired by Carl Chambers and included three members of the Board plus Emanuel R. (Manny) Piore from IBM.

The committee supported Heeschen's suggested move to Charlottesville, and there was a general consensus among the committee members and the Trustees to accept Heeschen's recommendation, although "some Trustees still voiced objection to making any move from Green Bank."[179] Of those that agreed to the move, many favored a location where the staff would find the intellectual stimulation that they felt was lacking in Green Bank. Princeton and Cambridge were touted as providing "all the social amenities and also the intellectual stimulation to be derived from close association with a major university maintaining a vigorous program in astronomy," and it was suggested that, by comparison, Charlottesville lacked "a congenial intellectual environment, ... and might seem a rather isolated site intellectually."[180] Rabi, who on his last day as AUI President, had appointed Heeschen as NRAO Director, was particularly adamant about the merits of Princeton or Columbia as the best location for the NRAO headquarters. Heeschen, while fully cognizant of the advantages of co-locating at a university, particularly one with vibrant physics and astronomy departments, was more concerned about how to keep the staff engaged in Green Bank activities, and looked for a home for the NRAO Headquarters closer to Green Bank. But some Trustees maintained that "modern transportation greatly diminishes the importance of geographical proximity."

Fed up with the debate and procrastination, Heeschen characteristically pointed out the need for an "immediate decision," but a motion and second to approve the move from Green Bank was withdrawn after further discussion. Instead the Executive Committee decided unanimously "that the question of establishment at Charlottesville, Virginia of a headquarters for the scientific staff of the Observatory and the approval of the Director's proposal for carrying out this plan be referred to the full AUI Board of Trustees at its annual meeting to be held on October 18, 1963."[181]

The full AUI Board continued the debate when they met in Green Bank on 18 October. Reynolds, Chambers and Manny Piore spoke in favor of the move, but others argued against the move, which they felt would be "a dissolution of the Observatory as an institution." Finally, the Trustees voted to move the headquarters to Charlottesville, but the minutes recorded that the decision was made only by majority vote, and unlike most AUI Board actions, was not unanimous.[182] According to Heeschen, Rabi was so upset with Heeschen's insistence on moving to Charlottesville rather than Princeton or Columbia that he walked out of the meeting before the critical vote to wander around the Observatory, and that Rabi did not speak to Heeschen or Piore again for two years. Heeschen also described his experience with AUI over the move, as the most disagreeable dealing he ever had with AUI during his 17 years as the NRAO Director.[183]

In trying to understand Rabi's dogmatic opposition to locating the Observatory Headquarters at Charlottesville's University of Virginia, it might be noted that at the time, the University of Virginia was essentially all white and all male. Moreover, only three years earlier, the city of Charlottesville, as part of Virginia's broader "Massive Resistance," had closed the city's public schools to protest the 1954 Brown v. Board of Education Supreme Court decision against school segregation. As seen by Rabi, a Jewish immigrant from

Eastern Europe, Virginia was part of the still-segregated South. The records do not indicate which of the other Trustees also spoke in opposition to Charlottesville, but many of the AUI Trustees of that period shared a similar background with Rabi.

Having persuaded AUI, Heeschen still had to deal with the NSF, which in turn had to deal with the WV Congressional delegation. Fortuitously, the NRAO budget included a $600,000 item to build a new laboratory building, and the Foundation agreed to a proposal from UVA to divert this sum toward the new Charlottesville building. UVA then built a 25,000 square foot building, which was leased to AUI at a yearly rental rate so that the construction cost was amortized over a period of five years, and that after this period, the rent was reduced to a level covering only maintenance. The new headquarters building, which became known at UVA as Stone Hall, was finished in December of 1965, and most of the tenured and tenure track scientific staff, along with many engineers, technicians, and administrative staff, moved from Green Bank to Charlottesville (Fig. 4.13). With the large staff increase associated with the VLA, and growing technical, administrative and human resources staff, NRAO soon had outgrown the new UVA building and in November 1972, NRAO had to rent additional space in Charlottesville to house the technical staff. Although the engineers were located only a few miles away, the sociological impact of the separation of the scientific and technical staff became an increasing concern. Historically, there had been close collaborations between the

Fig. 4.13 In this cartoon from the December 1955 *Observer*, the Green Bank telescopes are weeping and waving as the moving vans head east to the big city of Charlottesville. Credit: NRAO/AUI/NSF

NRAO scientists and engineers in developing and testing new instrumentation, and in the design of new telescopes. In April 2005, with the intention of bringing the engineers and scientists back together, the capacity of Stone Hall was doubled by adding a new wing. But soon after construction began in 2003, it was apparent that even the addition could not accommodate the growing number of new staff involved with Atacama Large Millimeter/submillimeter Array, and the technicians and engineers of the Central Development Laboratory, as it was then named, remained in a separate location.

NOTES

1. Sunderlin to diary, 23 January 1956, LOC-ATW, Box 26.
2. Bok to Pawsey, 31 July 1956, NAAustrl, C3830 Z3/1/VII. Miffed that he had been passed over as Director of the Harvard College Observatory, Bok had resigned his faculty position at Harvard (DeVorkin 2018). It isn't clear if his forceful "NUTS" reply to Waterman was based on the famous WW II reply of General Anthony McAuliffe to the German army general who demanded the American surrender in December 1944.
3. Berkner to Menzel, 21 September 1956, AUI-BOTXC, 21 September 1956. The other members of the committee suggested by Berkner were Ira Bowen, C. D. Shane, W. W. Morgan, Otto Struve, and J. B. Wiesner.
4. Menzel to Bowen, 3 October 1956, HLA-IB, Box 34.
5. Menzel to Search Committee, 18 October 1956, HLA-IB, Box 32.
6. Menzel to Search Committee, 22 October 1956, HLA-IB, Box 32.
7. Bok to Tuve, 3 July 1956, NAA-NRAO, Founding and Organization, Correspondence; CITA-JLG, Box 111, Folder 5.
8. Bok to Pawsey, 23 October 1956, NAAustrl, C3830 Z3/1/VII.
9. Report of the Ad Hoc Committee to Nominate the Director of the National Radio Astronomy Facility, 18 October 1956, HLA-IB, Box 34.
10. Goldberg to Berkner, 30 November 1956, HUA, Papers of Leo Goldberg, HUGFP 83.20.
11. Greenstein to Menzel, 6 March 1957, CITA-JLG, Box 23, Folder 4.
12. e.g., AUI-BOTXC, October 1958.
13. DSH, 31 July 1995, unpublished notes prepared for 140 Foot Birthday Symposium, September 1995, NAA-DSH, US Radio Astronomy History, Talks.
14. Minutes of AUI Board Meeting, 19 February 1960.
15. Struve to DSH, 9 February 1960, NAA-NRAO, Founding and Organization, Antenna Planning, Box 1.
16. The initial members were A.J. Deutch (MWPO), W.E. Gordon (Cornell), F.T. Haddock (Michigan), A.B. Meinel (NOO), J.B. Wiesner (MIT), E.L. McClain (NRL), G.C. McVittie (Illinois), D.H. Menzel (Harvard). Merle Tuve graciously declined to participate (Tuve to Berkner, 18 December 1956, CITA-JLG, Box 113, Folder 1).
17. Sullivan interview with Findlay, 14 August 1981, NAA-WTS, Individuals. https://science.nrao.edu/about/publications/open-skies#section-4
18. Jack Campbell only came to NRAO in 1972 as an Electronics Engineer to work on the VLA in Socorro. All of the other former Harvard students started

out on the NRAO Scientific Staff in Green Bank, but in 1965 most moved to the new NRAO headquarters in Charlottesville, VA.

19. KIK, ENB, and David Hogg interview with DSH, 31 May 2011, NAA-KIK, Oral Interviews. https://science.nrao.edu/about/publications/open-skies#section-4

20. Minutes of the Meeting of the AUI Advisory Committee on Radio Astronomy, 16–17 October 1956, NAA-NRAO, Founding and Organization, Meeting Minutes.

21. Both the official program for the dedication and the NSF press release included the words "National Radio Astronomy Observatory" in their titles.

22. DSH to Berkner, 12 November 1957, NAA-NRAO, Founding and Organization, Correspondence.

23. Undated 1957 report from DSH, "The Research Programs of the NRAO," NAA-NRAO, Founding and Organization, Planning Documents.

24. National Radio Astronomy Observatory, Site Development Program, December 1957, NAA-NRAO, Founding and Organization, Planning Documents.

25. Emberson to DSH, 21 April 1958, NAA-NRAO, Green Bank Operations, 85 Foot Tatel Telescope, Box 1.

26. Emberson to file, 8 July 1958, NAA-NRAO, Green Bank Operations, 85 Foot Tatel Telescope, Box 1.

27. DSH, NRAO Internal Symposium, 12 June 1992, NAA-DSH, US Radio Astronomy History, Talks.

28. DSH to Observers, 30 June 1960, NAA-NRAO, Green Bank Operations, 85 Foot Tatel Telescope, Box 2.

29. The possibility of detecting the Zeeman splitting of the 21 cm hydrogen line due to the galactic magnetic field was first discussed by Bolton and Wild (1957).

30. Technical and popular accounts of the planning and early history of the NRAO 140 Foot Radio Telescope are given by Emberson and Ashton (1958), by Emberson (1959), and by Small (1965). Personal accounts of the complex story surrounding the design and construction of the 140 Foot Telescope are given by Heeschen (1996, 2007a, b, 2008). Much of this section is adopted from Heeschen's 31 July 1995 unpublished notes prepared for 140 Foot Birthday Symposium, September 1995, NAA-DSH, US Radio Astronomy History, Talks. The story of the 140 Foot construction is portrayed in a movie which can be seen at https://vimeo.com/120155652

31. Letter written by Grote Reber, 7 October 1957, cited by Emberson and Ashton (1958).

32. DSH to Keller, 2 December 1957, NAA-DSH, Radio Astronomy History, NRAO.

33. DSH, 31 July 1995, op. cit.

34. AUI-BOTXC, 17 June 1955.

35. AUI-BOTXC, 14 July 1955.

36. These numbers are contained in Appendices A-1, A-2, and A-3 to the 11–12 July 1955 AUI Steering Committee report, AUI-BOTXC, 14 July 1955.

37. *Plan for a Radio Astronomy Observatory*, p. 34, NAA-NRAO, Founding and Organization, Planning Document. https://science.nrao.edu/about/publications/open-skies#section-4

38. Engineering Design Objectives for a Large Radio Telescope, Vol. 1 pg. 1, CAMROC, NAA-JWF, LFST.

39. A contract let with the MIT Servomechanisms Laboratory and inquiries to commercial suppliers such as Goodyear Aircraft did not provide any encouragement that an alt-az mount and mechanical coordinate converter could provide the desired precision.

40. Tuve to Seeger, 23 February 1956, CITA-JLG, Box 111, Folder 6.

41. AUI-BOTXC, 17 February 1956.

42. Greenstein to Tuve, 6 March 1956, CITA-JLG, Box 111, Folder 6.

43. Seeger to Berkner, 22 March 1956, LOC-ATW, Box 26.

44. Minutes of the AUI Advisory Committee Meeting, 26–27 March 1956, NAA-NRAO, Founding and Organization, Planning Documents.

45. It was widely suspected that the obvious ripples in the 250 foot antenna surface were caused by the heat generated during the welding process, which caused a thermal deformation of the dish surface.

46. The radio emission from planetary nebula and H II regions is due to thermal processes and is typically stronger at shorter wavelengths.

47. Emberson to Sheppard, 16 April 1956, NAA-NRAO, Founding and Organization, Correspondence.

48. Tuve to Emberson, 14 March 1956, NAA-NRAO, Founding and Organization, Correspondence.

49. KIK interview with Frank Drake, 14 September 2010, NAA-KIK, Oral Interviews. https://science.nrao.edu/about/publications/open-skies#section-4

50. Tuve to A. H. Jackson, Vice President, Blaw-Knox, 25 July 1956, NAA-NRAO, Founding and Organization, Correspondence; CITA-JLG, Box 111, Folder 5.

51. Bok to DSH, 8 June 1956, NAA-NRAO, Founding and Organization, Correspondence.

52. Emberson to Haddock, 1 August 1956, NAA-NRAO, Founding and Organization, Correspondence.

53. AUI Advisory Committee on Radio Astronomy, Minutes of 16–17 October 1956, NAA-NRAO, Founding and Organization, Meeting Minutes.

54. AUI-BOTXC, 18 October 1956.

55. Minutes of the AUI Advisory Committee on Radio Astronomy, 16–17 October 1956, NAA-NRAO, Founding and Organization, Meeting Minutes. See also Emberson to Ad Hoc Committee, 31 October 1956, NAA-DSH, Radio Astronomy History.

56. AUI-BOTXC, 16–17 October 1956.

57. DSH, 31 July 1995, op. cit.

58. KIK, ENB, and David Hogg interview with DSH, 31 May and July 2011, NAA-KIK, Oral Interviews. https://science.nrao.edu/about/publications/open-skies#section-4

59. AUI-BOTXC, 21 September 1956; Minutes of Meeting of the AUI Advisory Committee on Radio Astronomy, 16–17 October 1956, NAA-NRAO, Founding and Organization, Meeting Minutes.

60. Minutes of 28 June 1957 meeting at the NSF, NAA-NRAO, Founding and Organization, Meeting Minutes.

61. Estimated Costs of Construction, 3 January 1958, NAA-NRAO, Green Bank Operations, Site Selection and Development.

62. Ibid.
63. Minutes of Pre-Proposal Conference on 140 Foot Equatorial Radio Telescope, 25 September 1957, NAA-NRAO, Founding and Organization, Meeting Minutes.
64. AUI-BOTXC, 15 November 1957; See also Condensed History of the 140-Foot Radio Telescope Project, 5 January 1961, NAA-DSH, Radio Astronomy History, NRAO.
65. Berkner to Bronk, 12 February 1958, appended to AUI-BOTXC minutes, 14 February 1958.
66. Record of Meeting at the NSF, 13 March 1958, appended to minutes of AUI-BOT, 18 April 1958.
67. Findlay undated 1959 memo, NAA-NRAO, Founding and Organization, Correspondence.
68. AUI-BOTXC, 14–15 July 1960.
69. Struve memo, 12 September 1960, appended to AUI-BOTXC minutes, 23 September 1960.
70. Berkner memo to file, 2 September 1960, appended to AUI-BOTXC minutes, 23 September 1960.
71. Ibid.
72. As reported in the minutes of the 20–21 October meeting of the AUI-BOT, Berkner had announced his resignation at an executive session of the 28 September meeting of the AUI-BOTXC. As instructed by the Chairman, the Secretary did not record the results of the 28 September executive session.
73. AUI-BOT, 21 October 1960.
74. AUI-BOTXC, 23 September 1960.
75. Berkner to Burchill, 22 September 1960, appended to minutes of AUI-BOTXC, 23 September 1960.
76. AUI-BOTXC, 23 September 1960.
77. Luton to Burchill, 1 September 1960, appended to minutes of AUI-BOTXC, 23 September 1960. Luton was the NSF Assistant Director for Administration.
78. AUI-BOTXC, 23 September 1960.
79. Waterman to Berkner, 5 October 1960, appended to minutes of AUI-BOT, 21 October 1960; also found in NAA-DSH, Radio Astronomy History, NRAO.
80. Other members of the Committee included metallurgical experts from Chicago Bridge & Iron Co., Bethlehem Steel, Battelle Memorial Institute, and US Steel. Bruce Rule from Caltech later joined the committee after several members resigned following the first meeting.
81. AUI-BOT, 20–21 October 1960.
82. Ibid.
83. AUI-BOTXC, 18 November 1960.
84. Ibid.
85. DSH, 31 July 1995, unpublished notes prepared for 140 Foot Birthday Symposium held September 1995, NAA-DSH, Talks.
86. D.R. Lord to President Eisenhower's Science Advisor, G.B. Kistiakowski, 8 September 1960, DDE, Radar and Radio Astronomy, Box 5, Records of the U.S. President's Science Advisory Committee.
87. DSH, 31 July 1995, op. cit.
88. Berkner to Kinzel, 17 October 1960, appended to minutes of AUI-BOT, 21 October 1960; also found in NAA-DSH, Radio Astronomy History, NRAO.

89. Berkner to Haworth, 31 October 1960, appended to AUI-BOTXC, 18 November 1960.

90. Lindberg to Emberson, 14 November 1960, appended to AUI-BOTXC, 18 November 1960.

91. Lindberg to Haworth, 6 December 1960, appended to AUI-BOTXC, 6 December 1960.

92. AUI-BOTXC, 16 December 1960; 19 January 1961.

93. AUI-BOTXC, 17 February 1961.

94. Haworth to Waterman, 29 December 1960, appended to minutes of AUI-BOTXC, 19 January 1961.

95. AUI-BOTXC, 18 November 1960.

96. Ibid.

97. AUI-BOTXC, 17 February 1961; Luton to Haworth, 23 February 1961, appended to minutes of AUI-BOTXC, 17 March 1961.

98. AUI-BOTXC, 20 April 1961.

99. Reynolds to Lindberg, 6 April 1961, appended to minutes AUI-BOTXC, 20 April 1961.

100. AUI-BOTXC, 20 April 1961.

101. Design Responsibility 140 ft. Radio Telescope, N. Cleveland to C. Davis, 18 February 1963, NAA-DSH, Radio Astronomy History, NRAO.

102. AUI-BOTXC, 19 May 1961.

103. Drake to Green Bank staff, 2 May 1961, NAA-DSH, Radio Astronomy History, NRAO.

104. DSH, Frank Drake, Roger Lynds, Kochu Menon, and Campbell Wade, all signed the letter to Struve, 9 May 1961, NAA-DSH, Radio Astronomy History, NRAO.

105. Struve to Rabi, 10 May 1961, NAA-DSH, Radio Astronomy History, NRAO.

106. AUI-BOTXC, 19 May 1961.

107. Struve to Rabi, 10 May 1961, op. cit.; Struve to Chambers, 16 May 1961, NAA-DSH, Radio Astronomy History, NRAO.

108. DSH, 1995, op. cit.

109. AUI-BOTXC, 16 June 1961.

110. Ibid.

111. AUI-BOTXC, 20–21 July 1961.

112. Ibid.

113. AUI-BOTXC, 14 September 1961.

114. AUI-BOTXC, 26 October 1961.

115. AUI-BOTXC, 17 November 1961.

116. AUI-BOT, 27 October 1961.

117. AUI-BOTXC, 15 December 1961.

118. Ibid.

119. AUI-BOTXC, 16 February 1962.

120. AUI-BOTXC, 16 March 1962.

121. The sphere was poured by General Steel Industries. Westinghouse fabricated the polar and declination shafts and also machined the spherical bearing. The critical spring-actuated brakes were fabricated by the Goodyear Tire and Rubber Company, and gear segments by the Philadelphia Gear Company.

122. AUI-BOTXC, 21 September 1962.

123. AUI-BOTXC, 17 October 1963.

124. AUI-BOTXC, 16 January 1964; AUI-BOTXC, 21 February 1964.

125. *The 140 Foot Telescope Construction Story*, a 1965 film made for NRAO/AUI by Peter B. Good, describes the lift in detail. https://vimeo.com/120155652

126. The half power beamwidth of the 140 Foot Telescope is given approximately by θ (arcmin) ~ λ (cm) so at 1.3 the half power beamwidth was only a little over a minute of arc, and the pointing errors up to half a minute of arc made observations difficult.

127. Findlay notes for a 29 April 1988 invited talk at Ithaca, NY that apparently was not presented. NAA-NRAO, GB-GBT, Planning and Design, Box 1.

128. Burke to Vanden Bout, 21 July 1987, NAA-NRAO, Green Bank Operations, GBT Planning and Design.

129. Reber to Findlay, 2 April 1965, NAA-NRAO, Green Bank Operations, LFST, Box 7. https://science.nrao.edu/about/publications/open-skies#section-4

130. DSH after dinner talk on 12 January 1992, op. cit.

131. Drake to KIK, 4 May 2018, private communication.

132. Due to their limited angular resolution, especially for filled aperture instruments, the sensitivity of radio telescopes to detect weak sources can be limited due to the blending of separate sources contained within the antenna beam. This is commonly called "confusion." See Condon and Ransom (2016, p. 121) for more details on how confusion varies with angular resolution and observing frequency.

133. TRR to KIK, private communication.

134. Pauliny-Toth et al. (1978) were the first to exploit this previously unanticipated use of the 300 Foot antenna. A decade later the antenna collapsed while Jim Condon (2008) was using the same technique.

135. AUI-BOT, 21 October 1960.

136. Reynolds to Rabi, 23 August 1963, LOC-IIR, Box 36. Rabi at this time was Chair of the AUI Board.

137. KIK interview with DSH, 31 May 2011, NAA-KIK, Oral Interviews. https://science.nrao.edu/about/publications/open-skies#section-4

138. Report of the Trustees Committee on the National Radio Astronomy Observatory, 18 January 1961, appended to the minutes of AUI-BOTXC, 19 January 1961.

139. AUI-BOTXC, 19 January 1961.

140. AUI-BOTXC, 17 March 1961.

141. Bok to Pawsey, 16 June 1961, J.L Pawsey private papers, courtesy of Hastings Pawsey and Miller Goss.

142. Pawsey to Bok, 21 June 1961, J.L. Pawsey private papers, courtesy of Hastings Pawsey and Miller Goss.

143. Pawsey to Rabi, 5 October 1961, LOC-IIR, Box 36. When the Rabi papers at the LOC were first examined by one of the authors (KIK) in April 2011, this letter had been replaced by a pink sheet of paper stating that "The following item(s) have been removed from the collection because they contain security classified information." A handwritten note said "Confidential (Australia)." Nearly two years later, following a FOIA request and the intervention of Virginia Senator Mark Warner's office, the letter was finally declassified and released. Pawsey had probably typed the letter himself, as evidenced by the large number of typographical errors, and because he had made personal comments about the quality of various people, he had typed at the top of the letter

"<u>CONFIDENTIAL,</u>" underlined and in capital letters. Pawsey was travelling at the time, so he typed and mailed letter from the Australian Scientific Liaison Office in London, using their stationary with the official letterhead of the Office of the High Commissioner for Australia. Thirty years later, an LOC staff member processing Rabi's papers apparently saw the "CONFIDENTIAL" and the official Australian stationary and, knowing that Rabi dealt with highly classified material, unwittingly had the letter classified. It took nearly three years to get it unclassified.

144. AUI-BOTXC, 26 October, 1961.

145. Ibid. AUI Vice President T.E. Reynolds had been in contact with the NSF concerning the length of the contract renewal (Reynolds to Scherer, 30 September 1960, attached to the minutes).

146. AUI-BOT, 27 October 1961.

147. Struve to Rabi, 31 October 1961, Papers of Otto Struve, UCB Bancroft Library.

148. Rabi to Pawsey, 31 October 1961, J.L. Pawsey private papers, courtesy of Hastings Pawsey and Miller Goss.

149. Pawsey to Rabi, 9 November 1961, J.L. Pawsey private papers, courtesy of Hastings Pawsey and Miller Goss.

150. AUI-BOTXC, 17 November 1961, Report of the Executive Session.

151. Rabi to Struve, 17 November 1961, NAA-NRAO, Founding and Organization, Correspondence. https://science.nrao.edu/about/publications/open-skies#section-4

152. Rabi to NRAO Staff, 17 November 1961, NAA-NRAO, Founding and Organization, Correspondence. https://science.nrao.edu/about/publications/open-skies#section-4

153. White to Pawsey, 22 November 1961, White to Rabi, 28 November 1961, J.L. Pawsey private papers, courtesy of Hastings Pawsey and Miller Goss.

154. Pawsey to Rabi, 26 November 1961, NAA-NRAO, Founding and Organization, Antenna Planning. https://science.nrao.edu/about/publications/open-skies#section-4

155. Rabi to Pawsey, 5 December, 1961, NAAustrl, C3830 Z1/14/A/1.

156. DSH to Rabi, 21 December 1961, NAA-NRAO, Fiscal and Business Services, Budgets, Box 1.

157. Rabi to Bowen, 2 December 1961, NAAustrl, C3830 Z1/3/1/VIII; quoted in Robertson (1992, p. 202).

158. Pawsey to Struve, 25 September 1959, NAAustrl, C3830 Z3/1/IX.

159. KIK interview with C. Wade, 29 December 2003, NAA-KIK, Oral Interviews.

160. AUI-BOTXC, 19 April 1962.

161. AUI-BOTXC, 18 May 1962.

162. AUI-BOTXC, 15 June 1962.

163. Pawsey to Rabi, 27 June 1962, J.L. Pawsey private papers, courtesy of Hastings Pawsey and Miller Goss.

164. G. Tape, Notes on Pawsey for Discussions with Wilde [*sic*], 25 July 1962, NAA-NRAO, Founding and Organization, Antenna Planning, Box 2.

165. J.L. Pawsey: Notes of Future Program at Green Bank, 17 July 1962, NAA-NRAO, Founding and Organization, Antenna Planning, Box 2. https://science.nrao.edu/about/publications/open-skies#section-4

166. Ibid.

167. AUI-BOTXC, 19/20 July 1962.
168. Wild to Rabi, 31 July 1962, NAAustrl, C4660/1 Part 9.
169. Notes on the NRAO Directorship—for I. I. Rabi, Summary of Ideas from Conversations with Trustees and Discussions at the AUI Executive Committee Meeting on 7/20/62. G. Tape, NAA-NRAO, Founding and Organization, Antenna Planning, Box 2.
170. AUI-BOTXC, 19 February 1960.
171. KIK interview with DSH, 13 July 2011, NAA-KIK, Oral Interviews. https://science.nrao.edu/about/publications/open-skies#section-4
172. At the time, the NRAO scientific and technical staffs were essentially all male.
173. The name presumably derives from the large number of small children living in the NRAO housing complex.
174. KIK interview with DSH, 31 May 2011, op. cit.
175. KIK unrecorded interview with L. Fredrick, 1 May 2014; see also AIP Fredrick interview by David DeVorkin. https://www.aip.org/history-programs/niels-bohr-library/oral-histories/31235
176. AUI-BOTXC, 18–19 July 1963.
177. Dunbar to Reynolds, 8 July 1963, LOC-IIR, Box 36.
178. DSH to AUI Trustees, 18 July 1963, appended to minutes of AUI-BOTXC, 18–19 July 1963.
179. AUI-BOTXC, 15 October 1963.
180. Ibid.
181. Ibid.
182. AUI-BOT, 18 October 1963.
183. KIK interview with DSH, 31 May 2011, op. cit.

BIBLIOGRAPHY

REFERENCES

Bolton, J.G. and Wild, J.P. 1957, On the Possibility of Measuring Interstellar Magnetic Fields by 21-CM Zeeman Splitting, *ApJ*, **125**, 296
Condon, J.J. 2008, ZAPPED! … by Hostile Space Aliens! In ASPC **398**, *Frontiers of Astrophysics: A Celebration of NRAO 50th Anniversary*, ed. A.H. Bridle, J.J. Condon, and G.C. Hunt (San Francisco: ASP), 323
Condon, J.J. and Ransom, S.M. 2016, *Essential Radio Astronomy* (Princeton: Princeton University Press)
DeVorkin, D. 2000, Who speaks for Astronomy? How Astronomers Responded to Government Funding After World War II, *Historical Studies in the Physical and Biological Sciences*, **31**, 55
DeVorkin, D. 2018, *Fred Whipple's Empire* (Washington: Smithsonian Institution)
Drake, F.D. 1960, The Position-Determination Program at the National Radio Astronomy Observatory, *PASP*, **72**, 494
Emberson, R.M. and Ashton, N.L. 1958, The Telescope Program for the National Radio Astronomy Observatory at Green Bank West Virginia, *Proc. IRE*, **46**, 23
Emberson, R.M. 1959, National Radio Astronomy Observatory, *Science*, **130**, 1307
Evans, D.S. and Mulholland, J.D. 1986, *Big and Bright: A History of the McDonald Observatory* (Austin: University of Texas Press)

Heeschen, D.S. 1996, The Establishment and Early Years of NRAO, *BAAS*, **28**, 863 (abstract only), NAA-DSH, US Radio Astronomy History, Talks (full text of talk)

Heeschen, D.S. 2007a, The First Ten Years, 1955 to 1965, paper presented at the 140 foot Birthday Symposium (1995). In *But It Was Fun*, eds. F.J. Lockman, F.D. Ghigo, and D.S. Balser (Green Bank: NRAO), 265

Heeschen, D.S. 2007b, The 300 Foot Telescope and the National Center Concept, paper presented at the 300 Foot Birthday Symposium (1987). In *But It Was Fun*, eds. F.J. Lockman, F.D. Ghigo, and D.S. Balser (Green Bank: NRAO), 133

Heeschen, D.S. 2008, The Origins and Early Years of NRAO. In ASPC **398**, *Frontiers of Astrophysics: A Celebration of NRAO's 50th Anniversary*, ed. A.H. Bridle, J.J. Condon, and G.C. Hunt (San Francisco: ASP), 311

Keller, G. 1960, Report of the Advisory Panel on Radio Telescopes, *ApJ*, **134**, 927

Krisciunas, K. 1992, *Otto Struve 1897–1963, Biographical Memoir* (Washington D.C.: National Academy of Sciences)

Osterbrock, D.E. 1997, *Yerkes Observatory, 1892–1950* (Chicago: University of Chicago Press)

Pauliny-Toth, I.I.K., et al. 1978, The 5 GHz Strong Source Surveys. IV - Survey of the Area Between Declination 35 and 70 Degrees and Summary of Source Counts, Spectra and Optical Identifications, *AJ*, **83**, 451

Ramsey, N. 1993, *I. I. Rabi, a Biographical Memoir* (Washington: National Academy of Sciences)

Robertson, P. 1992, *Beyond Southern Skies – Radio Astronomy and the Parkes Telescope* (Cambridge: CUP)

Robertson, P. 2017, *Radio Astronomer – John Bolton and a New Window on the Universe* (Sydney: NewSouth Publishing)

Small, M. 1965, The New 140-foot Radio Telescope, *S&T*, **30**, 267

Struve, O. 1940, Cooperation in Astronomy, *SciMo*, **50**, 142

Struve, O. 1960, The National Radio Astronomy Observatory: The Outlook in 1960, *PASP*, **72**, 177

Struve, O. 1961, National Radio Astronomy Observatory Annual Report, *AJ*, **66**, 465

Weinreb, S. 1962a, An Attempt to Measure Zeeman Splitting of the Galactic 21-CM Hydrogen Line, *ApJ*. **136**, 1149

Weinreb, S. 1962b, A New Upper Limit to the Galactic Deuterium-to-Hydrogen Ratio, *Nature*, **195**, 367

FURTHER READING

Bok, B. 1956, A National Radio Observatory, *SciAm*, **195**, 56

Drake, F. and Sobel, D. 1992, *Is Anyone Out There?* (New York: Delacorte Press)

England, J.M. 1982, *A Patron for Pure Science* (Washington: National Science Foundation)

Findlay, J.W. 1963, The 300-Foot Radio Telescope at Green Bank, *S&T*, **25** (2), 2

Findlay, J.W. 1974, The National Radio Astronomy Observatory, *S&T*, **48** (6), 352

Kenwolf, L.G. 2010, *A Social and Political History of the National Radio Astronomy Observatory at Green Bank, WV* (Morgantown: West Virginia Libraries).

Lockman, F.J., Ghigo, F.D., and Balser, D.S. 2007, *But It Was Fun* (Green Bank: NRAO)

Malphus, B. 1996, *The History of Radio Astronomy and the National Radio Astronomy Observatory: Evolution toward Big Science* (Malabar, Fla.: Krieger Pub.)

Munns, D. 2003, If We Build It, Who Will Come? Radio Astronomy and the Limitations of 'National' Laboratories in Cold War America, *Historical Studies in the Physical and Biological Sciences*, **34** (1), 95

Munns, D. 2013, *A Single Sky: How an International Community Forged the Science of Radio Astronomy* (Cambridge: MIT Press)

Struve, O., Emberson, R.M., and Findlay, J.W 1960, The 140-Foot Radio Telescope of the National Radio Astronomy Observatory, *PASP*, **72**, 439

CHAPTER 5

Is Anyone Out There?

In the shadows of the struggles surrounding the construction of the 140 Foot Radio Telescope, Frank Drake, NRAO's newest and youngest scientist, carried out a small observing project to detect radio signals from nearby stars that would indicate the present of extraterrestrial intelligent life. Naming his program "Project Ozma," after the mythical princess of the Land of Oz, Drake observed two nearby stars, Tau Ceti and Epsilon Eridani. This project, the first modern Search for Extraterrestrial Intelligence (SETI), captured the imagination of the public and scientific communities alike. In the following decades, other investigators initiated a variety of SETI programs of ever-increasing capability at NRAO and elsewhere. NASA began a major SETI program, but it became mired in controversy over whether searching for intelligent life in the Universe is a proper scientific pursuit or should be relegated to the realm of science fiction. While no confirmable evidence for extraterrestrial intelligent life has yet been found, the discovery of the widespread existence of other planetary systems combined with the vastly improved sensitivity of radio telescopes has reinvigorated SETI research, and the exciting promise of detection continues to attract the attention of new generations of astronomers and the public.

5.1   Project Ozma

Frank Drake had long been fascinated by the possibility of life on other worlds and more generally by science and engineering. On a Navy ROTC[1] scholarship, he followed these interests by enrolling at Cornell University. Initially

---

The title of this chapter is taken from the book by Frank Drake and Dava Sobel 1992, *Is Anyone Out There? The Scientific Search for Extraterrestrial Intelligence* (New York, NY: Delacorte Press).

© The Author(s) 2020
K. I. Kellermann et al., *Open Skies*, Historical & Cultural Astronomy,
https://doi.org/10.1007/978-3-030-32345-5_5

interested in designing airplanes, Drake ultimately settled on a degree in engineering physics. Cornell also provided the opportunity for Drake's first forays into astronomy. He enrolled in an elementary astronomy course and was captivated by the lectures of Otto Struve, and he even built a small optical telescope. Although Drake wanted to pursue his interests in astronomy by enrolling in a graduate astronomy program, he first had to complete his required Naval service.

During his three years of active duty, Drake was a Naval electronics officer, learning skills that would soon greatly impact his future career. After his time in the Navy, he enrolled in 1955 in the graduate astronomy program at Harvard University. Bart Bok, chair of the Harvard astronomy program, quickly put Drake's electronics training to work by offering him a position with the Harvard radio astronomy project, where he further developed his technical abilities and became a skilled radio astronomer. After receiving his PhD from Harvard based on studies of neutral hydrogen in galactic clusters, Drake joined the NRAO. He arrived in Green Bank in April 1958, joining Dave Heeschen and John Findlay as the only other members of the scientific staff[2] (Fig. 5.1).

As described in Chap. 4, within a few months of Drake's arrival in Green Bank, NRAO acquired an 85 foot radio telescope. The new 85 foot Tatel Radio Telescope was essentially off the shelf, and did not have the collecting area or precision of the planned 140 Foot Radio Telescope. However, Drake realized that an 85 foot antenna had the capability to detect radio signals equivalent to Earth's most powerful transmissions from a distance of up to 10 to 20 light-years. Carefully choosing the right occasion, Drake softly suggested the first SETI project over lunch one day with Heeschen, Findlay, and Acting NRAO Director Lloyd Berkner. Drake proposed using the Tatel Telescope to search for signs of extraterrestrial life from nearby stars. To Drake's surprise and

Fig. 5.1 Frank Drake, 1962. Credit: NRAO/ AUI/NSF

delight, Berkner, whom Drake described as being an optimistic gambler in science, immediately gave his enthusiastic approval and encouragement. Drake dubbed his investigation Project Ozma after the fictional princess Ozma of the Land of Oz in L. Frank Baum's series of Oz books, suggesting a sense of "a land far away, difficult to reach, and populated by strange and exotic beings" (Drake and Sobel 1992). Project Ozma thus became the first modern Search for Extraterrestrial Intelligence (SETI)[3] and the standard by which future search programs have been measured.

Drake selected two nearby Sun-like stars, Tau Ceti and Epsilon Eridani, for his search. T. Kochu Menon, a fellow radio astronomy graduate from Harvard, had recently joined the NRAO scientific staff and was interested in studying magnetic fields in galactic hydrogen clouds using what is known as the Zeeman effect.[4] The instrumental requirements to search for the Zeeman effect were similar to those Drake would need for Project Ozma, and Menon and Drake went to work together to build the equipment needed for their projects, with the Zeeman experiment serving as a convenient front for Ozma (Fig. 5.2).

On 11 April 1960 Drake made the first observations for Project Ozma (Fig. 5.3). After six hours of observing Tau Ceti with no results, he turned the antenna toward Epsilon Eridani, the third closest star to the Sun located only about 10 light-years away.[5] Almost immediately Drake saw an off the scale

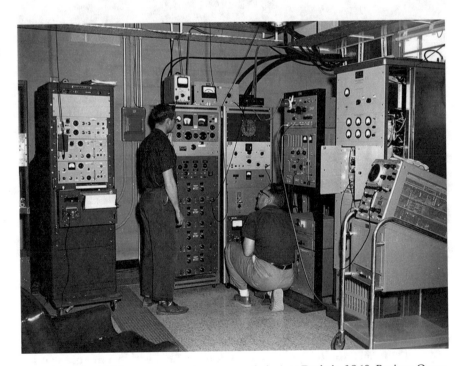

Fig. 5.2 Tatel Telescope control room used during Drake's 1960 Project Ozma. Credit: NRAO/AUI/NSF

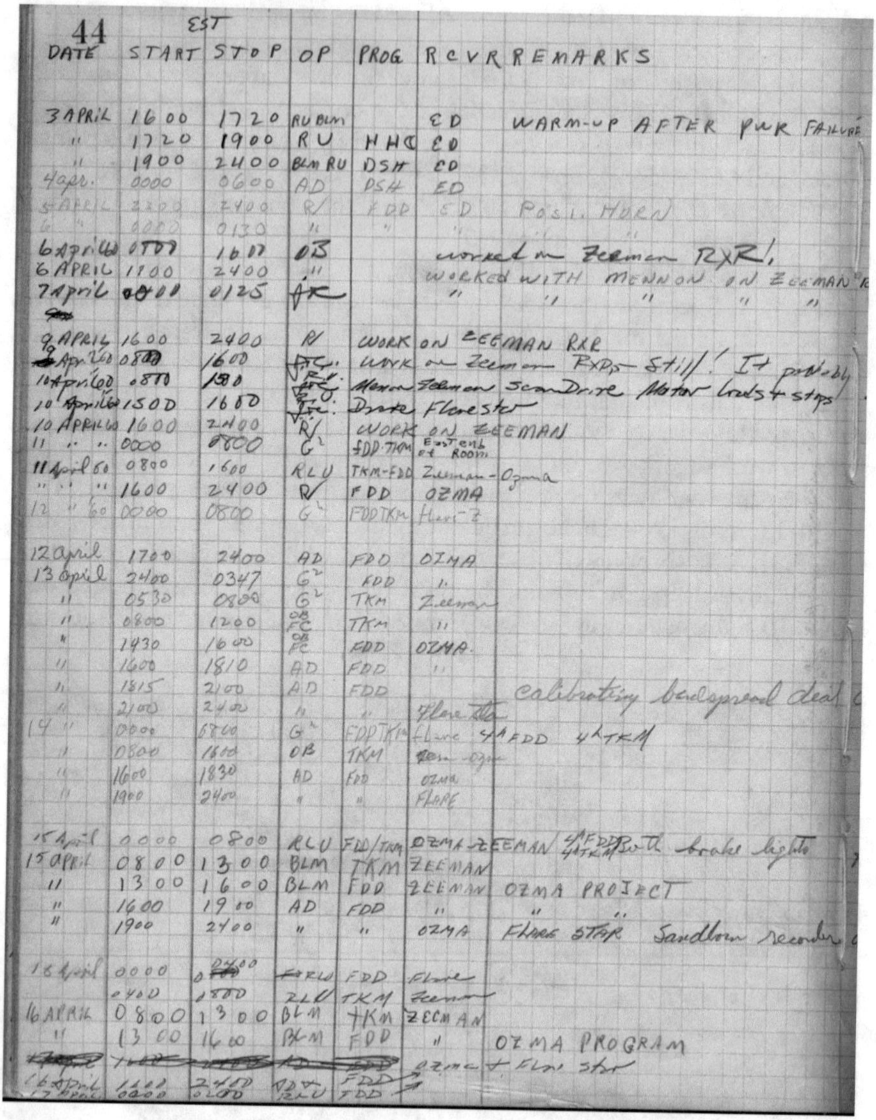

44 EST

DATE	START	STOP	OP	PROG	RCVR	REMARKS
3 APRIL	1600	1720	RU BLM		ED	WARM-UP AFTER PWR FAILURE
"	1720	1900	RU	HHC	ED	
"	1900	2400	BLM RU	DSH	ED	
4 apr.	0000	0600	AD	DSH	ED	
5 APRIL	2200	2400	R/	FDD	ED	POS'L HORN
6 "	0000	0130	"	"	"	"
6 April 60	0700	1607	OB			worked on Zeeman RXR!
6 APRIL	1700	2400	"			WORKED WITH MENNON ON ZEEMAN R
7 April	0000	0125	JK			" " " " "
9 APRIL	1600	2400	R/	WORK ON ZEEMAN RXR		
9 Apr 60	0800	1600	FC	Work on Zeeman RXRs still! It probably		
10 April 60	0800	1500	FC	Mennon Zeeman Scan Drive Motor loads + stops		
10 April 60	1500	1600	FC	Drake Flare star		
10 APRIL 60	1600	2400	R/	WORK ON ZEEMAN		
11 " "	0000	0800	G²	FDD TKM	East end of Room	
11 April 60	0800	1600	RLU	TKM-FDD	Zeeman - Ozma	
" " "	1600	2400	R/	FDD	OZMA	
12 " 60	0000	0800	G²	FDD TKM	Flare Z	
12 april	1700	2400	AD	FDD	OZMA	
13 April	2400	0347	G²	FDD	"	
"	0530	0800	G²	TKM	Zeeman	
"	0800	1200	OB FC	TKM	"	
"	1430	1600	OB FC	FDD	OZMA	
"	1600	1810	AD	FDD	"	
"	1815	2100	AD	FDD		calibrating bandspread dial
"	2100	2400	"	"	Flare Star	
14 "	0000	0800	G²	FDD TKM	Flare 4^A FDD 4^A TKM	
"	0800	1600	OB	TKM	Zeem-Ozma	
"	1600	1830	AD	FDD	OZMA	
"	1900	2400	"	"	FLARE	
15 April	0000	0800	RLU	FDD/TKM	OZMA-ZEEMAN	4^A FDD 4^A TKM Both brake lights
15 April	0800	1300	BLM	TKM	ZEEMAN	
"	1300	1600	BLM	FDD	ZEEMAN	OZMA PROJECT
"	1600	1900	AD	FDD	"	" "
"	1900	2400	"	"	OZMA	FLARE STAR Sanborn recorder
16 April	0000	0800	RLU	FDD	Flare	
	0400	0800	RLU	TKM	Zeeman	
16 APRIL	0800	1300	BLM	TKM	ZEEMAN	
"	1300	1600	BLM	FDD	"	OZMA PROGRAM
					ozma + Flare star	
16 April	1600	2400	AD	FDD		
17 April	0000	0800	RLU	FDD		

Fig. 5.3 Tatel Telescope log book for April 1960, with entries for Ozma observations on 11–16 April. Credit: NRAO/AUI/NSF

pulsed signal—8 pulses per second. An apparent successful detection of signals from another planet stunned Drake and his colleagues. Drake later recalled his experience, "When it happened, we were all dumbfounded. Could it be this easy? All you need to do is point to a random star and within a minute you see a signal that puts a receiver into overload? We were so surprised and so unprepared for it, we didn't know what to do. Everybody just looked at each other."[6]

But detecting signals from extraterrestrial intelligence was not to be that easy. The signal lasted for only a few minutes and Drake spent the next few days trying to confirm the apparent detection. To determine whether the signal was of terrestrial or extraterrestrial origin, he placed a small horn antenna, which was sensitive to radiation coming from a wide area of sky, outside the control building window. About ten days later they again saw the same signal. But both the Tatel Telescope and the horn antenna detected the signal with equal intensity, and it was clear that it originated from a terrestrial source, but one of unknown origin. Twenty-five years later at a gathering of the group involved in the original Ozma study, Drake commented, "We have never known what that was, but it had all the earmarks of being an electronics countermeasure system, probably airborne based on the timescale during which we heard it." (Kellermann and Seielstad 1986, p. 25).

After a little over a week of observing, there was a catastrophic receiver failure. Drake took a short break to get the receiver fixed while the telescope was used at another frequency for more conventional radio astronomy observations. When Drake resumed his observations, he continued to observe both Tau Ceti and Epsilon Eridani. But after 150 hours of observing spread over a total period of about two months between April until late June, there were no suggestions of any signals from either Tau Ceti or Epsilon Eridani.

The amount of data collected for Project Ozma motivated NRAO to implement new, digital methods of recording data. Initially all data for Project Ozma was recorded with a moving pen on strip charts, the standard technique for recording radio astronomy data at the time. However, Drake and his colleagues soon found that analyzing these chart recording was extremely tedious, and over the course of the project they developed the first digital data-recording system used at NRAO. At first they used a newly available digital voltmeter with a nixie tube display. This setup required the observer to press a button to freeze the display and then write down the numerical output of the receiver. Later, they attached a printer to the digital output to avoid the tedious task of writing down the numbers. Drake recalled the next leap forward in the process:

> The dramatic breakthrough that took place then was that [visiting] astronomer Gart Westerhout came and recognized that he could make life much more pleasant for himself if he attached a string to the button so that he wouldn't have to stand by the button. He rigged a string across the ceiling of the telescope control room and down from the ceiling right over an easy chair, in which he could sit. He attached a pull[ey] to this string and then he sat in the easy chair and whenever he wanted to take a reading he would pull the chain. (Kellermann and Seielstad 1986, p. 24)

Ultimately the system evolved to record the digital output of the receiver on punched paper tape to be analyzed later by NRAO's first computer, an IBM 610.

5.2   Cocconi and Morrison Paper

Concerned that the scientific community would view Project Ozma as science fiction and a waste of resources, Drake and newly appointed NRAO Director Otto Struve initially agreed not to publicize it. The fledging Observatory was already facing extensive criticism resulting from what many felt were the heavy-handed methods used by AUI in securing the NSF contract to operate NRAO, the delays and cost overruns of the 140 Foot construction, and envy over the generous way the Observatory was being funded compared to the university astronomy facilities. However, their hand was forced when Cornell physicists Giuseppe Cocconi and Philip Morrison published their September 1959 *Nature* paper, "Searching for Interstellar Communication" (Cocconi and Morrison 1959). Drake was pleased that his ideas had been independently discussed by two prominent scientists, in particular Morrison, who had earlier played a major role in the Manhattan Project at Los Alamos and who had personally assembled the atomic bomb dropped on Hiroshima. However, Struve was annoyed that Drake and NRAO were not receiving credit for their enterprising plans. It was Frank Drake who not only was the first to suggest searching for 21 cm radio signals from extraterrestrial intelligence, but who also took the initiative to build the necessary instrumentation, and planned to carry out the first meaningful search.

Despite that, Cocconi and Morrison were receiving all the attention from both scientists and the public. Struve and Drake wanted to ensure that NRAO would get the proper credit for their pioneering work. So in November 1959, reacting to the Cocconi and Morrison paper, Struve announced the existence of Project Ozma in his Karl Taylor Compton Lectures in Astronomy at MIT where he said[7]:

> Many of us have seen a report that thus far has come only in a mimeographed form from 2 American Scientists now in Europe. I believe their paper is soon to be in *Nature*. At Green Bank we are also thinking about this problem and there is underway a project which goes by the letters O-Z-M-A and I do not recall what these letters stand for, but those of us who forget what the name means, call it the project of the little green men.[8]

Curiously, in the published version of his lectures, written more than a year later, Struve (1962) made no mention of Project Ozma. Also, four months before his Project Ozma began, Drake discussed the problem of detecting transmissions from distant planetary systems in a January 1960 popular article in *Sky and Telescope* where he also described his plans for Project Ozma (Drake 1960). Both Drake and Struve were treading a fine line. They did not want NRAO to be ridiculed for spending money on what might be perceived as a science fiction project; but neither did they want to be scooped by others. So Drake wrote to the NSF Press Officer with an "On Demand Information Regarding NRAO Project Ozma" press release describing Project Ozma and saying, "We very much wish to withhold this information from the press until such time as the experiment is successful" and "if the press should become aware of the project ... or if another group also embarks on research in this field."[9] Nevertheless, as a result of Drake's article, Struve's lectures, and the Cocconi and Morrison paper, the search for extraterrestrial civilizations drew increased interest from both the scientific community and the public.[10]

Cocconi and Morrison had considered what type of signal extraterrestrial civilizations might use in an attempt to contact each other or to announce their existence. They argued that since hydrogen was the most abundant element in the Universe, a fact which would be known to any intelligent being, the best place to start searching for signals from extraterrestrial life would be the 1420.405 MHz (21 cm) radio line of neutral hydrogen. Curiously, Drake chose this same frequency for Project Ozma, but not for the same reasons as Cocconi and Morrison. By selecting the 21 cm line, Drake could then bury the $2000 development costs as part of Menon's project to detect the Zeeman effect (Drake and Sobel 1992, p. 28). Recalling the planning process for Project Ozma, Drake later explained, "We would build it and do the search at the 21 centimeter line.... It was a way to prevent criticism of the Observatory, and in a way, kill two birds with one stone." (Kellermann and Seielstad 1986, p. 19). Drake's plan for Project Ozma thus allowed him to meet his scientific goals for the project while limiting the criticism that could be leveled against NRAO.

As a result of the reluctantly released publicity, Microwave Associates gave Drake one of the first parametric amplifiers used in radio astronomy, which allowed a major improvement in sensitivity not only for Project Ozma, but for Menon's Zeeman experiment, as well as other 21 cm projects on the Tatel Telescope. Interestingly, with time some SETI researchers pointed out that 1420 MHz is the wrong place to search, since any intelligent civilization would want to keep that frequency quiet for radio astronomers.[11] Later SETI researchers focused on the region between the 21 cm hydrogen line and the 18 cm hydroxyl (OH) line which has been characterized as the "water (H_2O)[12] hole," an equally logical place for a water based civilization to transmit their presence (e.g., Oliver 1979).

5.3   REACTIONS TO SEARCHING FOR EXTRATERRESTRIALS

Cocconi, Morrison, and Drake all anticipated that their speculations would meet criticism from other scientists due to the widespread perception that extraterrestrial life was more of a plot device for science fiction novels than a serious topic of scientific discussion. Indeed, the Cocconi and Morrison paper created quite a stir in both the scientific and popular press. Their primarily theoretical paper resulted from follow-up discussions of a paper Morrison had published the previous year on gamma ray astronomy. In their discussions, Cocconi and Morrison realized that gamma rays were being artificially produced at Cornell, just a few floors below them as a by-product of the university's synchrotron experiments. They structured their paper in a way to limit criticism they expected to receive. By assuming that intelligent, communicative life existed on other planets, they concentrated on the best methods for attempted contact and their article focused on this issue. "We shall assume that they (an extraterrestrial civilization) established a channel of communication that would one day become known to us...What sort of channel would it be?" Though their initial discussions addressed artificially produced gamma rays, based on energy considerations and relative transparency of the interstellar medium and planetary atmospheres, they concluded that radio frequencies between 1 MHz and 30 GHz provided the best opportunity for establishing interstellar communication. Furthermore, they noted that an advanced alien civilization would use the most obvious and simplest wavelength to communicate with our relatively primitive civilization, and that would be the 21 cm (1420 MHz) line of neutral hydrogen, the most abundant element in the Universe. In their *Nature* paper, Cocconi and Morrison concluded, "We therefore feel that a discriminating search for signals deserves a considerable effort. The probability of success is difficult to estimate; but if we never search, the chance of success is zero."

In common with all subsequent SETI programs to date, Project Ozma found no evidence for any intelligent extraterrestrial civilizations, although some skeptics questioned whether there was intelligent life on Earth, or at least in Green Bank.

But Project Ozma did not go unnoticed. During this period, Green Bank was visited by West Virginia Governor Cecil Underwood, by Theodore Hesburgh, the new president of the University of Notre Dame, and by Bernard Oliver, the vice president of Hewlett Packard, who would later play a major role in the American program to detect radio signals from an extraterrestrial civilization. Representatives of various news media also visited Green Bank, and Drake was interviewed for television, complete with cue cards.

All NRAO telescopes were actually controlled by a professional telescope operator based on instructions from the scientist in charge of the program. During Governor Underwood's visit, the telescope operator noted the Governor's visit in the logbook by writing in parenthesis, "Republican fool." On another occasion, a newscaster quietly took the telescope operator aside to ask, "If you had really heard something, you would tell me; wouldn't you?" (Crews 1986, p. 29). But after examining hundreds of yards of chart paper, Project Ozma disclosed nothing but noise (Drake 1979, p. 13). While Ozma was not successful in the narrow sense of detecting extraterrestrial civilizations, in a broader sense Ozma brought SETI to the public arena and defined all subsequent work in the quest for extraterrestrial civilizations. Drake never expanded Project Ozma to look at other stars, but over the next half century he became the recognized senior statesman and spokesman for SETI. No doubt his growing stature was enhanced by his prematurely white hair, which conveyed an image of wisdom and authority even when he was a young Green Bank astronomer.

With no positive results to report, Drake never published the results of Ozma in the refereed scientific literature. However, in a 1961 paper in *Physics Today*, Drake discussed the arguments for searching for advanced extraterrestrial civilizations and gave a technical description of the Ozma instrumentation. In a very brief statement, he simply said, "A search for intelligent transmissions has been conducted in Green Bank. We looked at two stars, Tau Ceti and Epsilon Eridani, which are the nearest solar system type stars. After some 150 hours of observing, we obtained no evidence for strong signals from these stars." (Drake 1961).

In their paper, Cocconi and Morrison discussed the broader issue of whether searching for extraterrestrial intelligence should be considered a legitimate scientific activity and noted, "The reader may seek to consign these speculations wholly to the domain of science-fiction." However, Cocconi and Morrison argued that the search for extraterrestrial intelligence lay on firm theoretical grounds and should be considered a valid scientific inquiry and went on to say, "We submit, rather, that the presence of interstellar signals is entirely consistent with all we now know, and that if the signals are present the means of detecting them is now at hand."

Recalling the public reaction, Morrison (1990, p. 24) said in a later interview, "It got huge newspaper and media coverage, which we didn't anticipate...The media kept chasing me because I was going around the world. In every city I visited there would be messages from reporters wanting to talk to me...." The attention SETI received in the public media can be attributed to the booming interest in extraterrestrial life and the growing enthusiasm for space exploration. Project Ozma was no doubt connected in the public's mind with the widespread fascination with "flying saucers" that had developed following the Roswell UFO[13] incident a decade earlier and with the popular speculation that UFOs or flying saucers were spacecraft sent to Earth by aliens from another planet. But the nature of this attention was a double-edged sword. Increased general interest in SETI would be a key factor in receiving funding for future SETI projects; however, SETI scientists also wanted to ensure that their work was viewed as valid scientific research, not lumped together with UFO sightings.

Indeed, while the reaction of the public was generally positive, the scientific community had more mixed views. Drake recalled the attitude of his colleagues as, "...uniformly positive but not enthusiastic. Again I think that it was the fact that we weren't investing a great deal of resources... People didn't think it was worth a very careful analysis, but since it wasn't crazy they said: These guys want to spend two thousand dollars, let them do it." (Drake 1990, p. 69). Morrison's recollection of his colleagues' reactions echoed Drake's experience, "Most felt it was not a good idea, probably foolish, certainly completely speculative, and hardly worth discussing." (Morrison 1990, p. 24). Though reactions were mixed in the scientific community, the attention that Cocconi, Morrison, and Drake received helped to connect scientists who were interested in extraterrestrial intelligence. Drake later commented that finally, "People knew who they could write to find out who was interested" (Drake 1990, p. 60).

5.4   DEVELOPMENT OF THE SETI COMMUNITY

After his earlier hesitation, Drake did not shy away from public exposure. Encouraged by the increasing interest, if not support for SETI, Drake thought about holding a conference to discuss Project Ozma and the broader aspects of extraterrestrial life, in particular intelligent civilizations. Following a talk Drake gave at the Philosophical Society of Washington on the "Detection of Extraterrestrial Intelligent Life," he exchanged ideas with Peter Pearman, a staff officer of the National Academy of Science's (NAS) Space Studies Board (SSB).[14] After returning to Green Bank, and with the concurrence of Struve, Drake proposed a "quiet symposium on extraterrestrial life" to be held in Green Bank which the SSB agreed to sponsor with NRAO acting as the host.[15] Pearman defined the goals of the meeting as.[16]

a) The considerations which may lead to the expectation that intelligent transmitters are likely to be observable;
b) Whether or not it would be worthwhile to engage in further exploratory investigations with existing apparatus or whether the prospects of detecting an interesting event are, in fact, too small to be of interest;
c) If the consensus turns out to be largely negative, it may be that some suggestions can be derived for further investigations which could be made either to verify or to refute the null hypothesis or which might enable a re-assessment to be made.

According to Drake, Pearman was trying to build support in the government for the possibility of discovering life on other worlds (Drake 2010). It did not hurt that Lloyd Berkner, President of AUI, was the Chair of the SSB, although curiously, Berkner did not attend the Green Bank conference. Participation in the Green Bank conference was by invitation only and Struve requested that the invitees cooperate in conducting the conference "privately, without publicity or press coverage."[17] This became only the second scientific conference held in Green Bank, following the joint US-USSR radio astronomy meeting held six months earlier.

Held on 1–2 November 1961, the Green Bank Conference on Extraterrestrial Intelligent Life brought together a diverse group of scientists and engineers to discuss issues pertinent to communicative extraterrestrial life. Participants included Morrison and Cocconi; Su-Shu Huang, author of papers on planet formation; John C. Lilly, author of the book *Man and Dolphin* on dolphin intelligence; and chemist Melvin Calvin, who was notified during the conference that he had received the 1961 Nobel Prize in Chemistry, giving a celebratory atmosphere to the gathering. Carl Sagan and Bernard Oliver also attended, both of whom would later play key roles in future SETI activities. NRAO Director Otto Struve acted as Chair.

The Green Bank conference provided the opportunity to examine the assumptions Drake, Cocconi, and Morrison had previously made about extraterrestrial intelligent life. As previously discussed, these first publications were limited to technical issues in order to limit the criticism their projects would encounter. However, at the Green Bank conference the attendees expanded their analysis to include debating the possibility of the existence of extraterrestrial life based on current understanding of astronomy, planet formation, and evolution, the optimum frequencies for communication, the form of messages, and, even at this early stage, they speculated on when it would be appropriate to send our own messages.

In preparation for the conference Drake organized his thoughts into a format that would shape subsequent SETI investigations for the next half century. The Drake Equation estimated the number of communicative civilization in the Galaxy by accounting for factors necessary for intelligent life to develop. This included the number of stars with habitable planets in the Galaxy, the fraction of those planets that develop life, and most important, but also most uncertain, the mean lifetime of a technical, communicative civilization.

The Drake Equation is given by:

$$N = R_* \, f_p \, n_e \, f_l \, f_i \, f_c \, L$$

where

R_* = mean rate of star formation over galactic history

f_p = fraction of stars with planetary systems

n_e = number of planets per planetary system with conditions ecologically suitable for the origin and evolution of life

f_l = fraction of suitable planets on which life originates and evolves to more complex forms

f_i = fraction of life-bearing planets with intelligence possessed of manipulative capabilities

f_c = fraction of planets with intelligence that develops a technological phase during which there is the capability for and interest in interstellar communication

L = mean lifetime of a technological civilization

Throughout history humanity has often speculated on the possibility of extraterrestrial civilizations. In the early part of the twentieth century, there were even primitive attempts to make contact with Martians or civilizations elsewhere in the Galaxy[18] (See Dick 1993, 1998). But for the first time, the Drake equation put these speculations on a quantitative basis. Although none of the factors in the Drake equation could be reliably estimated at the time, the

conference participants could optimistically argue that each of the first six factors could be in the range of 0.1 to 1. The big uncertainty was the lifetime of
technically advanced civilizations, which might be as short as a few hundred
years or as long as hundreds of millions of years for those civilizations that did
not destroy themselves by war or by exhausting their resources.[19] Subsequent
work leading to the recognition of the widespread existence of exoplanets and
advances in evolutionary biology has mostly confirmed these early speculations, leaving L as the big uncertainty.

The camaraderie of Green Bank conference attendees was evident at the
meeting as they dubbed themselves the Order of the Dolphin, and Calvin had
dolphin pins made in honor of Lilly. The formation of the Order of the Dolphin
demonstrated the sense of group identity that developed during the conference. Although the Order never developed into an official organization in any
sense, after the meeting Carl Sagan wrote to J.B.S. Haldane inviting him to
join the Order. Haldane's response to Sagan's request illustrates the type of
organization the Order of the Dolphin was. Sagan recalled his response, "...he
(Haldane) wrote me that membership in an organization that had no dues, no
meetings, no responsibilities was the sort of organization he appreciated; he
promised to try hard to live up to the duties of membership." (Sagan 1973,
p. 168). For the next few decades, the Green Bank conference attendees
formed the core group of scientists involved in the search for extraterrestrial
intelligence.

Following the 1961 Green Bank meeting, many of the still surviving participants returned to Green Bank in 1985 to celebrate the 25th anniversary of
Project Ozma (Kellermann and Seielstad 1986) (Fig. 5.4). But at the 2010
50th anniversary workshop, only Frank Drake himself was able to participate.[20]
By the time of the 1985 conference, the search for extraterrestrial civilizations
had split between two strategies. One approach was to look for extraterrestrial
beacons consisting of high-powered transmitters and highly directional antennas that might be pointed toward the Earth. However, beacon research assumes
that the extraterrestrial civilization has some knowledge of the Earth as being
inhabited by a technical society. The other approach is to look for signals resulting from the analogue of Earth's entertainment broadcasting or the powerful
defense related radars—so-called "eavesdropping." The 1985 conference participants vigorously debated the relative merits of the two approaches, as well
as the effectiveness of looking outside the radio band, in particular at optical or
infrared wavelengths, where Harvard's Paul Horowitz was beginning an innovative search.

Unlike most previous SETI conferences, the 2010 Green Bank 50th workshop paid less attention to technical and implementation strategies, instead
discussing such things as the social, moral, religious, and legal implications of a
confirmed discovery of an extraterrestrial civilization, and the likely societal
reactions. The participants also reflected on the accomplishments and impact
of the previous half century of SETI and speculated on the possible consequences

Fig. 5.4 25th Project Ozma reunion, Green Bank, 1985. Front (left to right): Bob Viers, Dewey Ross, Bill Meredith, Troy Henderson, Bob Uphoff. Back: George Grove, Fred Crews, Omar Bowyer, Frank Drake, Kochu Menon. Credit: NRAO/AUI/NSF

of realizing that we may be alone in the Universe. Among the participants of the 2010 workshop was Judge David Tatel, the son of Howard Tatel who had designed the 85 foot Green Bank telescope for the Blaw-Knox Company when he was working at the Carnegie Institution's Department of Terrestrial Magnetism. His son, David, a distinguished jurist and member of the US Court of Appeals for the District of Columbia Circuit, shared his thoughts on the legal and moral implications of a confirmed detection of a transmission from an extraterrestrial civilization.

For years after Drake's short-lived observations, Project Ozma became one of the most highly discussed programs at NRAO and the focus of broader discussions about NRAO. Ozma, along with Cocconi and Morrison's famous paper and the 1961 SETI Conference, launched SETI into the popular vernacular (Drake and Sobel 1992). Criticism of SETI often focused on the whimsical nature of searching for signals from extraterrestrials and was often discussed in the same way as the widely debunked UFO research. Scientists such as Drake, Morrison, and Sagan, argued against that characterization of their work, claiming that the advent of the modern radio telescope provided the technology to conduct a sober, scientific search for interstellar signals of intelligent origin.

5.5   SETI After Project Ozma

In 1963, Drake left NRAO for the Jet Propulsion Laboratory (JPL) in Pasadena, California. Frustrated by the bureaucracy and paperwork associated with a NASA laboratory, he left JPL to become Director of the Cornell Arecibo Observatory after only a year. There he initiated a number of SETI programs, often in collaboration with Carl Sagan.[21] In the years following Project Ozma, NRAO accepted proposals to use the 300 Foot and 140 Foot Radio Telescopes in competition with other proposals for more conventional astronomical research, but only if the investigators agreed to publish in the normal literature and not solicit undue publicity. Gerrit Verschuur used the 140 Foot and 300 Foot Telescopes to search for intelligent signals from possible planets orbiting ten nearby stars (Verschuur 1973), but buried his SETI observations as part of a more extensive program to study galactic neutral hydrogen clouds. Only about a decade after Ozma, Verschuur estimated that his search was already about 30 times more sensitive than Drake's Ozma. Radio astronomers Pat Palmer and Ben Zuckerman convinced NRAO to spend 500 hours to observe 674 stars with the 300 Foot antenna which they dubbed Project Ozma II or Ozpa. But perhaps concerned that if they tried to publish in a peer reviewed astronomy journal they might damage their reputation as rapidly rising young radio astronomy stars, they reported their work first in an internal NRAO newsletter (Palmer and Zuckerman 1972) and later in the proceedings of a 1979 IAU conference (Zuckerman and Tarter 1980).

In 1974, Frank Drake again shook the scientific community with a dramatic and bold experiment. As part of the ceremonies marking the inauguration of the Arecibo radio telescope upgrade, Drake used Arecibo's powerful million-watt transmitter to broadcast a message to the globular cluster M13 located about 21,000 light years away. Drake's message consisted of a stream of 1679 bits of 1's and 0's which, he argued, any intelligent species would recognize as the product of the prime numbers 73 and 23. When arranged in a pattern of 73 lines by 23 rows, the 1's and 0's conveyed a simple picture of human life.

Until Drake's Arecibo message, all previous SETI research was passive. That is, powerful radio telescopes were used to try to receive transmissions originating from extraterrestrial civilizations. Drake reasoned that if everyone only listened, there would be no signals to receive. At a distance of 21,000 light years, it would be at least 42,000 years before one might expect a reply from M13. Nevertheless, Drake's message generated some controversy. In England, Martin Ryle argued that by sending a message, Drake was irresponsibly disclosing our presence to any alien civilization that might live on a planet orbiting a star in M13, and that there was no guarantee that such a civilization would be friendly. Of course, since the 1920s humans had begun broadcasting radio signals into space. And starting in the 1950s with the rise of powerful TV transmitters and the Distant Early Warning system (the DEW Line) radar, man-made radio signals from the Earth were already propagating through the Galaxy and could be detected by alien civilizations no more advanced than our own. In fact, due to these artificial radio transmissions at VHF and UHF frequencies, to a distant observer the Earth appears brighter than the natural radio emission from the Sun.

The increasing number of SETI programs, particularly in the United States, perhaps fueled by Drake's transmission to M13, raised numerous ethical, moral, legal, and political questions about how to react to the reception of any signal thought to come from an advanced extraterrestrial civilization. Should a successful detection be kept confidential or immediately made public, and if made public should national governments be consulted first? If consulted, would governments try to classify any relevant information about the detection? If made public how would countries (particularly rogue countries), religious or quasi-government groups, or even individuals be constrained from responding and perhaps misrepresenting our society and culture? Behind much of the debate is the uncertainty about whether alien civilizations would be benevolent or would be a threat, intentional or not, to emerging societies, and it was noted that the history of contact between terrestrial civilizations is not encouraging in this respect.[22]

It was becoming increasingly clear not only to scientists, but to the public and particularly politicians, that the detection of signals sent by extraterrestrials or aliens would have a profound impact on human life. At worst, it was feared that two-way contact with aggressive aliens with advanced weaponry might destroy our civilization, much as the colonial powers on Earth have destroyed the less technologically developed earthly societies with which they came into contact. At the other extreme was the possibility of obtaining advanced medical knowledge with the promise of curing famine and disease, and providing advanced technologies to support human activities as well as the immeasurable impact to human religious beliefs. SETI was becoming too important for the government to leave to scientists.

At the request of Don Fuqua, Chairman of the US House of Representatives Subcommittee on Space Science and Applications,[23] the Library of Congress Science Policy Research Division[24] issued a report on the "Possibility of Intelligent Life Elsewhere in the Universe." The report, which was compiled by Science and Technology Analyst Marcia Smith (Smith 1975), gave a definitive update on the status of SETI research in both the United States and the USSR. It served to legitimize SETI and served as a blueprint for future SETI programs. In 1982, the International Astronomical Union set up a new commission on "Bioastronomy: Search for Extraterrestrial Life." Further legitimization of SETI came from the 1980, 1991 and 2001 National Academy of Sciences Decade Reviews of astronomy which recognized the Search for Extraterrestrial Intelligence as an important and valid area of scientific research, and the special role played in SETI by radio astronomy (Field 1982, p. 150–151; Bahcall 1991a, p. 62; McKee and Taylor 2001, p. 131–132). Perhaps an even more influential endorsement came from Theodore Hesburgh, President of

Notre Dame, who wrote, "This proposed search for extraterrestrial intelligence (SETI) is also a search of knowing and understanding God." (Morrison et al. 1977, p. vii).

In view of the large uncertainty in the nature of the signals being sought, the National Academy reports stressed the need for a variety of approaches by independent researchers, and cautioned against too much reliance on highly organized and visible agency-directed expensive programs such as those being pursued by NASA. The NASA SETI observations began in the early 1970s when Jill Tarter, Jeff Cuzzi, and others used the Green Bank 300 Foot Radio Telescope to search for narrow band radio signals.[25] But by this time, it had become apparent that no existing or planned radio telescope probably had sufficient sensitivity to detect signals from any but the nearest stars. Under the leadership of Bernard Oliver, NASA convened a summer study to design a radio telescope with at least two orders of magnitude better sensitivity than any existing facility that would be capable of a meaningful SETI search (Oliver and Billingham 1973). Project Cyclops, as it was known, was to consist of an array of more than one thousand 100-meter antennas. The study team also identified a number of exciting radio astronomy applications, but the anticipated cost of approximately $10 billion far exceeded any conceivable funding.[26] Nevertheless, Oliver's bold vision served to ignite the scientific community, especially radio astronomers, to the prospects for SETI investigations, as well as to begin to smooth the way toward the construction in the US of a large multiple antenna array for more conventional radio astronomy programs. But the Cyclops study also had a negative impact. Although never intended to be a blueprint for an actual construction program, the high price tag left a long-lasting stigma that the search for extraterrestrial civilizations involved huge amounts of money that could be better spent, depending on one's outlook, on conventional astronomy or on addressing the nation's sociological problems.

SETI and NASA NASA's involvement in Project Cyclops led to a series of workshops organized by John Billingham at the NASA Ames Research Center. Billingham, who was a British military medical doctor and head of a small group called the Committee on Interstellar Communications within the Exobiology Office at Ames, had previously recruited Bernard Oliver to lead the Cyclops study. He was probably the first person to be head of a US government office with official responsibility for extraterrestrial intelligence (Billingham 1990). Billingham's committee was part of the NASA Life Science Division which was the result of NASA's interest in life beyond the Earth. The NASA SETI workshops, which were chaired by Morrison, were convened to systematically examine the fundamental basis of SETI, the preferred search approaches, and the social, environmental, and political impact of the success or failure of SETI. They concluded that SETI was timely and feasible, that a significant program would not require substantial resources, and that the US could take the lead in this intrinsically international endeavor.

The report of the NASA workshop on "The Search for Extraterrestrial Intelligence" (Morrison et al. 1977) laid out the blueprint for the American SETI program for the following decades, and by the close of the 1970s, SETI had developed a robust community of scientists and engineers, strong public support, and sets of detailed options for future work. Interestingly, although there has been no space based component of the US SETI program, starting with the Ames workshops, for several decades the US SETI effort was led by NASA rather than by the NSF which, at the time, had no interest in life beyond the Earth.

The end of the 1970s brought strong challenges to further SETI research in the United States. In February 1978, Senator William Proxmire (Wisconsin) awarded SETI his infamous "Golden Fleece Award," a dubious monthly honor meant to single out projects that the Senator felt wasted federal funds. In making his announcement of the "award," Proxmire suggested that NASA was "riding the wave of popular enthusiasm for *Star Wars* and *Close Encounters of the Third Kind*," and proposed that SETI should be "postponed a few million light-years."[27] While Proxmire was often criticized for being ill-informed about the recipients of his Golden Fleece Award, he regularly was able to terminate funding for the projects he honored. SETI continued to receive some NASA funding, but, not wanting to threaten the more important and much more expensive space programs, NASA kept SETI funding below the Congressional radar.

However, the Golden Fleece award sparked a decades-long battle over NASA and even NSF funding for SETI projects. While strong scientific and public support successfully rebutted Proxmire's legislative attacks on proposed funding, the continuing struggles over NASA funding prompted Tom Pierson and Jill Tarter to create the private SETI Institute and Bruce Murray and Carl Sagan to form the Planetary Society, both of which could operate at lower cost than a government agency such as NASA. By the mid-1980s, following the urging of Carl Sagan, Proxmire relented, apparently leaving SETI with no strong opponents in Washington (Drake and Sobel 1992, pp. 195–196), and encouraging NASA to develop the SETI Microwave Observing Project (MOP). Subsequently, NASA established the MOP much in the same way as other NASA missions, issuing a "Research Announcement" soliciting proposals for several facility instrumentation teams as well as the usual Interdisciplinary Investigators. A project office was set up at Ames and began negotiations with NRAO for NASA to take over the operation of the 140 Foot full time for SETI beginning in 1995 when the new 100 meter GBT was expected to be completed (See Chap. 10).

The Microwave Observing Project, which was later renamed the High Resolution Microwave Survey (HRMS) consisted of two complementary strategies: the NASA Ames based Targeted Search System (TSS) and the JPL based Sky Survey (Dick 1993). The TSS was planned to use primarily the Green Bank 140 Foot, the Parkes 210 foot, the Nançay 94 meter, and the 1000 foot Arecibo radio telescopes to examine 800 nearby stars between 1 and 3 GHz, while the Sky Survey planned to use the antennas of JPL's Deep Space Tracking Network in California, Australia, and Spain to search the entire sky between 1 and 10 GHz. Each project developed its own wide bandwidth spectrum analyzer with up to 30 million independent frequency channels having a resolution as narrow as about 1 Hz. The NASA HRMS was formally launched on 12 October 1992 (500 years after Columbus landed in the Bahama Islands) with simultaneous celebrations at the JPL Goldstone Deep Space Communications Complex in California and at the Arecibo Observatory in Puerto Rico. Ironically, although the engineers at Goldstone and Arecibo were looking for signals originating from many light-years away, they were unable to get the planned communications link between the two US sites to work.

After only a year of full funding, the HRMS and all US SETI research suffered a major setback. The HRMS project was projected to cost $108 million spread over about a decade, and was vigorously promoted in Washington by Jill Tarter and others. Congressional lobbying can sometimes be successful in raising funds for a pet project, and usually at worst is ignored. Typically, a ten million-dollar budget item in NASA's then seven billion-dollar annual budget would go unnoticed by Congress, or perhaps would be buried somewhere and not even appear as a line item. But the broad public interest in SETI along with the intense lobbying effort brought the HRMS to the attention of Democratic Senator Richard Bryan from Nevada. Bryan, the former governor of Nevada and a long-standing opponent of SETI, had unsuccessfully tried to kill the NASA SETI program for FY1992 and FY1993, perhaps in an attempt to get visibility for himself and the state of Nevada. On 20 September 1993, Bryan introduced a late amendment to the 1994 Housing and Urban Development appropriations bill to "prohibit the use of funds for" NASA's $12 million HRMS funding appropriation. Only the previous day, in a *Parade Magazine* article,[28] Carl Sagan had inadvertently provided fuel for Bryan's anti-SETI rhetoric. Sagan's article started out by specifically drawing attention to the year-old NASA SETI program and tried to minimize the cost by comparing it to the price of a military attack helicopter. Sagan's article was widely distributed as a supplement to the 19 September Sunday newspapers around the country, and almost certainly came to the attention of Bryan, who introduced his amendment the following day. Referring to SETI as "*The Great Martian Chase,*" Bryan went on to absurdly state, "As of today millions have been spent and we have yet to bag a single little green fellow. Not a single Martian has said take me to your leader, and not a single flying saucer has applied for FAA approval."[29]

Bryan argued that he was not against science, but pointed out that the previous year both the House and Senate had eliminated NASA's MOP. He was clearly annoyed that NASA had adopted the new HRMS name, which he felt was a weak attempt to hide SETI funding. He argued that even after decades, SETI had never detected any signs of intelligent life and that $12 million could send 9000 needy students to the University of Nevada.[30] Despite a valiant effort from Senators Barbara Mikulski (D-MD), Phil Graham (R-TX) and Jay

Rockefeller (D-WV), two days later by a voice vote, the Senate passed Bryan's amendment to save taxpayers money, and NASA's HRMS program died on the Senate floor.

Having just rebounded from the Hubble Space Telescope spherical aberration mirror fiasco,[31] NASA was not in a strong position to argue with Congress for SETI funding. Formally the funding bill applied only to NASA's 1994 budget, but NASA was not willing to incur Congressional wrath and fight over this relatively small budget item that was out of the mainstream of NASA programs. Much of the previous effort and expenditures of approximately $15 million at Ames and at JPL had gone into designing and prototyping several generations of high resolution multichannel spectrometers and other instrumentation. So there were few actual observations to use as a basis for soliciting additional funding. For more than a decade following Senator Bryan's intervention, SETI was an unpopular subject at NASA and at other government agencies such as the NSF, where SETI research was explicitly excluded from NSF grants until 2000. Nevertheless, NRAO continued to support modest SETI programs on the 140 Foot and 300 Foot Telescopes, provided they were not given undue publicity and the results would be published in the normal astronomical journals rather than the popular literature.

It was perhaps unfortunate that the national SETI program had become so entwined with NASA, primarily because some of the key people interested in SETI research happened to already work for NASA. Unlike the NSF, NASA has historically concentrated on big missions in space with well-defined realistic goals, and has not looked favorably on ground-based research. SETI as an ongoing research activity with constantly changing procedures and goals was probably more appropriate for the NSF than NASA. Ironically, SETI's association with NASA created the image within the astronomy community that SETI, like other NASA programs, had generous funding, at least in comparison with NSF funded research programs.

SETI Goes Private Jill Tarter and other SETI researchers have repeatedly pointed out SETI's awkward situation: SETI is one of those scientific endeavors that must justify additional expense for increased capability on the basis of previous failures (e.g., Garber 1999). Fortunately, Tarter and others have been able to exploit the extensive instrumentation development begun by the HRMS and prior NASA SETI programs. Supported largely by private funding, the SETI Institute initiated Project Phoenix to search the ~800 nearby stars originally specified by the HRMS Targeted Search. Phoenix used nearly 30 million simultaneous 1 Hz spectral channels and had the capability to detect signals in the range between 1 and 3 GHz which might have originated from transmitters comparable to the most powerful terrestrial radars such as those at Arecibo and Goldstone. Following an extensive negotiation with NRAO, the SETI Institute purchased time on the Green Bank 140 Foot Radio Telescope for Project Phoenix. But after nearly a decade of searching with the 140 Foot

Telescope, as well as the Parkes, Nançay, and Arecibo radio telescopes, no convincing signals from any extraterrestrial civilizations were detected.

Three other long running SETI programs need to be mentioned.[32] Using private funding, SERENDIP or the "Search for Extraterrestrial Radio Emissions from Nearby Developed Intelligent Populations," used a secondary feed for so-called commensal or "piggy-back" research, primarily at Arecibo and Green Bank, to examine random directions in the sky determined by the regularly scheduled astronomical program being pursued at the telescope. SERENDIP, which went through a series of four hardware and software upgrades, used as many as 168 million channels to cover a 200 MHz wide band.

At Ohio State University, following the completion of their all sky radio source survey in 1973, Robert Dixon and John Kraus initiated a SETI survey using their standing parabolic reflector which Kraus referred to as "Big Ear." Kraus also self-published a journal called *Cosmic Search*, which promoted a variety of SETI programs as well as provided a broad introduction to topics in radio astronomy.[33] Aside from a widely publicized but never confirmed "WOW" signal recorded on 15 August 1977, Big Ear found no signs of any signals originating from any extraterrestrial civilizations.[34] The program ran for many years using largely student and volunteer labor, until the telescope was finally shut down. In 1998, despite widespread popular protest, Big Ear was removed to build a golf course.

A particularly innovative SETI activity has been the SETI@home program developed by Dan Wertheimer and colleagues at the University of California, Berkeley. SETI@home uses data taken at a number of the world's most powerful radio telescopes which is then is distributed over the internet to more than 80 million personal computers in more than 200 countries around the world where the data is analyzed for evidence of signals not of natural origin and not due to terrestrial interference. The combined computing power of the distributed SETI@home network, which operates on the host computers as a screensaver, rivals that of the world's largest super computers and has resulted in tens of millions of amateurs involved in what must be the largest citizen science programs ever implemented. However, the amateur SETI effort has not been confined to data analysis, as a number of amateurs have built modest facilities which they have used to survey the sky for signals from extraterrestrial civilizations. Many of these amateur activities are coordinated by the SETI League which has more than one thousand members.

Naturally, there were many false alarms, perhaps the most dramatic being Jocelyn Bell's 1967 discovery of pulsars. After eliminating the possibility that the pulsating signals had a terrestrial origin, but noting that the pulsars had all the characteristics of terrestrial radar systems, Cambridge radio astronomers wistfully speculated that they might be observing interstellar beacons used to guide interstellar travel, and whimsically named them Little Green Men or LGMs (Bell 1984). Even earlier, Gennady Sholomitsky (1965) used a classified Soviet military facility in Crimea (Fig. 5.5) to observe the radio sources CTA 21 and CTA 102 which were known to have peculiar radio spectra (Kellermann

Fig. 5.5 The ADU-1000 antenna at the Evpatoria Deep Space Communication and Control Center in Crimea used by Gennady Sholomitsky in 1964 and 1965 to observe CTA 102. The array of eight 16 meter diameter antennas was built from parts of a Soviet battleship, a military railway bridge, and the hull of a captured Italian submarine. Credit: State Space Agency of Ukraine

et al. 1962). Sholomitsky, who was a student of the Russian astrophysicist Iosef Shklovsky, made the surprising discovery that over a period of only a few months, the strength of the radio source CTA 102 changed by about 30 percent (Sholomitsky 1965) (Fig. 5.6). It was difficult to explain such rapid variability in terms of what was then understood about synchrotron radiation from radio galaxies and quasars.[35] On 12 April 1965 Alexander Midler, a TASS (Soviet equivalent of the Associated Press) science reporter, overheard a discussion between Shklovsky and Nikolai Kardashev speculating that perhaps the observed radio emission from CTA 102 might have been generated by an extraterrestrial civilization, and later that day, TASS issued a "telegram" on the discovery by Soviet scientists of an artificial cosmic signal. At a large press conference widely attended by Soviet as well as foreign press, Shklovsky and Kardashev, playing with the assembled journalists, did not deny the possibility that the signals from CTA 102 were due to an extraterrestrial intelligence (Fig. 5.7). The press took it seriously, and the 14 April 1965 issue of *Pravda* reported that extraterrestrials were signaling the Earth. The startling news quickly spread to newspapers throughout the Soviet Union and around the

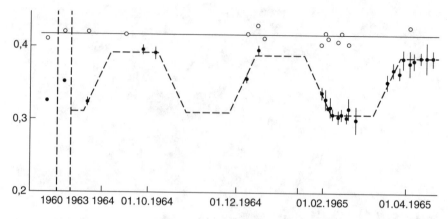

Fig. 5.6   Variability of CTA 102 from late 1963 to mid-1965 as it appeared in Pravda. Filled circles show the flux density of CTA 102 relative to 3C 48. The open circles are for CTA 21. Credit: Courtesy of the Sholomitsky family

Fig. 5.7   From left to right, on the steps of Moscow State University, Gennady Sholomitsky, Josef Shklovsky, and Nikolai Kardashev having a good laugh following the 12 April 1965 press conference. Credit: Courtesy of Sternberg Astronomical Institute, Moscow State University

world. The Byrds, a famous rock group, popularized a song about extraterrestrials from CTA 102 that even entered the peer reviewed scientific literature.[36]

Approaching the close of the twentieth century, the SETI Institute inaugurated a new study to review the status of SETI and plan for the future. Over an 18-month period in 1998 and 1999, the SETI Science and Technology

Working Group (SSTWG), led by Australian radio astronomer Ronald Ekers, met on four occasions to develop a strategy for both the near term as well as the long term opportunities for SETI. The Working Group report confirmed that searches in the electromagnetic spectrum should continue including optical as well as radio wavelengths; that both targeted and sky surveys are needed; that telescopes be developed with multiple beams; that searches include both narrow band signals as well as broad band pulsed emissions; that the focus should be on the detection of beacons, but although thought to be less likely, the eavesdropping scenario should not be excluded; and that multiple site detection systems be used to discriminate against terrestrial interference, thus greatly improving the credibility of the search (Ekers et al. 2002).

An important outcome of the SSTWG study, the impact of which went well beyond SETI, was the development of the Large-N-Small-D (LNSD) concept to build up a very large collecting area by using many small, possibly commercially available dishes and receivers, instead of a single large aperture. The SSTWG recommended the construction of the One hectare Radio Telescope (1hT) which later became the basis of the proposal to build the Allen Telescope Array. The LNSD concept was further developed by the US Square Kilometre Array (SKA) Consortium as part of a proposal from the United States to build the International SKA, consisting of thousands of 12 meter class dishes spread throughout the western United States. The United States later withdrew from the SKA project, but the International SKA mid-frequency design, which plans to include first hundreds then thousands of 15 meter (49 foot) diameter antennas to be erected at a remote site in South Africa, is also based on the LNSD concept developed by the SETI Institute's SSTWG (Sect. 11.7).

With the less than enthusiastic interest in SETI at NASA and at the NSF, US SETI activities have depended more and more on growing private philanthropy. In addition to the modest donations from thousands of individuals, people such as Bernard Oliver from Hewlett Packard, Microsoft co-founder Paul Allen, and Nathan Myhrvold, who made his fortune as Chief Technology Officer at Microsoft, generously supported the search for extraterrestrial civilizations. A potentially major advance to SETI research came from Paul Allen's gifts amounting to $30 million to construct the first radio telescope specifically dedicated to the Search for Extraterrestrial Intelligence. The Allen Telescope Array (ATA) was expected to include 350 six meter diameter dishes to obtain a total collecting area of about 10,000 square meters—roughly the equivalent of a 100 meter diameter steerable dish antenna. The monolithic parabolic reflecting surface of each antenna was designed to be stamped out from a single mold instead of the conventional construction method using a large number of small panels. However, the design and construction of the antenna structure and the associated electronics proved to be more complicated and more expensive than anticipated, so the available funds only permitted the construction of 42 of the originally planned 351 antennas or 12% of the planned collecting area.

The ATA, which is located at the Hat Creek Radio Observatory in northern California, began operation in 2007. As has been the case for many modern radio telescopes, especially in the United States, operating funds turned out to be even more difficult to raise than construction funds. Following a brief funding hiatus in 2011, the ATA has been operated by SRI International with support largely from the US Air Force for a variety of programs, including SETI as well as conventional radio astronomy, to complement several Defense Department programs.

5.6   SETI in the USSR

Interestingly, outside of the United States, it was primarily scientists in the Soviet Union, inspired by the imaginative astrophysicist Iosef Shklovsky, who have been seriously involved in SETI research.[37] Unlike the American programs which concentrated on high spectral resolution to look for very narrow band radio signals, the Soviet effort, largely led by Vsevolod Troitsky, considered broad band very short time duration pulses as well as monochromatic radio emissions.[38] (See Troitsky et al. 1971, 1974). On the theoretical side, Nikolai Kardashev's famous paper on Type I, Type II, and Type III civilizations focused on the development of civilizations based on their ability to harness the energy from their planet, their sun, and their galaxy respectively (Kardashev 1964). Kardashev also argued that the broad spectral region near the 1421 MHz hydrogen line had minimum noise and so was optimum for SETI research.

Characteristically, the Soviet SETI program was more structured than its American counterpart. Like much scientific research in the USSR, especially if related to electronics, computers, or other high tech instrumentation, there was a significant government component, generally through the Soviet Academy of Sciences. Drake even remarked,

> I believe … it had little to do, in my opinion, with a wide regard for the search enterprise itself, but was more of a reflection of the state of Soviet science in general, particularly the lack of peer review. There was also a political motive behind the government support for these activities: The authorities correctly perceived the search enterprise as an area where Soviets could compete with and possibly excel over American efforts (Drake and Sobel 1992, p. 96).

Following Project Ozma, in 1964 Soviet astronomers met at the Armenian Byurakan Astrophysical Observatory for the Soviet National Conference on Problems of Communication with Extraterrestrial Civilizations to set out a plan for future SETI research (Tovmasyan 1964). This early Soviet conference was a remarkable preview of SETI conferences that would be held over the course of the next half century and included speculations on the multiplicity of inhabited worlds, the existence of alien civilizations at various levels of development, the best means for establishing communication, how to distinguish an extraterrestrial signal sent by an advanced alien civilization from terrestrial interference, as well as the problems of linguistics. The conference concluded that it would be appropriate to begin a simultaneous program of transmission and reception of radio signals. Reacting to skepticism about transmitting, Shklovsky argued that starting to transmit was like burying a time capsule—only more expensive.

In 1972 the Soviet Academy considered and approved a national research program on "Communication with Extraterrestrial Civilizations" (Scientific Council of the USSR Academy of Sciences 1975). The intended Soviet program included targeted searches and all sky surveys at both radio and infrared wavelengths, searches for monochromatic signals as well as broad band pulses, and noted the need for special attention to develop the techniques that would be needed to decipher any received signals. Interestingly, however, there was no mention of transmitting in the 1972 plan.

In a noteworthy departure from the Cold War atmosphere of the time, in May 1971 scientists from the USSR and the US gathered at the Byurakan Observatory in Soviet Armenia for the first ever international conference about the search for extraterrestrial intelligence (Fig. 5.8). The Byurakan conference,

Fig. 5.8 Participants in the 1971 US-USSR SETI Symposium at the Armenian Byurakan Observatory. One of the present authors (KIK) is standing in the back 5th from the right wearing dark glasses. Standing in the front are Frank Drake (6th from left), Vitaly Ginzburg (7th from left), Carl Sagan (9th from left). Credit: KIK/NRAO/AUI/NSF

which was co-sponsored by the Soviet and US Academies of Science, was remarkable in bringing together scientists from a wide range of disciplines, including anthropology, linguistics, biology, and world history as well as astrophysics and SETI, to discuss the scientific basis for intelligent life elsewhere in the Universe; the technical and sociological challenges of SETI; and the moral, social, and legal implications of a successful detection. Critical to the meeting was Boris Belitsky, who served as the incredible bilingual nearly real-time translator. Belitsky was born in the United States and grew up in the Soviet Union with his parents who had emigrated in the 1930s. He was the English language science editor for Radio Moscow and had served as the translator at the 1960 trial of Francis Gary Powers after his U-2 spy plane was shot down by a Soviet missile over Soviet territory.[39]

The Byurakan conference confirmed that the search for extraterrestrial intelligence was technically feasible; that a successful detection would have a profound influence on the future of civilization on Earth; that a successful search would likely require a large expenditure of funds and resources, but that a modest start was feasible and was recommended. The group also made some specific recommendations for future work to search for civilizations not only at our own level of technical development, but also at a "level greatly surpassing our own" and discussed building "a decimeter radio telescope with an effective area ≥ 1 km^2." The recommendations included searches for narrow band signals, but also for strong impulsive signals from both ground and in space, and set the agenda for SETI research for the next decades.[40] Two later meetings of USSR and US SETI scientists were held in Tallinn, Estonia in 1981 (Sullivan 1982) and in Santa Cruz, California in 1991, although by the time of the Santa Cruz meeting SETI research had all but ceased in the rapidly crumbling USSR[41] (See Shostak 1993).

5.7   Continuing SETI Programs

The Soviet focus on the evolution of civilizations mirrored the development of SETI in the United States in the 1950s and 1960s. With time and the progress of astronomical research, the rate of star formation in the Galaxy, the fraction of stars with planets, and the number of habitable planets were reliably estimated, even measured, and is now known to be very large. Thus, the biggest uncertainty about the number of advanced communicative civilizations is still the formation and longevity of intelligent civilizations. But even as early as his 1961 lecture at the Philosophical Society of Washington, Drake commented,[42]

The number of civilizations with which we might communicate today is strongly affected by the average length of time during which technology and motivation allow a civilization to be contacted. We have a very poor estimate of this time at present, with a resultant large uncertainty in the number of civilizations we might contact.

Our increasing recognition of the ubiquity of planets around other stars in the Galaxy, combined with the improvement by many orders of magnitude of the sensitivity of SETI programs and the lack of any positive detections, has raised a concern for the future of humanity. As early as 1950, Enrico Fermi is said to have famously asked, "Where is Everybody?"[43] leading to many speculations ranging from the motivations of extraterrestrials to a worrisome short value of L due to the inevitable self-destruction of technological civilizations, whether from war or environmental negligence. The scientific debate over the lifetime of civilizations echoed the vociferous debates over nuclear weapons during the Cold War period. The threat of global annihilation particularly impacted US and Soviet SETI scientists, perhaps motivating them to better understand our place in the Universe and the lifetime of our own civilization. Indeed, at the Green Bank 25th anniversary Project Ozma celebration, Sebastian von Hoerner, who had served in the German army on the Russian front and later lived through the 1945 Allied destruction of Dresden, explained that the aftermath of Dresden was "much more gruesome" than any fictional apocalypse, and that it will be not the day after a nuclear war but "it is the year after, when all of the food has been eaten up and when the thin skin of our civilization comes peeling off in large chunks." (von Hoerner 1985, p. 3).

In 2016, Frank Drake and SETI returned to Green Bank. Yuri Milner, a Russian billionaire, has pledged $100 million over a ten-year period to support SETI investigations at the Green Bank Telescope, the Parkes 210 foot dish, the 500 Meter FAST fixed radio telescope in central China, and elsewhere. Milner's *Breakthrough Listen* project will search a million Galactic stars and 100 nearby galaxies. This project is led by former Air Force General Pete Worden, the former Director of the NASA Ames Research Center and one of the leaders of President Ronald Reagan's Strategic Defense Initiative (also known as Star Wars), with the support of Frank Drake, Sir Martin Rees, and the late Stephen Hawking. Milner started out studying physics at Moscow State University. Apparently his teacher, Andrei Sakharov, was not impressed with Milner's promise in physics, and suggested that he might consider a different career. This inspired Milner to study business at the University of Pennsylvania and to go on to a very successful career in finance. More than half a century after Frank Drake's Project Ozma, with Milner's generosity, Green Bank is again probing the skies for signals from extraterrestrial intelligent civilizations, but with more than a million times better sensitivity than Project Ozma. Thus *Breakthrough Listen* can detect in one second a signal that would have taken Project Ozma 100,000 years. On the downside, the radio spectrum is vastly more polluted now than it was more than half a century ago for Project Ozma. Today, SETI has to deal with the proliferation of interfering signals from commercial, military, and scientific satellites, as well as from a wide range of terrestrial transmissions, interference which will only get worse with time.

In the more than half-century since Project Ozma, a small group of dedicated enthusiasts such as Kardashev, Oliver, Drake, Bracewell, Kraus, Morrison, and later Tarter have remained passionate about the need to search for alien intelligence. Others such as University of Virginia's Robert Rood, Shklovsky, and von Hoerner later became disillusioned about the prospects for a successful detection, while yet others saw SETI as a backdoor approach to fund new large radio telescope systems, or to protect the radio spectrum from interference. Further, the continued lack of any observational evidence for extraterrestrial civilizations has led to serious objections from some about the justification for SETI research. Others, such as the SETI Institute's Jill Tarter (2010), point out that only a very small amount of the nine-dimensional "cosmic haystack" (3 directions, time, frequency, modulation, sensitivity, and 2 polarizations) has been searched for the illusive needle in the cosmic haystack. Sir Martin Rees famously pointed out, "The absence of evidence is not evidence of absence," (Oliver and Billingham 1973, p. 3) a remark arguably made even more famous by Carl Sagan in defense of his continued support for SETI. Starting with Drake's Project Ozma, NRAO has treated proposals for SETI the same as more conventional astronomical research programs. While there was no secrecy, neither was there any undue publicity associated with the NRAO SETI proposals. Even during the period when the NSF would not entertain grant proposals for SETI, NRAO continued to support modest SETI proposals on the 140 and 300 Foot Telescopes. Considering that the proposed NASA SETI program amounted to less than 0.1 percent of the NASA budget, one cannot help but wonder whether, if there had been less publicity and lobbying by the NASA SETI researchers, they might have escaped the scrutiny of Congress.

Probably no other subject in the history of science has had more conferences held or books published based on the absence of any experimental results. Indeed, it is hard to think of any other area of human inquiry where we know less about what we are looking for, where to look, how to look, or even if there is anything to look for. Astronomers know much more about the planets, stars, and galaxies than about the technology and motivation of extraterrestrials, but as discussed in Sect. 6.2, astronomers did not do a very good job of predicting cosmic masers, pulsars, quasars, solar and Jupiter radio storms, gravitational lenses, the cosmic microwave background, the rotation of Mercury and Venus, or indeed cosmic radio emission itself, until they stumbled across them in the course of other investigations, some of which were motivated by commercial and military goals or as demonstrations of national strength and prestige. SETI search strategy is an interesting intellectual exercise, but may not lead to anything. The first detection of an intelligent communicative extraterrestrial civilization may well come from a similar serendipitous discovery, perhaps even by an amateur, rather than a directed SETI investigation. Nevertheless, there is universal agreement, as argued in the National Academy of Science 1991 Decade Review of astronomy that, "A successful 'contact' would be one of the greatest events in the history of mankind." (Bahcall 1991b, p. I–13).

NOTES

1. ROTC is the Reserve Officers Training Corp, a university program to train military officers. ROTC students received a stipend to cover their education expenses toward a four year degree in return for a commitment to serve in the active military.
2. Drake was recruited by Dave Heeschen and joined NRAO on 1 January 1958 but spent the first few months finishing up his PhD thesis at Harvard and planning the 85 foot research program. For further information about Drake's childhood and education see Chap. 1 in Drake and Sobel (1992).
3. We will use SETI throughout this chapter to refer to both individual projects and the general field of research dealing with the detection of radio signals from extraterrestrial civilizations. The term SETI came into common use in the 1970s. Throughout the 1960s scientists referred to the various components of what became known as SETI by many names including interstellar signals, communication with extraterrestrial life, communicative civilizations. The term SETI can originally be traced back to the 1971 Communication with Extraterrestrial Intelligence (CETI) Conference held at the Byurakan Astrophysical Observatory in Armenia. At the conference, Drake, Kardashev, and Shklovsky acknowledged that you first have to find extraterrestrials before you can communicate with them. The phrase "search for extraterrestrial intelligence" first appears in Project Cyclops (Oliver and Billingham 1973), the 1972 NASA study on detecting extraterrestrial intelligent life. Another NASA study led by Phil Morrison in 1977 on the same topic was the first to use SETI as an acronym. SETI quickly became the standard terminology of the field.
4. The Zeeman effect results from a magnetic field which splits the 21 cm line of neutral hydrogen into multiple lines. Since the separation of the lines is proportional to the magnetic field strength, the Zeeman effect can be used to determine the magnetic field strength in hydrogen clouds.
5. Epsilon Eridani is now known to have at least one planet, known as Epsilon Eradani b, which orbits the star with a period of 6.85 years. However, Epsilon Eradani b has a mass about 1.5 times that of Jupiter and so may be unable to support any kind of life (Benedict et al. 2006).
6. Pulsars were unknown at the time, see Kellermann and Seielstad (1986), p. 25.
7. Otto Struve, Karl Taylor Compton Lecture Series, MIT Libraries, Institute Archives and Special Collections.
8. Actually, the Cocconi and Morrison paper was already published in the 19 September issue of *Nature*, but a copy had not yet reached Green Bank.
9. Drake to Paine (NSF), 19 October 1959, NAA-NRAO, DO, Conferences, Symposia, and Colloquia.
10. *New York Times*, 22 November 1960, p. E 11.
11. Indeed, a small band centered on the 1420.4 MHz hydrogen line is the only globally protected frequency on Earth.
12. Between the 1.4 GHz hydrogen line and the 1.7 GHz OH hydroxyl line is the water ($H + OH = H_2O$) hole.
13. Near Roswell, New Mexico, beginning on 7 July 1946, local ranchers found debris from a crash, and the US military quickly launched an effort to recover the debris. The official explanation of the event attributes this debris to the collapse of a military surveillance balloon. However, the more popular explanation

attributed the debris to a downed spaceship that contained alien life, which the US government covered up with their official version of the events. The Roswell UFO incident, as it was called, became the most famous supposed UFO encounter, though the 1950s and 1960s marked a high point in reports of such phenomena. For information on the Roswell Incident and encounters with extraterrestrial life, see Clark (1993).

14. Pearman to Villard, 13 March 1961, NAS-NRC-A, Organized Collections, SSB, Conferences, Extraterrestrial Intelligent Life.
15. Drake to Pearman, 13 March 1961, NAS-NRC-A, Organized Collections, SSB, Conferences, Extraterrestrial Intelligent Life.
16. Pearman to File, 9 June 1961, NAS-NRC-A, Organized Collections, SSB, Conferences, Extraterrestrial Intelligent Life.
17. Letter of invitation to the Green Bank conference from Otto Struve to approximately 20 invitees, NAA-NRAO, DO Conferences, Symposia, Colloquia.
18. People like Heinrich Hertz, Guglielmo Marconi, Nicola Tesla, Donald Menzel, and even Albert Einstein considered methods to communicate with extraterrestrials, particularly Martians.
19. See Pearman (1963) for a discussion of the 1961 Green Bank SETI Conference.
20. A webcast of the 2010 conference is available at https://vimeo.com/album/3095975
21. Sullivan interview of Drake, 27 April 1979, NAA-WTS, Individuals. https://science.nrao.edu/about/publications/open-skies#section-5
22. A non-binding protocol that recommends how the detection of extraterrestrial intelligence should be disseminated was authored by the International Academy of Astronautics and the International Institute of Space Law, and now has been adopted by a group of individuals and institutions participating in the search for extraterrestrial intelligence. http://www.seti.org/post-detection.html. However, like many other well-intentioned agreements, it was not signed by all potential signal recipients and, in any event, is not enforceable.
23. Part of the House Committee on Science and Technology.
24. The Library of Congress Science Policy Research Division, later known as the Congressional Research Service, provides bipartisan information to Congress.
25. Tarter et al. (1980) used the NRAO VLBI recording system (see Sect. 8.4) to digitally record a 360 kHz IF band which they subsequently analyzed in a CDC 7600 computer. In this way they were able to create an early form of post-observation Digital Signal Processing to implement a multichannel spectrometer with more channels and narrower bandwidths than then possible with conventional analogue hardware.
26. Probably the true cost of building Cyclops would have vastly exceeded the $10 billion estimate given in the Cyclops report, which was only $10 million per 100-m antenna exclusive of instrumentation. For the VLA, the cost of the instrumentation was about twice the cost of the antenna elements.
27. Press release from Senator William Proxmire, 16 February 1978.
28. "The Search for Signals from Space," *Parade Magazine*, 19 September 1993.
29. Senator Richard Bryan Press Release, 22 September 1993.
30. Congressional Record, 20 September 1993.
31. When launched in 1990 after more than a decade of development and huge cost overruns, the Hubble Space Telescope mirror was found to have been incorrectly ground, resulting in badly defocused images.

32. For a full list of SETI projects undertaken between 1959 and 1992 see Drake and Sobel (1992).
33. Between 1979 and 1982 John Kraus and his students published 4 volumes of *Cosmic Search*.
34. On 15 April 1977 Big Ear registered a short lived but strong signal, characteristic of a cosmic rather than terrestrial origin. When looking later at the recorded telescope output chart, Jerry Ehman wrote, "Wow." See Grey (2012).
35. The argument went that if a radio source varied on a time scale of 100 days it could not be more than 100 light days across, otherwise the signals from different parts of the source would arrive at different times and the variability would be smeared out. In such a small source, the density of relativistic electrons responsible for the observed synchrotron radiation would be so great that the radiation would be self-absorbed and not able to escape the source.
36. In 1968, after receiving the referee's report, Eugene Epstein of the Aerospace Corporation sneaked a note about the Byrds' CTA 102 song into their *Astrophysical Journal* paper about radio source variability. See Schorn et al. (1968). One of the present authors (KIK) was the referee of this otherwise serious publication.
37. In 1960, Shklovsky wrote a stimulating article in the Soviet journal *Priroda*, No. 7, 21 which was reprinted in *Interstellar Communication*, ed. A. G. W. Cameron (New York: W. A. Benjamin), p. 1.
38. Other related programs involved the Moscow Power Institute, the All-Union Electrical Engineering Institute of Communications, the Russian Language Institute, and the USSR Academy of Sciences.
39. Powers was flying over the USSR as part of a CIA mission to monitor Soviet nuclear weapons development. Unexpectedly, his U2 aircraft was shot down by a Soviet missile on 1 May 1960; Powers survived and was taken prisoner by the KGB. President Eisenhower, not knowing that Powers had survived, initially claimed that it was a weather plane that had drifted off course. Powers was tried and found guilty of spying, but was returned to the US in exchange for Soviet spy Rudolf Abel. Powers later died in a helicopter crash while reporting on Los Angeles traffic for a local TV station.
40. Reports of the 1971 US-USSR CETI Conference, which was attended by one of the authors (KIK), are given in a short report by the Organizing Committee, chaired by V.A. Ambartsumian (1972) and in a colorful article written by Freeman Dyson (1971).
41. *The Third Decennial US-USSR Conference on SETI* was held in August 1991, just a week prior to the ouster of the USSR leader Mikhail Gorbachev that led to the collapse of the USSR five months later.
42. Frank Drake, "Detection of Extraterrestrial Intelligent Life," Washington Philosophical Society, Smithsonian Institution Archives, Washington, DC. We are grateful to Steven Dick for providing us with copies of the records of Drake's talk.
43. This is known as "The Fermi Paradox."

BIBLIOGRAPHY

REFERENCES

Ambartsumian, V.A. 1972, First Soviet-American Conference on Communication with Extraterrestrial Intelligence (CETI), *Icarus*, **16**, 412

Bahcall, J. ed. 1991a, *The Decade of Discovery in Astronomy and Astrophysics: Report of the Radio Astronomy Panel* (Washington: National Academy Press)

Bahcall, J. ed. 1991b, *Working Papers: Astronomy and Astrophysics Panel Reports* (Washington: National Academy Press)

Bell Burnell, J. 1984, The Discovery of Pulsars. In *Serendipitous Discoveries in Radio Astronomy*, ed. K.I. Kellermann and B. Sheets (Green Bank: NRAO/AUI), 160 http://library.nrao.edu/public/collection/02000000000280.pdf

Benedict, G.F. et al. 2006, The Extrasolar Planet ε Eridani b: Orbit and Mass, *AJ*, **132**, 2206

Billingham, J. 1990. In *SETI Pioneers: Scientists Talk About Their Search for Extraterrestrial Intelligence*, ed. D.W. Swift (Tucson: University of Arizona Press), 246

Clark, J. 1993, *Extraordinary Encounters: An Encyclopedia of Extraterrestrials and Otherworldly Beings* (Detroit: Gale Research)

Cocconi, G. and Morrison, P. 1959, Searching for Interstellar Communications, *Nature*, **184**, 844.

Crews, F. 1986, Project OZMA – how it really was. In *The Search for Extraterrestrial Intelligence*, ed. K.I. Kellermann and G.A. Seielstad (Green Bank: NRAO/AUI), 27 http://library.nrao.edu/public/collection/02000000000301.pdf

Dick, S.J. 1993, The Search for Extraterrestrial Intelligence and the NASA HRMS: Historical Perspectives, *SSRv*, **64**, 93

Dick, S.J. 1998, *Life on Other Worlds: The 20th Century Extraterrestrial Life Debate* (Cambridge: CUP)

Drake, F.D. 1960, How Can We Detect Radio Transmissions from Distant Planetary Systems?, *S&T*, **19**, 140

Drake, F.D. 1961, Project Ozma, *PhT*, **14**, 40

Drake, F.D. 1979, A Reminiscence of Project Ozma, *CosSe*, **1** (1), 10

Drake, F.D. 1990. In *SETI Pioneers: Scientists Talk About Their Search for Extraterrestrial Intelligence*, ed. D.S. Swift (Tucson: University of Arizona Press), 54

Drake, F.D. 2010, *Astronomy Beat* (San Francisco: Astronomical Society of the Pacific)

Drake, F.D. and Sobel, D. 1992, *Is Anyone Out There? The Scientific Search for Extraterrestrial Intelligence* (New York: Delacorte Press)

Dyson, F. 1971, Letter from Armenia, *The New Yorker*, November 6, 126

Ekers, R.D. et al. eds. 2002, SETI 2020: A Roadmap for the Search for Extraterrestrial Intelligence (Mountain View, California: SETI Press)

Field, G. ed. 1982, *Astronomy and Astrophysics for the 1980s* (Washington: National Academy Press)

Garber, S.J. 1999, Searching for Good Science: The Cancellation of NASA's SETI Program, *JBIS*, **52**, 3

Grey, R.H. 2012, *The Elusive Wow: Searching for Extraterrestrial Intelligence* (Chicago: Palmer Square Press)

Kardashev, N.S. 1964, *AZh*, **41**, 282. English translation: Transmission of Information by Extraterrestrial Civilizations, *SAs*, **8**, 21

Kellermann, K.I. et al. 1962, A Correlation between the Spectra of Non-thermal Radio Sources and their Brightness Temperatures, *Nature*, **195**, 692

Kellermann, K.I. and Seielstad, G.A. eds. 1986, *The Search for Extraterrestrial Intelligence* (Green Bank: NRAO/AUI) http://library.nrao.edu/public/collection/02000000000301.pdf

McKee, R. and Taylor, J. eds. 2001, *Astronomy and Astrophysics for the New Millennium* (Washington: National Academy Press)

Morrison, P. 1990. In *SETI Pioneers: Scientists Talk About Their Search for Extraterrestrial Intelligence*, ed. D.S. Swift (Tucson: University of Arizona Press), 19

Morrison, P., Billingham, J., and Wolfe, J. eds. 1977, *The Search for Extraterrestrial Intelligence*, NASA SP-419 (Washington: NASA)

Oliver, B.M. 1979, Rationale for the Water Hole, *Acta Astronautica*, **6**, 71

Oliver, B.M. and Billingham, J. eds. 1973 revised, *Project Cyclops: A Design Study for a System for Detecting Extraterrestrial Intelligent Life*, NASA CR114445 (Originally published 1972, revised 1973, reprinted 1996 by the SETI League and the SETI Institute with additional material)

Palmer, P. and Zuckerman, B. 1972, Ozma Revisited, *NRAO Observer*, **13** (6), 26 https://science.nrao.edu/about/publications/open-skies#section-5

Pearman, J.P.T. 1963, Extraterrestrial Intelligent Life and Interstellar Communication: An Informal Discussion. In *Interstellar Communication*, ed. A.G.W. Cameron (New York: Benjamin), 287

Sagan, C. 1973, *The Cosmic Connection: An Extraterrestrial Perspective* (Garden City, N.Y.: Anchor Press)

Schorn, R.A. et al. 1968, Quasi-Stellar Radio Sources: 88-GHZ Flux Measurements, *ApJ*, **151**, L27

Scientific Council of the USSR Academy of Sciences 1975, *AZh.*, **51**, 112. English translation: The CETI Program, *SvA*, **18**, 669

Sholomitsky, G.B. 1965, *AZh*, **42**, 673, English translation: Fluctuations in the 32.5-cm Flux of CTA 102, *SvA*, **9**, 516

Shostak, S. ed. 1993, *The Third Decennial US-USSR Conference on SETI* (San Francisco: Astronomical Society of the Pacific)

Smith, M. 1975, *Possibility of Intelligent Life in the Universe*, Report prepared for the Committee on Science and Technology, US House of Representatives, Ninety-Fifth Congress (Washington: Government Printing Office), updated in 1977 to include new astrometric information and the status of the NASA SETI program

Struve, O. 1962, *The Universe* (Cambridge: MIT Press)

Sullivan, W. 1982, SETI Conference at Tallinn, *S&T*, **63**, 350

Tarter, J.C. et al. 1980, A High-Sensitivity Search for Extraterrestrial Intelligence at Lambda 18 Cm, *Icarus*, **42**, 136

Tarter, J.C. et al. 2010, SETI Turns 50: Five Decades of Progress in the Search for Extraterrestrial Intelligence, *Proc. SPIE 7819, Instruments, Methods, and Missions for Astrobiology XIII* doi:https://doi.org/10.1117/12.863128

Tovmasyan, G.M., ed. 1964, *Extraterrestrial Civilizations* (Yerevan: Armenian Academy of Sciences Press) in Russian; English translation for NASA and the NSF from the Israel Program for Scientific Translation, no. 1823 (Springfield, VA: U S. Dept. of Commerce)

Troitsky, V.S. et al. 1971, *AZh*, **48**, 645. English translation: Search for Monochromatic 927-MHz Radio Emission from Nearby Stars, *SvA*, **15**, 508

Troitsky, V.S. et al. 1974, *UsFiN*, **13**, 718. English translation: The Search for Sporadic Radio Emission from Space, *SvPhU*, **17**, 607, 1975

Verschuur, G. 1973, A Search for Narrow Band 21-cm Wavelength Signals from Ten Nearby Stars, *Icarus*, **19**, 329

von Hoerner, S. 1985, Life in Space and Humanity on Earth. In *The Search for Extraterrestrial Intelligence*, ed. K.I. Kellermann and G.A. Seielstad (Green Bank: NRAO/AUI) http://library.nrao.edu/public/collection/02000000000301.pdf

Zuckerman, B. and Tarter, J. 1980, Microwave Searches in the U.S.A. and Canada. In *Strategies for the Search for Life in the Universe*, ed. M.D. Papagiannis (Dordrecht: Reidel)

FURTHER READING

Bracewell, R. 1974, *The Galactic Club: Intelligent Life in Outer Space* (San Francisco: W. H. Freeman)

Cameron, A.G.W. ed. 1963, *Interstellar Communication* (New York: Benjamin)

Davies, P. 2010, *The Eerie Silence: Renewing Our Search for Alien Intelligence* (Boston: Houghton Mifflin Harcourt)

Dick, S.J. 1996, *The Biological Universe: The Twentieth-Century Extraterrestrial Life Debate and the Limits of Science* (Cambridge: CUP)

Dick, S.J. 2018, *Astrobiology, Discovery, and Societal Impact* (Cambridge: CUP)

Isaacson, H. et al. 2017, The Breakthrough Listen Search for Intelligent Life: Target Selection of Nearby Stars and Galaxies, *PASP*, **129**, 054501

Oliver, B.M. 1993, The Search for Extraterrestrial Intelligence, *Mercury*, **2** (2), 11

Sagan, C. and Drake, F.D. 1975, The Search for Extraterrestrial Intelligence, *Scientific American*, **232**, May, 80

Shklovsky, I.S. and Sagan, C. 1966, *Intelligent Life in the Universe* (San Francisco: Holden-Day)

SETI@50, Video presentation of talks at workshop held in Green Bank, 13–15 September 2010 on the occasion of the 50th anniversary of Project Ozma, http://library.nrao.edu/setitoc.shtml

Sullivan, W. 1966, *We Are Not Alone* (New York: McGraw Hill)

Swift, David S. 1990, *SETI Pioneers: Scientists Talk About Their Search for Extraterrestrial Intelligence* (Tucson: University of Arizona Press)

Tarter, J. 2001, The Search for Extraterrestrial Intelligence, *ARAA*, **39**, 511

Vakoch, D.A. and Dowd, M.F. 2015, *The Drake Equation: Estimating the Prevalence of Extraterrestrial Life through the Ages* (Cambridge: CUP)

CHAPTER 6

The Bar Is Open

With the successful completion of the 140 Foot Radio Telescope in 1965, and the increased use of the 300 Foot Telescope, NRAO finally began to serve its role as a national observatory. Any scientist with a good program had access to world class facilities without regard for institutional affiliation—a policy which later became known as "Open Skies."

These were exciting years for radio astronomy. In addition to NRAO, new powerful radio telescopes at Caltech, MIT, and Cornell in the US, and in Australia, Canada, the UK, and later the Netherlands and Germany, were making exciting new observations. During a period of only a few decades in the middle of the twentieth century, radio astronomers made a series of discoveries which fundamentally changed our understanding of our Universe and its constituents. Radio astronomy was at last recognized as a legitimate part of astronomy. NRAO flourished in this exciting environment and became the poster child for the National Science Foundation's (NSF) support for large expensive scientific facilities, an expansion of their traditional role of funding individual research grants.

6.1   NRAO REACHES MATURITY

With the dedication of the 140 Foot Telescope in 1965, NRAO was finally a true national observatory with multiple state-of-the art facilities. The 300 Foot Telescope had been in operation for three years, the new 140 Foot Telescope offered unprecedented opportunities for centimeter wavelength research, and the Green Bank Interferometer was about to begin operation (Fig. 6.1).

Administration Partly as a result of the separation of activities between Green Bank and Charlottesville, but due also to the increasing size of the NRAO operation and the resignation of the NRAO Business Manager Frank Callender in 1965, NRAO Director Dave Heeschen reorganized the administrative struc-

© The Author(s) 2020, corrected publication 2021
K. I. Kellermann et al., *Open Skies*, Historical & Cultural Astronomy,
https://doi.org/10.1007/978-3-030-32345-5_6

Fig. 6.1   Aerial view of the NRAO Green Bank site showing the three element inter-ferometer, the 300 Foot transit dish, and the 140 Foot Telescope in the background. Credit: NRAO/AUI/NSF

ture of the Observatory. Hein Hvatum and Ted Riffe (Fig. 6.2) were appointed as Assistant Directors of Technical Services and Administration, respectively, while Bill Howard became Assistant to the Director. Hvatum had responsibility for the electronics lab, the machine shops, and telescope operations. Riffe was in charge of business management and plant maintenance, and continued in his role as the NRAO Fiscal Officer. Hvatum moved to Charlottesville but the division heads who reported to him remained in Green Bank, creating an administrative challenge. Riffe remained in Green Bank, where he was responsible for the Green Bank site operations, but was replaced in 1968 by John Findlay as Assistant Director for Green Bank Operations. Fred Crews was formally in charge of only the Telescope Operations Division, but with time assumed more and more responsibilities for all operations in Green Bank. John Hungerbuhler, who had come to Green Bank to assist Max Small during the final years of the 140 Foot construction, was the Chief of the Engineering Division and Thomas Williams was in charge of Plant Maintenance. Heeschen ruled NRAO from Charlottesville with a firm (some would say iron) hand. Once, when describing the NRAO organizational structure, Heeschen drew two circles on a blackboard connected with a vertical line. In the upper circle he wrote "Me," and in the lower circle, "Everyone Else."

The commute between Green Bank and Charlottesville was not as easy one. In the first years after the move, before a series of road improvements, it would

Fig. 6.2 Ted Riffe served as the NRAO Assistant then Associate Director for Administration. Credit: NRAO/AUI/NSF

take a careful driver nearly three hours to make the trip between the two sites—more, if one were to get stuck behind a slow-moving logging truck. Several local speed traps along the way presented an additional challenge, but that did not inhibit the more ambitious drivers, including the NRAO Director, from an unofficial competition to see who could post the best time. Regular communication between Green Bank and Charlottesville was facilitated by a daily car shuttle and dedicated telephone lines. Each day cars departed from the two sites at precisely 9:00 a.m. carrying passengers, equipment, and data. The two cars met at a halfway point on a mountaintop which still shows the remains of a Confederate Civil War encampment (Fig. 6.3). The drivers exchanged cars so that each driver could return to his home, and the passengers and equipment continued to their destination, arriving at noon at the other site. Although visiting Green Bank observers could fly into more local airports in Elkins or Clarksburg, West Virginia, many chose to go through Charlottesville, where they could first meet with NRAO staff to discuss their observing programs. Green Bank staff returning from a trip would usually fly into one of the local airports where they would be met anytime of the day or night by an NRAO driver who would return them to Green Bank. The drivers were all long-time local residents who would help make the time pass by relating tales of local history, no doubt with some embellishment.

The split of operations between Green Bank and Charlottesville, and later Tucson and Socorro, was not without problems. Gone were the informal contacts between the scientists in Charlottesville and the engineers, who were mostly based at the observing sites, and the culture was akin to that of colonies

Fig. 6.3 Green Bank—Charlottesville daily shuttle cars meeting at the halfway point to exchange passengers, equipment, and data. Credit: NRAO/AUI/NSF

with all major decisions made at the Charlottesville Headquarters. Interestingly, after the Charlottesville office was opened in late 1965, it was mostly the US staff members who took advantage of the opportunity to move from Green Bank, leaving behind scientists who had come to NRAO from such varied places as England, Germany, Netherlands, Iraq, India, Poland, and Sweden. Often the only US member of the NRAO Scientific Staff found in Green Bank was one of the present authors, Kellermann. On one occasion, Sebastian von Hoerner's sons put up a sign on the road leading to the housing area reading, "You are now leaving the American Zone," a spoof on the then divided German city of Berlin.

In 1969, John Findlay was replaced by Mort Roberts as Assistant Director for Green Bank Operations, and in 1970, Dave Hogg returned to Green Bank as the Assistant Director. Hogg was succeeded by Bill Howard in 1974, when Hogg returned to Charlottesville to become Associate Director for NRAO Operations with broad responsibility for both the Tucson and Green Bank site operations. Reflecting the increased size and complexity of the Observatory, the Assistant Directors for Tucson and Green Bank Operations now reported to Hogg rather than Heeschen, who was increasingly turning his attention to the Very Large Array (VLA) (Chap. 7). Recognizing the increased responsibilities resulting from the growing administrative staff needed to deal with the VLA, as well as the increasingly complex NRAO operations, in 1976 Ted Riffe became the Associate Director for Administration and served in this capacity until he retired in 1987 and was succeeded by James (Jim) Desmond. Bill Howard left NRAO in late 1976 to become Assistant Director of the Astronomy Division at the NSF and was replaced by Ken Kellermann as Acting Assistant Director in Green Bank before Kellermann went to Germany to become director at the Max Planck Institut für Radio Astronomy in Bonn. Bob Brown became the Green Bank Assistant Director in June 1977, followed by Rick Fisher (1981), Martha Haynes (1982), and George Seielstad (1984). After Seielstad's departure to the more urban environment of Grand Forks, North Dakota, a number of technical and scientific staff cycled through as Green Bank Assistant Directors until 2008, when Karen O'Neill assumed that role.

Phyllis Jackson, who was born and educated in Marlinton, West Virginia, joined NRAO in April 1959 to provide secretarial support to the new Observatory in Green Bank. In 1965 Phyllis moved to Charlottesville with the Scientific Staff. For 38 years, until she retired in 1997, she ran the Director's office with great efficiency. Many believed that, in effect, Phyllis Jackson ran the organization. Among her numerous skills was her lightning fast typing speed, and she often typed scientific papers, correcting the spelling and English of young scientists lacking in those skills (Fig. 6.4).

The National Science Foundation Organizational changes also occurred at the NSF. Initially, NRAO reported to Randal (Randy) Robertson at the NSF, and was administered out of a separate division headed by Daniel Hunt, which

Fig. 6.4 Phyllis Jackson, longtime secretary to NRAO Directors, ca. 1983. Credit: NRAO/AUI/NSF

dealt with all of the National Facilities such as NRAO and the Kitt Peak National Observatory (KPNO). In this way, funding for grants to individual scientists, which was under the Division of Astronomy headed by Robert Fleisher, was kept separate from that of the National Centers. However, in 1976 the NSF placed the National Astronomy Centers under the Astronomy Division. This pitted NRAO funding against proposals for individual research grants, setting up a competition for funding that continues to this day.

Mark Price, who had previously been the first NSF Spectrum Manager, became the Acting Director of the new Astronomy Division until Bill Howard was appointed as the permanent Director in November 1976. Under Howard, Price remained in charge of the individual investigator grants program, and Ronald LaCount administered the facilities. Price left the NSF in 1979 to accept an appointment as Chair of the Physics Department at the University of New Mexico in Albuquerque, and was replaced by Morris Aizenman as head of the grants program. Throughout his tenure as AST Director, Howard strove to keep the centers and grants funding in the constant ratio of 2 to 1, but resigned in 1982 over the contentious issues surrounding the abandonment of the controversial NRAO 25 meter millimeter wave telescope (Sects. 8.7 and 10.3).

On 1 May 1986, the Astronomy Division was moved from the Directorate for Astronomy, Atmospheric, Earth, and Ocean Sciences (AAEO), sometimes referred to as the division for "Earth, Air, Fire, and Water," to the Directorate for Mathematical and Physical Sciences, and AAEO became simply the Directorate for Geosciences.

Scientific Staff The NRAO scientific staff was structured following the AUI policy originally put in place for Brookhaven. This typically consisted of a series of two- or three-year term appointments as Assistant, then Associate Scientist, followed by an "up-or-out" tenure decision and promotion to the rank of Scientist. Members of the NRAO Scientific Staff generally divided their time between independent research and the development of new facilities and instrumentation as well as providing support for visiting users. Although he delegated most operational responsibilities to others, throughout his tenure as the NRAO Director Dave Heeschen remained in direct charge of the scientific staff.

In addition to the regular scientific staff, there was a steady flow of Research Associate or postdoctoral appointments meant to give new PhDs full time opportunities for research. Research Associates generally had no observatory responsibilities and were expected to move on after a two- or three-year temporary appointment. To give the Research Associate appointments more prestige and to compete with the NASA Hubble Fellowship and various university prize fellowships, the NRAO Research Associates were later called Karl Jansky Fellows, and the doctoral students became Grote Reber Predoctoral Fellows.

Tenured appointments at NRAO were granted by AUI in recognition of outstanding research accomplishments or other intellectually creative activity appropriate to the mission of the Observatory, and could only be terminated by the AUI Board for financial exigency or what AUI called "moral turpitude," a concept which no one really understood, but which was apparently never tested. The meaning of tenure at an organization such as AUI is unclear, but was intended to offer some degree of security of continued employment. Tenured staff were expected to provide leadership in the planning, design, construction, and operation of the NRAO facilities as well as carry out a vigorous program of individual research. Tenured staff members had some freedom to direct their own efforts, but, starting in the 1970s, the design and operation of the VLA, and then the Very Long Baseline Array (VLBA), required more staff support than could be provided by the limited number of tenured staff, and NRAO adopted the practice of offering "continuing" or "indefinite" appointments that were neither tenured nor term appointments. Those scientists with continuing appointments reported directly to the relevant site director. However, with the increasing autonomy of the Green Bank, Tucson, and Socorro sites, the individual site directors complained that they had no management control over the tenured scientists at their site who reported only to the distant NRAO Director, and with time the tenured staff became more integrated into the site operational structure. This was especially true after 1999, when the new AUI President, Riccardo Giacconi, insisted that each staff member be assigned well-defined "functional responsibilities," a concept that he had introduced earlier as Director General of the European Southern Observatory (ESO).

During the 1960s and early 1970s many scientists rotated through NRAO, either after a Research Associate appointment, or following one or more term appointments on the scientific staff. They were generally able to find good academic positions in the post-Sputnik market, often with better professional opportunities than they might have had at NRAO. Many went on to distinguished careers as leaders in radio astronomy and started their own radio astronomy groups, and, in some cases, developed their own facilities or became Principal Investigators on NASA missions.

As discussed previously (Sect. 4.2), the early NRAO Scientific Staff was dominated by recent graduates from Harvard, but in 1959, Roger Lynds, who had received his PhD from the University of California, Berkeley, joined the NRAO staff after a one-year appointment in Canada. Following his Green Bank visit as a University Toronto graduate student, Dave Hogg joined the staff as an Assistant Scientist in 1961, followed by Frank Low from Texas Instruments in 1962 (Sect. 10.1), Peter Mezger from Germany in 1963, Mort Roberts in 1964, and by Barry Clark and Ken Kellermann from Caltech in 1964 and 1965 respectively. Unlike the other Harvard recruits, Roberts had received his PhD from the University of California, Berkeley, after which he became a Research Associate at Harvard before coming to NRAO.

In order to provide some theoretical support to the young scientific staff whose interests and expertise were mostly in technical and observational areas, in 1962 NRAO recruited the theoretical astrophysicist Sebastian von Hoerner from Germany. When the FBI called NRAO as a follow-up to von Hoerner's visa application, Ted Riffe received the call as the Head of Administration. Riffe responded to the routine questions confirming that von Hoerner would have a salary so would not become a welfare recipient and that he was not a known criminal. But when asked what he would be doing at NRAO, Riffe, not knowing anything about radio astronomy replied, "I can't tell you," to which the caller, assuming that NRAO was involved in some highly secret activity that was above his clearance level, responded, "Oh, I understand!"

At NRAO, von Hoerner initially did research on some cosmological problems, but infected by the Green Bank culture, he soon turned his attention to observing with the 140 Foot Telescope and to the design of large radio telescopes; he became a major player in the NRAO Largest Feasible Steerable Telescope program (Chap. 9). NRAO later recruited three young theoretical astrophysicists, Robert (Bob) Brown, Robert (Bob) Hjellming, and David DeYoung, but they too were seduced by the opportunities to make new discoveries by observing with the NRAO telescopes. Hjellming worked with Campbell Wade observing radio stars using the Green Bank Interferometer (Sect. 6.2) and later wrote the VLA Observing Guide (*The Green Book*) and was in charge of the VLA off-line software development (Sects. 7.6 and 7.7). Brown and DeYoung became dedicated observers, and both went on to assume responsible administrative positions. DeYoung left NRAO to join the Kitt Peak National Observatory, later the National Optical Astronomy Observatory (NOAO) where he rose to become the Associate Director. Bob Brown became a skilled observer, and after serving as Assistant Director for both Green Bank and Tucson Operations, became the Associate Director for Charlottesville Operations in 1985 and assumed responsibility for the Scientific Staff and NRAO Scientific Services. Starting in 1991, Brown oversaw the development of the NRAO Millimeter Array (MMA) and later ALMA, and became the NRAO Deputy Director in 2000. In 2003 Brown left NRAO to become Director of the Cornell National Astronomy and Ionospheric Center (NAIC) which oversaw operation of the Arecibo Observatory.

In order to get advice on Observatory priorities and policies, Heeschen instituted an Observatory Council, which met monthly and included all tenured members of the Scientific Staff as well as the key Division Heads and other administrative staff. In the first years following the completion of the 140 Foot Telescope, NRAO had a generous budget for new equipment. At least once a year the Council would hear from the Electronics Division about proposed new receivers or other instrumentation. Heated discussions followed. Each scientist vigorously defended personal priorities, but Weinreb generally decided what to do based more on technical considerations rather than astronomical whims. In those years only capital costs were considered, as it was assumed that the engineers and technicians were getting paid anyway. Invariably the

Electronics Division would take longer to deliver the promised instrumentation, as each engineer would be in charge of multiple projects. However, the engineering talent and motivation were exceptional, and soon the world looked to NRAO for expertise in radio astronomy instrumentation.

Publications In the early years of the Observatory, NRAO debated where to publish scientific and technical papers on radio astronomy. The *IRE Transactions of the Professional Group on Antennas and Propagation* was considered, but radio astronomy was more than just antennas. At the suggestion of AUI's Richard Emberson, there was broad agreement to use the *Astronomical Journal*, but that never materialized. Apparently the NRAO radio astronomers still did not consider their work to be appropriate for the *Astrophysical Journal*, and in 1961 NRAO started its own publication series as had been common at many optical observatories. By the mid-1960s, however, the NRAO staff was finally publishing largely in the peer-reviewed journals, and the *Publications of the National Radio Astronomy Observatory* was terminated after two years and 17 papers. In principle the Director was responsible for all Scientific Staff publications, but in order not to delay timely publications, Heeschen only asked that he be shown all papers; if he did not respond within 24 hours, that was to be considered approval to submit the paper. In order to allow publication of NRAO data by visiting scientists with no grant support, NRAO paid for publication costs, most of the cost of travel to Green Bank, and later to the other sites, as well as making the NRAO computing facilities available at no cost. MIT Professor Bernard Burke once explained that it was cheaper to pay his students to travel and to pay for Green Bank lodging than to pay the high costs of using the MIT computing facilities.

The Green Bank US-USSR Radio Astronomy Conference Radio astronomers, as a species, are prolific travelers. As we have noted (Sect. 2.4), both Frank Kerr from Australia and Henk van de Hulst from the Netherlands happened to be at Harvard when Ewen detected the 21 cm hydrogen line. Earlier van de Hulst had visited Grote Reber in Wheaton, Illinois. Due to their relative isolation, Australian radio astronomers, including Taffy Bowen,[1] Joe Pawsey, and John Bolton, made frequent round-the-world trips to learn about radio astronomy progress in Europe and the US. Cambridge was more secretive about their work, but Peter Scheuer did spend two years in Australia working with Bernie Mills. It was not uncommon for this closely knit international community to stay in the homes of colleagues when visiting other observatories, and many lifelong international friendships were established among the radio astronomers as well as among their spouses.

An exception during the Cold War period was the USSR. Travel to the USSR was difficult due to both visa restrictions and the language differences. It was even more difficult for Soviet scientists to travel to the West. Most Soviet papers were published in Russian, while the availability of Western journals was very limited within the USSR. To foster better contact between Soviet and American radio astronomers, in 1961 Otto Struve organized a joint US-USSR Symposium on radio astronomy[2] that was held under the exchange agreement between the US and Soviet Academies of Science (Fig. 6.5). After the opening session held in Washington on the afternoon of 15 May 1961, all the delegates proceeded by overnight train to White Sulphur Springs, West Virginia, and then by bus to Green Bank to continue the following afternoon. Six Soviet and 28 American scientists participated in what was the first of many international scientific meetings held in Green Bank. Viktor Vitkevich led the Soviet delegation, while Otto Struve was the host in Green Bank. Frank Drake reviewed the history and status of US radio astronomy and Vitkevitch did the same for the Soviet Union. Most of the talks dealt with reports of recent observations or theoretical considerations, but on the last day, both sides presented their ideas on the design and construction of new radio telescopes.

In addition to the official presentations, informal groups discussed source polarization, 21 cm line studies, discrete sources, and radio telescope design. NRAO staff hosted members of the Soviet delegation at dinner parties in their Green Bank homes. Following the conference, the Soviet delegation travelled to Washington to visit the White House and the Carnegie Institution Department of Terrestrial Magnetism, but their request to visit the Sugar Grove station was not approved.

Fig. 6.5 Attendees at the USA-USSR Radio Astronomy Symposium, Green Bank, May 1961, left to right: Row 1 (seated)—G. Getmantsev, F. Haddock, M. Wade, S. Edmundsen (interpreter), R. Minkowski, V. Vitkevitch, O. Struve, R. Sorochenko, J. Firor, G. Keller, A. Kuzmin, R. Bracewell, F. Drake. Row 2—C. Wade, E. McClain, V. Sanamyan, P. Kalachev, G. Stanley, A. Barrett, H. Weaver, G. Swenson, C. Mayer, D. Heeschen, J. Kraus. Row 3—G. Field, T. Menon, C. Seeger, L. Woltjer, A. Sandage, A. Lilley, A. Blaauw, F. Kahn, B. Burke. Absent—G. Burbidge, J. Findlay, C. Lynds. Credit: NRAO/AUI/NSF

6.2   First Scientific Studies

While Berkner, Greenstein, Menzel, and the others involved in setting up NRAO certainly were looking forward to the expected contributions of radio astronomy, probably no one really appreciated the full potential of this new window on the Universe. As early as 1954, John Hagen had already reflected that "while we have perhaps 'skimmed the cream' of the top of radio astronomy, I feel it has a great future."[3]

Indeed, following the earlier discoveries of the non-thermal galactic radio emission, the hot solar corona, solar radio bursts, radio storms on Jupiter, radio galaxies, and the radio supernovae remnants that we discussed in previous chapters, in less than a few decades, radio astronomers went on to make another series of remarkable discoveries. They reported the first evidence for cosmic evolution, discovered quasars, pulsars, the microwave background, interstellar molecules, radio recombination lines, cosmic masers, the greenhouse effect on Venus, Jupiter's radiation belts, the first extra-solar planets, made the most precise tests of general relativity, and detected the first observational evidence for gravity waves (see e.g. Kellermann and Sheets 1984; Wilkinson et al. 2004). During the same period, active radar experiments disclosed the unexpected rotation of Mercury, were able to measure the rotation of cloud-covered Venus, determined the Astronomical Unit with new precision, and conceived of and confirmed a new fourth test of general relativity. Finally, starting with a simple experiment in Green Bank, the human race began the first credible research for intelligent counterparts elsewhere in the Universe. Subsequently, observations in other regions of the spectrum, obscured by the Earth's atmosphere, were opened by space-based observatories. But the radio spectrum was the first to be explored outside the narrow optical window, and the discoveries rolled in. These were truly the golden years of radio astronomy, and astronomy, limited for millennia to optical observations, would never be the same (Fig. 6.6).

A discussion of all these discoveries is beyond the scope of this book, and so we confine our discussion below to the early research with the Green Bank facilities.[4]

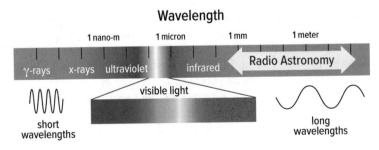

Fig. 6.6 Electromagnetic spectrum showing the range covered by radio astronomy. Credit: J. Hellerman/NRAO/AUI/NSF

The Tatel Telescope Observations with the new 85 foot Tatel Telescope were largely devoted to continuum studies of radio galaxies, supernovae remnants, and galactic ionized hydrogen (H II) regions and Solar System planets. Dave Heeschen (1961a) inaugurated a program of studying radio source spectra, with observations at four frequencies spanning the range from 440 MHz (68 cm) to 8 GHz (3.75 cm), and was able to confirm the reported decrease in the flux density of the Cas A supernova remnant of 1 to 2% per year at 1.4 GHz (Heeschen 1961b). Campbell Wade, Roger Lynds, and Hugh Johnson used the Tatel Telescope to study individual radio galaxies and the weak radio emission from the nearby normal galaxy M31. Dave Hogg, then a visiting graduate student from Canada and the first international observer at NRAO, used the Tatel Telescope to observe supernovae remnants and H II regions, while University of Indiana student Yervant Terzian studied planetary nebula, and Kochu Menon and Roger Lynds observed the thermal emission from the Orion Nebula and other H II regions. Frank Drake used observations at 3.75 cm to show the complex structure of the Galactic Center region.

Perhaps the most important observations made with the Tatel Telescope were Frank Drake's studies of radio emission from the planets that established the high temperature and greenhouse effect on Venus, the internal heating of Jupiter and Saturn, and the non-thermal radiation from Jupiter's Van Allen belt, all of which preceded confirmation by widely heralded NASA space-based observations.

The first spectroscopic observations at NRAO were by MIT student Sander (Sandy) Weinreb who brought his 21 channel digital autocorrelation receiver to the Tatel Telescope to search for the 327 MHz deuterium line and to try to detect the effect of Zeeman splitting of the 21 cm hydrogen line from interstellar magnetic fields. Although Weinreb did not detect either deuterium or the Zeeman effect after hundreds of hours of integration, his digital spectrometer (Weinreb 1961, 1963) was the forerunner of the very successful series of future NRAO autocorrelation spectrometers, as well as the implementation of similar devices at essentially every radio observatory in the world.

The 300 Foot Transit Telescope One of the first projects on the 300 Foot Telescope was a survey by Dave Heeschen (1964) of normal galaxies, while Ivan Pauliny-Toth et al. (1966) studied all sources in the 3C catalogue at 750 (40 cm) and 1400 MHz (21 cm) with the aim of determining accurate flux densities and positions free of interferometer lobe ambiguities. This work was later extended by Bridle et al. (1972), who used the 300 Foot Telescope to compile a complete sample of 234 sources above 2 Jy at 1.4 GHz. Dave Hogg (1966) observed a number of supernovae remnants, and Gart Westerhout showed up each summer with his family and his group of University of Maryland students to map the distribution of H I in the Galaxy (Fig. 6.7). At the same time, Mort Roberts used the 300 Foot Telescope to study H I in nearby galaxies. A particularly notable study was the investigation of the Andromeda galaxy (M31) by Roberts and Whitehurst (1975) which was able

Fig. 6.7 Gart Westerhout (left) and Frank Drake examining 300 Foot telescope output. Credit: NRAO/AUI/NSF

to probe the kinematics of the Galaxy way beyond its optical boundary, convincingly demonstrating the existence of dark matter, suggested decades earlier by Fritz Zwicky (1933).

The 140 Foot Telescope The first astronomical observations with the 140 Foot Telescope were made in March 1965, even before the formal acceptance and dedication of the telescope in October. Earlier, Sebastian von Hoerner (1964) developed a widely used technique for restoring radio source brightness distributions from observations of lunar occultations, and, as soon as he could, von Hoerner began using the 140 Foot to observe lunar occultations of radio sources to determine the positions and angular structure. He was assisted by Joe Taylor, then a graduate student at Harvard who was working with von Hoerner on his PhD thesis.

When the 140 Foot Telescope was completed in 1965, there was only one known radio spectral line, the 1420 MHz (21 cm) line of neutral hydrogen. Van de Hulst's classic 1945 paper only devoted a few paragraphs to the 21 cm hydrogen line (van de Hulst 1945). The rest of the paper was devoted to free-free emission from ionized hydrogen and the possibility of detecting high order radio recombination lines (RRL) at radio wavelengths corresponding to changes in an electron energy level. However, van de Hulst erroneously concluded that due to Stark broadening, RRLs would not be observable, and until Nikolai Kardashev (1959) showed that the effects of Stark broadening were previously overestimated, there were no serious attempts to observe RRLs. At the 1962 IAU General Assembly in Hamburg, two Soviet groups from Moscow and Leningrad reported the detection of RRLs at 8872.5 MHz (H90α)[5] and 5763 MHz (H104α) respectively. However, due in part to the limited information about the instrumentation permitted by Soviet authorities and the poor quality of the visual presentation, compounded with language difficulties, their results were not generally accepted by radio astronomers in Europe or the US.

Peter Mezger had joined the NRAO scientific staff from Germany and first tried to detect RRLs using the Tatel Telescope in late 1964, but the results were inconclusive, and Mezger eagerly awaited completion of the 140 Foot to search for RRLs. Even before the formal acceptance of the telescope by NRAO, Mezger and Bertil Höglund, a visiting scientist from Sweden, shared time with the painters working during the day, and on 9 July 1965, they were able to detect the 5009 MHz (6 cm) H109α RRL in the galactic H II regions M17 and Orion. Recognizing the importance of demonstrating that the 140 Foot was really working and had made a major discovery, Höglund and Mezger quickly sent off a letter to *Nature*. When they didn't hear from *Nature* after a few weeks, Mezger sent a telegram to the editor, demanding prompt action, but was told in response that "*Nature* will not be dictated to with respect to publication. Your paper is rejected." Dave Heeschen then pulled some strings and their paper was published in *Science* (Höglund and Mezger 1965) (Fig. 6.8).

Mezger, T. K. Menon, and others at NRAO, as well Ed Lilley at Harvard and Bernie Burke at MIT, were excited by the new research opportunities made possible by studying RLLs. Lilley sent two of his graduate students, Pat Palmer and Ben Zuckerman, to Green Bank to use the 140 Foot to study recombination lines from helium as well as hydrogen transitions with $\Delta n = 2$. Burke sent two of his students, Ted Rifenstein and Thomas Wilson, who worked with Mezger and Wilhelm Altenhoff on a variety of galactic continuum as well as recombination line studies.

Concurrently with the RRL studies, Kellermann and Pauliny-Toth (1968, 1969) initiated a long running program to study extragalactic radio source spectra and variability. Their 140 Foot data were supplemented by 750 and 1400 MHz data from the 300 Foot along with low frequency data from Cambridge (Kellermann et al. 1969) and demonstrated the nature of very compact radio sources. Exploiting the high frequency capability of the newly

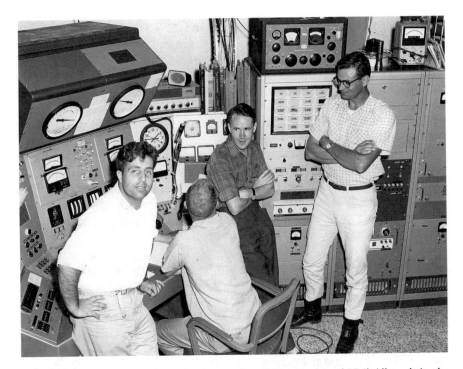

Fig. 6.8 Peter Mezger, Troy Henderson, Bertil Hoglund, and Neil Albaugh in the 140 Foot control room, 27 July 1965, a few weeks after their discovery of radio recombination lines. Credit: NRAO/AUI/NSF

completed 140 Foot Telescope, Pauliny-Toth and Kellermann (1966) followed the dramatic changes in the centimeter wavelength emission of radio galaxies and quasars which challenged existing radio source theories. During this same period, Heeschen (1968) used first the 140 Foot and then the 300 Foot and the Green Bank Interferometer (GBI) to detect the first active galactic nucleus in an Elliptical galaxy.

The biggest scientific impact of the 140 Foot Telescope was no doubt in the discovery and investigation of interstellar molecules. At the 1955 Jodrell Bank Symposium on Radio Astronomy, Charles (Charlie) Townes presented a classic paper speculating on the potential to detect lines at radio frequencies due to various atomic and molecular rotation transitions. Townes was an accomplished physicist who would share the 1964 Nobel Prize for Physics for the invention of the maser. In the published version of his Jodrell Bank paper, Townes (1957) derived microwave transition frequencies for a number of molecules including carbon monoxide (CO), ammonia (NH_3), water (H_2O), hydroxyl (OH), and hydrogen cyanide (HCN).

The calculation of the transition frequencies of ammonia and water go back even earlier than Townes' presentation at the Jodrell Bank Symposium. While working at Bell Labs following WWII, Townes (1946) and Townes and Merritt

(1946) published two seminal papers on the 22 GHz (1.3 cm) spectra of ammonia and water respectively that would later lead to the important discovery of these molecules at Hat Creek by Cheung et al. (1968, 1969) at 1.3 cm. The water lines turned out to be surprisingly strong due to maser action. But to the embarrassment of NRAO, an earlier proposal to detect the 1.3 cm water line by NRAO postdocs David Buhl and Lewis Snyder was turned down on the advice of a referee with distinguished theoretical credentials, but who had not considered the possibility of water masers.

Starting in 1967, the 140 Foot Telescope was used regularly for very long baseline interferometer high resolution studies of the continuum radiation from quasars and for OH, and H_2O interstellar maser emission (Sects. 8.1–8.5). The need to simultaneously schedule the 140 Foot together with an increasing number of other radio telescopes, each with their own scheduled maintenance, became a challenge to NRAO management.

The Green Bank Interferometer (GBI) Although the Green Bank Interferometer was built primarily to give the young NRAO staff experience in interferometry, it was a powerful research instrument as well. The GBI complemented the Caltech interferometer that operated at longer wavelengths and over a shorter baseline and so had less angular resolution. Moreover, the GBI was probably unique in not being laid out in an east-west or north-south baseline. Even though the 243 degree azimuth of the GBI was determined by the Observatory site geometry, it turned out to have some unanticipated advantages in making absolute position measurements (Wade 1970). Dave Hogg et al. (1969) mapped the brightness distribution of several bright radio galaxies and the Cas A supernova remnant. Heeschen (1968) used the GBI to show that the radio sources in NGC 1052 and NGC 4278 were small and confined to the galactic nucleus, in contrast to radio galaxies whose radio dimensions typically exceeded the optical boundaries of the galaxy. Later, using the GBI, Hjellming and Wade (1971) discovered the first true radio stars.

The addition of the remote antenna at Spencer's Ridge and Huntersville (Sect. 7.3) improved the angular resolution of the GBI by more than a factor of ten, opening up new research opportunities. Probably the most exciting observation was the discovery by Bruce Balick and Bob Brown of the small radio source at the Galactic Center which they showed was less than 0.1 arcsec (1000 AU) (Balick and Brown 1974). Brown later named this compact source Sgr A*. This was the first evidence for a massive black hole at the center of a normal galaxy. Later VLBI observations using the 140 Foot together with other radio telescopes showed that Sgr A* was less than 10 AU in extent (Kellermann et al. 1977). Perhaps the most important GBI program, however, was the series of elegant measurements of the gravitational bending of radio waves by the Sun which confirmed the general relativity prediction to within 1 percent precision (e.g., Fomalont and Sramek 1975), far better than the then best contemporary optical observations made at the time of solar eclipses.

Little Big Horn One of the outstanding problems facing radio astronomers in the 1960s was the need to establish an absolute flux density scale. It was straightforward, using any of the NRAO telescopes, to measure the relative intensity of radio sources. But putting the measurements on an absolute scale required knowledge of the effective area of the telescope, including the illumination efficiency as well as losses due to diffraction and surface imperfections. It is difficult to calculate these quantities for parabolic antennas, but relatively straightforward for a simple horn. Conventional horn antennas, however, were historically too small to give sufficient sensitivity for meaningful measurements.

In 1958, John Findlay decided to address this problem by building a very large calibration horn in Green Bank, which became known as "Little Big Horn." Little Big Horn was constructed on the side of a hill pointing toward a position where the strong radio source Cassiopeia A would pass once a day through the antenna beam. It was 120 feet (36.7 m) long and had an aperture 17.6 by 13.1 feet (5.4 by 4 m), and was probably the largest horn antenna ever built anywhere (Fig. 6.9). Construction was completed in September 1959, and the first observations began the following month and lasted for several decades. After careful calibration and correction for the contribution from the galactic background, Findlay (1972) was able to measure an absolute flux density of Cas A with a precision of 1.8 percent, and he also determined the rate of

Fig. 6.9 Little Big Horn calibration antenna designed and used by John Findlay to establish an absolute 1.4 GHz flux density scale. Credit: NRAO/AUI/NSF

decrease of flux density as (1.38 ± 0.15) percent per year in good agreement with other measurements and the earlier prediction by Shklovsky (1960).

6.3   THE CENTRAL DEVELOPMENT LABORATORY

The 140 Foot Telescope was smaller than the Jodrell Bank, Australian, Canadian (Sect. 6.6), and the later German (Sect. 9.2) dishes, and did not have the pointing accuracy or surface precision of the Haystack, Canadian or Australian antennas. Nevertheless, the 140 Foot was arguably more productive than any of the competing facilities due to the excellent instrumentation made available by the NRAO Central Development Laboratory (CDL) located in Charlottesville. This was particularly true for spectroscopic observations, due to the series of low noise receivers and multi-channel spectrometers built for Green Bank that were unequaled anywhere in the world. The CDL pioneered the development of actively cooled Field Effect Transistor (FET) and High Electron Mobility Transistor (HEMT) amplifiers for centimeter wavelengths, as well as cooled Schottky diode mixers and Superconductor-Insulator-Superconductor (SIS) devices for millimeter wavelengths.

The first receivers at NRAO were mostly based on commercial products from Ewen-Knight, Microwave Associates, or Long Island-based Airborne Instruments Laboratory (AIL). They used simple mixer receivers or, in some cases, parametric amplifiers, but their performance was not as good as at some other observatories, such as Caltech. In 1962 Hein Hvatum led an effort to purchase a 6 cm maser amplifier from AIL. It cost \$135,000, a lot of money at the time, but not too much for the then generously funded young observatory. Stability issues limited the sensitivity, and the maser was never used for any astronomical programs. NRAO clearly needed to improve its instrumental support to complement the growing collection of antennas, the Tatel Telescope, the 300 Foot Telescope, and the finally completed 140 Foot Telescope.

When Joe Pawsey visited Green Bank in 1961 (Sect. 4.6), he noted the weakness of the technical staff, but took note of the young MIT graduate student, Sandy Weinreb (Fig. 6.10), who was using the 85 foot Tatel Telescope in an attempt to detect deuterium and the Zeeman effect with his novel digital spectrometer. After arriving in England, Pawsey wrote to AUI President I. I. Rabi suggesting that AUI might try to hire Weinreb.[6] Four years later, Weinreb joined NRAO at the age of 28 to become head of the NRAO Electronics Division reporting to Hein Hvatum. Initially Weinreb lived in Green Bank, where he led the development of receivers for the 140 and 300 Foot Telescopes and the Green Bank three element interferometer, and was also responsible for the growing electronics research and development group in Charlottesville.

In 1968 Weinreb and his young family moved to Charlottesville, where he set up the NRAO Central Development Lab (CDL). Mike Balister, who had come to Green Bank in 1966 from Canadian Westinghouse where he worked on several classified military programs, became Associate Head of the NRAO

Fig. 6.10 Sandy Weinreb led the NRAO Electronics Division, built the first radio astronomy digital spectrometer, and provided much of the architectural design of the VLA. Credit: NRAO/ AUI/NSF

Electronics Division under Weinreb, as well as Head of the Green Bank Electronics Division, creating an awkward dual reporting line. Balister left Green Bank for Charlottesville to lead the low noise development group in 1972, and was replaced by Craig Moore as Head of the Green Bank Electronics group. In 1986 this group was placed under George Seielstad, then the Assistant Director for Green Bank Operations, and Weinreb became the NRAO Assistant Director for Technical Development. After a leave of absence from NRAO when he taught at the University of Virginia, Weinreb left NRAO in 1987 to join Martin Marietta. He later joined the faculty of the University of Massachusetts, then moved to California where he continued to develop low noise centimeter and millimeter wave amplifiers at the Caltech Jet Propulsion Laboratory (JPL). During his 23 years at NRAO, Weinreb pioneered the use of low-noise, cryogenically-cooled solid state amplifiers which greatly enhanced the sensitivity of the NRAO radio telescopes. He was also the architect for the electronic systems design for the VLA, and led the group which developed the VLA instrumentation. Weinreb was honored in 2008 with the Grote Reber Medal for his lifetime innovative contributions to radio astronomy and in 2011 with NRAO's Jansky Lecturership.

Following Weinreb's departure, Ballister became Acting Assistant Director for Technical Development as well as Head of the CDL. NRAO was unable to recruit a replacement for Weinreb, and in 1989 Balister became the permanent Assistant Director for Technical Development. Due to the rapid growth of the

NRAO staff resulting in part from the VLA program (Chap. 7), NRAO soon outgrew its new headquarters at the University of Virginia, and after 1972, the CDL was housed in rented space elsewhere in Charlottesville. Unlike the close collaboration between the scientists and engineers previously found in Green Bank, at CSIRO, and at many other radio observatories, the NRAO CDL and Basic Research staffs grew apart, primarily as a result of their physical separation, and there were few collaborative programs between NRAO scientists and engineers. Further, with the increasing independence of Green Bank, Tucson (Sect. 10.2), and Socorro (Sect. 7.7), new instrument construction and maintenance electronics work increasingly became the responsibility of the local electronics groups, which were more integrated into the local management structure, and there was less communication among the technical staffs at the different sites. The CDL increasingly concentrated on fundamental research and the production of low noise amplifiers for other observatories and for use in non-astronomical environments rather than in direct support of the instrumentation at NRAO facilities.

Low Noise Amplifiers and Mixers Probably the biggest success of the CDL was in the development of cryogenically-cooled centimeter wavelength low-noise HEMT and FET amplifiers under the leadership of Weinreb and Marian Pospieszalski (e.g., Pospieszalski et al. 1988). In addition to providing state of the art sensitivity for the NRAO telescopes, nearly every major radio observatory in the world used amplifiers designed and built at the NRAO CDL either as low noise front ends, or as IF amplifiers in millimeter wave receivers. The income from the sale of these amplifiers provided a valuable supplement to the eroding NSF funding, and supported the further research by the CDL staff.

NRAO cooled amplifiers were also used in support of several space missions, including the NASA Wilkinson Microwave Anisotropy Probe (WMAP) that measured the anisotropy of the cosmic microwave background with unprecedented precision and gave precise new constraints to the cosmological parameters (Bennett et al. 2003). The WMAP satellite contained 80 NRAO HEMT amplifiers in a dual-beam dual-polarization configuration (Jarosik et al. 2003) designed and built at the CDL by Pospieszalski. The NRAO amplifiers, which were passively cooled to about 90 K and covered the frequency range from 20 to 106 GHz in five bands, operated flawlessly for the nine year operational lifetime of the satellite. Later the CDL built 22 GHz (1.3 cm) amplifiers for use on the Russian RadioAstron space VLBI mission (Sect. 8.9). The RadioAstron amplifiers were similar in design to the lowest frequency WMAP amplifiers, and worked for the eight year duration of the mission, giving badly needed sensitivity at the shortest RadioAstron operating wavelength. The CDL also fabricated 8 GHz (3.75 cm) amplifiers for the VLA to support the 1989 NASA Voyager flybys to Uranus and Neptune (Sect. 7.7). The new 3.75 cm receivers also gave a badly needed boost to the VLA sensitivity and were used for a wide range of continuum observations.

Receivers for millimeter wavelengths were particularly challenging. When NRAO's Kitt Peak 36 Foot Millimeter-Wave Telescope first went on the air in early 1968, typical system temperatures were over 1000 K using simple mixer receivers. Weinreb and Kerr (1973) were able to achieve significant improvement by cooling Schottky diode mixers down to 15 K. By the 1980s, the introduction of SIS devices by Anthony (Tony) Kerr had replaced Schottky diodes as mixer elements for millimeter wave receivers. More recently, the frequency multipliers and wide bandwidth sideband-separating SIS mixers developed at the CDL along with Kerr's SIS mixers were critical to the technological success of ALMA (Sect. 10.7). Throughout this period, the millimeter-wave advancements made at the CDL were enhanced by a close collaboration with the University of Virginia Semiconductor Devices Laboratory led by Robert Mattauch.

All of the NRAO receivers used cooled radiometers. But commercial refrigerators were not fully reliable and were a constant source of downtime at the telescopes. To deal with the repeated breakdowns, and in anticipation of the coming needs of the 27 element VLA, Howard Brown and David Williams in Green Bank developed an innovative modification to the commercial refrigerators. Green Bank finally did obtain a successfully operating maser when Craig Moore spent time at JPL to fabricate a copy of the JPL 22 GHz (1.3 cm) traveling wave maser, and that operated on the 140 Foot Telescope for many years. The new maser gave unprecedented sensitivity for observing the water vapor (H_2O) and ammonia (NH_3) lines but was disappointing for continuum work due to gain fluctuations induced by the mechanical pump used to cool the liquid helium.

Spectrometers and Digital Back Ends Weinreb's MIT digital spectrometer was soon replaced by a series of more advanced systems designed and built at the CDL in Charlottesville. Even before Weinreb joined NRAO in 1965, NRAO had hired Art Shalloway from Cornell to develop digital spectrometers for Green Bank. Shalloway visited Weinreb, who was then working at MIT, to seek his guidance on building a digital spectrometer, and Weinreb continued to advise on the construction. The Model I autocorrelator, which had 100 spectral channels, was installed at the 300 Foot Telescope in 1964, and the Models II and III, with 384 spectral channels, on the 140 Foot and 300 Foot respectively in 1968 and 1972. The Model IV spectrometer, with 1024 channels, was based on the VLA custom chips (Sect. 7.7) and replaced the Model II spectrometer on the 140 Foot in 1980. In 1972, NRAO gave the original Model I correlator to Caltech to be used at Owens Valley Radio Observatory.

Digital spectrometers based on Weinreb's design were restricted by the limited sampling rate of existing digital hardware. To study the much broader spectral lines observed at millimeter wavelengths, Weinreb devised a hybrid system whereby an analogue filter bank system was used to divide the band into eight smaller bands, each of which could be analyzed by a digital spectrometer

(Gordon 2005, p. 104). Future digital systems at NRAO exploited the rapid developments in the speed of digital devices, enabling up to 262,536 spectral channels over a 1.6 GHz bandwidth on the GBT as well as capabilities for sophisticated analysis of pulsar data. Later digital correlator systems were built by Ray Escoffier for the VLA, and by Escoffier and Richard Lacasse for ALMA, all designed and built at the CDL. However, the JVLA WIDAR[7] correlator was designed and constructed by Brent Carlson and his group at the Dominion Astrophysical Observatory as part of the Canadian contribution to the JVLA.

Feed Systems Radio telescopes have historically used a single feed/receiver combination at the primary or secondary focus of the antenna, and so at any one time could observe only a single region in the sky. Soon after the 140 Foot antenna was completed, Jaap Baars (1966) introduced a dual beam system to reduce the effect of tropospheric emissions. Later multiple receiver/feed systems were built for use on the 300 Foot Telescope with up to seven dual polarization independent beams at 5 GHz.

Starting in the mid-1990s several radio observatories, including the NRAO CDL, began to develop multiple element phased array feeds (PAFs) (Fisher and Bradley 2000). Unlike the focal plane arrays, the signals from each interferometer pair of feed elements were correlated and used to form multiple beams in the sky in the same way that ground-based arrays are used. By proper adjustment of the weighting and phasing of the individual elements, PAFs can also correct for reflector surface errors. However, the construction and operation of PAFs is technically challenging since the elements need to be small and closely spaced, so mutual coupling among the elements makes it difficult to configure the system for optimum efficiency. In practice, the advantage of the multiple beams has been at least partially offset by the degradation in system temperature.

In order to address this shortfall, NRAO, in collaboration with Brigham Young University, the Green Bank Observatory, and West Virginia University, developed a 1.4 GHz cryogenically cooled PAF system. Using a 19-element array to form seven dual polarized beams, the system temperature is under 20 K, and the sensitivity of the seven synthesized beams is comparable to the best cooled single pixel systems, thus giving up to a factor of seven improvement in survey speed (Roshi et al. 2018). See Fig. 6.11.

6.4   Open Skies

Traditionally, astronomical observatories existed for their staff or for the faculty and students of a parent observatory or university. Observing time was allocated by the Director or by agreement among the staff members. Visitors were not uncommon, but generally they were either on long term sabbatical visits or were working with a host institution staff member. Often the larger and more

Fig. 6.11 NRAO 19 element dual polarized cryogenically cooled 1.4 GHz Phased Array Feed. Credit: J. Hellerman/NRAO/AUI/NSF

unique the telescope, the harder it was for outsiders to gain access.[8] NRAO was established to facilitate competition by American radio astronomers, especially those from the influential northeast universities, with their European and Australian counterparts. However, in October 1959, Heeschen wrote to the editors of *Astronomical Journal, Science,* and *Publications of the Astronomical Society of the Pacific* that "the facilities of the Observatory are open to any competent individual with a program in radio astronomy, regardless of institutional affiliation."[9] In practice, access to NRAO facilities was considered independent of not only institutional affiliation, but national affiliation as well. At the same time, NRAO declared that previous experience in the techniques of radio astronomy was not necessary to use the NRAO instruments.

This NRAO policy became known as "Open Skies" following the nomenclature adopted by the international airlines governing reciprocal landing rights. The NRAO Open Skies policy had a profound and long-lasting impact on how global astronomical research evolved. For the first time, any qualified scientist with a good idea could gain access to a world class facility, independent of his or her national or institutional affiliation. Starting with the 1969 Mansfield Amendment to the Military Authorization Act, which shifted the burden of federal support for the university radio observatories from the Defense Department to the NSF, the NSF required as a condition of receiving operating funds that a fraction, typically one-half, of the available observing time be made available to users from outside the university receiving the grant. With time, under pressure from the NSF and NASA, the Open Skies concept impacted all large American astronomy observatories, whether on the ground or in space. Ultimately, following the US model, other observatories, especially radio observatories worldwide, have adopted an Open Skies policy for at least a fraction of their observing time. Within the US, the Open Skies concept has become a matter of US policy applying to all federally operated research facilities.

Open Skies produces the best science because there is a broader pool of investigators, and the competition sharpens the US investigators. Interestingly, with the limited support available for individual investigator grants, Open Skies also means that other countries provide the labor to use and interpret the results from NRAO and other US facilities, but at no cost to the US taxpayer because salary and training are provided elsewhere. On the other hand, if the facility use becomes dominated by international users, it may be interpreted that US scientists are less capable and do not have the best ideas. Indeed, the NSF constantly draws attention to the high fraction of non-US scientists using NRAO facilities, and this may be reflected in the level of support received by NRAO from the NSF (Sect. 11.7).

By the mid-1970s, the NRAO instruments were becoming increasingly complex and a growing fraction of users did not have the experience needed to effectively use the NRAO radio telescopes. To provide added support for visiting scientists, a staff scientist was appointed as a "friend" of each telescope. The telescope friends were responsible for the calibration of each instrument, and for instructing visiting observers on how best to use the telescope, informing them of any new equipment or computer software, and alerting them to any known problems with the telescope or instrumentation. The concept of "telescope friend" has now spread to other radio observatories, both in the US and internationally.

Allocating Observing Time With the suite of world-class observing facilities provided by the 300 Foot and 140 Foot Telescopes, as well as the Green Bank Interferometer, and the corresponding increase in requests for observing time from visiting scientists, the informal allocation of observing time by consensus of the staff or by the Director was no longer tenable. Starting in the mid-1960s, each proposal, whether from staff or from visitors, was reviewed by outside referees. The allocation of observing time on all NRAO facilities has remained the responsibility of the NRAO Director, but has been generally delegated: initially to Bill Howard, Assistant to the Director; by the mid-1970s to the local site directors; then to an in-house committee; evolving to committees of mixed NRAO and outside members; and finally to a Time Allocation Committee (TAC) composed entirely of non-NRAO staff. The referee reports remain, in principle, only advisory, but with time have played a stronger and stronger role in the assignment of observing time. More recently, the TAC has been assisted by a series of discipline-related panels that provide priority ranking of proposals within the discipline, which are then assembled by the TAC into a master list of priorities for each NRAO telescope. NRAO staff receive no special treatment and there is no time reserved for scientific observing by NRAO staff, although time needed for commissioning of new equipment, calibration, or testing after a change of instrumentation between observers has been allocated as needed to NRAO support staff. See Hogg (2006) for more details on the NRAO telescope time allocation process.

Later, NRAO established new rules for so called "Large Proposals" in excess of 1000 hours that were subject to more extensive review, and required additional commitments from the observing team to make their results available in a timely manner. Also, in response to the increasing interest in multi-wavelength observations, NRAO implemented agreements with NASA-operated space-based facilities whereby small amounts of telescope time on these facilities or on NRAO facilities may be granted by the other TAC in response to a single joint proposals for the two facilities.

Data from all NRAO facilities are archived. Soon after the start of VLA operations, NRAO started to receive requests for data taken by other observers. Starting in August 1983, all observers were allowed an 18 month period where they had exclusive use of the data, after which time the data became available to the public. Approximately ten percent of the publications that use NRAO facilities are based on archival data rather than on new observations.

6.5   COMMUNITY INTERACTIONS

The strength of NRAO is derived from its user community. In order to receive scientific, technical, and management advice from the national as well as international community, both NRAO and AUI receive input from a variety of committees.

Advisory Committees Top level advice and support to NRAO has been derived from the AUI Board, which appoints the Observatory Director, approves all senior level appointments, grants tenure to qualified staff members, and is the legal entity entering into contracts and agreements with the NSF and other organizations. In 1958 the previous AUI Advisory Committee on Radio Astronomy morphed into the AUI Advisory Committee for the National Radio Astronomy Observatory. AUI had a long tradition of Visiting Committees to advise on their operation of the Brookhaven National Laboratory, and starting in 1961, the AUI Advisory Committee became the AUI Visiting Committee, with a mandate to report to AUI on the management of NRAO and how the Observatory was responding to national needs.

To complement the Visiting Committee, in 1965 Dave Heeschen convened a Users Committee reporting to the NRAO Director to evaluate the performance of the telescopes and advise on priorities for new instrumentation. The

User Committee meetings were often heated, especially when something was not working or a promised new receiver was late. On one occasion, after a particularly critical comment, Heeschen told the committee that if they weren't happy they could find a new Director, and he walked out. From time to time, in response to special needs such as the design or early construction of a new telescope, the Director would convene a special advisory committee. There was also a Computer Advisory Committee to advise the Director on the long-standing hardware and software issues that seemed to continually plague NRAO and frustrate observers. NRAO hosted many other meetings and conferences, including scientific symposia and workshops that brought together visiting scientists and members of the Scientific Staff. All of these meetings, including the Visiting and User Committee meetings, were opportunities for staff and colleagues to meet informally, and were generally highlighted by a nice dinner at a nearby restaurant or country club. Heeschen detested after dinner talks. Invariably at the conclusion of the dinner, he, as the host, would stand up, ceremoniously bang an eating utensil on a glass to get attention, and announce, "The bar is open," then sit down so the participants could continue their informal discussion.

The Jansky Lectures Starting in 1966, AUI initiated the Karl G. Jansky Lectureship as an honor recognizing outstanding contributions to the advancement of radio astronomy. Each recipient has presented an annual public lecture in Charlottesville and usually at one or more of the other NRAO sites in Green Bank, Socorro, or Tucson. In addition, the recipient often presents a professional colloquium on his or her research to the local staff and spends a few days interacting with staff members. Since the mid-1980s the Jansky Lecture has been attended by Karl's son David Jansky and other members of the Jansky family. The recipient is chosen each year by the NRAO Scientific Staff based on nominations received from the broader community. The first Jansky Lecture was given by John Bolton from Australia in 1966, followed by Jan Oort in 1967, Iosef Shklovsky in 1968, Fred Hoyle in 1969, and Robert Dicke in 1970. More than 50 scientists have been recipients of the Jansky Lectureship, many of whom were not themselves radio astronomers, but who, through their theoretical contributions or observations in other wavelength bands, contributed to the enhancement of radio astronomy.

Student Programs As part of its mandate to educate the next generation of radio astronomers and to provide for a scientifically literate public, NRAO has since 1959 brought in undergraduate students during the summer to work with staff members. John Findlay was the first administrator of the program, which was later expanded to include graduate students. Many former NRAO summer students later joined the NRAO scientific or engineering staff or went on to distinguished careers in astronomy and astrophysics or other areas of science and engineering or government service. Stephen Chu, a 1970 student, went on to win a Nobel Prize in Physics and later became Secretary of Energy in Barack Obama's administration. Stephen Hawley, who spent the summers of 1973 and 1974 in Green Bank, became a NASA astronaut who deployed the Hubble Space Telescope and participated in several of the HST servicing missions. In addition to the summer student program, each year a few graduate students are typically in residence working on their dissertation research under the joint supervision of an NRAO scientist or engineer and a faculty member from their home institution.

The NRAO-KPNO Exchange Visits Perhaps in response to the concerns that had been raised earlier by Tuve and others about the segregation between radio and optical astronomers and the intellectual isolation of the Green Bank staff, NRAO and the Kitt Peak National Observatory Directors in 1959 initiated a series of alternating annual exchange visits which were used to exchange ideas about running a national astronomy observatory for visitors as well as for scientific presentations on current research by the two staffs. These visits were greatly appreciated and enjoyed by both staffs, but died out in the late 1960s, ostensibly due to their expense and the concern that these visits did not directly help the Observatory's user communities.

6.6   Growing Competition

NRAO was not alone in benefitting from the Sputnik-inspired wave of support for science, especially for astronomy and space science, which provided generous financial support to NRAO, but also led to other new American radio astronomy facilities and a certain amount of competition for recognition and continued funding. New university facilities were funded by the Office of Naval Research (ONR) and the Air Force, as well as by the NSF. Other US radio astronomy facilities were expanded and new facilities initiated, also mostly with Department of Defense (DoD) funding. Meanwhile Australia, Britain, Canada, Germany, and the Netherlands were building on their earlier successes and embarking on new radio astronomy initiatives characteristic of the transition of radio astronomy to big science.

Under Ed Lilley's leadership, the Harvard radio astronomy program gradually shifted emphasis from the 60 foot telescope to the more powerful new 120 foot Haystack antenna. Additional competition to NRAO would come from the Caltech two-element interferometer and the Cornell 1000 foot spherical dish at Arecibo. In Australia, Taffy Bowen was building a 210 foot (64 m) steerable dish at Parkes, and in Canada, the National Research Council was building a 150 foot (46 m) radio telescope at Algonquin Park. In Cambridge, Martin Ryle was developing a series of innovative synthesis radio telescopes that were setting the benchmark for high resolution sensitive imaging of radio galaxies (Sect. 7.1).

Ohio State and Big Ear John Kraus' helical element array at Ohio State University (Sect. 2.5) had limited bandwidth, and it was not realistic to consider a significant increase in collecting area using an array of helices. With support from the NSF, the Air Force, and the University, Kraus built a novel fixed paraboloid 360 feet long by 70 feet high (109 × 21 m) oriented in the east-west direction. Declination adjustment was made with a 260 foot long by 100 foot (79 × 30 m) flat tiltable reflector that was illuminated by feeds covering the frequency range 20 MHz (15 m) to 2 GHz (15 cm). Although basically a transit instrument, the feed system was mounted on a movable carriage that allowed motion in hour angle up to 30 min from the meridian. The angular resolution ranged from 12.5 degrees by 50 degrees at 20 MHz to 7 by 28 arcmin at 2 GHz.

Plans to increase the collecting area by a factor two never materialized, but the Ohio State Radio Telescope was used over a period of six years, primarily for a discrete source survey covering 8.66 steradians at 1.4 GHz (21 cm) down to a limiting flux density of 0.25 Jy. The Ohio State catalog of nearly 20,000 radio sources was published in a series of papers in the *Astronomical Journal* between 1967 and 1975 and summarized by Rinsland et al. (1975). The Ohio State surveys uncovered a number of flat and peaked spectrum radio sources, many of which were optically identified with quasars and BL Lac Objects. However, the Ohio State survey was limited by confusion resulting from the large north-south fan-beam and side-lobe responses and was overshadowed by the emergence of more reliable source catalogues from Parkes and NRAO. Starting in late 1973, "Big Ear," as Kraus (1995) called it, began a search for narrow band 21 cm radio signals from extra-terrestrial intelligent civilizations (SETI). In 1998, after years of opposition by Kraus and many Big Ear supporters, the Ohio State radio telescope was demolished and the land turned into an 18-hole golf course.

The University of Illinois Vermilion River Observatory (VRO) The University of Illinois Astronomy Department was headed by George McVittie, a well-known theoretical cosmologist. Like John Kraus, McVittie was fascinated by the prospects of addressing cosmological problems by using radio sources, and in 1956 he recruited George Swenson from Michigan State University to begin a radio astronomy program. Swenson had a strong background in electronics, radio propagation, and acoustics, but no experience in astronomy. Following a global familiarization tour of radio observatories, and with support from ONR, Swenson built a radio telescope with sufficient sensitivity and resolution to detect radio sources an order of magnitude weaker than those that had been cataloged at Cambridge or Sydney.

To satisfy the requirements of collecting area, resolution, and low sidelobes, as well as the constraints of cost, Swenson (1986) constructed a fixed parabolic cylinder 600 by 400 feet (183 × 122 m), oriented in the north-south direction. The parabolic surface was lined with a one-inch steel wire mesh reflecting surface. The antenna was steered in declination by appropriate phasing of the 276 receivers and feeds located 153 feet (47 m) above the parabolic surface oriented along the north-south axis. No hour angle motion was provided, so the telescope operated in a transit mode in the unused 608–614 MHz (49 cm) Channel 37 TV band, which was later allocated world-wide as a protected band for radio astronomy. The initial receivers included electron beam parametric amplifiers, replaced later by Field-Effect-Transistor amplifiers.

Although Swenson had to deal with weather-induced degradation of the parabolic surface and parametric amplifier reliability, over a period of nearly a decade the VRO detected and catalogued over a thousand discrete radio sources, mostly radio galaxies and quasars. Notable among these was the radio source known as VRO 42.22.01 which was identified with what was presumed to be a long recognized star, BL Lacerte, or BL Lac (Schmitt 1968). However, BL Lac turned out to be a quasar-like object of the class of radio sources since referred to as BL Lac objects, which typically have weak or no optical emission lines and are highly polarized.

By 1969, ground erosion had altered the parabolic surface beyond repair and the telescope was abandoned for use. Its work was completed, but the 49 cm protected radio astronomy band remains as a legacy of the Vermilion River Observatory.

The University of Michigan 85 Foot Telescope During WWII Fred Haddock worked at the Naval Research Laboratory (NRL), where he developed a submarine mounted radar that was used to locate Japanese ships. After the war he remained at NRL, working primarily at short centimeter and millimeter wavelengths to study the Sun and thermal radio sources, and he became a strong advocate for pushing the NRAO 140 Foot Telescope to the shortest possible wavelengths. In 1956, Haddock left NRL to begin a radio astronomy program at the University of Michigan that was funded by ONR. Under Haddock's leadership, Michigan acquired a Blaw-Knox 85 foot antenna similar to the NRAO 85 Foot Tatel Telescope, but inexplicitly the surface accuracy and pointing were about a factor of two better than the Green Bank antenna. Haddock, who was a gregarious, outspoken scientist, later took pride in pointing out that his Michigan antenna was better than the one delivered to Green Bank. As a result, Michigan radio astronomers and students were able to push their observations to 4 cm and later 2 cm to study flat spectrum radio sources (Dent and Haddock 1965) and radio variability (Dent 1965) while operation of the Green Bank antennas was essentially limited to 4 cm minimum wavelength. The University of Michigan 85 foot radio telescope remained in operation for more than half a century, during which Hugh and Margo Aller led a program to monitor the variability of extragalactic radio sources at 1.3, 2, and 6 cm wavelengths.

The Naval Research Laboratory Following the construction of their first 25 meter-class radio telescope at their Maryland Point Observatory, ten years later

NRL purchased a new high-precision 85 foot radio telescope from the Rohr Corporation (McClain 1966). The new dish was designed by Robert Hall and was based on his earlier design of the 85 foot antennas that were built by Blaw-Knox for NRAO (Sect. 4.3) and the University of Michigan, but the NRL antenna was modified to allow good operation at wavelengths as short as 1 cm. It was used in 1967, together with the Green Bank Tatel Telescope, as part of the first successful NRAO independent-oscillator-tape recording interferometer (Sect. 8.1).

The Haystack 120 Foot (37 m) Precision Antenna In parallel to NRAO's construction of the 140 Foot radio telescope, the US Air Force was building a radome-enclosed 120 foot antenna. Like the Jodrell Bank Mark I, or the Arecibo 1000 foot dish, the Haystack antenna was not originally conceived to do radio astronomy, but rather as multi-purpose facility to operate as a ground station for satellite communication, for radio propagation experiments, and as a satellite tracking radar facility.[10] In fact, much of the motivation for building the Haystack facility was the military interest in evaluating the use of metal-space frame supported radomes to protect large precision antennas against potential deleterious environmental effects of wind, snow, ice, and uneven solar heating. More conventional types of air-supported or self-supported plastic radomes were not considered practical for such large antennas due to the significant loss that would occur from any dielectric surface strong enough to support its own weight. A 120 foot antenna was the largest that could fit inside a 150 foot radome that the US Air Force was considering for an Arctic environment, although a radome specifically designed for the more benign New England weather could probably have been built with less transmission loss.

Construction of the Haystack facility by North American Aviation, Inc. began in 1960 and was completed in 1964 as part of the Lincoln Laboratory, which was operated by MIT for a variety of defense related activities. Haystack construction was under the direction of Lincoln Laboratory scientist Herbert Weiss who had previously been in charge of the Defense Department's Distant Early Warning (DEW) Line radar facility. At the time, Haystack was the largest radome ever built. As the Haystack facility was motivated primarily by defense needs, there was little or no input from the conservative astronomers who insisted on an equatorial mount for the NRAO 140 Foot. Rather, Haystack pioneered the use of digital computers to control the alt-az mounted antenna as well as to evaluate the data. The pointing accuracy of the radome enclosed Haystack 120 foot antenna was not affected by the wind and the effects of solar heating were minimized. The pointing accuracy was much better than that of the NRAO 140 Foot and the alt-az mounted dish did not suffer from the same astigmatic deformation when the telescope was tipped in elevation. Early use of the Haystack antenna was for a variety of experiments in space communication and space situation awareness, but gradually more time was devoted to astronomy, particularly for radar studies of the Moon and planets. A particularly important and innovative experiment was the verification by Irwin Shapiro and colleagues (Shapiro et al. 1971) of the so-called Fourth Test of General Relativity.[11] Starting in 1967, with support from NASA, the Haystack antenna was also used in a series of Very Long Baseline Interferometry (VLBI) experiments, particularly involving high resolution studies of interstellar hydroxyl (OH) and water (H_2O) vapor masers, as well as innovative applications of VLBI to precise measurements of the rotation of the Earth with applications to global timekeeping and for studies of tectonic plate motions (continental drift). Much of the instrumentation and techniques used for VLBI, including several generations of the VLBA recording system, have been developed at the Haystack Observatory (Sect. 8.4).

Subsequent upgrades to both the 120 foot dish as well as the radome and radar capability have enabled effective operation to short millimeter wavelengths for both radio astronomy and for a variety of passive and active communication and space surveillance activities. Although initially funded by the Air Force and later by the Defense Department Advanced Research Project Agency (ARPA),[12] as a result of the 1969 Mansfield amendment to the 1970 Military Authorization Act, the Haystack Observatory was turned over to MIT to operate primarily for radio astronomy programs sponsored by the North East Radio Observatory Corporation (NEROC) with financial support from the NSF.[13] But starting in the late 1980s, with decreasing support from the NSF, the amount of time available for radio astronomy has been reduced.

The Arecibo 1000 Foot Dish[14] Meanwhile, further competition to NRAO was appearing on the Caribbean island of Puerto Rico where Cornell University was constructing a 1000 foot (305 m) fixed spherical antenna 10 km south of the city of Arecibo. Like the Haystack and Jodrell Bank antennas, the Cornell antenna was not initially designed for radio astronomy, but for studies of the ionosphere. The Arecibo antenna was conceived in 1958 by Cornell Professor of Electrical Engineering William Gordon to study the density and temperature distribution in the ionosphere using incoherent radar backscatter (Gordon et al. 1961). Gordon calculated that even using the most powerful radar transmitters then available, a dish diameter of 1000 feet would be required in order to obtain a decent return signal from the upper F-layer of the ionosphere. Marshall Cohen noted that with such a huge antenna, it would also be possible to detect radar reflections from the Sun and the planets. Further discussion at Cornell led to increasing interest in using the proposed dish for general radio astronomy.

After rejecting potential sites in the Philippines and Cuba, a suitable site centered on a sinkhole was located in the Puerto Rico karst district south of Arecibo, and funding was obtained from the new ARPA. There was considerable interest within the military establishment in better understanding the ionosphere, since it might contain traces of disturbances by Soviet ICBMs or satellites as well as by high altitude nuclear explosions. Moreover, the ionosphere plays an important role in the world-wide propagation of high frequency radio communications, and the International Geophysical Year (1957–1958) and the 1957 launch of the Russian Sputnik were added incentives to study the ionosphere.

However, Gordon's calculation of the required sensitivity was faulty. Gordon assumed that the returned signal would be Doppler broadened by the random motions of ionospheric electrons. But the radar scattering is mostly from electrons moving with the singly ionized oxygen ions and so the width is given by the Doppler broadening of the ions, not the electrons, and is about a factor of a hundred more narrow than that calculated by Gordon. Gordon's goals could have been attained with a substantially smaller dish than 1000 feet, but by time this was recognized both Cornell and ARPA had become enamored with the concept of a 1000 foot dish with its potential non-ionospheric applications, and there was apparently no consideration of reducing the size of the dish (Cohen 2009).

When completed in 1963, the 20-acre collecting area of the Arecibo antenna was by far the largest of any radio telescope in the world, and would not be exceeded for another half a century when the Chinese 500 meter dish was completed in 2017. The Arecibo dish had a spherical instead of parabolic surface so that the beam could be steered up to 20 degrees from the zenith by moving the feed structure.[15] However, with a spherical surface there is no single focal point; rather the incoming radiation is focused along a line, so that simple horn or dipole feeds used on conventional parabolic antennas had to be replaced with elaborate line feeds. Unfortunately, due to a design error the original line feeds, including the high-powered 440 MHz radar feed, had poor sensitivity and unacceptably high sidelobes, and it would take years before they were replaced by properly designed line feeds.

For many applications, including 21 cm spectroscopy, SETI, and pulsar research, the Arecibo telescope was more sensitive than any radio telescope in the world. Active radar programs also allowed the exploration of solar system objects out to the orbits of Jupiter and Saturn. However, in common with other filled aperture radio telescopes, continuum observations were limited by confusion and gain stability. For the first few years, the Arecibo Ionospheric Observatory (AIO) was operated as part of the Cornell-Sydney University Astronomy Center with funding from ARPA. In 1971, the AIO became a national observatory and part of the Cornell National Astronomy and Ionospheric Center (NAIC) with primary funding from the NSF, while planetary and ionospheric radar programs were supported by NASA. In 1973 the antenna was upgraded by replacing the original wire mesh surface with 38,788 perforated aluminum panels which permitted operation up to 3 GHz (10 cm), and a new 420 kW 2.3 GHz radar was installed. A second upgrade was completed in 1997 when the line feeds were replaced by a Gregorian subreflector, the surface was reset to increase the maximum observing frequency, and a more powerful 1 MW transmitter installed for planetary radar.

Soon after completion of the antenna, radar observations of Mercury showed that, contrary to all text book descriptions, Mercury was not in synchronous rotation with its 88 day period of revolution, but instead rotated at

2/3 of the 88 day period or 59 days (Pettengill and Dyce 1965). Later observations included the discovery of the first millisecond pulsars (Backer et al. 1982), the first exoplanets (Wolszczan and Frail 1992), and the first detection of the effects of gravitational radiation (Taylor et al. 1979). In 1966 Frank Drake became Director of the Arecibo Observatory and initiated a vigorous program of pulsar research. As discussed in Sect. 5.5, with its extraordinary collecting area and sensitivity, the Arecibo telescope became a home for SETI observations by Frank Drake, Carl Sagan, and others.

Caltech and the Owens Valley Radio Observatory As was discussed in Sect. 3.2, following the lengthy discussions among Caltech's Jesse Greenstein, Robert Bacher, and Lee DuBridge, Taffy Bowen from Australia, Mt. Wilson and Palomar director Ira Bowen, and Carnegie's Vannever Bush and Merle Tuve, Greenstein and Tuve organized the January 1954 Washington conference on radio astronomy which led to an unintended shift in US radio astronomy planning toward the East Coast establishment and ultimately in the formation of the NRAO. However, Bacher, DuBridge, and Greenstein had not lost interest in creating a radio astronomy program at Caltech, one that would have a unique relation with the Mt. Wilson and Palomar Observatories (MWPO). Shortly after the conclusion of the Washington conference, ONR Science Director Randal Robertson visited Caltech and began a discussion with DuBridge and Bacher about possible funding from ONR for a Caltech/ MWPO radio astronomy program. Although excited about the potential of radio astronomy, DuBridge hesitated, in part because of his uncertainty about how to proceed, his concern about whether the US should or could support two major facilities, and by the threat of competition from the planned Jodrell Bank 250 foot radio telescope or Taffy Bowen's planned large Australian radio telescope.

By July 1954, Greenstein had succeeded in convincing DuBridge of the need for Caltech to establish its own radio astronomy program, and DuBridge asked ONR Chief Scientist Emanuel (Manny) Piore "to keep a few dollars earmarked off in a corner for such a program."[16] DuBridge added that he had selected Caltech Professor Bill Pickering, then temporarily at JPL, to lead the new Caltech radio astronomy program. Robertson quickly responded that since his visit to Caltech in February, "I have had some funds mentally set aside for your project."[17] Within a few days, Elliot Montroll, the ONR Director of Physical Sciences, followed up, saying he wanted to meet with DuBridge in early August to "discuss the projected ONR program on radio astronomy."[18] Things were moving fast. Pickering accepted DuBridge's offer to take charge of the radio astronomy program starting 1 January 1955 and had already "begun to develop ideas in this direction."[19] Then JPL Director Luis Dunn suddenly and unexpectedly resigned to accept an industrial position at Ramo Woolridge, leaving DuBridge with the problem of finding a new JPL Director. JPL was a $13 million a year program, providing needed overhead funds to Caltech, and was a key component of the then burgeoning US space program.

DuBridge needed to appoint the best man as JPL director and that was Bill Pickering, leaving his radio astronomy program without a leader.[20]

A few months earlier, Grote Reber, who apparently had gotten wind of the developing Caltech interest in radio astronomy, had written DuBridge to inquire if there might be place for him in the Caltech program.[21] DuBridge wisely consulted Greenstein, who reported that while Reber was "extremely original and has done remarkable good work, ... I'm not sure he would be a suitable person to build an organization around."[22] DuBridge responded to Reber that "our plans for radio astronomy at Caltech are at a very elementary and formative stage and we do not know if it will be possible to get any work underway in the near future or not," and went on to say in the traditional way, "if we do find it possible to undertake a project in which you might be interested, I will get in touch with you."[23] There is no evidence, however, that Reber was ever considered for a position in the Caltech radio astronomy group, even when DuBridge was desperately in need of finding someone to replace Pickering.

The "Old Boy Network" then went into quick operation. While DuBridge was pondering how to find a new leader for his radio astronomy project, Taffy Bowen, who was in the US to raise money for his planned radio telescope, stopped at Caltech and met with DuBridge, Bacher, Greenstein, and Walter Baade from the Mt. Wilson and Palomar staff. Recognizing the lack of experienced American radio astronomers, DuBridge suggested to Bowen that one of his men be invited to Caltech to help start a radio astronomy program. Several possibilities were discussed, but they agreed that John Bolton would be the best person. Following his remarkable discoveries described in Sect. 2.1, Bolton had run into conflict with his Radiophysics boss, Joe Pawsey. To ease tensions, Bowen had transferred Bolton from radio astronomy to his own rain-making program, but recognized the opportunity afforded by Caltech to get Bolton back into radio astronomy and perhaps to get him out the difficult situation at the Radiophysics Lab.[24] Although Bowen claimed that he did not want to lose Bolton, he agreed to talk to Bolton and explore his interest in going to Caltech. Encouraged by his discussions with Bowen, DuBridge rather casually suggested to Piore that "possibly you could add say $50,000 to the contract at ONR for radio astronomy purposes."[25]

Only after Bowen had talked with Bolton and obtained in principle Bolton's interest in going "to Cal. Tech. to help you start a Radio Astronomy programme,"[26] did DuBridge finally contact Bolton with an offer of an appointment "as a Senior Research Fellow in Physics and Astronomy.... The initial appointment would be for a two-year period, beginning whenever you would find it possible to arrive. At the end of the two years, we would like to explore the question of your future with a definitive possibility in mind that it may be mutually agreeable for you to remain with us."[27] DuBridge, who was sensitive to the growing interest from AUI and other east coast scientists, was anxious to get started and expressed the hope that Bolton would start at the beginning of 1955.

John Bolton arrived at Caltech with his family in February 1955, initially on a two year leave of absence from CSIRO. At Bolton's urging, Caltech also hired his long-time friend and colleague Gordon Stanley, who arrived a few months later to provide expert technical support. To the surprise of DuBridge, Bolton announced that he did not intend to build another Mills Cross as DuBridge had discussed with Bowen, but rather had his own ideas about a variable spacing interferometer to measure accurate radio positions and morphology of discrete radio sources.

To carry out their planned radio astronomy program Bolton and Stanley needed to find a large, flat radio-quiet site. Stanley located an appropriate site in the Owens Valley nestled between the Sierra Nevada on the west and the Inyo Mountains to the east. Starting early in the twentieth century, the city of Los Angeles diverted water from the Owens Valley to the rapidly growing and water-starved city of Los Angeles, destroying farms and leaving the Owens Valley as a dry wasteland. With a diminishing source of livelihood, the population of the Owens Valley stagnated or even declined.[28] In order to minimize population growth in the area, in 1920, the Los Angeles Department of Water and Power (LADWP) purchased a large tract of land located near the small town of Big Pine about 250 miles north of Pasadena. Caltech seized the opportunity to rent for $1 per year several hundred acres of land from the LADWP about five miles from Big Pine where they could be assured of minimal population growth in the area and relative freedom from RFI.[29]

DuBridge had already laid the foundation with ONR for support of Bolton's proposal, and for the first few years the construction and operation of the OVRO was supported primarily by generous funding from ONR, with some private funding raised by Caltech for the buildings needed to house the control room, workshops, and living quarters for visiting staff. Arnold Shostak, the ONR program officer, did not bother with formal proposals or project reviews, but had an informal unmilitary-like seat-of-the pants approach to deciding who and what to fund and what not to fund. Aside from Caltech, Shostak provided ONR funds for George Swenson at the University of Illinois and Fred Haddock at the University of Michigan.[30] However, starting with the construction of the OVRO 130 foot (40 meter) antenna in 1964, OVRO, as well as the other ONR funded American radio observatories, depended more and more on NSF funding, fomenting an increasing level of competition between NRAO and the university-operated radio observatories.

Although much of the heavy construction of the OVRO interferometer was initially carried out by commercial contractors, as Bolton began to run out of ONR funds, the construction and operation at OVRO was increasingly done largely by his graduate students, who were mostly recruited from the Caltech physics rather than the astronomy departments. Under Bolton and Stanley's supervision, and especially Bolton's iron handed management, the students designed much of the instrumentation as well as ran bulldozers and tractors, dug trenches, laid cables, painted antennas, designed and built electronic systems, etc. The students were paid only a small fraction of what the commercial contractors were paid, which did not escape the attention of the local contractors. On one hot summer day, a local union organizer showed up while Bolton, along with a group of students, were laboring out in the desert sun and demanded to know if this was a union job. Bolton assured him that he was a member of the International Astronomical Union and that was the last time any labor union was involved at OVRO. For the students, who provided skilled, but cheap, labor, the experience was invaluable in their later careers, and many of Bolton's students and junior staff themselves went on to distinguished careers in radio astronomy. Robert Wilson received the 1978 Nobel Prize for Physics, along with Arno Penzias, for their discovery of the cosmic microwave background, which created a new field of precision observational cosmology from what previously had been a mathematical exercise. Six other former students later served as directors of radio observatories in the US, Europe, India, and Australia.

When the OVRO East-West interferometer went into operation at the end of 1959, NRAO had only a single 85 foot antenna and was struggling with what turned out to be only the beginning of extensive problems with the 140 Foot antenna. By the end of 1960, OVRO had a fully functional two-dimensional interferometer using two 90 foot (27.4 meter) antennas that could be moved on railroad track to various stations separated by up to 1600 feet in both the east-west and north-south directions. Young radio astronomers were recruited from around the world to join the OVRO staff. Jim Roberts and Kevin Westfold arrived from Australia and were soon followed by Venkataraman Radhakrishnan (known as "Rad") from India by way of Sweden, Per Maltby from Norway, and Dave Morris from Jodrell Bank. Tom Matthews, who had been Bolton's host while at a visit to Harvard, was the sole US-trained staff member.

At the end of his initial two-year appointment, Bolton resigned from CSIRO and accepted an indefinite appointment at Caltech, and on 1 January 1958 he became Director of the Owens Valley Radio Observatory (OVRO) and Professor of Radio Astronomy. To the surprise of many, citing Caltech bureaucracy, what he perceived as excessive overhead charges, pressures to teach, and family reasons, Bolton left Caltech and returned to Australia at the end of 1960 to take charge of the Parkes 210 foot radio telescope and to oversee the final stages of its construction.

Following Bolton's departure, Caltech tried to recruit Robert Hanbury Brown as OVRO director, but he had already committed to go to the University of Sydney to build a large optical intensity interferometer. It is not clear to what extent Caltech low-keyed the radio astronomy program for fear it would compete with either the ONR funded program at the Caltech synchrotron or Caltech/MWPO ambitions to build a southern hemisphere 200 inch telescope. Finally, in 1965 Caltech appointed Gordon Stanley as OVRO director. Stanley had even fewer academic qualifications than Bolton, and unlike Bolton was never given a professorial appointment. When Stanley retired in 1975,

Professor of Radio Astronomy Alan Moffet, who had been one of the earliest radio astronomy students at Caltech, became the OVRO director.

The early OVRO research program fully met Greenstein and DuBridge's ambitions to bring together radio and optical astronomy at Caltech. During his six years at Caltech, Bolton led the development of the OVRO which became in those years the most productive radio observatory in the world. The OVRO interferometer measured radio source positions with an accuracy of a few arc seconds (e.g., Read 1963; Fomalont et al. 1964) leading to secure identifications with radio galaxies at ever increasing distances (Minkowski 1960; Maltby et al. 1963). Caltech students and staff demonstrated the pervasive double nature of radio galaxies (Maltby and Moffet 1962), discovered the first quasars, made a number of important solar system discoveries (e.g., Clark and Kuzmin 1965), and pioneered interferometric radio spectroscopy (Clark et al. 1962; Clark 1965) and interferometric polarization studies (Morris et al. 1964). By the mid-1960s, the Owens Valley Radio Observatory with its modest funding from ONR was a clear success, in contrast to the much better NSF-funded NRAO, which was struggling with the 140 Foot antenna, and until the 1962 completion of the 300 Foot transit antenna, NRAO did not have a competitive radio telescope.

Tensions between NRAO and Caltech had existed since DuBridge had announced at Berkner's July 1953 meeting that he was going to begin a radio astronomy program at Caltech (Sect. 3.2). Bolton, and later Gordon Stanley, served on various NRAO advisory committees, and never missed an opportunity to criticize the NRAO operation. Stanley's mixer receivers were more sensitive than NRAO's expensive receivers using commercial parametric and maser amplifiers, and he was not hesitant to remind NRAO of their problems. One Caltech student, unenthusiastic about doing manual labor in the Owens Valley, asked Bolton if he could instead spend the summer at the NRAO summer student program, which offered an apparently better opportunity to be involved in research as well as a series of interesting lectures. Bolton flatly told him, "Sure, but, if you go to NRAO, you don't need to come back." That student never got his PhD.

The rivalry between NRAO and OVRO reached its peak in late 1960s and 1970s when both organizations proposed to build the array of dishes that had been recommended by the first decade review of astronomy (Whitford 1964) (Sect. 7.2) and later, to a lesser, extent the competing proposals for a dedicated Very Long Baseline Array (VLBA) (Sects. 8.6 and 8.7). Ironically, however, nearly 40 former Caltech students, staff, and faculty later worked at NRAO, including two NRAO directors as well as a number of NRAO assistant directors. After the early influx of Harvard graduates, it was the Caltech graduates and postdocs who played major roles in planning and designing the new instrumentation and software that kept NRAO at the forefront of radio astronomy over the next half century. In particular, Barry Clark, Ed Fomalont, Eric Greisen, and Richard Sramek were key players in the design, construction. and operation of the NRAO VLA, and Ron Ekers, who had been a Caltech postdoc, became the first director of the VLA (Sect. 7.7). Clark, Jon Romney, and Craig Walker, along with one of the present authors (KIK) were heavily involved in planning for the VLBA (Sects. 8.6 and 8.7) while Alwyn (Al) Wootten became the ALMA Project Scientist for North America (Sect. 10.7).

The Caltech plan to build an eight-element array led to the construction of the first 130 foot (40 m) prototype dish which was used in conjunction with the two 90 foot (27 m) antennas as a three element interferometer (Sect. 7.2). But following the failure of the Caltech Owens Valley Array proposal in favor of the NRAO VLA, Caltech turned its attention first to Very Long Baseline Interferometry (Sect. 8.1) using the 130 foot telescope and then to millimeter astronomy based on a novel 10 meter dish designed by Caltech physics professor Robert Leighton (Sect. 10.4).

Jodrell Bank American ambitions were dwarfed by Bernard Lovell's 250 foot (76 m) fully steerable radio telescope at Jodrell Bank, near Manchester in the UK. Lovell was familiar with Grote Reber's work and, indeed, during WWII Lovell had implemented some of Reber's receiver designs in developing British radar systems. As early as 1950, Lovell (1987) began talking about building a very large fully steerable dish at the University of Manchester. Interestingly, Lovell's plans were not motivated by radio astronomy, but by his interest in detecting radar echoes from ionized trails left by cosmic rays as they passed through the Earth's atmosphere. Lovell's 250 foot Jodrell Bank radio telescope was finally finished in 1957, and would remain an icon of radio astronomy for many years. However, the years of delay and cost escalation during the 250 foot construction were sadly prophetic of many of the other large steerable radio telescope projects which were to follow throughout the twentieth century.

After nearly a decade of continual operation, the Mark I telescope was beginning to show signs of wear resulting from stresses induced by the repeated elevation motion. The antenna needed a major overhaul, more than could be accomplished from the regular maintenance program. Lovell managed to raise £350,000 to not only overhaul the Mark I, but to upgrade it to what he called the Mark IA, which would work at frequencies up to several GHz. This was to be accomplished by overhauling the azimuth track, reducing the load on the elevation bearing, strengthening the backup structure, increasing the dish diameter by 15 feet (4.5 m), putting a new surface on the dish, strengthening the single feed support, and revamping the control system. But the cost estimate turned out to be over £500,000, and it was admitted by the designer, Charles Husband, that this was only an estimate, and that the price could not be determined without actual contractor bids. Once again, repeated delays and design changes ran the final cost of the overhaul and upgrade to £650,000, even including some descoping of the upgrade, including the decision to strengthen the single feed support structure rather than replace it with a tetrapod and abandoning any plan to increase the dish diameter. In 1987, the Jodrell Bank 250 foot Mark IA radio telescope was renamed the "Lovell

Telescope" and remains in operation more than 60 years after its completion in 1957.

In addition to extensive stand-alone radio astronomy research programs, the Jodrell Bank telescope was a key component of the very successful Manchester long baseline interferometer program, and later MERLIN, the Multi-Element Radio Linked Interferometer Network (Sect. 8.1).

The 250 foot's delays and cost increases were in part the result of upgraded performance specifications following the Harvard discovery in the US of the 21 cm hydrogen line. Lovell naturally wanted to exploit this exciting discovery, and introduced costly design changes which the University was unable or unwilling to meet. Following the accelerated space program resulting from the Sputnik launch by the USSR, the US found itself with inadequate facilities to track either Russian or American spacecraft, especially those in so-called "deep space," beyond Earth orbiting satellites. Rather embarrassingly to the US, NASA had to contract with the University of Manchester to use their still unfinished 250 foot antenna at Jodrell Bank to track the rocket that launched the Sputnik spacecraft and for subsequent tracking of the first US satellites. Probably only the fortuitous launch of Sputnik and the income from the NASA contract saved Lovell from going to prison for his alleged misuse of University funds. Lovell himself became perhaps the best-known figure in radio astronomy, and was rewarded by knighthood a year earlier than Martin Ryle, who headed the very innovative and successful, but contentious, radio astronomy program at Cambridge (Sect. 7.1).

The Australian 210 Foot Parkes Radio Telescope The events leading to the construction and early operation of the Parkes 210 foot radio telescope have been described by Robertson (1992). After first flirting with Caltech, Taffy Bowen succeeded in raising $500,000 from the US-based Carnegie and Rockefeller Foundations for his Giant Radio Telescope (GRT) in Australia. Although he was able to obtain matching funds from the Australian government, they were not sufficient to meet Bowen's ambitions for an antenna that would compete with Lovell's 250-footer. Moreover, the Australian government money came with strings attached. The other, still very productive, Radiophysics radio astronomy programs would need to be cut, so there was little support for Bowen's GRT among the Radiophysics staff, who saw it as competition to their own ambitious plans. Bernie Mills wanted to build an expanded version of his successful Mills Cross, the so-called "Super Cross." Chris Christiansen and Paul Wild each had plans for solar radio telescopes. Ultimately Mills and Christiansen would leave Radiophysics for the University of Sydney where they could pursue their goals. Wild almost accepted an attractive offer to go to Cornell University in New York, but with the help of Joe Pawsey, he was able to obtain additional funds to build his solar radioheliograph near Culgoora in northern New South Wales. Later, as discussed in Sect. 4.6, Joe Pawsey planned to leave to become Director of NRAO, but became ill and died before he could take up the NRAO position. John Bolton supported

Bowen's plans for building a GRT, but Bolton had gone off to Caltech with Gordon Stanley to start the Owens Valley Radio Observatory.

Bowen first recruited Barnes Wallis to help design his radio telescope[31] and then engaged Freeman Fox, who had designed the Sydney Harbor Bridge, for the detailed engineering design. Freeman Fox produced several designs for various sized telescopes. The available funds suggested a diameter of 210 foot (64 meter) if the elevation was limited to 30 degrees above the horizon and if the operating wind speed was limited to only 30 miles per hour. The shortest operating wavelength was specified as 10 cm (3 GHz). A site near the farming town of Parkes in the rural New South Wales Goobang Valley was chosen due to its isolation from local sources of radio interference, expected low winds, mild snow-free climate, geological stability, and relative proximity to Sydney. With a 210 foot diameter, the Parkes radio telescope was clearly too large to use a conventional equatorial (polar) mount. Instead Wallis developed a novel "Master Equatorial" consisting of a small equatorially mounted unit located at the intersection of the antenna's azimuth and elevation axis (Robertson 1992, p. 147). Using a light beam reflected off a mirror on the back of the dish, the pointing of the radio antenna was servo-controlled to the position of the Master Equatorial and provided remarkable stable and precise pointing. Although the Master Equatorial design was known to the AUI Planning Committee, it was never seriously considered for the Green Bank 140 Foot Telescope.

After he returned to Australia from Caltech at the end of 1960, John Bolton oversaw the completion of the Parkes radio telescope, which was opened on 31 October 1961, a full four years before the completion of the NRAO 140 Foot Telescope (Sect. 4.4). As with the Jodrell Bank 250 foot antenna and many subsequent large radio telescope systems built in the US and in other countries, there were many construction delays and cost increases. The final cost of the Parkes radio telescope was about US$2 million, considerably more than the initial budget estimates, but only about one-seventh the cost of the 140 Foot. Although smaller than the Jodrell Bank radio telescope, the Parkes dish had far better surface and pointing accuracy than Lovell's telescope, and twice the collecting area of the NRAO 140 Foot Telescope. At least initially it did not operate at the same short wavelengths as the 140 Foot antenna, but subsequent upgrades, resurfacing, and strengthening resulted in it being used up to 22 GHz (1.3 cm) with a sensitivity comparable to that of the 140 Foot Telescope.

During the first few years of operation, Bolton began a variety of research programs, including a systematic survey of the sky south of +20 degrees Declination. Following his close involvement with optical astronomers at Caltech and at the Mt. Wilson and Palomar observatories, Bolton initiated a very productive collaboration with astronomers and students from the Mt. Stromlo Observatory near Canberra, leading to the discovery of many new quasars and the determination of their redshifts. Later, Richard (Dick) Manchester led a vigorous Parkes pulsar program and the first Fast Radio Burst was discovered at Parkes in 2007 by Duncan Lorimer et al. (2007). Along with the Sydney Opera House and the Sydney Harbour Bridge, the Parkes radio telescope has become an Australian icon, and the role of the Parkes telescope in support of the Apollo 11 Moon landing was the subject of the well-known movie, *The Dish*.

The Canadian Algonquin Park 150 Foot Dish Canadian scientists had developed a modest program in radio astronomy, mostly centered at an attractive and relatively radio quiet site near Penticton, British Columbia. Research at Penticton was based on several simple and inexpensive arrays operating at relatively long wavelengths of 15 to 30 meters, as well as a small instrument to monitor solar radio noise at 10 cm. The daily solar monitoring program, which began in 1947 near Ottawa, continues to this day, providing important data on solar activity and its impact on short wave radio proportion. It is certainly the longest running uninterrupted program in radio astronomy, and arguably the longest running scientific study of any kind. However, with the growing interest in the exciting developments leading to the discoveries being made in in Europe and Australia, Canadian radio astronomers needed a competitive radio telescope that would operate at centimeter wavelengths.

In 1966 the Canadian National Research Council completed a 150 foot antenna at Algonquin Park in Ontario. The Algonquin Radio Observatory (ARO) antenna was designed by Freeman Fox, who built the Parkes radio telescope, and followed many of the same design concepts, including a Master Equatorial to control the antenna pointing. The ARO radio telescope had a more precise surface and better pointing than the NRAO 140 Foot, and, owing to the alt-az design, gravitational deflections were better understood than they were for the polar-mounted 140 Foot. However, there was limited funding in Canada for instrumentation or operation. Except for the early success of the Canadian VLBI program, the scientific impact of the Algonquin Park 150 foot radio telescope was limited. After an ambitious plan to upgrade the telescope for operation at short millimeter wavelengths was cancelled following an NRC review, further radio astronomy observations with the Algonquin Park radio telescope ceased.

The Giant Metrewave Radio Telescope (GMRT) in India Radio astronomy in India began later than in many other countries. Govind Swarup started his radio astronomy career at the CSIRO Radiophysics Laboratory under the tutelage of Joe Pawsey. After spending a year at Harvard's Fort Davis radio telescope, Swarup earned his PhD at Stanford working under Ron Bracewell. In 1961, T. Krishnan, M. R. Kundu, T. K. Menon, and Swarup proposed starting a radio astronomy program in India. But only Swarup actually returned to India, where he led the design and construction of a novel new radio telescope located near the small town of Ooty and intended to observe lunar occultations. During the 1970s, Swarup and his growing cadre of young scientists used the Ooty radio telescope to determine the positions and morphology of more than a thousand extragalactic radio sources. With the construction of the more powerful VLA and the WSRT (Chap. 7), however, the time for lunar occultations had passed.

Swarup then led his young group to design and build the GMRT, comprised of thirty parabolic dishes each 45 meters (148 feet) in diameter. The GMRT, which went into full operation in the year 2000, is 25 km in extent, is located about 100 km from the city of Mumbai (Bombay), and covers the frequency range from 40 to 1700 MHz. Each dish is based on Swarup's novel SMART (Stretched Mesh Attached to Rope Trusses) design intended to exploit the low labor costs in India. Following a series of upgrades, the GMRT remains one of the world's most powerful radio telescopes, especially in the frequency range below 1 GHz.

6.7   Grote Reber Challenges NRAO[32]

Since Karl Jansky's early work, and starting with Reber's activities at Wheaton, radio astronomers have steadily moved to ever-shorter wavelengths in the quest for better angular resolution and to study the multitude of molecular transitions that exist in the millimeter and submillimeter bands. In 1953 Reber was approached by Merle Tuve of the Carnegie Institution to return to Washington to help develop plans for building a large parabolic dish. Characteristically departing from conventional wisdom, Reber decided to concentrate instead on the extremely long hectometer wavelengths where he felt he could make a bigger impact. While he was working in Hawaii, he had access to ionospheric records which showed regions of minimum ionospheric attenuation between latitudes of 40 and 50 degrees in both hemispheres. Reber chose Tasmania, with its access to the rich southern sky and more favorable climate compared with Canada. In November 1954 he moved to Bothwell, Tasmania, where, except for visits to the United States and Canada, he lived and worked for almost the next 50 years.

In Tasmania, Reber designed and built a series of arrays to study galactic radio emission at wavelengths as long as 2.1 km. Reber hoped to exploit the fluctuations in the ionosphere which he claimed led to the creation of narrow holes which would form a high resolution window to the Universe. Observing at hectometer wavelengths, he noted that the sky is very bright everywhere, especially at the poles, and that the Milky Way appears as a dark absorption band which he correctly understood was due to free-free absorption by interstellar electrons (Reber and Ellis 1956). Unfortunately, increasing levels of ionospheric absorption and the increased density of the interplanetary plasma following the unusually low sunspot minimum of 1955, combined with increasing levels of broadcast interference, limited the results of Reber's hectometer observations. But Reber's 144 meter (2045 kHz) array, which he built in a local farmer's pasture, consisted of 192 dipoles covering a full square km and remains the largest "filled-aperture" radio telescope ever built (Reber and Ellis 1968).

To overcome ionospheric absorption, Reber conceived the idea of releasing liquid hydrogen into the ionosphere so it could recombine with free electrons and thus make the ionosphere temporarily transparent to wavelengths as long as 300 meters. For the last twenty years of his life, he relentlessly tried to obtain demobilized American ICBMs to carry a canister of hydrogen into space, but he was frustrated by the seemingly endless bureaucracy and the large cost associated with any rocket activity. Letters to colleagues (including one of us, KIK), friends, observatory directors, NASA laboratory heads, and Congressmen gave no encouragement, but he stubbornly refused to take "no" for an answer and continued to seek surplus rockets until shortly before his death. However, Reber, Ellis, and others did convince NASA to fire the engines for 16 seconds on the ill-fated Challenger space shuttle during a pass over Bothwell on the night of 15 August 1985. A quarter ton of fuel was released in an attempt to create a temporary hole in the ionosphere. But the results of these hectometer observations were inconclusive (Ellis et al. 1987).

In 1967, Reber applied for an NSF grant of $1.25 million to construct a large array in the northern hemisphere to operate at a wavelength of 144 meters, but was turned down. Reber did not easily accept the rejection of his proposal or the growing emphasis on short wavelengths. His suggestion that his array could be funded by "slight curtailment of routine operations at Green Bank," and that "N.R.A.O. should be gradually contracted in favor of scientifically more auspicious long wave programs at other places throughout the country"[33] was not well received in Washington or by NRAO. He researched previous NSF astronomy grants and challenged the legality of the national astronomy centers, which he argued were prestige institutions designed to impress the ignorant, which constituted "a mortgage on astronomy money," and said that the members of the National Science Board were stuffy old men.

Reber wrote letters to the NSF Director Leeland Haworth; the President of the National Academy of Sciences Frederick Seitz; various Congressmen, including Senator William Proxmire, as well as Ralph Nader and his Center for the Study of Responsive Law. He testified before Congress that too much NSF funding for radio astronomy was going to NRAO instead of individual investigators[34] and claimed that the "staff at the NSF are mostly clerks quite uninterested and incapable of imaginative scientific leadership."[35] From the National Academy of Sciences he asked for "positive leadership instead of basking in renown."[36] To establish his credentials so that that the recipients of his letters did not think they were coming from a crank, he would often start his letters referring the reader to the *Encyclopedia Britannica* article describing his accomplishments.

Reber also expressed great concern at the National Academy of Sciences report (Whitford 1964) on the future of ground-based astronomy in the US, in particular what he considered an overemphasis on "huge instruments." In a letter to *Science*, Reber (1966) argued against the construction of a large radio array [the VLA, Chap. 7] and commented that the plans for a 400–600 foot radome enclosed dish "displays an acute lack of imagination." He was especially critical about NRAO and about the present and future NRAO scientists, writing,

> Green Bank might as well be closed down. The best work likely to ever be done there has already been completed and published: namely my beans.[37] Some of the things that go on there in the name of administration shouldn't happen to a dog. The net effect is that of a mortgage on astronomy. Only the duller members of the new generation will try to find a place at these institutions.[38]

His prescient remarks about the need for a tracking multi-beam array made of simple wire elements preceded by more than a quarter of a century the international discussions about building a Square Kilometre Array (e.g., Ekers 2013), but his uninhibited comments about mainstream radio astronomy, especially concerning NRAO, left Reber as a bit of a pariah.

6.8   Changing Leadership

NRAO By 1977, Dave Heeschen was getting tired of dealing with the increasing NSF bureaucracy and announced his desire to step down as Director of NRAO and spend more time with his family, sailing on Chesapeake Bay, and in reactivating his research career. During his 17 years, first as NRAO Acting Director and then as Director, Heeschen had transformed NRAO from an organization on the brink of failure into a highly respected facility that was universally acknowledged as the leading radio observatory in the world. During his tenure as NRAO Director, Heeschen established NRAO as the national observatory that had been envisioned by Menzel, Bok, Stratton, Berkner, and others. He saw the completion, finally, of the 140 Foot Telescope, led the drive to build the VLA, established millimeter radio astronomy at NRAO, and enthusiastically supported the early VLBI program and later the initiative to build the VLBA. He recruited an outstanding scientific and engineering staff and oversaw NRAO's transition into a true visitor institution. He took particular interest in the professional growth of the scientific staff, and throughout his tenure as Director, he remained the head of the "Basic Research Division." Accustomed to the informal discussions with Randy Robertson in the early years, he was annoyed when the NSF started to record their discussions and rebelled against the NSF request to review and edit the lengthy transcript of one of their meetings. Instead, he sent it back with the terse note, "If that's what we said, then that's what we said."[39] Nevertheless, he was greatly respected at the NSF which considered NRAO to be the best run of the national centers.[40]

After his retirement, Heeschen broke his personal tradition and gave an after dinner talk at an NRAO internal scientific symposium where he outlined his "advice to directors and managers and to would-be-directors and managers:"[41] (1) Hire good people, then leave them alone; (2) Do as little managing as possible; (3) Use common sense; (4) Don't take yourself too seriously; and finally, (5) Have fun.

Throughout his tenure as Director, Heeschen expressed concern that NRAO was getting too large and that university operated facilities were not receiving adequate support. In a letter to then AUI President, Keith Glennan, Heeschen wrote,[42]

> For all its logical, quantitative aspects, I think the best science is a highly personal, individualistic activity—as much so as art or music or writing—and it needs the right kind of atmosphere. If NRAO, in one context, or AUI in another, gets so large that the atmosphere can't be maintained then we become just high-level technicians or facility operators, and while it may be necessary that someone do this, I don't think it is for us.

When Heeschen announced his retirement, AUI, under the direct leadership of President Gerry Tape, initiated a search for a new Director. But running a "user" oriented observatory does not necessarily appeal to the typical research scientist. After a national search, the top three candidates all turned down the NRAO Directorship, and AUI asked Mort Roberts to become the NRAO Director. Roberts was a long time member of the NRAO Scientific Staff and had previously served for a year as the Green Bank site director from 1969 to 1970. He had a distinguished research career primarily using 21 cm spectroscopy to study the kinematics of nearby galaxies which led to best evidence for the existence of dark matter.

Roberts became the Director of NRAO just as the construction of the VLA was being completed and NRAO was gearing up for the 25 meter millimeter radio telescope and the VLBA. He oversaw the completion of the VLA construction (Sect. 7.6) and the complex transition from VLA construction to VLA operations (Sect. 7.7), hired Ron Ekers, then working in the Netherlands, to become director of VLA operations, and helped procure the funding for the VLBA (Sect. 8.7). However, he was blamed, probably inappropriately, for the failure of the popular 25 meter millimeter-wave telescope proposal (Sects. 8.7 and 10.3).

Roberts strongly believed that NRAO should have a world class scientific staff, and tried to protect the research time of the staff. Young staff members, especially those on term appointments, were encouraged to concentrate on their research. As he put it, their job was "to get tenure." This led to some resentment from those staff members on continuing or indefinite appointments, as they had to absorb the full load of supporting Observatory operations, but, in return, they weren't subject to the threat of termination at the end of a few-year term appointment.

After the completion of his five year term as NRAO Director in 1984, Roberts returned to full time research. Hein Hvatum became the Interim Director for three months until Paul Vanden Bout arrived as NRAO Director from the University of Texas on 1 January 1985. Vanden Bout had built a solid reputation for making the University of Texas Millimeter Wave Observatory a major player in millimeter spectroscopy, and he had trained a new generation of millimeter wave astronomers. Before coming to NRAO, Vanden Bout had served on both the NRAO Visiting and Users Committees, and chaired the Advisory Committee charged with selecting the site for the VLBA Operations Center, so he was well known to the AUI Board. During his 17 years at the helm of NRAO, Vanden Bout oversaw the construction of the VLBA (Sects. 8.7 and 8.8); dealt with the unprecedented circumstances surrounding the collapse of the 300 Foot Telescope leading to funding and construction of the Green Bank Telescope (Sects. 9.6 and 9.7); began the Expanded Very Large Array project (Sect. 7.8); together with Bob Brown, initiated the design of the Millimeter Array (MMA); and was an important part of the complex negotiations that lead to the international partnership and the funding and construction of ALMA (Atacama Large Millimeter/submillimeter Array) (Sects. 10.6 and 10.7). Vanden Bout resigned as NRAO Director in June 2002 to become Interim Director of ALMA, and W. Miller Goss became the Interim NRAO Director.

AUI reached out to Kwok-Yung (Fred) Lo, then director of the Taiwan Institute of Astronomy and Astrophysics to fill the Director position. Earlier, Lo had been the Chair of the University of Illinois Department of Astronomy and had served on the faculty at Caltech. He received his PhD in physics from MIT in 1974 and had a wide range of research interests including star formation and VLBI. Lo's management style contrasted with that of Vanden Bout, who tried to rule by consensus, whereas Lo ruled more by edict. During his ten year, sometimes contentious, tenure as NRAO Director, Lo oversaw the complex transformation of the VLA to the JVLA (Sect. 7.8); made the tough decision to curtail the long running AIPS++ project; and played a crucial role in the establishment of ALMA and the North American ALMA Science Center at NRAO in Charlottesville. Lo stepped down as NRAO Director in May 2012 and died in 2016 after a 25-year-long struggle with cancer (Fig. 6.12).

The next Director of NRAO was Anthony (Tony) Beasley who arrived at NRAO with impressive credentials obtained as a result of his forceful management of some of the NSF's largest projects. Beasley first came to NRAO in 1991 on a postdoctoral appointment after receiving his PhD in Astrophysics from the University of Sydney. He rapidly rose through the ranks, becoming Deputy Assistant Director for VLA/VLBA Operations and for Computing in 1997, and Assistant Director in 1998. In 2000 Beasley left NRAO to become Project Manager for the Combined Array for Research in Millimeter-Wave Astronomy (CARMA) which combined the Caltech and BIMA millimeter arrays (Sect. 10.4) at a relatively dry site in the Inyo Mountains east of OVRO. In 2004 he returned to NRAO as an Assistant Director and International Project Manager for ALMA, under construction in Chile by NRAO and ESO (Sect. 10.7). His next challenge was as Chief Operating Officer and Project Manager for the NSF-funded National Ecological Observatory Network (NEON), a continental-scale ecological observatory designed to detect ecological change and enable forecasting of its impacts. In 2012 Beasley returned as Director of NRAO and in 2016 he also became AUI Vice President for Radio Astronomy Operations. He brought a new style of aggressive no-nonsense management to an Observatory faced with NSF-mandated changes and was committed to bringing under-represented groups to NRAO.

While open to input and criticism, Beasley responded decisively to each crisis situation facing NRAO. Hitting the ground running, he initiated the new VLA Sky Survey, began the effort to provide Science Ready Data Products to enhance the productivity of the increasingly complex NRAO telescopes, and initiated the early development of the next generation VLA (ngVLA) project[43] (Sect. 11.2) to provide the US response to the international SKA project (Ekers 2013).

AUI Following the rapid turnover in the AUI Presidency discussed in Sect. 4.6, Gerry Tape returned from his appointment as Ambassador to the International Atomic Energy Commission to become President of AUI from 1969 to 1980. Tape was succeeded by Cornell chemist Robert E. Hughes, who had previously served as the NSF Assistant Director for Math and Physical Sciences. Hughes retired in 1997 and was replaced by Lyle Schwartz from the

Fig. 6.12 Four NRAO Directors at the NRAO 50th anniversary symposium, June 2007. Left to right: Fred K. Y. Lo (5th Director), Paul A. Vanden Bout (4th Director), Morton S. Roberts (3rd Director), David S. Heeschen (2nd Director). Credit: NRAO/AUI/NSF

National Institute of Standards and Technology (formerly the National Bureau of Standards). Schwartz was immediately confronted with the Brookhaven controversy over the alleged leakage of radioactive tritium into the local drinking water (Sect. 9.7), resulting in the loss of the contract with the Department of Energy to operate Brookhaven.

AUI was in serious trouble. Faced with the loss of the lucrative Brookhaven contract which was an order of magnitude larger than the NRAO contract, Schwartz resigned just over a year after he arrived. Some of the AUI Trustees also resigned, and AUI was restructured to become a self-perpetuating not-for-profit manager of scientific facilities. No longer were the nine original universities represented on the Board by a university scientist and administrator, and AUI was governed by a new, more diverse Board of Trustees.

The new AUI Board asked long-time Trustee and Cornell Professor of Astronomy Martha Haynes to act as Interim AUI President until they could find a permanent replacement for Schwartz. Around this time Riccardo Giacconi was about to complete his term as Director of the European Southern Observatory (ESO) where he oversaw the construction of a suite of four 8 meter telescopes in Chile known as the Very Large Telescope. Giacconi had previously served as the first Director of the Space Telescope Institute in Baltimore, Maryland, where he defined how the Hubble Telescope would serve the astronomical community. Earlier, at the Harvard-Smithsonian Center for Astrophysics, he directed the effort leading to the Einstein Observatory and initiated what later became the Chandra X-ray Observatory. In 2002 Giacconi received the Nobel Prize for Physics for his "pioneering contributions to astrophysics, which have led to the discovery of cosmic X-ray sources."

Giacconi had a reputation both in the US and Europe as a strong, some felt too strong, leader who could return AUI to its former prominence, and on 1 July 1999, he became the first president of AUI with a background in astrophysics. Coupled with the fact that, for the first time, AUI had no responsibilities other than NRAO, it meant Giacconi brought a strong hand to his close management of NRAO and its Director Paul Vanden Bout.

At ESO, there were no permanent members of the Scientific Staff, and Giacconi questioned the appropriateness of tenure at NRAO. He expected that all of the scientists should have well-defined "functional" responsibilities, instituted strong project management, and full cost accounting for all NRAO projects. Until his retirement in 2004, Giaconni also played a strong role in the management of ALMA (Sect. 10.7), a project which he helped to initiate while still at ESO.

Giacconi was succeeded as AUI President by his long-time friend and colleague Ethan Schreier, who not only had a background in astronomy, but who had even been a user of the VLA. Schreier took an active role in the US participation in the planning for the international Square Kilometre Array (SKA) (Sect. 11.6), until it became apparent in 2011 that the US would not play a role in the project. In 2017 Adam Cohen, who had been the Deputy Under Secretary for Science and Energy at the US Department of Energy, became the new AUI President. Cohen greatly expanded the AUI corporate office and initiated a number of new scientific, technical, and business initiatives to broaden AUI's purview.

NOTES

1. Strictly speaking, Taffy Bowen was not a radio astronomer, but was effectively a member of the international radio astronomy community.
2. NAA-NRAO, DO, Conferences, Symposia, Colloquia.
3. Hagen to Greenstein, 16 August 1954, CITA-LAD, Box 35, Folder 2.
4. See Kellermann and Sheets (1984) and Wilkinson et al. (2004) for reviews of these discoveries.
5. Nomenclature to describe atomic RRLs is given by the name of the element followed by the lower electron energy, followed by Greek letter denoting the number of energy levels involved in the transition.
6. Pawsey to Rabi, 5 October 1961, LOC-IIR, Box 36.
7. Wideband Interferometric Digital Architecture.
8. Carl Borgman, Director of Science and Engineering at the Ford Foundation, reported that 97 to 98 percent of the time on the Palomar 200 inch was used by the Observatory staff, but only 25 percent of the 100 inch time (Edmondson 1997, p. 179).
9. NAA-NRAO, Founding and Organization, Correspondence. Heeschen's notice was published in *Science* (**130**, 1179) and the *AJ* (**64**, 1273), but we have found no record of any publication appearing in *PASP*.
10. The design and construction of the Haystack 120 foot antenna has been described by Weiss (1965). Much of the description given here is taken from this paper.
11. Albert Einstein proposed three tests of his General Theory of Relativity: the precession of Mercury's orbit; the bending of light by the Sun; and the shifting of spectral lines in a gravitational field. In 1965 MIT professor Irwin Shapiro proposed a fourth test resulting from the excess delay of the reflected radar signal from a planet as the signal passes close to the Sun (Shapiro 1964). Using Mercury and Venus as targets when they were near superior conjunction, Shapiro and his colleagues confirmed the delay calculated due to General Relativity (Shapiro et al. 1971).
12. The agency has been variously known as ARPA or DARPA.
13. Further background on the history of the Haystack facility can be found at http://www.haystack.mit.edu/hay/history.html
14. This section is adopted from Altschuler (2002) and Cohen (2009).
15. When using a parabolic reflector, the feed system must remain close to the focal point in order to minimize losses due to aberration.
16. DuBridge to Piore, 9 July 1954, CITA-LAD.
17. Robertson to DuBridge, 19 July 1954, CITA-LAD, Box 35, Folder 2.
18. Montroll to DuBridge, 23 July 1954, CITA-LAD, Box 35, Folder 2.
19. Bacher to Piore, 12 August 1954, CITA-LAD, Box 35, Folder 2.
20. DuBridge to Priori, 21 September 1954, CITA-LAD, Box 35, Folder 2. After the launch of Explorer I, just a few months after the launch of Sputnik, Pickering

led the JPL development of US missions to the Moon and planets for the next two decades.

21. GR to DuBridge, 8 April 1954, CITA-LAD, Box 35, Folder 2.

22. Greenstein to DuBridge, 8 April 1954, CITA-LAD, Box 35, Folder 2.

23. Dubridge to GR, 13 April 1954, CITA-LAD, Box 35, Folder 2.

24. John Bolton's early career at the CSIRO Radiophysics Laboratory, his exile to cloud physics research, his time at the Caltech Owens Valley Observatory, and his return to direct research with Bowen's 210 foot radio telescope are discussed in the book by Peter Robertson (2017).

25. DuBridge to Piore, 21 September1954, CITA-LAD, Box 35, Folder 2.

26. Bowen to DuBridge, 28 September, 1954, CITA-LAD, Box 35, Folder 2.

27. DuBridge to Bolton, 8 October 1954, CITA-LAD, Box 35, Folder 2.

28. See *The Story of Inyo* by W. A. Chalfant (1933) for a descriptive account of the impact to the Owens Valley by the city of Los Angeles.

29. Oral History, Robert Bacher, CITA.

30. Arnold Shostak's son, Seth, studied radio astronomy at Caltech and became one of the first people to recognize that 21 cm observations of the rotation of galaxies suggested presence of more matter than was visible to optical telescopes, which later became known as dark matter. Seth later went on to a prominent career in SETI research and as a well-known spokesman for understanding life in the Universe.

31. During WWII, Wallis had designed bombers and had developed innovative techniques for destroying German dams critical to the war effort. Wallis conceived the concept of an independently mounted small "master equatorial" unit located at the interion of the antenna axes and optically coupled to the telescope.

32. This section on Grote Reber is adapted from Kellermann (2004), with permission from the Astronomical Society of the Pacific.

33. GR to E. H. Hurlburt, 20 February 1967, NAA-GR, Correspondence, General Correspondence.

34. GR Testimony before US Senate Committee on Independent Offices and Department of Housing and Urban Development Appropriations for Fiscal Year 1970, 11 July 1969, NAA-GR, Correspondence, General Correspondence.

35. GR to Connecticut Representative E. Q. D'Addario, 14 July 1966, NAA-GR, Correspondence, General Correspondence.

36. Ibid.

37. In 1959 and 1960 Reber spent time in Green Bank during which he tried to increase the harvest of beans by forcing them to wind in the reverse direction (Reber 1964).

38. One of the current authors (KIK) joined the NRAO scientific staff in 1965 where he remained for 52 years.

39. WEH III to KIK, 2 April 2012, NAA-KIK, Open Skies, Chapter 6.

40. Ibid.

41. DSH, 12 June 1992, NRAO Internal Symposium, Green Bank, WV, NAA-DSH, Radio Astronomy History, US Radio Astronomy, Undated talks.

42. DSH to Keith Glennan, 22 November 1967, NRAO-DO, Organizational Charts and Memos, 1958–2012.

43. https://ngvla.nrao.edu/

BIBLIOGRAPHY

REFERENCES

Altschuler, D.R. 2002, The National Astronomy and Ionospheric Center's (NAIC) Arecibo Observatory in Puerto Rico. In ASPC **278**, *Single–Dish Radio Astronomy: Techniques and Applications*, ed. S. Stanimirovic et al. (San Francisco: ASP), 1

Baars, J.W.M. 1966, Reduction of Tropospheric Noise Fluctuations at Centimetre Wavelengths, *Nature*, **212**, 494

Backer, D.C. et al. 1982, A Millisecond Pulsar, *Nature*, **300**, 615

Balick, B. and Brown, R.L. 1974, Intense Sub-Arcsecond Structure in the Galactic Center, *ApJ*, **194**, 265

Bennett, C.L. et al. 2003, Nine-Year Wilkinson Microwave Anisotropy Probe (WMAP) Observations, *ApJS*, **208**, 20

Bridle, A.H. et al. 1972, Flux Densities, Positions, and Structures for a Complete Sample of Intense Radio Sources at 1400 MHz, *AJ*, **77**, 405

Chalfant, W.A. 1933, *The Story of Inyo* (Chicago: The Author)

Cheung, A.C. et al. 1968, Detection of NH^3 Molecules in the Interstellar Medium by Their Microwave Emission, *Phys. Rev. Lett.*, **21**, 1701

Cheung, A.C. et al. 1969, Detection of Water in Interstellar Regions by its Microwave Radiation, *Nature*, **221**, 626

Clark, B.G. 1965, An Interferometric Investigation of the 21-centimeter Hydrogen-Line Absorption, *ApJ*, **142**, 1398

Clark, B.G. et al. 1962, The Hydrogen Line in Absorption, *ApJ*, **135**, 151

Clark, B.G. and Kuzmin, A. 1965, The Measurement of the Polarization and Brightness Distribution of Venus at 10.6-CM Wavelength, *ApJ*, **142**, 23

Cohen, M.H. 2009, Genesis of the 1000-Foot Arecibo Dish, *JAHH*, **12**, 141

Dent, W.A. 1965, Variation in the Radio Emission of 3C273, *Science*, **148**, 1458

Dent, W.A. and Haddock, F.T. 1965, A New Class of Radio Source Spectra, *Nature*, **205**, 487

Ekers, R.D. 2013, The History of the Square Kilometre Array (SKA) Born Global, In *Resolving the Sky – Radio Interferometry: Past, Present, and Future*, ed. M.A. Garrett and J.C. Greenwood (Manchester: SKA Organization), 68

Edmondson, F. 1997, *AURA and Its US National Observatories* (Cambridge: CUP)

Ellis, G. et al. 1987, Low-Frequency Radioastronomical Observations During the Spacelab 2 Plasma Depletion Experiment, *Austr. J. Phys.*, **24**, 56

Findlay, J.W. 1972, The Flux Density of Cassiopeia A at 1440 MHz and its Rate of Decrease, *ApJ*, **174**, 527

Fisher, J.R. and Bradley, R.F. 2000, Full-Sampling Array Feeds for Radio Telescopes, *Proc. SPIE*, **4015**, 308

Fomalont, E.B. et al. 1964, Accurate Right Ascensions for 226 Radio Sources, *ApJ*, **69**, 772

Fomalont, E.B. and Sramek, R.A. 1975, A Confirmation of Einstein's General Theory of Relativity by Measuring the Bending of Microwave Radiation in the Gravitational Field of the Sun, *ApJ*, **199**, 749

Gordon, M.A. 2005, *Recollections of Tucson Operations* (Dordrecht: Springer)

Gordon, W.E. et al. 1961, The Design and Capabilities of an Ionospheric Radar Probe, *Trans. IRE*, **AP-9**, 17

314 K. I. KELLERMANN ET AL.

Heeschen, D.S. 1961a, Observations of Radio Sources at Four Frequencies, *ApJ*, **133**, 322

Heeschen, D.S. 1961b, Secular Variation of the Flux Density of the Radio Source Cassiopeia A, *Nature*, **190**, 705

Heeschen, D.S. 1964, A Radio Survey of Galaxies, *AJ*, **69**, 277

Heeschen, D.S. 1968, Radio Properties of Elliptical Galaxies NGC 1052 and NGC 4278, *ApJ*, **151**, L35

Hjellming, R. and Wade, C.M. 1971, Radio Stars, *Science*, **197**, 173

Hogg, D.E. 1966, Radio Emission from a Number of Possible Supernovae Remnants, *ApJ*, **144**, 819

Hogg, D.E. et al. 1969, Syntheses of Brightness Distribution in Radio Sources, *AJ*, **74**, 1206

Hogg, D.E. 2006, Selecting and Scheduling Observing Proposals at NRAO Telescopes. In *Organizations and Strategies in Astronomy Vol. 7*, ed. A. Heck (Dordrecht: Springer), 181

Höglund, B. and Mezger, P.G. 1965, Hydrogen Emission Line n_{110} to n_{109}: Detection at 5009 MHz in Galactic H II Regions, *Science*, **150**, 339

Jarosik, N. et al. 2003, Design, Implementation, and Testing of the MAP Radiometers, *ApJS*, **145**, 413

Kardashev, N.S. 1959, On the Possibility of Detection of Allowed Lines of Atomic Hydrogen in the Radio-Frequency Spectrum, *Astron. Zh.*, **36**, 838. English translation in *SvA*, **3**, 813

Kellermann, K.I. 2004, Grote Reber (1911–2002), *PASP*, **116**, 703 https://doi.org/10.1086/423436

Kellermann, K.I. and Pauliny-Toth, I.I.K. 1968, Variable Radio Sources, *ARAA*, **6**, 417

Kellermann, K.I. and Pauliny-Toth, I.I.K. 1969, The Spectra of Opaque Radio Sources, *ApJ*, **155**, L71

Kellermann, K.I. et al. 1969, The Spectra of Radio Sources in the Revised 3C Catalogue, *ApJ*, **157**, 1

Kellermann, K.I. et al. 1977, The Small Radio Source at the Galactic Center, *ApJL*, **214**, 61

Kellermann, K.I. and Sheets, B. eds. 1984, *Serendipitous Discoveries in Radio Astronomy:* (Green Bank: NRAO/AUI)

Kraus, J.D. 1995, *Big Ear Two* (Powell: Cygnus-Quasar Books)

Lorimer, D. et al. 2007, A Bright Millisecond Radio Burst of Extragalactic Origin, *Science*, **318**, 777

Lovell, A.C.B. 1987, *Voice of the Universe: Building the Jodrell Bank Telescope* (New York: Praeger)

Maltby, P. and Moffet, A.T. 1962, Brightness Distribution in Discrete Radio Sources. III. The Structure of the Sources, *ApJS*, 7, 141

Maltby, P., Matthews, T.A., and Moffet, A.T. 1963, Brightness Distribution in Discrete Radio Sources IV. A Discussion of 24 Identified Sources, *ApJ*, **137**, 153

McClain, E.F. 1966, A High-Precision 85-foot Radio Telescope, *S&T*, **32** (7), 4

Minkowski, R. 1960, A New Distant Cluster of Galaxies, *ApJ*, **132**, 908

Morris, D. et al. 1964, On the Measurement of Polarization Distribution over Radio Sources, *ApJ*, **139**, 55

Pauliny-Toth, I.I.K. et al. 1966, Positions and Flux Densities of Radio Sources, *ApJS*, **13**, 65

Pauliny-Toth I.I.K. and Kellermann, K.I. 1966, Variations in the Flux Densities of Radio Sources *AJ*, **71**, 1

Pettengill, G.H. and Dyce, R.B. 1965, A Radar Determination of the Rotation of the Planet Mercury, *Nature*, **206**, 1240

Pospieszalski, M.W. et al. 1988, FETs and HEMTs at Cryogenic Temperatures – Their Properties and Use in Low-Noise Amplifiers, *IEEE Tr. Microwave Theory and Tech*, **MTT-36**, 552

Read, R.B. 1963, Accurate Measurement of the Declinations of Radio Sources, *ApJ*, **138**, 1

Reber, G. 1964, Reversed Bean Vines, *Journal of Genetics*, **59**, 37

Reber, G. 1966, Ground Based Astronomy: The NAS 10-Year Program, *Science*, **152**, 150

Reber, G. and Ellis, G.R. 1956, Cosmic Radio-Frequency Radiation near One Megacycle, *J. Geophys. Res.*, **61**, 1

Reber, G. and Ellis, G.R. 1968, Cosmic Static at 144 Meters Wavelength, *J. Franklin Inst.*, **285**, 1

Rinsland, C.P., Dixon, R.S., and Kraus, J.D. 1975, Ohio Survey Supplement 2, *AJ*, **80**, 759

Roberts, M.S. and Whitehurst, R. 1975, The Rotation Curve and Geometry of M31 at Large Galactocentric Distances, *ApJ*, **201**, 327

Robertson, P. 1992, *Beyond Southern Skies – Radio Astronomy and the Parkes Telescope* (Cambridge: CUP)

Robertson, P. 2017, *Radio Astronomer – John Bolton and a New Window on the Universe* (Sydney: NewSouth Publishing)

Roshi, A. et al. 2018, Performance of a Highly Sensitive, 19-element, Dual-polarization, Cryogenic L-band phased-array Feed on the Green Bank Telescope, *AJ*, **155**, 202

Schmitt, J.L. 1968, BL Lac Identified as a Radio Source, *Nature*, **218**, 663

Shapiro, I.I. 1964, Fourth Test of General Relativity, *Phys. Rev. Lett.*, **13**, 789

Shapiro, I.I. et al. 1971, Fourth Test of General Relativity: New Radar Result, *Phys. Rev. Lett.*, **26**, 1132

Shklovsky, I.S. 1960, Secular Variation in the Flux and Intensity of Radio Emission from Discrete Sources, *AZh*, **37**, 256

Swenson, G.W. 1986, The Illinois 400-ft Radio Telescope, *IEEE Ant. Prop. Newsletter*, **28** (6), 13

Taylor, J.H. et al. 1979, Measurements of General Relativistic Effects in the Binary Pulsar PSR1913+16, *Nature*, **277**, 437

Townes, C.H. 1946, The Ammonia Spectrum and Line Shapes Near 1.25 cm Wave-Length, *Phys. Rev.*, **70**, 665

Townes, C.H. 1957, Microwave and Radio-Frequency Lines of Interest to Radio Astronomy. In *IAU Symposium No. 4, Radio Astronomy*, ed. H.C. van de Hulst (Cambridge: CUP), 92

Townes, C.H. and Merritt, F.R. 1946, Water Spectrum Near One-Centimeter Wave-Length, *Phys. Rev.*, **70**, 558

van de Hulst, H.C. 1945, Origin of Radio Waves from Space, *Nederlandsch Tijdschrift voor Natuurkunde*, **11**, 210. English translation in Sullivan, W.T. III 1982, *Classics in Radio Astronomy* (Cambridge: CUP), 302

von Hoerner, S. 1964, Lunar Occultations of Radio Sources, *ApJ*, **140**, 6

Wade, C.M. 1970, Precise Positions of Radio Sources. I. Radio Measurements, *ApJ*, **162**, 381

Weinreb, S. 1961, Digital Radiometer, *Proc. IRE*, **49**, 1099

Weinreb, S. 1963, *A Digital Spectral Analysis Technique and its Application to Radio Astronomy*, PhD Dissertation, Massachusetts Institute of Technology

Weinreb, S. and Kerr, A.R. 1973, Cryogenic Cooling of Mixers for Millimeter and Centimeter Wavelengths, *IEEE J. Solid-State Circuits*, **8**, 58

Weiss, H.G. 1965, The Haystack Microwave Research Facility, IEEE Spectrum, February, 50

Whitford, A. 1964, *Ground-based Astronomy: A Ten Year Program* (Washington: National Academy of Sciences)

Wilkinson, P.N. et al. 2004, The Exploration of the Unknown. In *Science with the Square Kilometre Array*, ed. C. Carilli and S. Rawlings (Amsterdam: Elsevier), 1551

Wolszczan, A. and Frail, D. 1992, A Planetary System Around the Millisecond Pulsar, *Nature*, **355**, 145

Zwicky, F. 1933, Die Rotverschiebung von Extragalaktischen Nebeln, *Acta Helvetica Physica*, **6**, 1102

FURTHER READING

Berge, G.L. and Seielstad, G.A. 1967, New Determinations of the Faraday Rotation for Extragalactic Radio Sources, *ApJ*, **148**, 367

Berge, G.L. and Seielstad, G.A. 1969, Polarisation Measurements of Extragalactic Radio Sources at 3.12 cm Wave-Length, *ApJ*, **157**, 35

Burke, B.F. 2003, Early Years of Radio Astronomy in the U.S. In *Radio Astronomy from Karl Jansky to MicroJansky*, ed. L.I. Gurvits et al. (France: EDP Sciences), 27

Cohen, M.H. 1994, The Owens Valley Radio Observatory: The Early Years, *Engineering and Science*, **57** (3), 8

Cohen, M.H. 2007, A History of OVRO: Part II, *Engineering and Science*, **70** (3), 33

Giacconi, R. 2008, *Secrets of the Hoary Deep: A Personal History of Modern Astronomy* (Baltimore: Johns Hopkins University Press)

Goddard, D.E. and Milne, D.K. eds. 1994, *Parkes: Thirty Years of Radio Astronomy* (Melbourne: CSIRO)

Graham-Smith, F. 2003, The Early History of Radio Astronomy in Europe. In *Radio Astronomy from Karl Jansky to MicroJansky*, ed. L.I. Gurvits et al. (France: EDP Sciences), 1

Kellermann, K.I. 1996, John Gatenby Bolton (1922-1993), *PASP*, **108**, 729

Kraus, J. 1963, The Large Radio Telescope at Ohio State University, *S&T*, **26** (1) 12

Leverington, D. 2017, *Observatories and Telescopes of Modern Times* (Cambridge: CUP)

Lockman, F.J., Ghigo, F.D., and Balser, D.S. eds. 2007, *But It Was Fun* (Green Bank: NRAO/AUI)

Lovell, A.C.B. 1985, *The Jodrell Bank Telescopes* (Oxford: Oxford University Press)

Lovell, A.C.B. 1990, *Astronomer by Chance* (New York: Basic Books)

Munns, D.P.D. 2013, *A Single Sky* (Cambridge: MIT Press)

CHAPTER 7

The Very Large Array

Starting in 1961, NRAO scientists began the process of designing a radio tele-scope that could make images with an angular resolution comparable to the best optical telescopes operating from a good mountain site. In 1967, the Observatory submitted a proposal to the National Science Foundation (NSF) for the construction of the Very Large Array (VLA). The VLA proposal was for 36, later reduced to 27, fully steerable 25 meter diameter antennas spread over an area some 35 km in diameter. However, there was a competing, much simpler and much cheaper proposal from Caltech for an 8 element array of 130 foot dishes. Several NSF review committees praised the VLA concept but indicated that it was too ambitious, and recommended that NRAO further study the VLA design, and that construction of the Caltech array should begin immediately. Following a confrontational battle among proponents of the NRAO and Caltech arrays, as well as a competing proposal for a 440 foot radome-enclosed antenna proposed by an MIT-Harvard led consortium, support of the VLA by the 1970 National Academy Decade Review of astronomy led to approval of its construction.

The 1973 oil crisis and the subsequent period of excessive inflation nearly killed the fixed budget project. But under the leadership of Dave Heeschen, NRAO brought the VLA project to completion in 1980, on schedule and close to the planned $78M budget appropriation. The VLA has been by far the most powerful and most successful radio telescope ever built.

7.1 Background

The 1957 Soviet launch of Sputnik created a widespread and frenzied concern that the US had fallen behind Russia in all matters scientific, especially in any-thing connected with space. In astronomy, the long tradition in optical astron-omy of building large telescopes on excellent mountain sites clearly established the United States as the world's leader in observational astronomy (see e.g.,

© The Author(s) 2020, corrected publication 2021
K. I. Kellermann et al., *Open Skies*, Historical & Cultural Astronomy,
https://doi.org/10.1007/978-3-030-32345-5_7

Florence 1994). Meanwhile, as discussed in Chap. 2, radio astronomers in the US, Europe, and Australia were reporting on exciting new discoveries ranging from solar system science to cosmology. The time was ripe to review the status of US astronomy and to plan for the future growth.

Radio vs Optical Resolution In spite of the dramatic advances and new discoveries made during the quarter century following Karl Jansky's pioneering work, by 1960 radio astronomers faced two challenges to further progress. First, the angular resolution of any optical or radio telescope is determined by the ratio of wavelength to size of the telescope. Because radio wavelengths are longer than optical wavelengths by a factor of about one hundred thousand, for many years it was assumed that the resolution of radio telescopes was fundamentally limited compared with the resolution of optical telescopes. Second, while optical telescopes can produce images of celestial objects with millions of independent pixels, conventional radio telescopes typically respond to the emission from only a single area in the sky. Thus, in order to map the area of interest, radio astronomers traditionally had to make a time-consuming raster scan. In this chapter, we describe how radio astronomers developed interferometric synthesis techniques to improve the angular resolution over what is possible from any filled aperture instrument, and discuss how NRAO was able to overcome considerable opposition and technical challenges to build the Very Large Array to make images of the radio sky with resolution comparable to that achieved by the best ground based optical telescopes. In the following chapter, we discuss how radio interferometry was extended to obtain angular resolutions hundreds to thousands of times better than the best optical telescopes on the best mountain sites or in space.

Early Radio Interferometry and Synthesis Imaging[1] The naive comparison between the resolution of radio and optical telescopes has turned out to be wrong for three important but not widely appreciated reasons. First, because radio wavelengths are long (indeed they are comparable with every day physical scales), it is possible to build diffraction-limited radio telescopes of essentially unlimited dimensions. Second, in practice, the resolution of ground based optical and infrared telescopes has been traditionally limited not by diffraction, but by turbulence in the Earth's troposphere known as "seeing."[2] Finally, while optical and infrared interferometers are feasible, their sensitivity is limited by the need to divide the incoming signal among two or more detectors, with a corresponding loss of sensitivity, whereas at radio wavelengths the signals can first be amplified before splitting with little loss of sensitivity.

Using their single antenna on a cliff overlooking Sydney Harbor,[3] McCready et al. (1947) were probably the first to recognize that the response of a simple two-element interferometer was one Fourier component of the sky brightness distribution. In their paper, McCready et al. famously noted, "It is possible in principle to determine the actual form of the [sky brightness] distribution in a complex case by Fourier synthesis using information derived from a large number of components." However, they went on to comment that varying the height of the cliff antenna "would be feasible but clumsy. A different interference method may be more practicable."[4]

Wilbur (Chris) Christiansen later used Earth-rotation synthesis imaging by combining the output of multiple one-dimensional scans of the Sun. Christiansen and Warburton (1955) first formed two orthogonal phased arrays, which they used to make multiple strip distribution scans across the Sun at different orientations as the Earth rotated on its axis. They then laboriously calculated the Fourier components of each strip distribution, followed by a two-dimensional Fourier inversion to obtain a two-dimensional image of the Sun.

It would be Martin Ryle and his group at Cambridge who were later able to exploit the full power of two-dimensional Fourier synthesis imaging, which is commonly referred to as "aperture synthesis" (e.g., Ryle and Hewish 1960; Ryle 1975). By combining data from a variable spacing interferometer and exploiting the rotation of the Earth to change the orientation of their east-west baseline, Ryle and Neville (1962) obtained a two-dimensional image of a 25 square degree region centered on the north celestial pole with a resolution of 4.5 arcmin.[5] This technique was informally referred to in Cambridge as "super-synthesis" and in Sydney as "Earth-rotation synthesis." However, super-synthesis using the meridian fixed parabolic cylinders was restricted to the north polar region.

The Cambridge group then went on to build the One-Mile Radio Telescope using two fixed and one moveable 60 foot steerable dishes located on a one-mile-long east-west baseline. The One-Mile Radio Telescope initially operated at 21 and 73 centimeters, exploiting the changing orientation of the array due to the rotation of the Earth to sample the Fourier transform (u,v) plane, giving resolutions of 25 arcsec and 1.5 arcmin respectively (Ryle 1962). Later, the resolution was improved to 12 and 6.5 arcsec with the installation of receivers for 11 and 6 cm respectively. The One-Mile Radio Telescope was followed a decade later by the 5-km Radio Telescope (Ryle 1972) using four fixed and four movable steerable antenna elements. Initially equipped to operate at 6 cm wavelength, and in 1974 at 2 cm, the 5-km Radio Telescope was able to make images with an initial resolution of only 2 arcsec and later better than 1 arcsec, or comparable to that of the best large ground based optical telescopes.

The Caltech Owens Valley Radio Observatory (OVRO) In Sect. 6.6 we described the development and the many spectacular successes of the Caltech twin 90 foot interferometer (Fig. 7.1). One may wonder why Caltech did not make better use of the capabilities of their two-element interferometer to do the same kind of full two-dimensional super-synthesis (Earth-rotation synthesis) pioneered by Martin Ryle and colleagues at Cambridge. Although Cambridge was characteristically secretive about their plans, a full description of the first super-synthesis array, the Cambridge One-Mile Radio Telescope was published in 1962 (Ryle 1962). At that time, Ryle had received funding, but

Fig. 7.1   The Owens Valley Interferometer. Credit: Caltech Archives

construction of the three-element array of 60 foot steerable dishes had not yet started. The OVRO two-element east-west interferometer was already complete and in operation by the end of 1959. By the end of 1960 the north-south baseline was also operational. This was six and five years respectively, before the first results from the Cambridge One-Mile Radio Telescope (Ryle et al. 1965). But the Caltech radio astronomers Per Maltby and Alan Moffet used Fourier inversion of their OVRO east-west and north-south OVRO transit observations of interferometer phase and amplitude to obtain only separate one-dimensional strip distributions which they then used to discuss the two-dimensional brightness distributions (Maltby and Moffet 1962). Maltby and Moffet did make some amplitude-only observations at different hour angles but these data were used only to constrain the model fitting, and they never made full use of Earth-rotation synthesis and two-dimensional Fourier inversions to derive two-dimensional source images as was done later by Ryle and colleagues at Cambridge. It seems that true synthesis imaging was not implemented at Caltech until nearly a decade later (Rogstad and Shostak 1971).

Alan Moffet later claimed that, unlike Cambridge, Caltech did not have access to sufficient computing power to do a complete two-dimensional Fourier inversion of data taken at many different hour angles. While this may have been partially true, it seems that both Caltech and Australian radio astronomers were slow in appreciating the full power of synthesis imaging and were unable, or at least unwilling, to take advantage of the large digital computers available to them. Australian radio astronomers continued to exploit hardware solutions (e.g., Frater et al. 2017) to build high resolution radio telescopes, while the Caltech radio astronomers continued to depend on model fitting techniques to analyze their data and never made the step to full synthesis imaging with the OVRO two-element interferometer system. Nevertheless, for the first half of the decade, OVRO was clearly the most productive US radio observatory. In fact, George Swenson suggested that it was just because of their success, especially the work leading to the discovery of quasars, that the Caltech radio astronomers did not see the need to pursue synthesis imaging.[6] Nevertheless, many of the students who were involved in the design and operation of the Owens Valley interferometer later went on to play major roles in the design, construction, and operation of the VLA.

The Westerbork Synthesis Radio Telescope[7] Based on a suggestion from US radio astronomer Charles Seeger,[8] Dutch radio astronomers, driven by Jan Oort, developed plans for a large radio telescope with dimensions of 3 to 5 kilometers and a resolution goal of 1 arcmin. Because of the anticipated large cost, Oort initiated talks with Belgium and Luxembourg for sharing the cost of what came to be known as the "Benelux Cross." At the December 1961 meeting of the European Organization for Economic Co-operation and Development (OECD) on Large Radio Telescopes, Oort (1961a, b, c) presented the scientific rationale for an antenna with a resolution of the order of an arcmin while Jan Högbom (1961) and others outlined a range of design concepts. Starting with a fairly conventional Mills Cross consisting of parabolic cylinder elements working at 75 cm (Christiansen and Högbom 1961), the Benelux Cross project went through a series of designs and evolved to a cross composed of a hundred or more 25 to 30 meter diameter parabolic dishes working at 21 cm (Christiansen et al. 1963a,b). The revised array also included a single 70 meter dish to provide the missing short spacings.[9] The change in operating wavelength was motivated partly by the desire to observe the 21 cm hydrogen line and partly because this frequency is protected by international agreement from RFI. Also, using steerable parabolic dishes instead of cylindrical parabolas opened the possibility of observing at wavelengths in addition to 21 cm, thus facilitating spectral and polarization studies. In each of these early designs, Leiden visitors Chris Christiansen from Australia, William (Bill) Erickson and Charles Seeger from the United States, and Jan Högbom from Sweden, together with Leiden's Lex Muller, provided the technical leadership for the Benelux Cross, which they planned to locate near the Belgian-Dutch Border.

All of these early designs were based primarily on phased arrays which formed multiple simultaneous beams, although Christiansen et al. (1963a) commented on the possibility of recording the interferometer amplitudes and phases from the different antenna spacings. Högbom had joined the Leiden-based Benelux Cross group after receiving his PhD with Ryle's group in Cambridge, so was fully familiar with the techniques of Earth-rotation aperture synthesis. In fact, Högbom's 1959 PhD thesis on "The Structure and Magnetic

Field of the Solar Corona" included the first detailed description of what came to be called "Earth-rotation synthesis." According to Högbom (2003), after writing up his work on the Sun, he was embarrassed that his thesis contained a mere 78 pages, far fewer than other Cambridge radio astronomy theses, which were all longer than 100 pages. So he "fattened" up his thesis with two additional chapters, one on "The Fundamental Relations of Aperture Synthesis" based on well-known ideas, and a completely original chapter on "Aperture Synthesis Using the Earth's Rotation."

Högbom's analysis of using the rotation of the Earth to cover the Fourier transform plane was based on using a transit interferometer with a broad primary beam to allow the necessary observations at large hour angles. It was not until Högbom saw Ryle's (1962) published description of the One-Mile Radio Telescope that he appreciated the possibility of doing Earth rotation synthesis using tracking antenna elements, and immediately applied these ideas to a more practical and cost effective implementation of the Benelux Cross. According to Raimond (1996), in 1963, Högbom proposed two possible 21 cm east-west arrays of 28 (34) parabolic dishes spread over 1600 (3000) meters based on Earth-rotation synthesis to give a resolution of 17 (10) arcsec. Apparently Luxembourg was never a serious participant, and by 1967, with no active radio astronomers and commitments to the newly established European Southern Observatory (ESO), Belgium had lost interest in the project, which then became a responsibility of the Netherlands Foundation for Research in Astronomy (NFRA). A new, more radio-quiet, site was chosen in the northern part of the Netherlands, and the Benelux Cross became the Westerbork Synthesis Radio Telescope (WSRT).

With the loss of the expected funding from Belgium, the design was further modified to contain only ten fixed 25 meter diameter equatorially-mounted antennas uniformly spaced over 1.6 km, plus an additional dish moveable along a 300 meter railroad track. However, when the bids came in, there were sufficient funds to build *two* moveable antennas. The WSRT (Fig. 7.2) was completed and went into operation in 1970, initially at only 21 cm. Originally, each of the ten fixed antennas were correlated each with only the two moveable antennas. Since the fixed antennas were uniformly spaced, all of the available Fourier components were still recovered. However, ignoring the data from the fixed interferometer pairs resulted in a loss of sensitivity by nearly a factor of two. Although not at the time anticipated by Högbom or anyone else, after the correlator was upgraded in 1977 to include all antenna pairs, the redundant spacings turned out to be a great advantage in removing the effects of ionospheric and tropospheric phase fluctuations.[10] This enhancement of conventional self-calibration techniques gave the WSRT greatly improved image dynamic range (Noordam and de Bruyn 1982).

Initially WSRT adopted a hands-off approach to observing and data reduction. The local staff supervised the observations, and the data were reduced by NFRA staff at the Leiden University computer center to produce the radio images.[11] Considering the long traditions of H I research by Oort and other

Fig. 7.2 The ten fixed antennas of the Westerbork Synthesis Radio Telescope. Credit: NFRA

Dutch radio astronomers, a spectroscopic capability was soon added to the initial continuum-only system. Over the next years the reliability, flexibility, resolution, and sensitivity of the WSRT continued to improve with the introduction of new low noise amplifiers, the addition of new observing bands at 3.6, 6, and 49 cm, and the migration to digital electronics. Two additional moveable antennas were added in 1976, and later the baseline length was doubled to 3 km to improve the resolution, and the telescope continued to be used by astronomers from around the world.

7.2   Origins of the Very Large Array and the Owens Valley Array

By the end of the 1950s, it was becoming increasingly clear from the exciting results coming from Sydney, Cambridge, Manchester, and Caltech that the major outstanding problems in radio astronomy required interferometers and arrays capable of arcsec resolution. As described in Sect. 3.2, as early as 1954, during the debate over the appropriate size antenna for the planned national radio astronomy facility, Bob Dicke's innovative suggestion to build an interferometer system was lost in the enthusiasm for building the largest possible antenna, but it would resurface nearly a decade later.

The Pierce Committee Following the establishment of NRAO in 1956 and the uncertain start on the construction of the Green Bank 140 Foot Radio Telescope, NRAO Director Otto Struve suggested that the AUI Advisory Committee on Radio Astronomy request that the NSF appoint a committee to review the scientific goals of radio astronomy and the instruments needed to address these goals. The NSF responded in December 1959 by appointing a committee under the Chairmanship of John Pierce, to "1) study the present and predictable needs of radio astronomers with regard to improved instrumentation; 2) study existing and proposed instruments with regard to improved instrumentation; and 3) advise the Foundation with regard to the desirability and feasibility of constructing more powerful instruments" (Keller 1961). Pierce, Executive Director of the Bell Labs Communication Sciences Division, was a well-known engineer and expert on information theory who had conceived and promoted the first communication satellites and the national network of microwave-linked telephone relay towers. Together with Claude Shannon and Barney Oliver, Pierce had developed the first concepts of speech digitization. He coined the term "transistor" for the revolutionary device which was developed under his direction, and wrote science fiction under the name of J.J. Coupling (David et al. 2004). The other Panel members were primarily radio astronomers.[12]

The Panel was impressed by the newly emerging very high resolution observations coming from Jodrell Bank (Sect. 8.1), the interferometric observations at Caltech (Maltby and Moffet 1962) and Nançay (Lequeux 1962), and the demonstration of the power of aperture synthesis by Ryle and Neville (1962). With great perception, they noted that "as the Manchester group has shown, the addition of a third element gives the phases without the need for calibration." The Panel discussed, in some detail, a proposal by John Bolton for 8-element and 16-element arrays of 200 foot dishes arranged in a Mills Cross configuration. But Bolton proposed a real-time phased array of parabolic elements, possibly making use of multiple spacings to reduce sidelobe levels, but not aperture synthesis and certainly not Earth-rotation (super) synthesis. Indeed, the later 1962 Caltech proposal for the Owens Valley Array followed the basic strategy that Bolton had presented to the Pierce Committee before he left Caltech at the end of 1960. The Panel also discussed the merits of Cornell Professor William (Bill) Gordon's plan to build a fixed spherical reflector in Puerto Rico, and anticipated the electrical and mechanical problems of designing suitable feed systems.

The Panel report, which, curiously, was published under the name of the NSF Assistant Director for Mathematical, Physical, and Engineering Sciences, Geoffrey Keller (1961), recognized the need for angular resolution of at least 1 arcmin and ultimately 1 arcsec. Although the Panel appreciated the power of aperture synthesis to obtain the needed high resolution, they correctly worried about how to achieve the instrumental and atmospheric phase stability needed to obtain 1 arcsec resolution. They recommended further "experimental and theoretical studies of aperture synthesis, antenna design, phase preservation, phase shifting, and stable low noise preamplifiers that that would lead to a detailed and practical plan for a radio telescope of high resolving power" and went on to suggest that "technical considerations of terrain, atmospheric and interference environment, and other scientific factors should govern the selection of the site, and convenience of access to particular institutions should be given only secondary consideration." But they also noted that "in the United States, the chief weakness of radio astronomy is not the lack of instruments or funds for instruments but a lack of radio astronomers," so the Panel recommended that universities support their most promising graduates and give postdoctoral fellowships using a combination of government and private funding. Unfortunately, it would be another decade before US radio astronomers could agree on what to build and who should build it, and yet a further decade before the Very Large Array would be completed and in operation. Meanwhile, in 1970, the WSRT began operating in the Netherlands, and in the UK, Martin Ryle and colleagues were doing exciting work with his One-Mile and, starting in 1971, the 5 km radio telescopes. In Australia, Paul Wild completed his 96-element solar heliograph array in 1967 and was making impressive movies of solar radio bursts.

Early NRAO Planning At the 1961 IAU General Assembly in Berkeley, CA, Commission 40 (Radio Astronomy) discussed high resolution radio telescopes. Campbell Wade later recalled that Dave Heeschen returned from the IAU meeting enthusiastic about the potential for building an array for high resolution radio observations, and led an impromptu discussion with Wade, Frank Drake, Roger Lynds, and Dave Hogg in the Green Bank cafeteria.[13] While there was apparently a lot of interest among the Green Bank staff, there was no one at NRAO with any experience in interferometry. As discussed in Chaps. 3, 4, and 9, in the early 1960s, NRAO was focused on building a very large filled-aperture radio telescope, using either a fully or partially steerable reflector or a large fixed reflector. They had even given a name to this hypothetical future project—the Very Large Antenna or VLA! Later this became the "VLAA" for Very Large Antenna Array, which finally reverted back to "VLA," but now meaning Very Large Array. NRAO and AUI were struggling to deal with the increasing problems surrounding the 140 Foot Telescope construction and the threats from Bliss to sue AUI for breach of contract regarding the 85 foot telescope project (Sect. 4.4), as well as completing the construction and commissioning of the 300 Foot Transit Telescope. Nevertheless, through the autumn of 1961 Heeschen, Wade, and others gave further thought to constructing an array of dishes. However, there was little dedicated effort until Heeschen brought things into focus on 5 March 1962 when he called a staff meeting for that same afternoon to discuss a draft development program for what he called "the very large telescope," declaring that the plan "will be submitted to NSF this week (probably tomorrow) as justification for our 1964 budget."[14]

Joe Pawsey had already been appointed to succeed Struve as the next NRAO Director, and was due to visit two weeks later. Although Heeschen was only serving as an Acting Director until Pawsey was to take over in October, he acted boldly and requested $3 million to be included in the FY1964 budget "for the first phase in the development of a very large radio telescope."[15] Heeschen proposed that Phase I begin by establishing the performance requirements and the antenna configuration needed to meet those requirements, followed by designing the antenna elements and electronics, studying of the effects of the atmosphere on phase stability, along with selecting a site and building a small number of antenna elements with electronics as a prototype. Regarding the site, Heeschen noted, in passing, that "Green Bank may not be suitable." Phase II, which Heeschen optimistically projected could begin in FY1965 or 1966, would be to "construct full telescope by expanding the portion built in Phase 1." As Heeschen noted in a 1991 handwritten note scribbled on a copy of his 5 March 1962 memo, "We never got the $3M – but this was the formal beginning of the VLA pjt [project] & in fact the pgm [program] outlined was generally carried out."[16]

Heeschen's ambitious plan was encouraged by Pawsey during his March 1962 visit to Green Bank, although a week after he arrived in Green Bank, Pawsey's trip was abruptly terminated by his illness (Sect. 4.6). Following his surgery in Boston, Pawsey described his "high resolution project" as part of his carefully considered plans for the future of NRAO.[17] However, the following day, AUI President Jerry Tape and Pawsey agreed that Pawsey would not take up the NRAO directorship, but that he would stay involved in NRAO programs. It is curious that, while Pawsey's report acknowledged his discussions with Drake and Heeschen about their proposed millimeter initiative (Chap. 10), he makes no reference to any discussions about an imaging antenna array. Earlier, Pawsey had written from his Massachusetts General Hospital bed to Bill Erickson to recruit Erickson to come to NRAO to be in charge of a project to develop the necessary "equipment capable of giving pictures ... of discrete sources in the sky with sufficient resolution to show all the significant physical features."[18] Pawsey sent a copy of his letter to Heeschen who distributed Pawsey's ideas about an imaging array to the NRAO Scientific Staff. Pawsey also contacted Peter Scheuer, expressing the hope that Scheuer and Henry Palmer might also come to Green Bank to examine the statistics of radio source interferometric measurements from Cambridge, Jodrell Bank, Sydney, and Caltech, and, in this way, define the needed instrument parameters.[19] Erickson responded that he had accepted a position at the University of Maryland and was unable to consider Pawsey's request, but remained interested and expressed willingness to help where feasible.[20] Palmer did spend a year at NRAO from October 1972 to October 1973 working with the Green Bank Interferometer.

Heeschen did not wait for Erickson or Scheuer, and asked Wade and others to assemble what was known about radio source structure and to investigate the various technical issues that would be needed to plan for the construction of the VLA.[21] Some guidance on the desired array parameters was already available to NRAO from the Pierce Advisory Panel, which led Heeschen to suggest a goal of "one arcminute beam at 21 cm and usability at 10 cm to later give even higher resolution." For the most part, Heeschen's 5 March plan was actually implemented, but the time scale would be much longer than he had anticipated, and it would take another decade before funding was approved and nearly two decades before the VLA became fully operational. NRAO never got the $3 million that Heeschen requested for the FY1964 budget, but by late 1962, Heeschen had been named as the NRAO Director and was ready to seriously address the construction of a large radio telescope array. In January 1963, he informed the AUI Board about the Green Bank discussions of an "array made up of 100' to 150' antennas"[22] and appointed Deputy Director John Findlay to initially lead the VLA development program.

Recognizing that a major fraction of the cost of the planned array would be in the individual antenna elements, NRAO needed "the best telescope for the lowest cost" and issued a Request for a Proposal to antenna companies, soliciting design studies for "a very large radio astronomy antenna system .... [that] will consist of a number of parabolic dish telescopes." Prospective bidders were asked to determine "the relative cost of the various choices that will have to be made by the Observatory staff," which included dish diameter, upper frequency limit, polar or alt/az mount, surface accuracy, and sky coverage.[23]

In 1962, NRAO had no experience in interferometry. Indeed, although the Pierce Committee enthusiastically endorsed an interferometric array to advance US radio astronomy, the only place doing serious interferometry in the US was Caltech, where John Bolton, his students, and post docs had built the Owens Valley two element interferometer. Bolton, and later Gordon Stanley, served on the AUI Visiting Committee for NRAO, so there were good opportunities for NRAO to learn from Caltech experience, and in October 1962, Cam Wade was dispatched to spend three weeks at Caltech to become more familiar with interferometric techniques. It was the first time Wade had actually seen an interferometer in operation. His main reaction on returning to Green Bank was that NRAO needed to find a better way of taking data than the pen and ink tracings on chart recorders used at Caltech. Wade also visited Stanford, as well as several industrial laboratories in Silicon Valley, where he learned about early developments of fiber optic technology, but concluded that while the technology was promising, it "had a hell of a long way to go," and concluded that NRAO should "stick with cable."[24]

By the end of 1963, Wade (Fig. 7.3) was able to put down on paper a basic description of the VLA.[25] Assuming that the proposed array would need to be able to observe a few hundred discrete radio sources with hundreds of pixels per source and that it should not take longer than a month to observe each source, Wade concluded that they needed to build an array several miles in extent, with at least twenty 80 foot diameter paraboloids able to be placed on 106 stations arranged on a Tee configuration. The VLA that was later built had little resemblance to Wade's early plan except perhaps for the antenna diameter, which Wade later admitted was based more on the commercial availability of 25 meter diameter antennas than on sensitivity arguments.[26] Dave Hogg later recollected that Wade's memo marked the real starting point of the

Fig. 7.3 Cam Wade wrote the 1963 memo that initiated the VLA project. Later, Wade led the study leading to the choice of the VLA site on the plains of St. Agustin, and became the first Director of VLA Operations. Credit: NRAO/AUI/NSF

VLA project.[27] A few months later, Heeschen brought the project to the attention of the AUI Executive Committee, where the acronym VLA appears to have been used for the first time for the Very Large Array and not for the Very Large Antenna, which was being pursued as a separate project now known as either the Largest Feasible Steerable Telescope or the Largest Feasible Steerable Paraboloid (Sect. 9.4).[28] Assuming that the VLA design would be completed in 1967 and 1968, the schedule called for "construction to begin in FY1969 and be completed in FY1971," and that "as the VLA approaches completion, the effort on the Largest Feasible Steerable Paraboloid will be augmented."[29]

The Owens Valley Array Hoping to build on their outstanding successes with the two-element OVRO interferometer, Caltech proposed adding four new 125 foot diameter antennas and an extension of the track to be used in various configurations of a phased or synthesis array. Knowing that operations support from the Office of Naval Research was becoming more difficult to obtain, Gordon Stanley sent the proposal to the NSF.[30] Although Stanley and others were aware of the growing problems with the NRAO 140 Foot Telescope, they boldly proposed a polar mount for the new OVRO antennas.[31] Working at a minimum wavelength of 10 cm, the proposed array would be able to synthesize a 1 arcmin beam or work as a simple one-dimensional interferometer with 10 arcsec resolution. The proposed cost was about $5 million. However, as early as October 1962, tensions started to build between NRAO and Caltech, when during a visit of Marc Vinokur to Pasadena, Caltech director Gordon Stanley expressed concern that the proposed Green Bank interferometer might compete with the Caltech proposal for enlarging the OVRO interferometer.[32]

The First Decade Review of Astronomy—The Whitford Report In late 1962, the National Academy of Sciences' Committee on Science and Public Policy convened what was to become the first of a series of Decade Reviews of astronomy. Albert Whitford, of the University of California's Lick Observatory, chaired the Panel on Astronomical Facilities, which was charged "to study the probable need for major new astronomical facilities in the United States during the next five to ten years, and to recommend guiding principles and estimates of cost in order that federal funds might be employed with maximum efficiency to promote advancement of astronomy in all of its branches."[33]

Recognizing the growing importance of radio astronomy that led to the discovery of "new and previously unsuspected phenomena," three of the six scientists on the seven-person panel were radio astronomers and three others were optical astronomers. Bruce Rule, a Caltech engineer who had been instrumental in the construction of both the Palomar 200 inch telescope as well as the two Owens Valley 90 foot radio telescopes, rounded out the committee. So although effectively half of the Panel represented radio astronomy interests, the NAS first convened a separate ad hoc committee "to guide the deliberations ... concerning the current and future needs in the field of radio astronomy."[34] Some 20 radio astronomers participated in the two-day meeting held at the NAS, which was attended by Frederick Seitz, NAS President, along with other representatives from the NAS Committee on Science and Public Policy (COSPUP), the NSF, ONR, and the President's Science Advisory Committee (PSAC). This was clearly a high level meeting reflecting the perceived importance of radio astronomy as a national priority.

The two days of discussion "developed a consensus of opinion" on the need for "the largest single undertaking," recommended by the group, which "would be the construction of a large array composed of about 100 separate parabolic antennas partially steerable, each of about 100 foot diameter, with surfaces good for 3 cm work,"[35] with an intended resolution of 1 arcmin. "A project of this magnitude," the group continued, "is obviously beyond the capabilities of a single university and naturally falls within the province of the NRAO, ... which would have the responsibility for the planning and the actual construction of the instrument." However, the report of the meeting also went on to discuss "a second, less expensive array with a resolution of 10 arcsec but with higher side lobes, designed primarily for the investigation of extragalactic sources that is already funded for construction at the Owens Valley Radio Observatory (OVRO)."[36] In fact the OVRO array had not been funded; it would require the next Decade Review to resolve the growing animosity between NRAO and Caltech over who would get NSF funding to build their array.

The Whitford Committee confined its deliberations to ground-based astronomy and called attention to the discoveries of "exploding galaxies" and quasars, the better picture of the rotation of our Galaxy, the distribution of galactic neutral hydrogen, along with the solar system studies leading to the measure of the magnetic field surrounding Jupiter, the structure and temperature of the invisible surface of Venus, and an improved measure of the Astronomical Unit. Citing the study of quasars as "excellent examples of the complementarity of radio and optical astronomy," and the faintness of quasars at all wavelengths, the committee drew attention to the need for better access to large facilities at both optical and radio wavelengths. In their 1964 Report, however, the committee tempered their evaluation of very large new optical telescopes with concerns about seeing, the cost effectiveness of large apertures and corresponding ancillary instrumentation, and "the need for access to large telescopes by a much larger number of astronomers." The report recommended the construction of three optical telescopes with apertures in the range 150 to 200 inches as well as "*four* general purpose telescopes of aperture range 60 to 84 inches" and "*eight* telescopes of 36 to 48-inch aperture." They then went on to suggest that only after the construction of three large telescopes was underway, "a representative study group be assembled to consider the problems of building a telescope of the largest feasible size." As would be the case in future Decade Reviews, reflecting the different emphasis by the radio and optical communities, the committee recommended constructing more modest-sized optical instruments in order to provide more observing time rather than instruments which would give new capabilities.

The report took a much more aggressive approach on radio astronomy. Noting that "the technical knowledge exists to build instruments that can reach beyond the thresholds of information now foreseen." the Whitford (1964, p. 19) Committee drew attention to:

- The major factor that limits the advance of radio astronomy is not particularly lack of observing time …but rather the lack of instruments of the proper design to meet problems now recognized.
- None of the proposed or existing instruments will provide the versatility, the speed, and particularly the resolution demanded for substantial progress.
- Contrary to the situation in optical astronomy, radio telescopes have not nearly approached the ultimate limitation on performance produced by inhomogeneities in the Earth's atmosphere.
- Clearly … no definitive knowledge of the radio sources throughout the universe can be obtained until the resolution of the order of seconds of arc is available for radio astronomers.

Although the Whitford Committee report recognized the need for "a group of lesser instruments useful in special problems and for student training," their highest recommendation for radio astronomy was for "a major high resolution instrument ... with a resolution of less than 10 arcsec at centimeter wave-lengths." The panel recommended "as the largest single undertaking in radio astronomy, the construction of a large array that would achieve these goals." As an example of the magnitude of the project, the Committee suggested an array of "about 100 separate parabolic antennas, each perhaps 85 feet in diameter" capable of operating "down to wavelengths as short as 3 cm" which they esti-mated could be built for a cost of about $40 million. Characteristic of the unwarranted confidence of many preliminary cost estimates, the panel noted that "the cost is fairly predictable," since 85 foot antennas were readily avail-able from several industrial suppliers, and that there was considerable experi-ence in interferometry. Although the considerations about the antenna cost were probably not far off, the panel failed to recognize the true complexity and corresponding costs involved in building and using an array that would meet the astronomical requirements. In the end the VLA cost about twice the Whitford Committee estimates, perhaps not so bad considering the significant inflation that would occur before the VLA was completed 15 years later.[37]

The Whitford Committee noted that the complexity of the array would place the project "beyond the capabilities of a single university." However, they fell short of a full endorsement of NRAO to construct the array, only remark-ing that the project "falls naturally into the category of instruments that should be constructed by NRAO," and went on to specify that "means should be provided for extensive participation by scientists who are not members of the NRAO staff in the planning and development of the instrument." (Whitford 1964, p. 52) Moreover, recognizing that it might take a decade to build the VLA, and prompted, no doubt, by Bruce Rule and knowledge of the growing ambitions at Caltech, the Committee not only recommended the funding of the "already-proposed extension" of the OVRO array to add four new 130 foot antennas along with an increase in the length of the interferometer, but suggested that "a further increase in the available equipment by a factor of two will allow useful resolutions of less than 10 seconds of arc." (Whitford 1964, p. 52) The estimated price tag for the enhanced OVRO array was only $10 million dollars, and the panel recommended that "construction should be commenced immediately." Comparison with the $40 million price tag and decade-long construction time estimated for the VLA positioned the Owens Valley Array and the VLA for a long and bitter conflict that would drag out for another decade, during which time nothing would be built in the United States.

Proposing the VLA Encouraged by the August 1964 Whitford Report, NRAO began serious planning for the VLA in the summer of 1964. Progress was greatly expedited by the arrival of Barry Clark on the NRAO staff only a few months later. Clark had just received his PhD from Caltech, where he had become an expert in radio interferometry and participated in the early design of the Owens Valley Array. For the next half a century, Clark, probably more than anyone else, was the intellectual force behind the VLA software, and argu-ably was the only person who understood all aspects of the VLA design. Dave

Heeschen formally established the VLA Project, and in September 1964 he hired George Swenson (Fig. 7.4), on leave from the University of Illinois, to come to NRAO to help design the VLA. On 20 October 1965, Sander (Sandy) Weinreb came to NRAO to lead the Green Bank Electronics Division and to be responsible for the conceptual design of the VLA hardware, including the front ends and correlator. The proposed NSF FY1967 budget contained $1 million for preliminary design of the VLA, and Heeschen later noted that this was "the first specific action taken by the NSF to allocate funds for this facility."[38]

Swenson, together with Cam Wade, investigated potential sites for the VLA; Wade also investigated the needed sensitivity and antenna size. David Hogg worked on the antenna configuration, Hein Hvatum on the antennas, Sandy Weinreb and Warren Tyler on the electronics, and Barry Clark on the computing system and data processing (Heeschen 1981, p. 16). Under Swenson's leadership, by the end of 1965 the Design Group had made sufficient progress "to solicit comments, criticisms, ideas, and assistance for further work." NRAO (1965) released a preliminary design report for the VLA that described in considerable detail the status of work on the development of the VLA and the desirable properties for an instrument to address the outstanding astronomical problems of the time.[39] These early specifications were:

(a) Wavelength: 10 cm
(b) Resolution: 10 arcsec
(c) Field of view: 5 arcmin
(d) Sensitivity: 4 mJy rms
(e) Versatility to address a wide variety of problems
(f) Expandability

Fig. 7.4 George Swenson served as VLA Project Manager while on leave from the University of Illinois. Credit: NRAO/AUI/NSF

After comparing the various approaches to obtaining high angular resolution, the VLA report concluded that a correlator array based on super-synthesis or Earth-rotation synthesis ideas was the only practical way to meet the desired goals of the VLA. Various configurations including a `Tee`, circle, cross, and a random configuration were investigated, but Leonard Chow, visiting from Waterloo University in Canada, came up with an innovative 3-arm `Wye` configuration with each of the arms separated by 120 degrees. The `Wye` has the same comparable coverage of the Fourier transform plane as a ring, but has the advantage that the antenna elements can transported on rails or road along a straight line so that the antenna spacing can be varied, and, if later desired, the array can be extended. The proposed configuration in the 1965 report had 12 antennas spaced along each of the three arms of the `Wye`, with one additional antenna placed at the center. It was suggested that initially each arm would be 2.4 km long, which gave the desired 10 arcsec resolution at 11 cm. Later, "as can be justified by the progress of observations ... and as funds become available, the arms of the `Wye` can be extended," and it was noted that, "In choosing a site, the requirement for 25 km arms will be considered ... to achieve 1" resolution."

The NRAO report made the point that the VLA would use point sources to calibrate the baselines, rather than the laborious precision survey used by Ryle for the Cambridge One-Mile Radio Telescope. The report left open the questions of antenna size, alt-az or equatorial mount, Cassegrain or prime focus feed, `Wye` or `Tee` configuration, railroad track or road, length of the arms, type of delay system, local oscillator distribution, and single or dual polarization. NRAO proposed that the VLA use the same basic system being used for the Green Bank Interferometer (GBI), with uncooled parametric amplifiers followed by a double sideband mixer, and with each antenna pair multiplied in a correlator. There was no discussion of possibly using digital delays or a digital correlator, although it was recognized that as in the GBI, the correlator output would be digitized and fed to a high speed computer for further processing. Cautiously, the report remarked, "Cooling the amplifier to achieve low noise temperature should be avoided."

The ambitious—many felt too ambitious—NRAO VLA Report No. 1 indeed generated a lot of community interest, but also generated controversy. As later described by Heeschen (1996), initially,

> The VLA did not enjoy much support, either in the US or in the rest of the world. The proposed instrument was considered to be unimaginative, undesirable, unneeded, technically unfeasible, far too costly, or some combination of these. It took a long time to convince the community and the NSF that was what they really wanted.

From the beginning, "the primary mission of the telescope" was considered to be "mapping of extra-galactic sources," so there was no provision for spectroscopy, which was dismissed with the remark that "the addition of line spectrometer equipment to the already formidable array of data processing apparatus required for seconds-of-arc angular resolution appears to increase the complexity of the whole system to a point not consistent with the present state of the electronic art" (NRAO 1965, p. 6). This raised a lot of objection from H I observers and the growing group of outspoken molecular and maser spectroscopists.

A major challenge to the feasibility of the VLA concept came from the UK, where Martin Ryle argued that atmospheric irregularities and turbulence would introduce phase fluctuations on interferometer baselines longer than a few kilometers. While the Cambridge development of aperture synthesis and later super-synthesis had an enormous impact on the future development of radio astronomy, and in particular on the NRAO proposal to build the VLA, ironically, Cambridge radio astronomers incorrectly argued that phase fluctuations due to tropospheric irregularities would fundamentally restrict the resolution of radio telescopes to about 1 arcsec (Hinder and Ryle 1971), or about the same seeing limit achieved by optical telescopes located on a good mountain site.

The instrumental phase of the Cambridge instruments was sufficiently stable that calibration observations were needed only at the start and end of each 12 hour run. In fact, there was no capability provided in the control or data reduction software to allow calibration data to be inserted during the continuous 12 hour track. Since, at least initially, the Caltech interferometer had poor instrumental phase and amplitude stability, it was standard practice to observe a calibration source several times an hour. Unlike the Cambridge One-Mile Radio Telescope which ran under computer control, at Caltech, the pointing of the telescope was manually controlled at all times, and data were reduced by hand from chart recordings. So, even when later instrumental improvements greatly reduced the instrumental instabilities, it was natural at Caltech to extend the same calibration technique to reduce the effect of tropospheric phase fluctuations. In this way it became possible to build radio telescopes with resolution better than the nominal seeing limit. With the later development of "self-calibration," radio telescopes were routinely able to achieve resolutions orders of magnitude better than optical telescopes.

In January 1966, Heeschen informed the AUI Board that he planned to allocate $1 million to the design of the VLA, and appointed a VLA Design Group of ten scientists and engineers with George Swenson as the Chair.[40] Over the following year, the Design Group under Swenson studied various antenna configurations, explored potential sites, and, with industrial contractors, studied various antenna designs. In January 1967, NRAO (1967) sent a formal proposal to the NSF to build the VLA. The proposal called for operation at 2.7 and 5.4 GHz (11 and 5.5 cm) with up to 1 arcsec resolution. To achieve the desired sensitivity (0.02 mJy rms at 2.7 GHz) and dynamic range (20 dB) NRAO proposed to use thirty-six 25 meter diameter antennas in a Wye configuration with each arm up to 21 km in length. Still, there was no spectroscopic capability planned, other than to note that the design "should not preclude the ultimate use of the instrument for line work." The estimated cost of construction, including 15% contingency and allowance for cost escalation over the planned four-year construction period was $51.9 million. Annual operating costs were projected to be $1.7 million. As noted earlier, the NRAO proposal was not without controversy. Many radio astronomers, both external to NRAO as well as within NRAO, felt that the VLA was too ambitious and too expensive. Moreover, with the increasing US budget deficits resulting from the escalating confrontation in Vietnam, there was little national interest in spending money on an expensive scientific enterprise of dubious national relevance. The NRAO budget request for FY1969 was reduced from $24.091 million to $6.4 million, and it was clear that there would be no VLA construction funds in 1969.[41]

7.3   THE GREEN BANK INTERFEROMETER (GBI)

As early as December 1958, while the Tatel Telescope (85-1) was still under construction, Dave Heeschen inquired of Blaw-Knox about the possibility of putting the antenna on a railway track so that it might be used later with the 140 Foot as part of a variable spacing interferometer. Cam Wade suggested starting instead with a small two-element interferometer to (a) test methods of local oscillator and IF signal transmission, and (b) develop methods of correlating the data. Wade recognized that these questions needed to be addressed before waiting for the completion of the first two VLA antennas, which he very optimistically stated "can hardly be finished sooner than 18 months from now."[42]

In January 1963, the AUI Board of Trustees approved Heeschen's request to obtain a second 85 foot antenna in order to "gain experience with interferometers."[43] The 85-2 antenna as it was called, was essentially a clone of 85-1, except that it was mounted on a set of 96 large truck tires and could be towed by two bulldozers along a roadway. For actual observing the antenna was lowered and bolted to stations with spacings that varied between 1200 meters and 2700 meters from 85-1, oriented along an azimuth of 243 degrees as restricted by the site geography.

The two-element Green Bank Interferometer was in operation by the middle of 1964 at 2695 MHz (11.3 cm) using a double sideband mixer with an IF band extending from 2 to 10 MHz.[44] A room temperature commercial parametric amplifier was used in front of each mixer to give a system temperature of about 125 K.[45] Initially all the GBI observations were recorded and reduced using strip chart recorders, but soon Wade discussed the techniques needed to find the amplitude and phase of digitally recorded interferometer data.[46] Then, in December 1964, only a month after arriving at NRAO, Barry Clark refined Wade's procedure for the digital reduction of GBI data[47] which was then implemented by Clark and Wade.[48]

The GBI had a resolution of about 10 arcsec. It met all of its design requirements and provided the interferometry experience the NRAO staff needed to pursue the VLA project. However, Caltech's Owens Valley two-element interferometer was still getting all the attention because of its exciting series of quasar identifications at larger and larger redshifts, new planetary results, pioneering observations of radio source polarization, and the ground-breaking investigations of radio galaxy structure (Sect. 6.6). Heeschen was anxious to get some visibility for the Green Bank Interferometer, and encouraged Wade to give a talk on his precision position measurements at the spring 1965 American Astronomical Society (AAS) meeting in Lexington, Kentucky.

Unfortunately, a large gulley located between 85-1 and the nearest 85-2 station precluded interferometer spacings less than 1200 meters. With no short spacings, the GBI had limited imaging capability, and in January 1966 Heeschen informed the AUI Board that NRAO needed a second moveable dish for the GBI.[49] As shown in Fig. 6.1, a third element, 85-3, allowing baselines as short as 100 meters, was added in 1967, along with a new interferometer control building and a new observing station. Interferometer control and data reduction were handled by a DDP-116 computer.[50] Starting in 1966, a portable 42 foot dish was placed at Spencer's Ridge, 11.3 km from Green Bank, to form the world's first phase stable radio interferometer with a baseline longer than a few kilometers (Fig. 7.5). The 2 to 12 MHz IF signal from the remote antenna was returned over a microwave radio link operating at 1347.5 MHz, which also provided the local oscillator synchronization. John Basart et al. (1970) ran a long series of observations to study the effect of atmospheric turbulence on interferometer phase.

Fig. 7.5 The 42-foot antenna components arrive at Bartow railway depot. George Grove standing at the far right with his ever present pipe. Credit: NRAO/AUI/NSF

In 1968, the GBI was further upgraded to operate at both 2695 and 8085 MHz (11.3 and 3.6 cm) with dual polarized front ends, and the portable 42 foot dish was replaced with a 45 foot dish having sufficient precision to operate at 8 GHz.[51] The 45 foot antenna was placed on a hilltop near the town of Huntersville, about 35 km from Green Bank, and was operated with an upgraded link. To avoid attenuation of the link signal from intervening foliage located in the direct line of sight, the radio link was bounced off a reflector mounted on a nearby Green Bank hill. In this way, NRAO was able to demonstrate the ability to maintain adequate phase stability over the longest baselines planned for the VLA and to make images with 1 arcsec resolution.

The 45 foot antenna was operated unattended with only a few-hour maintenance visit scheduled once a week. A decade later, the success of the remotely operated radio-linked antenna gave NRAO some confidence that it could successfully maintain and remotely operate the antennas of the proposed Very Long Baseline Array (Sect. 8.6).

The first GBI spectroscopic observation occurred in 1968, an unsuccessful attempt to detect the H134α radio recombination line near 2700 MHz. Following a meeting in Green Bank in August 1968, attended by 13 NRAO and university scientists, 21 cm single sideband front ends, a wider bandwidth delay system, and a digital correlator were added to permit H I spectroscopy.

By 1969, with no clear prospects for VLA funding, the NRAO staff began to discuss enhancements of the GBI to improve its imaging capability by adding a fourth 85 foot dish, three 13 meter dishes, two additional observing stations along the existing roadway, and a new baseline orthogonal to the existing one. NRAO made it clear that the proposed expansion of the GBI was not a substitute for the VLA, but rather, in view of the delay in funding the VLA, it was intended as a stopgap measure to permit the kind of research not possible with the existing GBI. As described below, following the recommendations from the Greenstein Committee, NRAO received the first VLA construction funds in late 1972, and so the proposed GBI expansion never happened.

The GBI served its intended purpose, giving the NRAO scientific and technical staff the experience needed to credibly design and build the VLA, as well as exposing the broader NRAO user community to the opportunities provided by synthesis imaging. Perhaps the most important contribution of the GBI was the demonstration, using the radio linked interferometer, that although phase fluctuations initially increase with antenna separation, beyond spacings of a few kilometers each antenna is looking through essentially independent atmospheres, and the interferometer phase fluctuations remain essentially unchanged as the separation is further increased (Basart et al. 1970). Specifically, with the 35 km spacing of the portable dish, NRAO was able to demonstrate that it would be possible to maintain phase coherence over interferometer scales comparable to those planned for the VLA, especially since the VLA would be located on a far better site than Green Bank. The concerns expressed by Martin Ryle, who of course had great influence, were shown to be unfounded: there were no natural constraints to achieving the stated goals of the NRAO proposed VLA. Another clear result of the experience with the GBI antenna transport and the limited lifetime of its tires was that the VLA antennas should be moveable on rails and not a roadway.

An unanticipated but far-reaching contribution of the GBI came, ironically, from the deficiencies of the GBI and not from its merits. Jan Högbom, when visiting from Holland in 1967, had observed some 60 radio galaxies and quasars, but was discouraged by the poor quality of the images resulting from the large gaps in the distribution of GBI antenna spacings. As he later described it (Högbom 2003),

> I found myself looking at 'dirty maps' of many sources including some calibration sources. It was then a small step to ask: if I subtract a full theoretical point source pattern, a suitably scaled and positioned 'dirty beam' from the map then there should be nothing left – unless of course there is something else out there. Often there was, and I went on subtracting. Returning to the map only the nice part – the central lobe – of each subtracted pattern was a temptation I couldn't resist and it actually seemed to work. ... So CLEAN had a very simple minded beginning but in the end it turned out to be more useful than I had ever expected.

By October 1978, the VLA was in operation in four frequency bands at 1.4–1.8 GHz, 5 GHz, 15 GHz, and 22 GHz, on baselines up to 12 km. However, the NSF provided little or no VLA operating funds at this time, so NRAO closed the GBI as an NSF funded user facility, not only to free up operating funds for the VLA, but to encourage staff and visitors to use and debug the VLA. This met with some resistance, since at this time the GBI was a smoothly operating and scientifically productive instrument, whereas the VLA, not unexpectedly for a new facility, was still under construction and not straightforward to use. Moreover, there was no overlap in frequency. The VLA did not operate in the GBI bands at 2.7 and 8.1 GHz, so observations begun on the GBI often could not be completed on the partially finished VLA. But Heeschen had no sympathy for complainers. He knew the only way to get the VLA debugged was to discontinue access to the GBI and force the staff and visitors to turn their attention to the VLA.

However, until 1996, NRAO continued to operate the GBI under contract to the US Naval Observatory for their program in Earth orientation and time keeping, together with their long-running project to monitor variable radio sources at 3.6 and 11 cm. Although the GBI was originally conceived of and was built to give the NRAO staff experience in interferometry and to prototype instrumentation for the VLA, it was an important research instrument as well for both NRAO staff and visitors. Chapter 6 discuses some of the key discoveries made with the GBI.

7.4 THE NRAO-OVRO WARS

Buoyed by the Whitford Committee report, and by NSF funding for the start of the first 130 foot antenna, Caltech quickly submitted a revised and enhanced proposal for the Owens Valley Array (OVA). The new proposal now included a total of eight 130 foot alt-az-mounted antennas to operate at wavelengths as short as 3 cm.[52] Caltech proposed completing the construction of the array by 1971 at a cost of nearly $15 million. The new OVA proposal apparently received "excellent reviews," but in an April 1967 visit to the NSF, Stanley was informed that there was no possibility of funding in FY1968 but that FY1969 looked more promising.[53] Stanley also suggested the possibility of funding only one additional antenna in 1969, a suggestion he later regretted when he learned that the OVA was already included in the NSF's planning for FY1969.[54]

The VLA and OVA proposals were very different. NRAO was proposing to build an elaborate national facility to be used by any qualified scientist with an appropriate program, and thus needed to be "flexible and versatile" (Heeschen 1981). This meant full sky coverage and ability to form images in one day or less. Caltech proposed a more modest instrument, with only limited public access. When first proposed, NRAO considered the VLA primarily as a continuum instrument, but recognized that spectroscopy was important, and that the design should not "preclude its future use for spectroscopy" (Heeschen 1981). The OVA put more emphasis on spectroscopy. In spite of the Whitford Committee recommendation to phase the construction of both instruments, it was clear that it would not be feasible to build both instruments, and until someone decided which would get built, nothing would get built. But, how would the decision be made? Who should decide?

The Dicke Committees By 1967 the NSF had been either unable or unwilling to fund either the OVA or the VLA. Moreover, there were other competing proposals: from Cornell for upgrading the Arecibo radio telescope to permit observations down to at least 10 cm wavelength, from Harvard/MIT for a large radome-enclosed radio telescope, and from a Caltech-Berkeley-Michigan consortium for a 100 meter fully steerable dish. These were all viable projects with persuasive scientific need and strong technical preparation. To consider these five major proposals, NSF convened an "Ad Hoc Advisory Panel for Large Radio Astronomy Facilities" with Princeton's Robert Dicke as the chair.[55] The Panel met in Washington DC for five days at the end of July 1967 to receive testimony from each of the five proposed projects and to make recommendation to the NSF. In addition to the eight members of the Panel, more than 40 representatives of the proposing organizations, government agencies, and all three military services participated in at least some of the deliberations.

The Panel report,[56] issued just over two weeks after their final meeting, recommended as its clear first priority that the Caltech proposal for the eight-element Owens Valley Array "be accepted in its entirety and funded as soon as possible, with an adequate operating budget." Secondly, the Panel urged that the Cornell proposal to upgrade the Arecibo telescope "also be accepted in its entirety, and funded as soon as possible." The MIT-Harvard proposal for the 440 foot dish was deferred pending the outcome of the Arecibo upgrade and further engineering studies. The Caltech-Berkeley-Michigan proposal was declined. Acting on the recommendations, the Arecibo upgrade was included in President Richard Nixon's proposed FY1970 budget, but was not approved by Congress. The NSF also took the first steps toward funding the OVA, and included funds for building the first 130 foot antenna at the Owens Valley. As it turned out, the rest of the OVA was never funded, but the single 130 foot telescope had a long successful history as part of the early US VLBI program (Chap. 8) and for single dish studies.

All of the Dicke Committee recommendations carried the proviso "that at least 50% of the time available for astronomy on such facilities should be made nationally available to qualified visitors,"—a clear endorsement of an "open skies" operating philosophy, but noticeably "open" only to US-based scientists. The Panel supported the VLA concept and the need for 1 arcsec resolution. However, they argued that more work was needed to demonstrate the advantage of the VLA proposal "in terms of economy of dishes and tracks, optimization of picture resolution elements, sky coverage, observation time, and flexibility," and only recommended continued study and actual measurements to "demonstrate the feasibility of interferometric techniques over very long baselines."

Disappointed and upset with the Dicke Committee report, which appeared to "damn the VLA with faint praise," NRAO had no choice but to continue the design effort as recommended by the Committee and demonstrate that the VLA would work as claimed. To reduce the cost, the number of antennas was decreased from 36 to 27. This resulted in an increase in the side lobe level from about one percent to about two percent. The updated design was issued in January 1969 as Volume III of the VLA proposal (NRAO 1969). Volume III included a discussion of prospective sites, a more detailed analysis of possible configurations, a conceptual design of the antenna elements and transportation system, along with the design of various components and subsystems, including the local oscillator, IF distribution, and delay systems. The proposal also reported on the successes of the GBI, including the demonstration that it is possible to maintain the required phase stability over baselines comparable to the extent of the VLA. The antenna and transporter studies, the design of the front end parametric amplifiers, the IF delay system, and the evaluation of computing requirements were contracted to industry. Most of the instrumental design work and planning for computing resources was done by NRAO scientists and engineers, to a large extent led by Weinreb and Clark respectively. As Cam Wade later explained "We took advantage of the delays to do things over again that we'd done in haste the first time."[57] The new cost estimate was now only just over $32 million.

Volume III of the VLA proposal made only brief mention of a possible future spectroscopic capability. In June 1969, Caltech countered with an update of their OVA proposal which focused on current spectroscopic observations at OVRO and spectroscopic applications of the OVA.[58] The new OVA report also discussed the possible expansion of the OVA within the Owens Valley and to adjacent valleys as well as stating that 50 percent of the observing time at OVRO was being made available to outside observers, thus addressing two of the NRAO criticisms of the OVA. The proposed cost of the OVA was close to $17 million.

Meanwhile, the NSF continued to be vague about VLA funding and asked that NRAO submit two construction plans, one for receipt of funds in FY1971 and the other in FY1974 or later. Heeschen advised the AUI Board that FY1971 "would not be unsatisfactory," but "if construction funds are postponed until 1974 or thereafter, it would be necessary to stop all design work until it was known precisely when construction funds would be available."[59] However, the NSF would not commit to the VLA or to the OVA without more explicit community endorsement.

Volume III brought the VLA design up to a point where NRAO felt the VLA was ready for final prototyping and construction. With no clear prospects for VLA construction, in 1969 Dave Heeschen dissolved the VLA Design Group and ceased further development work. With the uncertain prospects for VLA funding NRAO enthusiasm waned. George Swenson had no enthusiasm for continued fighting for the VLA and likened the situation to "scrubbing the decks on the Titanic."[60] Dave Heeschen had no patience for defeatism and suggested that it was time for Swenson to return to the University of Illinois.

By 1969 none of the Dicke Committee recommendations had been funded, but the Arecibo Observatory became part of the NSF-funded National Astronomy and Ionospheric Center (NAIC). The discovery during the previous two years of pulsars (neutron stars), atomic recombination lines and interstellar (organic) molecules, along with new precision tests of General Relativity and new observations of quasars and radio galaxies, had changed the landscape of radio astronomy, which the Committee "contrasted with the tragic standstill in the funding of new facilities." Meanwhile, the 100 meter Effelsberg antenna and the 12-element Westerbork Array were nearing completion, as were new radio telescopes in India (Ooty) and at Cambridge in the UK. In view of the changes since the 1967 Dicke Committee report, the NSF reconvened the Committee "to reconsider its former recommendations, ... and to reaffirm or alter the recommendations and priorities." There was no mandate to prioritize the recommendations. Accordingly, the Panel "reaffirmed its previous recommendation" that the Arecibo telescope be improved and that the Owens Valley Array be constructed," and recommended "with equal urgency the construction of the large radome-enclosed fully steerable dish and the Very Large Array." (Dicke 1969)[61] The Committee reaffirmed the recommendation that at least half of the observing time on these facilities be available to visitors and that there be sufficient operating funds to facilitate their use by non-expert observers. They also made a point of endorsing "the support of radio astronomy research and facilities at the universities." All the proposed projects received an enthusiastic excellent recommendation and were all deemed urgent. Such an unrealistic blanket endorsement wasn't really useful to the NSF. Only the Arecibo resurfacing would get a new start in FY1971, but there was no resolution of the VLA/OVA issue. NEROC tried an end run to fund the construction of the 440 foot dish through a special Congressional appropriation to the Smithsonian Institution (Sect. 9.5), but that plan failed in Congress.

In an attempt to resolve the stalemate, Heeschen and the NRAO staff held several discussions with Caltech to explore the possibility of jointly building an array. At an 18–19 September 1968 meeting in Charlottesville, Heeschen and OVRO Director Gordon Stanley discussed their views on some of the scientific, technical, and administrative issues facing a joint operation.[62] Apparently there was sufficient common ground to agree to extend the discussions with a visit by NRAO staff to the Owens Valley. On 11–12 November, Clark, Heeschen, Hogg, Hvatum, and Wade met with OVRO's Marshall Cohen, Alan Moffet, Duane Muhleman, and George Seielstad. While there was general agreement that the two groups needed to have close and continued contact on scientific and technical issues, both sides came away suspicious of the motives and commitment of the other. Curiously, Stanley could not, or chose not, to attend the meeting, but his report to the Caltech administration emphasized the disadvantages to Caltech of a joint program.[63] In early 1969, Stanley again met with NRAO scientists during visits to the potential VLA sites in Arizona and New Mexico, but returned claiming to have detected "the unshakable determination of the NRAO people to proceed with the VLA," and said that "the unanimous consensus of the [Caltech] radio astronomy group is that we do not proceed further with the attempt at cooperation on an array with the NRAO people."[64]

As later described by Hogg,[65] there were perhaps four areas of disagreement between the Caltech and NRAO concepts:

(a) Caltech argued for a smaller number of larger dishes to facilitate calibration and to minimize maintenance. NRAO argued for a larger number of smaller elements to improve the u,v coverage.

(b) The Owens Valley was too small in the east-west direction to accommodate the full extent of the NRAO VLA concept. At one point Heeschen offered to consider a joint project that would more closely follow the Caltech design, but only if the array were built on a site that allowed for future expansion.

(c) The NRAO scientists designing the VLA and using the GBI all had strong backgrounds in continuum research, while spectroscopists were mostly using the 140 Foot and 36 Foot. Noting that the H I work at Westerbork was clearly very productive, Caltech put more emphasis on spectroscopic observations. NRAO ultimately appreciated this deficiency of the VLA and adopted full spectroscopic capability for it.

(d) Caltech did not fully buy into the visiting user concept and envisioned an operating model more like that of the Owens Valley Observatory and not the NRAO national observatory model which emphasized user support.

Desperate to find funding for the VLA, in August 1969 Heeschen, Hvatum, and Wade met in Reno, Nevada, with officials from the University of Nevada to seek their possible support in obtaining VLA seed money from the Reno-based Max C. Fleischmann Foundation. The University expressed interest and offered office space, but could offer no help with persuading the Foundation to support the VLA. AUI President Gerry Tape's three-page proposal to the Fleischmann Foundation[66] was rebuffed with a curt response that the Fleischman Foundation was not interested in funding the VLA.[67]

The Greenstein Committee By early 1969, both NASA and the NSF, as well as the Bureau of the Budget (BOB), were becoming increasingly aware of the need to prioritize the many planned initiatives in both space and ground based astronomy. Following a "prospectus" prepared by the BOB,[68] both the NSF and NASA approached the National Academy of Science (NAS) to conduct "an independent study … which can assess the priorities of astronomy from the scientific point of view, specifically cutting across the lines of responsibility which may tend to bias the planning of individual agencies in favor of particular techniques."[69] It took the NAS four months to respond with a two page proposal to form a "main committee of approximately 12 experts … to undertake detailed planning of the study, to oversee the work of some 12 panels, and to prepare the final report" with oversight by the NAS Committee on Science and Public Policy (COSPUP) chaired by Harvey Brooks.[70] The NAS moves deliberately with their studies, and expected that the report would take two years and would be delivered in mid-1971. The NSF, perhaps surprisingly, was apparently optimistic about the prospects for early construction funding, and NSF Director Leeland Haworth responded that the summer of 1971 would be marginally late to address even the NSF FY1973 budget proposal. He expressed concern that "information on astronomy is urgently needed by the Federal agencies and the Executive offices to develop astronomy support plans for earlier fiscal years. Absence of this information might slow down the U.S. astronomy program."[71] Haworth then proceeded to request "an interim preliminary report … by early spring 1970 … [which would] make it possible for your study to have a real impact already on the FY1972 budget and prevent any undue delays."

Jesse Greenstein from Caltech was approached to lead the study, but was less than enthusiastic. Although he had devoted most of his career to optical spectroscopy, as discussed in Chaps. 1–3, Greenstein was involved in radio astronomy almost from its beginnings. He had organized the first major international conference on radio astronomy which ultimately led to the creation of the NRAO, had convinced DuBridge to begin a radio astronomy program at Caltech, and had played a major role at Caltech in the 1963 discovery of quasars, although his personal (and Caltech's?) goals were unmistakably for a southern hemisphere partner for the Palomar 200 inch optical telescope (Greenstein 1984a, b; Trimble 2003; Kraft 2005).[72]

Greenstein noted that none of the recommendations of the five-year-old Whitford report had been implemented, and anticipated that the Second Dicke Committee meeting scheduled for the following month would serve to set priorities for radio astronomy. He expressed doubt on the value of a new study without some broader indication from BOB, Congress, and the President's Science Advisory Committee (PSAC), of what they wanted from a new report, whether it was an appropriate time for a new report, and he asked whether "there is any point at all in proposing large sums of money for a physical science which is not notorious for its extensive contributions to the industrial welfare, to the inner-city, to pollution etc. which seems to be the major interest of the informed Congress, and of the current administration."[73]

Greenstein was ultimately persuaded to Chair the Astronomy Survey Committee and agreed to provide the requested interim report. He initially approached 21 colleagues to join the main Steering Committee, only one of whom, Bernard Burke from MIT, was a radio astronomer. At their first meeting on 11–12 October 1969, the Steering Committee heard from the NSF, NASA, BOB, and Congress about their plans and expected budget levels, discussed the final composition of the committee and membership of the panels, and reviewed previous recommendations, including the Whitford report and the recently issued, but inconclusive, second Dicke Committee report.

Rather than appoint panel members who would be perceived as neutral, as was done for the Dicke Committee, Greenstein populated the Radio Panel with representatives of all the competing proposals: Dave Heeschen for the NRAO VLA, Marshall Cohen for the Caltech OVA, Frank Drake for the Arecibo resurfacing, and Bernie Burke for the NEROC 440 foot dish, and he asked Heeschen to Chair the panel. The choice of Heeschen as panel chair was not without controversy, as some committee members felt that he would bias the panel toward the VLA.[74]

Greenstein promptly informed Heeschen and the Panel that, "the NSF has recently taken a very strong position in favor of a major expansion in radio astronomy," and he put the Radio Panel on a fast track so that the BOB could not use the existence of the Greenstein Committee as an excuse to delay.[75] Understanding that the Arecibo resurfacing would be in the NSF FY1971 budget, at their first meeting on 10 November 1969 the Radio Panel debated only the relative merits of the VLA, the OVA, and the NEROC dish.[76] But they were unable to reach any consensus. If they assumed that all three projects would be funded during the next decade, the Panel argued that the OVA should be built first, but if only one project were to be funded, then the Panel favored the VLA, with only Cohen and Burke dissenting, supporting instead the OVA and the NEROC dish respectively. With so little time to meaningfully address the long-unresolved issues of priority, the Panel report did little more than endorse the Dicke Committee report that all four proposed projects (including the Arecibo upgrade) were important and urgent and that they be started in FY1971, although they also added a fifth project, a millimeter wavelength dish also being proposed by NRAO (Chap. 10). The Radio Panel interim report,[77] dated 1 December 1969, was approved by Greenstein and Brooks without the normal lengthy Academy approval process, and was forwarded to the new NSF Director William McElroy on 16 December.[78] However, at the same time, the NAS President Philip Handler informed Greenstein, that

> the whole picture developed more rapidly than McElroy or DuBridge[79] expected, and the FY 1971 books are now closed. At this time nothing would be gained by dissemination of the report beyond Dr. McElroy within the Foundation ... but you can appreciate the privileged and sensitive nature of this paragraph."[80]

Greenstein could only reply, "We do what we can, in a rather rough world. I shall try to encourage our younger experts in the field of radio astronomy to plan for a realistic future."[81]

With the exception of the Radio Panel, all of the other panel chairs were members of the parent Survey Steering Committee, but the Radio Panel was represented only by Frank Drake and Bernie Burke, each of whom had their own priorities. Heeschen informed Greenstein that this was a problem, and threatened to resign if it wasn't fixed.[82] Whether Greenstein was trying to correct this imbalance or was reacting to NAS President Handler's criticism that there were no committee members from the South,[83] in March 1970, Greenstein belatedly asked Heeschen to join the Steering Committee.[84] Again, there was some concern raised, including by Heeschen himself, that this would give NRAO and the VLA an appearance of an inappropriate advantage, but Greenstein pointed out that Heeschen was sensitive to the issue of bias, that he was a member of the NAS, and that "he is viewed by radio astronomers of the country as one of the most well-balanced and fair-minded persons possible."[85] Still, having concerns that his "real or assumed bias toward NRAO could serve to work against radio astronomy in general and NRAO in particular," Heeschen only reluctantly accepted this increased responsibility.[86]

The four previous studies of radio astronomy priorities, the Pierce, Whitford, and two Dicke Committees had all endorsed the construction of a large radio array, but none set priorities among the modest sized university array proposed by Caltech, the more elaborate and more expensive national facility proposed by NRAO, a large steerable radio telescope of the type proposed by NEROC, or the proposed upgrade of the existing Arecibo fixed spherical reflector. Greenstein realized that the only way to get anything funded required making hard decisions about priorities, and Heeschen was determined that the NRAO VLA be the top priority.

As requested by the NSF, the Astronomy Survey Committee issued an interim report which was limited to ground based astronomy projects that might be started in FY1972 or FY1973. Having been told that there would not be more than \$3 to \$6 million a year for any new starts in these years,[87] which excluded even a start on the high price tag VLA or OVA, the Radio Panel could not agree about the relative merits of enlarging either the Owens Valley or Green Bank interferometers. After much debate, the Radio Panel endorsed the 65 meter millimeter wave dish also proposed by NRAO as its top priority (Sects. 8.7 and 10.3) and recommended that the NSF also take over the university radio astronomy projects that had been dropped by the Department of Defense (DoD) as a result of the Mansfield Amendment.[88] But after the meeting, Cohen wrote to Heeschen, "I am very concerned over the millimeter dish being put in front," and he argued instead for "two more telescopes at Owens Valley, with about 1½ miles of track," pointing out the "cost effectiveness of the Owens Valley Observatory" and that three Caltech OVRO graduates were on the NRAO senior staff.[89] The interim report of the Radio Panel also suggested that the NSF explore the possibility of increased cooperation with NASA for very long baseline interferometry (Chap. 8).

The interim report of the parent Survey Committee included the NRAO millimeter telescope as its first priority, NSF support for all former DoD astronomy facilities, and gave an honorable mention to expanding the GBI along with other modest optical and infrared opportunities. In approving the report, COSPUP stressed that "such interim measures should not be taken as implying any decreased importance of the various items in the list provided by the Dicke panel of the NSF [and that] delay in the Dicke program will permit the Europeans to move ahead of the U.S. in this important area."[90] This caveat seems to have escaped the notice of the BOB, as did the concern expressed by NAS President Handler about the eroding position of US radio astronomy and the "brain drain" of young American radio astronomers.[91]

In responding to the interim report, the NSF Director expressed his view that "In spite of the present fiscal stringencies, I am convinced that the U.S. must start on the VLA."[92] Although the OVA would have been the cheaper choice between the two array proposals, the NSF was reluctant to spend so much money on a single university facility, rather than at the NRAO where the array would serve the broader community. Moreover, the NRAO was the poster child of the NSF and NRAO had a direct link to relatively high levels at the NSF that university groups did not enjoy. The NSF was committed to making the national observatory a success and wanted to build the VLA, but they needed the endorsement of the community. This recognition that the NSF, as well as the White House Office of Science and Technology (OST), already favored the VLA helped to ultimately swing the Radio Panel to support the VLA.

However, the modest recommendations for radio astronomy contained in the interim report appeared inconsistent with the ambitious Dicke Committee recommendations and the earlier endorsement of the Radio Panel which claimed that the VLA, the OVA, the NEROC dish, and the Arecibo resurfacing were all important and were all urgent. The apparently unaggressive interim report was based on earlier information provided by BOB Director Hugh Lowerth and Philip Yeager from the House Committee on Science and

Astronautics "that there was no possibility at all of funding for any part of the Dicke program beyond resurfacing of Arecibo for at least the next two years."[93] But as Harvey Brooks reported to Handler, "I am now told by Bill McElroy that the information given to us … was wrong, and that the interim report and COSPUP letter have confused the issue within OMB [White House Office of Management and Budget],[94] and resulted in general confusion about the priorities within the Administration…. The NSF now feels that the VLA should be top priority but that the COSPUP letter undermined [their] case with OMB for the VLA," and that according to McElroy, "The difference between the interim report and the Dicke panel seems to have been deliberately used as an excuse for deferring any new starts in radio astronomy."[95] Indeed, the minutes of National Science Board (NSB) Executive Session for 3–4 September 1970 show that the NSF had already included the VLA in the NSF FY1972 budget request to OMB, although in view of the on-going Vietnam War and the then large budget deficit, it did not survive to get into the President's FY1972 budget request. Interestingly, this information was already known to NAS President Handler, since at the same time, Handler was also a member (and recent Chair) of the NSB, but he was not free to divulge this confidential information to the members of Greenstein Committee. In fact, the NSB minutes show that "The Chairman reminded the Board that all subjects discussed in Executive Session, particularly with respect to the budget, are to be treated as highly confidential."[96]

Having dispensed with the interim report for modest new starts in FY1971 and FY1972, the parent Survey Committee and the Radio Panel now had to address the serious issue of dealing with the major projects: the VLA, the OVA, and the NEROC dish. Since the Radio Panel interim report had included the NRAO newly proposed 65 meter millimeter wave radio telescope in its preliminary recommendation for a 1972 new start when they had thought that any new start had to be limited to $6 million, Heeschen was caught having to either appear to reverse that interim recommendation, or support a project that was competing with the VLA.

Greenstein stressed the need to prioritize the panel's recommendations and not just to present what might appear as a shopping list. Burke, Cohen, and Heeschen were committed to the large dish, the OVA, and the VLA respectively, to which they and their colleagues had already devoted many years of hard work and significant design funds. They were not in the mood to compromise. According to anecdotal reports, the Radio Panel deliberations were intense, with no holds barred, resulting in figurative "blood on the floor." On one occasion, when a panel member complained of the bias of the Chair toward the VLA, Heeschen walked out in disgust and threatened to resign.

After several Radio Panel meetings, it was clear that the non-committed panel members as a whole preferred one of the arrays over the NEROC dish, and it came down to choosing between the OVA and the VLA. Burke had participated in the VLA Design Group, and appreciated the potential power of the VLA. Perhaps more relevant, he was a member of the AUI Board of

Trustees, and was sympathetic to the role of a national observatory of which he was a major user and from which he received significant support. With the NEROC dish off the table, Burke cast his lot with the NRAO VLA, and the rest of the panel went along. But there was no real agreement on whether or not the Panel should report a prioritized list which might make it more likely that at least the top one would be funded, or an unranked list which might increase the chance for two or more projects to be supported. At the last meeting of the Radio Panel in San Francisco on 18–19 February 1971, the panel agreed to stress that the entire program was needed if the US was to be preeminent in radio astronomy, but, recognizing that they could not all start at the same time, that the first new start would be the VLA. Once the Radio Panel agreed to support the VLA as the first priority for the radio astronomy, things moved very fast. Even before the Steering Committee had issued its formal report, Greenstein, together with Heeschen, went to OST to make the case for the VLA.

The formal report of the Radio Panel (Heeschen 1973), which appeared much later than the report of the parent Survey Committee (Greenstein 1972), was broad and convincing, citing the exciting discoveries by radio telescopes over the past decade that so fundamentally changed our view of the Universe. While not mentioning any specific proposals, appropriately leaving that for the NSF, the Radio Panel recommendations were nevertheless clear and unambiguous. Recognizing that the Arecibo resurfacing had already been authorized for construction, the Radio Panel recommended in order of priority the construction of (1) a large aperture synthesis array, (2) a large fully steerable parabola, and (3) a large telescope for millimeter observations (Sect. 10.3) (Findlay and von Hoerner 1972). To balance the strong support given to the NRAO VLA and millimeter telescope projects, the panel also expressed strong support for a wide range of university activities by recommending that "construction of new instruments at university facilities should continue, … in some cases, where outstanding competence exists, major new university instruments should be provided," and said that "Support for new operations, new state-of-the-art equipment, and maintenance of existing university facilities must be maintained at a level that will allow effective research." In support of the VLA, the Radio Panel specifically noted, "One of the most active areas of radio astronomy is the study of non-thermal sources, including quasars and radio galaxies."

Concurrent with the Astronomy Survey, the NAS also ran a physics study, led by Alan Bromley from Yale. A panel on astrophysics, chaired by George Field from Harvard, was appointed to jointly support both the astronomy and physics surveys. One of the present authors, (Kellermann), represented radio astronomy interests on the Astrophysics and Relativity Panel, which recommended that "the Astronomy Survey Committee take into account the need for a large array that can synthesize a beam of the order of seconds of arc in a reasonable period of time, for study of extragalactic radio sources" (Field

1973a, b) so the VLA came to the parent Survey Committee blessed by two separate panels.

There were no other large "shovel ready" projects coming up from the other panels. The optical astronomers preferred more observing time and improved detectors over building more powerful new facilities, and were willing to support the VLA.[97] Acceptance of the VLA as the top project for the Committee was perhaps easier in the full Steering Committee than it was in the Radio Panel. Multiple straw ballots, each with different constraints and weighting criteria, put the VLA on top each time, usually by a wide margin. According to Heeschen,[98] Greenstein was initially very opposed to the VLA. Although he had played a prominent role in founding NRAO, Greenstein had come to see the relatively well-funded big national observatories as a threat to university-based, individually-driven scientific research.[99] He too threatened to resign from the Committee but realized that doing so would undermine the whole study.[100]

Heeschen knew that the VLA had strong supporters on the Committee, and thought it better that he did not attend the final Steering Committee meeting held in Boulder, and thus it was Burke who presented the case for the VLA. Greenstein himself later explained that he was finally sold on the VLA by its expected capability to resolve the long-standing radio source count controversy, and also by the expectation that it would contribute to the broader cosmological issues facing astronomy.[101] Also, Greenstein was never enthusiastic about the OVA. He came into the Survey hoping to get a copy of the Palomar 200 inch telescope in Chile. Moreover, he had little confidence in the OVRO management to construct and operate something of the magnitude of the OVA. At a higher level, within Caltech, there was more interest in enhancing the high energy physics program than in the radio astronomy program. Indeed, after John Bolton left Caltech at the end of 1960, there was only a token effort to bring in a new Director from the outside. The main competition to the VLA came from the NASA proposed series of High Energy Astronomy Observatories (HEAO) but since this was a NASA, not an NSF program, and since the cost of HEAO was an order of magnitude larger than that of the VLA, they were not really in any direct competition.

The final report of the Astronomy Survey Committee (Greenstein 1972) recommended as its top priority "A very large array, designed to attain a resolution equivalent to that of a single radio telescope 26 miles in diameter," but added, "this should be accompanied by increased support of smaller radio programs and facilities at the universities or other smaller research laboratories." A program to develop instrumentation for optical telescopes, support for the new field of infrared astronomy, a series of High Energy Astronomy Observatories, and the large millimeter-wavelength antenna received second through fifth priorities respectively. The NEROC proposal for "a large steerable radio telescope designed to operate efficiently at wavelengths of 1 cm and longer" was given only tenth priority and was never built. Radio astronomers would need to wait another 30 years before a large steerable radio telescope would be built in the US, and it only happened then as a result of a freak accident (Chap. 9).

The Survey Committee report probably ended up with a stronger endorsement of the VLA than intended by many members of the Radio Panel. Indeed after the report was released, Cohen again wrote to Heeschen that he was having second thoughts; that "the VLA will not touch on the exciting and fundamental problems: molecules and compact objects," and that he was getting a lot of negative comments from other radio astronomers.[102] Nevertheless, on 22 March 1971, Heeschen, together with Cohen and other members of the radio panel, met with Edward David, President Nixon's controversial new Science Advisor, at a meeting organized by Geoff Burbidge. David noted that the group's support for the VLA "reaffirms the budgetary proposal made last year by NSF," and indicated that no further discussion was needed on this topic.[103] However, as result of his abandonment of the NEROC dish and his public support of the VLA, Burke faced a formidable challenge at home from his MIT/Harvard colleagues, who accused him of something just short of treason.[104]

Sensing that the tide was shifting toward the VLA, in May 1970 Stanley wrote to the Caltech management suggesting that the time had passed for the OVA.[105] Trying to salvage something for OVRO, Caltech withdrew the OVA proposal and instead proposed a more modest Owens Valley Interferometer (OVI). The new Caltech proposal exploited a perceived weakness of the VLA proposal and emphasized the spectroscopic opportunities made possible by building only two new 130 foot antennas to operate together with the existing 130 foot and two 90 foot antennas. In an apparent about-face from their earlier position, the new proposal discussed the OVI as "a nationally-available facility." The proposal for $6.5 million was sent to both the NSF and NASA, but was never funded.[106] Gordon Stanley stepped down as OVRO Director in 1975 and was succeeded by Alan Moffet. Under the leadership of Marshall Cohen, Caltech shifted their emphasis to VLBI (Chap. 8).

Ironically, following the lengthy period of controversy between NRAO and Caltech, once the decision was made in favor of the VLA, it would be Caltech graduates such as Barry Clark, Edward Fomalont, Eric Greisen, and Richard (Dick) Sramek who played major roles in the final design, construction, and later the operation of the VLA. Two of the long-time VLA site directors, Ron Ekers and Miller Goss, had both worked at Caltech.

7.5   CHOOSING THE VLA SITE

There is no "best site" for a radio telescope, or for that matter for any telescope, as the quality of the site depends on many different criteria. The "best site" will depend on how the different criteria are weighted, and different people will weight them differently. The criteria adopted for choosing the VLA site included many of the criteria that went into choosing the Green Bank site. These included:

1. Freedom from radio frequency interference, which meant isolation from population centers and being surrounded by mountains to shield the array from radio transmissions;
2. Low latitude in order to observe the largest part of the sky, particularly the galactic center region, and in US territory;
3. Freedom from extreme weather conditions and earthquakes that might damage the instrument, and also low average wind speeds so as not to compromise the antenna pointing;
4. Availability of adequate power and water;
5. Proximity to a nearby town with adequate schools, cultural, and medical facilities, and access to reasonable surface and air transport.

As is the case for all radio telescopes, criteria 1 and 5 can be mutually exclusive, and the VLA had its own additional requirements. As was noted in the VLA Report No. 1, a large flat area of at least 20 miles in diameter was required to allow the individual antenna elements to be transported, and the land had to be available at reasonable cost. Moreover, it was becoming increasingly clear from experience with the GBI and at OVRO that to operate at centimeter wavelengths, clear dry skies were required, as atmospheric turbulence contributes to interferometer phase fluctuations. All the criteria suggested a site in the desert southwest. From examination of topographic maps, NRAO engineer Sidney Smith identified 14 potential sites, which he labeled Y1 through Y14. Wade later added the Plains of San Agustin in central New Mexico, which Smith had missed as it lay on the corners of four different topographic maps, but which clearly stood out as potentially an attractive site. It was labeled Y15.

Wade and Smith went to New Mexico in November 1965 to inspect the Plains of San Agustin both from the ground and the air, and to enquire about land availability. Wade was immediately impressed, and for the next five years, he considered this as the site to beat. In choosing the site for the VLA, NRAO had to consider not only the technical and logistical criteria, but a variety of social, economic, political, and environmental criteria as well. Everyone wanted to be involved—the local landowners, the politicians, concerned citizens, and of course the NSF. Although the Y15 site stood out from the beginning as being the most desirable, 33 other sites were considered. Some of these were quickly rejected. Two were active oil drilling fields; several others were Air Force bombing ranges (Heeschen 1981, p. 11). Wade and others investigated all 34 sites between 1965 and 1971. As Wade described it, "I got to eat in lots of backwoods restaurants."[107]

Much later, Wade recalled that he was troubled about discrepancies in the contour levels which described some of the potential sites that were located on different topographical maps, so before one of his trips to investigate prospective sites in the Southwest he purchased an altimeter that had been salvaged from a wrecked airplane to check site altitudes. Taking off from the Cleveland airport on the second leg of his flight back to Charlottesville from inspecting the Arizona site, Wade was playing with the altimeter to see if he could detect when the cabin pressure changed. Suddenly, the pilot announced that they were returning to the airport for an emergency landing. Like the other passengers, Wade, concerned about the emergency landing, hastened his exit through one of the emergency chutes, only to be taken aside by FBI agents for questioning. Apparently, spotting Wade fooling with his altimeter, another passenger notified the flight crew that Wade was about to set off a bomb. The plane set off again, but only after a long delay, and Wade was not the most popular passenger on the flight.[108]

To the extent possible, NRAO tried to keep the search process quiet to avoid possible political interference and land price speculation. Unlike in Green Bank, where the Observatory land was privately held, the area chosen for the VLA was mostly federal and state owned land, but was leased to private ranchers who were very protective of their grazing rights. Fortunately, Wade had grown up on a farm in Kentucky and knew how to talk to farmers and ranchers without alarming them and without the local politicians getting too involved. After multiple visits, Wade, often accompanied by NSF or AUI staff, managed to satisfy the ranchers that the VLA would not harm their ranching interests.

Of 34 potential VLA sites, NRAO let contracts to a civil engineering firm to study seven sites[109] for ground stability, drainage, the suitability of underground water for drinking, suitability of soil content for construction, etc. While Wade continued to prefer the Y15 site, there were strong arguments for the Y23 site which was close to Tucson and close to where NRAO was already operating its 36 Foot Telescope (Sect. 10.2) and also close to the Kitt Peak National Observatory. But according to Wade, the Y23 site was subject to flooding, had a large rattlesnake population, and was too close to potential interference from Tucson.[110] Y27, the site near Marfa, Texas and the McDonald Observatory, was considered by some to be attractive as it was in Texas, the home state of then President Lyndon Johnson. In 1967, the Plains of San Agustin became NRAO's proposed site, but it was kept quiet until 1971 when it was clear that further progress on the project depended on developing the site. This meant going public with disclosing the Plains of San Agustin as the preferred site, and NRAO (1971) submitted Volume IV of the VLA proposal to the NSF in December. Volume IV described the site selection process, discussed the relative merits of seven acceptable sites, the reasons for rejecting the remaining 27 sites, and the merits of the Plains of San Agustin as the preferred site for the VLA.

All astronomers think they are experts on telescope site selection, and nothing is ever more controversial in any big telescope project than choosing a site. The VLA was no exception. In submitting Vol. IV to the NSF, Heeschen's covering letter succinctly summarized the choice of Y15 in terms of its elevation, level ground, drainage, accessibility, and cost, concluding with "The site Y15 in the Plains of San Agustin is remarkable, and is perhaps uniquely suited to the requirements of large radio astronomy arrays."[111] Although Vol IV of the VLA Proposal probably gave more detail and more extensive justification for the VLA site selection than for any previous telescope project, the NSF needed reassurance before approving the selection of the site, and asked the

National Academy of Science to review Vol. IV, to advise on the adequacy of selection criteria, to review the analysis of the site selection data, and to assess the conclusions. The NAS sent Vol. IV, along with Heeschen's cover letter, to C. Mayer (NRL), E.M. Purcell (Harvard), J.R. Pierce (Caltech), R.B. Leighton (Caltech), and R.N. Bracewell (Stanford), requesting their advice. All responded positively endorsing the methodology and the choice of the Plains of San Agustin, but one reviewer could not resist the opportunity to question whether or not the VLA was worth the huge cost and suggested that some of the money could better be spent on VLBI (Chap. 8).[112]

Work on the site began in 1974. Even though most of the land was owned by the state or federal government, gaining access was not straightforward. It took more than seven years to complete the paper-work to give the NSF the right-of-way through one parcel of land owned by the Department of the Interior. Each of the ranchers who owned land near the ends of the three arms brought suit against the government condemnation of their land, and the suits had to be settled by a court appointed commission, costing the project another $200,000. One of the ranchers, who owned land near the end of the north arm, objected to the encroachment on a piece of land that he had developed for irrigation and farming. Unable to negotiate a mutually agreeable settlement, the north arm of the VLA was shortened to 19 km, and so is 2 km shorter than the other two arms. Only one square mile of land, housing the control building, cafeteria, dormitory buildings, and the Antenna Assembly Building, was actually purchased for the VLA; the three strips of land, 300 feet on each side of the baselines, are leased.

New Mexico state law requires that land used for any new project be inspected by a state-licensed archeologist for evidence of historical land use. A preliminary survey by a New Mexico State University archeologist disclosed evidence of ancient habitation near the end of the planned array's southwest arm. The site was on land owned by a local rancher who refused admittance to the site until a court order rejected his claim (Lancaster 1982). The state of New Mexico, the NSF, and the Department of the Interior all turned down applications to fund the required excavation, and the VLA project had to pay almost $100,000 for the archeological work, which uncovered more than 3,000 artifacts dating back as much as ten thousand years (Beckett 1980).

7.6   BUILDING THE VLA

Selling the VLA to the radio astronomy community, to the NSF, and to Congress took a decade, but this was only the beginning. It would take nearly another decade to address the multitude of managerial, funding, technical, and logistical challenges facing NRAO and the NSF. In the spring of 1971, when it first appeared that the VLA would be funded, Heeschen appointed Hein Hvatum, the NRAO Associate Director for Technical Services, as the new VLA Project Manager. Heeschen declared his own work finished and left for a well-earned six-month Caribbean sailing trip with his family, leaving Hvatum in charge. Hvatum then led a detailed review of the design of the antenna and other instrumentation. The 1960s were not only a period of rapid discoveries in radio astronomy, but one of major technical advances. Cryogenically cooled low noise amplifiers, largely developed at NRAO by Sandy Weinreb and his staff, greatly improved the sensitivity of radio telescopes, although there remained reliability issues. Digital signal processing had replaced chart recorders and analogue electronics, and astronomers were becoming more comfortable with large scale computing machines. The rapid scientific and technical advances represented both opportunities and challenges. In 1971, astronomers expected more from the VLA than they did when it was first discussed in the 1965 report, and even the 1969 proposal was technically obsolete. But no one had ever simultaneously operated 27 cryogenically cooled receivers. Indeed, it was often a challenge in Green Bank to keep a single cooled receiver operational for more than a few days at a time. The VLA goal of 10 arcsec resolution was replaced by 1 arcsec, but the dynamic range requirement was modestly set at only 50 to 1 corresponding to maximum sidelobe levels of the order of two percent, or comparable to that of a carefully illuminated parabolic dish.

Following the Congressional approval of the VLA project in August 1972, the NSF made $3 million available in November 1972 for VLA design and prototyping. The final antenna configuration was based on a total of 28 antennas, nine along each of the three arms plus a spare, so that at any given time one antenna could be scheduled for routine servicing and possible installation of new receiving equipment. Four configurations of the antennas were proposed to vary the resolution and field of view. The four antenna configurations provided maximum arm lengths of 600 m, 1.95 km, 6.4 km, and 21 km, and became known as the D, C, B, and A configurations respectively. Along each arm, the spacing of the nine antennas was concentrated toward the center and followed a power law distribution of spacing that minimized the total number of stations required.

Electronics Division head Sandy Weinreb did most of the system design for the VLA instrumentation. The Green Bank Interferometer (GBI) operated in only two bands with concentric feeds, and the initial proposal to build the VLA was based on a similar system. Due to satellite interference near the 11 cm band, the primary VLA band was shifted to 6 cm with additional bands at 18–21 cm, 2 cm, and 1.3 cm. Each receiver first-stage was mounted on separate circularly polarized feeds located on a 2 meter diameter ring centered on the vertex of the dish. An asymmetric secondary reflector at the Cassegrain focus was rotated to illuminate each feed and direct the beam along the electrical axis of the telescope. The 6 cm receiver was conventional and included a parametric amplifier followed by a mixer and IF system. For the 18–21 cm system, the signal was up-converted to 5 GHz (6 cm) and the 6 cm paramp used as the second stage. In order to properly illuminate the 18–21 cm subreflector and to keep the feed from being prohibitively large, a dielectric lens was placed in front of the feed. The 1.3 and 2 cm mixers also used the 6 cm paramp as the second stage. Although the addition of the 1.3 and 2 cm bands improved the resolution to 0.1 arcsec at the shortest wavelength, this cost effective arrangement resulted in relatively poor sensitivity at both 1.3 and 2 cm. At each band there were two independent receivers, one each for left and right hand circular polarization. For each polarization, the 100 MHz IF band was split into two 50 MHz bands, which was the largest that the digital electronics of the era could accommodate.

In order to optimize the aperture efficiency, the primary and Cassegrain secondary reflectors differed from their canonical parabolic and hyperbolic shape (Williams 1965). To evaluate the planned fixed position 4-feed system, the Green Bank 140 Foot Telescope was converted to a Cassegrain optics employing a rotating asymmetric secondary reflector. Unfortunately, the Green Bank prototype did not uncover a problem caused by the offset feed geometry, which resulted in the two circularly polarized beams being displaced by 0.06 beam widths (Napier and Gustincic 1977). By the time the problem was discovered, six systems had already been purchased, and it was decided not to implement any changes.

Although NRAO had experienced considerable issues with the reliability and stability of cryogenically cooled receivers on the Green Bank 140 Foot Telescope, Weinreb made the bold decision that in order to obtain the best sensitivity he needed to use cooled parametric amplifiers on the VLA front ends. There were initial reliability issues with the VLA front ends, but by the end of the construction project Lancaster reported that "reliability was no longer a problem" (Lancaster 1982). Ultimately the 5 GHz paramps and mixers were replaced by separate cooled Gallium Arsenide Field Effect Transistor (GaAsFET) amplifiers for each band which gave lower noise, better stability, and improved reliability.

One of the key technical innovations employed during the construction of the VLA, was the use of the newly-developed low loss TE_{01} mode circular waveguide to carry the IF signals back from each antenna. The same waveguide carried the common local oscillator reference signal from the central laboratory to each antenna and the extensive monitor and control signals to and from each antenna and the central control building.[113] In the 1967 VLA proposal, the signals were to be transmitted by conventional coaxial cable, but to compensate for the attenuation over the long baselines extending up to 27 km, expensive and perhaps unreliable amplifiers would be needed every few kilometers to maintain the needed signal strength. It was felt that optical fiber technology was not sufficiently well developed at the time to be used for the VLA. Weinreb became aware of the new low loss circular waveguide that had been developed to replace the microwave relay towers then in use to support the AT&T national telephone network. However, when they went to the NRAO business office to get approval to buy the waveguide, Weinreb and Hvatum had to admit that the waveguide was being manufactured only in Japan. Moreover, the Japanese plant where the waveguide was being fabricated was scheduled to be closed. In order to procure the waveguide needed for the VLA, NRAO only had a month to place the order. NSF approval was straightforward, but the US Department of Commerce was not so easily convinced why such a large government contract should go to Japan, especially without an open bidding process.

No one at NRAO had any experience in using the new circular waveguide. The loss resulting from inserting couplers at the many antenna stations in the inner few kilometers of the array meant that there would not be enough signal to reach the outer antennas.[114] Moreover, standing waves set up by the couplers threatened to generate spurious propagation modes in the waveguide. After a five-year effort, NRAO solved the problem with the invention of a new coupler with a low insertion loss (U.S. Patent No. 4025878). Also, as used by Bell Telephone, the waveguide was mounted on steel springs attached to an outer steel pipe. NRAO wanted to save money and directly buried a 1.25 km test section at the VLA site to evaluate the long-term stability. Although the attenuation of the waveguide increased significantly over the next few months, it ultimately stabilized and the decision was made to directly bury the waveguide at least one meter deep (~3 feet) along each of the VLA arms without using any protective enclosure (Fig. 7.6). According to Lancaster (1982), because the incremental funding prevented placing a single order and there was only one manufacturer, the Sumitomo Corporation, which was located in a foreign country, "the procurement of the waveguide was one of the most difficult actions of the VLA construction." Although the test section was obtained at a cost of $32 per meter, subsequent asking prices jumped to $79 per meter, but after negotiations were reduced to keep the waveguide procurement within budget.

Another important design change from the original proposal, made possible by the rapid development of digital signal processing technology, was to build a digital delay-multiplier system based on custom designed Application Specific Integrated Circuits (ASICs) to process the four 50 MHz IF bands, two in each circular polarization. To compensate for the delay of up to 140 microsecs in the differential path length from each antenna, the digital delay system needed to maintain an accuracy of better than two nanoseconds across the 50 MHz band.[115] This complex digital system included a test and replacement capability to automatically detect component failures and replace failed components with spare units. However, the actual implementation of all four IF bands had to be deferred as the initial computer system was not adequate to handle the full data load. Although a spectral line capability was not included in the original proposal, the VLA correlator that was finally built employed a technique known as recirculation, whereby an increased number of frequency channels could be obtained at the expense of limited bandwidth.[116] The VLA correlator was a large digital system operating at 100 MHz and included 13 racks of NRAO developed hardware containing 85,000 integrated circuits. It was made feasible by using two custom developed integrated circuits which reduced the number of multilayer circuit cards from 864 to 156.[117] In simple terms of megaflops,

Fig. 7.6 The TE_{01} mode circular waveguide was buried alongside the railway track to transfer the local oscillator and IF signals between the central control building and each antenna. Credit: NRAO/AUI/NSF

the VLA correlator rivaled the most powerful general purpose computers then available.

A particularly challenging area, the computing hardware and software, was divided into two areas. Real time control of the VLA, data acquisition, and data processing were assigned to the "synchronous" system comprised of a series of Modcomp II mini computers, and was designed to have minimal real time interaction with the VLA operator or the observing scientist. The "asynchronous" system included off-line data editing, calibration, and processing and imaging that was initially implemented in a Digital Equipment Corporation DEC 10 mainframe computer. Recognizing the enormous challenge that the VLA image processing presented, NRAO investigated the feasibility of using analogue optical imaging processing.[118] An internal committee appointed by Heeschen reported that the optical system could give better dynamic range for about the same price, but as the risk was higher, they fortunately recommended the digital processor, which was initially implemented in PDP 11/40 and PDP 11/70 mini-computers and an array processor. The original PDP 11s were later replaced by several more cost-effective and popular VAX 11-780 machines, including one in Charlottesville, and later by Convex C1 mini-supercomputers. Robert (Bob) Hjellming[119] led the development of the asynchronous software system and Barry Clark the synchronous system, but they were supported by a growing team of programmers along with a growing software budget. In order to take advantage of the rapid growth in computing power, the initial hardware acquisition was limited to that needed to handle only the 10 antenna continuum system, with the intention of acquiring the rest of the computing hardware closer to the end of the VLA construction. In the end, this approach gave the best computing power for the money, but severely limited the use of the partially completed VLA. In part, this was the result of the excellent VLA sensitivity and an antenna configuration that could give reasonable images even for short "snapshot" observations lasting only a few minutes rather than 8 to 12 hours. This meant one could observe a hundred or more sources per day instead of the planned two to three sources, with a corresponding increase in the computing load. Moreover, the VLA proposal assumed a relatively straightforward single data pass of gridding and Fourier transform, but the use of deconvolution techniques and self-calibration led to multiple passes and an interactive data reduction process. NRAO scientists initially assumed that due to the large number of interferometer baselines deconvolution would not be needed for VLA data. Attempts to develop a so-called "pipeline" combination of hardware and software lasted over a decade, but were never satisfactorily implemented.[120]

Finding the Money The FY1973 Congressional Appropriation Bill HR 15093 including initial funding for the VLA was signed by President Richard Nixon on 14 August 1972. Cam Wade recollected that when he learned Nixon had approved the VLA, Heeschen's mixed response was, "We've wanted this thing so long, and now we are getting it from a crook."[121] The original NRAO plan called for a one-year design phase followed by a four-year construction plan at a total cost of $63 million. However, the NSF wanted to limit funding to not more than $10 million per year to minimize the impact of the VLA construction on other NSF programs. A new construction plan was negotiated, with the first year for final engineering design and prototyping funded at $3 million, followed by a constant funding level of $10 million per year, which stretched the construction over a period of nearly eight years. This not only delayed the start of full VLA operations, but the extended production schedule increased the cost due to the then high level of inflation, the loss of quantity discounts for large purchases, as well as the need to maintain the administrative, scientific, and technical support structure over a longer period of time. With an assumed rate of inflation of 6% a year, the total cost projection increased to $76 million. As it developed, the extra time and increased funding turned out to be a blessing, as it allowed time for prototyping, testing, and, where necessary, design changes with a minimum of retrofitting. Although the VLA was technically state-of-the-art, because most of the construction cost was in straightforward areas such as antennas and railway track, it was felt that the budget plan was sound, and "NRAO was determined to build the VLA on schedule and within budget" (Heeschen 1981, p. 31). However, as described below, the actual rate of inflation became much higher, which resulted in continual modifications to the construction plan and threatened the successful completion of the VLA.

The FY1973 NSF budget passed by Congress included the requested $3 million for VLA design and prototyping, but funding for the first year of construction hit a roadblock in Congress.[122] The House of Representatives Science and Astronautics Authorization Committee included the planned $10 million for VLA construction as part of the NSF's total $610 million authorization bill for FY1974. However in the House Appropriations Sub-Committee for Housing, Urban Development, and Independent Agencies (HUD), Representative George Shipley (D-Illinois) commented, "The stars will still be shining in 20 or 30 years, but pollution is going to be a heck of a lot worse in 20 to 30 years."[123] Subsequently, the Appropriations committee report stated, "Although this committee approved the initial funding for this project in fiscal 1973, it now feels that that in view of general budget constraints and other earthbound National Science Foundation priorities, the VLA can be deferred," and VLA funding was eliminated from the House NSF Appropriations bill. Coincidently, as reported in the *Wall Street Journal*,[124] the Authorization and Appropriations bills reached the House floor and were each passed on the same afternoon. With no appropriation, the authorization was meaningless. The VLA was saved when New Mexico Governor Bruce King found himself on the same plane with Senator Joseph Montoya (D-New Mexico) where they discussed the VLA problem. Montoya was a member of the Senate Appropriations Committee and brought the VLA problem to the attention of the HUD sub-committee chair, Senator William Proxmire (D-Wisconsin) of Golden Fleece fame (Sect. 5.5), and other sub-committee members. With Proxmire's support, the Appropriations bill passed by the Senate included the full $10 million for the start of VLA construction. The House-Senate Conference Committee, as is typical in such situations, split the difference, and the final FY1974 Appropriations bills containing $5 million for the start of VLA construction passed both houses without discussion. But this reduced funding caused yet another redrafting of the construction plan and increase of the projected project cost to $78.2 million. At the request of the NSF, dozens of other funding arrangements were prepared over the course of the project, with 17 alone in FY1974.

But there was still another hurdle to overcome. The NSF had to be convinced that the VLA as designed was feasible and could be built for the planned $63 million, so they contracted with the Stanford Research Institute (SRI) to do a feasibility study of the proposed VLA. The charge to SRI was to (1) determine the proposed system feasibility and ability to meet specifications in the light of existing technology, (2) confirm the cost and time schedules for the construction, development, and operation of the system, and (3) evaluate the method for managing the VLA Project proposed by AUI. SRI convened an ad hoc committee that did not include any astronomers. The committee met five times over a period of three months. Their report[125] was very favorable, but they expressed concern about whether NRAO's informal management style would be appropriate to a complex construction project such the VLA. The report confirmed that (1) the VLA was technically feasible, (2) the cost had been accurately estimated by NRAO, (3) the time for construction could be as short as four years, (4) NRAO's technical competence was confirmed, and (5) the proposed project management "is generally good, but could be improved."

VLA Construction[126] Instead of bidding the entire project to build the VLA to a single contractor, NRAO decided to act as its own prime contractor to minimize the cost and to maintain tight control over the construction. Following the approval of the VLA by Congress in August 1972, Jack Lancaster joined NRAO two months later as the VLA Project Manager and NRAO Assistant Director (Fig. 7.7). Prior to coming to NRAO, Lancaster had been the Chief Project Engineer at Brookhaven, where he oversaw the construction of the major reactors and accelerators. In November 1972, the VLA Construction Project was organized as a Division of NRAO with Hein Hvatum retaining responsibility for the overall technical design (Heeschen 1981, p. 31). Groundbreaking on the Plains of San Agustin took place on 4 December 1972. In April 1973 Lancaster opened an office in Magdalena, New Mexico, about 20 miles from the center of the VLA site. Over the next eight years, Lancaster, Hvatum, and Heeschen expertly guided the VLA project to its successful completion in January 1981, on time and officially within the 1972 revised budget.

As with other radio arrays, the largest single cost item for the VLA was for the antennas. The Request for Proposals (RFP) for the antennas was based on an in-house design and cost estimate led by NRAO engineer Bill Horne. Reflecting the increasing interest in going to shorter wavelengths, the antennas were specified to have a surface accuracy of 0.75 mm rms and pointing accuracy of 2 arcsec, sufficient to permit observations at wavelengths as short as 1.3 cm. NRAO estimated that the antennas would cost $19.3 million. The bids ranged from a low of $16.8 million by E-Systems Inc. to a high of $31.8 million from the Collins Radio Company. Following evaluation of the business and technical aspects of the proposals and discussions with each of the potential vendors, NRAO received five "Best and Final" proposals. In October 1974 NRAO signed a contract for the fabrication and construction of 28 antennas

Fig. 7.7 Jack Lancaster, VLA Project Manager who oversaw the VLA construction. Credit: NRAO/AUI/NSF

with the Dallas, TX-based firm E-Systems, Inc. The E-Systems contract also included an Antenna Assembly Building to facilitate the construction of the antennas, the installation of instrumentation, as well as ongoing antenna maintenance and repair.[127]

Since the fabrication and erection of the antennas was planned to be stretched out over a number of years, the contract was complex, since there is no guarantee from year to year that Congress will appropriate the needed funds. During FY1974, E-Systems completed the engineering design, and the first two prototype antennas delivered in 1975 met all specifications. The antenna contract contained options for NRAO/AUI to purchase a predetermined number of antennas each year at a predetermined fixed-price that increased each year to allow for an anticipated six percent annual inflation. This lead to the first serious problem in the VLA construction program.

Due to the oil crisis resulting from the OPEC oil embargo following the 1973 Yom Kippur War, and the abandonment by Richard Nixon of the US Gold standard and subsequent dollar devaluation, the 1970s experienced a period of extreme inflation. Within eight months the cost of steel doubled. On 6 January 1975, E-Systems notified NRAO/AUI that they were no longer able to meet their contractual fixed-price obligations. It was apparent that any attempt to enforce the predetermined prices would result in bankruptcy, leaving NRAO with no path to secure the antennas. After lengthy negotiations, it was agreed that NRAO/AUI would advance the funds to purchase the steel for all of the remaining antennas, which would help mitigate the impact of the high national inflation, and that E-Systems would deliver the completed antennas at the previously agreed price, but on a faster schedule. However, this meant that by spending an unplanned large fraction of the limited NSF annual funding on the antennas, the instrumentation of the antennas was delayed and the antennas were not available for commissioning or scientific observations at the planned rate.[128]

The less than satisfactory experience with moving the GBI antennas along a roadway led to an early decision that the VLA antennas would move along two parallel railway lines. Rather than trying to push or pull the antennas using a bulldozer as was done at Green Bank and Caltech, antenna transporters were specially designed to reconfigure the VLA antennas. The initial plan called for three transporters, one working along each arm, but there was only enough money in the budget for two transporters that were christened "Hein's Trein" and "CamTrak" (Fig. 7.8). Also, as a result of the rapid period of inflation following the oil crisis, the cost of used railroad track increased from $90 per ton to $330 per ton. NRAO hired two retired railway track foremen who were able to locate 14,000 tons of US government surplus track at some 28 different locations around the country. As described by Heeschen (1981), with the aid of the NSF, some of this was declared surplus and made available to NRAO for the cost of shipping to the VLA site. Another 221 tons of new rail was purchased at near scrap prices after it had been rejected by the US Department of

Fig. 7.8 Hein's Trein, used to transport the VLA 25 meter antennas when reconfiguring the array. Credit: NRAO/AUI/NSF

Transportation for not meeting the required specifications for commercial rail lines (Figs. 7.9 and 7.10).

All of the instrumentation, the front ends, IF systems, digital delays, and the correlator, were designed by NRAO engineers, and for the most part, were fabricated in-house. Other instrumentation, including the monitor and control system, feeds, paramps, cryogenics, and the waveguide distribution system, were fabricated commercially. With a careful system of testing and noting failure rates, redesigns and retrofitting were kept to a minimum. Not unexpectedly, in view of the problems experienced in Green Bank, the cryogenic systems proved to be the least reliable component until a new manufacturer was found.

With the instrumentation of the first completed antenna, and the start of commissioning, the VLA project management, scientists and engineers moved from Charlottesville to temporary headquarters in Socorro, New Mexico, during the spring and summer of 1975. It was a one-hour bus ride, each way, from Socorro to the VLA site. Although on most days most of the staff were not normally needed at the construction site, Lancaster adopted the practice of having everyone—management and administrative personnel, scientists, engineers, and technicians—all ride together to the site on one of two buses. This practice enabled a high level of communication among the disparate groups, which many later agreed was crucial to the successful completion of the VLA.

Fig. 7.9 Unloading surplus rail in Socorro. Crane is unloading Crab Orchard rail from rail cars. The truck in the background is leaving with rail for the VLA site. Credit: NRAO/AUI/NSF

Fig. 7.10   Surplus rail from Holloman Air Force Base arriving at the VLA site. Credit: NRAO/AUI/NSF

The first antenna was completed and moved from the Antenna Assembly Building in July 1975 (Fig. 7.11), less than three years after the initial VLA funding was authorized. First fringes were detected between two antennas over a 1.24 km baseline in February 1976. By the beginning of 1977 five antennas were in operation using a 2 km baseline, and the first scientific results from the VLA were reported by Balick et al. (1977). By June 1978, there were ten antennas in operation using a 10.6 km baseline, and NRAO (1978) announced that the VLA was open for scientific proposals from the community. All 27 antennas were in operation by July 1980 and in use for scientific observations. The installation of all 122 km of railroad track was completed by the end of 1980. Under the tight management of Heeschen, Lancaster, and Hvatum the VLA was built close to the planned budget and completed on schedule. When completed in January 1981, the VLA met, or, in many cases exceeded, all of its performance goals.

Throughout the eight-year construction, Heeschen continually stressed that NRAO was committed to meet the agreed budget. If some item came in at higher than the planned price, something else had to go. Numerous such adjustments were made during the process; fortunately, many of the deleted items were restored in the later years. As a result of constantly changing budget projections from the NSF, OMB, and Congress, NRAO had prepared nearly 50 different funding schemes by the time the VLA was completed in 1980. The final VLA price tag was $78.578 million which was only three percent over

Fig. 7.11 First VLA antenna emerges from Antenna Assembly Building in July 1975. Credit: NRAO/AUI/NSF

the original March 1971 budget of $76 million. The increase in the Consumer Price Index over the same period was more than a factor of two. NRAO was able to keep the impact of the unprecedented high inflation modest, due in part to the procurement adjustments made to the fixed-price antenna contract, the more modest rate of inflation for electronic instrumentation, and level or even reduced prices for computing equipment. However, some of the NRAO scientific and administrative staff working on VLA planning and construction remained on the NRAO Operations budget throughout the project, so that the true cost of the VLA was actually somewhat higher than the official number.

Characteristically, Congress and the NSF were nervous throughout the construction period. The VLA was the most expensive NSF project ever attempted and they had to be convinced that it was all going well. In 1976, Congress initiated a review of the VLA project, followed by a Hearing at the House Subcommittee on Science, Research, and Technology.[129] The subsequent report of the House Science Committee stated, "The Committee is very pleased with the close agreement between the original and current budget and time schedules and commends the project's accomplishments to date." But they worried that the existing NRAO and AUI advisory committees might be concerned with only science and technology and not management, so they added that "The Committee strongly recommends that the Director of the Foundation establish an ad hoc advisory panel to examine the VLA management and technical plans and activities."[130] Meanwhile, in 1975 and 1976, several delegations of aides from the House Appropriations Committee, including the Chief of Staff, Richard Mallow,[131] and the House Committee on Science and Technology[132] visited the VLA. They were mostly concerned about how NRAO was reacting to the potential cost increases resulting from escalation, but also asked about how telescope time would be awarded, how many women were employed on the project, and interestingly, what additions to the array were foreseen for the 1980s. Reportedly, one of the visitors noticed Barry Clark's cluttered office and said, "If this is an example of how the VLA is being run, we're in trouble."[133] Congressional staff also participated in the annual NRAO/AUI presentation to the NSF on 14 February 1976.[134] This was followed by another hearing held on 30 September 1976 where Dave Heeschen was asked to testify.[135]

The NSF responded to the Congressional mandate by appointing a panel of representatives from industry, universities, and government agencies chaired by Cornell University Vice President Robert Matyas. There were no astronomers on the panel, which met five times during 1977. The panel report issued on 31 December 1977 noted that "The program has now progressed far enough to state with assurance that it will be both a technical and scientific success, ... but also has the potential for discovery in allied fields."[136] However, the panel criticized the NSF for creating difficulties with the stretched-out budget, and recommended that future "projects of this magnitude and complexity be planned and scheduled within a more optimum engineering construction time." The panel praised the project management and the dedication to "living within current budgets and schedules," but raised concerns about the low level of the remaining contingency and the level of effort devoted to software. Interestingly, the panel endorsed the plan "in which a basic publishable 'product' is provided by the VLA facility," a goal which took nearly another forty years to reach.[137] In addition, the NSF conducted a further audit to "render an opinion as to NSF management on the economy, efficiency and control with which the VLA Project is being administered within the Foundation and by NRAO." Interestingly, eight of the ten recommendations by the audit committee pertained to NSF and not to NRAO record keeping and financial statements.[138]

The official dedication, attended by some 600 staff and guests, was held on 10 October 1980 (Fig. 7.12). Ten years later, at an October 1990 conference sponsored by both the IAU and URSI, more than 220 scientists from 17 different countries gathered in Socorro to celebrate the 10th anniversary of the opening of the VLA. The conference included presentations on astronomy, instrumentation, history, and planning for the future (Cornwell and Perley 1991). The first discussions leading to the Square Kilometre Array began at this meeting (Chap. 11).

In 1973 Heeschen established an internal VLA Steering Committee to replace the defunct VLA Design Group. The Steering Committee met monthly to provide continuing advice on the various aspects of the VLA construction program. Also in 1973, he appointed an external VLA Advisory Committee

Fig. 7.12 VLA dedication, 10 October 1980; President's Science Advisor Frank Press standing at the podium. Credit: NRAO/AUI/NSF

"to help assure ... that the engineering and construction of the VLA are consistent with the performance goals, and to participate in the further general development of the concept and design of the instrument."[139] The VLA Advisory Committee generally met twice a year throughout the construction project and into the early operations phase. It contributed significantly to setting the final parameters of the VLA as well as addressing various technical problems as they arose. The Committee also drew attention to the lack of sufficient computer power to deal with making images from VLA data.

NRAO's computing difficulties were not confined to the VLA, so in early 1982, Roberts appointed a Computer Advisory Committee to "elicit advice from highly qualified experts in the field."[140] In appointing the Committee members, Roberts noted the "data explosion in the last half decade," and said, "The recent completion of the Very Large Array (VLA) ... particularly dramatize [*sic*] this problem." All but one of the members were computing experts from industry and academia, but some of the VLA Advisory Committee members,[141] such as Burke and Moffet, frequently took part in the meetings, along with relevant NRAO staff.

7.7   TRANSITION TO OPERATIONS

When completed in 1980, the VLA was not only the most powerful radio telescope in the world, it was, not surprisingly, the most complex radio telescope ever built. In recognition of its sophistication and complexity, VLA users needed extensive documentation which was initially provided by a widely used user manual known as *The Green Book* (Hjellming 1978). *The Green Book* was an indispensable reference source to observing with the VLA and included detailed instructions for post-observation data calibration and imaging. It was ultimately replaced by the *Astronomical Image Processing Software* (AIPS) and its associated *Cook Book*.

One of the advantages of array-type radio telescopes is that they are built in steps, and early observations can begin as soon as there is an interesting number of antennas. Also, being able to test many aspects of the final array after only a few antennas were completed and instrumented meant that debugging and the development of observational procedures could begin early. Even the partially completed VLA far exceeded the scientific capability of any other radio telescope, so the user community got a head start on using the VLA.

A key concept of the NRAO plan was, to the extent possible, to transfer project development personnel, especially the scientists and engineers, to operations, and so minimize turnover and exploit the expertise and experience of the development team for operations. When ten antennas became operational in 1978, it was becoming clear that VLA operations needed to be separated from the continuing construction activities, but during the several years of overlap, this meant that operations started more slowly than desired. Richard (Dick) Thompson, who had been the key systems engineer, was placed in charge of the operations phrase. Thompson had begun his career at Jodrell Bank as part of the team that developed radio-linked long baseline interferometry in the late 1950s (Sect. 8.1).

The operation of the VLA, with about 110 employees located in New Mexico, nearly doubled the size of the NRAO staff. The VLA needed a scientist, not an engineer, to coordinate VLA commissioning and operations, and Heeschen asked Cam Wade to serve as the initial Assistant Director for VLA Operations. Dave Heeschen had been the energetic NRAO leader for 18 years, and as the VLA approached completion he decided to step down to return to research and to be able to spend more time with his family. Following a national search, Mort Roberts was named as the NRAO Director, effective 1 October 1978. Roberts and Wade were longtime friends and colleagues, but they had different approaches to management. Things finally came to a breaking point over Wade's supervision of some of the local staff who lived in the small town of Datil about 20 miles west of the VLA site. According to Wade,[142] who was concerned that the local staff were underpaid, as a small gesture, he and Lancaster let them use an NRAO van to travel between their home and the VLA. When he learned of this practice, Roberts instructed them to stop. Incensed, at what he perceived as micromanagement, Wade responded by resigning as VLA Director, but he agreed to stay on until Roberts could find a replacement. However, in defiance of Charlottesville management, Wade informed one of the technicians that he was to be on 24-hour call, so it would be necessary for him to take the van to Datil each night, and that if anyone else wanted to ride with him that would be OK.[143]

Roberts and AUI President Jerry Tape succeeded in recruiting Ronald Ekers to serve as the first permanent director of VLA Operations. Ekers was then a Professor at the University of Groningen in The Netherlands, where he had established himself as an expert in radio interferometry. He had received his PhD in 1967 working at Parkes under John Bolton and Bart Bok, followed by three years at Caltech and then a year at the Cambridge Institute for Astrophysics with Fred Hoyle. Ekers had been a member of the NRAO VLA Advisory Committee and so was familiar with the VLA. Even a year earlier, he and Heeschen had discussed the possibility of Ekers coming to NRAO as the Director of VLA Operations. However, he had just returned from a year's sabbatical in Australia, and was obliged to return to Groningen, so was unable to accept without negotiating a "buy-out" with the University. Roberts was impatient to remove Wade, and after returning from vacation, Wade found that he had been replaced by Peter Napier as the new VLA Acting Director. NRAO and the University of Groningen finally did work out a deal by which Ekers would spend part of his time in Groningen. Starting in October 1980, Ekers served as the VLA Director of Operations until February 1988, during which time he defined the VLA operation style, much of which continues to this day.[144] After leaving NRAO, he became the first director of the Australia Telescope National Facility.

NRAO never received the level of funding planned for the effective operation of the VLA. As early as 1973, Ekers drew attention to the apparent inadequacy of the computer plans.[145] In the early years, this resulted in inadequate computing power to deal with the growing amount of data and the increasingly sophisticated techniques being developed for image processing. Although the power of CLEAN had already been demonstrated by Högbom (2003) in dealing with the poor u,v coverage from the GBI, it wasn't clear if CLEAN would continue to be effective with the initially superior images obtained from the VLA with its better u,v coverage. Many early opponents of the VLA, especially those at Caltech, argued, correctly as it turned out, that the VLA was over-designed to meet the stated goals. However, the use of CLEAN (Högbom 1974, 2003; Schwarz 1978; Clark 1980) and then self-calibration using the closure phase relations (Readhead and Wilkinson 1978; Cotton 1979), resulted in an instrument having very much greater power than originally planned. Moreover, the development of CLEAN, maximum-entropy, and self-calibration enabled the VLA to produce tens or even hundreds of images per day instead of the planned two or three. All of this combined to greatly increase the computing demands. Initially, the available computing power at the site was inadequate to fully exploit these computationally intensive algorithms, and NRAO was criticized for not providing adequate computing power to support the VLA. Heeschen later commented (Tucker and Tucker 1986, p. 31),

The expectations of what people would do with the VLA increased tremendously between the time we first submitted the proposal and the time we eventually began building it. … The techniques that are used today didn't exist at all in the sixties. ... These procedures are extremely valuable, but they increase the data-processing requirements by orders of magnitude. When we first designed the computer, we thought it was adequate for what we then thought the thing would do. But, in the interim, so much happened that we simply couldn't afford.

Heeschen resisted the temptation to ask for the additional funds that would have enabled the earlier full exploitation of the VLA, particularly for spectroscopic observations. In January 1980, in order to keep up with the flow of data, use of the VLA was reduced to 50 percent of the available observing time, and full time observing was not restored until April 1981. Adequate computing power was arguably the single biggest constraint to the VLA scientific productivity, and throughout the early 1980s NRAO struggled with the computing issue. The VLA Advisory Committee repeatedly urged NRAO to acquire more computing hardware and devote more resources to VLA software development. The NRAO Computer Advisory Committee recommended a "long range plan based on astronomical requirement ... flexible and growable computer architecture ... [requiring] a major new infusion of capital from the NSF."[146] Fortunately, however, the rapid development of relatively inexpensive powerful work stations and then personal computers, along with adoption of the user-friendly Astronomical Image Processing System (AIPS) (Greisen 1990, 1998)[147] and other on-line documentation,[148] mitigated the computing bottleneck and, ultimately, greatly contributed to the success of the VLA. Combined with the increasing availability of powerful, yet inexpensive work stations, AIPS made it possible for the users to reduce their VLA data at their home institution, thus relieving NRAO of the need to provide a major computing center.

Initially, computing resources were so limited that users were normally restricted to making images no larger than 256 x 256 pixels, and needed to specifically request permission if they wanted to make larger images. Although VLA image processing was initially restricted by the limited computing power available, if NRAO had instead chosen what many argued was a more cost effective optical image processing, it would have excluded the use of CLEAN and self-calibration, which would probably have restricted the VLA imaging capability to that proposed in 1967.

As mentioned, with the growing use of inexpensive powerful computers, the data processing was no longer an issue, but with time the VLA hardware became outdated. Other components of the VLA such as the power cables and surplus railroad ties were deteriorating at an alarming rate. Because the A-configuration makes up 70 percent of the VLA's baseline, the high cost of replacing so many railway ties and power cables threatened the continued operation in A-configuration.[149] Between 1984 and 1986, NRAO added simple uncooled 327 MHz (92 cm) receivers to the prime focus. But until the 2001 EVLA project (Sect. 7.8), the only major VLA improvement came about from an agreement with NASA to provide observing time on the VLA to download images from the Voyager spacecraft at the time of its encounter with Neptune in August 1989. To optimize the VLA sensitivity, NASA provided funds for NRAO to construct state-of-the-art receivers at 8.4 GHz, which for a long time remained the most sensitive VLA receiving band. Since support of the Voyager encounter required a higher level of reliability than normal radio astronomy observations, NRAO convinced NASA to provide sufficient funding to also replace the backup power generators, deteriorating railway ties, and power cables supporting the inner configurations.

As discussed in Chap. 8, the decision to co-locate the VLBA Operations Center in Socorro together with the VLA Operations necessitated moving many of the VLA activities to a new Array Operations Center on the New Mexico Tech Socorro campus, which opened in December 1987.

Impact of the VLA The VLA was not only completed on schedule and within budget, but in nearly every respect—sensitivity, resolution, image quality, speed, number of frequency bands, spectroscopic capability, field-of-view—it far exceeded its design goals. As described by Heeschen (1981), the VLA

> was motivated by a clearly perceived need, in the early 1960s, for an image forming instrument of the greatest feasible resolution, sensitivity and general versatility. While its designers were most strongly influenced by the opportunities and problems presented by extragalactic radio sources, the need for such an instrument was apparent in all areas of radio astronomy, and the VLA was in fact designed to be used for almost all kinds of radio astronomy studies.

He went on to point out that

> The VLA has 10 to 100 times greater resolution and sensitivity than any other existing radio telescope, and its resolution is comparable to or greater than that obtainable at optical and other wavelengths. Its speed, sky cover, ability to measure polarization, and ability to make high frequency, high resolution spectroscopic observations give it tremendous power and versatility for a wide variety of problems.

Perhaps the biggest change brought about by the VLA was that, prior to the VLA, radio astronomy was pretty much a "black belt" experience confined to those with training or experience in radio astronomy. NRAO users were expected to do their own calibration and analysis. Student use generally required attentive support from a faculty or in some cases an NRAO staff advisor. The power and complexity of the VLA led NRAO to provide more hands-on support for users, which, in turn, began to attract a broader group of astronomers who needed radio data to enhance their research program.[150]

To support the growing user community, and especially to train the new generation of scientists who would use the VLA, NRAO started the very successful series of synthesis imaging workshops. The first workshop held in Socorro in June 1982 was attended by 85 scientists and students (Thompson and D'Addario 1982) and was followed by similar workshops held every two or three years.

An unanticipated use of the VLA developed when NRAO received two separate proposals to use the VLA for sky surveys intended to detect and catalogue an unprecedented number of discrete radio sources. One proposal was from an internal group led by Jim Condon, who proposed an "all-sky" survey with a resolution of the order of an arcmin to detect the nearly two million radio sources stronger than 2.5 mJy. A competing proposal came from a group led by Robert Becker from the University of California, Davis, proposing a deeper, higher resolution survey that would reach sources as faint as 1 mJy, but only covered the limited area of the sky corresponding to the Sloan Digital Sky Survey (SDSS) (York et al. 2000). With the higher angular resolution, the proposed Becker survey would give more accurate source positions needed for optical identification with SDSS counterparts, as well as imaging the arcsec structure of detected radio sources. However, unlike the NRAO survey, it would be insensitive to larger scale radio emission. The two proposed surveys were not only in competition, but each wanted thousands of hours observing time. The NRAO Director, Paul Vanden Bout, polled the user community, who were mostly supportive of the two proposals, so he appointed a small internal committee chaired by Frazer Owen to make a recommendation on choosing which proposal to approve. Recognizing the complementarity of the two projects, Owen's committee recommended that they both be approved. But once the observing began, other users, including those who supported the idea of big projects, complained that they were taking up too much observing time.

The two projects were each completed, but to minimize the impact to other observers, each was stretched out over a number of years. Both the Condon et al. (1998) NRAO VLA Sky Survey (NVSS) and the Becker et al. (1995) Faint Images of the Radio Sky at Twenty Centimeter (FIRST) were enormously successful resources for the astronomy community, each receiving thousands of citations. To help resolve any future quandaries between the merits of large proposals and their impact to other programs, Vanden Bout convened a "Large Proposal Review Committee" chaired by NRAO staff member Alan Bridle.

The VLA has also helped to propel radio astronomy into the popular media, appearing in several movies and numerous TV ads. The popular movie *Contact* brought particular attention to the VLA, although it gave a misleading impression that the VLA is used to search for signals from extraterrestrial intelligent civilizations. Images of the VLA appear frequently in TV advertisements, although mostly for products and services unrelated to radio astronomy.

The impact of the VLA was not all positive. Faced with limited operating budgets, the NSF could not support the expensive VLA operations at the same time as the many more modest university-operated facilities. Funding was cut for the Owens Valley Observatory interferometer, where many of the techniques used at the VLA had been developed and where many of NRAO staff that built the VLA were trained. Only modest support remained for operating the OVRO 130 foot antennas as an element of the US VLBI Network (Chap. 8). The five-element interferometer built by Ron Bracewell at Stanford University with NSF funds was closed soon after it was completed. A novel low frequency array built by Bill Erickson in the Southern California desert lost most of its NSF support. American radio astronomers and their students were more and more driven to the national observatory facilities operated by NRAO and NAIC. The reduction of the once vibrant university radio astronomy groups restricted the training of the next generation of technically-skilled observers, further increasing the pressure on NRAO to provide a turn-key observing opportunity at all of its facilities.

7.8   The Karl G. Jansky Very Large Array (JVLA)

The VLA has arguably been the most productive ground-based telescope ever built, certainly the most productive radio telescope (Trimble and Ceja 2008), and has more than lived up to Dave Heeschen and NRAO's expectations. Although conceived primarily for extragalactic continuum research, the VLA has been used to study essentially all fields of astronomy from the Sun and Solar System bodies to a wide range of galactic and extragalactic phenomena. A particular surprise has been the large fraction of time spent studying the radio emission from stars, an area not even mentioned in the 1965 VLA Report No. 1. Spectroscopic studies, which were mentioned only in passing in the 1967 proposal, have typically represented about one-third of the VLA observing time. As noted by Trimble and Zaich (2006),

> The VLA [is] responsible for 22% of the papers and 27 percent of the citations [in radio astronomy]. The VLA is, therefore, proportionally even more influential in world radio astronomy than HST is in world optical astronomy.

When built in the 1970s, the VLA receivers, waveguide transmission, and digital correlator were all state-of-the-art. However, the limited funding made available for VLA operations, combined with the rapid advances in technology, meant that only two decades after its dedication the VLA instrumentation was becoming woefully obsolete. Moreover, the great changes in astronomy over the last few decades of the twentieth century led to new demands for better sensitivity and image quality as well as improved spectral and angular resolution.[151] Responding to these forces, in May 2000 NRAO/AUI submitted a proposal to the NSF to increase the sensitivity of the VLA by up to an order of magnitude, provide better frequency coverage, and greatly improved spectroscopic capability.[152] NRAO intended this to be just the first phase of the VLA upgrade, but the Phase II proposal[153] to "expand" the VLA by adding additional antennas, was never funded, although the name "Expanded Very Large Array" (EVLA) stuck, at least throughout the construction period. Following review by the NSF, the EVLA Phase I project was partially funded through an increment to the NRAO operating budget rather than through the Major Research Equipment and Facilities Construction (MREFC) budget, but about half of the $96 million cost of the EVLA was born by the base VLA operating budget. Additional funding came from in-kind contributions from the Consejo Nacional de Ciencia y Technologia (CONACyT)[154] of Mexico, and from the Canadian National Research Council for the WIDAR (Wide-band Interferometer Digital ARchitecture) correlator. The Memorandum of Understanding (MOU) with Canada was far reaching and established the North American Program in Radio Astronomy (NAPRA) which, in return for the Canadian WIDAR correlator, gave Canadian scientists the same access as US scientists to all NRAO facilities. The NAPRA agreement included Canadian participation in ALMA as well as joint NRAO-Canadian efforts toward the SKA development.[155]

Upgrading the VLA instrumentation began in 2001 and was completed a decade later. A particularly challenging aspect was the stated goal of keeping the VLA in operation throughout the decade-long EVLA construction period, which meant simultaneously operating combinations of old and upgraded antennas with first the old and then the new correlator. Except for a seven-week period in 2010 when the old VLA correlator was replaced by the new one, the VLA remained in operation during the entire 10 years of construction. The upgraded VLA, which was dedicated on 31 March 2012, was given a new name, the Karl G. Jansky Very Large Array (JVLA), replacing the term "EVLA," which has since been redefined to refer to the construction project and not the instrument (Fig. 7.13).

The JVLA has complete frequency coverage across the entire band from 1 to 50 GHz provided by eight feeds and receivers at the secondary focus. A ninth receiving band at the prime focus covers the range from 50 to 450 MHz with reduced sensitivity. The front end systems are based on cooled HEMT amplifiers and corrugated feed horns, while the local oscillator is derived from a hydrogen maser frequency standard, and distributed via fiber to each of the 27 antennas. The IF signals are digitized at the antenna and sent by fiber to the new WIDAR correlator, designed by Brent Carlson from Canada. The WIDAR combines the data on each of the 351 interferometer baselines at up to 10^{15} 32 bit operations per second to give up to four million frequency channels, or, for continuum studies, up to 8 GHz total bandwidth in each of two polarizations. More detailed descriptions of the JVLA are given by Perley et al. (2009, 2011). The basic performance parameters of the JVLA are shown in Table 7.1 compared with the 1967 proposal, what was achieved in 1980 at the end of VLA construction, and in 1986 after various hardware and software upgrades.

The JVLA reaches unprecedented levels of sensitivity extending to below 1 µJy in a 12 hour integration. But with the improved sensitivity and larger bandwidths came the need for better interference suppression and improved dynamic range in order to reach the thermal sensitivity limits of the array in long integrations. New imaging algorithms and new software needed to meet these challenges was implemented both within AIPS (Greisen 1990) as well as

Fig. 7.13   Karl G. Jansky Very Large Array rededication, 31 March 2012. Center: The JVLA in D array. Top: Attendees at the rededication ceremony. Lower right: NRAO Director Fred Lo (standing) initiates the start of first official Jansky VLA observation. Seated behind Lo are Ethan Schreier, AUI President, and James Ulvestad, NSF Astronomical Sciences Division Director. Lower left: Anne Moreau Jansky Parsons, daughter of Karl Jansky. Credit: D. Fiinley/NRAO/AUI/NSF

Table 7.1   VLA performance

Parameter	1967	1980	2000	2011 JVLA
Resolution (arcsec)	1	0.1	0.04	0.04
Sensitivity (μJy)	300	50	10	1.5
Dynamic Range	100	1000	100,000	200,000
Field of View (arcmin)	1 to 10	1 to 30	1 to 300	1 to 300
No. λ Bands	2	4	6	1–40 GHz continuous
Shortest Wavelength (cm)	11	1.3	1.3	0.7
Declination Range	−20° to +90°	−40° to +90°	−40° to +90°	−40° to +90°
Speed (images/day)	3	100	200	20,000
No. Freq. channels	none	256	512	4×10^6
Map Size (pixels)	100×100	512×512	4096×4096	4096×4096

Notes: The first three columns are adopted from Ekers' 1987 report on the First Seven Years of VLA Operations (Ekers 1987, op. cit.) and the final column from Perley et al. (2011). The values given for the sensitivity represent the 3σ detection level from a 12 hour synthesis image. The specification of the dynamic range is a bit subjective as it depends in part on the complexity of the field being imaged. The number of images per day given for the JVLA is based on the number of independent fields observed in the NRAO VLA Sky Survey using a scan rate of 3.31 arcmin/sec

CASA (Common Astronomical Software Application) (McMullin et al. 2007). Although data analysis, especially from the first few years of the JVLA, was rather labor intensive, the publication of a special issue of the *Astrophysical Journal Letters* in 2011 (Vol. 739), demonstrated the already wide range of possible research and the enormous potential of the JVLA for new discoveries.[156]

NOTES

1. Detailed reviews of synthesis imaging in radio astronomy are given by Taylor et al. (1999), Kellermann and Moran (2001), Thompson et al. (2017), and in references contained therein.

2. The recent implementation of adaptive optics has enabled some ground based telescopes to achieve diffraction limited imaging at near infrared wavelengths.

3. The so-called "sea interferometer" combined the signal received from reflection by the ocean with the directly received signal to form a two element interferometer, although only one antenna was used.

4. The authors do not elaborate, but presumably they are talking about a conventional 2-element interferometer. Apparently Pawsey had been the first author on this paper, but the author list was rearranged to be alphabetical.

5. Earlier, as part of Cambridge PhD work, Patrick O'Brien and Peter Scheuer (Scheuer 1984) and Högbom (1959, 2003) used Earth rotation synthesis to image the Sun in 1953 and 1958 respectively.

6. Swenson, G.W. Jr. 1977, A Case Study of the Decision to Construct a Large Radio Telescope, (Washington: National Academy of Sciences Joint Working Group on National Systems for the Stimulation of Fundamental Research), NAA-NRAO, NM Operations, VLA Design and Construction.

7. The design and construction of the Westerbork Synthesis Radio Telescope, including the earlier considerations of the Benelux Cross, are discussed by Raimond (1996).

8. Charles Seeger was the brother of the folk singer Pete Seeger.

9. In all radio arrays, data corresponding to interferometer spacings comparable to and smaller than the antenna diameter is missing and this results in the inability to observe structures whose angular scale is comparable with the primary beam of the individual elements.

10. Each of the redundant spacings from different antenna pairs sampled the same Fourier component of the sky, so any differences in the measured interferometer response were due to instrumental, tropospheric, or ionospheric effects.

11. Detailed technical descriptions of the WSRT are given by Baars et al. (1973), by Baars and Hooghoudt (1974), and by Högbom and Brouw (1974).

12. The other members of the panel included Ronald Bracewell (Stanford), Bernard Burke (Carnegie Institution of Washington), P. Chena (Purdue), David Heeschen (NRAO), Rudolph Minkowski (University of California Berkeley), L.J. Chu (MIT), Richard Emberson (AUI), William Gordon (Cornell), George Swenson (University of Illinois), and J.H. Trexler (Naval Research Laboratory). John Bolton, representing Caltech, served as a consultant to the Panel, but returned to Australia before the work of the Panel was

completed. NSF Astronomy Director Geoffrey Keller (1961) served as liaison between the Panel and the NSF and authored the report.

13. RH interview with CMW, 29 December 2003. https://science.nrao.edu/about/publications/open-skies#section-7

14. DSH Memo to Findlay, Drake, Hvatum, Wade, Hogg, Vinokur, Callendar, 5 March 1962, NAA-NRAO, Founding and Organization, Antenna Planning. https://science.nrao.edu/about/publications/open-skies#section-7

15. Ibid.

16. Ibid.

17. J. Pawsey, Notes of Future Program at Green Bank, 17 July 1962, NAA-NRAO, Founding and Organization, Antenna Planning. https://science.nrao.edu/about/publications/open-skies#section-7

18. Pawsey sent copies of his letter to Heeschen, Tape, and Rabi, with a note to Heeschen that "it is highly provisional and could well be modified after further discussion with you people in Green Bank." Pawsey to Erickson, 27 June 1962, NAA-NRAO, Founding and Organization, Antenna Planning. https://science.nrao.edu/about/publications/open-skies#section-7

19. Pawsey to Scheuer, 4 July 1962, NAA-NRAO, Founding and Organization, Antenna Planning. https://science.nrao.edu/about/publications/open-skies#section-7

20. Erickson to Pawsey, 28 August 1962, NAA-NRAO, Founding and Organization, Antenna Planning. https://science.nrao.edu/about/publications/open-skies#section-7

21. DSH to J. Findlay, Ivan Pauliny-Toth, M. Vinokur, C. Wade, and V. Venugopal, 7 September 1962, NAA-NRAO, NM Operations, VLA, Design and Construction, Box 1.

22. AUI-BOT, 18 January 1963.

23. Request for a Proposal for Antenna Design Study, 8 November 1962, NAA-NRAO, Founding & Organization Series, Antenna Planning.

24. KIK interview with CMW, 21 March 2015, NAA-KIK, Oral Interviews. The VLA did not use fiber optic links until the EVLA upgrade 40 years later. https://science.nrao.edu/about/publications/open-skies#section-7

25. CMW, November 1963, NAA-NRAO NM Operations, VLA, Design and Construction. https://science.nrao.edu/about/publications/open-skies#section-7

26. RH interview with CMW, op. cit.

27. KIK interview with DSH, 31 May 2011, NAA-KIK, Oral Interviews. https://science.nrao.edu/about/publications/open-skies#section-7

28. AUI-BOTXC, 15 February 1963.

29. AUI-BOTXC, 18 February 1966.

30. Caltech Proposal to the NSF, 1962, NAA-KIK, Open Skies.

31. Ironically, a few years later the Caltech 90 foot dishes were converted to alt-az mounts.

32. Vinokur to DSH and JWF, 30 October 1962, NAA-NRAO, Founding and Organization, Antenna Planning.

33. Other committee members included radio astronomers Ronald Bracewell, Frank Drake, and Fred Haddock along with optical astronomers William Liller, W.W. Morgan, and Allan Sandage. Whitford, A. 1964, Ground-based

Astronomy: A Ten Year Program, National Academy of Sciences, Washington, hereafter referred to as the Whitford Committee.

34. Summary Minutes of the Ad Hoc Meeting on Radio Astronomy, November 1 and 2, 1963, ANAS, Central File, COSPUP, Astronomical Facilities.
35. Ibid.
36. Ibid.
37. During this period the Consumer Price Index increased by a factor of 2.4.
38. AUI-BOTXC, 16 September 1965.
39. The main contributors to the design report were L.C. Chow, B.G. Clark, D.S. Heeschen, D.E. Hogg, H. Hvatum, G.W. Swenson, W.C. Tyler, and C.M. Wade. From time to time, a few external people, including Frank Drake and Bernie Burke, participated in the monthly design meetings.
40. AUI-BOT, 20 January 1966.
41. AUI-BOTXC, 17 August 1967.
42. CMW to DSH and JWF, 3 November 1962, NAA-NRAO, Founding and Organization, Antenna Planning.
43. AUI-BOT, 18 January 1963.
44. AUI-BOTXC, 16–17 July 1964.
45. NRAO, November 1969, GBI Memo No. 25, The NRAO Interferometer, A Development Program. http://library.nrao.edu/public/memos/gbi/GBI_025.pdf
46. CMW, November 1964, GBI Memo No. 4, A Method for Finding the Phase and Amplitude of Interferometer Fringe Patterns. http://library.nrao.edu/public/memos/gbi/GBI_004.pdf
47. BGC, December 1964, GBI Memo No. 7, On the Reduction of Digital Interferometer Records. http://library.nrao.edu/public/memos/gbi/GBI_007.pdf
48. Clark, B.G. and Wade, C.M., April 1965, GBI Memo No. 14, The Interferometer Fringe Reduction Program. http://library.nrao.edu/public/memos/gbi/GBI_014.pdf
49. AUI-BOT, 20 January 1966.
50. BGC, July 1968, GBI Memo No. 23, Programs for the DDP-116 Computer. http://library.nrao.edu/public/memos/gbi/GBI_023.pdf
51. NRAO, November 1969, The NRAO Interferometer, A Development Program, op. cit.
52. Caltech Proposal to the NSF, April 1966, NAA-KIK, Open Skies.
53. CITA-GJS, unprocessed papers.
54. Ibid.
55. Other committee members were Bart Bok (University of Arizona), Stirling Colgate (New Mexico Institute of Mining and Technology), Rudolph Kompfner (Bell Laboratories), William Morgan (Yerkes Observatory), Eugene Parker (University of Chicago), Merle Tuve (Carnegie Institution Department of Terrestrial Magnetism).
56. https://science.nrao.edu/about/publications/open-skies#section-7
57. KIK interview with CMW, 21 March 2015, op. cit.
58. The Owens Valley Array – 1969, A Report to the Second Session of the Dicke Committee on Facilities for Radio Astronomy, NAA-KIK, Open Skies.
59. AUI-BOT, 18 January 1968.

60. KIK interview with CMW, 21 March 2015, op. cit.
61. The Panel membership was retained from the first meeting, although Merle Tuve was unable to attend the second meeting of the Panel. https://science.nrao.edu/about/publications/open-skies#section-7
62. GJS to Anderson, R. Bacher, and L. Bonner, 20 November 1968a, CITA-GJS.
63. GJS to Anderson, R. Bacher, and L. Bonner, 20 November 1968b, CITA-GJS.
64. GJS to R. Bacher, 3 April 1969, CITA-GJS.
65. KIK interview with DEH, 26 March 2013, NAA-KIK, Oral Interviews.
66. Tape to J. Bergen, 12 January 1970, NAA-NRAO, New Mexico Operations, VLA, Design and Construction.
67. J. Bergen to Tape, 17 February 1970, NAA-NRAO, New Mexico Operations, VLA, Design and Construction.
68. Bureau of the Budget (BOB), Interagency Continuing Special Study: Federally Supported Astronomy, undated. NAS-NRC-A, Central File: Divisions of the NRC: Physical Sciences: Astronomy Survey Committee: 1970. The BOB prospectus contained six major astronomy projects, two of which, "a large radio astronomy array" and "a large steerable antenna," were for radio astronomy.
69. L. Haworth to H. Brooks, 22 January 1969, NAS-NRC-A, Central File, ADM: Committees and Boards: COSPUP Panels: Astronomical Facilities: Report: General: 1965. Brooks was the Chairman of the NAS Committee on Science and Public Policy (COSPUP).
70. Proposal for Support of the Study of Priorities in Astronomy June 1, 1969–September 30, 1971, NAS-NRC-A, Central File, ADM: Committees and Boards: COSPUP Panels: Astronomical Facilities: Report: General: 1965. The Committee on Science and Public Policy (COSPUP) was renamed in 1982 as the Committee on Science, Engineering, and Public Policy (COSEPUP), and in 2016 as the Committee on Science, Engineering, Medicine, and Public Policy (COSEMPUP).
71. Haworth to Brooks, 27 May 1969, NAS-NRC-A, Central File, ADM: Committees and Boards: COSPUP Panels: Astronomical Facilities: Report: General: 1965.
72. Greenstein has related his role in early radio astronomy in an unpublished memoir, *I Was There in the Early Years of Radio Astronomy*, NAA-KIK, Open Skies.
73. Greenstein to Brooks, 7 July 1969, NAS-NRC-A, Central File, ADM: Committees and Boards: COSPUP Panels: Astronomical Facilities: Report: General: 1965.
74. The other members of the radio panel were Geoffrey Burbidge (University of California San Diego), George Field (University of California Berkeley), Gordon Pettengill (MIT), James Warwick (University of Colorado), and Gart Westerhout (University of Maryland). All were radio astronomers except Field and Burbidge, who were theoreticians, although Field had had some experience as a radio observer at NRAO. Burbidge was also a member of the joint Astronomy/Physics Panel on Relativity and Astrophysics.
75. Greenstein to Radio Panel, 14 October 1969, NAA-NRAO, Director's Office, Professional Organizations/Committees, NAS Committees.

76. Minutes of the Radio Panel Meeting, 14 November 1969, NAA-NRAO, Director's Office, Professional Organizations/Committees, NAS Committees.
77. Report of the Radio Astronomy Panel of the Astronomy Survey Committee, 1 December 1969, NAA-NRAO, Director's Office, Professional Organizations/ Committees, NAS Committees.
78. Handler to McElroy, 16 December 1969, NAS-NRC-A, Central File: Divisions of the NRC: Physical Sciences: Astronomy Survey Committee: 1970.
79. President Nixon had appointed Caltech President Lee DuBridge as his Science Advisor and Chair of the Presidents Science Advisory Committee (PSAC).
80. Handler to JLG, 16 December 1969, NAS-NRC-A, Central File: Divisions of the NRC: Physical Sciences: Astronomy Survey Committee: 1970.
81. JLG to Handler, 22 December 1969, NAS-NRC-A, Central File: Divisions of the NRC: Physical Sciences: Astronomy Survey Committee: 1970.
82. DSH to JLG, 10 March 1970, NAA-NRAO, Director's Office, Professional Organizations/Committees, NAS Committees.
83. Handler to JLG, 20 November 1969, NAS-NRC-A, Central File: Divisions of the NRC: Physical Sciences: Astronomy Survey Committee: 1970.
84. JLG to DSH, 31 March 1970, NAA-NRAO, Director's Office, Professional Organizations/Committees, NAS Committees.
85. JLG to Handler, 1 June 1970, NAS-NRC-A, Central File: Divisions of the NRC: Physical Sciences: Astronomy Survey Committee: 1970.
86. DSH to JLG, 23 April 1970, NAA-NRAO, Director's Office, Professional Organizations/Committees, NAS Committees, Astronomy Study; CITA-JLG, Box 96, Folder 12.
87. Brooks to JLG, 16 March 1970, NAA-NRAO, Director's Office, Professional Organizations/Committees, NAS Committees.
88. Minutes of the Third Meeting of the Radio Panel, 9–10 April 1970, NAA-NRAO, Director's Office, Professional Organizations/Committees, NAS Committees.
89. They were Barry Clark, Ed Fomalont, and one of the present authors (KIK). MHC to DSH, 2 October 1970, CITA-GJS.
90. Brooks to Handler, 3 September 1970, NAS-NRC-A, Central File: Divisions of the NRC: Physical Sciences: Astronomy Survey Committee: 1970.
91. Handler to Brooks, Dicke, Greenstein, and Whitford, 15 December 1970, NAS-NRC-A, Central File: Divisions of the NRC: Physical Sciences: Astronomy Survey Committee: 1970.
92. McElroy to Handler, 30 October 1970, NAA-NRAO, Director's Office, Professional Organizations/Committees, NAS Committees; NAS-NRC-A, Central File: Divisions of the NRC: Physical Sciences: Astronomy Survey Committee: 1970.
93. Brooks to Handler, 21 December 1970, NAS-NRC-A, Central File: Divisions of the NRC: Physical Sciences: Astronomy Survey Committee: 1970.
94. In 1970, President Nixon replaced the Bureau of the Budget (BoB) with the Office of Management and Budget (OMB).
95. Brooks to Handler, 21 December 1970, op. cit.
96. Minutes of the NSB, 2–4 September 1970, NSB-70-265.
97. Greenstein was an exception. In 1969 at a meeting of the National Research Council Division of Physical Sciences, Greenstein emphasized the need for building bigger optical telescopes to address the big problems.

98. KIK Interview with DSH, 31 May 2011, op. cit.

99. JLG, unpublished memoir, *I Was There in the Early Years of Radio Astronomy,* op. cit.

100. Ibid.

101. KIK interview with JLG, January 1995.

102. Cohen to DSH, 23 April 1971, NAA-NRAO, Director's Office, Professional Organizations/Committees, NAS Committees.

103. David to McElroy, 12 April 1971, NAA-NRAO, Director's Office, Professional Organizations/Committees, NAS Committees.

104. KIK interview with BFB, 20 June 2012, NAA-KIK, Oral Interviews. https://science.nrao.edu/about/publications/open-skies#section-7

105. GJS to H. Brown, R. Bacher, C. Anderson, JLG, 11 May 1970, CITA-GJS.

106. GJS, et al. 1969, The Owens Valley Array – 1969; A Report to the Second Session of the Dicke Committee on Facilities for Radio Astronomy, NAA-KIK, Open Skies.

107. RH interview with CMW, op. cit.

108. CMW to KIK, 22 June 2018 telephone interview.

109. Limbaugh Engineers, Inc. studied two sites in New Mexico, two in Nevada, and one each in Arizona, Texas, and Utah. They submitted an interim report in August 1966 and a more detailed final report in November 1966.

110. RH interview with CMW, op. cit.

111. DSH to T. Owen, 7 February 1972, NAA-NRAO, NM Operations, VLA Site Selection, Procurement, and Development.

112. Very Large Array (VLA) Site Selection Review Panel: NAS-NRC-A, Central File: Divisions of the NRC: Physical Sciences: 1972.

113. The circular waveguide used the so-called TE_{01} transmission mode which has very low loss, but only if bends are avoided.

114. NRAO VLA Electronics Memos No. 105, 120, 128, 145, 146, 150–155, 179. http://library.nrao.edu/vlae.shtml

115. To maintain coherence, the signals being correlated need to be synchronized to an accuracy of the order of 1/bandwidth.

116. The VLA correlator allowed various combinations of frequency resolution and bandwidth ranging from 16 channels covering 50 MHz bandwidth (3.125 frequency resolution) to 256 channels covering 97 kHz (381 Hz resolution).

117. VLA Electronics Memo No. 207, The Correlator System for the Very Large Array, C. Broadwell and R. Escoffier, July 1982. http://library.nrao.edu/VLAE_207.shtml

118. See NRAO VLA Computing Memo No.122. http://library.nrao.edu/public/memos/vla/comp/VLAC_122.pdf

119. An initial software effort called CANDID led by Bob Hjellming was an almost complete failure. Sadly, Bob Hjellming died in 2000 after suffering a heart attack during a diving training class.

120. R. Duquet, NRAO VLA Computer Memorandum 172. http://library.nrao.edu/public/memos/vla/comp/VLAC_172.pdf. For more detailed information on the history of the implementation of the VLA computing hardware and software systems see the series of VLA computing memoranda at http://library.nrao.edu/vlac.shtml

121. RH interview with CMW, op. cit.

122. The tortuous path of the VLA funding bill through Congress was described in the *The Money Maze: How a Group of People Wanting a Big Telescope Fared in Washington, Wall Street Journal*, 1 November 1973, and by Gloria Lubkin (1975).
123. Ibid.
124. Ibid.
125. Stanford Research Institute, VLA Feasibility Study Final Report, February 1972 (pts. 1–2) and March 1972 (pt. 3), NRAO-NAA, New Mexico Operations, VLA, Design and Construction.
126. This section is based in part on the detailed report by Jack Lancaster (1982) on construction of the VLA, NAA-NRAO, NM Operations, VLA Operations. https://science.nrao.edu/about/publications/open-skies#section-7
127. NRAO RFP-VLA-01 *Antenna Procurement Contract Selection Committee – Final Report*, 30 July 1973, NAA-NRAO, NM Operations, VLA, Design and Construction.
128. The final price of the 28 antennas was $17,755,258. The Antenna Assembly Building cost $400,796. NAA-NRAO, NM Operations, VLA, Design and Construction
129. The hearing was held on 30 September 1975. NAA-NRAO, NM Operations, VLA, Design and Construction.
130. House of Representatives Committee on Science and Technology Authorizing Appropriation to the National Science Foundation, HR-94-930, 18 March 1976, NAA-NRAO, NM Operations, VLA, Design and Construction.
131. A.R. Thompson to DSH, DEH, Hvatum, and Lancaster, 31 December 1975, NAA-NRAO, NM Operations, VLA, Design and Construction.
132. A.R. Thompson to DSH, DEH, H. Hvatum, and J. Lancaster, 17 February 1976, NAA-NRAO, NM Operations, VLA, Design and Construction.
133. KIK interview with WEH, 20 September 2011, NAA-KIK, Oral Interviews. https://science.nrao.edu/about/publications/open-skies#section-7
134. Memo from Claude Kellett, NSF Program Officer for NRAO, to file, NAA-NRAO, NM Operations, VLA, Design and Construction.
135. NAA-NRAO, NM Operations, VLA, Design and Construction. This hearing was also attended by Robert Hughes, the NSF Assistant Director for Mathematical and Physical Sciences. Hughes later became the President of AUI, serving from 1980 to 1997.
136. Report of the Ad Hoc Advisory Panel for the Very Large Array (VLA), National Science Foundation, 31 December 1977, NAA-NRAO, NM Operations, VLA Advisory Committees, Box 2.
137. Only with the commencement of JVLA operations and the development of the VLA pipeline has it become possible to obtain a basic publishable image without a lot of labor-intensive data processing.
138. Report on the Audit Very Large Array Radio Telescope Project, NSF Office of Planning and Resources Management, Audit Office, 4 May 1977, NAA-NRAO, NM Operations, VLA, VLA Budget.
139. Initial members of the VLA Advisory Committee included B. Burke (MIT), J. Douglas (Univ. of Texas), F. Drake (Cornell), R. Ekers (Groningen, Netherlands), C. Heiles (Univ. Calif. Berkeley), M. Kundu (Univ. Maryland),

A. Moffet (Caltech), G. Swenson (Univ. Illinois), A. Rogers (MIT, Haystack Observatory), R. Wilson (Bell Labs). NAA-NRAO, NM Operations, VLA Advisory Committees, Box 1.

140. Letters from MSR to VLA Advisory Committee, 18 September 1981, NAA-NRAO, NM Operations, VLA Advisory Committees, Box 2.

141. Members of the VLA Computer Advisory Committee were A. Brenner (Fermi National Laboratory), W. Brouw (Westerbork), P. Green (IBM), K. King (Cornell), H. S. McDonald (Bell Labs), T. Paganelli (General Electric), G.S. Patterson (Cray Laboratories), S.F. Reddaway (IBM), and H. Schorr (IBM).

142. KIK interview with CMW, 21 March 2015, op. cit.

143. Ibid.

144. See Ekers, R. 1987, The First Seven Years of VLA Operations – VLA 1980-1987, NAA-NRAO, NM Operations, VLA Operations. https://science.nrao.edu/about/publications/open-skies#section-7

145. A. Moffet to DSH, 28 May 1973, NAA-NRAO, NM Operations, VLA Advisory Committees.

146. Ekers, 1987, op. cit.

147. See also AIPS Memos 61 and 87. http://library.nrao.edu/public/memos/aips/memos/AIPSM_061.pdf, http://library.nrao.edu/public/memos/aips/memos/AIPSM_087.pdf

148. VLA Observational Status Reports. http://library.nrao.edu/obsstat.shtml

149. While the use of used surplus railway ties allowed NRAO to finish the VLA within the planned budget, the later operating cost of replacing the ties was greater than the initial cost saving.

150. Even earlier the WSRT provided complete hands off images for the Dutch astronomers.

151. The VLA Development Plan, Proceedings of a Science Workshop Held in Socorro, NM, 13–15 January 1995, NAA-NRAO, NM Operations, EVLA, Box 1.

152. The VLA Expansion Project, Phase I – The Ultrasensitive Array, Proposal submitted to the NSF, May 2000, NAA-NRAO, NM Operations, EVLA.

153. The Expanded VLA, Phase II, Resolving Cosmic Evolution, Completing the EVLA, 15 April 2004, NAA-NRAO, NM Operations, EVLA, Box 2.

154. J. Martuscelli to P. Vanden Bout, 4 April 2000, NAA-NRAO, NM Operations, EVLA, Box 1. CONACyT pledged to contribute 20 million Mexican Pesos (about USD 1 million).

155. MOU between the Hertzberg Institute of Astrophysics and the NRAO to establish a North American Program in Radio Astronomy, September 2001. NAA-NRAO, NM Operations, EVLA, Box 1. The value of the WIDAR correlator was estimated to be about USD 20 million.

156. *ApJL,* 20 September 2011, **739.**

157. Draft of Heeschen's AAS talk is in NAA-DSH, US Radio Astronomy, History, Talks, Folder 2.

BIBLIOGRAPHY

REFERENCES

Baars, J.W.M. et al. 1973, The Synthesis Radio Telescope at Westerbork, *Proc IEEE*, **61**, 1258

Baars, J.W.M. and Hooghoudt, B.G. 1974, The Synthesis Radio Telescope at Westerbork. General Lay-out and Mechanical Aspects, *A&A*, **31**, 323

Balick, B., Bignell, C., and Hjellming, R.M. 1977, VLA Radio Maps of Four Planetary Nebulae, *BAAS*, **9**, 601

Basart, J.F., Miley, G.K., and Clark, B.G. 1970, Phase Measurements with an Interferometer Baseline of 11.3 km, *IEEE Trans AP*, **18**, 375

Becker, R.H. et al. 1995, The First Survey: Faint Images of the Radio Sky at Twenty Centimeters, *ApJ*, **112**, 407 http://sundog.stsci.edu/top.html

Beckett, P.H. 1980, *Dept. of Sociology and Anthropology Report 357* (Las Cruces: New Mexico State University)

Christiansen, W.N. and Warburton, J.A. 1955, The Distribution of Radio Brightness over the Solar Disk at a Wavelength of 21 cm. III. The Quiet Sun. Two Dimensional Observations, *Aust. J. Phys.*, **8**, 474

Christiansen, W.N. and Högbom, J.A. 1961, The Cross Antenna of the Proposed Benelux Radio Telescope, *Nature*, **191**, 215

Christiansen, W.N., Erickson, W.C., and Högbom, J.A. 1963a, Modification of the Design of the Proposed Benelux Radio Telescope, *Nature*, **197**, 940

Christiansen, W.N., Erickson, W.C., and Högbom, J.A. 1963b, The Benelux Cross Antenna Project, *Proc. IRE Aust.*, **24**, 219

Clark, B.G. 1980, An Efficient Implementation of the Algorithm 'CLEAN', *A&A*, **89**, 377

Condon, J. J. et al. 1998, The NRAO VLA Sky Survey, *AJ*, **115**, 1693

Cotton, W.D. 1979, A Method of Mapping Compact Structure in Radio Sources Using VLBI Observations, *AJ*, **84**, 1122

Cornwell, T.J. and Perley, R.A. eds. 1991, ASPC **19**, *Radio Interferometry: Theory, Techniques, and Applications, Proceedings of the 131st IAU Colloquium* (San Francisco: ASP)

David, E.E., Mathews, M.V., and Noll, A.M. 2004, John R. Pierce. In *Biographical Memoirs of the National Academy of Sciences*, **85** (Washington: National Academy Press) http://www.nasonline.org/member-directory/deceased-members/51643.html

Dicke, R. ed. 1967, *Report of the Ad Hoc Advisory Panel for Large Radio Astronomy Facilities* (Washington: National Science Foundation)

Dicke, R. ed. 1969, *Report of the Second Meeting of the Ad Hoc Advisory Panel for Large Radio Astronomy Facilities* (Washington: National Science Foundation)

Field, G. ed. 1973a, Report of the Astrophysics and Relativity Panel. In *Physics in Perspective, Vol. II, Part B* (Washington: National Academy of Sciences), 771

Field, G. ed., 1973b, Report of the Astrophysics and Relativity Panel. In *Astronomy and Astrophysics for the 1970s, Vol. 2* (Washington: National Academy of Sciences), 268

Findlay, J.W. and von Hoerner, S. 1972, *A 65-Meter Telescope for Millimeter Wavelengths* (Charlottesville: NRAO)

Florence, R. 1994, *The Perfect Machine* (New York: Harper Collins Publishers)

Frater, R.H., Goss, W.M., and Wendt, H.W. 2017, *Four Pillars of Radio Astronomy: Mills, Christiansen, Wild, Bracewell* (Giewerbestrasse, Switzerland: Springer Nature)

Greenstein, J.L. ed. 1972, *Astronomy and Astrophysics for the 1970s, Vol. 1* (Washington: National Academy of Sciences), 2

Greenstein, J.L. 1984a, Optical and Radio Astronomers in the Early Years. In *Serendipitous Discoveries in Radio Astronomy*, ed. K.I. Kellermann and B. Sheets (Green Bank: NRAO/AUI), 79

Greenstein, J.L. 1984b, Optical and Radio Astronomers in the Early Years. In *The Early Years of Radio Astronomy - Reflections Fifty Years after Jansky's Discovery*, ed. W.T. Sullivan III (Cambridge: CUP), 67

Greisen, E. 1990, The Astronomical Image Processing System. In *Seminar on Acquisition, Processing and Archiving of Astronomical Images*, ed. G. Longo and G. Sedmak (Naples: Observatorio Astronomico di Capodimonte)

Greisen, E. 1998, Recent Developments in Experimental AIPS. In ASPC **145**, *Astronomical Data Analysis Software and Systems VII*, ed. R. Albrecht, et al. (San Francisco: ASP), 204

Heeschen, D.S. ed. 1973, In *Astronomy and Astrophysics for the 1970s, Vol. 2*, ed. J. L. Greenstein (Washington: National Academy of Sciences), 2

Heeschen, D.S. 1981, The Very Large Array. In *Telescopes for the 1980s*, ed. G. Burbidge and A. Hewitt (Palo Alto: Annual Reviews, Inc.), 1

Heeschen, D.S. 1996, The Establishment and Early Years of NRAO, *BAAS*, **28**, 863[157]

Hinder, R. and Ryle, M. 1971, Atmospheric Limitations to the Angular Resolution of Aperture Synthesis Radio Telescopes, *MNRAS*, **154**, 229

Hjellming, R. ed. 1978, *An Introduction to the Very Large Array* (Socorro: NRAO)

Högbom, J.A. 1959, *The Structure and Magnetic Field of the Solar Corona*, PhD Dissertation, Cambridge University

Högbom, J.A. 1961, A Comparison between Two Different Designs for the Benelux Radio Telescope with Special Reference to Their Sensitivity. In *Large Radio Telescopes* (Paris: OECD), 131

Högbom, J.A. 1974, Aperture Synthesis with a Non-Regular Distribution of Interferometer Baselines, *A&AS*, **15**, 417

Högbom, J.A. 2003, Early Work in Imaging. In *ASPC* **300**, *Radio Astronomy at the Fringe*, ed. J.A. Zensus, M.H. Cohen, E. Ros (San Francisco: ASP), 17

Högbom, J.A. and Brouw, W.N. 1974, The Synthesis Radio Telescope at Westerbork. Principles of Operation, Performance and Data Reduction, *A&A*, **33**, 289

Keller, G. 1961, Report of the Advisory Panel on Radio Telescopes, *ApJ*, **134**, 927

Kraft, R.P. 2005, Jesse Greenstein, *National Academy of Sciences Biographical Memoirs*, **86** (Washington: National Academies Press), 1 https://www.nap.edu/read/11429/chapter/10

Lancaster, J. 1982, *Very Large Array Completion Report* (Socorro: NRAO) https://science.nrao.edu/about/publications/open-skies#section-7

Lequeux, J. 1962, Mesures Interférométriques à Haute Résolution du Diamètre et de la Structure des Principales Radio Sources à 1420 MHz, *An. Astrophys*, **25**, 221

Lubkin, G. B., 1975, *The Decision to build the Very Large Array* (Cambridge: Harvard University Press)

Maltby, P. and Moffet, A. 1962, Brightness Distribution in Discrete Radio Sources. III. The Structure of the Sources, *ApJS*, **7**, 141

McCready, L.L., Pawsey, J.L., and Payne-Scott, R. 1947, Solar Radiation at Radio Frequencies and Its Relation to Sunspots, *Proc. R. Soc. London A*, **190**, 357

McMullin, J.P. et al. 2007, CASA Architecture and Applications. In ASPC **376**, *Astronomical Data Analysis Software and Systems XVI* (San Francisco: ASP)

Napier, P.J. and Gustincic, J.J. 1977, Polarization Properties of a Cassegrain Antenna with Off-Axis Feeds and On-Axis Beam, *IEEE AP-S International Symposium,* 452

Noordam, J. E. and de Bruyn, A. G. 1982, High Dynamic Range Mapping of Strong Radio Sources with Application to 3C 84, *Nature,* **299**, 597

NRAO 1965, *VLA Design Report No. 1* (Green Bank: NRAO)

NRAO 1967, *A Proposal for a Very Large Array Radio Telescope Vol. I and II* (Green Bank: NRAO)

NRAO 1969, *A Proposal for a Very Large Array Radio Telescope Vol. III* (Green Bank: NRAO)

NRAO 1971, *A Proposal for a Very Large Array Radio Telescope Vol. IV* (Green Bank: NRAO)

NRAO 1978 Announcement, *BAAS,* **10** (1), 386

Oort, J.H. 1961a, Some Considerations Concerning the Study of the Universe by Means of Large Radio Telescopes. In *Large Radio-Telescopes* (Paris: OECD), 35

Oort, J.H. 1961b, Considerations Concerning the Minimum Resolving Power Required for 21-cm Line Observations with a Very Large Antenna. In *Large Radio-Telescopes* (Paris: OECD), 53

Oort, J.H. 1961c, Some Suggested Programs. In *Large Radio Telescopes* (Paris: OECD), 123

Perley, R. et al. 2009, The Expanded Very Large Array, *Proc. IEEE,* **97** (8), 1448

Perley, R. et al. 2011, The Expanded Very Large Array: A New Telescope for New Science, *ApJL,* **739**, 1

Raimond, E. 1996, Historical Notes: Four Decades of Dutch Radio Astronomy, Twenty-Five Years Westerbork Telescope. In *The Westerbork Observatory, Continuing Adventure in Radio Astronomy,* ed. E. Raimond and R. Genee (Dordrecht: Kluwer), 11

Readhead, A.C.S. and Wilkinson, P.N. 1978, The Mapping of Compact Radio Sources from VLBI Data, *ApJ,* **223**, 25

Rogstad, D.H. and Shostak, G.S. 1971, Aperture Synthesis Study of Neutral Hydrogen in the Galaxy M101: I. Observations, *A&A,* **13**, 99

Ryle, M. 1962, The New Cambridge Radio Telescope, *Nature,* **194**, 517

Ryle, M. 1972, The 5-km Radio Telescope at Cambridge, *Nature,* **239**, 435

Ryle, M. 1975, Radio Telescopes of Large Resolving Power, *Science,* **188**, 1071

Ryle, M. and Hewish, A. 1960, The Synthesis of Large Radio Telescopes, *MNRAS,* **120**, 220

Ryle, M. and Neville, A.C. 1962, A Radio Survey of the North Polar Region with a 4.5 Minute of arc Pencil-Beam System, *MNRAS,* **125**, 39

Ryle, M., Elsmore, B., and Neville, A.C. 1965, Observations of Radio Galaxies with the One-Mile Telescope at Cambridge, *Nature,* **194**, 517

Scheuer, P.A.G. 1984, The Development of Aperture Synthesis at Cambridge. In *The Early Years of Radio Astronomy,* ed. W.T. Sullivan III (Cambridge: CUP), 249

Schwarz, U.J. 1978, Mathematical-Statistical Description of the Iterative Beam Removing Technique (Method CLEAN), *A&A,* **65**, 345

Taylor, G.B., Carilli, C.L. and Perley, R.A. eds. 1999, *Synthesis Imaging in Radio Astronomy II,* ASPC **180**, (San Francisco: ASP)

Thompson, A.R. and D'Addario, L.A. 1982, *Synthesis Mapping, Proceedings of the NRAO-VLA Workshop Held in Socorro, NM, June 21-25* (Green Bank: NRAO)

Trimble, V. 2003, Jesse Leonard Greenstein (1909-2002), *PASP*, **115**, 890

Trimble, V. and Zaich, P. 2006, Productivity and Impact of Radio Telescopes, *PASP*, **118**, 933

Trimble, V. and Ceja, J.A. 2008, Productivity and Impact of Astronomical Facilities: Three Years of Publications and Citation Rates, *Astron Nachr*, **329**, 632

Tucker, W. and Tucker, K. 1986, The Very Large Array. In *The Cosmic Inquirers* (Cambridge: Harvard University Press)

Whitford, A. 1964, *Ground-Based Astronomy: A Ten Year Program* (Washington: National Academy of Sciences)

Williams, W.F. 1965, High Efficiency Antenna Reflectors, *Microwave J.*, **8**, 79

York, D.G. et al. 2000, The Sloan Digital Sky Survey: Technical Summary, *AJ*, **120**, 1579

FURTHER READING

Bracewell, R.N. 1958, Radio Interferometry of Discrete Sources, *Proc. IRE*, **46**, 97

Bracewell, R.N. and Roberts, J.A. 1954, Aerial Smoothing in Radio Astronomy, *Aust. J. Phys. 7*, 6

Elbers, A. 2017, *The Rise of Radio Astronomy in the Netherlands* (Cham, Switzerland: Springer)

Finley, D.G, and Goss, W.M. eds. 2000, *Radio Interferometry: The Saga and the Science: Proceedings of a Symposium Honoring Barry Clark at 60* (Washington: AUI)

Fomalont, E.B. and Wright, C.H. 1974, Interferometry and Aperture Synthesis. In *Galactic and Extragalactic Radio Astronomy*, ed. G. Verschuur and K.I. Kellermann (Berlin: Springer)

Heeschen, D.S. 1967, Radio Astronomy: A Large Antenna Array, *Science*, **158**, 75

Heeschen, D.S. 1975, The Very Large Array, *Sky and Telescope*, **49** (6), 344

Heeschen, D. 1991, Reminiscences of the Early Days of the VLA. In ASPC **19**, *Radio Interferometry: Theory, Techniques, and Applications; Proceedings of the 131st IAU Colloquium*, ed. T.J. Cornwell and R.A. Perley (San Francisco: ASP), 150

Hogg, D.E. 2000, The Green Bank Interferometer. In *Radio Interferometry: The Saga and the Science: Proceedings of a Symposium Honoring Barry Clark at 60*, ed. D.G. Finley and W. M. Goss (Washington: AUI), 20

Kellermann, K.I. 2014, David S. Heeschen. In *Biographical Memoirs of the National Academy of Sciences* (Washington: National Academy of Sciences) https://aas.org/obituaries/david-s-heeschen-1926-2012

Kellermann, K.I. and Moran, J.M. 2001, The Development of High-Resolution Imaging in Radio Astronomy, *ARA&A*, **29**, 457

Lancaster, J. 1991, The Design and Construction of the VLA: The Project Managers View. In ASPC **19**, *Radio Interferometry: Theory, Techniques, and Applications; Proceedings of the 131st IAU Colloquium*, ed. T.J. Cornwell and R.A. Perley (San Francisco: ASP), 165

Leverington, D. 2017, *Observatories and Telescopes of Modern Times* (Cambridge: CUP)

Napier, P.J. et al. 1983, The Very Large Array: Design and Performance of a Modern Synthesis Radio Telescope, *Proc. IEEE*, **71**, 1295

Sullivan, W.T. III 2009, *Cosmic Noise: A History of Early Radio Astronomy* (Cambridge: CUP)

Swenson, G. The Early Technical Development of the VLA. In ASPC **19**, *Radio Interferometry: Theory, Techniques, and Applications; Proceedings of the 131st IAU Colloquium*, ed. T.J. Cornwell and R.A. Perley (San Francisco: ASP), 160

Thompson, A.R. et al. 1980, The Very Large Array, *ApJS*, **44**, 151

Thompson, A.R., Moran, J.M. and Swenson, G.W. 2017, *Interferometry and Synthesis in Radio Astronomy* (Cham, Switzerland: Springer) https://www.springer.com/us/book/9783319444291

The VLA Takes Shape, *S&T*, **52** (5), 320

Weinreb, S. et al. 1977, Waveguide Systems for a Very Large Antenna Array, *Microwave J.*, **20** (3), 49

CHAPTER 8

VLBI and the Very Long Baseline Array

Beginning in the 1950s radio interferometers and arrays of antennas were connected by cable, waveguide, or radio links separated by up to a hundred kilometers or more. Starting in 1967, radio astronomers in the US and Canada began to experiment with independent local oscillators and broad band tape recorders to record data collected by widely separated antennas, a technique which came to be known as Very Long Baseline Interferometry or VLBI. Using radio telescopes spread throughout the United States, Australia, and Europe, VLBI baselines were increased to thousands of kilometers, and ultimately to space, with baselines ranging out to hundreds of thousands of kilometers.

Within the United States, the informal US VLBI Network initially managed the complex logistics of organizing simultaneous observations by many radio telescopes, each with their own management and their own scientific programs. European radio astronomers later organized the European VLBI Network (EVN). The non-optimum location of the antennas being used for VLBI and the difficulty of scheduling observations flexibly led NRAO to construct the Very Long Baseline Array (VLBA), consisting of ten 25 meter diameter antennas spread throughout the United States from St. Croix in the US Virgin Islands to Hawaii. With an angular resolution as good as 0.0001 arcsec, the VLBA was the highest angular resolution telescope in the world. Observations with the VLBA have revealed the nature of jets ejected from the supermassive black holes found in quasars, shown the structure of cosmic masers associated with the birth and death of stars, determined the expansion rate of the Universe independent of the traditional cosmic ladder, measured the rotation of the Milky Way, and determined with great precision the relativistic bending of radio waves.

© The Author(s) 2020, corrected publication 2021
K. I. Kellermann et al., *Open Skies*, Historical & Cultural Astronomy,
https://doi.org/10.1007/978-3-030-32345-5_8

8.1   INDEPENDENT-OSCILLATOR-
TAPE-RECORDING INTERFEROMETRY[1]

With its overall dimensions of 35 km, the Very Large Array (VLA) at the time represented about the longest practical interferometer baseline with direct electrical connections. As early as the 1950s and 1960s radio astronomers at Jodrell Bank, led by Henry Palmer, began to experiment with radio links to provide a common frequency reference and to return the data from a remote antenna to Jodrell Bank, where they were correlated with data from the Jodrell Bank 250 foot antenna. In a series of elegant observations, they gradually extended their interferometer out to baselines of 115 km to show that some radio sources were smaller than an arcsec (Allen et al. 1962). Later the Jodrell Bank radio astronomers teamed up with a group at the Malvern Royal Radar Establishment to link two antennas separated by 127 km. Observing at wavelengths as short as 6 cm, Palmer et al. (1967) were able to demonstrate that some radio sources were as small as 0.05 arcsec. The Jodrell Bank to Malvern radio link involved two repeater stations. Extension to longer baselines was impractical or at best would exceed radio observatory budgets.

However, motivated by the Jodrell Bank results, the rapid variability of radio quasars,[2] the observation of low frequency cutoffs in the quasar radio spectra,[3] and observations of interplanetary scintillations,[4] several radio astronomy groups around the world had begun to think about further extending interferometer baselines using atomic frequency standards as independent oscillators and high speed (broad bandwidth) tape recorders to record the data at each end of the interferometer for later playback and correlation.

Early VLBI Development The first serious discussions of independent-oscillator-tape-recorder-interferometry apparently took place in Moscow in early 1962 (Matveyenko et al. 1965). Realizing the potential applications of this powerful technique, the Russian scientists wanted to publish and patent their ideas, but were thwarted by Soviet bureaucracy and secrecy (Matveyenko 2013). The following year, during Jodrell Bank Director Bernard Lovell's visit to the USSR, Leonid Matveyenko discussed the possibility of doing interferometry between Jodrell Bank and the USSR. However, neither Jodrell Bank nor the Russians were able to develop or obtain the needed instrumentation, and nothing resulted from these discussions. Matveyenko, Nikolai Kardashev, and Gennady Sholomitsky (1965) were finally able to publish their paper, but they wrongly concluded that the sensitivity depended inversely on the interferometer baseline length because they incorrectly assumed that integration times were limited by the natural fringe rate.

VLBI became a practicality as a result of three key technical advances: precision atomic clocks to provide precise time and frequency,[5] high speed tape recorders capable of recording the broad bandwidths needed to obtain adequate interferometer sensitivity,[6] and fast digital computers to correlate the data, all of which became commercially available in the mid-1960s. Starting in

1965, unaware of the Soviet work, two groups, one in Canada and one in the United States, began to develop a VLBI capability. The Canadian group used analog-type tape recorders, which had just become popular in the TV industry, to record a 1 MHz IF bandwidth. The observing frequency was 448 MHz (67 cm), and time synchronization was facilitated by simultaneously recording timing data on the audio track. In order to compensate for any timing uncertainties, the speed of one of the playback systems was adjusted until the appearance of interference fringes signaled proper time alignment.[7]

The NRAO program was initiated by NRAO scientists Barry Clark and Ken Kellermann, who were joined by Professor Marshall Cohen from Cornell and David (Dave) Jauncey, a Cornell postdoc who had just arrived at Cornell from Australia as part of the new Cornell-Sydney University Astronomy Center. After informal discussions between Cohen and Kellermann in August 1965, the following month Kellermann approached NRAO Director David Heeschen about possible funding to develop a tape-recording independent-oscillator interferometer. When asked about the cost, Kellermann was caught off guard as he and Cohen had not really thought about cost, so he threw out a guesstimate of $100,000. After a brief pause to check the status of the NRAO budget, Heeschen responded with, "Will $50,000 be enough until the end of the year?" That was it! No proposal. No review committee. No debates within NRAO or discussions with the radio astronomy community. Work could begin immediately. A few days later, however, perhaps perceiving that this might become a big enterprise, or perhaps just wanting to cover himself in case it was a failure, Heeschen asked for a short written proposal which was quickly produced.[8] An equally brief proposal was submitted to Cornell which shared in the development costs.[9] Heeschen's formal request to the NSF to include $100,000 in the budget for "an independent local oscillator interferometer" came six months later.[10]

The NRAO group chose to use digital rather than analog recordings. The sensitivity of a digital interferometer depends on the number of bits recorded and correlated. This, in turn, is proportional to the product of the observing time and the bandwidth (bit rate). The length of the integration time is limited by the coherence of the independent local oscillators or the atmosphere and ionosphere. Since each bit is recorded at a precise time determined by an atomic clock, the digital recording is self-clocking which reduces the need for precise stability of the record-playback system. The data are then precisely aligned in time in the playback computer. By storing and shifting the data bits, it was easy to examine a range of time alignments to compensate for any uncertain timing or antenna location. In order to minimize development costs and to be able to do the correlation in a general purpose computer, NRAO used standard computer reel-to-reel tape drives to record one-bit digital data at 720 kilobits per second (kbps) appropriate to sampling a 360 kHz bandwidth at the Nyquist sampling rate.[11] Each 12-inch reel of tape lasted about three minutes, and it took about an hour to correlate each pair of tapes in the NRAO IBM 360/50 computer.

Fig. 8.1 Hewlett-Packard Model 5065A Rubidium frequency standard and power supply used to maintain time and to determine the local oscillator frequency for the early NRAO VLBI experiments. Credit: NRAO/AUI/NSF

To optimize the sensitivity and to compensate for the necessarily narrow bandwidth which was limited by the recording system, the NRAO-Cornell team planned to use the 1000 foot Arecibo radio telescope at one end of the interferometer baseline and the newly completed Green Bank 140 Foot Telescope at the other. Claude Bare from the NRAO Electronics Division joined the team to provide engineering support. The commercially available Hewlett Packard HP 5065A Rubidium standard (Fig. 8.1) was used as the time and frequency reference.[12] Considering the advertised frequency stability of one part in 10^{11}, NRAO felt that this would be adequate for operation at 611 MHz (50 cm) for integration times up to a few hundred seconds. At the time, this was also the highest frequency where the Arecibo telescope was operating.

Two approaches were considered to synchronize the separate clocks at each end of the interferometer. The most straightforward method was to first bring a running clock to Washington by car, where it was synchronized with the US master clock at the US Naval Observatory (USNO), and then driven to Green Bank where it would be synchronized with a second clock, which would then be carried by car or by commercial airline to the other end of the interferometer baseline. Since transporting clocks more than a few hundred miles was impractical, the NRAO group also used the 100 kHz LORAN C stations located along the east coast of the United States and at several offshore European sites. Each LORAN C station broadcast a characteristic time code that was synchronized with the USNO master clock.[13] At nighttime when ionospheric disturbances were low, the timing signals could be measured with an accuracy of a few microseconds.

Serious development work at NRAO began in late 1965 with the design of the recording hardware by Claude Bare and the correlator software by Barry Clark. A friendly rivalry developed between the NRAO and Canadian groups, with frequent telephone exchanges reporting successes and problems. As described by Norman Broten (1988),

> During all of this time Kellermann and I had stayed in touch. It was a remarkable friendly competition between the two teams striving to be the first to use successfully tape-linked interferometry. Scarcely a week would go by without the phone ringing, either in Ottawa or at Green Bank, to keep both teams abreast of each other's progress. ... Later in the experiment the response to a telephone call would be "Got any fringes yet?"

All of the US participants were, at the same time, involved in other projects. Bare was responsible for supporting other Green Bank digital systems; Clark, Cohen, Jauncey, and Kellermann all were pursuing various observational programs, and Clark was busy with the VLA design (Chap. 7). Probably the project did not proceed as fast as might have been possible, but following successful bench tests in October 1966, NRAO sent one of the recording terminals to Arecibo for the first observations. Jack Cochran, then an NRAO technician, traveled to Puerto Rico to install and operate the equipment. Somehow Pan American Airlines lost track of the shipment, which was finally inexplicably traced to a warehouse in Baltimore. Tired of waiting, Cochran had returned to Green Bank to spend the Thanksgiving holidays with his family.

Following Cochran's return to Puerto Rico, the first observations between Green Bank and Arecibo were finally made in January 1967, but were unsuccessful. There were no interference fringes observed. A second experiment in February was equally unproductive, and it was never clear what was wrong with either set of those first observations with Arecibo. All of the equipment was returned to Green Bank to be carefully checked. On the night of 5–6 March 1967, the NRAO group ran a successful test on a 650 meter baseline between the 140 Foot Telescope and one of the 85 foot antennas of the Green Bank Interferometer. Interference fringes were readily found the next day after correlation on the Charlottesville IBM 360/50 computer.

The first successful observation using well-separated antennas was on 8 May 1967, again using one of the Green Bank 85 foot antennas and the Naval Research Laboratory 85 foot antenna at Maryland Point (Fig. 8.2), a distance of 220 km or 460,000 wavelengths at 610 MHz (Bare et al. 1967). Three sources, 3C 273, 3C 286, and 3C 287, were unresolved and demonstrated that the system worked as expected. However, considering the relatively long wavelength, the resolution was no better than had been previously obtained by the Jodrell Bank-Malvern radio linked interferometer at 6 cm (Palmer et al. 1967).

Fig. 8.2 NRL 85 foot radio telescope at Maryland Point, used together with the Howard Tatel 85 Foot telescope in Green Bank for the first successful NRAO VLBI observation. Credit: NRL

Probably the first successful VLBI observations of real astrophysical interest was by the Canadian group using a 3074 km transcontinental baseline at 448 MHz (67 cm) between the Algonquin Park 150 foot radio telescope and a 25 meter antenna located at Penticton, British Columbia. These observations were made on 13 April 1967, nearly a month before the Green Bank-Maryland Point observations, but they were not able to successfully correlate the data until 21 May, a few weeks after the US data were correlated. The Canadian observations directly demonstrated that several quasars were less than a few hundredths of an arcsec in diameter. Both groups reported their results at the August 1967 URSI Radio Science Meeting held in Montreal.

Meanwhile, an MIT/Haystack group was using a radio-linked interferometer on a 13.4 km baseline between the Millstone Hill 84 foot antenna and the Agassiz 60 foot radio telescopes to show that OH maser sources appeared unresolved with angular dimensions less than a few arcsecs (Moran et al. 1967a). These maser sources were highly variable and so were expected to be very small. In June 1967, the MIT/Haystack group joined forces with the NRAO/Cornell group to observe both quasars and OH masers on a 845 km baseline between the Haystack 120 foot radio telescope and the Green Bank

140 Foot Telescope. These observations showed that both quasars (Clark et al. 1968a) and 1.7 GHz OH masers (Moran et al. 1967b) had angular structures less than a few hundredths of an arcsec.

In August 1967, both the MIT/Haystack and NRAO/Cornell groups used the 85 foot radio telescope at the University of California Hat Creek Observatory together with the Green Bank 140 Foot Telescope to extend the interferometer baseline to 3500 km to demonstrate structures on scales less than 0.01 arcsec (Clark et al. 1968b; Moran et al. 1968). Also in August, the Green Bank-Arecibo baseline finally gave results at 611 MHz (Jauncey et al. 1970). The next step was to go to shorter wavelengths, but negotiations with the University of California to install a 6 cm receiver on the Hat Creek Telescope fell through. Coincidentally, at about the same time, Olaf Rydbeck, head of the radio astronomy program at the Swedish Onsala Space Observatory, was asking his former student, Hein Hvatum, how they could get into VLBI. An observational program was quickly formulated, and recorders and clocks were shipped to Sweden for a January 1968 experiment at both 6 and 18 cm. The Green Bank-Sweden baseline was 6319 km long, and showed that some quasars and active galactic nuclei (AGN) were smaller than one thousandth of an arcsec. This was the highest angular resolution ever obtained for any astronomical observation; indeed, probably for any measurement. One thousandth of an arcsec is equivalent to reading ordinary newsprint at a distance of 100 miles.

These were exciting times for the NRAO VLBI group, who shipped or carried tape recorders, receivers, atomic clocks, and many pounds of magnetic tape to radio observatories around the world. Within a year, interferometer baselines had increased to intercontinental distances and the angular resolution to better than one thousandth of an arcsec. Although the principles were straight forward, there were enormous technical and logistical challenges to obtaining agreements to use antennas at observatories that may have been scheduled for other programs, shipping materials and supplies, synchronizing clocks, and dealing with failed atomic clock batteries, as well as building and installing new, often untested equipment at unfamiliar observatories. Anything that could go wrong, often did go wrong. Unlocked oscillators, time synchronization errors as large as one second, crossed polarization, incorrect wiring, and wrong frequency settings were not uncommon, and any one such error would lead to no fringes, which generally wasn't discovered until after the experiment was over.

The MK I[14] VLBI System was used extensively from 1967 to 1971. Observing was very labor intensive. An experienced person could record, rewind, dismount, and mount a new tape in 12 minutes, but two or three observers were needed at each telescope to support a multi-day observing session. On occasion, more tape would accidently end up on the floor than on the rewind reel. More than 100 tapes were recorded at each station in a single 24 hour observing session. A three station experiment meant three baselines had to be correlated, one at a time; a four station experiment, six baselines. So a single 24 hour experiment could mean 300–600 hours of playback at the NRAO IBM 360/50 in Charlottesville.

In 1966, Marshall Cohen left Cornell for the University of California San Diego, and two years later moved to Caltech, where he started a major VLBI program. Many of the later VLBI leaders were trained at Caltech as students or as postdoctoral fellows. Caltech operated a 360/75 machine that was able to correlate MK I VLBI tapes about five times faster than the NRAO 360/50, but the normal charges to use this facility to process VLBI tapes were prohibitive. Instead, Cohen was able to arrange to use time late at night at no charge, but only if he provided the personnel to run the machine and change tapes. Following the completion of the Caltech OVRO 130 foot radio telescope in 1968, when it was no longer part of the proposed Owens Valley Array (Sect. 7.2), the 130 foot became a workhorse for VLBI.

Meanwhile, the Canadian group continued their observations at 408 and 448 MHz with a total of ten successful single baseline observations extending across Canada and to Jodrell Bank (Broten et al. 1969; Clarke et al. 1969). Many of the experiences and logistical problems encountered by the NRAO group were experienced as well by the Canadian VLBI observers (Broten 1988). In particular, the Canadian group had replaced their studio TV recorders with the less reliable but more portable Ampex VR600 recorders (Sect. 8.4).

8.2   Penetrating the Iron Curtain

Following the 6 cm VLBI observations with baselines to Sweden (Kellermann et al. 1968), VLBI had quickly reached a resolution of about 0.001 arcsec (1 milli-arcsec). Many sources, particularly quasars, still had unresolved features. The longest baselines were already a significant fraction of the Earth's diameter, so it was clear that the only way to further improve the angular resolution would be to go to shorter wavelengths or to space, but the only radio telescopes outside the United States that could work at short centimeter wavelengths were located in the Soviet Union. Although this was during the depths of the Cold War, in February 1968, Marshall Cohen and one of the authors (KIK) boldly wrote to Viktor Vitkevich, a well-known leader of Soviet radio astronomy, suggesting a VLBI experiment between the NRAO 140 Foot Telescope and the Lebedev Physical Institute's precision 22 meter antenna located at the Puschino Observatory near Moscow. Although both Cohen and Kellermann had previously met Vitkevich, they did not realistically expect that a US-USSR VLBI experiment involving the exchange of highly sensitive atomic clocks and high speed tape recorders would be feasible, so were not surprised when they initially received no response to their letter.

But to their pleasant surprise, after five months Vitkevich responded by telegram, followed by a letter from his colleague Leonid Matveyenko, reporting that the proposed experiment had been approved by the Soviet Academy of Sciences. However, Matveyenko suggested that instead of using the 22 meter dish at Puschino, the program use the more precise 22 meter dish in Crimea.

Much later, NRAO learned that during the five month period before sending his response, Vitkevich, with the aid of the Soviet astrophysicist Iosef Shklovsky, had sought and gained approval not only from the USSR Academy of Sciences but also from the Soviet political and military authorities (Matveyenko 2013).

In spite of the Cold War tensions there were no objections from NRAO or the NSF to the proposed experiment, but NRAO first had to obtain an export license from the US Department of Commerce for all of the specialized equipment that would be temporarily sent to the USSR, including a commercial atomic clock and a high speed computer tape recorder.[15] There was an additional, potentially more serious military concern. VLBI observations are used by radio astronomers to investigate the size and structure of cosmic radio sources, but there are also a variety of terrestrial applications, including the precise determination of the Earth's axis of rotation as well as the distance between the two antennas that form the interferometer. It was just these two quantities that are needed for the precise delivery of ICBMs. While the Soviet government had similar concerns, there was a curious asymmetry. Since accurate US maps were publicly available, any American adversary could derive the distance from the Crimean radio telescope to any potential US target, such as the Pentagon, the White House, or a US based missile site. However, since maps of the USSR were not accessible to Americans, or to anyone outside of those who needed to know, there was no reciprocity. Indeed, it was widely recognized that even tourist maps of Moscow were deliberately distorted.

Clark and Kellermann were not surprised to receive a visit one day in Green Bank from two men who identified themselves as representing the US Defense Intelligence Agency. They wanted to know all the details of the proposed observations. It was clear from the questions asked that they were remarkably familiar with VLBI and what it could and could not do. They correctly noted that it would be easily possible for NRAO to corrupt the baseline and Earth rotation information without compromising the intended astronomical goals of the proposed experiment, but NRAO refused to take part in any such charade. In the end, it was agreed that since accurate global mapping was becoming widely available from satellite imaging, there were really no security concerns as long as either Clark or Kellermann were present with the equipment at all times. Rather, it appeared that the intelligence agencies on both sides recognized that there were no secrets in this business, but their goals were to not make it easy for the other side, and to not give away information.

Following a visit from two Russian scientists to Green Bank,[16] the first observations were scheduled for October 1969 at both 2.8 and 6 cm. The ensuing program turned out to be a logistical challenge. All of the NRAO instrumentation and 25 cartons of magnetic tape were sent by air to Moscow, where it was supposed to be sent on to Crimea. However, the Russians claimed that the tape recorder was too big to fit into the cargo hold of a Russian jet, so arrangements had to be made to send it by train. But the recorder was also over the train weight limit, and required an appeal for a waiver from the Soviet Academy of Sciences. Telephone or telex communications between the USSR and the US were at best unreliable. Calls had to be booked in advance, perhaps to give the intelligence agencies on both sides the opportunity to listen. Often one side or the other would be barely audible, and after a few minutes of yelling, the connection would be broken.

The biggest problem concerned the synchronization of the atomic clock in Crimea with the one in Green Bank.[17] Normally, VLBI observations made use of the extensive network of LORAN C stations established to facilitate navigation. Each LORAN C station broadcast accurate timing signals that were synchronized by periodic visits from USNO personnel carrying an atomic clock that had been synchronized to the master clock at the US Naval Observatory in Washington. However, the LORAN C station in Turkey, just across the Black Sea from the Crimean radio telescope, had not yet been synchronized with Washington. The backup plan was to synchronize the clock in Leningrad using a LORAN C station in the Baltic Sea and to carry the running clock by plane to Crimea. But the American LORAN C transmissions were blocked by a powerful Soviet imitation. The Russian radio astronomers denied any knowledge of any such Soviet transmission, but it later turned out that it was well known to the Swedish timekeeping service. Plan C involved shipping a running atomic clock from Sweden, recharging the batteries at the Leningrad Pulkova Observatory, and then flying it to Crimea. Since the batteries had died during the first flight from Leningrad to Crimea, the clock was returned to be resynchronized. The running clock was then placed on the floor of the commercial flight along with a backup car battery. The American and Russian radio astronomers, equipped with their voltmeter, ran back and forth from their seats to check on the health of the battery during the flight. With hindsight it is hard to believe that the Soviet authorities allowed such activities, which might be compared with having Russian scientists playing with a lot of sensitive equipment on an American flight from New York to Miami.

The scheduled observations were split into two parts, with a gap of several weeks to allow time for the tapes to be correlated in Charlottesville. A few of the first tapes recorded in Crimea were hand carried to Moscow, but when it turned out to be difficult to arrange for them to be shipped by air freight to New York, they were just given to a PanAm pilot at the airport with instructions that they would be picked up by a colleague at New York's Kennedy Airport. During the gap between the two observing sessions, the NRAO scientists were taken on an escorted trip to Armenia and Uzbekistan. Many years later, it was learned that as soon as the Americans left Crimea on their trip, KGB engineers arrived to take careful notes and photographs of the sensitive US instrumentation.

After a few false starts, characteristic of the early VLBI experiments, the observations were successfully completed over the 8035 km long baseline and resulted in the then record angular resolution of 0.0004 arcsec, or only a few light-years at the distance of the quasar 3C 273 (Broderick et al. 1970). Following the close of the marathon observing session, the Russian scientists insisted on celebrating the occasion with toasts to VLBI and to continued

Fig. 8.3 From left to right, Ivan Moiseev, John Payne, and Viktor Effanov celebrate the conclusion of the first US-USSR VLBI observations in October 1969. Seated is an unnamed member of the Crimean radio telescope staff. Credit: KIK/NRAO/AUI/ NSF

Russian-American friendship. While the Americans were able to deal with the Russian beer and vodka, it was a challenge to keep up with the Russian hosts who downed shots of (nearly) pure (190 proof) alcohol washed down with beer (Fig. 8.3).

Two years later, a second experiment, this time at wavelengths as short as 1.3 cm, using NRAO's new MK II VLBI recording system gave another factor of two improvement in resolution (Burke et al. 1972). In subsequent years the Russian radio astronomers, led by Leonid Matveyenko, built their own MK II compatible recording and playback systems. Russia became a regular participant in global VLBI observations using a variety of radio telescopes located throughout the USSR, and in 2011 launched the very successful space VLBI mission, RadioAstron (Sect. 8.9), extending VLBI baselines to more than 250,000 km in length.

It was only through the cooperation and support of the Soviet Academy of Sciences, Aeroflot, and governments on both sides that it was possible to rise above the pervading culture to carry out one of the few scientific collaborations of that Cold War period that involved the exchange of sensitive instrumentation. Although there had been a long-time exchange program between the US and the Soviet Academies of Science, it was typically shrouded on both sides in suspicion of the visiting scientists. The support of the Soviet Academy and the US government to grant the needed export license was crucial, as the USSR VLBI experiments were perhaps unique in that they were initiated and organized from the ground up by the participating scientists, with a minimum of government or even institutional involvement.

The good will established by the 1969 joint experiment nearly evaporated when the return shipment of the atomic clock, tape recorder, state-of-the-art digital instrumentation, as well as 25 cartons of recorded computer tape, apparently disappeared. As part of the agreement with the Commerce Department, NRAO had agreed that the Russians would return all of the American instrumentation immediately after the experiment was concluded. When the purported shipment apparently did not arrive in the US, NRAO followed with two weeks of frantic queries to PanAm and Aeroflot, along with a series of telegrams to the USSR with increasing concern and threats about the impact to future exchanges. To the embarrassment of NRAO, everything turned up in an Air France warehouse in New York where it had been sitting for two weeks. To avoid a repetition of this awkward situation after the 1971 experiment, the NRAO team was taken to the Moscow airport to witness the loading of the return shipment on an Aeroflot plane bound for New York. Sitting in the truck on the tarmac, as the plane left the ground, Matveyenko proudly informed the NRAO group, "Now it is your problem."

8.3   Faster than Light

In October 1970, an MIT/Haystack group observed the strong quasars 3C 273 and 3C 279 at 7840 MHz (3.8 cm) on a 3900 km baseline between the Haystack 120 foot antenna near Tyngsboro, MA and the NASA Deep Space Network Goldstone 210 foot dish near Barstow, CA (Knight et al. 1971). Both antennas used low noise maser amplifiers to give improved sensitivity over earlier observations, as well as hydrogen maser frequency standards for improved phase stability. These observations, colloquially referred to as "Goldstack," were not intended to study the structure of compact radio sources, but to measure the apparent change in the relative position of 3C 279 due to relativistic bending as it passed close to the Sun.[18] Nevertheless, for the first time, the observed fringe amplitudes were of sufficient quality to show unambiguously that both sources contained at least two distinct components. The separation of the two components was accurately determined as only 0.005 arcsec.

Four months later, in February 1971, both Whitney et al. (1971) and Cohen et al. (1971) repeated the October observations using the same equipment. Cohen et al. observed 31 sources including 3C 273 and 3C 279. Irwin Shapiro had made the October 1970 results available to the NRAO-Caltech group, who wanted to see if there had been any changes in their structure since the Knight et al. observations made four months earlier.

By this time, Cohen had moved to Caltech, and he assigned third-year graduate student David Shaffer[19] the task of analyzing the data which had been correlated on the Caltech IBM 360/75.[20] When Shaffer plotted the data, he was surprised to note that 3C 279 had changed in a manner reflecting an increase in the separation of the two components. Knowing the distance of 3C

279, Shaffer calculated the speed of separation to be three to four times the speed of light and burst into Cohen's office to pronounce his discovery. The results for 3C 273 were less clear, but also indicated component motion greater than the speed of light, a phenomenon which came to be referred to as "superluminal motion."

Although they were initially alarmed at this apparent violation of special relativity, Cohen et al. (1971) soon realized that superluminal motion had, in fact, been previously predicted by Martin Rees (1967). A long-standing problem of astrophysics was that the rapid variability of the powerful radio emission from quasars seemingly implied such small dimensions that they would rapidly self-destruct.[21] However, Rees pointed out that if the radio source was expanding or moving toward the observer at close to the velocity of light, the radiation would be focused along the direction of motion and so appear to be more luminous than it was if the radiation were assumed to be isotropic. Moreover, since the source of radiation was nearly catching up to its own radiation, any changes in luminosity seen by a distant observer located nearly along the direction of motion would appear to happen in a shorter time span than the intrinsic change. Thus, the apparent velocity could be arbitrarily large.

Whitney et al. (1971) also noticed the apparent superluminal motion when comparing their data taken four months apart. Both groups presented their results at the 13–14 April 1971 Rumford Symposium (Rogers and Morrison 1972) (Fig. 8.4). They considered alternate interpretations, including properly phased time variability in a set of stationary sources such as one observes on a

Fig. 8.4 Members of the NRAO-Cornell, MIT-Haystack, and Canadian VLBI teams gathered in Boston in April 1971 to receive the Rumford Medal of the American Academy of Arts and Sciences. From the NRAO-Cornell team, Dave Jauncey is seated in the front row 2nd from the left, Marshall Cohen, 7th from left, and Ken Kellermann, 9th from left. Barry Clark is standing in the rear center

movie marquee, or a searchlight effect where stationary material is excited by a shock front moving at an oblique angle. Also, because the Goldstack observations covered only a limited part of the Fourier Transform plane, it would be possible to reproduce the limited data with a time variable, more complex, but stationary morphology. Interestingly, the near equal double structure of 3C 279 observed in 1970/1971 has never been repeated. The actual structure of 3C 279 is indeed much more complex than a simple double, and the early interpretation of superluminal motion based on the limited data then available was probably premature, but nevertheless has been confirmed by later more detailed imaging of radio sources and their kinematics (e.g., Lister et al. 2016). However, perhaps the biggest challenge to superluminal motion came from the small but persistent group of scientists who argued that quasars are closer than indicated by their large redshifts, and so the observed angular motion would correspond to a much slower linear velocity (e.g., Burbidge 1978). These arguments continued for decades and only died when their proponents died.

8.4   Advanced VLBI Systems

The NRAO MK II VLBI System Although the NRAO digital VLBI system proved to be more reliable and easier to use than the Canadian analog system, there were several serious limitations. Tapes were expensive and only ran for three minutes, the tape drives were large and heavy and thus expensive to ship, and the narrow 360 kHz bandwidth limited the sensitivity. Processing time on general purpose computers was lengthy, which in some cases meant expensive. Nevertheless, compatible recording units were built at Haystack and used by CfA and NASA Goddard Space Flight Center (GSFC) for a series of geodetic studies. Soon after the first successful experiments, under the leadership of Barry Clark, NRAO began the design of an advanced recording system referred to as the MK II VLBI system. The NRAO MK II system used the same portable Ampex VR660C helical scan TV recorder then in use by the Canadian VLBI group, but recorded digital instead of analog data. Each reel of two-inch wide tape lasted for three hours instead of the three minute MK I tapes, weighed only about 10 pounds, and recorded a 2 MHz bandwidth with one-bit samples at 4 Mbps (Clark 1973). Initially seven record units were built by NRAO in cooperation with the Leach Corporation. Requiring only IF, 5 MHz, and a 1 pulse per second timing signals, the MK II units could be easily transported for temporary use at other observatories. By the end of 1976, 19 MK II units were in operation throughout the world either built by NRAO or built elsewhere following detailed designs made available by NRAO.

Allen Yen, one of the architects of the Canadian VLBI system, had advised NRAO against using the VR 660 recorder which the Canadians had found to be unreliable. Although the NRAO team anticipated that digital recordings would be more robust to timing irregularities or to imperfections in the tape itself, variations in the mechanical alignment among the different units used for

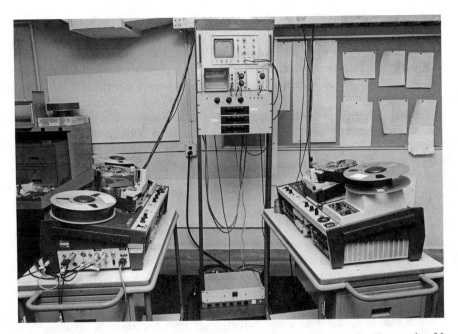

Fig. 8.5 Mark II VLBI correlator at NRAO offices in Charlottesville. Two reels of 2 inch wide tape are shown mounted on the VR660 video tape recorders. Credit: NRAO/AUI/NSF

recording and playback led to difficulties in playback, with losses in synchronization and unacceptably high error rates. An elaborate set of interactive mechanical adjustments in the playback units required considerable experience and skill to successfully play back MK II tapes (Fig. 8.5). Moreover, the recording problems were exacerbated by the use of government surplus tape, which turned out not be suitable for the VR 660 recorder. After years of frustration and unreliable observations, thousands of pounds of tape were buried in Green Bank. In 1976, at the suggestion of Yen, the record/playback units were replaced by the more reliable IVC 825 recorder that used one-inch wide tape. The IVC machines, manufactured by the International Video Corporation, proved to be more reliable than the VR 660s and required fewer adjustments on playback. But each reel of tape only lasted one hour, and playback errors were still a problem.

A major breakthrough occurred in the late 1970s with the introduction of the home Video Cassette Recorder (VCR). During the course of extended visits to NRAO, Caltech, and the Max-Planck-Institut für Radioastronomie (MPIfR) in Bonn, Germany, Yen developed remarkably inexpensive modifications which could be applied to consumer VCRs to record MK II compatible data. Each tape cost only a few dollars instead of a few hundred dollars, and could be inexpensively shipped by regular (customs-free) first class mail instead of the complex and costly air freight shipments required for the Ampex and

IVC tapes. With their simplicity, low cost, and wide-spread availability, dozens of MK II units were built and operated at radio observatories throughout the world, including Europe, Russia, China, South Africa, Brazil, and Australia. Global VLBI became a practicality, with some experiments using ten or more separate telescopes to obtain milli-arcsec images of unprecedented resolution and image quality.

Initially the MK II tapes were correlated one baseline at a time on a two-station correlator operated in Green Bank. Later the correlator was expanded to simultaneously play back tapes recorded at three telescopes and the spectroscopic capability was increased from 32 to 288 and finally 512 frequency channels. With the growing use of VLBI, the MK II correlator evolved from an experimental operation to an NRAO facility, and was moved from Green Bank to Charlottesville to enable easier access to visiting users. At first, VLBI investigators came to Charlottesville to operate the processor themselves, but with the growing demand for time and the complexity of the playback operation, trained operators provided support in the same manner as the NRAO telescope operators. As in the case for telescope users, NRAO helped to defray the cost of travel to Charlottesville, provided access to the NRAO computing facilities, and support for publication page charges. Other MK II correlators based on the NRAO processor were later built and put into operation in Germany, the USSR, and China for limited use by investigators in these countries.

Frustrated with the long delays at the NRAO correlator, Marshall Cohen and Arthur Niell of JPL built a two-station MK II processor at Caltech which was expanded to five stations in 1978. Later, in collaboration with JPL, Caltech built a large playback facility that allowed up to 16 MK II tapes to be simultaneously played back and correlated. The Caltech/JPL processor had no spectroscopic capability, but starting in 1986 it became the correlator of choice for multi-station MK II continuum observations (Cohen 2007).

The MIT/Haystack MK III VLBI System Not long after the first successful VLBI observations in 1967, NASA initiated an ambitious geodetic VLBI program initially using a MK I compatible recording system. In order to obtain increased sensitivity, a new broadband record system, known as MK III, was developed at the Haystack Observatory with NASA funding (Rogers et al. 1983). Like the NRAO MK I and MK II VLBI systems, the MK III system recorded 1-bit Nyquist sampled data but used a Honeywell Model 96 multi-track instrumentation recorder to simultaneously record up to 28 tracks each at up to 4 Mbps (2 MHz bandwidth) sustained rate, giving better than a five times improvement in sensitivity over the NRAO MK II system. Initially the tapes were all correlated in a special processor built at the Haystack Observatory, but other processors were later put into operation at the MPIfR and at Caltech/JPL. However, with a tape speed of 135 inches per second, each 9200-foot-long tape only held about 13 minutes worth of data. Since each reel of the one-inch wide tape cost about $250, the astronomy community continued primarily to use the less sensitive and less expensive MK II system. Although as early as

1975 there were fully developed plans within NRAO to build a large MK III correlator in Charlottesville, there were never sufficient funds to begin any construction. All MK III astronomy observations were correlated at Haystack, Caltech/JPL, or at the NASA Goddard Space Flight Center, and later at Bonn.

Real Time VLBI Although the advances in magnetic tape recordings allowed improvements in VLBI data rates and sensitivity, the recordings were often defective, resulting in large playback errors. Moreover, shipping the large quantities of magnetic tape was both expensive and logistically demanding. International experiments had additional complications requiring customs clearance, sometimes accompanied by demands for tariffs. On one occasion, when clearing tapes being returned to the US after being recorded in Australia, the Los Angeles Airport customs officer wanted to collect import taxes, even though the tapes were owned by the NSF and had only been sent to Australia a few weeks earlier. Apparently, as he explained, now that the tapes contained data they were more valuable, and so were subject to import taxes. As he tried to explain, Hollywood movies that were filmed abroad were taxed, but it was the movie content, not the film that was taxed. Fortunately, in that particular instance, whether it was because he was told that the tapes only contained noise and no information, or whether he just tired of arguing with astronomers who clearly had no money, he gave up and cleared the shipment.

Even more important than the cost, inconvenience, and reliability of magnetic tape recordings, as previously discussed, any one of a number of errors in timing, polarization, or other technical malfunctions could ruin observations, and it might be weeks or months before the tapes were correlated and the failure recognized. Several approaches to suitable long-distance, broad-bandwidth, real time data links were considered, including a series of microwave radio links, late nighttime use of the national television network, and communications satellites, but all appeared prohibitively expensive. In 1976 and 1977, a team of US and Canadian radio astronomers were able to obtain time on the Canadian Communications Technology Satellite (CTS) (later named Hermes) for a series of real time VLBI Observations, first between the Green Bank 140 Foot Telescope and the Algonquin Radio Observatory (ARO) 150 foot antenna, and later between OVRO and ARO (Yen et al. 1977). The CTS was a joint project of the Canadian Department of Communications and NASA, and was available at no cost for approved investigations.

The successful real-time CTS-based VLBI observations were made at 10.7 GHz (2.8 cm) using a 10 MHz wide IF bandwidth (20 Mbps data rate, five times that of the MK II tape recording system). The one-bit IF data streams were time stamped based on timing signals from independent hydrogen maser clocks at each end and sent from Green Bank, via the geostationary CTS transponder, to ARO where they were correlated in real time. Because there was an approximately 0.25 second delay in the signal from Green Bank arriving at ARO, Benno Rayher at NRAO built a 0.27 second 5.5 million-bit delay line for the 20 Mbps ARO data stream. Small fluctuations in the data path length were accommodated by the 64 channel (3.2 μsec) correlator built by Yen at the University of Toronto.

As with tape recording VLBI, the hydrogen masers also provided independent coherent local oscillator reference signals. A later series of observations used the ANIK-B synchronous satellite to synchronize the local oscillators at two radio telescopes located in British Columbia, and the NRL radio telescope at Maryland Point, Maryland (Knowles et al. 1981). These phase coherent observations were used to demonstrate the feasibility of geodetic, time synchronization, and Earth rotation VLBI observations.

Between 1977 and 1980, European radio astronomers studied a real time link using the European Large Satellite (L-SAT) to distribute both data and local oscillator signals to European radio telescopes. However, their ESA sponsored investigation never progressed beyond the Phase A Study.[22] Starting in 2006, the EVN began to accept proposals for real time VLBI whereby the data are sent by fiber from the observing stations to JIVE for correlation, with data transmission costs supported by the European Commission. A similar capability was demonstrated the following year in Australia. This so-called eVLBI has become increasingly popular for time-critical observations such as flaring AGN.

8.5   VLBI Networks

In order to determine radio source structure in any detail, simultaneous or near simultaneous observations over a range of baseline spacings and orientations are necessary. Under the leadership of Marshall Cohen, what had now become the NRAO-Caltech VLBI group arranged to use the 85 foot radio telescope at the Harvard Radio Astronomy Station near Fort Davis, TX to supplement the telescopes in Green Bank and at OVRO, as well as the telescopes in Sweden and later Effelsberg, Germany. In support of the Fort Davis VLBI observations, NRAO provided a MK II record system, frequency standard, and receivers for 2.8 and 6 centimeters.

The US VLBI Network As described in Sect. 8.1, the early VLBI observations were all organized by the scientific investigators. Each series of observations necessitated writing separate proposals to each observatory, innumerable phone calls to arrange for the common observing time at multiple telescopes, shipping equipment and magnetic tapes, and arranging for people to travel to the telescopes to change tapes, to run the experiment, and then travel to Charlottesville to oversee the correlation of tapes. After the discovery of superluminal motion, the pressure for repeated observations to monitor radio source kinematics strained the available resources and personnel. The three-station VLBI correlator in Charlottesville required multiple playback passes to deal with the increasing number of four and five or more station observations, and was hopelessly backed up. Moreover, the record-playback system proved to be unreliable. At best, the multiple interactive adjustments needed to play back tapes recorded on different recorders required skill and patience; often playback error rates were so bad that the data had to be discarded. At other times, it was not clear if a low measured amplitude was real or the result of record/playback errors. Other issues included the lack of an antenna in the Midwest to complement the cluster of radio telescopes in the Northeast and in California, and the lack of antennas that worked well at short centimeter wavelengths.

In April 1974 NRAO held a meeting in Charlottesville to confront the mounting VLBI problems and to plan for the future. The 25 participants agreed that in principle a new dedicated array of properly located modern antennas, complete with standard instrumentation including hydrogen maser frequency standards, and a large central processing facility were needed to meet the increasing requirements of the growing VLBI community. Such an ambitious program was clearly in the future, and a short term solution, even if not ideal, was needed to exploit the growing opportunities for VLBI research.

The group envisioned a three-phase program:

1. Given the limited funds available, and recognizing the independent management of existing US radio observatories, the existing radio telescopes should be organized to the extent possible.
2. A new antenna located in the Midwest should be built to fill in the "Midwest Gap" in baseline coverage.
3. NRAO should pursue the design of a new array of antennas dedicated for VLBI.

Over the next few years, a series of reports called "VLBI Network Studies" were written to address the three phases discussed in Green Bank.

I. A VLBI Network Using Existing Telescopes (Cohen 1977)
II. Interim Report on a New Antenna for the VLBI Network (Swenson et al. 1977)
III. An Intercontinental Very Long Baseline Array (Kellermann 1977)
IV. On the Geometry of the VLBI Network (Swenson 1977)

Interested scientists continued to meet to discuss the formation of a US VLBI Network in order to provide reliable, versatile, and convenient facilities for VLBI observations and to provide an organization to discuss VLBI problems of national interest (Cohen 1977). Cohen (2000) later recalled that he adopted the term "network" rather than "array" as the NSF objected to "array," since NRAO was already building the VLA and the NSF was concerned that Congress might wonder why radio astronomers needed two arrays. Five organizations, NRAO/Green Bank, MIT/Haystack, Harvard/Fort Davis, Caltech/OVRO, and the University of California/Hat Creek, each committed one week of coordinated observing time to VLBI every two months. The University of Illinois and the USNO agreed to make their telescopes available at the Vermillion River and Maryland Point Observatories respectively. A Network Users Group (NUG) was organized to provide a single source for receiving and refereeing proposals, to organize the distribution of magnetic tapes, and to coordinate the observations and correlation. The NUG, which included some 40 VLBI scientists, met regularly, usually in connection with the annual URSI meeting in Boulder, CO, and addressed many of the proposal and scheduling issues. The NUG also organized a Technical Committee to establish standards, but logistical and technical problems continued as the NUG had no power or funds to implement changes at the participating observatories, which each had their own priorities.

Volume IV (Swenson 1977) of the "VLBI Network Studies" series addressed the question of the so-called "Midwest Gap" in the array of existing antennas. Both the University of Illinois[23] and the University of Iowa proposed to construct a new Midwest antenna to plug the existing gap and to ultimately become the first antenna of a multi-element dedicated array. The proposals were supported by the NUG and were seriously considered by the NSF, but before funding became available the motivation for the Midwest telescope was overtaken by the VLBA.

In 1981, six groups, Caltech, Harvard-Smithsonian, MIT, University of California at Berkeley, University of Illinois, and University of Iowa, signed an MOU to form a VLBI Consortium with the goal of increasing the effectiveness of the Network by making observations more convenient and more reliable. For legal reasons connected with NRAO's status as a national facility, NRAO did not initially join the Consortium, although at least one NRAO scientist participated in each Consortium meeting as an at-large member, and NRAO participated in all Network organized activities. In 1986 this arrangement was formalized when NRAO became an Associate Member of the VLBI Network Consortium. Each member contributed $2000 a year to the Consortium, which could then purchase recording media and arrange for their distribution among the observatories and the correlators at Caltech or NRAO, set technical standards, and handle proposal reviewing and scheduling. Each observatory appointed a VLBI friend to help support in-absentia observing. No longer was it necessary for the observers to provide personnel at each telescope. With a single proposal, a small group, or even one person with a good scientific project, could now get simultaneous observing time at all the radio telescopes operated by Consortium members.

The European VLBI Network (EVN) Starting with the 1968 and 1969 VLBI observations with Sweden and the Soviet Union, European radio observatories became increasingly involved in VLBI observations, generally using NRAO MK II recorders but on occasion borrowed MK III systems. The data were mostly correlated in Charlottesville at the NRAO MK II processor. In 1977 the MPIfR began an ambitious VLBI program with the construction of first a MK II and later a MK III playback system and, in August 1978, the MPIfR hosted a major international VLBI Symposium in Heidelberg, Germany, attended by about 100 scientists.

Inspired by the US VLBI Network, and following a series of informal meetings, in March 1980 the directors of five European radio observatories[24] met in Bonn and agreed to create the European VLBI Network (EVN), and then in 1984 created the more formal Consortium of European Radio Astronomy Institutes for Very Long Baseline Interferometry. The EVN formed a Program Committee to review proposals three times a year and to schedule VLBI observations every two months. A Technical Working Group (later Technical and Operations Group) specified standard observing frequencies, polarization standards, and data formats. As in the US, the EVN accepted proposals from any qualified scientists, including those with no European affiliation, and provided local support at the individual telescopes. Italy built two new radio telescopes dedicated to VLBI, one near Bologna and one in Sicily, and other radio telescopes in UK were added to the EVN.

As the number of antennas involved in EVN observations increased, the three-station MPIfR MK II processor became oversubscribed due to the multiple passes required for each observation. Some of the larger MK II experiments were processed at NRAO or Caltech, or, in the case of MK III observations, at Haystack. MPIfR built a three station MK III processor which was later expanded to five stations, but even this was inadequate to handle the growing number of multiple antenna observations, and priority was given to experiments involving MPIfR staff.

Following several unsuccessful attempts to fund a large European VLBI processor, in 1993 the Dutch government established the Joint Institute for VLBI in Europe (JIVE) in Dwingeloo to build and operate a 16 station MK III processor to support European VLBI observations. Richard Schilizzi was appointed the Director of JIVE. With the support of the EVN observatories, EVN observations have, since 1999, been processed at JIVE, which has developed into the center of VLBI research in Europe. In addition to its role in processing EVN observations, JIVE archives the correlated EVN data.

The EVN ultimately grew to more than 20 telescopes at 15 institutes in 12 countries, including some in Africa, Asia, and North America. A unique feature of the EVN has been the series of well attended EVN Symposia and the EVN User Committee meetings, which started in 1993 and have been held every two years since 1994. These symposia, which are usually held in conjunction with one of the regular EVN Director's meetings, have served to coalesce the European VLBI community and specifically to introduce new young scientists to VLBI opportunities. More detailed discussions of the EVN are given by Porcas (2010), Booth (2013, 2015), and Schilizzi (2015).

Global VLBI To accommodate transatlantic observations, the EVN and NUG combined to schedule "Global" VLBI observations, which often also included radio telescopes elsewhere in the world. As many as 18 different antennas (e.g., Reid et al. 1989) were included in some of these global observations. In Japan, scientists began a VLBI program, partly motivated by geodetic interests relevant to potential earthquake prediction. Other VLBI networks were initiated in Australia, Korea, and China. In South Africa, a former NASA tracking antenna at Hartebeesthoek was given to the South African hosts and instrumented for VLBI. Initially, most global VLBI observatories used the simple and economical MK II record system, which was gradually replaced by MK III and later VLBA-compatible recording systems, although for a while Canada, Jodrell Bank, Australia, and Japan each continued to use their own incompatible recording systems.

8.6   Planning the VLBA

The first discussions about building a dedicated Very Long Baseline Array (VLBA) began at NRAO in the summer of 1973.[25] Shortly after the 1974 meeting in Charlottesville, Dave Heeschen set up a "VLBA Design Group to continue the development of the concept of a dedicated Intercontinental Array and to help upgrade the present activities in VLB Interferometry."[26] Following the Charlottesville meeting, Swenson and Kellermann (1975) discussed the status of VLBI and some early ideas about a dedicated VLBA. A more complete description of a ten element dedicated Intercontinental Very Long Baseline Array based on tape-recording interferometry was prepared at NRAO as a collaborative effort of NRAO and external scientists and engineers, especially those from Caltech and Haystack Observatory (Kellermann 1977). The NRAO report noted that in addition to addressing a wide range of astrophysical problems, the proposed array could be used for precise tests of General Relativity and for interplanetary spacecraft navigation, and would have applications to a variety of terrestrial phenomena including the measurement of Earth tides and continental drift, accurate global clock synchronization, and the possibility of earthquake prediction. Like other NRAO facilities, the VLBA would operate as a single instrument available to all scientists based on competitive proposals. The proposed VLBA consisted of ten 25 meter diameter antennas, at least eight of which were in the continental United States. Placing the other two antennas in Hawaii and Spain or the Azores would increase the angular resolution to be about the largest possible on the surface of the Earth. Locations in Alaska, Iceland, Mexico, and Easter Island were also discussed to improve the north-south resolution. Unlike the VLBI Network antennas, which were originally built for other purposes, the proposed Intercontinental Very Long Baseline Array antenna elements would be more optimally placed, but with consideration of road access and the availability of power. Each antenna would be supported by a small staff to maintain the instrumentation and the antenna, change tapes, and arrange for their transportation to and from the central processor. However, overseeing the observations, pointing the antennas, changing frequency, etc. would be under the control of a remote central operator.

The proposed VLBA was based on demonstrated technology: VLA type antennas and receivers and MK II or MK III recording and playback systems. The biggest challenge was to provide sufficient staff to change tapes as often as four times an hour, devising a robot tape changer, or developing a cost effective real-time satellite data link. Another challenge was the lack at the time of a commercial source of the hydrogen masers needed to provide the necessary frequency stability for operation at wavelengths as short as 1 cm. NRAO estimated the VLBA construction cost to be about $26 million, and the operation plan called for a staff of 53 including two people at each antenna site to oversee the observations, change tapes, and provide basic technical support.

At this time NRAO was in the midst of the VLA construction (Chap. 7) and was also being pressured to build the 25 meter millimeter radio telescope (Chap. 9) which had received considerable support from the Greenstein Decade Review Committee (Sect. 7.4). Support for the VLBA was divided, even among practicing VLBI scientists. More meetings were held, but NRAO had inadequate resources to pursue serious engineering work on the VLBA, and little progress was made.

In 1979, a conceptually similar idea, called the Canadian Long Baseline Array (CLBA), was proposed by Canadian radio astronomers "to serve the needs and interests of Canadian astronomers." The proposed CLBA contained eight antennas located in Canada and one in France for both scientific and political reasons.[27] The Canadians proposed using larger antennas for better sensitivity, but the VLBA operated at a shorter wavelength so would have better angular resolution.[28] The US and Canadian groups discussed a possible collaboration, but the Canadian group thought that their government would be more receptive to a purely Canadian project. Perhaps reflecting their true fears, the Canadian radio astronomers also expressed concern that any joint effort with the US would be dominated by the Americans, and felt that the CLBA was a chance for Canada to take the lead. At a meeting of the Canadian Astronomical Society, Ernie Seaquist, Chair of the CLBA Planning Committee, claimed that "the CLBA is currently funded," and that the funding prospects were poorer in the US, so that the CLBA would be built before any US instrument.[29] Instead of reacting to the challenge of competition from Canada, William E. (Bill) Howard III, then the NSF Director of Astronomical Sciences and a former NRAO Assistant Director, saw the CLBA as his solution to the growing competition between the VLBA and the NRAO millimeter telescope and asked, "Why not let the Canadians do it?" By 1983, the positions on both sides had softened, but only slightly. Both the Canadians and Americans wanted to go ahead with both the CLBA and VLBA proposals. At an April 1983 meeting in Charlottesville, the two groups discussed ways to collaborate in the unlikely possibility that both arrays were built.[30] There were many compatibility issues, particularly regarding the recording systems used. All of the VLBA technical, scientific, and management memos were available to the Canadian group and some Canadians participated in the VLBA Working Groups, but little information flowed in reverse. The CLBA was chosen by the National Research Council of Canada over several other big science projects. However, with increasing CLBA cost estimates and the deteriorating economic situation in Canada, funding never materialized; the CLBA initiative eventually died, and Canadian radio astronomers turned their attention to other directions.

Although there was as yet no formal proposal requesting NSF funds for the VLBA, informal exchanges between NRAO and the NSF led to the inclusion of the VLBA in the Astronomy Division's planning for new starts, but after the NRAO 25 meter millimeter wave telescope, and in competition with the planned KPNO NTT 15 meter optical telescope, along with one or two 10 meter submillimeter telescopes, as well as a wide range of upgrades, instrumentation, and support for existing telescopes at all wavelengths.

Uncertain about the commitment of NRAO and concerned about input to the upcoming Decade Review of astronomy, Marshall Cohen, a member of the Decade Review radio panel, began a semi-independent design effort at Caltech in collaboration with JPL, although many of the same people, including NRAO scientists and engineers, participated in both the NRAO and Caltech design programs. The broad goals which would motivate the design of both the NRAO and Caltech arrays were discussed in a meeting held in January 1980 at Caltech, which was attended by about 30 scientists and engineers. The participants agreed that ten antennas would be a reasonable compromise between cost and imaging quality. Caltech issued the results of their design study in September (Cohen et al. 1980), in time for consideration by the Decade Review Committee. The existence of two institutions pushing for a VLBA added credibility to the project, and the item by item comparison of the independent cost estimates ultimately led to a better understanding of the VLBA construction costs.

Encouraged by a specific request from the NSF for a "Conceptual Proposal" and by the apparent increase in community support, NRAO responded by resuming its VLBA design program. In May 1980, NRAO Director Morton Roberts appointed a formal VLBA Design Group to re-examine the scientific motivation and technical feasibility of the VLBA.[31] Working Groups on antennas, configuration, correlators, data transfer, electronics, feeds, front ends, local oscillator, recording systems, monitor and control, operations, post processing, science, sites, and management focused on preparing a new report, and continued throughout the VLBA construction period to oversee the design and provide engineering support. NRAO staff were joined by Working Group members from the US and Canadian VLBI communities, particularly members from Haystack and Caltech, who led the recorder and correlator groups respectively. From 16–18 September 1980, a group of some 70 astronomers, geodesists, and engineers from Canada, Germany, the Netherlands, and Italy, as well as from the US, met in Green Bank to iron out the differences between the NRAO and Caltech studies. The first day was devoted to scientific presentations highlighting recent VLBI research, followed by a VLBI NUG meeting. The second day began with overviews of the Caltech-JPL, NRAO, and Canadian design concepts, followed by discussions of the array configuration, the antennas, front end, record, local oscillator, and playback designs. The Green Bank meeting consolidated the main concepts behind the VLBA and served to unite its supporters to urge action from NRAO.[32]

Following the Green Bank meeting, the 1977 NRAO report was updated to reflect the scientific and technical progress over the preceding three years. The February 1981 "Very Long Baseline Array Design Study" included only antennas located in the United States, but also considered the use of real time satellite links to replace the expensive and commercially unavailable hydrogen masers, and also an upgraded MK III system that could allow an order of magnitude increase in capacity of each tape, allowing a single tape to last for up to six hours instead of 13 minutes (Kellermann 1981). The proposed construction and annual operating cost of the VLBA were given as $39.1 million and $3.8 million respectively. In order to help prepare a formal proposal to the NSF and to consolidate the community, in November 1981 Roberts appointed a VLBA Planning Group of external scientists to advise on the preparation of an actual proposal to the NSF for VLBA construction.[33] Although the Planning Group was initially expected to function for only a few months, it remained in existence until June 1985 when it was dissolved by Paul Vanden Bout.

8.7   Funding the VLBA

In December 1978, the National Academy of Sciences began its next Decade Review of Astronomy and Astrophysics. Patrick Thaddeus from Columbia chaired the Radio Astronomy Panel. VLBI interests were represented by Marshall Cohen from Caltech and Bernard (Bernie) Burke from MIT.[34] Both the NRAO and Caltech reports had demonstrated the potential scientific impact of a dedicated VLBI system, that such an array was technically feasible, and that the cost could be reliably estimated. However, the VLBA was competing with a proposed high altitude 10 meter submillimeter telescope and the long overdue 100 meter class fully steerable short centimeter wavelength dish, as well as the need for maintaining and upgrading the VLA and Arecibo. Although early in their deliberations Thaddeus and the Radio Panel placed the VLBA as the highest priority new start for radio astronomy in the coming decade, there were still important undecided issues. Who would build it? What was the right tradeoff between the number of antennas and cost? What was the best recording system?

To address these and other questions, Thaddeus took the unusual step of bringing the key VLBA protagonists to his country cabin in rural New York to thrash out the issues. Although he expressed enthusiasm for the VLBA project, Thaddeus was concerned that in order to sell it to the parent Survey Committee, and later to the NSF and Congress, it would be necessary to keep the cost below $30 million. But $30 million was insufficient to build a ten element array. This generated vigorous discussion about the minimum number of antennas needed, whether some existing antennas could be used instead of building all new ones, the broad issue of international participation, and the optimum funding schedule. Unlike the strict confidentiality of later Decade Reviews, both the NSF and NRAO personnel had the opportunity to comment on and contribute to successive drafts of the reports of both the panels and the main committee.

As expected, the final report of the Radio Panel listed the VLBA as its clear first priority for new ground-based construction (Thaddeus 1983).[35] Based on the NRAO and Caltech design studies, the Committee suggested a total construction cost of $35 million. The Radio Panel report also listed space VLBI as the first priority for space radio astronomy and gave strong endorsements to the construction of a 100 meter class centimeter wavelength telescope, primarily for the support of space VLBI. At the same time, the Ultraviolet, Optical, and Infrared (UVOIR) Panel was divided among those who argued for a state-of-the-art large ground-based OIR telescope and those who wanted more intermediate class instruments. Unable to reach a consensus on priority, the UVOIR Panel recommended as their first priority both a "national New Technology Telescope (NTT) of the 15-m class for optical and infrared observation," and "the construction at good sites of several smaller national telescopes with apertures between 2.5 and 5 m" (Wampler 1983, pp. 98–99).

The parent survey steering committee, chaired by Harvard's George Field, endorsed the construction of the VLBA as the first priority for all of ground-based astronomy in the 1980s with a price tag of $50 million that included ten years of operation funding (Field 1982).[36] The Committee also noted, "The 25-Meter Millimeter-Wave-Radio Telescope, which was recommended in an earlier form in the Greenstein report, has not yet been implemented" but left unanswered any recommended priority between the VLBA and the 25 meter telescope, both proposed NRAO projects. Moreover, the Panel made no recommendation about who would build or who would manage the VLBA once built. As with the proposed VLA and the OVRO Array a decade earlier, Caltech and NRAO were again competing for NSF support.

The second priority for ground-based astronomy was the 15 meter New Technology [optical] Telescope (NTT) which was deemed to be of equal scientific merit to the VLBA, but not yet technically ready for construction. So the report recommended only that the NTT "*design studies … are of the highest priority and should be undertaken immediately*" (Field 1982, p. 16). The National Optical Astronomy Observatory undertook design studies for a 15 meter OIR telescope, but as of 2019 the largest optical-infrared telescopes operating in the United States are only in the 8–11 meter class. A 24 meter equivalent Giant Magellan Telescope (GMT) is under construction in Chile, and a Thirty Meter Telescope (TMT) is planned pending the identification of a suitable site.[37] Meanwhile a 39 meter equivalent diameter telescope, known as the Extremely Large Telescope (ELT) is under construction in Chile by ESO.

During the Field Committee deliberations, the NSF National Science Board discussed "Big Science Policies and Procedures." The NSB recognized "Big Science" projects not only cost a lot to build, but would have a continuing operating cost that with level or even declining budgets could adversely impact a broad range of other programs in the field. The Board wisely set a high bar for supporting big projects at the NSF and defined a demanding procedure for funding "big science."[38]

The VLBA Versus the 25 Meter Millimeter Wave Telescope After years of planning and development, in 1977 NRAO had sent a proposal to the NSF to build a 25 meter millimeter wave telescope on a high altitude site chosen for low water vapor (Sect. 10.3). In January 1980, President Jimmy Carter's FY1981 budget proposal included $1.7 million for the engineering design of the 25 meter millimeter telescope, but a few months later the proposal was withdrawn by the NSF following a $70 million cut in the NSF budget. Carter's budget proposal for FY1982 again included a start for the 25 meter telescope, now on a proposed three instead of four year construction schedule. But with the nation dealing with high unemployment and inflation resulting from the unprecedented gasoline prices brought about by the Iran Oil Crisis, incoming President Ronald Reagan's economic recovery plan froze all FY1982 new starts, specifically mentioning the 25 meter telescope,[39] and the 25 meter telescope did not appear in the FY1983 budget request. However, the Decade Review Radio Panel had made an early decision, probably in 1980, not to re-evaluate "ongoing programs approved by previous advisory committees." Specifically, their report noted that "the most important such project in radio astronomy is the 25-m millimeter-wave telescope proposed by the NRAO." Assuming that better times lay ahead following the end of the Iran Hostage Crisis, the Radio Panel Report stated that, "The present report of the Panel on Radio Astronomy is predicated on the assumption that the 25-m telescope will be constructed during the early or middle years of the 1980's" (Thaddeus 1983, p. 212) [original underlining]. Similarly, the parent Survey Committee emphasized "the importance of approved, continuing, and previously recommended programs," and specifically noted that "The 25-Meter Millimeter-Wave-Radio-Telescope … has not yet been implemented" and would permit "the United States to maintain its leadership in this exciting and highly productive field" (Field 1982, pp. 13–14, 120). Both the Radio Panel and the parent Survey Committee stopped short of making the difficult decision of how to proceed if the 25 meter was not funded.

Meanwhile work on both projects continued at a low level at NRAO. With the new strong Field Committee recommendation for the VLBA, as well as a letter writing campaign from the VLBI community, and the long outstanding but still unfunded 25 meter project, Roberts was faced with a dilemma. NRAO was just bringing the VLA into operation with inadequate staff and an insufficient operating budget, and was dealing with conflicting pressures from the VLBI and millimeter communities for another new start. These were two fields, millimeter astronomy and VLBI, that had been started at NRAO, but where US leadership was being threatened. The NRAO scientific staff itself was split over the two projects (Gordon 2005, pp. 140–145), and there was growing concern from both communities that nothing was happening.

To help decide on the best approach, Roberts convened an ad-hoc committee to adjudicate between the two projects.[40] The committee met in the Washington offices of AUI on 25 January 1982. Their views ranged from

"25 meter with enthusiasm; the VLBA not yet ready; stay with mm telescope;" to "VLBA new, more attractive and more saleable; VLBA to maintain credibility of Field Report; 25 meter no longer attractive."[41] As Roberts later reported to the NSF, the committee felt that the scientific case for both projects was "equally strong" and that "a majority favored seeking funds in the FY1984 budget for the 25 meter telescope."[42] Following a discussions within NRAO, and realizing the long term impact "to the astronomy community in general and on NRAO in particular," Roberts wrote to Francis Johnson, NSF Assistant Director for Astronomy, Atmospheres, Earth and Oceans, to "urge the NSF to include the 25-meter telescope in its plans for FY1984." He also told Johnson, "We will complete, as rapidly as possible, the preparation of a VLBA proposal for submission to the NSF."[43]

This ambiguous NRAO position, along with the lack of any clear recommendation from the Field Committee, left the NSF in a quandary. They could not fund both NRAO radio astronomy projects in the coming decade and sought the advice of their own Astronomy Advisory Committee (AAC)[44] which met on 5–6 April 1982 at the NSF in Washington. Mort Roberts made the good suggestion to the NRAO and Caltech VLBA advocates that "it would be completely inappropriate to use that occasion to push for one's own proposal. Not only inappropriate but ineffectual, for the details of approving one proposal versus another will not be left to the AAC, but will be based on peer reviews and internal gyrations within the NSF."[45]

The presentations for the millimeter telescope went first, but ran overtime, leaving little time for the VLBA and for probing questions about the CLBA, the differences between the Caltech and NRAO plans, the use of existing radio telescopes instead of building new ones, and other delicate concerns. During the discussion following the presentations, the committee noted that the 25 meter proposal was clearly getting old; Japan had just completed the 45 meter Nobeyama millimeter wave telescope, and the 30 meter IRAM millimeter telescope on Pico Veleta, in the Spanish Sierra Nevada, was already under construction. On the other hand, the 25 meter concept was mature, while there was no real proposal or engineering design for the VLBA. The 25 meter was "shovel ready;" the VLBA was not.

Nevertheless, soft spoken committee member Richard McCray unexpectedly suggested that the time had passed for the millimeter telescope and that the VLBA represented a new opportunity to extend US leadership in this important area which had broad applications beyond astrophysics. McCray was a respected theoretical astrophysicist and had been a member of the Field Committee, so his comments were taken seriously, and the committee subsequently recommended unanimously that the NSF pursue funding for the VLBA, and at the same time voted seven to three to not go ahead with the 25 meter proposal.[46] Two weeks later, Roberts wrote to the NSF Director, John Slaughter, to "request that the NSF set aside our request for funds to construct [the 25-meter millimeter wave telescope]"[47] This was the effective end of the NRAO 25 meter project. Although Roberts' controversial action left a bitter taste among the millimeter astronomers, including those at NRAO, it would eventually lead to a major new US initiative in millimeter astronomy (Sects. 10.6 and 10.7) as well as to the funding and construction of the VLBA.

At its 19 October 1982 meeting, the AAC revisited the VLBA and unanimously passed the following resolution:[48]

> The Very Long Baseline Array radio telescope was recommended by the Astronomy Survey Committee as the highest priority new facility for ground-based astronomy. The Astronomy Advisory Committee recommends that the NSF seek the necessary funds to construct this facility as soon as possible.

Interestingly, another major previously approved but still unfunded project was the Space Infrared Telescope Facility (SIRTF). But unlike the NRAO 25 meter telescope, the infrared community and NASA did not abandon SIRTF, which was finally realized with the launch of the Spitzer Space Telescope in 2003, and its subsequent very successful mission.[49]

Multidisciplinary Use of the Very Long Baseline Array The response of any radio interferometer depends not only on the structure of the radio source, but also on the interferometer baseline length and its orientation, the coordinates of the radio source, and the local oscillator frequencies. In practice, the data are correlated using a range of probable fringe rates and time delays which are then examined in a computer to find the fringe rate and delay that gives the maximum interference fringe amplitude. Working backward, the observed fringe rate and delay can then be used to determine with great accuracy the geometry of the interferometer baseline and the relative time offset between the two antennas. It was, therefore, immediately clear from the first 1967 VLBI observations that in addition to astronomy and astrophysics, VLBI techniques also had a variety of important terrestrial applications, including the measurement of Earth rotation (UT1), polar motion, Earth tides, continental drift, and possibly earthquake prediction (e.g., Gold 1967; Cohen et al. 1968). Since radio source coordinates can be determined with great precision, VLBI is also used for precise tests of General Relativistic gravitational bending, for spacecraft navigation, and to locate lunar and planetary exploration vehicles.

Although NRAO and the VLBI community had promoted the VLBA to the NSF and the Field Committee based partly on the geodetic and other non-astronomical applications, the design discussions had not really responded to the needs of these other applications. It was clear that support of the geodetic community as well as the astronomy community was needed before the NSF would fund the construction of the VLBA. At Bernie Burke's initiative, the National Research Council held a two-day workshop at the National Academy of Sciences in Washington on 6–7 April 1983.

During the course of the Workshop on Multidisciplinary Use of the Very Long Baseline Array (NAS 1983), it became increasingly clear that although the geodetic community needed and wanted the VLBA, their support was predicated on NRAO being more sensitive to the needs of geodesy in designing the array, calibration procedures, providing auxiliary instrumentation, and in dealing with proposals and scheduling. To better accommodate the wide range of non-astronomical observations discussed at the workshop, NRAO agreed to increase the elevation range of the antennas, to increase their slew speed, to provide for simultaneous observations in the 4 and 13 cm bands commonly used for geodetic studies, and to increase the number of IF channels. All of these modifications added to the construction cost. From the viewpoint of the astronomy user, perhaps the most serious compromise was the choice of the IF system and some frequency bands to be compatible with existing geodetic VLBI systems rather than the VLA. Interestingly, the NAS meeting was not only funded by the NSF, NASA, and the NOAA National Geodetic Survey, but also by the Defense Advanced Research Projects Agency (DARPA) and the Defense Mapping Agency.

The NSF and Congress As with the VLA (Chap. 7) and the GBT (Chap. 9), obtaining VLBA construction funding was complex, but in each case for different reasons. Following the NSF Astronomy Advisory Committee decision in April 1982 and the President's Office of Science and Technology Policy (OSTP) briefing, things moved very fast, at least at first. The following month, Roberts sent the formal proposal for a Very Long Baseline Array (Kellermann 1982) to the NSF Director, John Slaughter, requesting "funds for such construction and operation."[50] In accordance with the agreement with Caltech, the front page of the proposal noted that it had been "prepared in collaboration with the California Institute of Technology." The NSF was still remembering the 25 meter situation and was unsure of NRAO's intentions. Roberts wrote to NSF Assistant Director Francis Johnson in July, saying, "*We* [NRAO] *conclude that we must now ask that the VLBA proposal be accorded the highest priority on the Foundation's agenda, ...* [and] we look forward to vigorous support of the VLBA project by the NSF and the NSB and hope that significant planning, design, and development funding will be available in FY 84." [italics in original].[51]

By this time Bill Howard had left the NSF over the disruption resulting from the reversal of support for the nearly funded 25 meter telescope in favor of the VLBA,[52] and had been replaced by Laura (Pat) Bautz as the NSF Astronomy Division Director. Larry Randall, the NSF Program Officer for NRAO and later Head of the Astronomy Centers Section, generously interacted with the NRAO Project Manager, (KIK), over the preparation of the VLBA proposal and the development of a budget plan for its construction, as did Kurt Weiler, who had specific responsibility for the VLBA at the NSF. During the final preparation of the proposal, some concern was raised that "VLBA" was not a very inspiring name, and that it might be easily confused with the VLA. Other names considered included "Trans American Radio Array (TARA)", "Trans American Radio Telescope (TART)", and "Trans American Telescope (TAT)". NRAO decided to adopt the appealing Caltech name,

Transcontinental Radio Telescope, or *TRT*, much to the dismay of the NRAO secretary who had to retype the proposal. But noting that they were already discussing the VLBA with the Office of Management and Budget (OMB) and Congress, Larry Randall rejected the name change, and the proposal had to be retyped yet again using "VLBA."

The 1982 proposal submitted to the NSF specified a construction budget of $50.729 million and an annual operating budget of $4.15 million. Characteristically, the NSF and the community focused their attention on the capital cost with little consideration paid to the operating cost. Indeed, NRAO never received the full incremental operating funds needed for the VLBA, and in particular, the annual $500,000 requested to upgrade the instrumentation never materialized. NRAO proposed an optimum five year construction plan to start in FY1984 that included a first year of engineering design along with the construction and corresponding site acquisition for the first antenna prototype. The plan initially called for $2.5 million for engineering design and to procure the first antenna, with the rest of the construction spread out over the next three years.

By this time, the price of oil and the rate of inflation had stabilized and Jay Keyworth, President Reagan's Science Advisor, announced that he was looking for "high leverage" areas where modest investments would have a "high impact" with regard to the 1984 budget. In response to Keyworth's request, the NAS appointed seven Briefing Panels to identify "those research areas within the field which are most likely to return the highest scientific dividends as a result of additional federal investment."[53] One of the authors (KIK) was a member of the "Briefing Panel on Astronomy and Astrophysics," which was chaired by George Field. Unlike most NAS studies and reports, this one was remarkably fast. It was just over a month between the appointment of the panel and the final report and presentation to OSTP, and "the Briefing panel quickly and unanimously identified the VLBA as the number one priority to bring to the attention of OSTP for a 1984 new start."[54] Field and NAS President Frank Press presented the report of the Briefing Panel on Astronomy and Astrophysics to Keyworth and other OSTP staff on 15 October 1982. Apparently Keyworth had already read the report, and told Field that "the briefing had revealed nothing new to them," and the one-hour presentation was dominated by the broader issues of the space station and NASA's emphasis on technology over science. In response to Field's "concern that the NSB looks unfavorably on large projects such as the VLBA," Keyworth reassured him that "OSTP, not the NSB, decides such issues," and that "several of the items were already being taken care of."[55]

The National Science Board, nervous about continuing operational requirements of big projects and their impact on grant support, only approved $0.5 million funding for the VLBA in FY1984. However, OSTP and OMB restored the full $2.5 million in President Reagan's budget proposal, which was included in the Congressional FY1984 NSF Appropriation as part of a large increase in the administration's support for science.

Although the VLBA had not yet gone through the NSF review process, Reagan's FY1985 budget request included $15 million for the first year of a four year VLBA construction project planned at approximately $15 million per year. Nevertheless, the NSF apparently still wanted to review the proposal and obtain endorsement from the National Science Board. In preparation for a presentation to the NSB, the NSF sent the NRAO proposal out for peer review, and convened a "blue-ribbon panel" led by Joseph Taylor from Princeton to make recommendations on the overall soundness of the project, management, technical specifications, staffing, timing, and costing.[56] In preparation for the September NSB meeting, the Review Panel met at the NSF on 30–31 May 1984. Their report concluded that the VLBA was scientifically important, the specifications were appropriate and attainable, the staffing and management plans were adequate, construction and operating costs credible, and the time scale realistic. But the panel suggested that the contingency be increased to 15 percent.[57]

However, in spite of the support from OMB and OSTP, the VLBA ran into an unforeseen snag in Congress. Massachusetts Representative Edward Boland was the powerful Chair of the House Appropriations Sub-Committee on Housing, Urban Development (HUD), and Independent Agencies that had jurisdiction over the NSF appropriations. Boland had little regard for astronomy or astronomers, apparently because of an earlier battle with supporters of the Hubble Space Telescope, and he pushed to include more funds to support supercomputing and for science education at the NSF. Five years earlier, at the dedication of the Five College Radio Astronomy Observatory 14 meter millimeter-wave radio telescope, Boland told the gathered astronomers that hard times were coming and not to expect money for more new radio telescopes, and he decided to hold the VLBA hostage to achieve his goal of increasing support for science education.

Boland's Chief of Staff, Richard Mallow, was also a formidable adversary. He had just authored a report supporting increased funding for supercomputing and cautioning against funding the VLBA and other Field Committee recommendations in view of the projected increasing costs of the Space Telescope and other ground-based optical telescopes in Arizona, Hawaii, and Chile. According to another Congressional staff member, "Dick Mallow tries to run the committee, but the other staff members do not always let him have his way."[58] Armed with Mallow's report, Boland argued that "it is more important that the NSF put more money into science education than into VLBA." To complicate the situation, Boland and Mallow's interest and support for computing was not entirely unwelcome at NRAO, since the Observatory was also interested in obtaining a super computer to deal with the growing volume of data from the VLA.

George Field and others wrote Boland a strong letter of support for the VLBA, explaining that the federal government spent a lot more on science education than was found in the NSF budget, and that the VLBA correlator was itself at the frontiers of computing technology.[59] In response, Boland remarked,[60] "This nation's future does not and will not depend on building the VLBA. It does depend, however, on how adequately we educate our children—particularly in science and mathematics." Field, Kellermann, and others tried to defend the VLBA with visits to key Congressional Offices,[61] but were unable to see either Boland or Mallow.

The "old boy network" then went to work to save the VLBA. MIT Professor Bernie Burke, a long-time supporter of VLBI and the VLBA, brought the VLBA problem to the attention of the MIT Dean of Science John Deutch and past MIT President Jerry Wiesner. Wiesner was a friend of Speaker of the House Tip O'Neill, who represented the old 8th Congressional District which included Cambridge. O'Neill previously shared living accommodations in Washington with Boland and reportedly persuaded Boland not to kill the VLBA in the House appropriations bill.

Following the community pressure, Boland allowed the VLBA to remain in the FY1985 House appropriations bill at the requested $15 million, but the bill stipulated that this money could not be spent until the NSF FY1986 budget request included at least 8.5 percent for science education.[62] The Senate appropriations bill also included the requested $15 million and, with the help of New Mexico Senator Pete Domenici and Jake Garn, Chairman of the Senate Appropriations Committee, eliminated the House education rider.

With $15 million for the VLBA included in both the House and Senate appropriations bills, the prospects looked encouraging. But to everyone's apparent surprise, when the House-Senate Conference Committee met on 26 June 1984, they "compromised" and appropriated only $9 million for FY1985 and retained the proviso that no money could be obligated until 1 April 1985, six months into the fiscal year,[63] and then only if Boland's education requirement were met in the NSF's proposed FY1986 budget. Boland had suggested a 4th quarter (July 1985) start for the VLBA, apparently a widely used tactic to avoid implementing an approved program, and the Senate had countered with a February start. April 1 was a compromise, but the final bill also specified that NRAO could not issue a Request for Proposals (RFP) for the antennas until after this date. When Mallow found out that an RFP had already been issued on 9 March, he reportedly went "non-linear."[64] Further, when Boland's committee agreed to include $15 million in the House Bill for the VLBA, they reduced the overall appropriation for the NSF AAEO Division by $12 million in order to minimize the impact to the federal deficit. But when the VLBA funding was reduced to $9 million by the Conference Committee, the AAEO cut remained, resulting a net loss of $3 million. The NSF was not happy about this, and suggested that community pressure was not necessarily useful and might even be counterproductive.[65]

When Kellermann met with Senator Domenici staffer George Ramonas in the Senator's office on 16 August 1984, he was assured that "As long as Pete Domenici is in the Senate, the VLBA will be protected, particularly if he remains as part of the majority party," but Ramonas acknowledged that Boland would probably try to remove the VLBA from the 1985 budget at the time of the FY1986 Appropriation Committee hearings in the spring of 1985.[66] Although the House and Senate HUD Appropriations sub-committees had to deal with 13 separate agencies and a total of $59 billion of appropriations, the VLBA had become a pawn in the OMB-House-Senate-NSF relationship. Reportedly most of the discussion at the House-Senate Conference for the FY1985 appropriation was devoted to the VLBA. Ramonas was a valuable contact in following the NSF-Congressional debates until he left Domenici's office in early 1985. When George Field later spoke with Ramonas's replacement, Joseph Trujillo, Trujillo indicated that he had never heard of the VLBA.[67]

The FY1985 VLBA construction funding was finally released to NRAO on 15 May 1985. By this time, reflecting the increased level of contingency suggested by the Taylor Committee and the projected inflation over a proposed project stretch-out, the expected cost of the VLBA had risen to $68.2 million.

For FY1986, the NSF and OMB requested $11.5 million for continued construction of the VLBA, but again asked for only $51 million for science education, and in defiance of previous instructions from Boland, they had deferred spending $31 million from the prior year's appropriation. At the 1986 budget hearings on 26 March 1985, Representative Boland was incensed at the NSF and the administration and informed Keyworth that, "The Administration doesn't seem to have any trouble finding money for VLBA. But it can't help with science programs for children." Keyworth responded by pointing out that the VLBA was the highest priority in astronomy and had gone through extensive peer review, adding, "I cannot think of a scientist in America who is a recognized authority in astronomy who questions the utility and viability of the VLBA," to which Boland retorted "Outside of astronomy, do you find any enthusiasts?" Under pressure from Boland, NSF Director Erich Bloch agreed to the further VLBA funding delay until May 15. There was more at stake than the FY1986 funding level.[68] Since Boland had cleverly delayed obligating the FY1985 funds, NRAO feared that if the FY1986 VLBA funding was zeroed out by the appropriations committee, Boland would then contrive to reverse the FY1985 VLBA appropriation, possibly leading to the same fate as the 25 meter millimeter dish. Fortunately for NRAO, the discussion at the hearing drifted away from the VLBA to the relative merits of HST and the Keck 10 meter telescope.

Two days later, in the Senate hearings, Domenici chastised Bloch for delaying the NSF science education program and spoke in strong support of the VLBA. Bloch responded that he could not fund the VLBA unless the rest of the NSF budget was preserved, but Domenici reminded Bloch that Congress, not the NSF Director, makes these decisions.[69] Again, George Field, Maarten Schmidt (AAS President) and Peter Boyce (AAS Executive Officer) led a letter writing campaign to Representatives and Senators involved in the appropriations process. As finally passed, the 1986 HUD-Independent Agencies Appropriations (P.L. 99–160) included $9 million for the VLBA, and this amount became the de-facto basis for the more or less level funding in subsequent years. The construction budget went first from three years at $20 million each, to four years at $15 million, and finally eight years at $9–$11 million a year. However, as it developed, it would have been challenging to have completed the VLBA on the original schedule. The design and production of the record and playback systems would prove to be more difficult and time consuming than originally anticipated, and it was fortunate that the NSF not only stretched out the funding, but added funds to account for inflation and increased management costs over the additional years.

8.8   Building the VLBA

The 1982 VLBA proposal noted the considerable technical progress made since the 1977 NRAO Design Study. In particular, Readhead and Wilkinson (1978) and Cotton (1979) had demonstrated how to recover most of the phase information from VLBI observations to produce full synthesis images with milli-arcsec or better resolution. At the same time recording data rates had increased, allowing bandwidths comparable to that of the VLA. The order of magnitude improvement in tape storage density offered a comparable reduction in consumption of tape, with a corresponding decrease in the cost of shipping tape and the ability to operate for many hours without human intervention. Meanwhile, the progress made with low noise cooled FET amplifiers offered both a cost saving and improved reliability over parametric and maser amplifiers.

The Antenna Configuration The far-flung location of the VLBA antennas presented a new paradigm for NRAO. Starting with the construction phase, in addition to Arizona, New Mexico, Virginia, and West Virginia, NRAO had to become licensed to do business and obtain legal counsel in eight additional states. The configuration of the array presented another challenge. The ten antenna sites would ideally be located to give the best imaging capability, but there were practical aspects to consider. The antennas had to be on land and it was important to avoid areas of high tropospheric water vapor such as the southeast or northwest parts of the country. Southern locations were preferred over northern locations to maximize access to the southern sky. Locations near the VLA and Socorro were desired both to exploit the finite size of the VLA when used with the VLBA, and to provide a convenient center for maintenance. Other considerations included availability of water, power, and communications, road access, freedom from RFI, low winds, proximity to transportation services for shipping magnetic tapes, security, and the ease of acquiring the land. Location at or near another radio observatory was considered attractive as a source of logistical support, but this turned out to be naïve, as the staff at most observatories did not have the expertise, training, or special skills needed to support the unique VLBA instrumentation. As pointed out by Napier (2000) each site had its own logistical, legal, and technical challenges— a bankrupt contractor at Ford Davis, a contract award protest at Owens Valley, DOE bureaucracy at Los Alamos, and environmental concerns at Hancock.

NRAO assumed that access to government-owned land would be more straightforward than private or institutionally-owned land. In fact, the opposite was true. The small plot of land needed for a VLBA antenna could be readily purchased or leased from private owners. But it was a bureaucratic nightmare to transfer land from one federal agency to another agency, and long term agreements were subject to changing agency personnel and changing priorities.[70]

Everyone agreed that each antenna should probably be on US soil, although some overtures were made about locating one antenna in Mexico to improve the north-south resolution.[71] There was also discussion about possibly placing one element at the Canadian radio astronomy observatory in Penticton, BC, instead of Washington, but this was discouraged by the NSF. In order to maximize the resolution of the VLBA, two of eight antenna sites were chosen to lie outside of the continental United States, but still on US territory. The antenna site in Hawaii presented a unique challenge. Except for the high altitude locations on Mauna Loa, Mauna Kea, and Haleakala, the water vapor content over the rest of the Island state was judged to be unattractive for radio astronomy. Extensive radio transmissions from an Air Force Laboratory on Haleakala rendered it unacceptable. The National Atmospheric Laboratory on Mauna Loa, which provides the important historical records of carbon dioxide in the atmosphere, did not want radio astronomers running around and possibly contaminating their measurements. Moreover, Mauna Loa is an active volcano and would have required a dyke to protect against possible lava flow. Mauna Kea, of course, was the home of many optical telescopes and offered good supporting infrastructure. The summit of Mauna Kea within the so-called "science reserve" was unattractive due to icing[72] and the prevailing high winds at the summit. So a site was chosen at an 11,800 foot location, but because it was outside of the "science reserve" long negotiations with the local governments and the University of Hawaii were necessary. Normally, the University of Hawaii requires a "guaranteed entitlement of UH scientists to a specified amount of observing time" at astronomy facilities located on Mauna Kea. The University of Hawaii waived this requirement "in view of the vital role of a Hawaiian VLBA antenna," but in return, the University asked for 100% of the single dish observing time on the Mauna Kea antenna. This was unacceptable, but NRAO did agree "to carry out tasks related to maintaining the radio frequency properties of astronomical sites in Hawaii."[73] And finally, NRAO agreed to give the UH astronomers some access to single dish observing, but this capability was never implemented, and the issue has been long forgotten.

The most eastern site was first planned to be in Puerto Rico. A location near the Arecibo Observatory where it could be supported by Arecibo personnel was interesting, but was rejected due to concerns about interference from the powerful ionospheric and planetary radar systems used at the Observatory. NRAO staff found another site at an about to be abandoned CIA communications station on Puerto Rico's southern coast that was shielded by a mountain range from the Arecibo radar. However, that location was threatened by a planned Voice of America powerful transmitting station right next to the intended VLBA site. At Frank Drake's suggestion, NRAO found a site on the island of Saint Croix. Being located near sea-level on a Caribbean island, the Saint Croix site not only suffered from the high precipitable water vapor content, but the salty damp air meant that the antenna had to regularly be repainted with a special corrosion-resistant paint. Moreover, dealing with the local legal system and a developer who objected to having the view of the ocean obscured by the 25 meter dish became a continuing issue. It wasn't until 1998, five years after the completion of the VLBA, that NRAO was finally given the approvals needed to make the erection of the VLBA antenna on Saint Croix legal. Construction of the Saint Croix antenna had just begun and only the concrete foundation existed at the time of Hurricane Hugo in 1989, but damage to the rest of the island delayed the completion of the antenna. Finally, Hurricane Maria in 2017, which caused widespread destruction on the island, did not do major damage to the antenna, but the impact to communication and transportation limited the operation of the antenna for many months.[74]

A different type of controversy arose over the location of the northeastern antenna site. While needed for good imaging quality, sites anywhere in the North East were subject to potential RFI due to the large population density throughout the region and the poor tropospheric conditions resulting from the large cloud cover and water vapor content prevalent throughout the area. Sites at the Five College Radio Observatory in central Massachusetts, near the University of Rochester in New York, and even in Canada were all considered. Craig Walker, who was responsible for optimizing the antenna configuration, argued for a location in northern New England, and Cam Wade located an attractive site in New Hampshire only about 50 miles from the MIT Haystack Observatory. But George Seielstad, the NRAO Assistant Director for Green Bank, argued that it would be more cost effective to place the VLBA antenna in Green Bank, where it could be supported by the existing Green Bank staff at no increased cost and with little impact to the array imaging capability. The arguments between optimizing the Array configuration and supporting Green Bank with a new antenna became very divisive within NRAO, but were finally decided in favor of the New Hampshire location. Figure 8.6 shows the final configuration adopted for the location of the VLBA antennas.

Construction The expected VLBA construction cost at the time of the proposal was $50.7 million, including an inventory of spare parts and 13 percent contingency. The annual operating costs were estimated to be about $4 million. After President Reagan signed the NSF budget in July 1983, which included $2.5 million for the VLBA design, NRAO rented additional office space in Charlottesville, and began the process of developing a staffing plan, completing a detailed work schedule, and establishing annual budgets. Hein Hvatum was appointed VLBA Project Manager and Kellermann became the Project Scientist. When Hvatum retired in 1987, Peter Napier,[75] who had been the Deputy Project Manager, took over as Project Manager.

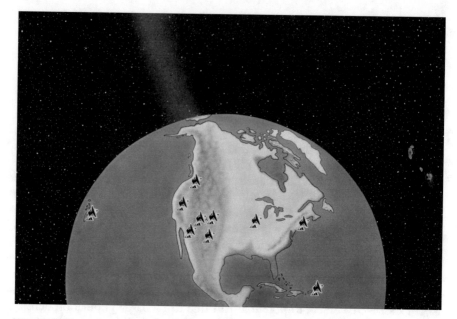

Fig. 8.6 Locations of the VLBA antennas: St. Croix, VI; Hancock, NH; North Liberty, IA; Fort Davis, TX; Los Alamos National Laboratory, NM; Pietown, NM; Kitt Peak, AZ; Owens Valley, CA; Brewster, WA; and Mauna Kea, HI. Credit: NRAO/AUI/NSF

The antenna elements were the most expensive part of the VLBA. In response to their RFP, NRAO received three bids to construct the ten antennas. TIW Systems Inc. and Radiation Systems Inc. (RSI) each proposed a conventional wheel and track antenna, while the Electronic Space Systems Corporation (ESSCO) proposed a pedestal mounted antenna enclosed in a radome. Following an independent analysis of the technical and business aspects of each proposal, NRAO chose RSI as the contractor for the ten antennas. The contract signed with RSI on 19 December 1984 for $19.61 million called for a five-phase approach, with each phase subject to authorization pending the availability of funds. In order to minimize state taxes, the contract was in two parts, one for the design, fabrication, and delivery of each antenna, and the other for the assembly and testing on site. But due to subsequent reductions in the expected NSF rate of funding, the contract with RSI was later renegotiated to deliver only two instead of three antennas a year.

While the Field Committee had recommended the construction of a VLBA, they were properly silent on who should build and operate the array, correctly leaving that as an NSF decision to be based on proposals and peer review. Although there had been acrimonious conflicts over the VLA and Owens Valley arrays (Sect. 7.2), NRAO and Caltech had worked together in developing plans for the VLBA and in advocating support from the community, even though their roles in the construction and operation of the VLBA were not clarified by the Field Committee. While it was becoming clear to Cohen that the construction and operation of a facility of the scope of the VLBA was probably beyond their interests and capability, Caltech still wanted to preserve some significant involvement. Maybe Caltech could build and operate the processor? But NRAO would not accept the responsibility for operating the VLBA without control over the processor, and several options were discussed. Maybe the processor could be located in Pasadena and be operated by NRAO. Maybe there should be two processors, one in Pasadena and one at NRAO.[76] NRAO needed Caltech's continued support if the VLBA was to be built. Perhaps more importantly, Caltech and the Jet Propulsion Lab (JPL) had a lot of experience and expertise on antennas, correlators, and imaging software that was vitally important for building the VLBA.

NRAO and Caltech staff met in Albuquerque, NM on 1 October 1981 to discuss how to best collaborate on the VLBA project.[77] Although Caltech recognized that NRAO would be the lead organization, they wanted to be "co-proposer," sharing decision-making responsibility through a joint "steering committee," but this was not acceptable to NRAO. Discussions and exchanges of seven draft MOUs continued for more than a year. NRAO stressed its need to maintain control, while Caltech stressed the value of its expertise and support.[78]

NRAO and Caltech finally agreed that Caltech would design and build the VLBA playback processor or correlator, but that when completed it would be moved to the NRAO. But there was no agreement about where NRAO should locate the processor or the VLBA Operations Center, and this became a matter of serious contention. The VLBI consortium leaders, especially from Caltech and MIT, continued to argue for a location near a major university to facilitate interaction with the broad astronomical community, while NRAO was more concerned about the logistics of VLBA operations and coordination with VLA operations. Even within NRAO, the VLBA debate triggered discussions about possibly shifting VLA operations to Socorro or Albuquerque from the array site on the Plains of San Agustin and possibly relocating the NRAO Headquarters. Four options were considered for the VLBA Operations Center: (a) Socorro, to facilitate coordination and to share resources with the VLA, (b) Charlottesville, where the NRAO Headquarters was located, (c) Albuquerque, which would provide some of the advantages of locating in Socorro, but perhaps provide more attractive living conditions for the VLBA staff and conceivably even VLA staff, and (d) co-location with VLA Operations on the Plains.

At the request of AUI, in September 1983 Mort Roberts appointed a committee to "review and advise on NRAO's selection of a site for the VLBA Operations Center."[79] Paul Vanden Bout, from the University of Texas and Chair of the NRAO Visiting Committee, was appointed as the committee chair.[80] The Vanden Bout committee was informed by a detailed report of the VLBA Operations Working Group, chaired by Carl Bignell, which examined the advantages and disadvantages of each option along with the potential impact on VLBA operations.[81] Vanden Bout's committee concluded that the control of the array operations and correlation of array data should be done at a common site, and that the Array Operations Center should be located near the VLA, specifically in Socorro or Albuquerque.[82] The committee, however, declined to make a recommendation on the location of the NRAO central offices, commenting only that "this issue depends on the future development of NRAO's activities," and that they found no connection between recommending a site for VLBA operations and the question of moving the NRAO headquarters to a new site.[83]

The discussions about moving the NRAO Headquarters slowly died away, but the decision to co-locate the VLBA and VLA operations had a profound impact on both facilities. The VLBA construction plan included funds for a VLBA Operations Center. Senator Pete Domenici was able to convince the New Mexico State Legislature to issue a $3 million bond that allowed New Mexico Tech to construct an Array Operations Center (AOC) which housed both the VLA and the VLBA operations staff. As a result, the VLA scientific, engineering, and business staff were able to move from the VLA site to Socorro, saving a two-hour daily commute. Locating all VLA personnel at the site had served well during the construction period, but by the 1990s, the operation had become sufficiently mature and most staff were not needed each day at the site, especially when the daily commute had some adverse impact on both the VLA operations and on staff morale. Ground breaking for the new combined Array Operations Center took place on 26 June 1987, and the AOC was opened for business on 8 December 1988. In 2008 the AOC was renamed the Pete V. Domenici Science Operations Center (DSOC) recognizing Domenici's "strong and effective support for science," and his role in securing Congressional support for the VLA as well as the VLBA along with the New Mexico legislature's support for the AOC (Fig. 8.7).

Perhaps the biggest technical challenge facing the VLBA was the choice of the recording system (Rogers 2000). The NRAO MK II VCR based system was reliable, relatively inexpensive, and could record for up to three hours on a single tape costing only a few dollars. But the MK II VCRs only recorded at 4 Mbps, limiting the bandwidth to 2 MHz. NRAO proposed to implement an upgrade based on work by Allen Yen at Toronto that would allow VCRs to record at 12.5 Mbps. The Haystack MK III system had a demonstrated recording rate (bandwidth) of 112 Mbps, or 28 times greater than the MK II system. But the MK III Honeywell Model 96 tape transport was very expensive; a single tape cost about $250 and lasted for only 13 minutes. The NRAO proposal suggested using a bank of eight upgraded MK II VCRs with a robot tape changer that would allow 24 hours of unattended recording at 100 Mbps (Fig. 8.8). Haystack proposed replacing the standard 28 track MK III headstack with a newly designed moveable 36 narrow track headstack that would allow multiple 128 Mbps passes on a single half-inch tape.

Fig. 8.7 The VLA-VLBA Array Operations Center, later named the Pete V. Domenici Science Operations Center (DSOC) in Socorro, New Mexico. Credit: NRAO/AUI/ NSF

Neither the upgraded NRAO MK II based system nor the upgraded Haystack MK III based system had been demonstrated, and considerable design work was still needed in each case. While there was some technical preference within NRAO for the MK II based system, MIT wanted to stay involved,[84] and NRAO agreed that MIT/Haystack would develop the VLBA recording system based on the upgraded MK III system. As the Haystack record system pushed the state of the art for magnetic tape recordings, its development suffered from continually increasing costs and corresponding delays. This led to constant tension between NRAO and Haystack/MIT. As a nonprofit university, neither Caltech nor Haystack/MIT could accept fixed priced contracts, and Hein Hvatum liked to complain that the Caltech/MIT definition of a deliverable was a new proposal asking for more money. As with Caltech, the increasing tensions between NRAO and Haystack and what NRAO called the Haystack/MIT/Caltech "grant mentality" resulted in NRAO assuming responsibility for the production of much of the record/ playback system. However, responsibility for the design of the challenging tape recorder upgrade remained with Haystack, since they had the unique expertise and experience to engineer the recorder to demanding specifications.

Though there were delays at Haystack in meeting the specified performance, the resulting VLBA recording system was a remarkable technical achievement. It recorded over 20 million bits of information on a square inch of magnetic

Fig. 8.8 Artist's
conception of the
proposed bank of eight
modified consumer TV
Video Cassette Recorders
using a robot cassette
changer to allow up to 24
hours of unattended
VLBA recording at a 100
megabit per second data
rate. The VCR concept
was abandoned in favor a
MK III based recording
system. Credit: NRAO/
AUI/NSF

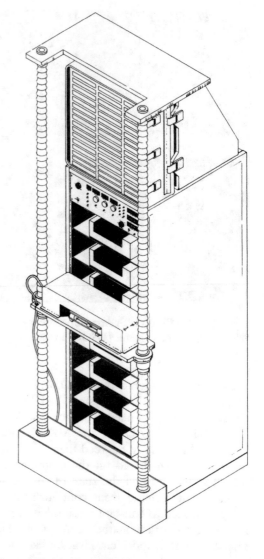

tape. Each 3.4 mile-long 38 micron-wide track of data deviated from a straight line by less than 0.001 inch. With 14 passes on each tape, a 14-inch reel of tape lasted for 10.5 hours at the nominal 128 Mbps recording rate, so that by using two recorders the tapes needed to be changed only once every day. For special experiments requiring higher sensitivity, the tapes could be run at twice the nominal speed to record at 256 Mpbs (128 MHz bandwidth) and on occasion the two tape drives were run in parallel, each at twice the nominal speed, to record at 512 Mbps (256 MHz bandwidth). In order to facilitate the use of other radio telescopes with the VLBA, VLBA-compatible record systems were fabricated and delivered at cost to radio observatories around the world.

Although the VLBA record system met the design goals, it pushed the state-of-the-art bit density, and recordings were sensitive to environmental conditions. Prior to recording, tapes needed to be stored in a room with carefully controlled temperature and humidity, and were easily damaged by friction heating as the tape rubbed against the transport tape edge guides at 140 inches per second. Before recording, each tape had to undergo a "pre-pass" to relax strains introduced during shipping. Nevertheless, recordings were not always error free, and the lifetime of the expensive headstacks was limited. A major improvement in the sensitivity, reliability, and operational ease of the VLBA occurred in 2007 when the tapes were replaced by commercial computer disk drives. Later advances in disk recording technology led to recording rates up to 2 Gbps, resulting in a factor of four increase in sensitivity for continuum observations over the original VLBA 128 Mbps tape recording system. By 2010, some 40 years after the first NRAO software correlator that ran on IBM 360, computers had been replaced by more powerful hardware correlators, the original VLBA hardware correlator was replaced by a cluster of commercial computers running a program known as DiFX (Deller et al. 2007).

As happened with the VLA construction, the annual NSF budget allocations were in a constant state of negotiation with OMB and Congress, resulting in continuing adjustments of the VLBA construction funding. By 1985, more than 30 separate budget scenarios had been prepared in response to constantly changing NSF requests. Probably the most serious funding impact was a result of the gap introduced in the FY1985 Congressional appropriations bill and the cut from $15 to $9 million. In order to purchase the long-lead times for all ten antennas, the NSF Astronomy Division considered supplementing the VLBA Congressional appropriation with a few million dollars of Division funds. But the NSF director was reportedly too "terrified of Boland" to reprogram NSF funds for the VLBA,[85] and it became necessary to delay work on the receivers, masers, record system, and processor, as well as renegotiate the antenna contract with RSI. Meanwhile, there was pressure from the user community, represented by the VLBI consortium, to fully instrument each antenna as it was completed and provide a correlator in order to begin observing programs, resulting in continual tension between NRAO and the VLBI community.

For two reasons, NRAO wanted to maintain the antenna construction schedule to the extent possible. First, if the contractual arrangements with RSI were not maintained, the cost of the antennas was likely to increase. Second, there was the residual concern that if the NSF or Congress were to cut off VLBA funding, it was important to at least complete the construction of all ten antennas, assuming that one way or another the instrumentation would somehow get built. However, maintaining adequate funding to complete the antennas on schedule meant that funding for other parts of the project would need to be deferred. In particular, this meant delaying the playback processer work at Caltech.

The VLBA playback processor was in a sense the brain and heart of the VLBA (Romney 2000). This was where the tapes from the ten remote sites were returned to be simultaneously played back and the data correlated. Using time stamps from a hydrogen maser atomic clock encoded at each antenna at record time, the tapes were synchronized at playback time to a small fraction of a microsecond and the signals from each antenna were correlated with the signals from each of the other nine antennas. Caltech did not have the resources to keep the correlator design team during a several year delay in funding, and agreed to terminate the correlator contract, which shifted back to NRAO, a move which was not completely unwelcome at NRAO. Jonathon (Jon) Romney, who had originally been hired at NRAO to work with the Caltech group, assumed responsibility for finishing the VLBA playback system in Charlottesville. Exploiting the delay resulting from the reduced level of funding, Romney and his group decided on a then unconventional approach to the design of the correlator, an idea originally suggested by Marty Ewing at Caltech and based on a concept proposed by the Japanese scientist Yoshihiro Chikada.[86] The VLBA correlator used a custom designed "FX Chip" which itself turned out to be a challenge, and had to undergo several rounds of prototyping before a satisfactory version was fabricated.[87]

When completed in 1992 and moved to Socorro, the VLBA correlator was able to execute nearly a trillion (10^{12}) multiplications a second, and supported up to 20 simultaneous playback systems, allowing the use of up to ten external antennas as well as the ten VLBA elements (Romney 2000). Alternatively the 20 playback drives could be used to support a double data rate (512 Mbps) if two drives were simultaneously used to record at double the sustainable rate (256 Mbps). The VLBA divided the data into 16 IF bands, each 8 MHz wide, and was not fully compatible with the MK III systems being used elsewhere which divided the data into 64 bands, each 2 MHz wide. But with time, VLBI systems at radio observatories around the world were modified to conform with the VLBA standard.

As a result of the constant budgetary concerns and the need to defer aspects of the VLBA construction, along with a confident, but naïve belief that most of the needed post-processing software was already available, NRAO did not devote sufficient resources to post-processing software development. As a result, VLBA users depended on the widely used Caltech Difmap VLBI package and on the Astronomical Image Processing System (AIPS), which had been developed for VLA data analysis. It would be several years after completion of the VLBA hardware before the VLBA could be considered fully operational and could be used by the non-expert observer. But the monitor and control software also lagged the hardware, in part due to a late start, and this impacted testing of the correlator (Walker 2000).

The last antenna at Mauna Kea was completed in April 1993, and was followed by the formal VLBA dedication on 20 August 1993 (Fig. 8.9). Pete Domenici, who played such a vital role in funding both the VLBA and AOC construction, was the keynote speaker. As Paul Vanden Bout later recollected, Domenici's staff and others were amused when, despite having it spelled out in large letters on a strip taped to the top of the podium, Domenici kept referring

Fig. 8.9 VLBA dedication on 20 August 1993. US Representative Joe Skeen (left) and NRAO Director Paul Vanden Bout (right) watch as Senator Pete Domenici scans the bar code below the word "Start" to initiate observations of the galactic hydroxyl (OH) gas cloud known as W3OH. The bar code triggered lights for each station on the displayed map, sequencing from east to west, and put a message on operators' screens prompting them to manually start the actual pointing sequence. Credit: NRAO/AUI/NSF

to the "Very Large Big Array." The final VLBA construction cost was $85 million, considerably more than the proposed $50.7 million. Much of the increase was the result of the stretched out, nearly level NSF funding and the consequential incremental purchases, and the need to maintain the standing army for the seven year construction period. Nevertheless, it would still be some years after the 1993 dedication before the VLBA became fully operational, including the specialized data reduction software needed to transform the data into high quality astronomical images.

The VLBA was arguably unique in having the broad involvement of the potential user community in specifying the design and contributing to its development. The Working Groups met regularly by teleconference and occasionally in person to discuss the many issues as they arose. However, the VLBA had an unexpected unfortunate consequence for VLBI research in the United States. It was understood by everyone that the construction of the VLBA would likely lead to the termination of NSF funding support for the operation of the OVRO, Iowa, Illinois, Fort Davis, Texas, and Haystack antennas that were being used in support of VLBI Network observing. The first casualty of the VLBA was the critically located antenna at Fort Davis following the rejection of Harvard's 1986 proposal to the NSF to operate the antenna through 1989, although the VLBA construction had barely begun. This was followed by the gradual but premature termination of NSF funding for VLBI support at OVRO, Haystack, Iowa, Hat Creek, and Illinois, which limited their participation in Network VLBI activities using the partially completed VLBA.

The closing of these facilities for VLBI itself was not a surprise, nor necessarily a disappointment, to their faculties and students, since it meant the end of their responsibilities to support VLBI observations in which they had no scientific involvement. But unexpectedly, the faculty and staff at these university radio observatories also lost their research funding, which had previously been packaged as part of the observatory operations grants. This had a long-ranging impact, specifically to the VLBA and more broadly to US radio astronomy. Without research support, it was just those university scientists that had developed VLBI techniques, including their active participation in the design of the VLBA and in supporting the proposal to build the VLBA, that were forced to turn their attention elsewhere. At Caltech, former VLBI scientists migrated to millimeter and optical astronomy, went to JPL to work on NASA missions, or left radio astronomy. Readhead and Tim Pearson devoted the next years to building an interferometer in Chile to investigate the small scale structure in the cosmic microwave background. At Haystack, the VLBI group focused their activities on NASA-supported geodetic research. At Berkeley, Don Backer became increasingly involved in pulsar, SETI, and Epoch of Reionization (EoR) research. Perhaps more important in the long term was the loss of students who had to follow the money. As a result, following its completion in 1993, the VLBA was used primarily by scientists from Harvard/Smithsonian, USNO, and NRL, with funding from Smithsonian and DoD respectively, as well as NRAO and foreign-based scientists. Indeed, much to the irritation of the NSF, about half of the available VLBA observing time has been used by non-US-based scientists, largely from Europe, but increasingly from China, Japan, and Korea, and ironically this probably contributed to the later NSF decision to divest from the VLBA (Sect. 8.10).

Transition to Operations[88] VLBA operations began as early as 1987, starting with the first completed VLBA antenna at Pie Town which was used to supplement the existing VLBI Consortium antennas. Additional VLBA antennas were added to the VLBI network antennas as they were completed. Initially, proposal review and scheduling were handled by the existing VLBI Consortium scheduling procedures, but starting in 1992, these activities were assumed by VLBA Operations. To support the Consortium observations, pending completion of the VLBA correlator, MK II terminals were installed at the first seven VLBA antennas, and the data continued to be correlated at the Caltech/JPL processor. Data obtained using the broad band VLBA recorders were initially processed at Haystack, Caltech, Goddard, or Bonn as appropriate. For these observations with the Consortium antennas, the capabilities at the stations included recording on narrow track recorders using either thick or thin tapes, or recording with the older wide track recorders. Maintaining capabilities to process all combinations stressed the operational capabilities of the processors.

During this period of "interim" VLBA operations, the NSF was unsympathetic to requests for interim "pre-operating" funds. Even after the completion of the VLBA, NRAO never received the planned additional $7 million annual operating funds. This impacted other NRAO operations as well as slowing upgrades to the VLBA.

8.9   Orbiting VLBI (OVLBI)

From the time of the first VLBI experiments, radio astronomers appreciated that there were no theoretical limits to the resolution of radio interferometers. Interferometers the size of the Earth were easily and quickly implemented. By going to space, baselines could be extended without limit, and the possibility of Earth-Space VLBI, commonly referred to as "Space VLBI," or "Orbiting VLBI" (OVLBI) was recognized as early as the Field (1982) Report. Its Radio Astronomy Panel Report (Thaddeus 1983) boldly suggested that a space VLBI mission required no new technology, and recommended that the VLBA be supplemented with a 25 meter orbiting radio telescope with compatible instrumentation, including IF data transmission to the Earth via the NASA Tracking and Data Relay Satellite System (TDRSS).

Probably the first serious Orbiting VLBI proposal was made in 1976 by Burke (MIT), Kellermann, and others, who suggested putting a 4 meter antenna on SpaceLab. The proposal was almost successful, but was beaten out by an infrared mission that apparently had broader engineering and surveillance applications. Burke's team proposed again in 1978 to orbit a 30 meter diameter antenna, but NASA later withdrew from the SpaceLab program. In 1979, Burke suggested that the NASA Venus Orbiting Imaging Radar antenna could be used as a variable spacing interferometer during its voyage from Earth to Venus. But after initial approval, NASA concerns about being able to stow the antenna before going into orbit around Venus killed Burke's ambitious VLBI project. Next, Burke and Frank Jordon (JPL) led an unsuccessful effort to fly a VLBI mission on the Space Shuttle.

The first demonstration of the practical feasibility of doing radio interferometry from an orbiting spacecraft came not from a mission designed for the purpose, but from the NASA TDRSS. In 1986 and 1987, a team of scientists from the US, Japan, and Australia, led by Gerry Levy from JPL, used a 4.9 meter antenna onboard the first NASA TDRSS antenna at 2.3 GHz (13 cm) together with 64 meter antennas in Australia and Japan to demonstrate the feasibility of Earth to space VLBI (Levy et al. 1986; Levy 1989). The TDRSS spacecraft are in geostationary orbit, and operational restrictions allowed only a restricted range of observations to give a maximum projected baseline up to 2.15 Earth diameters (27,400 km). These OVLBI observations demonstrated, for the first time, that some radio sources had brightness temperatures as high as a few times 10^{12} K, at or above the traditional Inverse Compton Limit supporting the existence of bulk relativistic motion (Linfield et al. 1989).

Starting in the early 1980s and continuing for the next three decades, US and European radio astronomers, sometimes separately, sometimes collaboratively, proposed a number of Earth to space interferometers including QUASAT,[89] the International VLBI Satellite (IVS),[90] and the Advanced Radio Interferometry between Space and Earth (ARISE) mission.[91] With primary support from ESA, NASA, and the US National Academy of Sciences, numerous reports were written and meetings held in Gross Enzersdorf (Austria), Budapest, Bologna, Noordwijk (Netherlands), Paris, and Tokyo, and in the US at NRAO (Green Bank and Charlottesville), JPL (Pasadena), NASA (Cape Canaveral, Florida), and at the NAS (Washington DC). Burke and Frank Jordon (JPL) led the effort in the US, and Richard Schilizzi in Europe. Kellermann, and later Bob Brown and Larry D'Addario, represented NRAO at these meetings. To coordinate these efforts, COSPAR set up an ad-hoc Committee on Space VLBI under the leadership of Graham-Smith of the UK. Meanwhile, the four space agencies from the US (NASA), Europe (ESA), Japan (ISAS), and the USSR (Intercosmos) set up their own Inter-Agency-Consultative-Group to exchange information on international OVLBI planning. The Global VLBI Working Group (GVWG) was organized at the 1990 URSI General Assembly in Prague at the suggestion of NRAO Director Paul Vanden Bout and URSI Commission V Chair Ron Ekers to coordinate both space and ground-based observing and tape management. Many of the same people from the small OVLBI community served on these multiple committees.

Considerable development work went into the studies, but none of the proposed US or European missions ever reached the launch pad. Launching large radio telescopes into Earth orbit is very expensive, and radio astronomy was doing very well from the ground. Within both Europe and the United States, radio astronomers were only looking to space to enhance their resolution, and they could not compete with the many proposals for infrared and high energy astrophysics missions where the science and the scientists were completely dependent on opportunities to observe from above the Earth's obscuring atmosphere. Moreover, except for spectroscopic observations or a few specialized observations relating to the maximum brightness temperature of synchrotron sources,[92] improved resolution can be obtained more easily and more cheaply by simply observing at shorter wavelengths. As a result, the relatively small radio astronomy community was unable to convince NASA or ESA to support a space VLBI program. However, they were more successful in Russia and Japan.

OVLBI also presents another challenge. Unlike other space astrophysics programs, OVLBI requires a network of ground radio telescopes to form the ground-based ends of Earth-Space interferometers. Moreover, typical space programs share the use of ground stations to send their data back, and OVLBI requires the full time use of at least two ground stations to receive the broad-band spacecraft data on a continuous basis. The need for NASA and ESA to team up with the ground community and surrender their control of the mission may explain their reluctance to become involved in OVLBI. In the US, the separation of funding for ground- and space-based astronomy between the NSF and NASA complicated the funding situation.

The 1984 meeting in Gross Enzendorf not only provided a focus for the proposed QUASAT mission, but western scientists heard, for the first time, about the proposed Japanese VSOP and Soviet RadioAstron OVLBI missions from Masaki Morimoto and Roald Sagdeev respectively. Morimoto was well known to US and European radio astronomers, not only for his role in building the Japanese 45 meter radio telescope at Nobeyama, but also for his bois-terous, alcohol-enhanced after dinner performances at numerous scientific conferences. Sagdeev, by contrast, was the prominent director of the Soviet Space Research Institute (IKI) who was an advisor and confidant of Mikhail Gorbachev, but at the time, was not known personally to the US or European radio astronomers.[93] Each of the two missions established their own international advisory committees—the RadioAstron International Science Council (RISC) for RadioAstron and the VSOP International Steering Committee (VISC) for VSOP. Both VSOP and RadioAstron were identified in the 1991 Decade Review of Astronomy and Astrophysics (Bahcall 1991) as excellent opportunities for international collaboration in astronomy, and recommended by the Radio Panel (Kellermann 1991) for NASA support for US participation in both missions.

VSOP The Japanese VLBI Space Observatory Programme (VSOP) was approved as an experimental mission by ISAS and was launched on 12 February 1997 aboard the first test flight of the Japanese Space Agency M-V rocket. It was widely assumed by the participants that the acronym VSOP was chosen by Morimoto after his favorite beverage. However, after launch, the spacecraft was renamed Highly Advanced Laboratory for Communications and Astronomy (HALCA). HALCA carried an 8 meter diameter dish into an ellip-tical orbit with a 21,400 km apogee, and was instrumented with receivers for 22 GHz (1.3 cm), 4.85 GHz (6 cm), and 1.66 GHz (18 cm). Unfortunately, the 1.3 cm system was damaged at launch. Without 1.3 cm, the resolution of VSOP/HALCA at the shortest wavelength (6 cm) was no better than the VLBA at 2 cm, and the opportunity to study H_2O maser emission was lost. VSOP remained in operation for six years, and was used primarily to study quasars at both 6 and 18 cm, and also made observations of pulsars and OH masers (Hirabayashi et al. 2000a). During the six-year lifetime of the mission, the VISC oversaw the proposal and scheduling process. NRAO played several important roles supporting VSOP operations. Starting in 1997, after modifica-tions under Jon Romney's leadership to accommodate Earth to Space base-lines, NRAO processed data from VSOP using the VLBA correlator. Larry D'Addario was successful in obtaining funds from NASA to build and operate a ground station in Green Bank using the old 15 meter antenna previously used as the remote station of the GBI. Ed Fomalont spent time at ISAS providing support for planning observations and analyzing data after it was correlated.

A later Japanese initiative, tentatively named VSOP2, proposed to use a 9 meter diameter antenna with cooled receivers at 5, 22, and 43 GHz in an elliptical orbit ranging from 1000 to 25,000 miles. To provide advisory support, the VISC was reconstituted as VISC-2. JPL, in collaboration with NRAO and US radio astronomers, requested NASA support for US supporting activities. The SAMURAI (Science of AGNs and Masers with Unprecedented Resolution in Astronomical Imaging) proposal requested NASA funding for a VSOP-2 tracking station, along with operational support for data analysis, and use of the VLBA and GBT. However, although VSOP 2 was initially approved by ISAS, they subsequently canceled the VSOP2 program due to technical problems and escalating costs.

RadioAstron Discussions of space-based interferometer systems in the USSR go back to the 1960s, but the details of the early planning have been lost to Soviet era secrecy. RadioAstron, also known as Spectrum-R, was one of three planned Soviet space astrophysics missions developed at the Cosmic Research Institute (IKI), the others being Spectrum-UV and Spectrum-X-gamma to work in the ultraviolet and high energy parts of the spectrum respectively. Each of the planned Soviet missions was led by an influential and respected Soviet academician—Nikolai Kardashev for RadioAstron, Alexander Boyarchuk for Spectrum-UV, and Rashid Sunyaev for Spectrum-X-γ, who vigorously competed for scarce resources. For years, the claimed priority for the first launch of the Spectrum series of satellites seemed to depend on who you were talking to.

Unlike VSOP, which was only 22,000 km above the Earth, Kardashev planned that RadioAstron would go out to 100,000 km, which he later extended to 350,000 km, close to the distance to the Moon. Like VSOP, RadioAstron required international participation, partly to provide access to large ground radio telescopes, partly to obtain the advanced VLBI recording technology not available in the Soviet Union, along with the need to have a global tracking network. From the beginning, the Western RISC members, as well as prominent scientists in the Soviet Union, argued against the high orbit proposed by Kardashev, first on the grounds that due to inverse Compton scattering, there would be no radio sources so small that they could be detected on such long interferometer baselines. Second, they argued that even if such small sources existed, interstellar scattering, at least for observations at the longer wavelengths, would likely broaden the source size, also rendering it unobservable with the high resolution corresponding to such long interferometer baselines.

It seemed for years that launch of RadioAstron was always scheduled to be five years from the date of inquiry, possibly reflecting the need to keep the project within the rolling Soviet five-year plan. When Kardashev lost his bid to become director of IKI he moved his whole team to the newly formed Astro Space Center (ASC), part of the well-known Lebedev Physical Institute, but due to space limitations at Lebedev, the ASC physically remained in the IKI building. The dissolution of the Soviet Union in 1991 and the ensuing deterioration of the Russian economy further delayed the mission. Kardashev managed to keep the RadioAstron team intact, but for at least a decade there was little progress toward a launch.

In 1989, Soviet Academicians Andrei Sakharov and Vitaly Ginzburg wrote to NASA Administrator Admiral Richard Truly asking for NASA support to provide tracking and data acquisition for RadioAstron and for funding for NRAO to build VLBA terminals for recoding the downlinked data in the USSR. The letter was signed by Sakharov only two weeks before he died.[94] Three months later, Vice President Dan Quayle, who headed the National Space Council, informed the Soviet ambassador to the US and issued a press release announcing that the US would participate in RadioAstron.[95] NASA set up a "Joint Working Group" specifically to deal with US-Soviet collaboration on astrophysics space missions.[96] US scientists, particularly from JPL and NRAO (Brown, D'Addario, Kellermann, and Weinreb), met frequently to develop plans for NRAO participation in RadioAstron. However, with the ensuing delays and uncertain status on the Russian side, as well the widely held skepticism about the choice of the orbit, NASA never got involved in RadioAstron, in spite of the NAS Decade Review which recommended "moderate" support from NASA for both VSOP and RadioAstron (Field 1982). Later, at the request of Kardashev, NRAO did build two low noise 1.3 cm FET amplifiers for RadioAstron which were sold to the ASC at cost after obtaining the necessary export license. The 1.3 wavelength receiver was particularly important for the success of RadioAstron, as it provided the highest resolution for continuum sources and was also needed for observations of the 1.3 cm H_2O maser sources.

RadioAstron was finally launched successfully from the Baikonur Cosmodrome in Kazakhstan on 18 July 2011. The spacecraft contained a 10 meter diameter antenna and receivers for the 1.3, 6, 18, and 92 centimeter bands. At the time of the launch Russia had only one ground station at Puschino, near Moscow, to receive the IF data from the spacecraft. A second ground station was badly needed to support observations when the satellite was not in view of Puschino. Since the high orbit extending out to 350,000 km, a high gain antenna was needed, and the retired NRAO 140 Foot was an obvious choice. But the 140 Foot antenna had been mothballed years earlier, and considerable work was needed before it could be restored to operational status. As NASA funding to support these activities never materialized, and OVLBI was beyond the purview of NRAO's NSF funding, shortly after the launch of RadioAstron NRAO and the Astro Space Center executed an MOU whereby the ASC provided the funds needed to refurbish the 140 foot antenna and to operate it as a downlink for RadioAstron. The Astro Space Center built a copy of the instrumentation used at Puschino and brought a team to Green Bank to install the equipment and to train the NRAO staff in its operation. Under a series of further MOUs, the 140 Foot antenna continued to downlink data from RadioAstron for later correlation in Bonn or in Moscow. After six years of operation, the on-board hydrogen maser that provided the local oscillator reference signal finally died, and starting in July 2017, both Green Bank and Puschino have transmitted to the spacecraft a real time local oscillator link referenced to ground-based masers. Following the loss of communication with the spacecraft, scientific observations with RadioAstron ceased in early 2019.

Starting in 2012 RadioAstron was used with a variety of ground-based radio telescopes to study quasars, OH and H_2O masers, pulsars, and the ISM, as well as doing tests of General Relativity with angular resolution as fine as 10 micro-arcsec. The RadioAstron scientific program was based on annual open calls for proposals which were reviewed by an international Program Review Committee.[97] For observations requiring the highest sensitivity, the GBT was used as the ground end of the Earth-Space interferometer. Much to the pleasant surprise of Western colleagues, RadioAstron observations showed fringes out to more than 200,000 kilometers, demonstrating brightness temperatures more than 10^{13} K, or several orders of magnitude greater than the Inverse Compton Limit for stationary sources (e.g., Kovalev et al. 2016; Pilipenko et al. 2018). The observation of fringes at 18 and even 92 cm on surprisingly long baselines has led to a new understanding of turbulence in the ISM and the nature of refractive scintillations (e.g., Johnson et al. 2016).

8.10   REFLECTIONS

The extraordinary milli-arcsec angular resolution of images made with the VLBA has enabled a wide range of galactic and extragalactic astronomy observations as well as important geodetic studies of continental drift and Earth rotation. As anticipated in the 1982 VLBA proposal, continuing observations of AGN jets have been a large part of VLBA observing programs, with data on individual sources now extending to as much as 25 years. Although much has been learned about the shapes (e.g., Pushkarev et al. 2017), kinematics (e.g., Cohen et al. 2007; Kellermann et al. 2007; Lister et al. 2016; Jorstad et al. 2017), and polarization (Homan et al. 2018) of AGN jets, there is still much unknown about how the jets are launched, collimated, and accelerated to nearly the speed of light.

Phase referencing, only briefly mentioned in the proposal, has become an important and routine part of the VLBA.[98] Precision VLBA astrometric measurements at unprecedented levels (Reid and Honma 2014) have been a pleasant surprise, more than meeting the proposal promises, and have enabled the determination of parallaxes (distances) to radio source throughout the Galaxy and the better delineation of its spiral arms (Reid et al. 2016) and overall structure, including size and rotational velocity. One of the important successes of the parallax measurements was the resolution of the distance controversy to the Pleiades star cluster (Melis et al. 2014).

Probably the single biggest impact of the VLBA, one of critical importance to cosmology, has come from the direct geometric measurement of the distance to the galaxy NGC 4258 to an accuracy of 3 percent through precise temporal monitoring of the motions of water masers in Keplerian orbits about the galaxy's center (Herrnstein et al. 1999, 2005; Miyoshi et al. 1995). This has provided an accurate anchor for the Cepheid distance scale (Riess 2016). This work has led to the Megamaser Cosmology Project that determined the Hubble Constant, based on maser distances alone, to an accuracy of 5 percent (Reid et al. 2013). These measurements also led to the best evidence for the existence of a supermassive black hole (10^8 solar masses) in another galaxy.

Throughout this period, the VLBA has also contributed to studies of Earth orientation and plate tectonics (e.g. Petrov et al. 2009), tests of general relativity (e.g., Fomalont et al. 2009), and interplanetary spacecraft navigation (e.g., Jones et al. 2011). The ongoing USNO program makes daily VLBA measurements to provide Earth orientation and rotation parameters needed for precision navigation. However, there have been some duds as well. Observations of stimulated radio recombination lines, which was claimed to be of "particular interest" in the 1982 proposal, never materialized.

Many VLBA observing programs have involved other radio telescopes, mostly in Europe, but also in Australia, Japan, China, Korea, and South Africa. More recently, the Large Millimeter Telescope (LMT) in Mexico and ALMA in Chile have been used to supplement millimeter VLBI. The use of these external antennas improves the image quality over that of the VLBA alone, but introduces compatibility and operational complexities of the kind that existed before the VLBA and that the VLBA was intended to eliminate. A particularly attractive mode of operation has been the High Sensitivity Array (HSA) which adds two or more of the large radio telescopes at Green Bank, Bonn, Arecibo, and the VLA[99] to the VLBA.

By the end of the 20th century, the VLBA had to an extent become a victim of its uniqueness. Because telescopes at other wavelengths do not have the resolution comparable to that of the VLBA, the range of VLBA scientific investigations has had little overlap with the interests of the broader American scientific community. Quasars, AGN, cosmic masers, and radio stars are point sources to OIR, X-ray, and γ-ray telescopes. Moreover, the US VLBI community never fully recovered from the loss of funding resulting from the VLBA construction and the termination of university based VLBI grant support. At the same time, VLBI has thrived in the rest of the world. Modest VLBI Networks were created in Australia, Russia (KVASAR Network),[100] China, Korea, and Japan to complement the broader East Asian and Asia-Pacific VLBI Networks. As part of the African SKA program, Africa has begun an ambitious program to repurpose redundant communication dishes for VLBI. Within Europe, VLBI has received strong national support, as well as generous funding from the EU, perhaps as a relatively non-controversial and relatively inexpensive means of promoting European unity. The EVN has expanded to include observatories in Africa, China, and even the US Arecibo Observatory.

Unlike the user community at other NRAO facilities, only about half of the VLBA users have been from US-based institutions, many from NRL, USNO, SAO, and the NRAO staff, rather than from the university community. Faced with limited operating funds and in anticipation of increased demands for operating funds for the planned Large Synoptic Survey Telescope (LSST), DKIST, and ALMA (Sect. 10.7), the VLBA became a likely target for decreased NSF funding. Claiming that future NSF budgets would grow no faster than inflation, in 2005, the NSF charged a "Senior Review Committee" to "examine the impact and the gains that would result by redistributing ~$30 million of annual spending from [Astronomy] Division funds."[101] As a boundary condition of the study, the NSF specified, "we will not use resources from unrestricted grants programs (AAG) to address the challenges of facility operations or the design and development costs for new facilities of the scale of LSST, GSMT, SKA, etc." AUI was asked to make "the case for and priority of each component of NRAO (VLA, VLBA, GBT, ALMA operations, etc.), along with a defensible cost for each." In addition, the NSF asked that AUI provide "as realistic an estimate as possible of the cost and timescale that would be associated with divestiture of each component."[102]

In its report, the Senior Review Committee recommended that

The Radio-Millimeter-Submillimeter base program should comprise the Atacama Large Millimeter Array, The Green Bank Telescope, and the Expanded Very Large Array [JVLA], operations together with support for University Radio Observatories and technology research and development through the Advanced Technologies and Instrumentation Program.[103]

The Committee went on to recommend that

The National Astronomy and Ionosphere Center and the National Radio Astronomy Observatory, ... should seek partners who will contribute to personnel or financial support to the operation of Arecibo and the Very Long Baseline Array respectively by 2011 or else these facilities should be closed.

Unless additional non-NSF sponsors could be found, the VLBA was clearly in trouble. Over the next few years, NRAO did reduce VLBA operating costs, but at the expense of reduced user support and poorer reliability. An agreement was reached with USNO by which USNO helped to support the VLBA in order to carry out their time measurements. Additional support to keep the VLBA operating was provided by the Universidad Nacional Autonoma de México (UNAM) in Mexico, MPIfR, and the European Radio Net. This external support helped, but was not sufficient to keep the VLBA operating. "In order to assess the most promising scientific areas for the VLBA, as well as review the options for new operational models and explore opportunities for additional support of VLBA operations," NRAO Director Fred Lo invited national and international observatory directors, NSF staff, and VLBI leaders to participate in a "Workshop on the Future of the VLBA."[104]

More than 60 scientists from 12 countries attended the Charlottesville Workshop held on 27–28 January 2011 (Fig. 8.10). Unfortunately, the start of the workshop was delayed by a major snow and ice storm which swept the East Coast on 26 January. Many participants spent the night at various airports or were on the road for up to nine hours to drive the 110 miles from Dulles Airport to Charlottesville. One participant obtained refuge in the back seat of a police vehicle when his rental car became stuck in the road. Following a series of talks on the major VLBA observing programs, the status of the various international VLBI networks, and discussions about recent and planned technical improvements, the participants agreed that the VLBA should emphasize key science and other large projects that involved less support from NRAO staff. The workshop participants also pledged sufficient external support that, combined with further cost saving measures, would enable NRAO to continue to operate the VLBA. In return NRAO would recognize the contributions of subscribers by awarding them a larger fraction of observing time, meaning less time for Open Skies proposals, even from US-based observers.

However, even this tough approach proved to be inadequate. The 2010 Astronomy Decade Review, "New Worlds and New Horizons" (Blandford 2010), provided an ambitious new agenda for the NSF Astronomy Division, which now faced potential additional operating funds for the highly recommended OIR, GSMT,[105] and LSST projects as well as for a variety of moderate programs. The NSF projected astronomy budgets were unable to support

Fig. 8.10 MPIfR Director Anton Zensus (right) confers with USNO Scientific Director Kenneth Johnston at the January 2011 Charlottesville VLBA Workshop. Credit: KIK/NRAO/AUI/NSF

these new initiatives as well as all of the existing facilities. James (Jim) Ulvestad,[106] the NSF Astronomy Division Director, convened a new "Portfolio Review Committee, Advancing Astronomy in the Current Decade: Opportunities and Challenges," that was charged with recommending the "AST portfolio best suited to achieving the decadal survey goals" under several budget scenarios. The Committee, chaired by Daniel Eisenstein[107] from Harvard, considered the whole AST portfolio of new and existing facilities and recommended that "AST divest from [the VLBA and GBT] before FY17," and that

> Within the context of open skies, the NSF should look to leverage its assets to maximize the ability of U.S. astronomers to access non U.S. capabilities or to obtain contributions toward operations and maintenance costs for U.S. facilities with high fractions of foreign users.[108]

In response to the Portfolio Review Committee report, when the AUI Cooperative Agreement to operate NRAO was due to expire in 2015, the NSF issued a competitive program solicitation for proposals to operate only the NRAO Jansky Very Large Array (JVLA), the North American share of ALMA, and the Charlottesville Central Development Lab.[109] It was the first time in the 60 year history of AUI management of NRAO that the NSF did not renew the NRAO five-year contract or Cooperative Agreement based on a non-competitive proposal. This time, a competing proposal was submitted to manage NRAO by the Southeastern Universities Research Association (SURA). Following a lengthy and detailed evaluation and review process, the NSF awarded AUI two new ten-year Cooperative Agreements, one to manage the North American share of ALMA and the other for NRAO operation of the JVLA, the Charlottesville Headquarters, and the Central Development Lab, effective 1 October 2016. Management of the Green Bank Observatory (GBO) and the VLBA under the Long Baseline Observatory (LBO) continued under an extension of the previous Cooperative Agreement, but with reduced funding for operations. Moreover, the LBO and GBO were established as new independent observatories, reporting directly to AUI and not as part of NRAO.[110] However, AUI appointed NRAO Director Tony Beasley as the AUI Vice President for Radio Astronomy, with direct responsibility for the NRAO, GBO, and LBO. Walter Brisken, a long-time member of the Socorro staff, was named as the LBO Director reporting to Beasley.

As planned when the decision was made to locate the VLBA operations in Socorro, the long-time operation of the VLBA jointly with the VLA as part of NRAO's New Mexico Operations was very effective. Many of the scientific, technical, computing, and administrative staff seamlessly supported both instruments. The new split, mandated by the NSF, added an extra layer of administration. The LBO did not have sufficient staff or resources to manage proposal review, human resources, or other administrative responsibilities, and depended on NRAO for these tasks, and it continued to use the nrao.edu email server. As mandated by the NSF, the LBO reimbursed NRAO for the cost of providing these various services. Considerable effort by AUI, NRAO, LBO, and the NSF was devoted to preparing the guidelines by which the LBO would operate as an "independent observatory" which was not really independent, and the NSF provided a one-time $1.5 million budget increment to set up the needed administrative framework.

Under the leadership of Brisken and Beasley, the LBO concluded an arrangement by which the USNO paid for half of the cost of the VLBA operations and development in return for half of the observing time to conduct observations to determine UT1 and other Earth rotation parameters. Smaller agreements with Australia, China, the MPIfR, the New York University in Abu Dhabi (UAE), and DoD provided additional financial support in return for observing time, enough that the NSF was satisfied that NRAO had created a sustainable operations model for VLBA. Following a non-competitive AUI proposal requested by the NSF, the once-threatened VLBA was reintegrated back into NRAO effective 23 October 2018. While providing less Open Skies observing, especially for small individual investigator projects, the long term stability of the VLBA was assured.

NOTES

1. Discussions of high resolution imaging in radio astronomy and the development of the NRAO-Cornell independent-oscillator-tape-recording interferometry system are given in Burke (1969), Kellermann and Cohen (1988), Moran (1998, 2000), and Kellermann and Moran (2001). The development of the Canadian long baseline interferometer system was reviewed by Gush (1988), Broten (1988), and Galt (1988). Section 8.1 is based, in part, on these papers.
2. Assuming that the variability time scale cannot be shorter than the light travel time across the source and knowing the distance to the quasars, the rapid variability suggested that the angular dimensions of variable radio sources was probably ≤0.001 arcsec.
3. For many quasars, the radio spectrum shows a sharp cutoff at low frequencies thought to be due to synchrotron self-absorption which is only important for very small dense radio sources.
4. Very small diameter radio sources scintillate or "twinkle" in the turbulent interplanetary medium in the same way that stars twinkle due to atmospheric turbulence.
5. In directly connected or radio linked interferometers, a common local oscillator (LO) signal is sent to each antenna where it is mixed with the incoming radio frequency (RF) signal to produce an intermediate frequency (IF) baseband signal. In VLBI systems the common LO is replaced by separate oscillators that are stabilized by atomic frequency standards that are sufficiently stable that they maintain coherence for the integration period—typically a few minutes to tens of minutes. The required stability is of the order of the reciprocal of the observing frequency. The atomic frequency standards are also used as atomic clocks, to provide synchronization of the recorded signals. The required

stability is of the order of the reciprocal IF bandwidth. For these early VLBI systems this was of the order of 1 microsec. Modern VLBI systems are generally stabilized by hydrogen maser frequency standards, but due to their greater cost and the lack of commercial sources of hydrogen masers, many of the earlier VLBI systems used the simpler and less stable commercially available Rubidium standards.

6. The sensitivity of radio telescopes depends inversely on the square root of the instantaneous bandwidth.

7. The Canadian group consisted of N.W. Broten, T.H. Legg, J.L. Locke, C.W. McLeish, R.S. Richards from the Canadian National Research Council; R.M. Chisholm from Queens University; H.P. Gush and J.L. (Allen) Yen from the University of Toronto; and J.S. Galt from the Dominion Radio Astrophysical Observatory.

8. B.G. Clark and K.I. Kellermann, 3 November 1965, General Considerations for a Very Long Baseline Interferometer, NAA-KIK, VLBI, Box 1.

9. Cohen et al., 22 November 1965, Some Considerations for a Very Long Baseline Interferometer between the Arecibo Ionospheric Observatory and NRAO, NAA-KIK, VLBA, History and Development.

10. DSH to R.M. Robertson, NSF Associate Director for Research, 15 April 1966, appended to the AUI-BOTXC minutes, 20 May 1966.

11. The NRAO MK I and later MK II VLBI systems used 1-bit samples of the digital data following a scheme developed by Sander Weinreb (1963) as part of his MIT PhD thesis. The data were sampled at twice the reciprocal bandwith, known as the Nyquist sampling rate. Harry Nyquist was a member of the Bell Laboratories staff and a contemporary of Karl Jansky. The correlation of 1-bit data suffers a loss of sensitivity of by a factor of $\pi/2 = 1.57$ compared with analog data, but is technically straightforward and is insensitive to gain fluctuations. The VLBA can use either 1-bit or 2-bit digitizing of the baseband data. With the bit rate limited by the recording technology, 2-bit digitizing at the Nyquist rate can cover only half of the bandwidth, but the sensitivity is about the same as 1-bit digitizing at the Nyquist rate.

12. Rubidium frequency standards made use of the hyperfine transition of rubidium-87 atoms at 6834682610.904 Hz.

13. The LORAN C (LOng RAnge Navigation) was used to locate the position of US naval ships. By comparing the time of arrival of transmissions from different LORAN C stations, ships could accurately determine their position without the need to depend on clear weather for traditional celestial navigation. When used for VLBI, the location of the observatory was known from conventional surveying techniques, and so knowing the distance to each LORAN C station, and thus the propagation time and the time that signals were transmitted, gave the accurate time at the observatory.

14. The nomenclature Mark I or MK I, II, III etc. was adopted by Barry Clark following the tradition of designating generations of naval equipment.

15. The Hewlett Packard HP 5065A Rubidium clock and 556 bits per inch (720 kilobits/sec) computer tape recorders were controlled items with potential military application.

16. One of the visitors, Dr. Leonid Matveyenko from the Lebedev Physical Institute, was a student of Shklvoskii and had been involved in the earlier discussions with Lovell. Matveyenko was accompanied on this initial trip by Dr.

Ivan Mossiev, who was in charge of the radio observatory in Crimea. For the actual observations one of the present authors (KIK), along with NRAO engineer John Payne, traveled to the USSR to supervise the installation and operation of the NRAO instrumentation.

17. In order to carry out a successful VLBI observation, the clocks at the two ends need to be synchronized to about an accuracy of the order of the reciprocal bandwidth. With the MK I system in use at the time, this corresponded to about 1 microsecond.

18. The classical test of relativistic light bending was first made during a solar eclipse in 1919. Sir Arthur Eddington barely measured the bending by an amount close to the predicted 1.75 arcsec at the limb of the Sun, which was widely acclaimed as confirmation of Einstein's theory of General Relativity. In later years Eddington's results were questioned. Radio measurements improved the precision to about ten percent, but the advent of VLBI opened an opportunity to greatly improve the accuracy.

19. Shaffer had begun his radio astronomy career as an NRAO summer student in 1966 through 1969. After receiving his PhD at Caltech in 1974, he spent a year at Yale, returned to NRAO as a member of the scientific staff for four years, and then spent the rest of his career at Radiometrics Inc. providing support for the MIT/NASA geodetic VLBI program.

20. Correlation of a pair of 3 minute tapes on the IBM 360/75 was about ten times faster than on the NRAO 360/50 computer.

21. When the energy density in a synchrotron radiation field exceeds the energy in the magnetic field, the relativistic elections lose energy by the Inverse Compton effect which produces X-rays, further enhancing the Inverse Compton losses.

22. Very Long Baseline Radio Interferometry Using a Geostationary Satellite, ESA Phase A Study, 1980, SCI (80) 1; ESA Study of the Ground Segment, 1981, SCI (81) 5. Although NRAO was not directly involved in this activity, Kellermann was then on leave from NRAO as a Director at the MPIfR, and participated in the study.

23. A New Midwest Antenna for the VLBI Network. A proposal to the NSF by G.W. Swenson, PI, June 1978, NAA-KIK, VLBA, History and Development.

24. Bologna, Jodrell Bank, MPIfR, Onsala, and Westerbork.

25. K. I. Kellermann, 1973, Some Thoughts on the Construction of an Intercontinental Very Long Baseline Array (VLBA), NRAO Internal Memo, NAA-KIK, VLBA, History and Development.

26. DSH to J. Broderick (VPI), B. Burke (MIT), T. Clark (Goddard), M. Cohen (Caltech), T. Clark (Goddard), W. Erickson (Maryland), M. Ewing (Caltech), S. Knowles (NRL), J. Moran (Harvard), A. Rogers (MIT), D. Shaffer (Interferometrics Inc.), G. Swenson (Ill), I. Shapiro (Harvard), 18 July 1974, NAA-KIK, VLBA, History and Development.

27. Report of the 13th meeting of the Canadian Astronomical Society, 3 June 1982, NAA-KIK, VLBA, History and Development.

28. The CLBA initially proposed to use 25 meter antennas, but later increased the size to 32 meter for greater sensitivity. The shortest wavelength of the 32 meter antennas was 1.3 cm compared with the 7 mm limit of the VLBA 25 meter antennas. "A Proposal for a Canadian Very-Long-Baseline Array," NAA-NRAO, NM Operations, VLBA. See also NAA-AHB, Canadian Long Baseline Array for more details of the CLBA.

29. Report of the 13th meeting of the Canadian Astronomical Society, op. cit.

30. Seaquist to KIK, 28 July 1983, NAA-KIK, VLBA, History and Development; A. Bridle and C. Walker, VLBA Memo No. 237. http://library.nrao.edu/vlba/main/VLBA_237.pdf

31. KIK, VLB Array Memo No 1, 22 May 1980, NAA-KIK, VLBA, History and Development. http://library.nrao.edu/vlba/main/VLBA_01.pdf

32. NAA-KIK, VLBA History and Development, Box 2.

33. Membership included D. Backer (UC Berkeley), B. Burke (MIT), M. Ewing (Caltech), K. Johnston (NRL), R. Mutel (Iowa), A. Rogers (Haystack), I. Shapiro (MIT), J. Welch (UC Berkeley).

34. Other Radio Panel Members were F. Drake (Cornell), M. Roberts (NRAO), J. Taylor (Princeton), J. Welch (University of California, Berkeley), and R. Wilson (Bell Labs).

35. Although the reports of the panels were published as Volume 2, and appeared in print only in 1983 after the Volume I report of the main committee, the reports of the panels were made available earlier to the main committee as input to their deliberations.

36. The first overall priority was AXAF, the Advanced X-Ray Astrophysics Facility, which was finally launched in 1999 and given the name "Chandra X-ray Observatory."

37. A site near the summit of Mauna Kea had been selected, but legal challenges from local groups have resulted in years of uncertainty and delay.

38. Big Science Policies and Procedures, 203rd meeting of the NSB Appendix D, 19 January 1979, NAA-KIK, VLBA, History and Development.

39. *Washington Post*, 11 February 1981.

40. Committee members were B. Burke (MIT), R. Dicke (Princeton), G. Field (Harvard), H. Friedman (NRL), D. Hogg (NRAO), J. Taylor (Princeton), P. Thaddeus (Harvard), and R. Wilson (Bell Labs).

41. KIK notes of 25 January 1982 meeting, NAA-KIK, VLBA, History and Development.

42. MSR to Slaughter, NSF Assistant Director, AAEO, 1 February 1982, NAA-NRAO, Director's Office, NSF Correspondence.

43. Ibid.

44. AAC membership at the time was J. Beckers (Chair-Arizona), E. Becklin (Hawaii), B. Burke (MIT), R. Giacconi (STScI), F. Gillet (KPNO), D. Hogg (NRAO), R. Humphreys (Minnesota), R. McCray (Colorado), D. Osterbrock (Santa Cruz), P. Pesch (Warner and Swasey), J. Taylor (Princeton), and A. Wolfe (Pittsburgh).

45. MSR to KIK, 18 February 1981, NAA-KIK, VLBA, History and Development.

46. Bautz to MSR, 30 June 1980, NAA-NRAO, Director's Office, NSF Correspondence.

47. MSR to Slaughter, 19 April 1982, NAA-NRAO, Director's Office, NSF Correspondence.

48. Kurt Weiler draft notes, 5 January 1983, NAA-KIK, VLBA, History and Development. Weiler was the NSF AST program manager for the VLBA.

49. SIRTF was initially approved by NASA for launch and return to Earth after the completion of the mission by the space shuttle, but concerns about contamination from the shuttle led to a redesign as a free-flyer.

50. MSR to Slaughter, 14 May 1982, NAA-NRAO, Director's Office, NSF Correspondence.
51. MSR to Johnson, 30 July 1982, NAA-NRAO, Director's Office, NSF Correspondence.
52. KIK interview with WEH III, 23 September 2011, NAA-KIK, Oral Interviews. https://science.nrao.edu/about/publications/open-skies#section-8
53. G. Low to KIK, 9 September 1982, NAA-KIK, VLBA, History and Development. Low was the Chair of COSEPUP.
54. KIK to MSR, 17 September 1982, NAA-KIK, VLBA, History and Development.
55. Field to Briefing Panel, 29 October 1982, NAA-KIK, VLBA, History and Development.
56. K. Weiler to Taylor, 30 March 1984, NAA-KIK, VLBA, History and Development. Other members of the Panel were B. Chrisman (Yale), R. Neal (SLAC), I. Shapiro (CfA), J. Welch (Berkeley), and R. Wilson (Cornell).
57. Report of the NSF VLBA Review Panel, 2 July 1984, NAA-KIK, VLBA, History and Development.
58. Nancy McGeown (from Representative Lindsey Boggs staff) to KIK, 16 August 1984, KIK to MSR, Hvatum, Hughes, 22 August 1984, NAA-KIK, VLBA, History and Development.
59. The letter was signed by G. Field (Harvard), J. Bahcall (Princeton), B. Burke (MIT), A. Code (AAS President), B. Oliver (Hewlett-Packard), C. Sagan (Cornell), M. Schmidt (AAS President-elect), and P. Thaddeus (Harvard). Additional letters of support were written by Haystack Director Joe Salah and Ed Ney from the University of Minnesota. NAA-NRAO, NM Operations, VLBA.
60. Boland to Field, 5 June 1984, NAA-KIK, VLBA, History and Development.
61. On 1 May 1984, Kellermann, Field, Thaddeus, and Peter Boyce, American Astronomical Society Executive Officer, met with a number of key Senate and House members and their staffs.
62. VLBA: A Congressman's Victory over NSF Project, *Physics Today*, October 1984, p. 56.
63. NSF to MSR, 28 June 1984, NAA-NRAO, NM Operations, VLBA.
64. Joel Widder (NSF Legislative Affairs) to KIK, KIK to Roberts, Hvatum, Hughes, 2 August 1984, NAA-KIK, VLBA, History and Development.
65. Bloch to NSB, 18 September 1984, NAA-NRAO, NM Operations, VLBA.
66. Ramonas to KIK, 16 August 1984, KIK Memo to Roberts, Hvatum, Hughes, NAA-KIK, VLBA, History and Development.
67. PVB to KIK, 7 March 1985, NAA-KIK, VLBA, History and Development.
68. House Appropriations Sub-committee on HUD minutes for 26 March 1985, NAA-KIK, VLBA, History and Development.
69. Senate Appropriations Committee hearing, 28 March 1985, NAA-NRAO, NM Operations, VLBA.
70. Only 15 years after the erection of the antenna, the Los Alamos National Laboratories (LANL) informed NRAO that for undisclosed security reasons, they wanted to remove the VLBA antenna situated in a remote part of the Laboratory site. Fortunately for NRAO, LANL management, and apparently their priorities, changed, and following some exploratory discussions, they did

not press the case. W. Press to F. Lo, 6 November 2002, NAA-KIK, VLBA, History and Development.

71. MSR to L. Rodriguez, 18 May 1982, NAA-KIK, VLBA, History and Development.

72. In 1993, 17 surface panels on the Mauna Kea antenna were damaged by ice falling from the feed support legs and had to be replaced.

73. R. Hall to PVB, 11 April 1989, NAA-NRAO, NM Operations, VLBA.

74. The Saint Croix VLBA antenna is located only 250 meters from the sea.

75. Napier (2000) has discussed the planning and construction of the VLBA.

76. KIK to Roberts, 2 November 1981, NAA-KIK, VLBA, History and Development.

77. NRAO was represented by Roberts, Clark, Hvatum, and Kellermann; Caltech by Cohen, Moffet, Readhead, and Rochus Vogt (OVRO Director).

78. Correspondence and internal memos among Ken Kellermann and Mort Roberts (NRAO), Marshall Cohen, and R. Vogt (Caltech); and Bernie Burke and John Evans (MIT), 1981–1982, NAA-KIK, VLBA, History and Development.

79. PVB to MSR, 20 December 1983, NAA-NRAO, NM Operations, VLBA, Box 3A.

80. Other members of the committee were: B. Burke (MIT), A. Davidson (Johns Hopkins), R. Dicke (Princeton), A. Hogg (Lowell), D. Hogg (NRAO), G. Preston (MWPO), M. Reid (CfA), and J. Taylor (Princeton).

81. VLBA Array Operations Center Site Selection Report, C. Bignell (Chair), December 1983, NAA-KIK, VLBA, History and Development.

82. Report of the VLBA Advisory Committee, 20 December 1983, NAA-KIK, VLBA, History and Development.

83. Ibid.

84. J. Evans to MSR, 1 December 1981, NAA-KIK, VLBA, History and Development. Evans was the Haystack Director.

85. PVB to KIK, 10 January 1985, NAA-KIK, VLBA, History and Development.

86. In a conventional radio array such the VLA, the data are first correlated with many different delays or lags introduced in each baseline pair, and then Fourier Transformed to obtain a spectrum of the fringe visibility. The VLBA uses a so-called FX correlator by which the data from each antenna, after digitizing, are first Fourier Transformed and then multiplied with the other antennas to give the visibility spectrum on each baseline. As the largest FX type correlator ever built, and the first one built outside Japan, it involved considerable risk and development time, but in the end was less expensive to fabricate. An earlier FX correlator was built for the Nobeyama 5-element millimeter array in Japan.

87. The custom designed VLBA correlator chips were later made available for a VLBI processor built at the Shanghai Astronomical Observatory and for a radio telescope in Mauritius.

88. Walker (2000) discusses early VLBA operations.

89. QUASAT *(Quasar Satellite)—A VLBI Observatory in Space*, 1984, in Proceedings of a Workshop held at Gross Enzersdorf, Austria (Noordwijk: ESA), NAA-KIK, VLBI, Space VLBI, Box 1; QUASAT A Space VLBI Satellite Assessment Study, ESA SCI (85) 5, NAA-KIK, VLBI, Space VLBI, Box 1; QUASAT, a VLBI Observatory in Space, a Proposal to NASA from JPL, NAA-KIK, VLBI, Space VLBI, Box 2. QUASAT received high marks for scientific

merit and technical feasibility, but the projected budget was greater than the ESA ceiling, and it was disqualified only a few days before the competitive review in October 1988 held in Paris.

90. IVS—An International Orbiting Radio Telescope, 1991, ESA SCI (91) 2, NAA-KIK, VLBI, Space VLBI, Box 2. The IVS proposal was based on a 20 meter diameter dish which was to be launched on the ill-fated Soviet Energia space shuttle and included scientists from the USSR, as well as from Europe and the US.

91. IVS and later ARISE proposed a 25 meter class antenna operating down to 3 mm wavelength in an elliptical orbit reaching up to 50,000 km. ARISE was recommended by the 2001 Decade Review, Astronomy and Astrophysics in the New Millennium (Taylor and McKee 2001), but NASA never provided funds to support the proposed US activities.

92. Because the resolution of an interferometer is given by $\theta = (\lambda/D)$ and the brightness temperature T is proportional to $S\lambda^2/\theta^2$, the maximum brightness temperature that can be observed depends only on the flux density and the square of the interferometer baseline, D, and is independent of wavelength.

93. Sagdeev later married Susan Eisenhower, daughter of the former US President Dwight Eisenhower, and immigrated to the United States, where he joined the faculty at the University of Maryland.

94. Ginzburg and Sakharov to Truly, 2 December 1989, NAA-BFB, Space VLBI.

95. Press release from the Office of the Vice President, 8 March 1989.

96. Kellermann and Kardashev (IKI) represented RadioAstron in the Joint Working Group, which was chaired by Charles Pellerin (NASA) and Rashid Sunyaev (IKI); Sunyaev was the project leader for the competing Spectrum-X-γ mission.

97. See http://www.asc.rssi.ru/radioastron/index.html for further details about RadioAstron and a list of relevant publications.

98. Phase referencing is a technique by which the antennas are rapidly switched between a reference calibration source and the target source. Phase referencing was in common use with connected element interferometers to improve the phase distortions due to tropospheric or instrumental instabilities. The adoption of phase referencing for VLBI with independent local oscillators increased the effective integration (averaging) time thus greatly improving the sensitivity.

99. Although the VLA and Westerbork telescopes are themselves arrays, all of the antennas can be electrically connected to operate as a single telescope with collecting areas equivalent to a 135 and 94 meter diameter dish respectively.

100. The Russian KVASAR (QUASAR) network consists of three 32 meter dishes originally built for precision measurements of Earth rotation and geodynamics by the Russian Institute of Applied Astronomy.

101. Report of the NSF AST Senior Review Committee, From the Ground Up: Balancing the NSF Astronomy Program, 22 October 2006, NAA-KIK, VLBA, History and Development. https://www.nsf.gov/mps/ast/ast_senior_review. jsp Committee members were T. Ayres (Colorado), D. Backer (Berkeley), R. Blandford, chair (Stanford), J. Carlstrom (Chicago), K. Gebhardt (Texas), L. Hillenbrand (Caltech), C. Hogan (Washington), J. Huchra (Harvard-Smithsonian), E. Lada (Florida), M. Longair (Cambridge), J.P. Looney (Brookhaven), B. Partridge (Haverford), V. Rubin (DTM).

102. W. Van Citters (NSF Division of Astronomical Sciences Director) to Ethan Schreier (AUI President), 7 April 2005, NAA-KIK, VLBA, History and Development.

103. Ibid.

104. KYL to recipients, 25 October 2010, NAA-KIK, VLBA, History and Development.

105. GSMT (Giant Segmented Mirror Telescope) was the generic name given to the proposed Thirty Meter Telescope (TMT) and the 20 meter Giant Magellan Telescope (GMT).

106. Ulvestad had come to the NSF from NRAO, where he had been the Assistant Director for New Mexico Operations and later head of the NRAO New Initiatives Office. Before coming to NRAO, Ulvestad was at JPL where he played a prominent role in space VLBI programs.

107. Other Portfolio Review Committee members were J. Miller (Lick, Vice-Chair), M. Agueros (Columbia), G. Bernstein (Penn), G. Blake (Caltech), J. Feldmeier (Youngstown), D. Fischer (Yale), C. Impey (Arizona), C. Lang (Iowa), A. Lovell (Agnes Scott), M. McGrath (NASA), M. Norman (UCSD), A. Olinto (Chicago), M. Skrutskie (Virginia), K. Schrijver (Lockheed Martin), J. Toomre (Colorado), R. Walterbos (New Mexico).

108. Advancing Astronomy in the Coming Decade: Opportunities and Challenges. Report of the National Science Foundation Division of Astronomical Sciences Portfolio Review Committee, 14 August 2012, NAA-NRAO, NSF Portfolio Review. https://www.nsf.gov/mps/ast/ast_portfolio_review.jsp

109. NSF Program Solicitation NSF 14-568, 25 November 2014.

110. The GBO management is discussed further in Sect. 11.9.

BIBLIOGRAPHY

REFERENCES

Allen, L.R. et al. 1962, Observations of 384 Radio Sources at a Frequency of 158 Mc/s with a Long Baseline Interferometer, *MNRAS*, **124**, 477

Bahcall, J.N. ed. 1991, *The Decade of Discovery in Astronomy and Astrophysics* (Washington: National Academy Press)

Bare, C. et al. 1967, Interferometer Experiment with Independent Local Oscillators, *Science*, **157**, 189

Blandford, R. ed. 2010, *New Worlds, New Horizons in Astronomy and Astrophysics* (Washington: National Academy Press)

Booth, R. 2013, The Origins of the EVN and JIVE: Early VLBI in Europe. In *Resolving the Sky - Radio Interferometry: Past, Present, and Future*, ed. M.A. Garrett and J.C. Greenwood (Manchester: SKAO), 51

Booth, R. 2015, *The Origins of the EVN and JIVE: Early VLBI Developments in Europe* http://www.jive.eu/jive-eric-symposium

Broderick, J. et al. 1970, Observations of Compact Radio Sources with a Radio Interferometer Having a Green Bank-Crimea Baseline, *AZh*, **47**, 784. English translation: 1971, *SvA*, **14**, 627

Broten, N.W. et al. 1969, Long Baseline Interferometer Observations at 408 and 448 MHz I. The Observations, *MNRAS*, **146**, 313

Broten, N.W. 1988, Early Days of Canadian Long-Baseline Interferometry: Reflections and Reminiscences, *JRASC*, **82**, 233

Burbidge, G. 1978, Physical Problems Associated with BL Lac Objects and QSOs, *Quasars and Active Galactic Nuclei, Physica Scripta*, **17** (3), 281

Burke, B.F. 1969, Long Baseline Interferometry, *Physics Today*, **22** (7), 54

Burke, B.F. et al. 1972, Observations of Maser Radio Sources with an Angular Resolution of 0.0002, *AZh*, **49**, 465. English translation: *SvA*, **16**, 379

Clark, B.G. et al. 1968a, High-Resolution Observations of Small-Diameter Radio Sources at 18-Centimeter Wavelength, *ApJ*, **153**, 705

Clark, B.G. et al. 1968b, Radio Interferometry Using a Base Line of 20 Million Wavelengths, *ApJ*, **153**, 67

Clark, B.G. 1973, The NRAO Tape-Recorder Interferometer System, *Proc. IEEE*, **61**, 1242

Clarke, R.W. et al. 1969, Long Baseline Interferometer Observations at 408 and 448 MHz II. The Interpretation of the Observations, *MNRAS*, **146**, 381

Cohen, M.H. et al. 1968, Radio Interferometry at One-Thousandth of a Second of Arc, *Science*, **162**, 88

Cohen, M.H. et al. 1971, The Small-Scale Structure of Radio Galaxies and Quasi-Stellar Sources at 3.8 Centimeters, *ApJ*, **170**, 270

Cohen, M.H. ed. 1977, *VLBI Network Studies I, A VLBI Network Using Existing Telescopes* (Pasadena: California Institute of Technology). Originally issued in 1975 by the Network Users Group

Cohen, M.H. et al. ed. 1980, *A Transcontinental Radio Telescope* (Pasadena: California Institute of Technology)

Cohen, M.H. 2000, Early Days of VLBI. In *Radio Interferometry: The Saga and the Science*, ed. D.G. Finley and W.M. Goss (Washington: AUI), 17

Cohen, M.H. 2007, A History of OVRO, *Engineering & Science*, **3**, 33

Cohen, M.H. et al. 2007, Relativistic Beaming and the Intrinsic Properties of Extragalactic Radio Jets, *ApJ*, **658**, 232

Cotton, W.D. 1979, A Method of Mapping Compact Structure in Radio Sources Using VLBI Observations, *AJ*, **84**, 1122

Deller, A.T. et al. 2007, DiFX: A Software Correlator for Very Long Baseline Interferometry Using Multiprocessor Computing Environments, *PASP*, **119**, 318

Field, G. 1982, *Astronomy and Astrophysics for the 1980's*, Vol. 1 (Washington: National Academy of Sciences)

Fomalont, E.B. et al. 2009, Progress in Measurements of the Gravitational Bending of Radio Waves Using the VLBA, *ApJ*, **699**, 1395

Galt, J. 1988, Beginnings of Long-Baseline Interferometry in Canada: A Perspective from Penticton, *JRASC*, **82**, 242

Gold, T. 1967, Radio Method for the Precise Measurement of the Rotation Period of the Earth, *Science*, **157**, 302

Gordon, M.A. 2005, *Recollections of Tucson Operations, The Millimeter-Wave Observatory of the National Radio Astronomy Observatory* (Dordrecht: Springer)

Gush, H.P. 1988, Beginnings of VLBI in Canada, *JRASC*, **82**, 221

Herrnstein, J.R. et al. 1999, A Geometric Distance to the Galaxy NGC4258 from Orbital Motions in a Nuclear Gas Disk, *Nature*, **400**, 539

Herrnstein, J.R. et al. 2005, The Geometry of and Mass Accretion Rate through the Maser Accretion Disk in NGC 4258, *ApJ*, **629**, 719

Hirabayashi, H., Edwards, P.G., and Murphy D.W. eds. 2000a, *Astrophysical Phenomena Revealed by Space VLBI* (Sagamihara: ISAS)

Homan, D.C. et al. 2018, Constraints on Particles and Fields from Full Stokes Observations of AGN, *Galaxies*, **6**, 17

Jauncey, D. et al. 1970, High-Resolution Radio Interferometry at 610 MHz, *ApJ*, **160**, 337

Johnson, M.D. et al. 2016, Extreme Brightness Temperatures and Refractive Substructure in 3C273 with RadioAstron, *ApJL*, **820**, L10

Jones, D. et al. 2011, Very Long Baseline Array Astrometric Observations of the Cassini Spacecraft at Saturn, *AJ*, **14**, 29

Jorstad, S.G. et al. 2017, Kinematics of Parsec-Scale Jets of Gamma-Ray Blazars at 43 GHz within the VLBA-BU-BLAZAR Program, *ApJ*, **846**, 98

Kellermann, K.I. et al. 1968, High-Resolution Interferometry of Small Radio Sources Using Intercontinental Base Lines, *ApJ*, **153**, 209

Kellermann, K.I. ed. 1977, *VLBI Network Studies III, An Intercontinental Very Long Baseline Array* (Green Bank: NRAO/AUI)

Kellermann, K.I. ed. 1981, *The Very Long Baseline Array, Design Study* (Green Bank: NRAO/AUI)

Kellermann, K.I. ed. 1982, *Proposal for The Very Long Baseline Array* (Green Bank: NRAO/AUI)

Kellermann, K.I., and Cohen, M.H. 1988, The Origin and Evolution of the NRAO-Cornell VLBI System, *JRASC*, **82**, 248

Kellermann, K.I. ed. 1991, Report of the Radio Astronomy Panel. In *Working Papers of the Astronomy and Astrophysics Survey Committee* (Washington: National Academy Press)

Kellermann, K.I. et al. 2007, Doppler Boosting, Superluminal Motion, and the Kinematics of AGN Jets, *Astrophys. Space Sci.*, **311**, 231

Knight, C.A. et al. 1971, Quasars: Millisecond-of-Arc Structure Revealed by Very-Long-Baseline Interferometry, *Science*, **172**, 52

Knowles, S.H. et al. 1981, Phase-Coherent Link between VLBI Stations via Synchronous Satellite, *BAAS*, **13**, 899

Kovalev, Y.Y. et al. 2016, RadioAstron Observations of the Quasar 3C273: A Challenge to the Brightness Temperature Limit, *ApJL*, **820**, L9

Levy, G.S. et al. 1986, Very Long Baseline Interferometric Observations Made with an Orbiting Radio Telescope, *Science*, **234**, 187

Levy, G.S. 1989, VLBI Using a Telescope in Earth Orbit. I. The Observations, *ApJ*, **336**, 1098

Linfield, R.P. et al. 1989, VLBI Using a Telescope in Earth Orbit. II. Brightness Temperatures Exceeding the Inverse Compton Limit, *ApJ*, **336**, 1105

Lister, M.L. et al. 2016, MOJAVE XIII. Parsec-Scale AGN Jet Kinematics Analysis Based on 19 years of VLBA Observations at 15 GHz, *AJ*, **152**, 12

Matveyenko, L.I., Kardashev, N.S., and Sholomitsky, G.V. 1965, Large Baseline Radio Interferometers, *Radiophysica*, **8**, 651. English translation: 1966, *SovRadiophys*, **8**, 461

Matveyenko, L.I. 2013, Early VLBI in the USSR. In *Resolving the Sky - Radio Interferometry: Past, Present and Future*, ed. M.A. Garrett and J.C. Greenwood (Manchester: SKAO), 43

Melis, C. et al. 2014, A VLBI Resolution of the Pleiades Distance Controversy, *Science*, **345**, 1029

Miyoshi, M. et al. 1995, Evidence for a Black Hole from High Rotation Velocities in a Sub-Parsec Region of NGC4258, *Nature*, **373**, 127

Moran, J.M. et al. 1967a, Observations of OH Emission in the H II Region W3 with a 74400λ Interferometer, *ApJ*, **148**, L69

Moran, J.M. et al. 1967b, Spectral Line Interferometry with Independent Time Standards at Stations Separated by 845 Kilometers, *Science*, **157**, 676

Moran, J.M. et al. 1968, The Structure of the OH Source in W3, *ApJ*, **152**, 97

Moran, J.M. 1998, Thirty Years of VLBI: Early Days, Successes, and Future. In ASPC **144**, *IAU Colloquium 164: Radio Emission from Galactic and Extragalactic Compact Sources*, ed. J.A. Zensus, G.B. Taylor, and J.M. Wrobel (San Francisco: ASP), 1

Moran, J.M. 2000, The Early Days of VLBI. In *Radio Interferometry: The Saga and the Science*, ed. D.G. Finley and W.M. Goss (Washington: AUI), 184

Napier, P.J. 2000, The VLBA – Planning and Construction, In *Radio Interferometry: The Saga and the Science*, ed. D.G. Finley and W.M. Goss (Washington: AUI), 198

National Academy of Sciences 1983, *Multidisciplinary Use of the Very Long Baseline Array*, https://nehrpsearch.nist.gov/static/files/NSF/PB84163690.pdf

Palmer, H.P. et al. 1967, Radio Diameter Measurements with Interferometer Baselines of One Million and Two Million Wavelengths, *Nature*, **213**, 78

Petrov, L. et al. 2009, Precise Geodesy with the Very Long Baseline Array, *Journal of Geodesy*, **83**, 859

Pilipenko, S.V. et al. 2018, The High Brightness Temperature of B0529+483 Revealed by RadioAstron and Implications for Interstellar Scattering, *MNRAS*, **474**, 3523

Porcas, R.W. 2010, A History of the EVN: 20 Years of Fringes, *Proceedings of Science*, 125 https://doi.org/10.22323/1.125.0011

Pushkarev, A.B. et al. 2017, MOJAVE XIV. Shapes and Opening Angles of AGN Jets, *MNRAS*, **468**, 4992

Readhead, A.C.S., and Wilkinson, P.N. 1978, The Mapping of Compact Radio Sources from VLBI Data, *ApJ*, **223**, 25

Rees, M. 1967, Studies in Radio Source Structure. I. A Relativistically Expanding Model for Variable Quasi-Stellar Radio Sources, *MNRAS*, **135**, 34

Reid, M.J. et al. 1989, Subluminal Motion and Limb Brightening in the Nuclear Jet of M87, *ApJ*, **336**, 112

Reid, M.J. et al. 2013, The Megamaser Cosmology Project. IV. A Direct Measurement of the Hubble Constant from UGC 3789, *ApJ*, **767**, 154

Reid, M.J. and Honma, M. 2014, Microarcsecond Radio Astrometry, *ARAA*, **52**, 339

Reid, M.J. et al. 2016, A Parallax-Based Distance Estimator for Spiral Arm Sources, *ApJ*, **823**, 77

Riess, A.G. 2016, A 2.4% Determination of the Local Value of the Hubble Constant, *ApJ*, **826**, 56

Rogers, A.E.E. and Morrison, P. 1972, Long-Baseline Interferometry, *Science*, **175**, 218

Rogers, A.E.E. et al. 1983, Very-Long-Baseline Radio Interferometry - The Mark III System for Geodesy, Astrometry, and Aperture Synthesis, *Science*, **219**, 51

Rogers, A.E.E. 2000, Challenges Facing the Recording System. In *Radio Interferometry: The Saga and the Science*, ed. D.G. Finley and W.M. Goss (Washington: AUI), 218

Romney, J.D. 2000, The Challenges of the VLBA Correlator. In *Radio Interferometry: The Saga and the Science*, ed. D.G. Finley and W.M. Goss (Washington: AUI), 227

Schilizzi, R. 2015, *What's in a Name: The Early Days of JIVE* http://www.jive.eu/jive-eric-symposium

Swenson, G.W. et al. 1977, *VLBI Network Studies II, Interim Report on a New Antenna for the VLBI Network* (Urbana: University of Illinois)

Swenson, G.W. 1977, *VLBI Network Studies IV, On the Geometry of the VLBI Network* (Urbana: University of Illinois)

Swenson, G.W. and Kellermann, K.I. 1975, An Intercontinental Array - A Next-Generation Radio Telescope, *Science*, **188**, 1263

Taylor, J. and McKee, C. eds. 2001, *Astronomy and Astrophysics in the New Millennium* (Washington: National Academy Press)

Thaddeus, P., ed. 1983, Panel on Radio Astronomy. In *Astronomy and Astrophysics for the 1980's* (Washington: National Academy Press), 211

Walker, R.C. 2000, Early VLBA Operations. In *Radio Interferometry: The Saga and the Science*, ed. D.G. Finley and W.M. Goss (Washington: AUI), 205

Wampler, J. ed. 1983, Panel on Ultraviolet, Optical, and Infrared Astronomy. In *Astronomy and Astrophysics for the 1980's* (Washington: National Academy Press), 98

Weinreb, S. 1963, A Digital Spectral Analysis Technique and its Application to Radio Astronomy, MIT PhD Dissertation and RLE Technical Report 42

Whitney, A.R. et al. 1971, Quasars Revisited: Rapid Time Variations Observed via Very-Long-Baseline Interferometry, *Science*, **173**, 225

Yen, J.L. et al. 1977, Real-Time Very Long Baseline Interferometry Based on the Use of a Communications Satellite, *Science*, **198**, 289

FURTHER READING

Clark, B. 1968, Radio Interferometers of Intermediate Type, *IEEE Trans. Ant & Prop.*, **AP-16**, 143

Clark, B. 2007, Travels with Charlie. In *But It Was Fun*, ed. F.J. Lockman, F.D. Ghigo, and D.S. Balser (Green Bank: NRAO/AUI), 563

Cohen, M.H. 1969, High Resolution Observations of Radio Sources, *ARAA*, 7, 619

Cohen, M.H. 1973, Introduction to Very-Long-Baseline-Interferometry, *Proc. IEEE*, **61**, 1192

Hirabayashi, H., Preston, R.A., and Gurvits, L.I. eds. 2000b, VSOP Results and the Future of Space VLBI, *Advances in Space Research*, **26**, No. 4

Kellermann, K.I. 1971, Joint Soviet-American Interferometry, *S&T*, **42**, 132

Kellermann, K.I. and Thompson, A.R. 1985, The Very Long Baseline Array, *Science*, **229**, 123

Kellermann, K.I. and Thompson, A.R. 1988, The Very Long Baseline Array, *SciAm*, **258**, 54

Kellermann, K.I. and Moran, J.M. 2001, The Development of High-Resolution Imaging in Radio Astronomy, *ARAA*, **39**, 457

Kellermann, K.I. 2007, First VLBI with the Soviets. In *But It was Fun*, ed. F.J. Lockman, F.D. Ghigo, and D.S. Balser (Green Bank: NRAO/AUI), 541

Minh, C. ed. 2003, MM VLBI between NRAO 14m and NRO 45m. In ASPC **306**, *New Technologies in VLBI*, ed. Y.C. Minh (San Francisco: ASP), 46

Napier, P. 1991, The Very Long Baseline Array. In ASPC **19**, *Radio Interferometer: Theory, Techniques, and Applications*, ed. T.J. Cornwell and R.A. Perley (San Francisco: ASP), 330

Napier, P. 1994, The Very Long Baseline Array. In *Very High Angular Resolution Imaging*, ed. J.G. Robertson and T.W.J. Tango (Dordrecht: Kluwer), 117

Napier, P. et al. 1994, The Very Long Baseline Array, *Proc. IEEE*, **82**, 658

Napier, P.J. 1995, VLBA Design. In ASPC **82**, *Very Long Baseline Interferometry and the VLBA*, ed. J.A. Zensus et al. (San Francisco: ASP), 658

Romney, J.D. and Reid, M.J. eds. 2005, ASPC **340**, *Future Directions in High Resolution Astronomy: The 10th Anniversary of the VLBA* (San Francisco: ASP)

Schilizzi, R. 2013, A Short History of Space VLBI. In *Resolving the Sky - Radio Interferometry: Past, Present, and Future*, ed. M.A. Garrett and J.C. Greenwood (Manchester: SKAO), 99

Shapiro, I.I. and Knight, C.A. 1970, Geophysical Applications of Long-Baseline Radio Interferometry in Earthquake Displacement Field and the Rotation of the Earth. In *Astrophysics and Space Science Library* **15**, ed. L Mansinha, D.E. Smylie, and A.E. Beck (Dordrecht: D. Reidel), 284

Shapiro, I.I. et al. 1974, Transcontinental Baselines and the Rotation of the Earth Measured by Radio Interferometry, *Science*, **186**, 920

Thompson, R., Moran, J.M., and Swenson, G.W. 2017, *Interferometry and Synthesis in Radio Astronomy* (Gewerbestrasse: Springer Nature) https://link.springer.com/book/10.1007%2F978-3-319-44431-4

VLBA Memo Series No. 1-698 http://library.nrao.edu/vlba.shtml

Zensus, J.A, Diamond, P.J. and Napier, P.J. ed. 1995, *Very Long Baseline Interferometry and the VLBA*. ASPC **82** (San Francisco: ASP)

CHAPTER 9

The Largest Feasible Steerable Telescope

From the very earliest stages, planning for NRAO included the construction of a very large fully steerable radio telescope with a diameter up to 1000 feet. However, following the 140 Foot debacle, there was no support for funding such an ambitious and risky construction program. After the construction of the 300 Foot Transit Telescope with its limited capabilities, NRAO initiated the Largest Fully Steerable Telescope (LFST) program to design and potentially construct a very large fully steerable radio telescope. The LFST team produced a series of designs for a 300 foot antenna capable of working at 1 cm wavelength, a 64 meter antenna working to 3 mm wavelength, and finally a 25 meter telescope working to 1 mm wavelength, but none of them were ever built. Although every review of the needs of radio astronomy supported the construction of a large fully steerable radio telescope, there was always a higher priority—the VLA, the VLBA, and most recently ALMA. In 1988, an NSF review committee recommended that the 27-year-old NRAO 300 Foot Transit Telescope be closed in order to provide funds for operating other new astronomical facilities. However, when the 300 Foot Telescope unexpectedly collapsed in November 1988, it was reported in the media as a national disaster for U.S. astronomy. West Virginia's Senator Robert Byrd demanded that the telescope be replaced. Although the NSF had other plans, Byrd included $75M in the 1989 Emergency Supplemental Appropriations Bill. The new 100 meter Green Bank Telescope would not be completed until the year 2000, and only after contentious litigation as to who was responsible for the delays and nearly factor of two increase in cost.

9.1 Early Discussions

Probably the first use of a parabolic dish for radio astronomy was in 1933 by John Kraus and Arthur Adel, who used a 1 meter diameter search-light mirror to try to detect the Sun at 20 GHz (1.5 cm) (Kraus 1984, p. 59). Although

© The Author(s) 2020, corrected publication 2021
K. I. Kellermann et al., *Open Skies*, Historical & Cultural Astronomy,
https://doi.org/10.1007/978-3-030-32345-5_9

they correctly speculated that sunspots might be regions of enhanced radio emission, they were unsuccessful due to the poor sensitivity of their receiver. As reported in a series of articles in the popular magazine *Radio-Craft*, even earlier, small parabolic dishes had been used for both transmitting and receiving radio waves in a variety of laboratory experiments and for communications over some tens of miles.[1] These pioneering programs were made at what were then called "ultra-short wavelengths" below 1 meter. In 1928, Fredrick Kolster of Palo Alto, California applied for a patent for a radio beacon to be used to guide airplanes to safe landings during periods of poor visibility. Kolster proposed using a small antenna at the focal point of a paraboloid to concentrate the radiation into a relatively small beam.[2] Even earlier, in 1888, during his pioneering experiments to demonstrate the existence of the electromagnetic waves predicted by James Maxwell, Heinrich Hertz used cylindrical parabolic reflectors to both transmit and receive radio signals generated by a spark gap.

When Grote Reber decided to follow up on Karl Jansky's discovery of cosmic radio emission at 20 MHz (15 meters), he recognized that a large parabolic dish would provide the most flexible opportunities, including the ease of changing frequency bands (Reber 1958). Reber's home-built dish became a prototype for later generations of antenna designs ranging from the familiar small consumer TV receiving dishes to the Jodrell Bank 250 foot telescope and German 100 meter steerable dishes to the ill-fated Sugar Grove 600 foot antenna. Starting with the 1964 Whitford Report (Whitford 1964), all of the National Academy of Sciences reviews of astronomy (Greenstein 1973; Field 1983), as well as the two NSF Dicke Committees (Dicke 1967, 1969), recognized the need for a large general purpose fully steerable parabolic dish for radio astronomy. But there was always another higher priority radio astronomy project that took precedence, and it would take a freak 1988 accident, a determined Green Bank scientist, and an influential, strong-minded US Senator before American astronomers would have a large fully steerable dish for radio astronomy.

Although Grote Reber's 32 foot radio telescope, described in Chap. 1, was not the first use of a parabolic radio reflector, in 1937 Reber's telescope was the largest parabolic antenna ever built. During WWII German engineers went on to build thousands of 3 meter (9 foot) diameter Würzburg antennas and hundreds of the so-called "Giant Wurzburg," 7.4 meter (23 foot) radar dishes, many of which found a home doing radio astronomy after the end of war hostilities (Sullivan 2009, p. 78). However, it would not be until 1951 that a larger purpose-built radio telescope would be erected on top of the Naval Research Laboratory building overlooking the Potomac River. The NRL 50 foot fully steerable dish had a very precise surface and made some of the first radio astronomy observations at millimeter wavelengths, although at the time, the limited sensitivity of millimeter wave receivers restricted millimeter observations to the thermal radiation from a few planets and H II regions.

Motivated primarily by the need for better angular resolution, as early as 1946 Grote Reber conceived an ambitious project to build a 200 foot diameter steerable antenna essentially based on his Wheaton design. Realizing the advantages of a fully steerable antenna, but also recognizing the complexity and cost of a 200 foot equatorially mounted telescope, he suggested using an alt-azimuth (alt-az) mount and an innovative analogue coordinate converter to provide a capability for tracking celestial sources. Reber assumed a maximum frequency of 3 GHz (10 cm) limited by electronics that might be available in the "visible future." In a letter to Otto Struve, Reber estimated that a 200 foot antenna could be built for $100,000.[3]

Nearly a decade later, Reber prepared a more detailed design of a 220 foot steerable antenna with a surface accuracy of about 3 mm, including sketches of all joints, a complete parts list, and a small model. By this time the cost had risen to $650,000 plus the unspecified cost of the drive system. Reber also outlined how he would extend the design to apertures up to 500 foot or more with a corresponding decrease in surface accuracy, and sketched the design of a 750 foot fixed spherical reflector mounted in a natural hole in the ground, such as Meteor Crater near Winslow, Arizona or Crater Elegante in Mexico.[4] This concept was later developed by Bill Gordon for the 1000 foot dish near Arecibo, Puerto Rico (Sect. 6.6). A year later, Reber argued that instead of the intermediate sized 140 Foot Radio Telescope, AUI should build a 600 foot diameter fully steerable antenna, which he estimated could be built for $10 million.[5]

With great prescience Reber sketched out many concepts for a fully steerable paraboloid that were rediscovered by others only much later. This included the realization that a structure that is strong enough not to bend will not fall; that due to turbulence, a wire mesh dish will not survive wind speeds greater than 20 mph any better than a solid surface; and that some antenna bending under gravity is not a problem, provided that the dish structure maintains a parabolic shape. Reber also made the innovative suggestion to mechanically adjust the dish structure using what he called "equalizers" to compensate for gravitational deformations, a concept that would not be successfully implemented for nearly another half century. He also suggested locating the antenna on a cliff looking to the south. In this way, he argued, the effect of sidelobes seeing the ground would be kept to a minimum when the antenna was pointed at the center of the Galaxy which would be low in the sky toward the southern horizon.

Reber unsuccessfully tried to interest the Carnegie Institution's Department of Terrestrial Magnetism, Harvard, MIT, the Office of Naval Research, and the NSF, as well as the nascent NRAO, in building his 220 foot radio telescope design. But Reber felt that his presentations were not taken seriously, no doubt, at least in part, due to his reluctance to follow what had by that time become fairly routine formal procedures to apply for NSF grant support. Only the New York-based Research Corporation found Reber's ideas of sufficient interest to provide modest support amounting to less than $250,000 over a 30-year period starting in 1951. However, the Research Corporation was interested in funding people, not big expensive facilities such as a large steerable radio telescope.

9.2   INTERNATIONAL CHALLENGES

Elsewhere in the world, radio astronomers were actively planning to build ever larger radio telescopes.

Bernard Lovell's Ambitious Plans for Jodrell Bank Following the successful completion of his 250 foot Mark I radio telescope in 1957, Bernard Lovell ambitiously began to think about building an even larger radio telescope. The Mark IV radio telescope was conceived as a fixed parabaloid, perhaps 1,500 to 15,000 feet across and up to 500 feet high. The 125 by 83 foot Mark II and Mark III telescopes were built as prototypes of the planned Mark IV instrument, and were used together with the Mark I as part of the very effective Jodrell Bank radio interferometry programs described in Chap. 2. But faced with increased competition for funds from Martin Ryle's group at Cambridge and Stanley Hey's group at Malvern, as well as growing government interest in participating in an international radio astronomy program such as the Benelux Cross, the funding for the ambitious Mark IV design study was repeatedly delayed, and never materialized. However, encouraged by hopes of funding from the United States, Lovell and Charles Husband conceived plans for a radio telescope "at least half the size of the visionary Mark IV," which Lovell named the "Mark V." Lovell's clearly stated goal was to build "the maximum possible size of dish for the money available," although this would mean compromising the accuracy and thus the shortest operating wavelength (Lovell 1985, p. 37).

By 1965, Lovell and Husband had a conceptual design for a 400 foot diameter telescope, which Husband estimated could be built for just over £4 million. Further engineering studies were developed by both Husband & Company and by Freeman Fox, who had designed both the Australian 210 foot antenna and the Canadian 150 foot radio telescope at Algonquin Park. During a visit to Harvard and MIT, Lovell became aware of the CAMROC design for a large radome-enclosed radio telescope (Sect. 9.5). Although the American scientists and engineers argued that an enclosed antenna could be built for much lower cost than one open to the environment, the CAMROC cost estimate was four times larger than the estimates for Lovell's Mark V antenna. The large discrepancy worried Lovell, but he was reassured by a meeting with the director of the National Science Foundation, after which Lord Francis Fleck said, "The NSF freely admit their dearth of genius by contrast which has led to expenditure far in excess of the British for far less results." (Lovell 1985, p. 27).

With the increasing emphasis on shorter wavelengths, especially by the emerging cadre of young radio astronomers interested in molecular spectroscopy, the original Mark V design goal of full efficiency at 21 cm no longer seemed adequate. However, the construction of a radio telescope of the proposed Mark V dimensions and capable of operation at such short centimeter wavelengths at a cost within the expected ceiling of £4.5 million, seemed to be an insurmountable challenge. Lovell apparently seemed unaware until 1968 of the developments in the homologous design concept, and of the already well advanced plans of German radio astronomers to construct a 100 meter radio telescope near Bonn to operate at wavelengths as short as one centimeter (see below). The projected price of only DM 32 million (equivalent then to about £3.4 million or $8 million) dismayed Lovell and his Jodrell Bank colleagues, who were skeptical of the German claims. When he later learned that the German telescope performed as expected, Lovell claimed that following the UK drive to join the European Common market, the pressure to collaborate with Germany killed his Mark V ambitions. Perhaps as a result of keeping his cards close to his chest and his failure to maintain usual scientific contact with his international colleagues, Lovell appeared to be surprisingly naïve about the German plans until he read of their completion in *Nature*.[6] The most optimistic Mark V scenario called for an antenna that was 2.5 to 5 times less accurate than the German telescope, depending on elevation and wind, and would not be operational until at least five years after the Bonn telescope. As related by Lovell, even the Jodrell staff were "rebellious" and recognized the futility of pursuing the Mark V concept (Lovell 1985, p. 104).

By late 1970, there was a further increase in the estimated (but acknowledged by Husband as not firm) cost of the Mark V radio telescope to nearly £8 million. Increased costs, the change of the UK government, Lovell's impending retirement, and decreasing prospects of support for new scientific projects, led to the abandonment of the Mark V project and its subsequent resurrection as a smaller Mark VA radio telescope. Over the next few years, Lovell advocated building a 375 foot diameter radio telescope, but by 1974, it appeared that even this smaller radio telescope would cost at least £17–20 million. Finally, nearly 15 years after Lovell had first proposed constructing a very large radio telescope, the UK Science and Research Council informed the University of Manchester that it would be unable to fund the proposed Mark VA radio telescope. Lovell responded by soliciting support from the respective radio astronomy commissions of the IAU and URSI. Citing long standing tradition, both of these international scientific unions declined to get involved in any political funding issues. Feelers from Germany offering time on the Effelsberg telescope were interpreted by Lovell as intending to "delay and destroy his Mark VA project" and "prevent us from building a larger instrument than the 100-m Bonn telescope." (Lovell 1985, p. 170) Krupp offered to build a copy of the Bonn telescope for half of the estimated Mark VA cost, but this was rejected by Lovell, who did not want to abandon the Husband design and start a protracted new study. Moreover, he argued that the UK should not "inject so much money in another economy," and that British engineers would not accept such a radical proposal.

The Effelsberg 100 Meter Radio Telescope (Fig. 9.1) The MPIfR 100 meter Effelsberg Radio Telescope[7] that so frustrated Lovell arose as a result of the visionary ambitions of the German scientist Otto Hachenberg, a generous gift from the German Volkswagen Foundation, and the recognition by the German

Fig. 9.1 MPIfR 100 meter radio telescope at Effelsberg, Germany designed by Otto Hachenberg. Elements of the International LOFAR Array can be seen in the foreground. Credit: Norbert Tacken, MPIfR

Max Planck Gesellschaft (MPG) of the growing opportunities in radio astronomy. Starting from the 1888 generation and detection of radio waves by Heinrich Hertz, through the WWII development of sophisticated radar systems, German radio research had a long and distinguished history. However, due to restrictions on all radio research imposed by the occupying American, French, and British military forces, radio astronomy was slow to develop in Germany. Among the major players in German wartime radar research were Leo Brandt and Otto Hachenberg, who had both worked at the Berlin-based Telefunken Company. After the war, Hachenberg became director of the East Berlin Heinrich Hertz Institute, and commuted between his home in West Berlin and his work in the East. But following the erection of the Berlin Wall on 13 August 1961, Hachenberg was unable to get to work and found himself without a job.

After the radio limitations were lifted in the 1950s, Bonn University built a 25 meter radio telescope on the nearby Stockert Mountain and in 1962 invited

Hachenberg to become the director. Being on top of a mountain in one of the heaviest industrial areas in the world, the high level of RFI (radio frequency interference) limited the effectiveness of the Stockert radio telescope. Hachenberg initially began to develop plans for building a 65 meter dish, comparable to the Parkes radio telescope, to provide the university with a competitive research facility. Although he was unable to find the DM 8 million needed to construct the 65 meter antenna, Hachenberg went on to design an 80 meter, then 90 meter, and finally a 100 meter radio telescope. Then, teaming up with other Bonn University colleagues, Hachenberg finally received half of a DM 32 million ($8 million) grant from the German Volkswagen Foundation to help build a 90 meter radio telescope at the University of Bonn.

By good fortune, Hachenberg's friend, Leo Brandt, had become a high ranking official in the German state of Nordrhein-Westfalen. Due to the Allied embargo, Brandt was unable after the War to find work in science and engineering, and began a career in politics. As Minister of Economy and Transport and later Secretary of State of Nordrhein-Westfalen, Brandt introduced the first speed limits within German cities. He was also able to help his old wartime friend Hachenberg by providing additional funds from the state and from the German Ministry for Research and Education to allow Hachenberg to build a 100 meter rather than 90 meter dish. Brandt also arranged to make a small plot of land available to build a radio telescope in a valley located near the small village of Effelsberg in the Eiffel Mountains about a one-hour drive from Bonn. A small river which marked the boundary between Nordrhein-Westfalen and the Rheinland-Pfalz had to be relocated to make room for the telescope. As a result, the state of Nordrhein-Westfalen became larger by about 2000 square feet.

Meanwhile, the MPG had become interested in the exciting new field of radio astronomy and invited Sebastian von Hoerner to become the director of the new institute for radio astronomy, located in Tübingen. As a young man, von Hoerner was drafted into the German army and sent to the Eastern Front where he participated in the German siege of Leningrad. After losing an eye at Leningrad, von Hoerner spent the rest of the War back in Germany in a research laboratory. In February 1945 he narrowly escaped the ravages of the Allied bombing of Dresden. Life in Germany did not get much easier after the War. Von Hoerner survived by collecting old tire tubes discarded by the occupying forces which he used to fabricate into rain coats that he then sold back to the US soldiers. Having no money, he then worked his way through his doctoral studies at Universität Göttingen by harvesting farm crops, in return for which he was allowed to take enough food to eat and was given a place to sleep. Initially he studied theoretical physics under Carl von Weizsäcker who guided him to problems in astrophysics. After receiving his PhD, von Hoerner moved to Heidelberg. In 1960, he came to NRAO as a one-year visitor at the invitation of Otto Struve, who was looking to broaden the scientific perspectives of his staff, which was heavily oriented toward radio problems. After returning to Germany for a few years, von Hoerner went back to Green Bank as a permanent

Fig. 9.2 Sebastian von
Hoerner, 1960. As a
member of the NRAO
Scientific Staff, he laid the
analytical foundations for
the homologous design of
radio telescopes, and,
along with John Findlay,
was an active participant
in the NRAO LFST
project. Credit: NRAO/
AUI/NSF

member of the NRAO scientific staff, only to be invited back to Germany to become co-director of the new Max-Planck-Institut für Radioastronomie (MPIfR), which was to be located near Tübingen (Fig. 9.2).

Although von Hoerner planned to go Tübingen, the Volkswagen Foundation awarded funds to Universität Bonn, and under pressure from Leo Brandt the MPG agreed to locate the new MPIfR in Bonn. As a result, von Hoerner declined the appointment at the MPIfR and remained at NRAO where, as described below, he went on to design, together with John Findlay, a series of large radio telescopes which were never built. The conflict between Bonn and Tübingen was further exacerbated by two factors. Von Hoerner and Hachenberg had mostly, but not entirely independently, developed the concept of homology whereby instead of trying to design a very rigid structure, the dish structure is allowed to deform to a new parabola as it is tipped to different elevations.[8] Von Hoerner used an analytical approach compared to Hachenberg's more empirical approach in designing the 100 meter Effelsberg radio telescope (Hachenberg 1970; Hachenberg et al. 1973). But each felt that he alone was responsible for developing the homology concept. Also, von Hoerner and Hachenberg had competed for a fixed level of Volkswagen funding which was initially split between them. However, when von Hoerner decided to remain at NRAO, the full DM 32 million was made available to Hachenberg and the Bonn group, sufficient to plan for a 100 meter size radio telescope.

The Effelsberg telescope was constructed by a consortium of the German firms Krupp and MAN, and has been in operation since 1972. Hachenberg became the founding director of the MPIfR and was shortly joined by Peter Mezger and Richard Wielebinski. Hachenberg brought with him scientists from Bonn University while Mezger, a German scientist who had earlier worked at the Stockert radio telescope, brought back a number of young radio astronomers who had been part of his team at NRAO. Following several upgrades to the surface and pointing, the Effelsberg radio telescope operates well today at 1.3 cm and is even used at wavelengths as short as 3.5 mm. Until the dedication of the Green Bank Telescope in 2000, the MPIfR 100 meter radio telescope remained the largest fully steerable radio telescope in the world.

Interestingly, many of the designers and builders of the large radio telescopes constructed in the 1950s and 60s made their reputation building bridges. Sir Charles Husband, who designed the Jodrell Bank 250 foot antenna, later went on to design the Britannia Bridge connecting the island of Anglesey with the Welsh mainland as well as the bridge featured in the movie *The Bridge Over the River Kwai*. The CSIRO 210 foot radio telescope and the Canadian 150 foot telescope were designed by Freeman Fox and Partners who had previously designed the Sydney Harbour Bridge. Ned Ashton, who built the NRL 50 foot radio telescope and later designed the 140 Foot Radio Telescope, had built several bridges over the Mississippi River. The 140 Foot project was initially contracted to the General Dynamics Electric Boat Company, the contractor for most of the US Navy's submarines, which may explain the windowless control room and submarine appearance of the room containing the declination bearing.

9.3   THE SUGAR GROVE FIASCO

While AUI, the NSF, and US radio astronomers were still debating the size of the NRAO telescope, scientists at the Naval Research Laboratory, under the leadership of James Trexler, conceived a plan to build a 600 foot diameter fully steerable antenna that would be both a powerful tool for radio astronomy, as well as for a variety of intelligence gathering applications. A few years earlier, Trexler had led the NRL program which successfully bounced the first voice signals off the Moon (Trexler 1958), suggesting interesting possibilities for surveillance of Russian radio transmissions. Discussions between the NRL Radio Counter Measurers Branch led by Trexler and the Radio Astronomy Branch led by John Hagen and Ed McClain resulted in a proposal to build a 600 foot fully steerable parabolic antenna. The concept was promoted simultaneously in the scientific and popular media as a radio telescope and in the halls of the Pentagon and Congress as a tool for military surveillance (van Keuren 2001; Greenberg 1964). Funds for the 600 foot antenna became available in 1958, and due to a perceived military expediency during the height of the Cold War, construction started immediately. The publicly stated use of the proposed 600 foot antenna was to monitor Soviet radio communications and radar reflected off the Moon's surface and for radio astronomy. Coincidently, the Navy antenna was to be located near Sugar Grove, WV, only 30 miles from

NRAO in Green Bank but a mountain range away. Subsequently this apparently led to considerable confusion among the public, if not the Washington bureaucracy, about the purpose of the Sugar Grove antenna and the distinction between Sugar Grove and Green Bank facilities. Indeed, in 1962, NSF Director Alan Waterman was chastised by Congressman Albert Thomas about "the real bear …. down there in West Virginia," (DeVorkin 2000, p. 56) and Harvard radio astronomer Edward Lilley referred to the Sugar Grove 600 foot antenna as "a radio telescope fiasco."[9] In part, the use of the term *radio telescope* for the Sugar Grove facility, whether intended or not, contributed to the confusion and misunderstanding.

The Sugar Grove antenna specifications were first laid out in December 1957 and were revised several times up until October 1959. The antenna was to be built from 30,000 tons of steel, 14,000 cubic yards of concrete, 600 tons of aluminum, and would stand 665 feet high. The movable reflecting surface was equivalent to two football fields in diameter. All of this was to be accomplished without the aid of modern computer-based finite element analysis.[10] Initial operation was anticipated for July 1962 (Figs. 9.3 and 9.4).

Fig. 9.3 Artist's conception of the US Navy's 600 foot Sugar Grove antenna. Credit: NRAO/AUI/NSF

Fig. 9.4 Partially built 600 foot surface backup structures lie on the ground at Sugar Grove in mid-1965. Credit: KIK/NRAO/AUI/NSF

The terms "radio telescope" and "radio antenna" were used interchangeably in Navy documentation. The coordinate conversion and steering of the telescope were based on an inertial guidance system under computer control using punched cards as the normal means of inputting instruction and punched paper tape to provide a continuous record of the antenna positioning. However, there was also a provision for allowing manual keyboard or "digital dial" input. The antenna surface was specified to consist of "no more than 210" individual panels. In order to maintain the precise surface and orientation of the dish in the sky under wind and thermal deformations, servo-controlled hydraulic jacks were designed to keep each panel within a planned 0.7 inch tolerance. Each panel was specified to be made of expanded, unrolled (non-flattened) aluminum alloy with openings not to exceed 0.625 by 1 inch, suggesting that operational frequency would not be greater than about 1 GHz. Provision was made to mount a four-inch optical telescope with a TV camera readout to point within 10 arcsec of the radio axis, presumably to support radio astronomy research. The antenna was to be painted with a white paint approved by the Navy's Bureau of Ships or BuShips for "top side" ship surfaces.

Although initially much of the activity surrounding the design and construction was classified, the existence of the project was publicly acknowledged, in fact even surprisingly well advertised (McClain 1960). The 1957/1959 antenna specifications document only carried the lowest classification level of "CONFIDENTIAL," which doesn't suggest a serious security concern, and was declassified in September 1962. A more detailed technical specification for the drive and control system, dated July 1962, was not classified. By this time, it was described only as a "radio telescope." A *New York Times* article stated that the Sugar Grove "telescope will be the largest ever built to tune in on the radio signals created in the stars and planets .... [and] will be able to look into space nineteen times further than the 200-inch optical telescope on Mount

Palomar."[11] When construction began in early 1959, the Navy apparently acted with typical military expediency to begin construction without waiting for detailed design, engineering, and cost evaluation or with any outside review or oversight.[12] Rather than contract with a commercial organization with experience in building large antennas, responsibility for the construction was given to the Navy Bureau of Yards and Docks (BuDocks). According to Admiral Frank Johnston, the design and construction were to proceed concurrently in order to save three or four years for what he described as an important military facility.

The construction cost was initially estimated to be $20 million, and it was expected that it would require about 30 people to operate the facility. A computer aided analysis of the structure made in 1959–1960 showed that the original design was faulty and had to be scrapped. Curiously, the 12,000 square-foot two-story antenna control and operations building was built underground, 500 feet away from the antenna. It was surrounded by two foot thick concrete shielding, ostensibly to provide RFI protection, although a simple Faraday Cage constructed of wire would have sufficed. One can only speculate about the true purpose of this underground concrete bunker. Interestingly, RFI specifications went down to 15 kHz probably in recognition of other activities on the site such as communication with the US Naval submarine fleet.[13]

The 600 foot cost estimates continued to grow. By the middle of 1962, there was considerable doubt about the feasibility of constructing such a large fully steerable dish, and the projected cost had ballooned to $300 million, while the estimated number of personnel that would be needed to operate the facility had increased by a factor of 30 to more than 1,000. Nearly $50 million had already been spent and another $50 million had been committed.

As early as September 1960, George Kistiakowski, Science Advisor to President Eisenhower, received a memorandum from his staff discussing the growing costs of the NRAO Green Bank Observatory and the Sugar Grove Naval Research Station.[14] In Green Bank, the projected cost of the 140 Foot Radio Telescope had risen to $5.5 million or nearly twice the original cost. At the same time, $134.6 million had been authorized to complete the Sugar Grove project, roughly 4.5 times the original estimate, but Kistiakowski was advised, "nobody believes it will" be completed. While the 140 Foot Antenna was expected to be exclusively used for radio astronomy, the Navy started to talk about Sugar Grove being available half of the time for radio astronomy, the remainder being for Navy operational requirements, including bouncing signals off the Moon, communication and electronic intelligence, and deep space probe communications. As late as October 1961, a *New York Times* article reported that McClain and Trexler had stated that "the major construction problems for the instrument had been solved," and that "construction is expected to stay within the latest authorized figure of $135 million."[15] But the weight of the moveable structure had increased to 32,000 tons, about equivalent to a naval battleship, in order to maintain the surface accuracy needed to support the now claimed maximum observing frequency of 2.3 GHz.

When Secretary of Defense Robert McNamara finally cancelled the Sugar Grove project in July 1962, the final cost to complete the construction was predicted to be $230 million. McNamara only mentioned in passing the technical problems and escalating costs, and stated that satellites could now provide the intelligence that the 600 foot antenna was to have obtained. Curiously, that same month the Navy issued an amendment to the technical specifications for the drive and control systems, by which time the Sugar Grove antenna was being increasingly described as a "radio telescope" that "could spot the edges of the Universe." The amount of scientific use of the Sugar Grove antenna was variously described as 25 to 50 percent. Otto Struve had been recruited to head a committee to review proposals for astronomical observing time on the Sugar Grove antenna, and although it was speculated that any astronomical observing would be highly coordinated and under the control of NRL, AUI was concerned that the advertised existence of astronomical studies at Sugar Grove would compromise NRAO's own goals of building a very large antenna.[16] As it turned out, the financial and technical embarrassment resulting from widely publicized cancellation of the ambitious 600 foot project, coupled with the apparent mismanagement, delays, and cost escalation of the 140 Foot project, became a serious black mark on US radio astronomy, and in particular on NRAO, that would take years to erase. The 17,000 cubic yard concrete foundation and 550 ton main bearing remain on the Sugar Grove site as a challenge to some future archeologists (Greenberg 1964).

Locally, residents of this small remote Appalachian Sugar Grove community had anticipated that the Navy's antenna project would bring in new employment opportunities with corresponding increases in land values. The resultant real estate speculation was premature and the ultimate cancellation of the 600 foot antenna project impacted the local economy as well as the image of US radio astronomy, especially at Green Bank.

Even after the 600 foot project was abandoned by the Navy, West Virginia Congressman Ken Hechler, a member of the House Science and Aeronautics Committee, asked the Navy to delay its "dismantling of the facilities" so that the NSF could consider a proposal by North American Aviation to use the already fabricated steel to build a less expensive 600 foot transit telescope at Sugar Grove for NRAO at a cost of only $20 million.[17] By this time the radio astronomers were wary of any involvement by the military, and NRAO declined to get involved. But West Virginia Senator Robert Byrd appealed to then President John Kennedy to maintain a presence at Sugar Grove, which continued until 2016 as the Navy Information Operations Command (NIOC), to conduct communications research and development for the Department of Defense. A quarter of a century later Senator Byrd would again intervene, this time to support radio astronomy in West Virginia.

Various antenna arrays, a 150 foot fully steerable antenna,[18] as well as smaller antennas and arrays were later built at Sugar Grove to support the Naval operation.[19] The Sugar Grove 150 foot antenna began as an investigation by John Findlay at NRAO to turn the NRAO 300 foot transit design into a fully steerable instrument, and was later pursued by Austin Yeomans at NRL who wanted to build a 300 foot antenna at Sugar Grove. Yeomans' planned 300 foot antenna never got built, but Yeomans later teamed up with Trexler and Edward Faelten to build a scaled down 150 foot version for Sugar Grove. Although built primarily for intelligence surveillance, in the 1960s and 1970s the 150 foot antenna was used infrequently for radio astronomy and was included in a number of VLBI programs.

Following the terrorist attacks on New York World Trade Center and the Pentagon on 11 September 2001, the Sugar Grove facility remained in operation by the National Security Agency for various intelligence surveillance applications. The continued presence of the NSA facility at the center of the National Radio Quiet Zone (NRQZ) has been important in preserving the NRQZ for radio astronomy in Green Bank. Half a century later, the contradiction between the classified nature of the project and the broad public dissemination of information about the project, along with the perhaps deliberately leaked confusion over the contrasting intelligence and radio astronomy goals, remains a mystery. After the project was canceled, a spokesman for the Defense Department acknowledged that "some of the capabilities from the beginning of the project had been overstated," and that "certain statements .... were scientifically inaccurate." An obvious limitation of using reflections from the Moon to warn of an impending Soviet rocket attack is that the Moon is visible at both Sugar Grove and Russia for only a few hours a day. One wonders about the real purpose of the Sugar Grove 600 foot antenna project and the massively shielded so-called telescope control building.[20]

By 2013 the NSA no longer had any need for the extensive domestic living facilities that had been built to support the large operational staff anticipated for the 600 foot antenna and other NIOC instruments. The General Services Administration announced the sale of the Sugar Grove Station including 80 single family homes, a 45 thousand square foot administration building, daycare and community centers, athletic facilities, and "much more."[21] Bids started at $1 million and on 25 July 2016 the NIOC was ultimately sold to an anonymous bidder for $11.2 million.[22]

9.4   THE LARGEST FEASIBLE STEERABLE TELESCOPE PROJECT

Although AUI wisely decided not to initially stretch its technical and financial resources, the construction of a very large fully steerable radio telescope was clearly on the agenda for the new radio observatory. As early as October 1957, the NRAO Advisory Committee met in conjunction with a broader group of astronomers, radio astronomers, and physicists to discuss future research programs at the Observatory and approved the following statement:

> The NRAO must continually anticipate the needs of and future developments in radio astronomy, and act promptly and decisively to provide for these needs. Because of the great time lag in the development of major instrumentation, the

NRAO should through its scientific advisors and staff, look now at the general direction of radio astronomy development in coming years, and commence planning for the next stage of development beyond the 140-ft.[23]

Work on the 140 Foot Telescope had hardly started when Dave Heeschen requested $250,000 from the NSF for "engineering studies and design of a very large antenna system." In this time period the acronym "VLA" was used to refer to the planned very large antenna rather than the Very Large Array, which came later (Chap. 7). In his proposal to the NSF,[24] which ran just over two pages, Heeschen stated that while the 140 Foot "will solve many of the current problems in radio astronomy, [it will] undoubtedly make many new discoveries and open new fields for investigation, ... many of which will require still more powerful instruments for their study. This is the way science works." Heeschen went on to argue the virtues of an antenna with 10^6 square feet of collecting area (equivalent to a 560 foot diameter dish). But faced with the construction of the first 85 foot antenna, the delays in the 140 Foot Telescope, and the construction of the 300 Foot Transit Telescope (Chap. 4), for the next five years, NRAO was not able to pursue these ambitious goals. In 1959, Heeschen again outlined the wide range of motivations for a very large antenna (VLA) ranging from the Sun and Solar System bodies to problems of galactic structure and dynamics and the structure and evolution of distant galaxies.[25] Meanwhile, the NSF Advisory Panel on Radio Telescopes emphasized the emerging technique of aperture synthesis over the construction of a single large aperture antenna.[26] (See Sect. 7.2). The Panel met three times. John Findlay and John Bolton were present as guests at the third meeting, which was held at the AUI office in Washington. Although Emberson was unable to attend, based on what he was told by others he informed the AUI Board that "John Bolton argued vigorously against the need for a Very Large Antenna, and that there seemed to be very little enthusiasm for the construction of such an instrument."[27] Undeterred, Heeschen and Findlay prepared a report for the NRAO Director arguing for a very large antenna or VLA at the NRAO.[28] Only with a VLA, Findlay and Heeschen claimed, could NRAO meet the instrumental needs of radio astronomy and support the research programs of staff and visiting observers alike. Findlay and Heeschen laid out a plan following the completion of the 300 Foot Transit Antenna to form a VLA design team consisting initially of 10 to 15 scientists and engineers starting in 1962.

In January 1963, following the successful completion and initial operation of the 300 Foot Transit Radio Telescope, Findlay and Robert Hall of Rohr Corporation discussed the possibility of constructing a 400 foot transit antenna with greater declination coverage and the capability to operate at shorter wavelengths than the existing 300 Foot structure. A month later, NRAO authorized the first phase of a design study by Rohr. Under Hall's leadership, Rohr completed their design work by June, but in view of the continuing issues with the 140 Foot Telescope, the estimated cost of more than $3.5 million was more than NRAO could realistically hope to find. Only a few months later, however, NRAO entered into a new contract with Rohr, for a feasibility study of a 100 meter fully steerable radio telescope capable of operating up to 3 GHz (10 cm). The Rohr study started with the design of the successful CSIRO 210 foot radio telescope which was the basis for the Rohr design of the 210 foot dish for the JPL Deep Space Network. Also under consideration was a 100 meter concept being developed by Harold Weaver at the University of California Berkeley.

As its first and second priorities for radio astronomy, the 1964 Whitford Report recommended the construction of the NRAO Very Large Array and the expansion of the Caltech Owens Valley Array (Sect. 7.2). However, as the third and fourth priorities, the Committee recommended the construction of two fully steerable 300 foot paraboloids at a cost of $8 million dollars each, as well as "approximately 15 ... smaller special-purpose instruments." The panel also supported the need for the very large fully steerable telescope and recommended $1 million for an engineering study for "the largest feasible steerable paraboloid" (Whitford 1964).

So even before the completion of the 140 Foot Telescope, US radio astronomers were looking past the 140 Foot to new possibilities. Succumbing to the pressure for observing time on the existing 300 Foot transit telescope, and buoyed by the Whitford recommendation, Heeschen sent a two-page letter to the NSF requesting that $2 million be added to the NRAO 1965 budget to build a second 300 foot transit dish, one that would work to 10 cm wavelength and have a larger declination range than the existing 300 foot dish.[29]

In October 1964, shortly after the release of the Whitford Report, a group of 14 American radio astronomers met in Green Bank "to discuss the question of whether this is the time to start thinking about a design study for a very large steerable telescope."[30] Although Heeschen, Director of NRAO, stated up front that NRAO's first priority was now the Very Large Array, he said that NRAO considered a large steerable radio telescope to be a second priority, and a millimeter wave telescope a third priority. The debate focused around a larger transit dish optimized for continuum work at centimeter wavelengths, and a smaller fully steerable antenna for spectroscopic studies working at somewhat shorter wavelengths. The group agreed at the end that radio astronomy needed a fully steerable telescope having a circular beam working down to at least 18 cm and preferably to 10 cm.

Although modest about the wavelength requirements, the group was more ambitious about the possible size, and discussed the construction of antennas with dimensions up to 600 feet (183 meters) in diameter. Perhaps unrealistically encouraged by the rapid funding and construction of the successful 300 Foot Transit Telescope, Heeschen suggested that a 400 foot (123 meter) transit telescope could be in operation by the end of 1966, while "a 100-m fully steerable dish could take about two years longer."[31] In their final report, the committee concluded that "the meeting gave the NRAO a mandate to undertake a feasibility study of a steerable instrument with a circular beam, a diameter of at least 600 ft. useful down to 18 cm and hopefully down to 10 cm."[32]

The following January, NRAO Deputy Director John Findlay documented the scientific and technical case for using NRAO resources to investigate design concepts for a large fully steerable radio telescope to be located in Green Bank.[33] Heeschen appointed a working group under Findlay (1965) to investigate the feasibility and options for constructing a very large fully steerable radio telescope as envisioned in the original NRAO planning discussions.[34] Initially the group began as "The Largest Feasible Steerable Paraboloid Working Group," but morphed into the "Largest Feasible Steerable Telescope" or "LFST" Group when they realized that configurations other than a paraboloid might be feasible and more cost effective. The group met for the first time on 2 April 1965. Over a period extending until 1972, the working group issued 57 reports, 42 of which formed a numbered series.[35] Engineers from interested industrial firms often attended the meetings.

A major result of these studies came from the work of von Hoerner, who investigated the fundamental principles and constraints for the design of large antennas. In an important paper, von Hoerner defined a "stress limit," a "gravitational limit," and a "thermal limit." Von Hoerner showed that for any given diameter there is a corresponding minimum wavelength, and stated that "any design that did not reach this limit was a waste of resources." (von Hoerner 1967). The stress limit, argued von Hoerner, is important only for diameters greater than 600 meters. Antennas larger than 40 meters are gravitationally limited, that is, they will deform under their own weight, while smaller ones are thermally limited due to temperature gradients across the structure, although thermal effects can be reduced at night, on cloudy days, or if the antenna is enclosed in a radome. However, he concluded that radomes are of no value for antennas larger than about 50 meters since they are limited by gravity, and even for smaller antennas, the value of a radome is "doubtful" (Fig. 9.5).

Another outcome of the LFST study, although one which was already subjectively understood by experienced observational radio astronomers, is that when pushed to shorter wavelengths, the performance of a radio telescope does not degrade as fast as theoretically predicted, assuming that the surface deviations from an ideal paraboloid are random. Moreover, it was recognized that, in practice, the effects of changing gravitational deformation result in a shift of the focal point of the dish as it folds up under gravity, and that the subsequent loss of gain experienced when the antenna is tipped in elevation can be partially mitigated by moving the antenna feed vertically to the new focal point. An extension of these ideas led to von Hoerner's "homologous deformation" concept described earlier. Traditional antenna designs required that the structure be as rigid as possible to minimize the impact of gravitational deformations. In homologous designs, the structure is allowed to bend, but in a carefully controlled way, so that under gravity, as the antenna is tipped, the surface bends into a new parabola, and the loss of gain can be recovered by merely moving the feed to the new focal point. But the price paid for a homologous design is that each structural member must be of a precise weight and

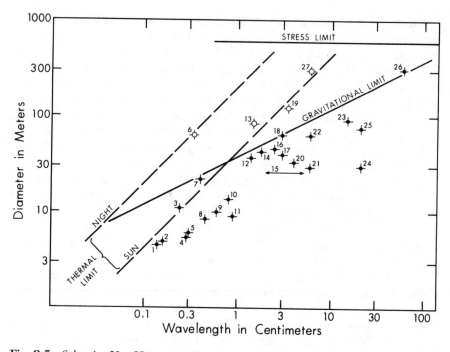

Fig. 9.5 Sebastian Von Hoerner's diagram showing the natural limiting size of antennas as a function of diameter and operating frequency due to gravitational and thermal distortions. The solid points show existing telescopes in 1972, while the open circles reflect planned new facilities. The proposed NRAO 65 meter millimeter-wave telescope shown in Fig. 9.7 is number 6 on the plot. Other major telescopes are Effelsberg (13), Arecibo before and after planned resurfacing (26 and 27), and the NEROC design (19). Credit: NRAO/AUI/NSF

cross section, so the use of standard structural steel is mostly precluded, thus increasing the cost.

The group also considered "ground supported antennas" such as the Arecibo 1000 foot antenna, as well as parabolic designs up to 200 meter in diameter with limited elevation movement which they argued could be built for about $60 million. One of the first concepts studied was for a 650 foot diameter floating concrete sphere structure that would support a 600 foot diameter reflector that would be driven in azimuth while a moving feed allowed celestial sources to be tracked up to one or two hours.[36] The motivating idea behind this concept was that in conventional radio telescopes, the limiting factor is the gravitational deflection as the dish structure is tipped. With a fixed elevation structure, gravitational deflections would not be relevant. Also, such a structure would have minimum wind and thermal loads as well as minimum maintenance and friction due to the lack of a mechanical bearing. Considerable effort was expended in developing the fixed spherical reflector concept with several commercial consultants as well as with University of Virginia civil engi-

Fig. 9.6 Floating spherical antenna concept, North American Aviation. Credit: NRAO/AUI/NSF

neers contributing to the design. Construction cost was estimated to be between $34 and $50 million to produce a structure capable of operating at wavelengths as short as 10 cm. However, the massive structure suggested unusual construction and drive problems requiring "unusual solutions," such as the large amount of antifreeze that would be needed to prevent freezing of the supporting liquid (Fig. 9.6).

In the end, it was clear that a conventional alt-az mounted fully steerable dish represented the only practical means to build a large radio telescope that would meet the scientific requirements of full sky coverage and flexibility. Over a period of seven years, the LFST group produced reasonably advanced designs for three telescopes that were thought to meet the requirements for a scientifically strategic antenna. The first was for a 200 meter dish based on the Navy's Sugar Grove design. Although the LFST design was for a dish ten percent larger than the Sugar Grove antenna, had a solid instead of mesh surface, and a considerably more accurate surface and pointing accuracy, the LFST group estimated the construction cost of a 200 meter fully steerable radio telescope as $105 million, or less than half of the estimated completion cost of the Sugar Grove 600 foot antenna.

Nevertheless, it was soon realized that a 200 meter antenna would be too heavy, too complicated, and too costly. The group went on to consider a 300 foot homologous dish that would operate with good performance down to 1 or 2 cm wavelength and could be built for only $8 million.[37] This was NRAO's answer to the radome enclosed 450 foot CAMROC antenna. However, the discovery of a variety of interstellar molecular species with transitions below 1 cm wavelength quickly made the 300 foot design appear obsolete, and the group turned its attention to the design of a radio telescope that could operate at millimeter wavelengths to exploit the new opportunities presented for studying interstellar chemistry.[38] It would be another four years before the group would produce a fully homologous 65 meter radio telescope design with good performance down to 3.6 mm, but only under "benign conditions" of temperature and wind (Findlay and von Hoerner 1972). The cost of constructing the 65 meter antenna was estimated to be just under $10 million (Fig. 9.7). The 65 meter antenna was never built, and in fact was never even formally proposed, but for a short time it competed with the Very Large Array in the deliberations of Greenstein's (1973) Astronomy Survey Committee Radio Panel (Sect. 7.4). With the increased emphasis on shorter wavelengths, the LFST group turned their attention to a 25 meter radio telescope which was later proposed by NRAO for operation at wavelengths as short as 1 mm wavelength. Following review by the NSF, the 25 meter millimeter telescope actually made it into President Carter's FY1981 and 1982 budget requests, but was ultimately dropped in favor the VLBA (Chaps. 8 and 10).

Years later, von Hoerner commented that LFST should have been the Largest *Fundable* Steerable Telescope, noting that the project began with 200

Fig. 9.7 Artist's conception of the 65 meter millimeter wave telescope. Credit: NRAO/AUI/NSF

meter diameter, then went down to 100, to 65, to 25 meters, and finally converged to zero. Although meant in jest, there was a real message here that ambitions must be carefully weighed against funding realities or, more accurately, perceived funding realities, because in practice it is so difficult to realistically predict funding scenarios even a few years downstream.

As late as 1979, NRAO Director Morton Roberts asked Findlay for an updated cost estimate to construct a large, fully steerable antenna for future planning. However, there is no record that anything further developed as a result of this initiative.

9.5   CHALLENGES FROM CALIFORNIA AND CAMBRIDGE

Planning for the next generation of large steerable radio dishes was not confined to NRAO. In 1965, around the same time as the NRAO LFSP/LFST program, two other independent initiatives appeared on the scene in apparent competition with the NRAO project as well as with each other. In the East, the MIT Lincoln Laboratory and the Smithsonian Astrophysical Observatory at Harvard University formed the Cambridge Radio Observatory Committee (CAMROC) to study designs for a large steerable radio/radar telescope to be located in the northeastern part of the United States.[39] Around the same time, OVRO Director Gordon Stanley wrote to Harold Weaver at U.C. Berkeley and Ron Bracewell at Stanford suggesting collaboration among the three universities for the construction of a 100 meter antenna. Caltech, Stanford, the University of California Berkeley, and later the University of Michigan joined to form a consortium known as the Associates for Research in Astronomy (ARA) to develop a "western regional radio astronomy facility." In 1965, the newly formed Western Regional Facility submitted a proposal to the NSF to design a 100 meter dish good to 10 cm for a cost of $12.3 million.[40]

The ARA antenna was to be located at the Owens Valley Radio Observatory where it could operate together with the proposed Caltech synthesis array (Chap. 7). Although it seemed expeditious to involve other universities in the proposal, Caltech was firm that they would maintain control. In particular, the OVRO site director, George Seielstad, questioned the added value of the consortium and urged construction and operation by Caltech alone, although he conceded that the antenna should be available for everyone with a "proposal to be judged solely on the basis of scientific merit" by a committee including representatives from outside Caltech but not confined to Stanford, Berkeley, and Michigan.[41] However, the ARA proposal ran head up against the competing MIT-Harvard proposal for a 440 foot radome enclosed radio telescope. Just as Lovell found in the UK with his proposed Mark V antenna, the NSF was unwilling to fund a $350,000 design study without some expectation of funding the construction, but they were also unwilling to consider construction funding without a detailed design and cost estimate. Moreover, both the NSF and Caltech were concerned that the proposed ARA 100 meter radio telescope would be in competition with the proposed expansion of the OVRO interferometer which had already received good reviews, and so the plan for the proposed Western Regional Facility in Radio Astronomy did not attract much national interest. It was unceremoniously killed in August 1967 by the first Dicke Committee "because of the more revolutionary possibilities inherent in the Arecibo and NEROC concepts"[42] (Sect. 7.4).

While it was noted that the US needed a large steerable radio telescope to keep up with the Jodrell Bank and Parkes radio telescopes, initially the motivation of the Harvard-MIT group was apparently to provide a competitive facility for Harvard, MIT, and other New England universities located close to home. Concerns about the extreme RFI environment in the New England area were dismissed as being no worse than anywhere else due to the increasing proliferation of RFI from aircraft and satellites which occur essentially everywhere.

Following the success of the Haystack 120 foot radome-enclosed antenna, the Lincoln Laboratory commissioned a study of even larger radomes. The report issued by Ammann & Whitney, Consulting Engineers, concluded in 1965 that radomes in the range 550 feet to 1100 feet were feasible at costs ranging from $7.75 million to $46 million respectively, and that such structures would result in antenna cost savings greater than the cost of the radome.[43] The CAMROC initial planning was for a 400 foot radome-enclosed structure working to 3 GHz (10 cm) wavelength. The actual CAMROC proposal, submitted two years later, discussed in two volumes the scientific merits of a large filled-aperture steerable dish, and the preliminary design of a 400 foot radome-enclosed dish with a surface good for operation up to 6 GHz (5 cm).[44]

As has been characteristic of the radio astronomy community, at the same time as there was a certain level of competition between the CAMROC and NRAO LFST groups, there was good technical cooperation, with members of each group serving on the "competition's" design committees and a number of common commercial consultants serving both projects. Indeed, when the Smithsonian Astrophysical Observatory (SAO) initiated a meeting in 1968 to solicit community support for their antenna project, SAO Director Fred Whipple asked NRAO's John Findlay to chair the meeting held at the Smithsonian Museum of History and Technology located on the Washington Mall.[45] The 31 meeting participants included representatives of Harvard, Smithsonian, MIT, Cornell, Ohio State, JPL, NRL, Caltech, the Universities of California and Maryland, NSF, and NASA. Whipple informed the group that given the "approval and endorsement" of the participants, the Smithsonian Institution would "attempt to take steps on their behalf to bring into being a national radio and radar filled-aperture telescope." According to Findlay, the Smithsonian was "ready to serve the national users."[46]

However, the participants largely had their own priorities and gave less than enthusiastic endorsement to the proposed CAMROC radome-enclosed large radio telescope. Instead they prefaced their report by pointing out the need for both "large arrays and large dishes," and urged the timely completion of the Arecibo upgrade. The report did point out the "urgent need for a large filled aperture radio-radar telescope," but recommended that it be located at a "site

selected primarily on the basis of scientific and technical criteria," thus effectively rejecting the Smithsonian arguments for a New England site. They also
agreed to endorse the Smithsonian design as the basis for the final design,
encouraged the SAO to submit a proposal to an unspecified "appropriate
agency of the Federal Government," and said that the Smithsonian should
"carry the general responsibility for the funding, design, construction, and
operation" of the antenna as a national facility for radio-radar astronomy.[47]

In 1967, thirteen northeastern institutions, including the Smithsonian
Astrophysical Observatory, Harvard, and MIT, formed the Northeast Radio
Observatory Corporation (NEROC), a nonprofit consortium to plan an
advanced radio/radar facility. In an attempt to circumvent the normal but
lengthy NSF proposal review process, NEROC sought to have the telescope
funding included in the FY1970 Smithsonian budget. The bill died in Congress,
partly as the result of what was considered only a lukewarm endorsement from
the two Dicke Committee reports (1967, 1969) (Sect. 7.4). In June 1970,
NEROC submitted a new proposal to the NSF for the design and construction
of a radome-enclosed 440 foot diameter fully steerable antenna capable of
operating to wavelengths as short as 1.2 cm (25 GHz).[48] But this time, after a
heated battle within the NAS Greenstein Committee reviewing the needs for
astronomy in the 1970s, the ambitious NEROC proposal to finally provide
American radio astronomers with the long planned large steerable radio telescope lost out to the NRAO proposal to build the Very Large Array (Sect. 7.4)
(Greenstein 1973).[49]

9.6   A National Disaster Leads to a New Radio Telescope

By the mid-1980s, the UK's Jodrell Bank 250 foot antenna had been upgraded
several times with a new, more precise surface and a new precision pointing
system which allowed operation at wavelengths as short as 6 cm. Meanwhile in
Australia, the 210 foot dish had also been upgraded, in part with funds received
from NASA to support the Apollo lunar program. Australian scientists were
using the inner part of the dish at 1.3 cm wavelength to study interstellar water
vapor and ammonia. In Canada, the National Research Council was operating
a 150 foot fully steerable alt-az mounted dish which had a better surface than
the 140 Foot Telescope. The most visible competition came from the Effelsberg
100 meter fully steerable antenna operated by the Max-Planck-Institut für
Radio Astronomie in Bonn, Germany.

The New Large Steerable Radio Telescope Study In September 1987, users of
the 300 Foot Telescope and other radio astronomers gathered in Green Bank
to celebrate 25 years of discoveries and to plan for future research programs.
But with the still-unfinished VLBA construction and the growing interest at
NRAO in millimeter astronomy, there appeared little prospect for the long-

desired US 100 meter class fully steerable antenna of the kind that was envisioned when NRAO was formed in the 1950s.

At the suggestion of one of the present authors (KIK), MIT Professor Bernard Burke, a longtime NRAO user, AUI Board member, and supporter of NRAO, wrote to NRAO Director Paul Vanden Bout, suggesting that NRAO explore the possibility of replacing the aging 140 Foot Telescope with its "antique" equatorial mount, inferior surface, poor pointing, and high maintenance cost. In a prescient remark, Burke noted that, "It would certainly be prudent to have the plans in readiness as soon as possible, should the proper occasion arise on short notice." Burke suggested the formation of a working group at NRAO to examine the scientific motivation, size-wavelength trade-offs, and cost of a 140 Foot replacement telescope.[50] The most likely funding source identified by Burke was support of the space VLBI missions planned by Japan and the USSR. Unable to convince NASA to support an American-led space VLBI mission (Chap. 8), Burke and others speculated that NASA might be willing to fund the construction of the large ground-based radio telescope that would be needed to complement the necessarily small orbiting space antenna, but at a small fraction of the cost needed to deploy a space-based radio telescope for VLBI.

Vanden Bout responded by appointing a committee to formulate a scientific justification for a new large antenna, to address the tradeoffs between size and short wavelength limit, and to consider concepts that would reduce the construction and operating cost. Since NRAO was still building the VLBA and was committed to the construction of the Millimeter Array (Chaps. 8 and 10) as the next NSF-funded major NRAO project, Vanden Bout instructed the committee to consider sources of funding other than the National Science Foundation.[51] The committee was charged with reporting by the end of 1988. Unable to come up with a reasonable descriptive name for a telescope that had not yet been designed, as a spoof on the LFST project, the prospective new telescope was provisionally named the NLSRT, the *New Large Steerable Radio Telescope*, "because that is so bad that there is no danger that it will, by default, become the final name."[52]

The NLSRT Committee concentrated on antennas in the range of 70 meters to 120 meters that would operate with short wavelength limits from a few millimeters to a few centimeters wavelength with estimated costs of $5 to $50 million. By November 1988, in anticipation of meeting the Director's end of the year deadline, a draft report was in hand discussing two broad classes of radio telescopes: (a) a general purpose 70 to 100 meter class instrument operating with "full efficiency" up to 22 GHz (1.3 cm), but with good efficiency up to 43 GHz (7 mm) and with limited performance up to 86 GHz (3.5 mm), and (b) a larger 100 meter to 150 meter diameter antenna capable of working only up to 3 GHz (10 cm). Although it was appreciated that the Green Bank site had limited capabilities at millimeter wavelengths, with the continuing increase in RFI, the location in the National Radio Quiet Zone offered unique opportunities at longer wavelengths.

Based on estimates received from various manufacturers and by scaling the cost of recently constructed radio telescopes[53] including the effects of inflation, the construction costs of the proposed "Very Large Dish (VLD)" were estimated to be in the ballpark of $50 million.[54] Suggested cost saving measures included choosing a site with low wind speed, restricting the slewing velocity and acceleration, limiting the elevation to 10 or 15 degrees above the horizon, avoiding complex joints and hard to fabricate pieces, and using a simple, easy to control computer such as the IBM PC-AT.[55] Understanding the Director's charge to the NLSRT committee that the next major NSF-funded NRAO facility was to be a millimeter array, the committee suggested that in view of their interest in SETI and in space VLBI, NASA might be the appropriate agency to support construction of the VLD. But a lot of work was still needed to understand the size/wavelength and conventional symmetric/unblocked aperture configuration trade-offs and their impact on the construction cost. In particular, the illumination efficiency and polarization properties of unblocked apertures, especially at the longer wavelengths, was not well understood. Community interest in the study was minimal. One prominent scientist wrote, "To be honest, I cannot justify spending any time on this now; the prospects for a positive outcome just seem too bleak."[56]

Meanwhile, during the course of the NLSRT study, an added incentive to consider the next generation of steerable radio telescopes developed as a result of a new threat to the continued operation of the Green Bank facility, especially the 300 Foot Transit Telescope. By 1988, in addition to the NRAO facilities, the National Science Foundation was operating a number of radio telescopes throughout the country. In recognition of its increased use for astronomy and the corresponding decrease in Department of Defense funding resulting from the 1969 Mansfield amendment, funding for astronomy at Haystack transferred to the NSF, although an important Haystack VLBI program remained largely supported by NASA. In 1969, ownership of the Haystack antennas was turned over to MIT, and management of the Haystack Observatory was assumed by NEROC, although MIT continued to provide administrative support and the Haystack Observatory staff remained as MIT employees. Similarly, responsibility for the Arecibo Observatory was transferred from the Air Force to the NSF. Cornell established the National Atmospheric and Ionospheric Center to administer the Observatory operation as a national observatory, joining NRAO and NOAO as the third national facility for astronomy, but placing increased burden on the NSF operating budget. Elsewhere in the US, e.g. at Caltech, Michigan, and Ohio State, other radio astronomy programs were transferred from ONR or AFOSR to the NSF. It is always difficult to determine in such circumstances whether or not new money was actually added to the budget, since one never knows what the budget would have been in the absence of the new initiatives. Nevertheless, by this time there were a total of 12 radio astronomy facilities in the US which were receiving funds from the NSF, and there were no comparable facilities which were operated solely with private or state support.

Faced with the prospect of inadequate funds to operate two national radio astronomy observatories, as well as a growing number of university operated radio astronomy facilities, the NSF Astronomy Division Director Laura (Pat) Bautz convened a sub-committee of their standing Advisory Committee for Astronomical Sciences to identify those NSF funded radio facilities "having the highest scientific priority so that they could be supported at levels sufficient to exploit their capabilities with the resources available."[57] The sub-committee was asked to "recommend relative priorities" with the implication that less productive instruments could be closed with minimal loss to astronomy. Donald Langenberg, a former Deputy Director of the NSF, was appointed sub-committee chair.[58] On 21–23 April 1988, representatives from all of the major US radio astronomy facilities gathered in Chicago at the Rosemont O'Hare Exposition Center to convince the Langenberg Committee, as the sub-committee was known, of the merits of continued operation of their facilities.

The Langenberg Committee weighed matters of frequency coverage, spatial resolution, versatility, future potential, ongoing research programs, impact to other disciplines, technology development, role in training students, and community access. Surprisingly, the Committee claimed it did not consider budgets and gave little weight to costs. The Committee reviewed and classified 12 US radio telescopes. Apparently not wanting to offend their radio astronomy colleagues and appear to be conspiring with the NSF to close radio telescopes, the Committee said nice words about each facility that they were asked to review and stated that "all currently funded facilities merit continued support." However, given the likely prospect of extremely limited funds for at least another year, and to be responsive to their charge, the Committee divided the NSF funded radio telescopes into three priority categories.

Group A contained those "deemed absolutely essential to the continued health of astronomy;" Group B facilities were "highly recommended … under all but truly disastrous funding levels;" while Group C telescopes were declared to be marginally less competitive than those in Group B. Group A facilities included the VLA, VLBA, the Berkeley-Illinois-Maryland Array, the OVRO mm array, and the Caltech Submillimeter Observatory on Mauna Kea. Group B contained NRAO's 140 Foot Antenna and its 12 Meter Millimeter Wavelength Telescope on Kitt Peak, the 14 meter millimeter wave dish at the Five College Radio Astronomy Observatory (FCRAO), and the Arecibo Observatory. The NRAO 300 Foot, the Haystack 120 foot, and the OVRO 40 meter antennas were deemed "less competitive" and placed in Group C. In the case of the 300 Foot Telescope, the Langenberg Committee drew attention to the original limited goals and the subsequent upgrades which enabled a wide range of important contributions. "However," the Committee concluded, "if the NSF finds that it cannot support even the current minimal complement of radio telescopes, the Committee reluctantly recommends diverting resources from the 300 Foot to adequately support higher priority facilities." In particular, NRAO and the NSF were about to be faced with the appreciable costs required to operate the VLBA, which was still under construction. As a result of the anticipated budget limitations and given the mandate from the Langenberg Committee, NSF funding was withdrawn from both Haystack and FCRAO. Almost surely NSF funds for the 300 Foot, and perhaps the 140 Foot as well, would have been terminated within a few years. Not only were the prospects for building a new large steerable antenna bleak, but the long term prospects for the continued operation of the existing Green Bank telescopes were not encouraging. Without the 300 Foot and 140 Foot Antennas, the future of Green Bank and the National Radio Quiet Zone was in doubt.

Collapse of the 300 Foot Telescope All that suddenly changed on the night of 15 November 1988. At about 10:30 pm, George Seielstad (Fig. 9.8), the NRAO Assistant Director for Green Bank Operations, received a telephone call from the head of telescope operations, Fred Crews.[59] "George," Crews reported, "You have a telescope down." Seielstad was not too alarmed. "The telescope is down," was standard radio astronomy-speak, generally meaning that telescope wasn't working properly because of some receiver or antenna malfunction, a not uncommon occurrence. However, Crews had a more serious message. At 9:43 pm that evening, the entire 300 Foot antenna structure had completely collapsed; all that remained was a tangle of steel members (Fig. 9.9). A later analysis reported that a large gusset plate which joined several members high up in the dish backup structure had cracked due to metal fatigue. That caused added stress on the adjoining members which then broke, spreading additional force to the surrounding structure, leading to the collapse of the whole antenna. Greg Monk, the 300 Foot Telescope operator that night, was working in the control building and noticed falling ceiling tiles and a steel member that had

Fig. 9.8 Green Bank site director, George Seielstad, played a major role in securing the support of West Virginia Senator Robert Byrd for the replacement of the 300 Foot telescope. Credit: NRAO/AUI/NSF

Fig. 9.9 Left: 300 Foot Radio Telescope photographed on 15 November 1988. Right: Remains of the 300 Foot Telescope on 16 November 1988, the morning after the collapse. Credit: Courtesy of Richard Porcas

crashed through the roof. Fortunately, no one was killed or injured in the accident. Only five months after the Langenberg Committee had sentenced the 300 Foot to death, the NSF was spared the administrative burden of withdrawing funds for the operation of the telescope, which they knew would mean dealing with West Virginia Senator Robert Byrd, who had been a consistent and staunch supporter of the NRAO Green Bank operation. Indeed, in previous years when NRAO wanted to consolidate administrative activities at the Charlottesville headquarters by moving the small NRAO fiscal division from Green Bank, Senator Byrd made clear that he did not approve.

In the 26 years since it went into operation, more than 1000 scientists had used the 300 Foot Telescope for 178,830 hours and published 429 peer reviewed scientific papers. Fifty-three students had used the telescope for at least part of their PhD dissertation research. About one-fourth of all known pulsars, including the Crab Nebula pulsar, had been discovered with the 300 Foot, and more than 100,000 discrete radio sources had been cataloged. This was more than all previously detected radio sources from all the radio observatories in the world. A total of $3.6 million had been spent on 300 Foot construction, upgrades, and ancillary equipment. When the telescope collapsed on the night of 15 November 1988, the estimated replacement value was nearly $10 million.

Like other US government facilities, NRAO carried no insurance, and the prospects for replacement appeared nil. All that might be recovered would be the salvage value of the nearly 500 tons of aluminum and steel lying in a West Virginia field. However, even that was not to be. The steel girders lay in a twisted mess under great tension. Releasing one member might release a dangerous spring that could cause serious, possibly fatal, injury. Indeed, over the next months, the interlocking pile of steel girders continued to flex and move. Salvage was dangerous. The best deal NRAO could get was from the Elkins

Iron and Metal company, who agreed to take away the debris at no cost. Other companies wanted to be paid to clean up the mess. During the cleanup nine months later, an unanticipated danger surfaced when a four foot rattlesnake emerged after being evicted from its home within the tangled jumble of steel girders.

Just four days after the collapse of the 300 Foot Telescope, the NSF and AUI commissioned an extensive formal investigation of the accident. A blue ribbon Technical Assessment Panel was appointed to determine the cause of the telescope failure. The panel was chaired by former Cornell Vice President Robert Matyas.[60] As frequently happens with such incidents, the NSF and AUI wanted to know if there was someone to blame. Contrary to some statements made immediately following the collapse, the 300 Foot Telescope was not built to physically last only 5 or 10 years, although the expected scientific lifetime was thought to be of that order before it would be superseded by a newer, more powerful instrument. However, the anticipated 100 meter class fully steerable telescope was never built. With the various improvements to receiver sensitivity, the two upgrades to the dish surface, the construction of the tracking feed, and the implementation of fast elevation scans, the 300 Foot Telescope remained scientifically productive until its collapse in 1988. There were no known structural compromises made at the time of construction that might have led to its collapse after more than 25 years, although a computer finite element analysis, which was not available a quarter of a century earlier, clearly indicated that the structure had been over-stressed. Specifically, the panel investigation showed that the gusset plate, which joined members of the back-up structure to the elevation bearing mounted on one of the two towers supporting the structure, had what were called "micro-fractures" that may have been introduced during the telescope construction, and which slowly expanded under repeated stress.[61]

At the time of the collapse, the 300 Foot Antenna was being used by James Condon to survey the sky at 6 cm wavelength for new radio sources. Due to the great improvements in receiver sensitivity since the telescope was first designed, it was no longer necessary to wait for the sky to slowly drift through the antenna beams as the Earth rotated. Instead, in order to speed up their observations, Condon and his colleagues were rapidly scanning the telescope in elevation, which likely contributed to the ultimate failure of the structure. By an unfortunate coincidence, at the time, the NRAO computing staff were "upgrading" the software needed to analyze the telescope data. So, although Condon was nearing the end of a month long observing program, he had been unable to analyze his data. Months later, when he was able to examine the data, he realized that even early in the previous month, there was a large hysteresis between the apparent positions measured when the telescope was driving up and when it was driving down. "Even more ominous," Condon (2008) later said, "during the final week before the collapse, the north-south beamwidth had increased from 3 arcmin to 4 or 5 arcmin." With the benefit of hindsight, Condon later realized that this was an early indication of the failing structure.

The review panel determined that the gusset plate which had suffered from metal fatigue "had been cracking for years and finally ran out of cross section." They also praised the original design and low cost construction of the antenna, as well as the quality of NRAO's continued inspections and maintenance, and concluded that there was no human error involved which could have prevented the incident. Nevertheless, with the benefit of hindsight, the stresses on a large number of structural members were perhaps as much as a factor of two higher than would be permitted by existing codes at the time the structure collapsed, and therefore the "structure was marginal with respect to structural failures." The panel concluded that the "failure of the telescope structure was not the result of inadequate maintenance or inappropriate operation of the telescope," and noted that there were no "unfavorable implications about the current ability to engineer future telescopes of this or larger size." These conclusions were met with a sigh of relief, not only at NRAO, but also over at the Naval Radio Station in Sugar Grove, where the staff was alarmed by the collapse of the 300 Foot Antenna as their 150 foot antenna was based on a similar design, and they worried that it might share whatever structural deficiency caused the destruction of the Green Bank 300 Foot.

Considering that the 1988 Langenberg Committee had declared the 300 Foot Telescope to be scientifically "less competitive," the response of the media and the scientific community was surprising if not startling. Washington newspapers sent helicopters to Green Bank to photograph the remains of the collapsed telescope. On his daily radio broadcast, Paul Harvey[62] announced, "The science of astronomy has suffered a devastating setback!" News media from around the world reported on "a major blow to world astronomy." In a front page picture caption, *The New York Times* stated that "the 300 foot radio telescope was one of the most powerful instruments in the world."[63] Several newspapers, including one in South Africa, blamed the collapse on "hostile space aliens" who wanted to stop earthlings from eavesdropping on their activities. A front page headline declared "Space Aliens Destroyed Radio Telescope" (Fig. 9.10).

A Controversial Congressional Earmark A more serious and a more sober response came the day after the telescope collapse when West Virginia's Senators Robert Byrd (Democrat, West Virginia) and Jay Rockefeller (Democrat, West Virginia) contacted the NSF about replacing the telescope. At the time, then NSF Director Erich Bloch was in Antarctica. Bloch, the former IBM Vice President and winner of the 1985 National Medal of Technology for his contributions to the development of the IBM System/360 series of computers, was the first NSF director to come from industry and to not have a PhD. He considered development of economic competitiveness to be a strong priority of NSF-sponsored basic research.[64] On November 28, just after he returned from Antarctica, Bloch was summoned to an evening meeting in Senator Byrd's office to explain what he planned to do to replace the 300 Foot Telescope. Byrd, who was at the time the Senate Majority Leader, was joined by Senator Rockefeller who was a member of the Senate Committee on Science,

NEWS

January 31, 1989   50SAT   70¢

BRIDE IN CHAINS!
Hubby and landlady made
her a slave, say police

**America's most powerful radio
telescope IS . . .**

ZAPPED!

. . . by hostile space aliens!

THE Green Bank radio telescope in West Virginia after
the alien attack. At left is how
it looked after construction.

BEFORE ▲   AFTER ▶

Space aliens zapped the enormous radio telescope at Green Bank, W. Va., with a powerful laser to keep scientists from monitoring their activities in the northern hemisphere!

That's the claim of Swiss astronomer Peter Voisard, who says the destruction of the 300-foot instrument on November 15 qualifies as the boldest act of extraterrestrial aggression in the history of the world.

"We know that extraterrestrials have shot down planes and abducted people but this is the first time they have been brazen enough to destroy a government research facility," the expert told newsmen

By RAGAN DUNN

in Geneva. "The Americans must have been learning too much about them," he continued. "There is no other rational explanation for such a strike as this. In some circles, it might be considered an act of war."

Dr. Voisard's report sent shockwaves throughout the world's scientific community. But the handful of men who can talk about the Green Bank disaster with authority refused to describe the incident as anything more than "a mystery."

There is no doubt that the instrument — which could detect objects almost two-thirds of the way to the edge of the visible universe — was essential to monitoring extraterrestrial activity in the northern skies.

And its devastation by a destructive beam stands as compelling proof that it was

zapped by extraterrestrials who wanted to mask their activity from mankind, Dr. Voisard said.

"Any other explanation defies logic," he continued. "The telescope had been in operation since 1962 and was solid as a rock.

"Suddenly it collapses in the dead of night. What are we supposed to think? That the telescope just fell apart?"

French radio astronomer Marc Kraemer was inclined to agree with Dr. Voisard but warned against jumping to conclusions.

"I firmly believe that we can

prove extraterrestrials toppled the telescope at Green Bank," he said. "But let's wait until all the evidence is in.

"Then we can take whatever steps are necessary to prevent things like this from happening in the future."

Brave dog

A German shepherd survived for three days after being buried by an avalanche near Berne, Switzerland.

When rescued by police the dog wagged its tail and appeared to be hungry but in good health.

Fig. 9.10 Front page of *Weekly World News*, 31 January 1989, declaring the revenge of the aliens. Credit: Weekly World News, 31 January 1989

Space, and Technology, NRAO Director Paul Vanden Bout, AUI President Robert Hughes, and George Seielstad, NRAO Assistant Director for Green Bank Operations, who, significantly, was the only resident of West Virginia from the NRAO delegation. However, Bloch, who was joined by Ray Bye, Director of the NSF Office of Legislative Affairs, was not intimidated by Senator Byrd. Vanden Bout explained that NRAO planned to propose to the NSF the construction of a new radio telescope to replace the collapsed 300 Foot Dish, but Bloch pointed out that any proposal to the NSF would need to be evaluated within the overall context of national needs and NSF priorities, and that such an evaluation would take considerable time. The Senators countered that they hoped for a "firmer commitment and a definite timetable." Clearly, more specifics were needed from NRAO including the cost and timescale for building a radio telescope to replace the fallen 300 Foot Antenna (Fig. 9.11).

Fig. 9.11 Top: Senator Robert C. Byrd, (D-WV); Senator Jay Rockefeller, (D-WV). Bottom: NSF Director Erich Bloch; NRAO Director, Paul Vanden Bout. Credit: Rockefeller and Byrd—US GPO; Bloch—NSF; Vanden Bout—NRAO/AUI/NSF

All NRAO had to offer, however, was the report of the NLSRT study with a "bench-mark" design for a 70 meter antenna operating to millimeter wavelengths, which was not intended to replace the 300 Foot Antenna but, as Burke had urged a year earlier, to replace the 140 Foot Radio Telescope, with its "obsolete equatorial mounting, its excessive gravitational deformations, non-repeatable pointing errors, and poor surface accuracy." The NLSRT report, which was rushed to completion within a few weeks of the 300 Foot collapse, reflected an emphasis on the shorter wavelengths and correspondingly smaller aperture than the 300 Foot.[65] At least another year of engineering work was still needed before a Request for Proposals (RFP) could be prepared, and a ballpark figure of $50 million was the only cost estimate available. Radiation Systems Inc. (RSI) president Richard Thomas saw an opportunity for new business, and within a few weeks of the 300 foot collapse, Thomas submitted an estimate of $9.6 million to replace the 300 Foot with a fully steerable antenna. In a series of letters to members of Congress from states where RSI had manufacturing facilities, Thomas actively lobbied to replace the fallen 300 Foot antenna. Thomas did not give any detailed specifications, but implied that aside from steerability, the RSI antenna would have the same performance as the 300 Foot Transit Antenna.[66] A re-evaluation of costs by the NLSRT committee using a wider range of data obtained from various manufacturers confirmed the NLSRT cost estimate of $50 million, although numbers ranged from less than $10 million (RSI) to nearly $100 million (JPL/Ford Aerospace).[67]

Just two weeks after the 300 Foot collapse, MIT Professor Bernard Burke sent a memo to the American Astronomical Society's Committee on Astronomy and Public Policy to alert them to the circumstances surrounding the telescope collapse and the strong congressional interest in providing a replacement.[68] Burke, who to a large extent had initiated the NRAO NLSRT study just a year earlier, wrote to reassure the AAS members that "powerful forces are at work" to fund a new radio telescope in Green Bank, but that it would be "disastrous" if the construction costs of a new telescope were to come from NRAO's limited operating budget. Burke pointed out that Senate Minority Leader (and former Senate Appropriations Committee chair) Pete Domenici (R-NM) would surely object if the Green Bank construction funds came at the expense of VLA operations in New Mexico or the then ongoing construction of the VLBA. Moreover, contended Burke, it was "absolutely essential" that Green Bank construction funds "not be borne by the NSF by taking resources from other areas of astronomy or physics without a supplement to the budget."

At the time, the next large NSF construction project was anticipated to be LIGO, the Laser Interferometer Gravity Wave Observatory, but LIGO construction was not yet funded. The NSF funding request for FY1990 only contained a nominal sum for continued LIGO research and development. LIGO was a very controversial project designed to detect the gravitational waves expected to be generated from the final stages of collapsing orbiting binary neutron stars. Although conceived by physicists, the label "Observatory" suggested to many that it was another expensive astronomy project. As a joint project of MIT and Caltech, LIGO had powerful support, but also strong opposition from many in the physics community who argued that LIGO would not have sufficient sensitivity to detect gravitational waves. There was also opposition from the astronomy community which had other priorities for new observatories. This was one instance where radio and optical astronomers were united against a perceived joint rival.[69] There was considerable support, however, for LIGO from NSF Astronomy Section Director Pat Bautz, who perhaps saw that replacing the 300 Foot Telescope with LIGO was a way of preserving her astronomy priorities.

As planned, LIGO consisted of a complex system of mirrors spaced 4 km apart located at the ends of two orthogonal excavated tunnels. The mirror separation was monitored by a system of lasers, and calculations showed that a passing gravitational wave was expected to change the mirror separation by only about 10^{-15} mm. In order to discriminate between a gravitational wave and disturbances from local vehicular traffic or seismic activity, the NSF planned to build two widely separated complexes, one of which was to be near Columbia, Maine. With the encouragement of the NSF, Caltech investigated the possibility of replacing the Maine site with Green Bank. In a report dated 31 January 1989, LIGO Principal Investigator Rochus (Robbi) Vogt reported that "it is technically feasible to build a LIGO installation" in Green Bank, but that due to the more difficult "topographical complexity" there would be a $7 million to $18 million increase in the cost compared with the alternate site in Maine.[70]

So when confronted by the senators, Bloch stubbornly ignored Rockefeller's offer to help, explained that the NSF had other priorities for astronomy, such as LIGO, and that, considering the Langenberg report, he did not plan to replace the Green Bank radio telescope. This apparently enraged Byrd who, reportedly red-faced, pointed his finger at Bloch claiming that in all his years in Washington he had never encountered such an uncooperative agency head. According to Vanden Bout, sensing that the situation was getting out of hand, Rockefeller then leaned over the table and, towering over the seated Bloch, said, "Leader is about to become Chair of Appropriations. He will have his finger on every dime of the Federal budget. Now, are you prepared to let us help you?"[71] AUI President Hughes, recognizing the need for at least the semblance of peer review, offered to write a proposal. Apparently all this made an impression on the NSF Director, who responded that he "could work with the Senator," to which Byrd replied that he had "been waiting all evening to hear that."[72]

An accomplished fiddler, former butcher, welder, and in his youth a Ku Klux Klan organizer, Robert C. Byrd was first elected to the US Senate in 1959 after serving three terms in the House of Representatives, and became the longest serving member of the US Congress. Since 1977, Byrd had served either as the influential Senate Majority or Minority Leader, but with the new 101st Congress, starting in January 1989, Byrd became the powerful Chair of the Senate Appropriations Committee, a position which he used to bring billions of dollars in federal funds to West Virginia and, wherever he could, to protect every West Virginian job. Recognizing that both attrition and lay-offs can reduce the scope of an organization, Byrd engineered the 1988 NSF Authorization Bill to explicitly forbid the elimination of a handful of positions at Green Bank which NRAO had proposed to transfer to Charlottesville.

George Seielstad, the Green Bank site director, had received his PhD from Caltech in 1963 for his research on radio source polarization at the Owens Valley Radio Observatory (OVRO). Following his graduate work, Seielstad spent a year on the faculty of the University of Alaska, and then returned to Caltech as a member of the OVRO staff and to serve as OVRO site manager. Living with his young family in the nearby small town of Bishop, Seielstad became involved in local politics. In 1974 he took a leave of absence from Caltech to run for Congress as a Democratic candidate in California's 18th Congressional District, representing the sparsely populated Inyo, Mono, and Alpine Counties. With what was probably a record low campaign budget, Seielstad easily won the Democratic primary, but lost the general election to the Republican incumbent, William Ketchum. Seielstad garnered 47% of the vote in this traditionally Republican district, and drew the attention of the Democratic National Committee. However, having spent significant personal funds to support his primary and general election campaigns, Seielstad gave up his political career to return to radio astronomy. But his service to the Democratic Party was not to be forgotten.

Both Senator Byrd and NSF Director Bloch were strong-minded individuals who normally got their way; neither wanted to be manipulated or, even worse, appear to be manipulated. Byrd was probably one of the most influential people to have served in Congress, having held all of the senior appointments in the Senate. Seielstad later described Bloch as an "acerbic, combative tough-guy personality."[73] Byrd was fascinated by astronomy and what astronomers knew about the existence of God. He was primarily motivated, however, by the opportunity to enhance the economy of West Virginia by bringing jobs to the state and more broadly raising the profile of West Virginia, which was widely perceived as an Appalachian backwater. He did not really care whether it was going to be a new radio telescope or LIGO that would be built in Green Bank, as long as it employed a lot of people and brought visibility to the state and to himself.[74]

However, before it could build a new radio telescope, NRAO and the radio astronomy community faced a long period of planning and construction. Characteristic of the solidarity shared by the global community of radio astronomers, Peter Mezger, Director of the MPIfR and former NRAO staff member, kindly offered to make time available to 300 Foot users on the Institute's 100 meter antenna, and the NSF made grants available to American users of the MPIfR telescope. This was a good deal for American radio astronomers, as well as for the NSF. Even if a different investigator were to fly to Germany each day to use the MPIfR telescope, it would only cost less than half a million dollars a year in plane fares and travel costs, or about an order of magnitude less than the annual operating cost of the German 100 meter telescope.

Meanwhile, encouraged by the apparent support of the West Virginia Senators, NRAO moved rapidly to present a credible plan to the NSF. Typically, planning for new telescopes, whether radio or optical, involves many years of design and engineering, years of study to choose the optimum location, as well as years of "selling" the facility, first to the astronomical community, then to the funding agencies, the Administration, and finally Congress. In this case the location was clear: Green Bank, WV. Ironically, Congress, if not the NSF, appeared to be supportive. Yet there was no telescope design, nor even a consensus of what kind of instrument to build to exploit the apparent funding opportunity. All that was available was the hastily completed 28 November 1988 NGLSRT report.

On 2–3 December, NRAO convened the first of several meetings between NRAO staff and members of the NRAO user community to reach a consensus on a replacement for the ill-fated 300 Foot Telescope. Fifty-six individuals from 15 separate institutions, plus NRAO staff, participated in this hurriedly arranged meeting to discuss priorities for the replacement telescope.[75] Many who could not attend, as well as those who did participate, wrote letters to NRAO and to the NSF presenting a variety of arguments supporting their particular scientific interest. Interestingly, strong support for the radio telescope came from Joseph Weber, who was the pioneer in developing instrumentation for the detection of gravity waves, but who saw LIGO as competition to his own search for gravity waves.[76] Others wrote opposing the construction of any new NRAO facility at a time of great need for correcting the diminishing support for American university radio astronomy facilities, or opposing the apparent use of "pork-barrel" funding.

The meeting participants were able to inspect the remains of the collapsed telescope and heard a report from former NRAO Director Dave Heeschen on the investigation already underway to find the cause of the collapse. But the presentations and discussion quickly turned to understanding the scientific drivers and to reviewing what was learned from the NLSRT study of a Very Large Dish (VLD). Considering the interest of the West Virginia Senators, suggestions for a southern hemisphere site or a drier site near the VLA, or for an array of small dishes, were quickly dispensed with as not being realistic. Recognition that the areas of scientific opportunity years in the future could not be defined, the participants argued for maximum flexibility. The discussions quickly led to a large steerable antenna with full sky coverage working to relatively short wavelengths. A tentative schedule at the December meeting called for fixing the telescope characteristics by the end of 1988, one month away, developing a conceptual design by the end of 1989, and an engineering design by the end of 1990. This would allow construction to begin in 1992 with a very optimistic completion date by the end of 1993.

Nevertheless, there remained several controversial areas to be settled immediately. For a given cost, there is a tradeoff between antenna size and the limiting operating wavelength. Based on established scaling laws, a 100 meter diameter telescope was expected to cost about six to seven times more than a

50 meter telescope with the same performance specifications. For a fixed size, a telescope built to operate at 3 mm wavelength might be expected to cost about twice as much as one designed for only 1 cm operation. Reflecting the rapidly evolving scientific interests driven by the existing discoveries in molecular spectroscopy, the series of radio telescopes designed by the earlier LFST group had successively decreased operating wavelength and compensating smaller diameters to mitigate the cost, and this argued for a telescope capable of operation to about 100 GHz (3 mm wavelength), where the Green Bank atmosphere becomes noisy and unstable. But other areas of research, mainly pulsar studies and 21 cm hydrogen research, argued for the largest possible size. Achieving a high operating frequency for pulsars was not important, as pulsars are strongest at lower frequencies, and it was considered fairly straightforward to meet the relatively easy performance specifications needed for pulsar studies. Scaling from the costs of existing antennas in size or wavelength limit, as well as estimates received from various manufacturers, cost estimates for antennas in the range of 70 to 100 meter diameter and operating at wavelengths as short as 3 mm ranged from less than $50 million to more than $100 million.

Following the 2–3 December Green Bank meeting, there was little agreement among members of the NRAO user community. Many argued for a very large dish operating at wavelengths of a few centimeters to replace the fallen 300 Foot for VLBI (especially in conjunction with a space-borne antenna), H I (21 cm) and OH (18 cm), pulsar, SETI, and studies of radio galaxy and quasar radio source distributions, luminosity functions, variability, and evolution. Others favored a smaller aperture that would work well at the shorter wavelengths needed for the rapidly growing field of molecular spectroscopy and searching for cosmic microwave background anisotropies. The proponents of the "big dish" argued that Green Bank was a terrible site for millimeter wave observations. The counter argument noted that important millimeter observations were being made at the FCRAO in central Massachusetts, where the weather conditions were comparable to those found in Green Bank. Spectroscopists and 21 cm workers argued for a clear aperture offset design to minimize reflections and sidelobes; others pointed out that no large offset antenna had ever been built and that it would be unreasonably costly to pursue this concept. Tor Hagfors, Director of the Arecibo Observatory, expressed concern about the impact that a new fully steerable radio telescope might have on the Arecibo Observatory, but noted that if the Arecibo Observatory were to be upgraded, the impact might be minimized.[77]

Yet others, such as Richard McCray from the University of Colorado and NOAO Director Sydney Wolfe, worried that any funds spent on a new NRAO antenna might come at the expense of other high priority programs or their own pet project, and argued against any 300 Foot replacement. Among the broader astronomical community, it had not escaped notice that the last two major NSF astronomy projects were for radio telescopes: the VLA and the VLBA which was still under construction. Optical astronomers had been waiting for the start of a national 8 meter telescope and wanted assurance that the "Byrd Antenna" would be "coupled with an overall improvement of funding for ground-based astronomy."[78]

There were also clear applications of a large radio telescope to spacecraft tracking and to various military uses, with implications for possible broader funding support. To address some of the concerns of the astronomical community, NRAO actively solicited interest in supporting the construction of a new radio telescope from NASA, the US Naval Observatory, and the Jet Propulsion Laboratory (JPL). However, this raised the possibility that with reduced NSF funding, Green Bank might ultimately perish as a radio astronomy facility, a specter that would resurface several decades later. Nevertheless, following the Green Bank meeting, Vanden Bout wrote to the NSF Director apprising him of the strong scientific support for replacing the 300 Foot Antenna with a modern 100 meter class fully steerable radio telescope. Noting the "important role in the missions of other agencies," such as NASA and the US Naval Observatory, Vanden Bout suggested that it might be appropriate to share the construction cost and use of the telescope.[79]

Meanwhile, the NRAO scientific staff and external radio astronomers continued to debate the design options of a new radio telescope. In a 12 December 1988 staff meeting, the New Mexico VLA staff argued that a compact array of smaller dishes would be more powerful than a single large antenna. But Vanden Bout quickly dispensed with the array concept on the largely non-technical grounds of higher operating costs and the need for NRAO to maintain excellence in both arrays and single dishes.[80] A particularly controversial topic was the relative merits of an unblocked aperture with its greater cost and potentially poorer pointing precision but lower sidelobes, better aperture efficiency, and relative immunity to interference versus cheaper and better understood, more conventional designs. Conventional radio telescopes have the feed[81] or subreflector mounted at the center of the dish supported by two, three, or four supporting legs. But, the feed or subreflector, as well as the feed supporting structures, partially block the aperture, decreasing the gain or sensitivity. Even more important, with conventional radio telescopes, the blocking structures set up reflections, or standing waves, between the dish and subreflector, which cause frequency ripples that greatly compromise spectroscopic observations. Moreover, the blockage generates antenna sidelobes that extend well outside of the main beam. Typically, about one third of the power received by a conventional radio telescope comes in through the sidelobes. This seriously impacts studies of widely distributed radiation, such as 21 cm studies of galactic hydrogen. Additionally, with lower sidelobes of an off axis feed/subreflector system, the telescope would be less subject to interference from satellites and aircraft. Off axis antennas had been built and were widely used in the telecommunications industry and for consumer satellite TV reception, but in 1989, the largest known antenna ever built with an unblocked aperture was only 7.5 meters in diameter. The construction of a large unblocked aperture antenna was a formidable challenge, as it required an asymmetric dish design, and many predicted a large but uncertain cost impact and likely reduction of antenna pointing accuracy due to inadequate stability of the feed-subreflector support structure.

A significant question was whether it was more appropriate for NRAO, the national radio observatory, to build a low cost specialized antenna or a general purpose antenna that would satisfy the needs of the large and diverse NRAO user community. Some argued that a general purpose instrument involves compromise so that it is not optimum for anything, while special purpose antennas, such as the old 300 Foot antenna, can have applications that often extend beyond the original design goals. Many letters were written to NRAO and to the NSF arguing one way or another. The arguments boiled down to a large dish of the order of 100 meters or larger but operating only at centimeter and longer wavelengths vs. a smaller but more precise antenna capable of operating at millimeter wavelengths. Green Bank astronomer and later NRAO Assistant Director for Green Bank operations, Jay Lockman, was probably the most vocal supporter of what he called a "Big Floppy Dish" (BFD) with an unblocked aperture and low sidelobes that would, among other advantages, be a unique instrument to study galactic H I, since all existing instruments suffered from stray radiation that contaminates 21 cm spectra. Burke protested that this was contrary to the consensus of the 2–3 December Green Bank meeting. With considerable prescience, he pointed out that a clear aperture asymmetric structure would introduce many design problems, cost risks, more complex feeds, and a delay in completion, which he argued NRAO and the radio astronomy community could ill afford.[82]

No matter how well designed, the effectiveness of optical telescopes is in practice limited by the environment, clouds, and "seeing,"[83] while the performance of radio telescopes depends on locally generated radio frequency interference (RFI) and increasingly at shorter wavelengths to atmospheric water vapor. Although the location of the proposed new dish was in a sense a non-issue if NRAO wanted to exploit the enthusiasm of the West Virginia Senators, this did not dissuade the purists. While there was general support for locating the new dish at one of the existing NRAO sites to minimize site development and operating costs and to exploit the presence of a highly trained technical staff, many radio astronomers argued for locating the antenna at the 7,000 foot elevation VLA site, with its clear skies and low water vapor content needed for effective operation at millimeter and short centimeter wavelengths. On the other hand, due to its location in the National Radio Quiet Zone, the Green Bank site was probably the best place in the US for doing radio astronomy at longer wavelengths where protection from RFI is an issue. It was also realized that if the 300 Foot replacement were to be built in New Mexico, this would likely expedite the migration of NRAO activities to New Mexico and thus lead to the closing of the Green Bank facility and the subsequent loss of the unique capabilities of the National Radio Quiet Zone. However, with an average cloud cover rivaling the Northwest and upper Great Lakes areas, it was argued that the Appalachian Mountain area is a poor location for millimeter observations. George Seielstad mounted a vigorous defense of the Green Bank site, pointing out that on a clear cold winter night, the short wavelength observing conditions can be excellent. Although it was understood that there are not too many clear cold winter nights each year that could support short millimeter wavelength observations, realistically the choice was between a radio telescope in Green Bank or LIGO.

NRAO Director Paul Vanden Bout actively solicited support from American radio astronomers, support which ranged from enthusiastic to lukewarm. There was a wide range of opinions about how much the replacement telescope should cost and about the relative priorities of wavelength coverage and size. In a hastily convened presentation to the NSF on 21 December 1988, just five weeks after the collapse of the 300 Foot Antenna, and less than three weeks after the Green Bank meeting to define the proposed telescope, Vanden Bout discussed the long history of planning for a large steerable radio telescope and presented a conceptual plan for a 70 meter diameter radio telescope at an anticipated cost of about $50 million dollars. In an effort to recognize the interests of millimeter astronomers, Vanden Bout suggested that the proposed antenna could have useful performance at millimeter wavelengths. Vanden Bout's strategy was to make the NSF comfortable with the Senate initiative. Bautz and others were clearly nervous about the appearance of pork and the expected resistance from the optical astronomy community, and from radio astronomers more interested in millimeter astronomy or arrays of smaller antennas, along with those concerned about the growing concentration of US radio telescopes at NRAO. In particular, the proposed 300 Foot replacement might threaten the planned Millimeter Array, also proposed by NRAO, or the planned upgrades of the Haystack and Arecibo radio telescopes. The latter was a particular concern, as Bloch and Bautz did not want to cause trouble with Cornell President Frank Rhodes, who was a member of the NSF National Science Board.[84] Closer to home, even the AUI Board of Trustees was less than enthusiastic about exploiting Congressional interest in a non-peer reviewed project that might impact other planned astronomy or physics programs.

A week later Byrd and Rockefeller announced that they wanted to see a proposal for the replacement of the destroyed telescope by January.[85] At the same time, NRAO learned that the NSF was seriously considering how to best exploit Senator Byrd's interest in Green Bank to satisfy the Foundation's own goal of building LIGO in Green Bank and perhaps closing down the radio astronomy operation. Vanden Bout was summoned by Byrd to appear in the Senate Appropriations Committee Room on 5 January 1989, along with NSF personnel. To their chagrin, Seielstad and AUI President Bob Hughes were initially not invited, but both managed to lobby for inclusion and ultimately did participate in the meeting. While waiting for Senator Rockefeller's late arrival, Senator Byrd explained to the NRAO/AUI participants how he had risen to such a powerful position in the Senate.[86] He made it clear that he expected to get whatever he wanted in Congress and from the NSF.

When the meeting began, the NSF presented their proposal to build LIGO in Green Bank instead of a new radio telescope. Vanden Bout produced a letter from Tony Tyson, a well-known astrophysicist from Bell Labs and long-time opponent of LIGO, who argued that LIGO was poorly conceived, but otherwise NRAO refrained from speaking against LIGO.[87] NRAO's position was difficult, as previously Seielstad had used his Caltech connections to try to convince Caltech's LIGO Director and former Provost and Vice President for Research, Robbi Vogt, to bring LIGO to Green Bank. But this was before the 300 Foot collapse and the threat that LIGO would compete with the proposed 300 Foot replacement radio telescope. Before the NRAO/AUI contingent was excused, Seielstad declared his preference for the new radio telescope over building LIGO in Green Bank. Nevertheless, following the meeting, Byrd's staff leaked that "NRAO blew it," by not coming down hard on LIGO.[88] Subsequent discussions with Byrd's Director of the Senate Appropriations Committee, Terry Sauvain, indicated Sauvain's strong preference for LIGO, and that he had discussed with the NSF's Ray Bye how to best achieve their goal. However, Carol Mitchell of Byrd's personal staff leaned toward the radio telescope and kept Seielstad and Byrd informed of the behind the scenes maneuvering by Sauvain and Bye, all of which infuriated Byrd, who did not tolerate any disagreements among his staff.

Having second thoughts about not having been more critical of LIGO, and not informing anyone else, Seielstad arranged a private meeting with Byrd to explain that he had been too meek because NRAO did not want to offend Caltech. Byrd, in return, explained that he didn't have the knowledge to discriminate between the value of LIGO and a radio telescope, and depended on the scientists.[89] Meanwhile, NRAO was instructed by Bloch to cooperate with a planned Caltech visit to evaluate the feasibility of placing LIGO in Green Bank. Not only was the proposed 300 Foot replacement telescope in danger, but Vanden Bout worried that NRAO would be "slaughtered" by Bloch in future NSF budgets. Another meeting was scheduled for 23 February, but no one was clear what that would achieve.

Hearing about the NSF proposal to "forego the building of a successor radio telescope to the lost 300 Foot Antenna at Green Bank, and instead to site one of the elements of the LIGO gravity experiment there," MIT's Bernard Burke, flexing his muscle as a new member of the National Science Board, wrote to Bloch addressing the unique importance of the National Radio Quiet Zone. Burke argued that the continuation of the NRQZ would be uncertain without a replacement telescope as well as important scientific contributions that would result from constructing a state-of-the-art 100 meter telescope in Green Bank.[90] Seielstad and Vanden Bout were regularly kept informed by Byrd staffers Terry Sauvain and Carol Mitchell about the continuing discussions between Byrd and Bloch on the merits of building LIGO in Green Bank. Aside from producing the letter from Tyson, NRAO refrained from criticizing the Caltech/MIT LIGO project, but Bloch and the NSF continued to resist the idea of building a new radio telescope in Green Bank. NSF head of Math and Physical Sciences, Richard Nicholson suggested to Byrd that LIGO would produce a Nobel Prize for West Virginia. Bloch and Nicholson were certainly not oblivious to the fact that Byrd's counterpart Chair of the House Appropriations Committee was from Livingston Parish in Louisiana where one of the LIGO elements was to be sited. Following his visit to Green Bank, Vogt was impressed with the infrastructure although concerned about access, and reported back to Bloch that Green Bank was indeed a suitable site for LIGO.

Input from the Congressional Research Service The US Congressional Research Service (CRS) was originally organized in 1914 as a special reference unit within the Library of Congress. In 1946 it was renamed the Legislative Reference Service and in 1970, it received its present name. The CRS offers bipartisan confidential research assistance to Members of Congress and their staffs. Their reports, while not classified, are not public unless the requesting Member of Congress chooses to make them public. According to the CRS web site,[91] the CRS is available to Congress 24/7 to offer authoritative, confidential, and objective analysis of current policies and present the impact of proposed policy alternatives.

Richard Rowberg was the Chief of CRS Science Policy Division when, on a Saturday night, he received a telephone call at his home from Terry Sauvain. Senator Byrd, explained Sauvain, wanted the CRS to tell him which was better for West Virginia, LIGO or a radio telescope.[92] Rowberg's background was in plasma physics, but he knew who was who in physics and astronomy. To respond to Byrd's request, he talked to a lot of people, including Tor Hagfors, director of the Arecibo Observatory, as well as scientists at the Lawrence Berkeley Laboratory in California.

On 17 February Rowberg sent Senators Byrd and Rockefeller a memorandum addressing the question of whether LIGO or a radio telescope would best benefit West Virginia.[93] The CRS report made clear that the issue was not about the scientific merits of building LIGO, but the benefits to West Virginia and the nation's scientific enterprise, and argued that "LIGO is likely to be built in any case, so the principal scientific question centers on the consequences to radio astronomy of not replacing the 300 Foot Telescope." Rowberg went on to discuss the issues of the number of personnel that would be involved in each project; the attention and scientific prestige that each project would bring to Green Bank and West Virginia; the impact to astronomy of not replacing the 300 Foot; the number of scientific users; and the potential for including West Virginia University in collaborative research. Describing LIGO as "a high risk experiment," the report noted that the successful detection of gravity waves would be "a major step in physics," but that it "will require a substantial advance to the limits of current technology."

Meanwhile, Caltech's LIGO Director Robbi Vogt was not going to let an opportunity for a Congressional earmark slip past. In a 24 February telephone call with Rowberg, Vogt suggested that perhaps their previous concern about ground noise in Green Bank was unfounded and that one of the LIGO elements could be built on the Green Bank site. Since only signals detected by *both* LIGO elements would be considered as due to gravity waves, Vogt suggested that any ground noise generated by only one element of the observatory would be unimportant. The Green Bank site was attractive to Vogt since there appeared to be local opposition to constructing LIGO in Maine. Moreover, argued Vogt, it would save the NSF a lot of money if LIGO could make use of the infrastructure that would be provided by the Green Bank radio observatory. But he also worried that unless a new radio telescope were to be built, the Green Bank site might likely be closed. In a 27 February memo to Byrd and Rockefeller's staff, the CRS reported that for this reason Vogt "hopes that a replacement telescope is built" in Green Bank. However, others argued that ground noise must be reduced as much as possible, and that activities surrounding the operation of the radio telescope would have significant impact to the effectiveness of LIGO, and so even the 140 Foot would need to be closed if LIGO were located in Green Bank.[94]

As initially planned by Sauvain, a 23 February meeting, presumably to discuss the input from the CRS, was to exclude NRAO. The meeting was postponed to 6 March, and at Byrd's insistence Hughes, Seielstad and Vanden Bout did finally attend, along with Bloch and Ray Bye, head of the NSF Legislative Affairs staff. This meeting was held not in Byrd's Office, but in the luxurious Senate Appropriations Hearing Room. Again Rockefeller was late, which gave Byrd time to tell more stories about the Presidents he had worked with. As expected, Bloch made a strong push for LIGO, referring to potential prestige and increased jobs for West Virginia. NRAO countered with the important science anticipated from the new radio telescope and the many prestigious discoveries already made by radio astronomers. Vanden Bout played hardball, pointing out that without a new radio telescope in Green Bank, in the face of declining budgets and commitments to the VLA and VLBA, the Observatory would be forced to leave Green Bank. The meeting closed with the NRAO representatives being excused while the Senators and their staffs continued to speak with the NSF. Apparently it didn't go well for Bloch, because he stormed out of the meeting room ignoring Vanden Bout and Seielstad on the way out.

On 7 March, the Senators issued a joint statement extolling the virtues of both LIGO and a replacement radio telescope, but argued that the collapse of the 300 Foot "radio telescope created an emergency situation ... that requires replacement at the earliest possible time." Working from the CRS report, they went on to point out that "replacing the telescope is also important to West Virginia from the standpoint of jobs, payroll, education, tourism and scientific prestige." Byrd was quoted as saying that he intended to "aggressively pursue funding" for the telescope.[95] Two days later the West Virginia Legislature unanimously passed a joint resolution urging Congress and the NSF to provide funding for a state-of-the-art fully steerable 100 meter diameter telescope to replace the collapsed 300 Foot. Discussion of the proposed bill on the floors of the WV House of Delegates and the Senate emphasized the economic impact to West Virginia and the additional jobs that the new telescope would bring to Pocahontas County.

On 14 March, the House Committee on Science, Space, and Technology held hearings on the NSF's 1990 budget request. Four radio astronomers, including Arecibo Director Tor Hagfors, testified on astronomy issues. There was agreement on the importance of replacing the 300 Foot, but concern that there were other higher priorities in astronomy that were already in the budget request. However, if new money were to be added to the FY1990 budget, the group felt that this would not be considered as circumventing the peer review process.[96] A week later, in a private note to Terry Sauvain, Rowberg pointed out that the NSF FY1990 budget request included $10 million for safety and environmental upgrades to the US Antarctic facility, as well as $250,000 for the start of repairs to the 25 year old Upper Atmospheres Facilities, neither of which had undergone formal peer review.

Although Robert Byrd ruled the Senate with an iron fist and usually got his way, he had two potential challengers to his plans for Green Bank: Representative James Whitten (D-MS), Byrd's counterpart as Chair of the House Appropriations Committee, and NSF director Erich Bloch. While the NRAO director and AUI president were kept in the dark about where Byrd and Whitten stood on the issue of LIGO vs. the radio telescope, George Seielstad claimed only he, Byrd, and an unnamed informant knew that the decision had already been made in favor of the radio telescope. But Seielstad was worried that due to the lack of a public statement about the future of Green Bank, his staff were leaving, and he was anxious to get started on constructing the new telescope.

Each year the US Congress passes an emergency supplemental funding bill which normally covers the cost of repair and recovery from things like tornadoes, earthquakes, floods, and fires that had occurred during the previous year, but is also used as a catch-all for a variety of other funding issues of special interest to Members of Congress. *The Dire Emergency Supplemental Appropriations and Transfers, Urgent Supplementals, and Correcting Enrollment Errors Act of 1989* or HR-2402 included, among other things, a prohibition on the use of Department of the Interior funds to place the Al Capone House in Chicago, Illinois, on the National Register of Historic Places.

Based on the NLSRT report, Seielstad and Vanden Bout had stated that a new telescope would be 70 meters (230 feet) in diameter and would cost about $50 million. Either Byrd misunderstood, or deliberately chose to appear that he misunderstood, and with the agreement of White House Office of Management and Budget Director, Richard Darman, Robert Byrd inserted $75 million into the FY1989 Senate Dire Emergency Act for the replacement of the NRAO 300 Foot Radio Telescope. In the House of Representatives, the radio telescope was competing with a pet project of the House Appropriations Committee Chair Jamie Whitten and a White House initiative for the war on drugs.

In a compromise with the House of Representatives, the final bill, which became Public Law 101-45, spread funds for the radio telescope over two years. A similar amount was included for Whitten's project in Mississippi, as well as funds for the war on drugs. On 23 June 1989, the House bill passed by a vote of 316-8 and the Senate approved it by a voice vote.[97] A week later, it was signed into law by President George H. W. Bush. However in 1990, as a consequence of the Gramm-Rudman-Hollings sequestration for FY1990, the GBT construction funds were reduced to $74,490,000.

Trying to avoid the stigma of apparent "pork" that the NSF and especially Erich Bloch were so strongly opposed to, Byrd noted when introducing the bill in Congress the priority placed by radio astronomers in replacing the fallen telescope, which he described as a "calamity." In fact, replacing the 300 Foot was not the highest priority in radio astronomy, even at NRAO, which was trying to obtain funds to construct a new array that would give astronomers an unprecedented opportunity to study the Universe at millimeter wavelengths. Meanwhile, the NSF was faced with a quandary. To oppose the powerful Senators might risk future budget allocations, but to agree to building a new telescope apparently meant compromising their stated priority for LIGO as the next major NSF construction project, as well as their commitment to the merits of peer review and their opposition to Congressional "earmarks."

Unlike NASA, the NSF for years did not have a standing budget line item for new facilities. However, by unwritten agreement, the 1989/90 GBT funding became the first funding for what later became the NSF's Major Research Equipment (later called Major Research Equipment and Facilities Construction or MREFC) budget, which later funded LIGO, the Atacama Large Millimeter Array (ALMA), the Next Generation Solar Telescope (NGST), and most recently the Large Synoptic Survey Telescope (LSST). At first the MRE/MREFC budget line was confined to the NSF's Directorate for Mathematical and Physical Sciences (MPS), but with time soon covered all of the NSF. Nevertheless, the NSF has always been on record as being opposed to Congressional earmarks, and subsequent declining NRAO budgets and consideration of closing the Green Bank facility as well as the Very Long Baseline Array (VLBA) may have reflected the Foundation's resentment of the perceived political pressure brought to bear in funding the new Green Bank Telescope.

9.7 BUILDING THE GREEN BANK TELESCOPE (GBT)

Typically, before a large new scientific instrument is built, it takes years of proposals and committee reviews, accompanied by forceful lobbying of lawmakers, as well as getting other scientists on board, each of whom have their own priorities. Especially in the case of proposed national observatory projects by NRAO and NOAO, university-based scientists are particularly skeptical and concerned that the national observatory project may come at the expense of their individual research grants. Although NRAO had managed to finesse funding in what was surely a record time of only about six months, its problems were only beginning. By April 1989, with input from several prospective manufacturers, NRAO had completed a re-evaluation of the technical options for 100 meter antennas with both blocked and unblocked apertures. The goal was to have good performance at 7 mm wavelength, and perhaps at 3 mm wavelength under the most benign conditions of wind and solar illumination. Recognizing that the largest unblocked aperture antenna which had ever been built was only 7.5 meters effective diameter, the NRAO study group recommended that a more detailed structural analysis be carried out. A homologous design did not appear feasible due to uncertainties in the computer modeling, fabrication tolerances, and the high cost of fabricating the large number of different-sized back-up structure members.

In late April 1989, Vanden Bout appointed a small team under the leadership of Seielstad to prepare a proposal. On 30 June, the same day that Congress authorized GBT funding, AUI submitted a hastily prepared formal proposal to the NSF for the construction of a fully steerable radio telescope with an "aperture of at least 100 meters," two arcsec pointing accuracy under good conditions, an active surface control, and operating wavelengths from 3 millimeters to meter wavelengths.[98] The proposal discussed, as an option, an unblocked aperture with an offset feed. It was no secret that, as a result of Senator Byrd's deft maneuvers in Congress, a total of $74.5 million was available for the project. NRAO engineers estimated that the construction costs would be about $58 million leaving somewhat less than the $20 million needed for project management and to build state-of-the-art receivers and other instrumentation for the new radio telescope. But it was unusual that a project that was expected to take years to build would be funded over just two years. This meant that inflation would eat into the budget, especially if the design and construction were to take longer than anticipated, so there was some pressure to avoid delay in the subsequent procurement and construction process.

Although it was clear that the NSF would support the project and funds had already been appropriated by Congress, the NSF went through the motions of a formal peer review and appointed a review panel that met in Washington on 31 July–1 August 1989. Panel members noted the potential strong scientific impact, the value of the NRQZ, and the unequaled potential of a 100 meter unblocked aperture working to 3 mm wavelength. Both the mail reviews and the reports of review panel members gave strong endorsement to the GBT program. Indeed, none of the 15 reviewers who ranked the proposal gave it a grade below that of "excellent." Knowing that the project was already funded, the reviewers concentrated on technical tradeoffs such as frequency coverage, size, and the nature of the optics. Interestingly, nearly all reviewers argued for an unblocked aperture, despite the current existence of nothing with an unblocked aperture larger than the Bell Labs 7.5 meter mm antenna (Fig. 9.12).

Just six weeks after the proposal submission, the NSF added $500,000 to NRAO's 1988 budget to "allow NRAO to begin a preliminary design study for the Green Bank Telescope."[99] However, it wasn't until 13 October 1989

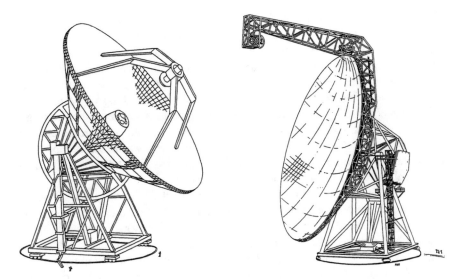

Fig. 9.12 100-meter telescope design concepts. Left: axially symmetrical design; Right: off-set reflector design. Credit: NRAO/AUI/NSF

that the NSF National Science Board got around to approving the funding for the preliminary design phase, after which a further $4.4 million was added to the NRAO budget to complete the design phase of the GBT. A year later, the NSB unanimously approved the GBT construction for an amount not to exceed $69,590,000 for a period of five years. Still nervous about the 300 Foot collapse, they requested "a complete design and structural review ... before manufacture and assembly of the telescope begins."[100]

NRAO did not want to merely duplicate existing radio telescopes, so proposed to build a novel state-of-the-art instrument that would have many years of productive life to deal with the wide range of scientific topics of interest to potential users. NRAO planned that the surface panels be controlled by a set of motorized actuators which would adjust the surface to compensate for the changing gravitational deflections as the telescope was moved in elevation, and ultimately to also compensate for the effects of wind and of thermal effects due to solar heating. In this way, it was argued the new telescope could be large and also have the precision to work at short wavelengths. More controversial was the issue of a well-understood conventional feed-sub-reflector support mechanism versus an asymmetric unblocked aperture. The asymmetric unblocked aperture offered reduced interference, improved spectroscopic capability, and higher efficiency, but at the expense of increased cost and complexity, as well as uncertain stability of the offset support structure which could introduce pointing uncertainties. The default was for a conventional symmetric structure, which was shown on the cover of the proposal to the NSF, but the possibility of an unblocked aperture and offset feed and asymmetric dish structure was also presented. It was estimated that an unblocked structure would cost about 12 percent more than for a conventional antenna of the same effective dimensions, but this would turn out to be far too optimistic. The main message from the astronomical community was to provide good high-frequency (i.e. 3 mm) performance. Although a radome-enclosed structure was not seriously considered, Herbert Weiss, who played a major role in the design of the earlier CAMROC/NEROC 440 foot enclosed antenna, later argued that a radome-enclosed antenna with the same effective sensitivity would be as much as a factor of four less expensive. Weiss's argument, however, was based on an overly conservative estimate of receiver sensitivity resulting in a unrealistically small fractional increase in system noise from the radome.[101]

Due to the fast-track funding of the Green Bank Telescope, NRAO was in the probably unique but unenviable position where the funding was ahead of the design. Vanden Bout appointed a GBT Specifications Working Group, chaired by Seielstad, which met every two weeks. But it would be a year before NRAO was able to submit a call for proposals to construct a 100 meter equivalent projected area clear aperture telescope with an actively controlled surface. The manufacturer was held responsible for achieving an overall surface accuracy better than 1.25 mm rms. NRAO was to be responsible for the active controlled surface which had a goal of overall accuracy of 0.4 mm rms. Under perceived pressure to complete the construction and start a research program, the RFP called for the unrealistic completion of construction and the start of operations in 1995.[102] During the debates about the antenna design parameters, the cumbersome term "NLSRT" was gradually replaced by reference to the generic "Green Bank telescope." When it came time to solicit proposals, George Seielstad declared, for the lack of a better name, that it be called the "Green Bank Telescope" or GBT, and that name stuck.

Former NRAO Director Dave Heeschen became the interim GBT project manager and stayed on as a consultant and advisor throughout the construction project. Heeschen made two difficult and far reaching decisions. Arguing that NRAO should not build just another telescope, he decided to go for the controversial off-axis asymmetric structure to give an unblocked aperture, and also to install an active surface that could be adjusted to compensate for gravity and ultimately the effect of solar heating induced effects of thermal gradients. Although there were some advantages to having the feed arm located at the bottom, such as improved accessibility and reduced spillover noise, the anticipated added cost of about $3 million argued for placing the feed arm at the top of the structure. The additional complexity, weight, and cost suggested that a fully homologous structure be dropped, as the active surface could compensate for gravitational deflections to form a best-fit parabola on a partially homologous structure.

Faced with demanding antenna specifications and a fixed budget, NRAO needed an experienced person to oversee the GBT construction. Out of 40 applicants, NRAO appointed Robert Hall as the GBT Project Manager. Hall, who had been an infantry commander during WWII, had previously designed the NRAO 85 foot antennas and consulted on the design of the 140 Foot Telescope while working at Blaw-Knox, the NRAO 36 Foot mm Radio Telescope while working at Rohr Corporation, and contributed to the design of NRAO 300 Foot Transit Radio Telescope. Hall had also overseen the construction of a number of antennas of the JPL Deep Space Network, including the 210 foot antennas at Goldstone, California; Tidbinbilla, Australia; and Madrid, Spain. As was done for the VLA and VLBA antennas, NRAO, under the leadership of the Observatory's structural engineer, Lee King, and with support from JPL, prepared an engineering design which the prospective manufacturer was free to use. However, as was the case for the VLA and VLBA antennas, the request for proposals specified that the manufacturer would be responsible for all aspects of the design and for meeting the performance specifications. This aspect of the GBT proposal process was later controversial and part of the basis for claims against NRAO/AUI and the NSF for payments of nearly $30 million, resulting from alleged changes in the GBT design made by NRAO after the construction contract had been signed.

On 1 June 1990, NRAO mailed a Request for Proposals (RFP) to prospective bidders, and three weeks later held a Preproposal Conference in Charlottesville. By this time, NRAO had become convinced that an unblocked clear aperture antenna was highly desirable and was feasible within the anticipated $55 million construction budget. Some 60 individuals with interest in some aspect of the GBT construction participated in the Charlottesville conference. Proposals were due on 1 October. NRAO received three bids for the construction of the GBT: from Brown & Root Services Corp., from the Fru-Con Corp, and from Radiation Systems Inc. (RSI), which had contracted with Ted Riffe, the retired NRAO Associate Director for Administration, to help prepare their bid. The bids ranged from a low of $57 million from RSI, to a high of $103 million from the Fru-Con Construction company. Fru-Con had teamed with the German MAN and Krupp consortium that had designed and built the MPIfR 100 meter Effelsberg radio telescope, while RSI partnered with Ford Aerospace and Electrospace Industries. The third proposal came from another consortium led by the Brown & Root Services Corporation. Brown & Root were joined by TIW Systems and the Vertex Communications Corporation to propose building the GBT for $83 million. The large spread in bids raised flags, but with no prospects for an increased Congressional appropriation, the three "prime proposers" were asked to reconsider their bids and to suggest appropriate cost savings by 9 November. Although some changes were subsequently made, only the RSI bid was within the available funding, although in many areas, one or other of the competing but more expensive proposals were judged to be superior.[103]

Under the leadership of its dynamic and ambitious president Raymond (Dick) Thomas, RSI had considerable experience in constructing radio telescopes and had been involved in the 1970 resurfacing of the Green Bank 300 Foot and the 1973 resurfacing of the 1000 foot Arecibo radio telescope. RSI also constructed the panels for the 28 VLA antennas and was the prime contractor for the ten VLBA antennas. RSI itself was composed of a number of fiscally independent subsidiaries located in Texas, Illinois, Florida, New Jersey, Nevada, and Georgia, as well as at their headquarters in Sterling, Virginia, and an office in the UK. Antenna fabrication was done primarily at the Universal Antennas Division, which was licensed in Nevada but had its construction facility in Richardson, Texas. Interestingly, Thomas was first inclined to bid for the full $75 million, which was widely known to be NRAO's budgeted amount, but Riffe pointed out to Thomas that the $75 million would need to include instrumentation for the telescope as well as NRAO's project management costs, so Thomas reduced his bid first to $57 million, and then under pressure from NRAO to $55 million.[104]

On 6 December 1990, the NSF added $65 million to the AUI cooperative agreement for the Green Bank Telescope project,[105] and on 19 December, AUI signed a contract with RSI for a firm fixed price of $55 million with an aggressive completion date of 31 August 1994.[106] To keep within the $55 million cost constraint, NRAO and RSI agreed to divide the project into two parts. First, a basic telescope that would be built by RSI using well understood engineering and construction practices and would operate up to 15 GHz, and second, a supplementary system designed and built by NRAO to enable operation first to 43 GHz then to 100 GHz. To achieve the required performance, NRAO proposed to use a system of actuators to adjust the surface to compensate for gravitational deformations as the telescope was tilted. A second phase would compensate for thermally induced deformations. The even more demanding task of accurately pointing the huge structure at the desired position in the sky and maintaining that position as the Earth rotates was also divided into two phases; the first using "conventional techniques" to be implemented by the antenna manufacturer, and the second to be developed by NRAO, using an elaborate system of lasers and retroreflectors to achieve the "precision pointing" and surface accuracy required for operation at the shorter wavelengths. It later turned out the RSI concept for the GBT did not meet the required surface accuracy by an order of magnitude, and RSI had underestimated by a significant amount the weight of the antenna.[107] Recalling the embarrassing collapse of the 300 Foot Telescope, members of the NSF's National Science Board Committee on Programs and Plans expressed concern about the structural integrity of the GBT design and called for a design and structural review that was held at the NSF on 10 October 1991. In order to provide scientific support to the GBT project, Jay Lockman was appointed as GBT Project Scientist, but was later replaced by David Hogg after Lockman replaced Seielstad as Green Bank Site Director in 1993.

Behind Schedule, Over Budget Groundbreaking for the ambitious GBT project occurred on 1 May 1991, and construction work began a few weeks later (Fig. 9.13). However, it would be ten years before the GBT was completed, and even longer before the challenging project was fully instrumented and operational at the planned shortest wavelengths. Not having adequate in-house

Fig. 9.13 Groundbreaking for the Green Bank Telescope on 1 May 1991. From left to right are Senator Byrd, NSF Director Walter Massey, and AUI President Robert Hughes. Credit: NRAO/AUI/NSF

engineering expertise, RSI sub-contracted the design to the California-based Loral Aerospace Corporation (previously Ford Aerospace), which had extensive experience in the design of radio telescopes. The Loral design, according to NRAO, did not meet the contract performance specifications, and NRAO offered its own optimized design.

In early 1991, Dave Heeschen expressed concern about "various suggestions to modify and/or expand some of the telescope specs" and cautioned against trying to evaluate or optimize the RSI design.[108] In 1992, NRAO called attention to a number of apparent deficiencies in the RSI design, and expressed "concern about the marginal aspects of the designs presented."[109] Already in

1992, in view of the optimistic completion date in 1994, Green Bank staff were building receivers and other instrumentation, and optimistically planned to return the NRAO GBT project team back to Green Bank operations by the end of 1994. But by now problems began to appear with the telescope construction. The actuators needed to correct the surface for deformations due to gravity were found to be causing radio frequency interference (RFI) which if not corrected would limit the performance of the radio telescope. Also, testing of the actuators indicated that they would not meet the lifetime specifications due to excessive internal wear. In mid-1992, Loral discovered that some of the tipping structure members would collide with the fixed alidade structure, requiring an altered geometry of the tipping structure, which in turn led to overstress in some structural members. By mid-1993, the design of the tipping structure by Loral was still not complete, and in June, RSI informed NRAO that the GBT completion would be delayed by a year until the end of 1994. In response, NRAO/AUI sent RSI a notice that RSI had defaulted on the contract, to which RSI reacted by sending a similar notice of default on the design contract to Loral. In an effort to mitigate the Loral design deficiencies, NRAO, with the assistance of JPL, offered an optimized design which met all specifications and was incorporated by Loral/RSI, ostensibly with RSI assuming responsibility for the design. Later this proved to be contentious as RSI interpreted the NRAO/JPL optimized design as a change order that they argued increased the construction costs and should be at the expense of NRAO/ AUI. Within NRAO, the impending delay was not considered all that bad, as it would give NRAO more time to develop, fabricate, and test all the electronics systems including the active surface and precision pointing system.

Even with the projected delays, it was not expected that these in-house tasks would be completed by the antenna delivery time, although as noted by GBT business manager William Porter, "they will be more mature than if there were no delay."[110] Nor was the delay completely unexpected, since as early as 1989, following discussions with TIW's Louis Becker, NRAO Engineer Larry D'Addario had called attention to the unrealistically short planned construction schedule. He then predicted that it would be late 1995 before the antenna would be completed. In 1994, the antenna contractor reported a new completion date in late 1996, which was later further delayed to March 1997. Following a series of review meetings, the re-baselined schedule placed the GBT delivery date at 15 December 1996, which later slipped to April 1998, and then due to safety issues with the derrick needed to lift the structural members "no sooner than the end of 1998." By April 1998, RSI reported that the GBT would be completed in October 1999, which became the end of 1999. These delays naturally resulted in increased personnel costs which NRAO had to absorb outside the fixed GBT project budget. In spite of the numerous delays in completing the antenna structure by the contractor, NRAO realized that due to poor NRAO management, it would be difficult to have the necessary software ready in time to support the commissioning of the GBT.[111] Moreover, development of receivers, spectrometers, and other ancillary instru-

mentation was delayed, as the small Green Bank engineering staff was continually called upon to support existing and ongoing Green Bank telescope
operations at the expense of developing GBT instruments. The GBT was not
handed over by RSI to NRAO/AUI until the end of 2000, some months after
the formal dedication, whose schedule was apparently set by Senator Byrd's
reelection campaign.

With the new design, the weight of the moving structure was only slightly
over 9 million pounds, roughly 15 percent less than the Loral design, and was
estimated to save RSI more than $1 million in the cost of steel. But a problem
appeared following the analysis of the dynamic pointing model which indicated
that following the repositioning of the antenna, the off axis feed arm would
vibrate for up to one minute with an unacceptably large amplitude that would
limit the performance of the telescope, particularly at high frequencies where
the primary beamwidth is as small as ten arcseconds. It would be years before
this problem was adequately resolved. As described by Dave Heeschen, "new
problems arise as fast as older ones get resolved, and it is not at all clear that any
real progress is being made." With great prescience, Heeschen added, "in the
case of RSI we need a paper trail in case we ever get into a legal hassle
with them."[112]

Meanwhile, the AUI Board of Trustees requested a written report from the
NRAO director to explain the "causes and consequences of the delay in the
GBT schedule."[113] The Board also noted with some unease the use in the GBT
construction project of personnel paid from NRAO operating funds. A routine
audit of the project by the Government Accounting Office (GAO) expressed
alarm about statements in a letter from Hall to RSI "regarding the unaccept-
ability of portions of the design and NRAO's concern about the overall safety
and performance of the structure."[114] Hall defended his letter by telling Vanden
Bout that "such statements are characteristic of tough contract management
and should not be over interpreted by those outside the project," and concluded that "the final design will show a structure which is safe and will perform to specification."[115]

The construction of the unconventional complex structure presented many
unanticipated challenges. Two-thirds of the welders applying for jobs were
deemed not qualified for the task. During the course of construction, an unfortunate accident led to the tragic death of one of the RSI ironworkers. On 16
November 1993, as several workers were lowered from the structure, they lost
communication with the crane operator, apparently due to the failure of the
batteries in their radio. While changing batteries, the basket carrying the workers apparently hung up on a rope safety line. The basket tilted, and one of the
workers fell 120 feet to the ground and was killed. The subsequent investigation by OSHA uncovered a variety of alleged safety violations, mostly unrelated
to the accident, but which incurred various monetary penalties.

In a rush to demonstrate that they were on schedule for the proposed 1994
completion date, RSI hastily began work by building the telescope foundation,
just four months after the start of the Loral design work and well before the

design was finalized. The foundation design was based on an estimated moving weight of about 12 million pounds. However, the foundation was only marginally adequate for a 12 million-pound load, and by the time the design was finished another 5 million pounds had been added to the structure, in the form of welding, paint, and the additional members needed to strengthen the structure to meet performance specifications. Pressured to meet the ambitious construction schedule, joints and backup structure beams were fabricated before the design was optimized by NRAO, creating a challenge to find appropriate locations to place the prefabricated members. This led to later litigation about whether NRAO's introduction of the optimized design constituted a change order leading to increased costs. In a further effort to speed up the construction, RSI did not accept Bob Hall's urging to trial-erect some of the antenna substructures at their plant in Mexia, Texas, and as predicted by Hall, this resulted in schedule delays and further increased costs. Tensions between NRAO and RSI staff, as well as among Green Bank Site Director Jay Lockman, Project Scientist Harvey Liszt, and Project Manager Bob Hall became a further challenge to completing the radio telescope and meeting the design specifications. Meanwhile, a heated debate arose within NRAO about the location of the GBT control building. Some argued for a location well removed from the telescope to minimize interference by people coming and going; others, led by Jay Lockman, argued for a more conventional location close to the telescope, where scientists and engineers could best interact with the instrumentation. In the end, a committee appointed by Vanden Bout recognized that there were no funds for a separate control building, and opted to place all the control in the new wing of the Jansky Laboratory, a decision which led to the resignation of Lockman as the Green Bank Site Director.

In June 1993, when RSI acknowledged that the GBT delivery date would be delayed by one year to 31 December 1995 and hinted that even this might be optimistic, NRAO refused to approve the delay, telling RSI to find a way within two weeks to explain what measures RSI would take to recover the schedule. RSI responded that they had already taken all possible options to reduce the completion time, and that the end of 1995 represented a reasonable schedule. Already, NRAO and RSI were setting the stage for what would prove to be a lengthy and costly litigation.

Faced with schedule slippage and escalating costs, in June 1994 RSI was acquired by the Communications Satellite Corporation (COMSAT) which formed COMSAT RSI Technical Products (CRSI or CRSI-TP) to handle their antenna business, including the completion of the GBT. But in July 1996, COMSAT CEO Bruce Crockett was ousted by the controversial Betty Alewine, who had her own agenda for COMSAT. Within a year, half of the 3,000 member COMSAT staff had left. Faced with their own financial troubles, and unable to find buyers for its shares in the Denver Nuggets basketball and Colorado Avalanche hockey teams, COMSAT sold CRSI to a subsidiary of TBG Industries Inc. But COMSAT was unable to divest the GBT contract which was then transferred to a newly formed COMSAT subsidiary, COMSAT

Radiotelescopes Inc.[116] However, CRSI-TP still retained the subcontract for the 2004 precision surface panels as well as for the drive motors and servo systems. COMSAT Radiotelescopes Inc. set up a new office in Herndon, VA, close to the CRSI facility in Sterling, whose sole responsibility was to oversee the completion of the GBT. John Evans, the former director of the MIT Haystack Observatory and one of the pioneers of radar astronomy, became the COMSAT Vice President and Chief Technology Officer and was given direct responsibility for the GBT construction until its completion in 2000. In 1998 COMSAT became part of Lockheed-Martin Global Telecommunications which had previously absorbed Loral, the company that did the complex GBT engineering design for RSI.

Claims and Counterclaims In the autumn of 1995, CRSI lodged a claim that, as a result of multiple continuing design changes, NRAO/AUI was responsible for a significant cost overrun of $14 million. Later, CRSI increased their claim to $29 million. Not receiving any resolution of their claim after more than a year, the new CRSI president Raymond Thomas (no relation to the former RSI President Dick Thomas) first threatened to stop work on the telescope, but then wrote to NSF Director Neal Lane, suggesting that the NSF provide additional funding for the GBT project, and that COMSAT would not be able to complete the telescope without additional funding.[117] A few months later, Thomas followed up with a letter to Senator Byrd arguing that NRAO had modified the specifications and "requesting your support for increased funding of approximately $29,000,000 … for the GBT."[118] The COMSAT claim was based on alleged unnecessary design work, an unreasonable life cycle specification, and inappropriate wind-load requirements. In response, NRAO/AUI argued that the design changes were necessary to meet the performance specifications as outlined in the construction contract. Moreover, parts were shipped to Green Bank in the wrong order, and poor workmanship resulted in time consuming repeated welds, structural elements that needed to be returned to the factory for reworking, or work that had to be done in the air after beams or joints had been erected. A particularly contested item was the number of expected antenna cycles and the corresponding impact to metal fatigue. NRAO/AUI countered with claims of $12 million for six years of increased project management costs, lost research time, the cost of operating the 140 Foot Radio Telescope for an additional six years, and the impact to science and NRAO's reputation.[119] Dave Heeschen described what he referred to as CRSI "bungles," which led to a 30 percent increase in weight, which in turn led to a greatly increased cost and a "telescope dangerously close to its survival and performance limits."[120]

Nevertheless, NRAO/AUI realized that the legal costs associated with a protracted dispute could well approach $5 million, and suggested that it might consider settling at a level of "something more than one million dollars," but not at "the thirty odd million dollars sought," and then only if it led to an early completion of the telescope construction.[121] In August 1997, Vanden Bout offered to settle for $4.5 million, but this was rejected by COMSAT, which made a counter-offer to settle for $15 million.[122] NRAO later offered to settle for $9 million, but it was rejected by COMSAT as being insulting.[123] To support their position, NRAO/AUI contracted with the accounting firm of Ernst & Young to audit CRSI's records to determine the merit of the claimed increased GBT construction costs. A later dispute arose when AUI accused Ernst & Young of excessive charges. Meanwhile, CRSI was involved in a similar dispute with Cornell University. CRSI had contracted with Cornell to upgrade the Arecibo radio telescope, but claimed that their cost overrun of $7 million was because Cornell had not fully disclosed "complete and accurate" information about the upgrade project and associated site limitations.[124]

During this same period, there was a major upheaval at AUI. For decades NRAO had enjoyed the valuable stewardship of AUI, which also managed the much larger Brookhaven National Laboratory under a contract with the Department of Energy. Brookhaven had about ten times as many employees as NRAO and a corresponding budget that was an order of magnitude larger than the NRAO budget. Not surprisingly, the membership of the AUI Board was dominated by scientists with interests and experience in nuclear physics, and many of the activities of the AUI Board were devoted to Brookhaven affairs. But traditionally, the AUI Board members, and especially the AUI President, were always available to help with particular issues that might arise at NRAO. Since 1980, Robert Hughes had served as AUI President working effectively with the NSF and Department of Energy (DOE) to help both Brookhaven and NRAO. Prior to assuming his position at AUI, Hughes had been a Professor of Chemistry at Cornell University, and in 1975 he became NSF Assistant Director for Astronomical, Atmospheric, Earth and Ocean Science, until he returned to Cornell in 1977. Hughes stepped down as AUI President in 1996, and Lyle Schwartz, who was formerly at the National Institute of Standards and Technology and the University of Maryland, became the new AUI President in March 1997.

By this time, rumors were circulating around Long Island that Brookhaven was dumping radioactive tritium which was contaminating the local drinking water. New York Senator Jacob D'Amato took up the war against Brookhaven, which argued that even if one drank a bathtub full of the local water every day, the radiation exposure would be less than that of a dental x-ray. But D'Amato persevered and following an investigation, the new DOE Secretary Frederico Pena, acting under pressure from the Senator, unilaterally dismissed AUI as the manager of Brookhaven, citing careless handling of a 12 year leak of radioactive tritium into the local ground (drinking) water.[125] The loss of 90 percent of its financial basis and embarrassing discredit raised questions at the NSF as to whether AUI would be able to continue to perform its obligations to NRAO.[126] The subsequent defections of a number of AUI Board members led to a restructuring of the AUI corporate structure as a self-perpetuating not-for-profit corporation. No longer were Board members representing their home universities, but rather were independent scientists and administrators. With only NRAO left to manage, the Board became more dominated by astronomers, some of whom had their own agendas, and after the Brookhaven experience, AUI naturally took a heavier hand in managing NRAO. Within a year, Schwartz resigned as AUI President, and Cornell Professor and AUI Trustee Martha Haynes became Interim President in April 1998, in the midst of the NRAO/AUI-CRSI dispute. Following a national search, Riccardo Giacconi, a pioneer of x-ray astronomy and later winner of the 2002 Nobel Prize in Physics for his pioneering work leading to the discovery of cosmic x-ray sources, was appointed AUI President in July 1999. Giacconi was well known as a strong-willed, no-nonsense individual who had previously served as the first director of the Space Telescope Science Institute and later the director of European Southern Observatory.

In case of dispute, the NRAO-RSI contract called for binding arbitration in lieu of a lawsuit. AUI had hired William (Randy) Squires, who later joined the Seattle based Summit Law Group, to represent NRAO/AUI. By the end of 1997, CRSI had submitted a demand for arbitration to settle their claim of $29 million. NRAO/AUI denied responsibility for any increased CRSI costs and submitted a counter claim for $3.8 million that CRSI moved to dismiss and which Squires described as "bereft of legal gunpowder."[127] Paul Vanden Bout recognized the need to keep the litigation issues from impacting ongoing construction work, so he appointed Dave Heeschen to lead a separate litigation team, which included NRAO scientists Dave Hogg and Harvey Liszt. They worked for several years with the legal team attorneys and staff to gather all relevant materials to reconstruct the decade-long record of design changes, delays, and communications between the manufacturer and subcontractors as well as between NRAO and the manufacturer. This included internal communications within NRAO and within RSI/COMSAT. Altogether NRAO/AUI spent over $5 million to prepare their legal defense.

Jacob Pankowski, of McKenna & Cuneo L.L.P., was the primary attorney for COMSAT, but at various times COMSAT used two other Washington, DC-based law firms to develop their case. COMSAT argued that "the requirement for 400,000 antenna cycles was unreasonable and unprecedented and that the impact to the design greatly increased the weight of the structure and extended the schedule," thus adding to the cost. Moreover, COMSAT claimed that the 400,000 cycles requirement was not specified in the request for proposals, that they were given inadequate guidance on how to calculate wind loads on the structure, and that after the design was nearly complete, AUI imposed an additional optimization process that was not a contract requirement and caused additional design effort which stretched out the program. COMSAT also rejected AUI's claim for damages suffered as a result of additional management costs associated with the delay and for the costs of using the less effective 140 Foot Telescope on the grounds that AUI would be reimbursed for these costs by the NSF.[128]

AUI contended that based on "excessive pride and self-confidence RSI had aggressively sought the contract to design and build the GBT, although they appeared to lack the understanding of the project's requirements and the capability to complete the design and construction of the GBT. Specifically, AUI argued that RSI/COMSAT's claims were "afterthoughts, dreamed up" to permit COMSAT to recoup their losses and argued that "RSI burdened its ill-conceived concept with a combination of poor or non-existent planning, lengthy and ineffective lines of communication and inexperienced managers." Furthermore, contended AUI, "RSI was hamstrung by the fact that its various subsidiaries utilized different, and apparently irreconcilable cost accounting systems that prevented project management from receiving accurate project fiscal performance information," and that "RSI did not recognize the magnitude of its overruns as they occurred."[129]

NRAO's Associate Director for Administration, James Desmond, summed up AUI's position:

> It is not often that the complexities of construction resolve in a way that permits a bona fide argument that a particular claims [sic] should be denied in its entirety [sic]. This is one of those unusual cases. Under the circumstances, COMSAT cannot be blamed for hoping that the size of its loss would overcome the paucity of its proof. It has failed to make the required showing, however, and the claims should be rejected.[130]

COMSAT responded, "Only now, without a shred of evidentiary or other support, does AUI make its mean-spirited and bizarre attack."[131]

As part of the "discovery" process, all records, notes, correspondence, technical calculations, etc. relevant to the GBT construction at both COMSAT and NRAO/AUI were made known to the other side. This involved copying costs at NRAO amounting to more than $100,000. More than 100 boxes of papers lined the halls at NRAO's Charlottesville headquarters waiting for CRSI staff and their attorneys to review and copy as needed. Following some administrative reshuffling within the American Arbitration Association, the AUI-CRSI case was moved from the Washington Regional Headquarters to a new Case Management Center in Atlanta, Georgia, and placed on the "Large Complex Case Track." In January 1998, Alan Kent, who was an experienced government procurement attorney, was appointed by the American Arbitration Association as the Arbitrator. The arbitration hearing, which was scheduled to last only four weeks, was originally scheduled to begin on 18 January 1999 (later realized to be a federal holiday), but was repeatedly delayed at the request of the COMSAT attorneys. The hearing finally began on 23 October 1999 at a Hyatt hotel in Reston, Virginia, and did not conclude until late January 2000. During the hearing, it was revealed that Judge Kent and AUI Trustee Claude Canizares had been college roommates, which almost resulted in a mistrial. Perhaps more important, Judge Kent was a WWII history buff and relished Bob Hall's tales of how he had served as an infantry officer under General George Patton.

NRAO/AUI presented nearly 50 depositions taken from various experts and non-experts. Preparation of post-hearing briefs and responses to the post-hearing briefs took nearly another six months, and review and deliberations by the Arbitrator yet a further six months. Finally, on 8 February 2001, Kent awarded COMSAT $6.62 million for its claim and NRAO $2.55 million for its counter claim. The Arbitrator recognized AUI's claim of the additional project management costs due to the COMSAT delays, but denied the claim of lost scientific data. Although the net cost to NRAO/AUI was only $4.07 million, the real cost to CRSI for building the GBT was independently estimated by both NRAO and COMSAT to be about $120 million, or $65 million over the contract value. COMSAT received only the $55 million contracted construction fee plus the $4 million arbitration award. Vanden Bout noted that $4 million amounted to only 5.5% of the total project cost and it was just that amount that he had offered COMSAT to settle in lieu of arbitration. But NRAO/AUI had also spent over $5M in preparing for the defense, most of which was in legal costs. The fee for the arbitration alone was $230,000 and was equally shared by AUI and COMSAT. The NSF took a hard line, and refused Vanden Bout's request for supplemental funding in FY98 and FY99 to cover the litigation expenses. While AUI agreed to loan NRAO $750,000 in FY98 to cover some of the litigation expenses, even more of a concern at that time was the possibility of an unfavorable judgment against AUI/NRAO of as much as $29 million.

Following the relatively modest adverse judgement, in early 2001 NRAO/AUI still faced bills totaling more than $9 million. Aside from issues of whether or not the cost of settling the claim was allowable under the terms of the AUI-NSF Cooperative Agreement, there were no funds available within the NSF Astronomy Division to cover such a large unplanned cost. Nor could budget funds be moved from other NSF divisions without express approval from the cognizant Congressional appropriation committees. There was a real possibility that NRAO/AUI would need to find the $9 million within the NRAO annual operating budget or from AUI corporate funds, and Vanden Bout started to implement a number of NRAO budget adjustments to at least cover the litigation expenses. However, following extensive strategy discussions, the NSF deftly adjusted the 2002 NRAO fiscal year to begin on 1 October 2001 instead of 1 January 2002, so in calendar year 2001, NRAO received 15 months of funding or an effective budget supplement of 25%. This was enough to pay the litigation costs, as well as to support other long overdue activities at the Observatory. Much to the chagrin of Che Kim, the powerful Clerk of the Senate Appropriations Committee, who was already at odds with NSF Director Rita Colwell, the NSF had cleverly and legally maneuvered a budget change without the required Congressional approval.

To their credit, even during the ongoing lengthy and sometimes bitter litigation process, COMSAT and NRAO engineers continued to work together to finally bring the GBT construction to a satisfactory completion by the end of 2000. Vanden Bout wisely allowed the project team to focus on completing the antenna, while enlisting others to support the lengthy legal proceedings. When finally dedicated in August 2000, the GBT became the largest movable structure on the surface of the Earth, weighing 17 million pounds and extending 100 x 110 meters across.[132] It has an unblocked aperture containing 2004 surface panels positioned by 2209 remotely controlled actuators or jack screws to constantly adjust the surface to compensate for thermal and gravitational distortions and keep the surface sufficiently accurate to a few tenths of a millimeter to allow operation at 86 GHz (3.5 mm). The planned innovative precision pointing system and adaptive surface, based on the use of lasers to measure the path length from various parts of the structure to fixed points on the ground, was never perfected. However, the finite element analysis gave such a good description of the structural behavior that the dish distortions under the effects of gravity are effectively removed by the active surface and a straightforward look-up table. Precision pointing is achieved by the use of tilt sensors located at strategic points in the structure, and feeding this information back into the pointing equations achieves a pointing accuracy of about one second of arc, equivalent to the thickness of a human hair at a distance of 15 feet. Paradoxically, when rejecting an application for funds to make a documentary film about the GBT construction, the NSF responded that, "unlike an optical telescope, a radio telescope is not very visual."[133] (Fig. 9.14).

Due to the various design changes implemented during the construction process, and the non-negligible weight of the weldings that had been neglected in the early weight calculation, the GBT, as delivered from the manufacturer, weighed between 17.0 and 17.5 million pounds, or about 30% more than the original design weight. Although the weight was thought to be within the required safety margins at the mid-span of each track segment, there had been no consideration of stresses at the joints or dynamic loading effects. Owing to the excess weight, some of the azimuth track plates began to slip and show excessive wear shortly after the completion of the GBT; numerous hold down bolts were shearing off, and gaps in the grout were filling with water and draining off grout particulate. As a result, the azimuth wheels were tilting and causing even more excessive wear of the track. NRAO engineers were concerned that, without repairs, the rate of track degradation would lead to a shutdown of GBT operations in 6–12 months.

Following a series of reviews of the extent of the damage, AUI submitted a claim to Lockheed Martin for $9,053,126.35 to cover their accrued costs and the expected cost of repairs. Lockheed Martin responded that AUI had been the "windfall beneficiary" of a telescope that cost $110 million to build and yet for which AUI paid only approximately $55 million but "regrettably took a 'throw in the kitchen sink' approach and that the AUI was unreasonable."[134] Lockheed claimed that the proposed AUI repairs were actually an upgrade of the original specifications for which they were not responsible.

Noting that the proposed upgrades were necessary to meet the original 20-year warranty, AUI was unwilling to settle for the $1.5 million offered by Lockheed. But following a visit to Lockheed by AUI Vice President Pat

Fig. 9.14   Completed Green Bank Telescope. Credit: NRAO/AUI/NSF

Donahue, Green Bank Assistant Director Phil Jewell, and the AUI attorney Randy Squires, AUI accepted Lockheed's check for $4 million to cover the cost of the track repairs.

Epilogue After the completion of the project, NRAO held a postmortem to evaluate how and why the GBT problems occurred.[135] Basically, it was agreed that nothing like this had ever been done before, and no one at NRAO, AUI, RSI, or the NSF, nor the distinguished members of the advisory committees, had ever built a 100 meter clear aperture structure with an active surface and with the GBT's exacting specifications. Clearly the initial cost estimates and schedule agreed to by both NRAO and RSI were overly optimistic. But the bidders proposed to meet the schedule because that is what the RFP demanded. Only RSI came even close to meeting the NRAO budget allocation, whether in ignorance or naivety, or perhaps in expectation of later negotiating a new cost. It was clear from the size of the Brown & Root Services Corp. and from the Fru-Con Corp. bids that the RSI bid was unrealistically low. However, NRAO's hands were tied. They could have rejected the RSI bid and redesigned the antenna, but RSI wanted the job and had apparently submitted a respon-sible bid consistent with the publically known budget and NRAO's own esti-mate. Had NRAO rejected the RSI bid, RSI would likely have protested. Alternately, NRAO could have gone back to the NSF for more funds, but considering the history of the funding process, that was not a viable choice. So NRAO chose to go ahead with RSI, which had a good reputation and whose bid was close to the NRAO budget estimate. On several occasions, NRAO threatened to hold RSI in default, but this was never a real option, since an alternate contractor would have required at least as much money, and the appropriated funds were already largely gone. NRAO took a calculated risk that RSI would not just walk off the job, as they were dependent on other existing and future government contracts. GBT Business Manager Bill Porter speculated that had there been time for a proper Design and Development phase, it would have been realized that the construction cost would be much higher than the RSI bid, but, he added, "had the real price been known, we might never have built the GBT."[136]

It took nearly half a century of discussion and debate, and numerous NSF and National Academy committees, but in the end it was a freak accident, coupled with the ambitions of a powerful Senator, a fiercely competitive radio astronomer with political connections, and a hungry, possibly naïve or unscru-pulous contractor, to finally build the largest and most powerful fully steerable radio telescope in the world.

Why did it take half a century before the United States could finally build a large fully steerable radio telescope? As pointed out by John Findlay in April 1988, it was not for the lack of design effort nor the lack of skilled people either in industry or academia.[137] But the NRAO 140 Foot and the 600 foot Sugar Grove fiascos were both embarrassments to the US radio astronomy commu-nity, from which it would take decades to recover. The 140 Foot itself was smaller than the Jodrell Bank or Parkes radio telescopes, both of which had been in operation for several years before the 140 Foot was finally completed, and the 140 Foot structure turned out to have serious limitations which impacted its short wavelength performance. The proposals for building a large fully steerable telescope in the US were led, or were perceived to be led, by engineers, not by an astronomer prepared to put his or her reputation on the line. By the mid-1960s, planning for the VLA was already dominating discus-sions at NRAO and AUI, and the LFST project was on the back burner. Finally, the scientific returns from the interferometric arrays at Cambridge, then OVRO, Westerbork, and later the VLA dwarfed the productivity of the large fully steerable radio telescopes at Jodrell Bank, Haystack, Effelsberg, and Algonquin Park, although the 210 foot Parkes antenna has been widely recog-nized as highly productive. Within both NRAO/AUI and the broader US radio astronomy community, the top priority was first the VLA, then the VLBA, and finally a millimeter array. A large fully steerable radio telescope for centimeter wavelengths remained a high priority for over a half a century, but never rose to first priority, normally a necessary, but by no means sufficient condition for obtaining federal funding for constructing a new scientific facility.

It is perhaps interesting to speculate that if Jim Condon had been able to inspect his 300 Foot survey data on a daily basis, he would have spotted the changing performance of the 300 Foot Telescope, which might have immediately been recognized as due to the increasing deformations of the structure. Further observations would have been halted; the 300 Foot Telescope, which was already earmarked for closure, would probably not have collapsed, but would have been closed for lack of funding and dismantled; Senators Byrd and Rockefeller would not have been alarmed; and the GBT would have never been built.

NOTES

1. *Radio-Craft*: January 1930, 312; October 1934, 213; May 1934, 648; *Short Wave Craft*: June-July 1931, 10; October-November 1930, 204; May 1934, 8; September 1935, 262; December 1935, 466; June 1936, 70. We are grateful to Bill Liles for bring these publications to our attention.

2. Patent Application filed on 23 June 1928 and granted US Patent No. 1,831,011 on 10 November 1931. We are grateful to Bill Liles for bringing this to our attention.

3. Reber to Struve, 16 July 1946, NAA-GR, General Correspondence I. https://science.nrao.edu/about/publications/open-skies#section-9

4. Reber, 11 March 1955, Large Mirror Design, NAA-GR, Notes and Papers. https://science.nrao.edu/about/publications/open-skies#section-9

5. AUI Advisory committee on Radio Astronomy, Minutes of 16–17 October 1956, NAA-NRAO, Founding and Organization, Meeting Minutes.

6. Lovell relates in *The Jodrell Bank Telescopes* his surprise in reading the article by Wielebinski, R. 1970, 100 m Radio Telescope in Germany, *Nature*, **228**, 507.

7. Material in this section describing the German 100 meter radio telescope is based on Wielebinski, R., Junkes, N., Grahl, B.H. 2011, The Effelsberg 100-m Radio Telescope: Construction and Forty Years of Radio Astronomy, *JAHH*, **14** 3; Hachenberg, O., Grahl, B.H., Wielebinski, R. 1973, The 100-meter radio telescope at Effelsberg, *IEEEP*, **61**, 1288; Wielebinski, R. 1970, 100 m Radio Telescope in Germany, *Nature*, **228**, 507; Wielebinski, R. 2003, The new era of large paraboloid antennas: the life of Prof. Dr. Otto Hachenberg, *AdRS*, **1**, 321; Wielebinski, R. 2007, Fifty years of the Stockert Radio Telescope and what came afterwards, *AN*, **328**, 388; and also on the 22 February 1973 Sullivan interview of Otto Hachenberg (NAA-WTS). https://science.nrao.edu/about/publications/open-skies#section-9

8. Based on experience with existing radio telescopes, many observational radio astronomers became aware that while an antenna is tipped the parabolic surface deforms, changing the focal point, so that by moving the feed, some of the loss of gain with elevation can be recovered. Hachenberg's 100 meter design was based on a series of iterative computer simulations intended to optimize the structure so that it maintains a precise paraboloid when subject to gravitational deformations. Von Hoerner showed analytically that with the right choice of structural members, a so-called homologous solution can be found.

Hachenberg and von Hoerner certainly met each other over a period of years in connection with the proposed new Max Planck Institute that they planned to jointly lead, but it is not clear if they ever discussed homologous antenna designs or to what extent either man consciously or unconsciously influenced the other.

9. Quoted by Butrica (1996, pp. 67–68).
10. Finite element analysis is a computer-based technique by which complex structures are broken down into a number of finite elements for subsequent analysis.
11. *New York Times*, 19 June 1959.
12. *Charleston Gazette-Mail*, 23 August 1964.
13. At the time, the Navy used very low frequency (VLF) radio to communicate with submarines around the globe.
14. D.R. Lord to President Eisenhower's Science Advisor, G.B. Kistiakowski, 8 September 1960, DDE, Radar and Radio Astronomy, Box 5, Records of the U.S. President's Science Advisory Committee.
15. *New York Times*, 15 October 1961.
16. AUI-BOTXC, January 1960.
17. Newspaper article of unknown source, 3 October 1963, NAA-NRAO, GB-LSFT, Box 6.
18. The Sugar Grove 150 foot steerable dish was designed by Ed Faelten who was also responsible for the design of the NRAO 300 Foot transit dish in Green Bank and who was later an active participant in the NRAO LFST program.
19. http://coldwar-c4i.net/Sugar_Grove/history.html
20. The politics of the Sugar Grove 600 foot antenna program are discussed in more detail in a publication by James Bamford (1983), *The Puzzle Palace*.
21. *Washington Post*, 22 May 2016.
22. *Charleston Gazette-Mail*, 26 July 2016.
23. DSH, Proposal for the development of a Large Antenna, 16 December 1957, NAA-NRAO, Founding and Organization, Planning Documents. https://science.nrao.edu/about/publications/open-skies#section-9
24. Ibid.
25. DSH, Comments on a Very Large Antenna, 21 August 1959, NAA-NRAO, Founding and Organization, Antenna Planning. https://science.nrao.edu/about/publications/open-skies#section-9
26. G. Keller, Report of the NSF Advisory Panel for Radio Telescopes (1960).
27. AUI-BOTXC, 19 February 1960.
28. J. Findlay and D. Heeschen, The National Radio Astronomy Observatory and a Very Large Antenna, 27 May 1960, NAA-NRAO, Founding and Organization, Antenna Planning.
     https://science.nrao.edu/about/publications/open-skies#section-9
29. DSH to Wilson, 17 December 1963, NAA-NRAO, Green Bank Operations, LFST, Box 4.
30. Gart Westerhout from the University of Maryland chaired the meeting which included Ron Bracewell from Stanford, Bernard Burke (see NAA-BFB) from DTM, John Kraus (see NAA-JDK) from Ohio State, Edward Lilley from Harvard, Richard Read from Caltech, George Swenson from Illinois, and Harold Weaver from the University of California at Berkeley. Frank Drake,

David Heeschen, John Findlay, and Morton Roberts represented NRAO, and Geoffrey Keller represented the National Science Foundation.

31. Wade notes from Green Bank Meeting to Discuss Next Large Paraboloid for NRAO, NAA-NRAO, Green Bank Operations, LSFT, Box 4.

32. Report on Ad Hoc Meeting of Radio Astronomers: Largest Feasible Steerable Filled-Aperture Telescope, Green Bank, 30 October 1964, NAA-NRAO, Green Bank Operations, LFST, Box 4.

33. J. Findlay, The Largest Feasible Steerable Filled-Aperture Telescope, 14 January 1965, LFSP/JWF/1. NAA-NRAO, Green Bank Operations, LFST, Box 9. https://science.nrao.edu/about/publications/open-skies#section-9

34. Other members included E. R. Faelten and Otto Heine, consulting engineers; NRAO engineers John Hungerbuhler and Max Small, and NRAO Scientist, Sebastian von Hoerner; and Richard L. Jennings a civil engineer from the University of Virginia.

35. LSFP, LFST and 65 Meter Reports. http://library.nrao.edu/65r.shtml See also the series of LFSP, LFST, and 65 Meter Unnumbered Reports. http://library.nrao.edu/65u.shtml

36. NAA-NRAO, Green Bank Operations, LFST, Boxes 8–12.

37. A 300 Foot High Precision Radio Telescope, January 1969, NRAO report. See also Youmans (1969).

38. A 25-Meter Radio Telescope for Millimeter Wavelengths, vol. 1, NRAO, September 1975.

39. *Harvard Crimson*, 17 January 1966.

40. Papers of Alan Moffet (unprocessed), CITA.

41. Stanley to Bonner, 27 December 1966, Papers of Gordon Stanley, CITA.

42. National Science Foundation Report of the Ad Hoc Advisory Panel for Large Radio Astronomy Facilities, 14 August 1967.

43. Large Diameter Radomes, a report prepared for MIT Lincoln Laboratory by Ammann & Whitney Engineering Consultants, April 1965 [Appendix IV to CAMROC Engineering Design Objectives for a Large Radio Telescope, October 1965], NAA-JWF, LFST.

44. A Large Radio/Radar Telescope, CAMROC Design Concepts, Vol I and Vol II, Cambridge Radio Observatory Committee, 1967.

45. Findlay later testified in Congress in support of the NEROC proposal which he claimed was "very unpopular within the NRAO management," JWF notes to KIK from April 1988, KIK, Open Skies, LFST.

46. Minutes of the 30 November–1 December 1968 Radio and Radar Astronomers meeting, NAA-JDK, Notes and Papers, US Radio Astronomy, Smithsonian Large Telescope.

47. Ibid.

48. "A Large Radio-Radar Telescope Proposal for a Research Facility" submitted to the National Science Foundation on 30 June 1970 by the Northeast Radio Observatory Corporation.

49. The saga of the CAMEROC/NEROC attempt to construct a large steerable radio telescope is discussed in more detail in Butrica (1996), 69–83.

50. Burke to PVB, 21 July 1987, NAA-NRAO, Green Bank Operations, GBT Planning and Design, Box 1.

51. PVB to KIK, 25 November 1987, NAA-NRAO, Green Bank Operations, GBT Planning and Design, Box 1.

52. NLSRT Memo No. 1, New Large Steerable Radio Telescope (NLSRT), NAA-NRAO, Green Bank Operations, GBT Planning and Design, Box 1. http://library.nrao.edu/public/memos/nlsrt/NLSRT_001.pdf

53. Experience from building a variety of fully steerable radio telescopes suggests that the cost varies approximately as $(Diameter)^{2.7}$ and as the (limiting wavelength)$^{-0.7}$.

54. NLSRT Memo No. 39, A Very Large Dish (VLD) Radio Telescope, 28 November 1988, NAA-NRAO, Green Bank Operations, GBT Planning and Design, Box 1. http://library.nrao.edu/public/memos/nlsrt/NLSRT_039.pdf

55. The IBM PC-AT was at the time an advanced personal computer using an 80286 processor running at 8 MHz with 18 MB of memory, and 20 MB hard disk drive.

56. Shapiro to KIK, 10 February 1988, NAA-KIK, Open Skies.

57. Langenberg, 1988. Report of Subcommittee on NSF Radio Astronomy Facilities, Advisory Committee for Astronomical Sciences, NAA-NRAO, NSF-Advisory Committees.

58. Other committee members were J. Gallagher (Univ. of Wisconsin), P. Kronberg (Univ. of Toronto), R. Mutel (Univ. of Iowa), J. Piper (Univ. of Rochester), B. Rickett (Univ. California San Diego), R. Wagoner (Stanford) and R. Wilson (Bell Labs).

59. The following section reporting on the impact of the collapse of the 300 Foot antenna and the subsequent negotiations between NRAO and the NSF is based in part on unpublished recollections of George Seielstad and Paul Vanden Bout, as well as on materials in NAA-NRAO, Green Bank Operations, GBT Planning and Design.

60. Matyas was a construction and management consultant. Other panel members were Edward Cohen, managing partner with Amman and Whitney, a structural engineering firm, and George Mechlin, retired Vice President for Research and Development of the Westinghouse Electric Corporation.

61. Report of the Technical Assessment Panel for the 300 Foot Radio Telescope at Green Bank WV, March 1989, NAA-NRAO, Green Bank Operations, 300 Foot, 300 Foot Collapse, Box 2. http://library.nrao.edu/public/memos/nlsrt/NLSRT_053.pdf

62. Paul Harvey broadcast a daily "News and Comment" show on the ABC radio network which was carried by more than 1200 national radio stations and 300 newspapers.

63. *New York Times*, 17 November 1988.

64. National Science Foundation. http://www.nsf.gov/about/history/bloch_bio.jsp

65. NLSRT Memo 39, op. cit.

66. NLSRT Memo 37, Richard Thomas to Richard Fleming, NAA-NRAO, Green Bank Operations, GBT Planning and Design, Box 1. http://library.nrao.edu/public/memos/nlsrt/NLSRT_037.pdf

67. NLSRT Memo 36, Typical Antenna Costs, NAA-NRAO, Green Bank Operations, GBT Planning and Design, Box 1. http://library.nrao.edu/public/memos/nlsrt/NLSRT_036.pdf

68. Burke to AAS CAPP, 1 December 1988, NAA-NRAO, Green Bank Operations, GBT Planning and Design, Box 2.

69. Many prominent scientists argued against LIGO as being too expensive and with minimal chances of success.

70. Report of R. E. Vogt to the NSF, 31 January 1989, NAA-NRAO, Green Bank Operations, GBT Proposals and Contracts, (Fact Sheets folder).

71. KIK interview with Vanden Bout, December 2016.

72. Unpublished notes of George Seielstad, NAA-KIK, Open Skies.

73. Ibid.

74. Even after the GBT was approved, Byrd continued to pressure the NSF to build LIGO in Green Bank. NRAO went on to prepare a pro-forma proposal which purposely contained the fatal flaw that if LIGO interfered with the operation of Green Bank radio telescopes then NRAO could shut down LIGO.

75. NLSRT Memo 32, Proceedings of a Green Bank Workshop, December 2–3, 1988, NAA-NRAO, Green Bank Operations, GBT Planning and Design, Box 1. http://library.nrao.edu/public/memos/nlsrt/NLSRT_032.pdf

76. Weber to Vanden Bout, 2 March 1989, NAA-NRAO, Green Bank Operations, GBT Planning and Design, Box 1.

77. Hagfors to Bloch, 30 December 1988, NAA-NRAO, Green Bank Operations, GBT Planning and Design, Box 1.

78. Schmidt to KIK, 21 December 1988, NAA-KIK, Open Skies.

79. Vanden Bout to Bloch, 19 December 1988, NAA-NRAO, Green Bank Operations, GBT Planning and Design, Box 1.

80. Vanden Bout, 1988, NAA-PVB, Journals and Calendars.

81. Traditional filled aperture antennas have a receiver connected to a simple antenna such as a horn or dipoles located at the focal point of a parabolic reflector. The expression "feed" comes from the early use of parabolic dishes for radar. Although a radio telescope works in reverse, the radar terminology is still in common use. Many modern radio telescopes as well as communication and satellite TV dishes use a so-called Cassegrain or Gregorian focus which uses a sub-reflector located near the focal point to reflect the incoming signal to the receiver and feed located near the dish surface. A Cassegrain system utilizes a convex hyperbolic sub-reflector located in front of the focal-point, while a Gregorian system uses a concave hyperbolic reflector located outside the focal point. Cassegrain and Gregorian configurations have been used in optical telescopes since the seventeenth century.

82. Burke to KIK, 27 March 1989, NLSRT Memo No. 56, NAA-NRAO, Green Bank Operations, GBT Planning and Design, Box 1. http://library.nrao.edu/public/memos/nlsrt/NLSRT_056.pdf

83. The resolution of optical telescopes is typically limited not by diffraction, but by tropospheric turbulence that is called "seeing."

84. NAA-NRAO, Green Bank Operations, GBT Planning and Design, Box 1.

85. Charleston Gazette, 30 December 1989.

86. Vanden Bout 1989, NAA-PVB, Journals and Calendar.

87. The original letter was handed to Bloch; no copies have been found in either the NRAO files nor in Tyson's personal files.

88. Vanden Bout, 1989, NAA-PVB, Journals and Calendars.

89. KIK interview with Seielstad, 17 February 2017.

90. Burke to Bloch, 26 January 1989. Copies were sent to Senators Byrd and Rockefeller as well as to NRAO. NAA-NRAO, Green Bank Operations, GBT Design and Construction, Box 1.

91. http://www.loc.gov/crsinfo/about/

92. Rowberg interview with Kellermann and Brandt, 14 November 2013.

93. The CRS report, which was co-authored by CRS member James Mielke, was addressed to both Byrd and Rockefeller with copies to Carol Mitchell and Amy Berger from Byrd's and Rockefeller's staff respectively. NAA-NRAO, Green Bank Operatioins, GBT Planning and Design, Box 2.

94. 27 February 1989 memo from the CRS to Carol Mitchel and Amy Berger, NAA-NRAO, Green Bank Operations, GBT Planning and Design, Box 2.

95. *Pocahontas Times*, 9 March 1989.

96. Rowberg to Carol Mitchell and Amy Berger, 14 March 1989, NAA-NRAO, Green Bank Operations, GBT Planning and Design, Box 2.

97. The following year, Congress appropriated the first funding for LIGO. In 1992, sites at Hanford, Washington, and Livingston, Louisiana, were selected. Following 15 years of cost escalation, schedule delays, and management upheavals the project had grown to include more than 1000 scientists from 75 institutions in 15 countries. On 11 February 2016, after more than 25 years of development and construction costing more than one billion dollars, the LIGO collaboration announced the unambiguous detection of gravity waves on 24 September 2015, from a pair of coalescing massive black holes. Abbot et al. 2016, *Phys. Rev. Lett.* **116**, 241103. Barry Barish, Kip Thorne, and Ray Weiss later received the 2017 Nobel Prize in Physics for their contributions to LIGO and the observation of gravitational waves.

98. A Radio Telescope for the Twenty-First Century, June 1989, NAA-NRAO, Green Bank Operations, GBT Proposals and Contracts.

99. Asrael to Hughes, 17 July 1989, NAA-NRAO, Green Bank Operations, GBT Construction, Box 1.

100. 297th Meeting of the National Science Board, 16 November 1990, NAA-NRAO, Green Bank Operations, GBT Construction, Box 1.

101. Weiss to Vanden Bout, 4 November 1989, 12 November 1989, NAA-NRAO, Green Bank Operations, GBT General, Box 1.

102. NRAO/AUI Request for Proposals, Green Bank Telescope, 1 June 1990, NAA-NRAO, Green Bank Operations, GBT Proposals and Contracts.

103. AUI Post-Hearing Brief to the American Arbitration Association, 16 May 2000, p. 17, NAA-NRAO, Green Bank Operations, GBT Litigation, Box 4.

104. KIK interview with Riffe, 7 November 2016.

105. Asrael to Hughes, 6 December 1990, NAA-NRAO, Green Bank Operations, GBT Construction, Box 1. This was later supplemented by an additional $4.56 million bringing the total amount close to the $75 million promised by Senator Byrd.

106. Contract No. AUI-1059, 19 December 1990, NAA-NRAO, Green Bank Operations, GBT Proposals and Contracts.

107. KIK and ENB interview with Liszt, 6 July 2016, NAA-KIK, Oral Interviews.

108. DSH to Hall, 26 February 1991, NAA-NRAO, Green Bank Operations, GBT Proposals and Contracts.

109. Hall to Hawkins, 9 October 1992, NAA-NRAO, Green Bank Operations, GBT Construction, Box 1.

110. Porter, Undated internal memorandum, NAA-NRAO, Green Bank Operations, GBT General, Box 1.

111. GBT Software Report of the Review Committee, NAA-NRAO, Green Bank Operations, GBT General, Box 2.
112. DSH to Hall, 7 June 1993, NAA-NRAO, Green Bank Operations, GBT Construction, Box 1.
113. NAA-NRAO, Green Bank Operations, GBT Construction, Box 1.
114. NAA-NRAO, Green Bank Operations, GBT Construction, Box 1.
115. Hall to Vanden Bout, 18 February 1993, NAA-NRAO, Green Bank Operations, GBT Construction, Box 1.
116. Letter from Hawkins, COMSAT RSI, to Porter, GBT Business Manager, 1 April 1998, NAA-NRAO, Green Bank Operations, GBT Litigation, Box 1.
117. Letter from Raymond Thomas (CRSI) to Lane, 16 May 1997, NAA-NRAO, Green Bank Operations, GBT Litigation, Box 1.
118. Letter from Raymond Thomas to Byrd, 5 February 1997, NAA-NRAO, Green Bank Operations, GBT Litigation, Box 1.
119. AUI Post-Hearing Brief to the American Arbitration Association, op. cit.
120. Memo from DSH to Squires, 9 January 1997, NAA-NRAO, Green Bank Operations, GBT Litigation, Box 1.
121. Squires to Pankowski, 17 June 1997, NAA-NRAO, Green Bank Operations, GBT Litigation, Box 1.
122. Vanden Bout to Dickman, 5 August 1997, NAA-NRAO, Green Bank Operations, GBT Litigation, Box 1.
123. KIK interview with Vanden Bout, 22 July 2016.
124. COMSAT RSI Inc. v. Cornell University, United States District Court, N.Y., 20 June 1996, NAA-NRAO, Green Bank Operations, GBT Litigation, Box 1.
125. Government Accounting Office report to Congressional Requesters on the Department of Energy Information on the Tritium Leak and Contractor Dismissal at the Brookhaven National Laboratory, 1997. Interestingly, only a year earlier DOE gave AUI an "excellent" rating for "environmental compliance and reactor safety."
126. Schwartz to Lane, 18 June 1997, NAA-NRAO, Green Bank Operations, GBT General, Box 2.
127. Respondents Memorandum in Opposition to Claimant's Motion to Respondent's Dismiss Second Amended Counterclaim, p.2, 13 April 1998, NAA-NRAO, Green Bank Operations, GBT Litigation, Box 1.
128. COMSAT Post-Hearing Brief to the American Arbitration Association, 15 May 2000, NAA-NRAO, Green Bank Operations, GBT Litigation, Box 4.
129. AUI Post-Hearing Brief to the American Arbitration Association, 15 May 2000, NAA-NRAO, Green Bank Operations, GBT Litigation, Box 4.
130. AUI's Post-Hearing Reply Brief, 30 July 2000, NAA-NRAO, Green Bank Operations, GBT Litigation, Box 4.
131. COMSAT's Claimant's Reply Brief, 30 June 2000, NAA-NRAO, Green Bank Operations, GBT Litigation, Box 4.
132. Since that time, the $1.5 billion Chernobyl Arch, built to confine the remains of the Chernobyl number 4 reactor, became the largest movable structure on earth, but it was only moved once to cover the destroyed reactor.
133. Summary of reviews, appended to J. Vanski to M. Haynes, undated (fax receipt date 20 July 1993), NAA-NRAO, Green Bank Operations, GBT General, Box 1.
134. O'Dea to Squires, 20 July 2006, NAA-NRAO, Green Bank Operations, GBT General, Box 3.

135. Vanden Bout notes, NAA-NRAO, Green Bank Operations, GBT General, Box 3.
136. Porter notes, NAA-NRAO, Green Bank Operations, GBT General, Box 1.
137. Findlay notes for a 29 April 1988 invited talk at Ithaca, NY that apparently was not presented. NAA-NRAO, Green Bank Operations, GBT Planning and Design, Box 1.

BIBLIOGRAPHY

REFERENCES

Bamford, J. 1983, *The Puzzle Palace*. (New York: Penguin)
Butrica, A.J. 1996, *To See the Unseen: A History of Planetary Radar Astronomy*, NASA SP 4218 (Washington DC: NASA)
Condon, J.J. 2008, ZAPPED! ... by Hostile Space Aliens! In ASPC **398**, *Frontiers of Astrophysics: A Celebration of NRAO 50th Anniversary*, ed. A.H. Bridle, J.J. Condon, and G.C. Hunt (San Francisco: ASP), 323
DeVorkin, D.H. 2000, Who Speaks for Astronomy, *Historical Studies in the Physical and Biological Sciences,* **31** (1), 2000
Dicke, R.H. 1967, *Report of the Ad Hoc Advisory Panel for Large Radio Astronomy Facilities* (Washington DC: National Science Foundation)
Dicke, R.H. 1969, *Report of the Second Meeting of Ad Hoc Advisory Panel for Large Radio Astronomy Facilities* (Washington: National Science Foundation)
Field, G. 1983, *Astronomy and Astrophysics for the 1980's, Report of the Astronomy Survey Committee* (Washington DC: National Research Council)
Findlay, J.W. 11 February 1965, *The Largest Feasible Steerable Paraboloid*, NRAO internal report https://library.nrao.edu/public/memos/65/65U/65U_100.pdf
Findlay, J.W. and von Hoerner, S. 1972, *A 65-Meter Telescope for Millimeter Wavelengths* (Charlottesville: NRAO)
Greenberg, D.S. 1964, Big Dish: How Haste and Secrecy Helped Navy Waste $63 Million in a Race to Build a Huge Telescope, *Sci,* **144**, 1111
Greenstein, J.L 1973, *Astronomy and Astrophysics for the 1970's, Report of the Astronomy Survey Committee, Vol 2* (Washington DC: National Research Council)
Hachenberg, O. 1970, The New Bonn 100-Meter Radio Telescope, *S&T,* **40**, 338
Hachenberg, O., Grahl, B.H., Wielebinski, R. 1973, The 100-meter radio telescope at Effelsberg, *IEEEP,* **61**, 1288
Keller, G. 1960, Report of the Advisory Panel on Radio Telescopes, *ApJ,* **134**, 927
Kraus, J.D. 1984, Karl Guthe Jansky's Serendipity, Its Impact on Astronomy and Its Lesson for the Future. In *Serendipitous Discoveries in Radio Astronomy*, ed. K.I. Kellermann and B. Sheets (Green Bank: NRAO/AUI), 59
Lovell, A.C.B. 1985, *The Jodrell Bank Telescopes* (Oxford: Oxford University Press)
McClain, E. 1960, The 600-Foot Radio Telescope, *SciAm,* **202** (1), 45
Reber, G. 1958, Early Radio Astronomy in Wheaton, Illinois, *Proc. IRE,* **46** (1), 15
Sullivan, W.T. III 2009, *Cosmic Noise, A History of Early Radio Astronomy* (Cambridge: CUP)
Trexler, J. H. 1958, Lunar Radio Echoes, *Proc. IRE,* **46**, 286
van Keuren, D. K. 2001, Cold War Science in Black and White: US Intelligence Gathering and its Scientific Cover at the Naval Research Laboratory 1948-1962, *Social Studies of Science,* **31** (2) 207

von Hoerner, S. 1967, Design of Large Steerable Antennas, *AJ*, **72**, 35

Whitford, A. 1964, *Ground-Based Astronomy: A Ten Year Program* (Washington DC: National Academy of Sciences)

Youmans, A.B. 1969, The Design of a 300-ft Research Antenna. In *Structures Technology for Large Radio and Radar Telescope Systems*, eds. J.W. Mar and H. Liebowitz (Cambridge: MIT Press), 5

FURTHER READING

Prestage, R.M. 2006, The Green Bank Telescope, *Proc. SPIE*, **6267** (12), 1.

Prestage, R.M. et al. 2009, The Green Bank Telescope, *Proc. IEEE*, **97**, 1382

Lovell, A.C.B. 1968, *The Story of Jodrell Bank* (New York: Harper & Row)

Lovell, A.C.B. 1974, *Out of the Zenith* (New York: Harper & Row)

Lovell, A.C.B. 1987, *Voice of the Universe* (New York: Praeger)

Saward, D. 1984, *Bernard Lovell: A Biography* (London: Robert Hale)

CHAPTER 10

Exploring the Millimeter Sky

In 1962, Frank Drake recruited Texas Instruments physicist Frank Low to come to Green Bank to develop bolometer receiver systems for use at millimeter wavelengths. Under Low's leadership, NRAO contracted with the Rohr Corporation to manufacture a 36 Foot Telescope designed for use at wavelengths as short as 1 mm. To minimize the effects of tropospheric water vapor, NRAO located the telescope at the Kitt Peak National Observatory near Tucson, Arizona. Fabrication errors led to long delays, and before the 36 Foot Telescope was finished, Low left NRAO to join the University of Arizona, where he could pursue his interests in infrared astronomy. Low's bolometers never reached the anticipated sensitivity at 1 mm, and manufacturing errors limited the performance of the 36 Foot dish. However, the unanticipated discovery of powerful 2.6 mm radio emission from interstellar carbon monoxide (CO), and later from other molecular species, led to a greatly increased interest in millimeter astronomy. Despite many technical and administrative concerns, the 36 Foot Telescope became the most oversubscribed NRAO telescope. In 1983, NRAO replaced the faulty 36 Foot dish with a more precise 12 Meter surface. Arguably, the 36 Foot/12 Meter Telescope became the most productive instrument in the world for millimeter spectroscopy until it was eclipsed by more powerful facilities both in the US and abroad.

An ambitious plan to build a 25 meter millimeter wave telescope on Mauna Kea in Hawaii was never funded, and it would be another quarter of a century before the NRAO would return to the forefront of millimeter wave radio astronomy with the completion of the Atacama Large Millimeter/submillimeter Array (ALMA) as a joint NRAO-ESO-NAOJ facility in northern Chile.

© The Author(s) 2020, corrected publication 2021
K. I. Kellermann et al., *Open Skies*, Historical & Cultural Astronomy,
https://doi.org/10.1007/978-3-030-32345-5_10

10.1   FIRST ATTEMPTS

Although the 1961 Pierce Panel report (Keller 1961) emphasized high resolution radio imaging, the Panel also drew attention to the potential opportunities at millimeter wavelengths noting that "the exploitation of wavelengths from 3 cm down through the millimeter range should be encouraged and supported." They also pointed out that "Such work can best be carried out at altitudes above 13,000 feet[1] with highly accurate dishes of moderate size (less than 100 feet)." But the Pierce Panel was primarily motivated by the drive for higher angular resolution, which they argued could be achieved with relatively small and therefore inexpensive dishes operating at millimeter wavelengths. Indeed, the highest resolution filled aperture radio telescope at the time was the Naval Research Laboratory's (NRL) 50 foot dish, which had a 3 arcmin beam at 8 mm wavelength.

As a physicist working for Texas Instruments, Frank Low (Fig. 10.1) developed sensitive liquid helium cooled germanium bolometer detectors that promised greatly improved sensitivity at infrared and short millimeter wavelengths (Low 1961). Since bolometer systems respond to all incoming radiation, including the warm radiation from the ground and atmosphere, the challenge was to develop effective filters that could isolate the desired waveband and attenuate everything outside the reception band by at least a factor of a million, while, at the same time, not introducing significant noise. This meant that the filters as well as the bolometer needed to be cooled to liquid helium temperatures. Frank Drake became aware of Low's work,[2,3] and

Fig. 10.1 Frank Low came to NRAO from Texas Instruments in 1962 to begin a millimeter astronomy program in Green Bank. Credit: NRAO/AUI/ NSF

recruited Low to come to Green Bank to develop millimeter wavelength receiver systems. Following his short visit to Green Bank in March 1962, Joe Pawsey warmly endorsed Drake and Low's millimeter initiative and also noted that Low's bolometer was "an ideal instrument for infra-red spectroscopy."[4]

After arriving in Green Bank in 1962, Low worked on 1.3 mm and infrared bolometer systems. He and Drake set up a 5 foot plastic dish with a gold plated surface and began the first astronomical observations in the 1.3 mm band (Low and Davidson 1965). This was NRAO's first experience with liquid helium cooled receivers, and the bolometer contract with Texas Instruments included two weeks of training in cryogenic techniques for Observatory personnel.[5] However, with their limited sensitivity, all Low and Drake could observe at 1.3 mm was the Moon, and they began to develop plans to build a larger antenna on a mountain site to minimize the absorption due to atmospheric water vapor.

Dave Heeschen shared their enthusiasm, and expressed the opinion that "Millimeter wavelength observations constitute a vast unexplored region of radio astronomy," and said he did not believe the Observatory should leave the millimeter wavelength field to others because this work could be done effectively only by a strong balanced group such as was available at Green Bank.[6] Although still struggling to complete the 140 Foot construction, Heeschen boldly proclaimed that the millimeter wavelength telescope ranked third in priority for NRAO "after the very large dish [LFST] ... and the interferometer array [VLA]."

10.2   The NRAO 36 Foot Millimeter Wave Telescope

As part of its 1964 budget submission to the NSF, NRAO included a request for $600,000, later increased to $800,000, to obtain a 36 foot diameter antenna designed to work at wavelengths as short as 1.3 mm. Frank Drake later recalled that it was a last minute thought, and that he added just a few paragraphs of explanation to NRAO's annual budget submission to support the request for a new millimeter wave telescope.[7] Prior to starting the 36 Foot project, Low and Drake embarked on a development program to build a series of smaller antennas ranging up to 12 feet in diameter (Fig. 10.2).[8] Although Heeschen was successful in the getting the NSF funds, it soon became clear that it was going to be a challenge to achieve the required surface accuracy of 0.002 inches (0.05 mm), less than the thickness of a sheet of paper, and 2 arcsec pointing accuracy, about the angle subtended by newsprint seen across a football field.[9] Moreover, in 1964, "in view of the relatively small initial cost and the scale of the operation," the AUI Board raised questions about whether or not a millimeter wave telescope was more appropriate for a university than for NRAO.[10] But Heeschen claimed that there were no universities prepared to invest in the technology development needed for observing at millimeter wavelengths.

Fig. 10.2   Frank Low (on the left) supervises the installation of his 1 mm bolometer on a 12 meter diameter dish behind the Green Bank Jansky Laboratory. Credit: NRAO/ AUI/NSF

In order to accommodate his 1.2 mm bolometer system, Low argued for an unusually long feed support structure. This made the optics for the conventional heterodyne receivers used at longer wavelengths more complex than would be the case for a more conventional f/D ratio of about 0.4, and was "the subject of prolonged discussion" within NRAO.[11] Also, as argued by Peter Mezger, the longer feed support legs were more subject to wind and thermal effects which compromised the pointing accuracy.[12] The controversy was finally resolved by adopting a compromise geometry, but as Mezger had anticipated, this still created problems in using the telescope at longer wavelengths.

After recruiting Low to start a millimeter program at NRAO, Drake left NRAO in 1963 to join the Jet Propulsion Laboratory. Two years later, after getting NRAO to agree to these unusual antenna specifications, Low left to join the University of Arizona Lunar and Planetary Laboratory. Low was already spending a lot of time in Tucson pursuing his interests in infrared astronomy, and suggested setting up an NRAO laboratory in Tucson to support the 36 Foot operation. Apparent in Low's request was his interest in remaining in Tucson instead of living in Green Bank or Charlottesville. Responding to Low, Heeschen firmly replied, "We do not intend to set up, instrument, and staff a lab in Tucson. This is a firm decision and applies to you and everyone else on the NRAO staff."[13] Later Heeschen added, "the Tucson site will always be—for the NRAO—purely an observing site. We will not have any appreciable staff there, no development lab there, and no scientists permanently in residence there…. It will not be possible for us to indefinitely maintain you in Tucson. At some time you should return to Charlottesville or affiliate with some other organization."[14] Low elected, instead, to join the University of Arizona, where he went on to have a very distinguished career as one of the pioneers of infrared astronomy. He also formed his own company to build and market infrared detectors for astronomical, industrial, and military use, and he liked to tell stories of dark-suited customers who would pay for bolometer systems with thousands of dollars in cash.

The departure of Drake and Low left NRAO with a novel but challenging millimeter wave telescope project, but without the two scientists who had initiated it. John Findlay took over as the project director, but he left in 1965 for a year's leave-of-absence to become director of the Arecibo Observatory in Puerto Rico, leaving Hein Hvatum, NRAO Assistant Director for Technical Services, in charge of the millimeter telescope effort. Peter Mezger, who was on the Green Bank Scientific Staff, assumed the role as the scientific leader of NRAO millimeter wave astronomy. In a thoughtful report,[15] Mezger noted that the few sources likely to be strong enough to study with the planned 36 Foot Telescope included the Sun, the Moon, and some of the planets. Although he noted that "there is some evidence of radio sources of very small apparent diameters with flat or increasing spectra which may become 'visible' at very short wavelengths," he commented, "it seems to be very doubtful if observations at 3 mm wavelength or shorter can contribute anything to the radio astronomy of galactic and extragalactic sources." He also went on to speculate on the possibility of observing atomic Radio Recombination Lines from high order electron transitions. In the same report, Mezger compared the short wavelength capabilities of the 36 Foot antenna with other facilities and reviewed the range of available millimeter wave amplifiers.

NRAO solicited proposals from eight potential suppliers and received three firm bids to construct the complete telescope.[16] Following evaluation of the proposals under Findlay's leadership, including visits to the three finalists' plants, NRAO chose the Rohr Corporation in Chula Vista, California to build the 36 Foot telescope. This was not NRAO's first experience with the Rohr Corporation. In 1963, when NRAO and Rohr were discussing a possible design contract for a 400 foot transit radio telescope, Rohr engineer Bob Hall had casually remarked that Rohr had recently completed a 15 foot dish designed for operation up to 140 GHz.[17] In order to protect the antenna from the weather, Rohr proposed enclosing the telescope in a 95 foot diameter rotating astrodome, allowing observations to be made through a 40 foot slit, much in the manner typical of optical telescopes.

Although Low and Drake had succeeded in making millimeter observations in Green Bank, the 36 Foot Telescope clearly needed to be located at a better site with less atmospheric water vapor. Two sites near Tucson, Arizona were considered: one on Kitt Peak Mountain, home of the Kitt Peak National Observatory (KPNO) and located about 50 miles to the west of Tucson, the other on Mount Lemmon, northwest of Tucson, home of the University of Arizona optical and infrared telescopes. Other sites near Climax, Colorado and at the Los Alamos National Laboratory in New Mexico were also discussed, the latter being pushed by some of the AUI Trustees with their atomic physics backgrounds. The Mount Lemmon site was located at an altitude of 9,000 feet, about 2,000 feet higher than the Kitt Peak site, but there were powerful radio transmitters, as well as other activities, on Mount Lemmon, which were a potential source of interference to millimeter astronomy. NRAO chose Kitt Peak, as KPNO agreed to provide logistical and administrative support for NRAO's millimeter telescope. Interestingly, there was no attempt made to evaluate any of the other potential sites, as Heeschen argued that "the difference between a so called 'good' site and a somewhat better one from the water vapor point of view is so much less than the difference between a good site and a bad site [Green Bank] as to make it unnecessary in his judgment to embark on detailed studies."[18] However, in spite of Heeschen and Findlay's reassurances about the adequacy of the Kitt Peak site and the attraction of collaborating with the optical astronomers at KPNO, the AUI Board continued to press the issue of seeking a more favorable site. On the other hand, the NSF Director Leeland Haworth cautioned Heeschen about the difficulties of operating such a distant site but otherwise supported the project.[19] Heeschen acknowledged that "a split operation presents real difficulties," but pointed out that the planned large array would also involve an additional site for NRAO.

Construction Challenges Construction of the dish itself, which took place at the Rohr plant in Chula Vista, was a challenge. In order to meet the 0.002 inch rms accuracy required for operation at 1.3 mm, Rohr decided to fabricate the surface in one piece rather than use multiple panels as for the Green Bank antennas, and machined the surface from welded sections of aluminum plate. The precision cutting procedure was so sensitive to vibrations that to avoid the effect of passing trucks and the effects of ocean tides, the cutting could only be done at night and at low tide. Moreover, the welding process distorted the structure, so that to achieve the desired parabolic shape, the thickness of the dish surface would then be less than the "desired minimum surface thickness." To correct for this, Rohr engineers sprayed additional metal to the low areas of the reflector surface. But the sprayed areas contained contaminants that dam-

Fig. 10.3 The 36 Foot dish was transported by road from the Chula Vista factory to the base of Kitt Peak. Credit: John Hungerbuhler/NRAO/AUI/NSF

aged the cutting tool and probably contributed to the resultant poor thermal characteristics of the dish.[20]

The dish surface was finally complete in late February 1966 and the 13,000 pound 36 foot dish was transported by road from the Chula Vista factory to the base of Kitt Peak (Fig. 10.3), accompanied by California and Arizona State Police escorts. The 425-mile trip took ten days and the entire Rohr convoy included eight truckloads of telescope components. Due to problems and delays in completing the dome, concerns about the impact of inclement weather conditions on top of the mountain, and ongoing repairs to the road up the mountain, the dish structure remained at the bottom of Kitt Peak for many months—under guard lest it be stolen or used as target practice by Arizona locals. Even after the dish was mounted, difficulties with the drive system, the azimuth bearing, control of the dome motion, and the on-line computer control delayed the start of telescope operations for another year. The 36 Foot Telescope (Fig. 10.4) was finally turned over to NRAO in April 1967, although a variety of problems remained, including the flexure of the bi-pod feed support, which could be mitigated by tightening the cables securing the legs, but there was a concern that this would lead to dish distortions. On one occasion, the cables were inadvertently loosened and the feed legs crashed into the dish, but fortunately there was no serious damage to either the telescope or personnel, except perhaps for the great embarrassment of the senior engineer who caused the accident.[21]

Fig. 10.4 The completed 36 Foot Telescope in its rotating dome enclosure on Kitt Peak

The 36 Foot millimeter wave telescope was novel for the time, as it was the first NRAO telescope to be designed from the start to be operated under the control of a digital computer. However, programing the computer introduced new challenges. Although the 85 foot, the 140 Foot, and the 300 Foot antennas all ended up being computer-controlled, it was only after years of experience with an operator interacting with analogue control systems. The 85 foot and 140 Foot telescopes were equatorially mounted and so needed no coordinate conversion, while the 300 foot was a simple transit telescope. The 36 Foot was NRAO's first alt-az telescope that required coordinate conversion between celestial right ascension-declination and altitude-azimuth, and indeed the concerns raised a decade earlier in the debates surrounding the design of the 140 Foot Telescope resurfaced (Sect. 4.4). Other alt-az radio telescopes such as those at NRL, Dwingeloo, Jodrell Bank, and Parkes used an analogue conversion system. The 36 Foot was one of the first telescopes anywhere to use a digital computer for the coordinate conversion. As a new experience, and due to errors in the operating system along with faulty hardware interfaces, it took several trips to Tucson by Green Bank engineers, and a new programmer, before the telescope was able to accurately point and track a celestial target. An interesting by-product of the later attempts to improve the computer control of the 36 Foot Telescope and real-time data analysis was the introduction of the FORTH[22] language developed by Charles (Chuck) Moore. After bringing FORTH to both Green Bank and Tucson, Moore left NRAO to form FORTH

Inc., which developed FORTH applications for a wide variety of end users including the space shuttle, medicine, oceanography, engineering, music, the San Francisco BART metro system, and the Boeing 777 avionics system.

Getting Going in Tucson The original scientific justification for the 36 foot millimeter wave telescope was marginal. It was not built to solve any specific scientific problem or to investigate any known phenomena, but rather to explore the opportunities for new discoveries that might be possible by working in this almost unexplored region of the electromagnetic spectrum, and in particular to exploit Frank Low's 1.3 mm bolometer system. A realistic estimate of what one might expect to observe with Low's bolometer and the 36 Foot telescope would have included the Sun, the Moon, the thermal emission from a few planets, and a few H II regions with thermal spectra. Only a few extragalactic sources were known to have spectra that when extrapolated to millimeter wavelengths might be detected with the 36 Foot. The class of compact "flat spectrum" radio sources were still unknown, and, ironically, there was no consideration of any spectroscopic observations.

Reflecting the anticipated nature of the 36 Foot operation as an experimental instrument, and perhaps realizing that the 36 Foot Telescope appeared to have limited attraction for outside users, Heeschen planned to keep the Tucson-based NRAO support staff to a minimum and the operation informal. After the telescope went into operation, George Grove, who had served in a variety of roles in Green Bank, transferred to be Head of Tucson Operations in August 1967 to support the 36 Foot operation and to provide some observing assistance. Initially, unlike at Green Bank, observers were for the most part expected to run the telescope and take care of the instrumentation themselves, but in 1968 Don Cardarella, who had been a Green Bank 300 Foot operator, moved to Tucson and became the first 36 Foot telescope operator.

When the telescope was finally placed in operation in the summer of 1967, there was no immediate rush of observers waiting to use the instrument. Drake and Low, who had started the project with great enthusiasm, were gone. Unlike the 140 Foot and 300 Foot Green Bank telescopes, the 36 Foot was conceived of as an experimental instrument and not a user facility. The telescope was scheduled informally, first by Heeschen and later by Bill Howard, then Assistant to the NRAO Director, with large blocks of time going to individuals or to small teams who would be in residence in Tucson for several weeks at a time. Much of the early observing was devoted to calibrating the pointing and learning how the focus and gain changed with elevation and temperature. However, the poor sensitivity resulting from the small antenna size, low efficiency, and high system temperatures made calibration challenging. Only the Sun and the Moon, and two planets, Venus and Jupiter, gave sufficient signal-to-noise ratio, and solar heating limited observations to the nighttime. When under computer control, the telescope moved very slowly, so large azimuth motions resulted in a lot of lost observing time. An adventurous and courageous observer knew how to unlock the computer and manually drive the telescope at high speed to a new position, hoping that the brakes would work, and risking tearing off the connecting cables if the telescope were not stopped in time.

Although Low continued to attempt 1.3 mm bolometer observations, the uncertain antenna pointing, poor aperture efficiency, and thermal distortions of the dish limited results. Scientific observing by NRAO staff, as well as by visitors, at 3 mm and 9 mm wavelength using simple mixer continuum radiometers to study extragalactic radio sources and thermal emission from compact H II regions were not productive. It was clear that the 36 Foot was not going to meet its design specifications, and, already, Heeschen was contemplating replacing the dish structure.[23] But AUI first called for a review to explain the increased cost, the delay in completion, and the failure to meet the anticipated specifications.[24]

The local oscillator systems for both the 3 and 9 mm receivers, were derived from klystron oscillators. Not only did they have limited lifetimes, but they were expensive, and not all of the klystrons lasted for their full 500-hour advertised lifetime. Since the klystrons were manufactured by the Canadian branch of the Varian Corporation, replacements were delayed by the need to get exemptions from the Buy American Act. More than one replacement klystron oscillator was hand carried across the border to minimize bureaucratic delays. Mark Gordon (2005, pp. 99–100) recalled the Charlottesville attempts to build a cooled parametric amplifier for millimeter spectroscopy. When finally delivered after years of development, it only worked over a narrow band around 49 GHz, where there were no spectral lines of interest, and where it was uncomfortably close to the atmospheric O_2 absorption feature. Gordon estimated that NRAO probably spent at least \$500,000 on the amplifier project. NRAO also obtained a new bolometer system that was fabricated at the University of Oregon and designed to operate at 1, 2, and 3 mm by using different filters. The new bolometer also had limited success.[25]

By 1968, there were a few external users observing the Sun and planets as well as bright H II regions and the Crab Nebula. In spite of its limitations, the 36 Foot Telescope was probably the most productive millimeter wave radio telescope in existence. Even with its low 10–15 percent efficiency at 1.3 mm, the 36 Foot had more collecting area, and better resolution than the Palomar 200 inch. However, problems operating the telescope with limited staff, inclement weather, and power failures continued to plague millimeter observers. A not uncommon visitor experience was, "Our run was pretty frustrating but not entirely unproductive."[26] Another observer asked for "some reasonable imitation of a working system."[27] In October 1968, Heeschen decided to stop scientific observing and give priority to long neglected repairs and better calibration of the efficiency and pointing. As he informed one potential observer, "The 36-ft has many problems associated with it: pointing calibration is difficult, other calibrations are difficult, the receivers have been unreliable, there have been mechanical problems with the dome, dish parameters—focus, pointing, gain—are unknown functions of temperature."[28] He finally realized that some of the problems in Tucson were the result of trying to manage the program from a distance and, in order to provide more effective local management, he hired Edward (Ned) Conklin in October 1969 as the first Tucson resident member of the NRAO scientific staff to act as the Tucson site manager. Conklin had received his PhD in electrical engineering at Stanford working with Ron Bracewell, and brought a new level of technical expertise to the Tucson group.

Local logistical support for NRAO's Tucson operations was provided by KPNO, which rented Tucson office and laboratory space and allocated several rooms in the KPNO mountain dormitory to NRAO. NRAO staff and observers on the mountain ate their meals at the KPNO cafeteria, which provided a pleasant opportunity to informally interact with observers using the KPNO optical telescopes, especially when poor weather prevented both radio and optical observing. However, with the introduction of spectroscopic capability in late 1969, there was growing pressure to use the telescope, even in the daytime, even if the gain and pointing were uncertain due to thermal deformations of the structure. As a result, the radio and optical astronomers kept different hours, and each complained of noise generated by those working at the other end of the spectrum. NRAO installed trailers, later upgraded to permanent buildings near the telescope itself, where the operators and observers were able to sleep in quiet, but instead of the noise, they then had to deal with the local scorpion and skunk population.

When Ned Conklin left NRAO in 1973 to join the Arecibo Observatory, he was replaced by Mark Gordon, who served as the first NRAO Assistant Director for Tucson Operations, with a charge to convert the 36 Foot from an experimental facility to a more user-friendly facility of the kind NRAO observers were familiar with in Green Bank. Gordon, who had spent a winter in Antarctica as part of the US Antarctic Research Project, brought a dynamic new leadership perspective to the 36 Foot operations. Soon the NRAO support staff in Tucson had grown to 20 people, including a full complement of telescope operators. However, the limited size of the Tucson Electronics Division, which perhaps reflected its original development as an experimental rather than a user facility, meant staff felt overworked maintaining the telescope and cryogenics, as well as the receiver instrumentation.[29] New receivers were built in Green Bank or Charlottesville, and a common complaint was that they would arrive untested only days before being scheduled on the telescope. However, the engineers in Charlottesville saw it differently, and complained about the misuse of their receivers by the Tucson engineers. The long commute between Tucson and Kitt Peak, especially in response to nighttime callouts, added to the low morale and likely contributed to the heavy turnover in the NRAO technical staff in Tucson.

Faced with growing tensions with KPNO and lack of adequate space resulting from the increased level of NRAO operations, Gordon moved the NRAO Tucson staff from their downtown KPNO offices to a free-standing facility in an industrial office complex some five miles away. But in October 1984, by agreement with the University of Arizona, the NRAO Tucson operations moved back to the university campus to occupy an upper floor of the new Steward Observatory building.

Interstellar Carbon Monoxide and Molecular Spectroscopy The possibility of observing narrow band radio emission from atomic and molecular transitions was discussed as early as 1955 by Charles (Charlie) Townes at the Jodrell Bank Symposium on Radio Astronomy (Townes 1957, p. 92). A few interstellar molecules, e.g., hydroxyl (OH), formaldehyde (H_2CO), water (H_2O), and ammonia (NH_3) had been detected at centimeter wavelengths (Sect. 6.2), but the transition probability of typical interstellar molecules increases rapidly toward higher rotational energy levels which occur primarily at millimeter wavelengths. Although there was no discussion of any spectroscopic capability when planning for the 36 Foot Telescope, by 1970 NRAO had installed a 40 channel spectrometer. However, in addition to the long-standing reliability and gain stability issues, spectroscopic observers were limited by the lack of any local data reduction capability at the telescope, and they had to wait until the next day to learn if they had discovered anything.

In February 1969, Arno Penzias from Bell Laboratories wrote to NRAO requesting eight weeks of observing time to search for "in descending order of our interest CN [cyanide], CO [carbon monoxide], and HCN [hydrogen cyanide]."[30] The CN observations were justified by the well-known detection of optical absorption lines (Field and Hitchcock 1966) but Penzias added that, "although CO has a much smaller dipole moment than CN, it is probably worth looking for." Heeschen granted Penzias only four weeks, but indicated that the other four weeks would likely follow. The Bell Labs group fully anticipated that any molecular lines would be weak and require long integration times to detect any signal. So in addition to bringing their own low noise front end to Kitt Peak, they also brought their own computer in order to average and display the results at the telescope. By this time, their main interests had shifted from CN to CO, but everyone was surprised when they pointed the telescope toward the Orion Nebula and saw a very strong CO signal in real time on the chart recorder. Robert Wilson et al. (1970) then went on to detect CO from a total of eight other Galactic sources including the Galactic Center.

The surprisingly strong CO emission discovered by Wilson et al. opened the door to the discovery of many other molecular species. Suddenly the 36 Foot Telescope was in heavy demand, and the competition to be the first to detect a new molecule or isotopic species was intense and not entirely cleanly fought. By this time Gordon had taken over the difficult task of scheduling observers with competing proposals. Although each proposal was sent to multiple referees for review, the referees were not always consistent in their comments, and there was considerable overlap between the referee pool and the observers. Complaints of unfairness or referee incompetence were not uncommon. With an oversubscription rate of about 5:1, only a small fraction of the proposals could be scheduled, but every proposer felt that their proposal was well above average. One group even threatened to go elsewhere to make their observations and thus deprive NRAO of the discovery of interstellar glycine (NH_2CH_2COOH).[31] But as Kellermann wrote to Heeschen, "the situation would be much worse if the available observing time <u>exceeded</u> the requested time by a factor of five."[32]

In order to search for a new molecule, observers needed to know the frequency, which could be calculated with some uncertainty, or in some cases determined from laboratory spectroscopy. Competing observers maneuvered to establish collaborations with theoreticians or laboratory spectroscopists to learn the correct frequencies needed to search for their favorite molecule; some then kept their search frequencies secret from other observers or even leaked false information. In principle, observers were supposed to follow their approved observing program, but some strayed into territory which had been staked out by other observers. Sometimes the frequency of a new line would become public, or at least known to competing groups. Whoever was the next observer could "discover" a new line. At least one observer was known to purposely enter an incorrect frequency in the telescope logbook in order to misdirect the next competing observer. Others misstated the sources they were observing or claimed that the reason they were observing a source not in their proposal was to use it as a calibrator. A particularly divisive situation arose in connection with the first detection of extragalactic CO. Competing observers argued among themselves and with NRAO that the other group had been approved for a different program and had acted unethically. Another observer recalled that, suspecting that someone was going through his desk at night, he invented a false molecule and a bogus observing proposal. Another observatory apparently spent considerable time looking for this molecule. When multiple groups discovered new molecules or new isotopes, there was a rush to publish before the other group, independent of who had actually made the first detection. These were arguably the most exciting times for millimeter astronomy but also perhaps the darkest days for millimeter astronomers.

As Mark Gordon (2005, p. vii) later wrote, molecular spectroscopy at Kitt Peak "revolutionized our understanding of the nature of interstellar gas, chemistry at extremely low temperatures, and how stars form and galaxies evolve." A whole new field of astrochemistry was largely born at the NRAO Kitt Peak millimeter wave telescope. Over one hundred different molecular and isotopic species had been detected, and NRAO was under considerable pressure to exploit this rapidly developing new field of astronomy. Moreover, millimeter astronomy was also being used to study the thermal radio emission from planets and other solar system bodies, as well as from stars and the energetic millimeter wave bursts from quasars. The 36 Foot Telescope had more than met the modest expectations of Drake and Low and was probably the most oversubscribed telescope in the world.

10.3   Replacing the 36 Foot Telescope

Although the 36 Foot Telescope had been responsible for many important discoveries, and, arguably, defined millimeter astronomy, it still had limited performance. The technical troubles remained, and observer complaints continued—along with a steady flow of advice on how to improve the 36 Foot operation. Not only did the surface distort due to differential thermal heating, but the pointing was erratic and non-reproducible, in part due to thermal distortions, but also to problems with the servo system. This made the telescope nearly useless for daytime observations. Even during nighttime, it was difficult to get quantitative results, although the competing astronomers were more interested in discovering a new molecule than in quantitative results that depended on accurately knowing things like the antenna gain and pointing. Various innovative attempts to shield the telescope from daytime heating proved less than effective (Gordon 2005, pp. 119–124). In 1973, the 36 Foot Telescope was converted to Cassegrain operation in order to facilitate the use of large cryogenically-cooled receivers and to permit beam switching using a nutating sub-reflector, but this did not address the more fundamental problems of telescope performance and safety. Perhaps the most serious issue arose in July 1972, when the 40 foot dome door jammed, driving the chief telescope operator to "declare the 95-foot radome housing the NRAO 36-foot radiotelescope [*sic*] condemned," and "you can consider this my formal resignation if the situation described herein is not corrected to my satisfaction."[33]

The 65 Meter Millimeter Wave Telescope In Sect. 9.4, we discussed how the growing interest in millimeter molecular spectroscopy led the LFST project to converge to a 65 meter antenna good to 3 mm under favorable observing conditions (Findlay and von Hoerner 1972). The proposed 65 meter telescope was designed to be homologous, although it was otherwise a conventional symmetric alt-az structure (See Fig. 9.7). Support for a major NRAO initiative in millimeter astronomy got a boost from the Greenstein (1973) Decade Review Committee, when it briefly appeared that there might be funds in the NSF FY1972 budget for a new millimeter wave radio telescope, but, as it turned out, the budget information was incorrect. Moreover, by this time, interest had moved to even shorter millimeter wavelengths, and there was no further attempt to fund the NRAO 65 meter telescope.

The Rise and Fall of the 25 Meter Millimeter Telescope Responding to the growing interest in the new field of astrochemistry, in 1974 Dave Heeschen established an NRAO committee to consider options for replacing the 36 Foot Tucson Telescope. He appointed Barry Turner as Project Scientist and chair of the committee, which included Findlay and von Hoerner as well as Mark Gordon and one of the present authors (KIK). Following a series of meetings, Turner's committee concluded that NRAO should build a 25 meter diameter telescope capable of operating down to 1 mm wavelength.[34] Heeschen then

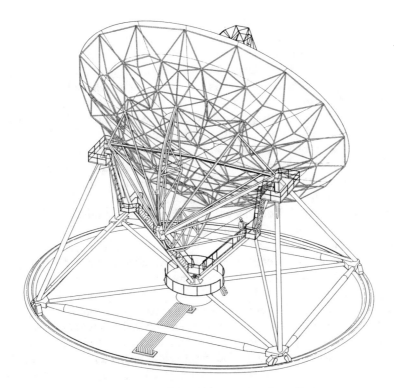

Fig. 10.5 Artist's conception of the 25 meter millimeter wave telescope. Credit: NRAO/AUI/NSF

established an external committee to give advice and to provide support for the new NRAO initiative.

By this time VLA construction was well underway, and in September 1975 NRAO submitted a formal proposal to the NSF to build and operate a 25 meter radome-enclosed telescope that would be good to wavelengths of 1 mm and shorter (Fig. 10.5). The main scientific motivations were for molecular spectroscopy to study star formation, the physical conditions (temperature, density) in interstellar clouds and in the atmospheres of cool stars, as well as tracers of galactic structure free of optical obscuration. Turner decided that his job was done and that he wanted to return to his research, and Mark Gordon replaced Turner as the 25 meter Project Manager and committee chair.

Discussions about where to site the 25 meter telescope became very controversial. Many committee members argued for a high altitude site with low water vapor content, important for the short millimeter and submillimeter wavelengths. The NRAO committee considered mountain sites in the continental US as well as the summit of Mauna Kea in Hawaii at nearly 14,000 feet elevation. A 12,470 foot high site on White Mountain, in the California Inyo Mountains near the OVRO appeared attractive, but access was limited, especially in the winter. The 9,000 foot high Mount Lemmon Observatory was a convenient, excellent observing site where the University of Arizona had many optical and infrared telescopes, but the radio and TV transmitters were a potential source of RFI. Mauna Kea has clear skies, low water vapor, and offered the best access to the galactic center and the southern hemisphere, but the projected construction and operations costs were much greater than for a continental site. One distinguished NRAO user commented that since the 2.6 mm CO transition was more important than the 1.3 mm band, a very high altitude site was not so important, and so it would be more cost effective to go to a less expensive site in Arizona.[35] However, citing the importance of low water vapor content to best exploit the capabilities of the telescope for observing near 1 mm, Gordon argued for Mauna Kea, and was supported by the NRAO Director, Mort Roberts, who felt that NRAO should provide the best possible instrument for the community.

After another two years of further design studies of the surface panels and the dome, as well as further analysis of potential sites, in 1977 NRAO submitted a revised proposal to the NSF to locate the telescope near the summit of Mauna Kea on the big island of Hawaii. The 1977 proposal differed from the earlier one in that NRAO now proposed an astrodome configuration, similar to that used for the 36 Foot or traditional optical telescopes, instead of a radome. The telescope would thus be protected from winds and inclement weather, but not suffer from the absorption characteristic of completely radome-enclosed telescopes such as Haystack (Sect. 6.6) and FCRAO (Sect. 10.4). But no rotating dome with an open aperture large enough to house a 25 meter diameter telescope had ever been built, and this added considerably to the estimated construction cost of $12.5 million and the annual operating cost of $1.35 million. The 25 meter antenna differed in an important way from all previous NRAO antennas which were "design and build" contracts based on performance specifications. Since the 25 meter specifications were so tight and difficult to measure, NRAO accepted the responsibility for the design and overall performance of the telescope, although William (Bill) Horne later noted that "while [NRAO] may possess the engineering capability, [it] certainly does not possess the engineering capacity ... for the required design work."[36] Six months later, Horne became the Project Manager for the 25 meter construction project.

As was the practice for all telescopes on Mauna Kea, the University of Hawaii, which operated the Mauna Kea site, expected that ten percent of all observing time would be given to University astronomers. However, there were no radio astronomers at the University of Hawaii, and NRAO was unwilling to compromise its Open Skies policy, especially in this very competitive field of millimeter spectroscopy. Instead, NRAO agreed to provide a one-time contribution toward a buried power line to the summit, as well as an annual contribution equivalent to the salary of a University of Hawaii Associate Astronomer to support the mountain astronomical infrastructure. This added another $2 million to the already expensive proposal. Later, when the University appeared to renege on the deal reached between Gordon and the U of H Institute for Astronomy Director, John Jefferies, Gordon threatened to re-open negotiations to locate the telescope in Arizona.[37] To complicate the situation, local environmental and cultural advocates, who by then opposed *all* astronomical facilities on Mauna Kea, especially objected to the 25 meter telescope because of its very large size and visibility compared with the Mauna Kea optical telescopes.

As described by Gordon (2005, p. 140), the university millimeter astronomy community was somewhat ambivalent about the 25 meter telescope. On the one hand, it promised a powerful new observing opportunity. But unlike the situation at longer wavelengths where there were no viable university facilities, university millimeter astronomers perhaps saw the proposed NRAO 25 meter telescope as competition to existing and planned university facilities at Berkeley, Harvard-Smithsonian, Caltech, and the Universities of Arizona, Massachusetts, and Texas. Typical of the university astronomers, Peter Strittmatter, Director of the University of Arizona Steward Observatory and Chair of the NSF Astronomy Advisory Committee, wrote, "I also believe that the committee will need to discuss how long the 25 m should remain as astronomy's No 1 new start priority if it is effectively blocked. Should other smaller projects be slipped in ahead of it?"[38]

Then, in 1979, Cornell's Frank Drake proposed a low cost 35 meter fixed spherical reflector alternative to the NRAO 25 meter telescope. By this time the 25 meter cost had risen to between $22 and $27 million, depending on the funding schedule. To advise them on deciding between the NRAO and Cornell proposals, the NSF appointed a sub-committee chaired by Alan Barrett to review the two projects. Following their meeting on 16 and 17 July 1979, the sub-committee unanimously and unambiguously "recommended without reservation that the NSF fund immediately the 25-meter millimeter wave telescope, as proposed by NRAO," and that "It is the unanimous judgment of the committee that the 35-meter fixed spherical telescope ... is not a realistic alternative to the 25-meter fully steerable telescope."[39]

The 25 meter project was saved, but it would only be a temporary reprieve. Bill Howard, now at the NSF as Astronomy Division (AST) Director, anticipated that with the ending of the VLA construction in 1980, the VLA funding level of about $10 million per year would remain in the AST budget and he could use these funds for the 25 meter telescope. Unfortunately, AST did not get to keep the VLA funding level, so AST proposed the 25 m telescope as a new start with new money in FY1981. As was described in Sect. 8.7, following the OMB cut to the proposed FY1981 NSF budget, the NSF director dropped the 25 meter telescope. It was included again in Jimmy Carter's final FY1982 budget proposal, but was killed when Ronald Reagan became president and froze all new starts for the new fiscal year. The 25 meter millimeter wave radio telescope then fell victim to the VLBA following the selection of the VLBA over the 25 meter telescope, first by the Decade Review Field (1982) Committee and then by the NSF Astronomy Advisory Committee. Lew Snyder at the

University of Illinois initiated a last desperate effort to save the telescope by sending the NSF a petition signed by many of the prominent workers in the field.[40]

It is a matter of speculation whether or not the telescope might have actually been built, if, instead of opting for the best site, NRAO had chosen one of the less expensive and more accessible sites in Arizona or New Mexico. Mark Gordon (2005, p. 146) later made a valiant effort to resurrect the 25 meter telescope by suggesting a less expensive surface structure and dome, and siting the telescope in the Santa Catalina mountain range near Tucson instead of on Mauna Kea. But it was too late; the millimeter astronomers had moved on to consider arrays. In stark contrast to the easy funding of the 36 Foot Telescope in the early 1960s, the level of effort that went into the 25 meter project was enormous. Over a decade of time NRAO staff had prepared dozens of funding plans and more than 150 internal reports dealing with everything from structural analysis to the electromagnetic properties of various paints, as well as detailed site and tropospheric water vapor studies; numerous contracts were negotiated but never implemented. Doing business at the NSF as well as at NRAO had changed, and would become even more complex during the long and difficult international negotiations leading to the construction of ALMA (Sect. 10.7).

The 12 Meter Upgrade Discouraged by the lagging progress with the proposed 25 meter telescope and the anticipated competition from the new millimeter telescopes being constructed by Caltech, Harvard-CfA, UC Berkeley, the University of Massachusetts, the Nobeyama Observatory, and IRAM, Gordon urged that the 36 Foot dish be replaced with a better reflector.[41] As Gordon (1984) later noted, "the popularity of the 36-foot was being killed by its very success."

Following a hastily called meeting in Charlottesville,[42] John Findlay was given the responsibility of replacing the 36 Foot dish structure. Instead of just matching the size of the existing 36 Foot (11.0 meter) diameter dish, Findlay elected to increase the size to 12 meters (39.4 feet), which he felt was the largest size compatible with the 12.5 meter (40 foot) dome slit (Fig. 10.6). As it turned out, this was probably a bad decision, as the telescope sidelobes did "see" the dome structure, limiting the performance, especially for sensitive continuum observations. Perhaps more important than the increase of 19 percent in collecting area, the focal ratio was changed to the more conventional value of 0.42, greatly facilitating the design of high efficiency feeds for millimeter wavelengths. After rejecting the possibility of obtaining a dish from Caltech (Sect. 10.4), NRAO solicited bids from 16 commercial sources, and awarded a contract to Central Fabricators, Inc. for $71,145 to fabricate a new reflector structure.[43] At the same time, the feed/subreflector bipod support was replaced by a quadripod to give better pointing stability. Unlike the solid 36 Foot dish, the 12 Meter reflector consisted of 72 aluminum petal-shaped panels manufactured by the ESSCO Corporation. When the new 12 Meter telescope went into operation in early 1983, it finally met the specifications originally set out

Fig. 10.6 John Findlay (right) and John Payne (left) discuss using the template to fabricate accurate surface panels for the 12 Meter Telescope

for the 36 Foot telescope and gave NRAO a badly needed competitive telescope with a surface that could support observations down to wavelengths as short as 0.8 mm.

Gordon stepped down as head of Tucson Operations in 1984 to return to full time research. He was followed as Tucson site director by Bob Brown, then Dave Hogg in 1985, and Darrel Emerson in 1986. But now a new cloud appeared on the horizon. For five consecutive years, NRAO had absorbed budget cuts and many of the new costs of operating the VLA by applying the budget cuts uniformly across all parts of NRAO. Everyone suffered and everyone complained, leading the new NRAO Director, Paul Vanden Bout, to announce that NRAO could no longer continue to do everything and that he might need to close the NRAO 12 Meter Telescope. Nearly fifty letters of protest from the user community as well as from NRAO staff were fired off to the NSF and to Vanden Bout.[44] Graduate students complained that their dissertation research was being arbitrarily terminated midway. But no one suggested any viable money-saving alternative, although some NRAO users suggested that there would be no loss if the Charlottesville headquarters were to be closed.

Fortunately, as described later by Gordon (2005, p. 187), Gordon was able to convince Arizona Senator Dennis DeConcini to use his influence to get sufficient funds restored to the NRAO budget. The NSF was not pleased with this political intervention, but as a result NRAO was able to keep the 12 Meter in operation until July 2000 when it was turned over to the University of Arizona to be used in their radio astronomy program. Some members of the NRAO

Tucson staff relocated to Charlottesville and become involved in the planning for ALMA; others retired, joined the University of Arizona program, or left NRAO to pursue other opportunities.

10.4   US Industrial and University Millimeter Wave Astronomy Programs

The NRAO 36 Foot/12 Meter telescope had opened up the new area of millimeter astronomy with its rich content of molecular transitions. Unlike other areas of radio astronomy, which were dominated by the large expensive telescopes and arrays, millimeter spectroscopy was much like optical spectroscopy, and limited more by observing time than by access to the most powerful facilities. It attracted not only traditional radio astronomers, who wanted to get away from their dependence on national facilities, but also laboratory spectroscopists such as Charlie Townes and Patrick (Pat) Thaddeus who saw opportunities to apply their skills in new ways.

Unlike centimeter to meter wavelength radio astronomy, states such as California, Illinois, Maryland, Massachusetts, and Texas, and at least two industrial groups, contributed to the construction and operation of a wide range of millimeter dishes and arrays, sometimes with additional support from the NSF. Perhaps motivated by their inability to fund the proposed NRAO 25 meter millimeter telescope and the recognition that the US was falling behind Europe in this emerging new area of astronomy, the NSF was particularly generous in supplementing both private and state funding for millimeter astronomy. But the NSF support came at a price: up to half of the observing time had to be made available to outside users.

Aerospace Corporation 15 Foot Millimeter Wave Antenna One of the first millimeter wave radio telescopes in the US was the 15 foot diameter dish operated by the Space Radio Systems Facility of the Aerospace Corporation. The antenna was located on top of the Aerospace building at the Los Angeles Air Force Station a few miles from the Los Angeles Airport and the Pacific Ocean, and had a surface accuracy of 0.09 mm rms. William Wilson, Robert (Bob) Dickman, and other Aerospace staff designed and built both continuum and spectroscopic receivers. In spite of the less than optimum location for millimeter observing, they, along with Eugene Epstein, used the telescope over a number of years for some of the first millimeter observations of quasars and planets, as well as for observing CO in the interstellar medium. (Stacey and Epstein 1964; Epstein 1977; Sargent 1979).

Bell Laboratories 7 Meter Millimeter Wave Antenna The Bell Labs Crawford Hill 7.5 meter millimeter wave antenna was built for propagation studies using the COMSTAR satellite beacons and for radio astronomy at frequencies up to 300 GHz. It was used over a period of years by Tony Stark, John Bally, and others, including many visitors. Highlights were studies of the Galaxy including a ^{13}CO survey of the plane (e.g., Stark et al. 1988), the Galactic Center region, and also studies of the structure and chemistry of molecular clouds. At the time it was probably the largest off-axis antenna ever built and remained so until the construction of the GBT (Chu et al. 1978).

University of Texas Millimeter Wave Observatory (MWO) The University of Texas 16 foot millimeter wave telescope was built in the early 1960s primarily for continuum studies of the planets and bright radio sources such as the Crab Nebula, the galactic center, and the Orion Nebula (Tolbert and Straiton 1965; Tolbert et al. 1965; Tolbert 1966). Although originally erected on the University of Texas campus in Austin in 1971, the antenna was moved to a better site at 2070 meters on Mt. Locke, the site of the University's McDonald Observatory. Motivated by the discovery of CO by Wilson et al. (1970), Paul Vanden Bout led an effort to bring the resources of the University of Texas, Harvard-CfA, Bell Laboratories, and the Columbia University/Goddard Institute for Space Studies to the MWO, and with support from NASA and the NSF, the MWO became a major player in millimeter wave spectroscopy. As described by Vanden Bout et al. (2012), "the amicable relations at the MWO stood in contrast to the NRAO 36-ft Radio Telescope where astronomers engaged in a vigorous competition to gain what was typically a few days of observing time, often to search for a new interstellar molecule." Rather than try to compete with NRAO observers in the race to discover new molecules, the MWO observing programs were largely devoted to using the strongest molecular lines to probe the physical conditions of their environment and to address questions posed by the discovery of an entirely new phase of the interstellar medium, including the nature of molecular clouds. Many of the future leaders of millimeter astronomy in both the US and Europe were trained at the MWO either as students or postdoctoral workers. In 1985, Vanden Bout left Texas to become the Director of NRAO, where he oversaw the construction of the VLBA and the GBT, and then spearheaded the US participation in ALMA.

Columbia University/Goddard Institute for Space Studies Shortly after the discovery of interstellar carbon monoxide (CO) by Wilson et al. (1970), Columbia University Professor Pat Thaddeus built a small 1.2 meter radio telescope which he placed on the roof of the Columbia physics building. In 1982 he installed a second telescope at the Cerro Tololo Observatory in Chile. When Thaddeus moved to Harvard-CfA in 1986, he took the Columbia instrument with him and placed it on top of a Harvard building. Over a period of many years, Thaddeus, together with numerous colleagues and students, used these two small radio telescopes to map out the 2.6 mm CO emission in the entire Galactic plane (e.g., Dame et al. 2001).

University of Arizona Steward Observatory Millimeter wave astronomy at the University of Arizona started with Frank Low's 1965 move from NRAO to Tucson, but stagnated as Low turned his attention toward infrared astronomy. In 1978, Peter Strittmatter, the Director of the UofA's Steward Observatory, spent a year at the MPIfR and began a discussion with Peter Mezger about millimeter and submillimeter wavelength astronomy. Mezger had hoped to build a submillimeter telescope on the summit of Pico Veleta above the 30 meter IRAM telescope, but access to the summit was hazardous, and Mezger was unable to gain permission for a summit site. Moreover, the Max Planck Gesselschaft (MPG) made it clear that they would not provide the additional annual funds which would be needed to operate the telescope. Strittmatter and Mezger then agreed to build and operate a 10 meter diameter submillimeter wavelength telescope at an altitude of 3180 meters on Mount Graham in eastern Arizona. The MPIfR—UofA Submillimeter Telescope (SMT), later renamed the Heinrich Hertz Telescope (HHT), has a surface accuracy of 0.015 mm rms and tracks to better than 1 arcsec. The HHT pioneered the use of carbon fiber reinforced plastic (CFRP) to minimize thermal effects in precision telescope structures (Baars et al. 1999) and operates at wavelengths as short as 0.35 mm.

Five College Radio Astronomy Observatory (FCRAO)[45] In 1976, FCRAO inaugurated a 14 meter diameter radome-enclosed antenna built by the ESSCO Corporation on the shores of the Quabbin Reservoir in central Massachusetts. Support for the construction and operation of the telescope came from a combination of NSF, private, and Commonwealth of Massachusetts funding. Until it closed for lack of operating funds in the spring of 2006, the FCRAO antenna was one of the largest millimeter wave telescopes in the US. During this period FCRAO scientists, engineers, and students designed and built a variety of innovative instrumentation, including a 16-element (QUARRY), then 32-element (SEQUOIA), MMIC arrays. The FCRAO receivers were among the best in the world, at times perhaps a factor two more sensitive than NRAO's 36 Foot/12 Meter receivers, and for many years FCRAO had a near-monopoly on structural studies of nearby galaxies. Many of the subsequent leaders in US millimeter wave science and instrumentation worked at or were trained at the FCRAO, and went on to distinguished careers in radio astronomy. Starting in the late 1990s, the FCRAO staff devoted their efforts toward building the LMT in Mexico. A 144 element bolometer 1.1 and 2.1 mm array known as AzTEC was developed at FCRAO in collaboration with others, and was used first on the James Clerk Maxwell Telescope (JCMT), then on the Japanese ASTE 10 meter submillimeter wave telescope in Chile, before being installed on the LMT in Mexico.

The Large Millimeter Telescope (LMT)[46] Planning for the 50 meter (164 foot) Large Millimeter Wave Telescope (LMT) on a high altitude site in Mexico as a joint effort between the University of Massachusetts and Mexico was already underway at the time of the 1990 Decade Review of Astronomy, but there were a number of competing proposals for what were considered "Moderate Programs." NRAO had proposed to fill the gap between the VLA and VLBA by constructing four new antennas in New Mexico. The Bahcall Committee Radio Panel was sensitive to the need to maintain viable university-based radio astronomy facilities in the US, and reluctant to allocate too much of the NSF's limited resources to NRAO, so it identified "A Large Millimeter Radio Telescope Working to at Least 230 GHz" as the highest priority for moderate sized projects. The expected federal share of the LMT cost was claimed to be only $15 million dollars, representing about half of the total cost (Kellermann 1991, p. I-9).

Normally, the parent committee of a Decade Review is tasked with interweaving the recommendations coming from the various wavelength panels, and it is rare for the parent committee to overturn the panel's ordered recommendations. However, in this case, the parent committee was apparently not impressed by the proposed LMT. The VLA extension proposed by NRAO appeared sixth and last among the recommended "Moderate Programs" (Bahcall 1991, p. 17), but the LMT was not mentioned at all in the Bahcall Committee report. The proposed VLA expansion was never funded, although an extensive refurbishing and modernization of the aging infrastructure, correlator, and other VLA instrumentation was later supported by the NSF, leading to the upgraded Karl G. Jansky Very Large Array (Sect. 7.8).

The reports of the Astronomy and Astrophysics Survey Committee Wavelength Panels have no formal status as recommendations of the NAS. However, based on the Radio Panel Report (Kellermann 1991), the University of Massachusetts working with the Mexican Instituto Nacional de Astrofísica, Óptica y Electrónica (INAOE) was able to obtain funding from Massachusetts and Mexican resources to build the LMT on the summit of Volcán Sierra Negra at an altitude of 4,600 meters (15,000 feet) in the Mexican state of Puebla. Following a series of technical and administrative disputes, compounded by funding delays, the LMT was finally completed in 2018 (Baars 2013). It operates at wavelengths as short as 0.85 mm on an excellent site. It is the world's largest filled aperture steerable telescope operating at such short millimeter wavelengths and is the largest, most complex, and most expensive scientific instrument ever built in Mexico.

Harvard-Smithsonian Sub-Millimeter Array (SMA) Planning for the SMA began in 1983. Motivated in part by the Field (1982) Committee recommendation, the new CfA Director Irwin Shapiro appointed a committee to study the feasibility of submillimeter interferometry. The committee, chaired by James (Jim) Moran, recommended the construction of an array of six 6 meter diameter dishes on a high dry site (Moran et al. 1984), but there were technical challenges in developing low noise receivers and movable antennas with sufficient precision to operate at submillimeter wavelengths. In 1987, CfA set up a laboratory for the development of submillimeter receiver technology and investigated potential sites in Arizona, Chile, and Hawaii. Under Moran's leadership, construction of a six element array of 6 meter diameter dishes near the summit of Mauna Kea at 13,350 feet elevation began in 1999. Two additional elements were added by the Taiwan Academia Sinica Institute of Astronomy and Astrophysics (Ho et al. 2004), and the eight element SMA with up to 172,000 spectral channels, 2 GHz of continuum bandwidth, and angular resolution up to 0.1 arcsec was completed in 2003. In 2008, the SMA was linked with the JCMT and California Submillimeter Observatory (CSO) to form a 10 element interferometer with baselines up to about 800 meters.

The SMA was the first imaging array to operate at sub-millimeter wavelengths. It made the first resolved radio images of the thermal emission of the Pluto-Charon system, of CO and HCN in the atmosphere of Titan, of the unscattered polarized continuum emission from Sgr A*, and of the extremely high velocity and low velocity collimated SiO outflows from a low luminosity proto star (Ho et al. 2004; Moran 2006). Unlike many of the other millimeter facilities, the SMA follows an Open Skies policy and observing time is available to all qualified scientists based on peer-reviewed proposals.

The Hat Creek Radio Observatory and the Berkley-Illinois-Maryland Association Millimeter Array (BIMA) Starting in the 1970s, University of California Professor Jack Welch built what was probably the world's first millimeter wave interferometer at the Hat Creek Radio Observatory in northern California. Under Welch's leadership, the initial two-element variable spacing interferometer was first expanded to three, then six dishes, each 20 feet in diameter. Starting in 1987, the array was operated by the Berkeley-Illinois-Maryland Association (BIMA). In 1993, the 85 foot diameter telescope at the Observatory collapsed during a violent wind storm. Instead of replacing the 85 foot telescope, Welch used the University insurance money to build new 20 foot diameter antennas to form a ten-element array that could be reconfigured to give angular resolutions up to 0.4 arcsec at 100 GHz (Welch et al. 1996). BIMA used cooled SIS mixers to operate up to 270 GHz or 1.1 mm. Data analysis was based on the MIRIAD (Multichannel Image Reduction, Image Analysis, and Display) software package developed by the BIMA group (Sault et al. 1995). Financial support for BIMA came from the states of California, Illinois, and Maryland, as well as from the National Science Foundation and the Taiwan based Academia Sinica Institute for Astronomy and Astrophysics (ASIAA). Thirty percent of the observing time at BIMA was made available on an Open Skies basis to users from outside the BIMA collaboration.

The Hat Creek interferometer and later BIMA were used to study H_2O masers, the HCN emission surrounding the galactic center, SiO masers in Orion, for a survey of CO in normal galaxies, for observations of the Sunyaev-Zelodvich effect, and the first millimeter VLBI observations (Plambeck 2006). BIMA also pioneered the use of mosaicked observations where observations based on hundreds of array pointings were combined to image a large extended area. In 2004, the BIMA antennas were moved to Cedar Flats to form part of CARMA (see below).

Caltech Submillimeter Observatory (CSO) and the Owens Valley Millimeter Array When it became clear in the early 1980s that they would not have a major role in the construction or operation of the VLBA, Caltech turned its attention to millimeter and submillimeter astronomy. Led by Professors Robert (Bob) Leighton, Alan Moffet, and Thomas (Tom) Phillips, Caltech developed two major facilities for millimeter/submillimeter astronomy. The Caltech program in millimeter wave astronomy was based on a novel antenna design by Leighton used to construct a series of 10.4 meter diameter dishes. Leighton's dishes were fabricated using 84 hexagonal aluminum honeycomb tiles which were figured after mounting on a steel backup structure using a custom-designed cutting machine installed at the same facility that was used to grind the Palomar 200 inch mirror. After surfacing, the dishes could be disassembled and reassembled in the field with a typical accuracy better than 0.035 mm rms.

A total of seven dishes were fabricated by Leighton and his colleagues. The most precise dish was the basis of the CSO located just below the summit of Mauna Kea at an altitude of 13,350 feet. The Mauna Kea antenna was mounted in a rotatable dome to provide protection from wind and weather. Under the direction of Phillips (2007), the CSO went into operation in 1987. It was used at wavelengths as short as 0.35 mm using a variety of bolometer arrays and coherent SIS mixer receivers to exploit the relatively high transition probability of molecules at higher frequencies, as well as the increased thermal emission from cold dust which peaks at infrared and submillimeter wavelengths. Due to a lack of operating funds from the NSF, the CSO was closed in 2015.

The other six dishes were erected in the Owens Valley to form a versatile imaging millimeter array operating in the 1.3 and 2.6 mm bands using cooled SIS mixers with an angular resolution of 1 arcsec at the shorter wavelength (Scoville et al. 1994). The Owens Valley millimeter array was used for a variety of spectroscopic observations ranging from studies of planetary atmospheres, evolved protostars, protoplanetary disks, nuclear starbursts, and luminous and ultraluminous high redshift galaxies, and was later absorbed into CARMA.

Combined Array for Research in Millimeter-Wave Astronomy (CARMA) It had been clear for some time that the BIMA and Caltech Millimeter Arrays would be much more powerful if they were combined into a single array, but both Caltech and BIMA resisted any change which threatened their independence. However, threatened by the potential loss of NSF funding, Caltech and the three BIMA institutions finally agreed to combine their facilities to form a more powerful 15-element array consisting of the six OVRO antennas plus nine BIMA antennas. Tony Beasley was recruited from NRAO to serve as the project manager to build CARMA at Cedar Flats in the Inyo Mountains east of the Caltech Owens Valley site at an altitude of 7,200 feet. An innovative aspect of CARMA was its use of the eight 3.5 meter antennas of the former Sunyaev-

Zeldovich Array, which were placed close to the CARMA antennas and used to simultaneously observe phase calibration sources. Beginning in 2007, CARMA provided a powerful northern hemisphere complement to ALMA, but it was closed in 2015 as the NSF concentrated its support for millimeter wave astronomy on ALMA.

10.5   International Challenges

The James Clerk Maxwell Telescope (JCMT)   Starting in 1983, the UK, together with the Netherlands and Canada, built a 15 meter diameter antenna near the summit of Mauna Kea close to the CSO. The JCMT is enclosed in a rotatable dome and observes through a slit covered with a membrane that is nearly transparent at millimeter and submillimeter wavelengths. With a surface accuracy about 0.025 mm rms, the JCMT had good efficiency at wavelengths as short as 0.3 mm. For many years the main instrument on the JCMT was the powerful Submillimetre Common-User Bolometer Array (SCUBA) which gave high sensitivity in both the 0.45 and 1.3 mm atmospheric windows with arrays of 91 and 37 pixels respectively, and was supplemented by single pixel bolometers at 1.1, 1.3, and 2 mm for photometry (Holland et al. 1999). SCUBA was used for both deep imaging and wide field mapping, and discovered the important new population of star forming galaxies (Barger et al. 1998). A 16-pixel SIS heterodyne receiver array was used for spectroscopy at 350 GHz. In 2011, SCUBA was replaced by SCUBA-2, with a 10,000-pixel bolometer camera cooled to 0.1 K and operating at the same 0.45 and 0.85 mm atmospheric windows as SCUBA (Holland et al. 2013). SCUBA-2 was able to map the sky about 100 times faster than its SCUBA predecessor.

SEST and APEX   When Roy Booth became Director of the Onsala Space Observatory in Sweden, he combined forces with IRAM millimeter wave astronomers and Peter Shaver at ESO to build the Sweden ESO Submillimetre Telescope (SEST). SEST was an open air 15 meter diameter telescope located at the ESO Observatory on La Silla in northern Chile at an altitude of 7,550 feet. The Cassegrain telescope was similar to the IRAM interferometer antennas, but was mounted on a fixed base, and was designed and built by IRAM in collaboration with French and German industrial partners (Booth et al. 1989). The antenna had a surface accuracy of only 0.07 mm rms and pointing accuracy of 3 arcsec. The construction and operating costs were shared equally by ESO and Onsala. Onsala was responsible for the technical operation and provision of the receivers and other instrumentation, while the operation on La Silla was managed by ESO along with their optical telescopes on the mountain. Starting in 1988, the SEST telescope was used primarily in the 1.3, 2.6, and 3.5 mm bands for spectroscopic observations of extragalactic interstellar molecules, especially CO, as well as for continuum observations of quasars. Observing time was shared equally between Swedish astronomers and ESO's European user community. The operation of SEST was ESO's first involvement with millimeter astronomy, and opened the door for ESO's later participation as a partner in ALMA. SEST was closed in 2003 when it was superseded by the Atacama Pathfinder Experimental Telescope (APEX).

APEX is a 12 meter diameter modified North American ALMA (Sect. 10.7) prototype antenna located at 16,500 feet altitude on the Chilean Atacama desert on the site of the ALMA telescope (Güsten et al. 2006). With its more precise surface of 0.017 mm rms and high altitude location, it replaced SEST, and operates primarily in the wavelength range between 0.2 and 1.5 mm, or between the radio and infrared parts of the electromagnetic spectrum. APEX was built as joint collaboration of the MPIfR, the Onsala Observatory, and ESO and, like SEST, is operated by ESO. NRAO was invited to join APEX, but declined due to the need to concentrate its limited resources on the MMA. Due to the high altitude location, the antenna is routinely operated from San Pedro de Atacama via a radio link. A particularly notable feature of APEX is its 295-element 345 GHz liquid helium cooled bolometer array known as LABOCA (Large Apex Bolometer Camera) (Siringo et al. 2009). LABOCA is the latest in a series of bolometer cameras developed by the MPIfR radio astronomer Ernst Kreysa, and has been used primarily to investigate star formation in the Milky Way Galaxy and in nearby galaxies.

The Institut de Radio Astronomie Millimétrique (IRAM) As early as the mid-1960s, Emile Blum began in France to develop plans for a millimeter wavelength interferometer (Encrenaz et al. 2011). During his 1967 visit to NRAO, he met Peter Mezger, then still on the NRAO scientific staff, but about to leave to become a Director at the MPIfR, where he would be in charge of the new Effelsberg 100 meter telescope. Perhaps based on his early exposure to the embryonic attempts in Green Bank by Frank Drake and Frank Low to experiment with millimeter wavelength astronomy in the early 1960s, Mezger had a long-time ambition to build a precise radio telescope to work at short millimeter wavelengths. With the support of the French Centre National de la Recherche Scientifique (CNRS) Director, Bernard Gregory, and the MPG President Reimar Lüst, Mezger, and Blum respectively, developed plans for a 30 meter antenna and a multi-element interferometer as parts of a joint observatory known as SAGMA.[47] The MPIfR group found an attractive site on Pico Veleta in the Spanish Sierra Nevada for their 30 meter antenna, while Blum and colleagues located a flat site on the Plateau du Bure in the French Alps suitable for an interferometer. But the Plateau de Bure site was only at an altitude of 2,550 meters, 300 meters lower than the proposed 30 meter site on Pico Veleta, and, more important, was further north, limiting access to the Galactic center.

Mezger was unyielding, arguing that the Plateau de Bure site was unacceptable for the 30 meter telescope, while Blum was equally firm that Pico Veleta could not accommodate an interferometer, especially if the baseline were to be expanded. Meanwhile, Gregory and Lüst were adamant that there would be no funding from CNRS or the MPG unless the two instruments were built as part of a joint French-German project. The issue was not so much a matter of saving money by a joint project, but a strong desire on the part of both CNRS and the MPG Max Planck Gesselschaft (MPG) in this post-World War II era to establish firm evidence for French-German collaboration. A radio astronomy project was perceived to be more straightforward than, for example, an agreement on agricultural subsidies. But Blum and especially Mezger were obstinate and held firm to their positions. Mauna Kea, on the big Island of Hawaii, was mutually acceptable to both Mezger and Blum, but was considered logistically unreasonable unless NRAO joined the project to provide local support and if significant funding came from the US.[48] The NRAO staff debated the idea of joining SAGMA on Hawaii and concluded that it would compromise its own plans for the 25 meter millimeter wave antenna (Sect. 10.3), and rejected the European proposal.[49]

Faced with an impasse, in early 1977 Gregory and Lüst convened an international committee of three so-called "wise-men" to adjudicate the siting issue. One of the present authors (KIK) served on the committee, along with Bernard Burke from MIT and Paul Wild from CSIRO in Australia. After meeting with MPG and CNRS and visiting the proposed sites, the committee met at the Paris CNRS headquarters to prepare their report and to deliver it to Gregory. Recognizing that the Plateau de Bure was by far the better of the two sites to locate the interferometer even though the latitude was higher than ideal, and that the Pico Veleta was by far the better of the two sites to locate the 30 m telescope, the committee noted that it would be inappropriate to favor one site over the other and recommended that the common center of the cooperative program should be located in an observatory headquarters in Grenoble, France. Recognizing the need to ease tensions and maintain the delicate balance between the French and German interests, the committee refrained from suggesting that the Director of the new joint observatory be German and that the chef be French. Following the Paris meeting, Burke and Kellermann traded in their first class plane tickets plus $50 each to purchase tickets for an unforgettable flight to Washington on the Air France Concorde.

The MPG Max Planck Gesselschaft (MPG) and CNRS accepted the recommendation, which led to the formation of IRAM, with headquarters in Grenoble, the three- (later expanded to six-) element interferometer on the Plateau de Bure and the 30 meter telescope on Pico Veleta (Baars et al. 1987). Peter de Jonge became the first director of IRAM, and established IRAM as a more independent and self-standing organization than either Mezger or Blum anticipated or found comfortable. In 1990 Spain became a full member of IRAM. Unlike in the US, where the NSF was not able to provide operating funds for CARMA at the same time as ALMA, IRAM, starting in 2014, constructed four more antennas with a goal of reaching a total of 12 by 2020 as part of the NOrthern Extended Millimeter Array (NOEMA), thus providing a powerful northern hemisphere complement to ALMA.

Both the 30 meter telescope and the Plateau de Bure interferometer operate up to 350 GHz (0.85 mm). The 30 meter is equipped with both multifeed spectrometer and bolometer cameras and, due to careful thermal control, operates well even in the daytime. A fatal accident with the cable car to the Plateau in 1999 killed 20 people, limiting access to the plateau to foot or helicopter, but was followed six months later by a helicopter crash which took the lives of another five people. Since then the rebuilt lift has been used only for transporting equipment, while IRAM staff and observers use a newly built road.

Japanese Millimeter Astronomy Under the leadership of Masaki Morimoto, Japanese radio astronomers built two world-class facilities at their Nobeyama Observatory, 150 km from Tokyo. The 45 meter dish is used at wavelengths down to 2.6 mm, and until the completion of the GBT (Sect. 9.7) was the largest telescope in the world operating at short millimeter wavelengths. A broad band 16,000 channel acoustical optical spectrograph was the heart of the 45 meter spectroscopic system. The millimeter interferometer, which contained six 10 meter diameter movable dishes operating between 1.2 mm and 1.3 cm, was closed for astronomical observing in 2007 as Japan devoted its resources to ALMA. The performance of both the 45 meter dish and the interferometer was limited by the modest 1,350 foot elevation and correspondingly high water vapor content.

As a prototype for their Large Millimeter and Submillimeter Array (Sect. 10.7), Japanese radio astronomers have also built the precision 10 meter diameter Atacama Submillimeter Telescope Experiment (ASTE) located near the ALMA site on the Atacama Desert in northern Chile. With its 0.02 mm rms precision surface and excellent site, ASTE is used (Kohno 2005) at frequencies up to 850 GHz (0.35 mm). Until it was moved to the LMT, ASTE used the FCRAO 144-element AzTEC bolometer for 1.1 mm continuum observations.

10.6   THE NRAO MILLIMETER ARRAY (MMA)

The millimeter wavelength facilities described in Sect. 10.4 brought new life to the US university radio astronomy programs, but the developing ambitions in Europe and Japan threatened US leadership in millimeter and submillimeter astronomy. Although the NRAO 36 Foot dish on Kitt Peak may have opened the field of millimeter astronomy, even after the 12 Meter upgrade it was no longer competitive with many of the other emerging millimeter wave facilities, which were larger, worked to shorter wavelengths, and were located on better sites. With the demise of the 25 meter project, NRAO was no longer a major player in this rapidly developing and promising field of millimeter astronomy.

The 1982 Astronomy Survey Committee (Field 1982) had assumed that the 25 meter telescope would be built, so they did not make any recommendations for any other major millimeter facility. Thus, following the April 1982 NSF Astronomy Advisory Committee decision to abandon the 25 meter telescope, there were no US plans to exploit this rapidly growing area of astronomy that had been pioneered in the US. The US millimeter astronomy community was not happy with NRAO's leadership, or perceived lack thereof, in selling the 25 meter to the NSF or to the broader astronomical community. Although many millimeter wave astronomers had gotten their start as a result of NRAO's pioneering efforts in millimeter wave astronomy, they now held NRAO responsible for the fall of the 25 meter telescope, in part due to what some felt was a stubborn insistence on sticking to the expensive Mauna Kea site, and in part for apparently abandoning the millimeter wave telescope in favor of the VLBA.

The Barrett Report In order to develop a strategy for moving forward after the collapse of the 25 meter project, Robert (Bob) Wilson (Bell Labs), Phil Solomon (Stony Brook), and Lewis Snyder (Illinois) convened a small meeting at the Crawford, New Jersey offices of Bell Laboratories on 28–29 October 1982 "to discuss future U.S. national instruments for mm-wave astronomy."[50] No one from NRAO was invited. The meeting participants acknowledged that the 25 meter telescope "would have been a world leading instrument when first proposed," but in view of "similar large instruments being built overseas in Europe and Japan, the time for the 25 meter telescope had passed."[51] The eighteen participants all signed a strong letter to NSF AST Director Pat Bautz and NSF Assistant Director for AAEO Frank Johnson presenting the case for building a "millimeter wave aperture synthesis instrument" based on a scaled down VLA and consisting of about 30 roughly 6 meter-sized antennas with a maximum baseline less than 3 km.[52] There was no mention in the letter of who should build the array or where it should be located.

    Previously unaware of the Bell Labs meeting, and also concerned about the future of US millimeter astronomy, Bautz "convened a Subcommittee of the NSF Astronomy Advisory Committee [chaired by MIT's Alan Barrett] to advise on the future needs of millimeter and of submillimeter wavelength astronomy."[53] Barrett called an open meeting of the Subcommittee at the NSF on 3 December 1982. The Subcommittee heard reviews of existing mm wavelength interferometry and single dish facilities, as well as possibilities for future developments. The Bell Labs and NSF groups agreed to work together and met again at Bell Labs on 9–10 February 1983, primarily to review and formulate the scientific case for millimeter interferometery.[54]

    The April 1983 report of the NSF Subcommittee, which became known as the "Barrett Report," recognized the advanced millimeter wave facilities already operating at IRAM and Nobeyama, and noted that the more modest interferometers at Caltech and Hat Creek were "unsuitable for general visitor use by a large segment of the mm-wave astronomers."[55] The first recommendation of the committee was the initiation of a design study of a millimeter wavelength aperture synthesis array with a minimum useable wavelength of 1 mm, an angular resolution of 1 arcsec or better at a wavelength of 2.6 mm, and a total geometric collecting area of 1,000–2,000 square meters. The NSF Subcommittee also did not specify who should build the array, but noted, "A project of this magnitude would be a national facility," and that, "It may well be that such an instrument would be situated with the present VLA in New Mexico in order to take advantage of the great expertise of the VLA staff."

Planning for the Millimeter Array The first serious discussions about building a millimeter array at NRAO took place at an internal workshop on future instrumentation held in Green Bank in October 1982. As input to the workshop, Frazer Owen prepared a memo calling attention to the millimeter wave dishes and arrays around the world "in the late planning or the construction stage," arguing that "the single dishes being planned seem likely to supersede the capabilities of the NRAO 12 meter fairly quickly," and that the time had come for NRAO to take the initiative.[56] Owen argued that the infrastructure already available at the VLA and the moderately high and fairly flat VLA site made it an ideal location for millimeter interferometry. He also pointed out that, in addition to the obvious drivers for spectroscopic imaging, interferometers were more effective than large single dish telescopes in suppressing the effects of ground and tropospheric emissions. Mort Roberts was impressed by Owen's presentation, and after the workshop asked Owen to form a small internal committee to review the scientific justification for millimeter interferometry,[57] but Owen was more concerned about the technical challenges of the array configuration.[58]

Encouraged by the Barrett report, Roberts formed a series of technical review committees to examine the configuration, siting, and antenna structures for a millimeter array. Potential sites in Antarctica, Arizona, Chile, Colorado, Hawaii, and Utah, were studied, along with the existing Owens Valley, Hat Creek, and VLA sites, as well as the nearby South Baldy site in New Mexico's Magdalena Mountains at 10,600 feet. A regular Millimeter Array Newsletter was issued, with Frazer Owen as Editor, and separate Millimeter Array technical and scientific memo series were begun.[59] NRAO Scientist Edward (Ed) Fomalont spent six months at the Nobeyama Observatory to implement AIPS on their interferometer and also to bring back to NRAO experience learned from working with the Nobeyama millimeter array. Following traditional NRAO procedure, a Millimeter Array Technical Advisory Committee, chaired by Bob Wilson, was established to solidify support from the university community.

Millimeter Array design work continued throughout the 1980s, with NRAO Associate Director Bob Brown as MMA Project Director, and included site testing in New Mexico, Arizona, and Hawaii. During this period, NRAO held a series of scientific and technical workshops to address a variety of technical issues and to tighten the scientific case for a Millimeter Array.[60] Interestingly, when asked about South America as a potential site, the MMA Advisory Committee responded, "This is not an attractive idea."[61] However, Mark Gordon expressed concern that in view of planned expansions of existing millimeter arrays in Japan, at IRAM, Caltech, and Hat Creek, the proposed NRAO Millimeter Array would not be sufficiently unique, and would be more attractive if located in Chile close to the CTIO facilities near La Serena.[62] NRAO staff also met with a group from the Smithsonian Institute to discuss possible collaboration between the MMA and the SMA. Joint Working Groups dealing with science, antennas, site selection, receivers, and management were formed. However, other than NRAO support for testing the Mauna Kea SMA site, the proposed collaboration did not materialize.

In January 1988, NRAO issued a two volume MMA Design Study. Volume I, *Science with a Millimeter Array* (Wootten and Schwab 1988), contained the Proceedings of the Green Bank Workshop held in October 1985 to define the scientific goals which a millimeter array might address. Volume II, *MMA Design Study* (Brown and Schwab 1988) discussed the design principle for a forty-element array using 7.5 meter antennas, instrumentation, and computing requirements. The estimated construction cost, including a 20% contingency, was $66 million.

Following another workshop, held in Socorro from 15 to 18 January 1989, to assess the scientific progress in millimeter wave astronomy, and after six years of planning, design, and prototyping, in July 1990, AUI/NRAO finally submitted a proposal to the NSF for the construction of a Millimeter Array (Brown 1990). The proposed MMA consisted of 40 transportable dishes, each 8 meters in diameter, to give a total collecting area of about 2,000 m^2 (Fig. 10.7). The planned frequency bands were 30–50 GHz, 70–115 GHz, 120–170 GHz, and 200–350 GHz. At its highest frequency of 350 GHz (0.85 mm) and in the largest 3 km diameter configuration the resolution was 0.06 arcsec. No site was specified, but the proposal reviewed the search for a suitable high dry site sufficiently large to hold the 3 km sized array. The attraction of a site in Chile was

Fig. 10.7 Artist's conception of the Millimeter Array, with 40 transportable 8 meter dishes. Credit: NRAO/AUI/NSF

discussed but, due to the much greater construction and operating costs that would be involved, it was dismissed in favor of sites in Arizona and one in New Mexico, close to the VLA.

By this time, the anticipated MMA construction price had risen to $120 million, including 15% contingency. Annual operating costs were estimated as $6.5 million. However, there were still many unanswered questions. The NRAO's MMA proposal to NSF was reviewed by 20 US and foreign scientists, who recommended that NRAO proceed with the MMA, but raised concerns about the site selection process and the estimated costs of construction and operation. Several reviewers noted that NRAO had no experience in millimeter interferometry.[63] In April 1991 the NSF brought a committee to Socorro to assess the project. The proposal reviews, the site visit, and the long range planning committee of the Advisory Committee for the NSF Division of Astronomical Sciences (ACAST) all "overwhelmingly endorsed the NRAO Millimeter Array,"[64] but ACAST raised concerns about where the operating funds would come from.[65]

The Bahcall Committee Responding to the growing threat from IRAM, the MMA also received the important blessing of the 1990 Decade Review of Astronomy (Bahcall 1991) in order to "recapture the once dominant position of the United States in millimeter astronomy."[66] There were no other "large" radio astronomy proposals competing with the MMA, so, unlike the earlier bitter battles over the VLA and VLBA which occurred in the 1970s and 1980s Decade Reviews, the 1990s Radio Panel quickly reached a consensus to recommend "as the highest priority for new construction a Millimeter Wave Array with sub-arcsecond resolution, comparable to that of the VLA, and having good image quality, a sensitivity adequate to study faint continuum and line emission, and a flexible spectroscopic capability in all of the millimeter wavelength windows between 30 GHz and 350 GHz." (Kellermann 1991, p. I-9).

In the parent Survey Committee, the MMA faced competition from the two 8 meter optical telescopes recommended by the OIR Panel, one located on Mauna Kea, optimized for infrared astronomy, and the other to be built in the Southern Hemisphere, optimized for optical and near ultraviolet wavelengths. Although there was a broad consensus that after the VLA, VLBA, and then the GBT, it was time to support other wavelengths, after vigorous debate within the Committee, the MMA was still given second priority, following the Mauna Kea infrared telescope, but far ahead of the Southern Hemisphere telescope (Bahcall 1991, p. 11). However, even during the Committee deliberations, before any decisions had been reached, NSF Director Erich Bloch reported that he had negotiated a deal with the UK Science and Technologies Facilities Council (STFC) to jointly build both of the 8 meter telescopes with the US and the UK each paying for half the cost. John Bahcall, the Survey Committee Chair, was incensed and argued that since the Committee had not yet reached any conclusions about the relative priorities of the various projects under consideration by the Committee, such an agreement was premature. But Bloch, not to be intimidated, retorted that if he had to wait for committees to decide anything, nothing would ever get built.[67]

NSF Approval In response to the issues raised by the reviewers, NRAO submitted a new proposal for a "Millimeter Array Design and Development Plan," requesting $22.3 million over three years to continue site evaluation, to provide final engineering design for the antennas and instrumentation, and for algorithm development.[68] As usual, things moved slowly in Washington, and it was not until November 1994, at the request of NSF Director Neal Lane, that the National Science Board (NSB) approved a project development plan for the MMA. In May 1995, the NSB authorized the expenditure of $26 million for a three-year MMA design and development program. To jump start the development program, NSF AST Director Hugh Van Horn added $1 million to the NRAO 1995 budget from AST funds to begin site studies and further planning. The three-year MMA Design and Development Program began in 1996 and included a prototype antenna, configuration studies, SIS mixer, and HFET amplifier design. By this time, the antenna concept for the MMA had evolved from a conventional on axis design to an offset configuration with an unblocked aperture constructed of carbon fiber instead of steel to reduce the effects of thermal deformations.[69] An additional requirement, which was to lead to increased cost, was the need to be able to quickly slew the antennas between the region under study and a nearby reference source.

As with the 25 meter telescope, community support was ambivalent, particularly from Caltech, Berkeley, Illinois, and Maryland, who saw the MMA as not only an exciting scientific opportunity, but also as a threat to their own ambitions. Recognizing the concern about the lack of millimeter interferometer experience at NRAO, and the opportunity to better engage the university community, Vanden Bout invited Caltech and BIMA radio astronomers to join an MMA Development Consortium (MDC). This was perhaps NRAO's first use of an embedded acronym. Meanwhile, in 1998 the NSF established their own MMA Oversight Committee (MMAOC) to provide further advice and oversight of the MMA project. NRAO scientists and engineers were spending more time writing reports and attending meetings than they were in designing the MMA, but it would get worse.

Just like the previous radio telescope projects we have discussed, a particularly challenging and controversial aspect of the MMA was choosing a site. With the growing interest in sub-millimeter wavelengths, the primary criterion was for a high dry site, but like the VLA, the MMA required a large flat area within which the antennas could be moved to different array configurations. The most desirable locations appeared to be in the Atacama Desert in northern Chile, which seemingly offered unmatched opportunity for low water vapor content, large flat areas, and unrivaled views of the southern sky. In fact, it was claimed that the Atacama Desert had the lowest precipitable water vapor of anywhere in the world. Unsubstantiated (and untrue) stories circulated that the proposed site received only 1 cm of precipitation each century.

NRAO recognized that a Chilean site would come with many practical logistical challenges and at a greater cost of construction, and especially operation. However, following an exhaustive study by Mark Gordon, and as the MMA development progressed, there was increasing interest in locating the MMA in Chile. NRAO needed to convince the NSF that Chile was worth the additional cost as well as the added administrative burden involved in spending federal funds in another country. In 1994, Paul Vanden Bout escorted NSF AST Director Hugh Van Horn and MPS Assistant Director William Harris on a visit to potential sites in Chile in the hope that they would be sufficiently impressed to ignore the negatives (Fig. 10.8). Apparently they were, but working in Chile turned out to be more expensive and more difficult than anyone anticipated.

Fig. 10.8 NSF MPS Assistant Director William Harris, NSF AST Director Hugh Van Horn, and NRAO Director Paul Vanden Bout standing on level ground at the 16,500 foot MMA—later ALMA—site, with 20,000 foot mountains rising in the background. Credit: NRAO/AUI/NSF

10.7   THE ATACAMA LARGE MILLIMETER/SUBMILLIMETER ARRAY (ALMA)

Although strongly endorsed by every review committee and enthusiastically supported by the NSF Astronomy Division, the MMA first needed the additional blessing of the National Science Board before it could be considered by the Administration or by Congress for construction funding. Largely as a result of the Congressional initiative to fund the GBT, the NSF had been able to establish the new Major Research Equipment (MRE) funding line to fund the construction of large new projects without impacting the operation of ongoing programs or grants to Individual Investigators.[70] However, by the time the MMA Development Plan was presented to the National Science Board in 1994, "the Federal funding landscape [had] changed substantially. In particular, Congress [had] made it increasingly clear that the viability of projects as large as the MMA may depend on the extent to which they are based on international partnerships."[71] International projects presented both opportunities and challenges. Vanden Bout declared at the start that NRAO was not interested in establishing an international partnership for the MMA just to save money or for the sake of satisfying the perceived Congressional wishes, but would do so only if a joint international program were to result in a more powerful capability. Conveniently, at the same time that NRAO was planning for the MMA, both European and Japanese scientists were developing their own plans for millimeter and submillimeter wave arrays. Both projects, as well as the MMA, were looking at potential sites in the Atacama Desert in northern Chile.

Japanese radio astronomers were discussing a Large Millimeter and Submillimeter Array (LSMA) to consist of fifty 10 meter antennas operating in six bands up to 500 GHz or 0.6 mm wavelength (Ishiguro et al. 1994). In order to explore a possible joint effort with Japan, Brown and Vanden Bout met with members of the Nobeyama Observatory and they "agreed on a memorandum to explore the possibility of a collaboration."[72] Initially, the two observatories discussed only the separate construction and operation of the MMA and LMSA on the same site, but perhaps with periodic joint operation of the 90 element array to give increased sensitivity, resolution, and image quality.[73]

There was also some earlier discussion of a Dutch participation in the US MMA.[74] But the potential Dutch collaboration became tied to an additional contribution to CARMA and conflicted with the Dutch aspirations for the 1hT (later called the SKA), and so the prospects for a Dutch collaboration evaporated.[75] Nevertheless, with a then estimated cost of $175 million, NRAO set out in a confidential memo the terms and conditions under which partners who contributed to the capital and operating costs could become MMA Associates with appropriate prorated shares of the observing time. Notably, the long standing NRAO Open Skies policy was being threatened by the statement that "aside from [Associates], U.S. observing time will not be available to non--U.S. observers."[76]

Meanwhile in Europe, IRAM, ESO, Sweden, and the Netherlands were developing their own plans for the Large Southern Array (LSA) to have a collecting area of 10,000 m², about five times that of the MMA, and to work to wavelengths only as short as 3 mm (Downes 1995). Early strawman concepts were for an array of fifty 16 meter dishes or one hundred 11 meter dishes with an estimated cost of 270 million Ecu ($360 million).[77] The European planning program was initially led by IRAM, and then later by ESO under its charismatic strong-minded Director, Riccardo Giacconi.

Collaboration between NRAO and ESO appeared attractive to both sides, and in June 1997, Giacconi (for ESO) and Vanden Bout (for NRAO) signed a resolution agreeing to an LSA/MMA Feasibility Study. Three Joint Working Groups, Science, Technology, and Management, were established to continue the design and planning. At NRAO, the design work continued using the $26 million that had been authorized for the MMA design. Both sides understood the complexities and delays of their ambitious plans that would be introduced by a joint project, and agreed they would insist on a project that was more powerful than either the MMA or the LSA. Ironically, at the end of his term as ESO director in 1999, Giacconi returned to the US where he accepted a position as President of AUI (Sect. 6.8), during which time he oversaw the AUI/NRAO negotiations with ESO to establish the governing structure of the joint facility in Chile.

In early 1999, Bob Brown ran a competition asking for ideas to name the new facility and received 33 suggestions. Following a ballot sent to about 100 individuals to choose among the 33 names, Brown presented the top eight candidates to a 30 March 1999 meeting of NSF, ESO, and PPARC representatives in Garching. The name ALMA appeared only sixth on the list, but according to Brown, Ian Corbett from PPARC, declared "ALMA is the best. I like acronyms I can pronounce." The entire room mumbled in agreement and the committee went on to the next agenda topic.[78] ALMA also means "soul" or "spirit" in Spanish.

In June 1999, just weeks before Giacconi joined AUI, the NSF signed a Memorandum of Understanding with European institutions for a joint "design and development phase of a large aperture mm/sub-mm array to be known as the Atacama Large Millimeter Array (ALMA)." ALMA joined the MMA (40 eight meter dishes working to 1.3 mm) with the European Large Southern Array (50 sixteen meter dishes working to 3 mm) to build an array containing 64 twelve meter diameter antennas with an angular resolution up to 0.005 arcsec at the shortest operating wavelength of 0.35 mm.[79] On the European side, the MOU was signed by Giacconi for ESO, CNRS, the MPG, the Netherlands Foundation for Research in Astronomy (NFRA), and the British Particle Physics and Astronomy Research Council (PPARC). But not everyone was enthusiastic. IRAM in Europe and CARMA in the US were building powerful millimeter wave arrays of their own that they knew would be threatened by the proposed plan for ALMA.

The first construction funding in the US for the joint ALMA project was approved by Congress in November 2001. Initially an ALMA Executive Committee (AEC), with Bob Brown as chair, was established to coordinate the ESO and NRAO activities. But establishing ALMA as an international project was not straightforward. ESO, NRAO, and later Japan, all came to the table with different goals. The ESO LSA emphasized a large collecting area to enable extragalactic spectroscopy, while the NRAO MMA stressed image quality, and the Japanese LSMA highlighted submillimeter wavelengths. Agreements among the five European partners, on one hand, and among the North American partners, US and Canada, on the other hand, and between the US and Europe were followed by separate agreements between the ALMA partners and the government of Chile and Consejo Nacional de Ciencia y Technologia (CONACyT). As part of the NAPRA (North American Program in Radio Astronomy) agreement, Canada agreed to work with the NSF in funding and supporting ALMA. But NRAO/AUI and the NSF first had to establish their own rules of engagement.

For various reasons, including the ongoing funding of the Japanese 8.2 meter Subaru optical telescope, Japan was unwilling or unable to enter into a firm agreement to participate in a combined telescope on the same time scale as Europe and the US. Japan, Europe, the US, and Canada agreed on a resolution expressing the intent to jointly construct ALMA and starting as early as 1999, a US-European-Japanese ALMA Liaison Group had met regularly to exchange technical progress and to establish the foundations for an "Enhanced ALMA." Bob Brown and Peter Napier represented NRAO at these discussions.[80] Japan did not formally join the ALMA project until funds became available in 2004 when Japan proposed to build a "Compact Array, a new correlator and new receivers, as well as to contribute to the infrastructure and operation."[81] By this time, the NSF/AUI/NRAO and ESO had already agreed on the legal structure and the basic parameters of the Array.

ESO and the NSF had different legal status with the government of Chile for the operation of ESO and CTIO respectively, and these agreements had to be respected with the joint ALMA project. In 1998, NRAO/AUI hired Eduardo Hardy, a native of Argentina, as the AUI representative in Chile, and in 2006, Hardy became the NRAO Assistant Director for Chilean Affairs. The 2006 Management Agreement for the construction and early operation of ALMA was a bilateral agreement between ESO and AUI acting as the NSF Executive. The management of the ALMA project was first overseen by an ALMA Coordinating Committee (ACC), and since 2017 by the ALMA Board, which includes representatives from the NSF, NRAO/AUI, the Canadian NRC, ESO, NOAJ, Chile, and ASIAA (Taiwan) as well as at-large members from Europe and Japan. A Joint ALMA Office (JAO) was established in Chile and managed by the ALMA Director to oversee the construction, commissioning, and operation of ALMA. Within Europe, the partners had their own European Coordinating Committee.

Due to the high 5,000 meter altitude of the ALMA site, it was clear that, to the extent possible, supporting activities should take place a lower altitude. An Operations Support Facility (OSF) was established not far from the town of San Pedro de Atacama at 9,500 feet elevation. Ground-breaking for the OSF took place in November 2003 (Fig. 10.9), and ground-breaking at the 16,500 feet Array Operations Site occurred in October 2005.

Paul Vanden Bout became the first ALMA Director and stepped down from his role as NRAO Director to concentrate on building ALMA. He served as ALMA Director between June 2002 and March 2003, during which time he led the negotiations between ALMA and the Chilean partners (Chilean Government and CONICYT). Massimo Tarenghi from ESO was the ALMA Project Scientist, and he succeeded Vanden Bout as ALMA Director in 2003. Tarenghi was followed by Thijs de Graauw from the Netherlands, Pierre Cox from France, and Sean Dougherty from Canada. In 2004, Anthony (Tony) Beasley returned to NRAO as the ALMA Project Manager. Beasley had previously been an Assistant Director at NRAO, after which he went to California to be Project Manager for CARMA. As the ALMA Project Manager, he inher-

Fig. 10.9 Groundbreaking for the ALMA Operations Support Facility. NSF's Bob Dickman is pouring Chilean wine in a tribute to the earth goddess Pachamama. Standing from left to right are Eduardo Hardy (AUI), Fred Lo (NRAO Director), Massimo Tarenghi (ESO), Catherine Cesarsky (ESO Director General) and Daniel Hofstadt (ESO). Credit: I. Dickman/NRAO/AUI/NSF

ited a project that was headed for a major cost overrun due to the unforeseen large cost of the antenna elements, the unanticipated complexity and cost of the international partnership, the unappreciated cost of building on the remote and challenging site, and unfavorable changes in the value of the Chilean peso. Following an agreement to "re-baseline," in December 2004, ALMA was downsized to 50 antennas, 25 each to be provided by NRAO/AUI and by ESO, and the number of frequency bands was reduced. Some of these were later restored when Japan joined the project.

Building a state of the art scientific instrument at this altitude was a challenge. At 16,500 feet elevation, the air density is only about half of that at sea level. Aside from the well-known impact to human performance, many electronic components do not function properly at this elevation. The ALMA correlator is among the faster supercomputers in the world, operating at about 2×10^{13} operations per seconds. Like all large computing systems, it needs to be cooled, but at 16,500 feet it takes twice as much cooling as it would at sea level. Ordinary computer disk drives do not work at the ALMA site, so solid state disks are used on all computers. Many other electronic components, such as electrolytic capacitors, are not rated for these altitudes. Another continuing problem is so called, "Single Event Upsets" (SEUs) which are the random flipping of a bit (zero to one, or one to zero) when a chip is hit by a cosmic ray particle. These SEUs are more common at ALMA than at sea level.

It was clear that the biggest challenge, and certainly the biggest technical risk facing ALMA, was in meeting the exacting specifications for the construction of the high precision antenna elements. In order to obtain a competitive design and cost, AUI/NRAO and ESO agreed to procure two separate prototypes for evaluation. ESO contracted with the French Alcatel-EIE consortium for their prototype, while AUI/NRAO chose the California based Vertex Antenna Systems, LLC for the design and construction of their prototype. Alcatel-EIE was the result of a complex series of mergers including the US Lucent Technologies, the successor of AT&T Bell Laboratories. Later Alcatel-Lucent became part of the Finnish Nokia Networks, and more recently was sold to a Chinese consortium. Vertex Antenna Systems was formed from mergers including TIW (Toronto Iron Works) and RSI, which had been involved in earlier NRAO antenna projects. The design and much of the fabrication of the AUI/NRAO prototype antenna actually came from the German based Vertex Antennentechnik subsidiary of Vertex Antenna Systems which had its origins in the Krupp group that was involved in building the Effelsberg, Pico Veleta, and HHT telescopes.

The two prototype 12 meter diameter antennas were erected at the VLA site where their performance was evaluated by a joint NRAO/ESO Antenna Evaluation Group (AEG), led by Jeff Mangum from NRAO. Although, at the time, Japan had not yet formally joined the ALMA project, a third prototype was built by the Japanese Mitsubishi Electric Company and was also erected at the VLA site, but was independently evaluated by a Japanese team. Owing to the late delivery of both the Alcatel and Vertex antennas to the VLA site, the evaluation (Mangum et al. 2006), especially for the Alcatel antenna, was not fully complete by the time the two partners had agreed to try to select a single contractor for the production antennas.[82]

Nevertheless, NRAO/AUI and ESO each issued a separate Request for Proposals, anticipating that the evaluation of the two prototypes would be completed in time to make a coordinated decision on the contractor. Both requests had common performance specifications but different business terms and considerations which were necessitated by their respective procurement policies. NRAO/AUI received bids from both Vertex RSI, which was later acquired by General Dynamics during the procurement process, and from Alcatel, which now included the German MAN, the former partner of Krupp in building the Effelesberg antenna. ESO received bids from Alcatel, the German based Vertex Antennentechnik as well as from the Italian contractor, Alenia Aerospace. Based on cost and performance, NRAO/AUI chose General Dynamics. In order to meet the July 2005 deadline before both of the General Dynamics pricings expired, NRAO/AUI signed a contract for $169 million for the construction of 25 antennas, fully anticipating that ESO would contract with Vertex for the other 25 antennas.[83] Previously, ESO had selected Vertex Antennentechnik, even before the NRAO/AUI/NSF decision to choose Vertex RSI (General Dynamics), but a last minute revised bid from the newly reorganized European-led Alcatel was lower than the Vertex bid, and ESO signed a separate contract with Alcatel for the other 25 ALMA antennas. Although the engineering design of the NRAO/AUI production antennas was led by Vertex Antennentechnik in Germany, fabrication actually took place in many countries. The project was managed by General Dynamics C4 Systems in Texas, where each antenna was first assembled and tested, then broken into sub-assemblies for shipment by boat from Houston to Chile.

Neither ESO nor NRAO/AUI management, scientists, and engineers were pleased with the separate contracts which involved different designs including different drive systems and a different sub-reflector support structure. Each side blamed the other for ending up with two different antennas. ESO considered that NRAO/AUI had acted prematurely in signing a contract with General Dynamics before the prototype evaluation was fully complete, while NRAO/AUI suspected that Alcatel had lowered their price below the Vertex price to make their bid more attractive to ESO. One can only speculate whether or not the Vertex bid was leaked to Alcatel, allowing them to undercut the Vertex price.

As it has turned out, many of the concerns about the operation and performance of two different antenna structures were unfounded, although the Vertex and Alcatel antennas do require different maintenance procedures. To complicate the situation, when Japan formally entered the project, they contracted with the Japanese Mitsubishi Electric Company to build four more 12 meter diameter antennas, as well as the twelve 7 meter antennas for the so-called "Compact Array" to provide critical short baselines. The Mitsubishi 12 meter dishes are yet a different design from either the ESO or NRAO/AUI 12 meter antennas. Although all of the Japanese antennas are primarily intended for use in the separate "Compact Array," in principle they can be used together with the AUI and ESO antennas as part of a single 54 or even 64 element array.

ALMA scientific observations officially started on 30 September 2011, and on 13 March 2013 ALMA was formally inaugurated after nearly three decades of planning, engineering, and construction at NRAO (Fig. 10.10), as well as in Europe and Japan. ALMA operates at wavelengths from 0.32 mm to 1 cm in configurations ranging from 150 meters to 16 km. By agreement, scientists from North America (US, Canada, and Taiwan) and the 14 ESO member states each get 33.75 percent of the observing time; East Asia (Japan, Korea and Taiwan) 22.5 percent; and Chile 10 percent. Taiwan participates in ALMA, not only through the Japanese East Asia group, but is also part of the North American group along with the US and Canada.

Interestingly, ALMA, which was the most expensive ground based telescope facility ever built, was itself never proposed to the NSF or reviewed in competition with other facilities by a US Decade Review Committee. Rather, it was only the more modest MMA that was recommended by the Bahcall (1991) committee at a projected construction cost of $115 million. The final cost of constructing ALMA was about $1.4 billion, with ESO and the NSF each paying 37.5 percent and Japan the other 25 percent.[84] The official cost to the NSF was $499 million.

Fig. 10.10 The completed Atacama Large Millimeter-Submillimeter Array (ALMA) shown in a compact configuration. Credit: NRAO/AUI/NSF

The operation of ALMA is perhaps unique among astronomical observatories. Instead of proposing for a specific amount of observing time, ALMA users propose to achieve a certain sensitivity, resolution, image quality, etc., and the ALMA staff determines the appropriate amount of observing time and the antenna configuration needed to meet the observer's requirements. In this way, observers (if one can still use that name) are not adversely impacted by bad weather or instrumental failures traditional to conventional telescope scheduling. Another innovative aspect of ALMA is that instead of raw data, the observers are given essentially science ready data products. ALMA Regional Centers (ARCs) were established to handle proposal review, scheduling, data reduction, and analysis, as well as archive support. The North American ALMA Science Center (NAASC), located at the NRAO in Charlottesville, is the North American ALMA Regional Center (ARC). In Europe, ALMA is supported by a central node at ESO in Garching, Germany, as well as eight ALMA regional nodes and centers of expertise. Other ARCs are located in Chile, Japan, and Taiwan.

Notes

1. Millimeter waves are absorbed by oxygen and especially water vapor in the lower atmosphere. The amount of absorption is determined by what is called the precipitable water vapor (PWV) content, which decreases with altitude and with decreasing temperature.
2. FDD, Report of Visit to Texas Instruments, Inc., 2 February 1962, NAA-NRAO, Tucson Operations, 36 Foot Telescope, Box 2.
3. FDD to F. Low, 27 March 1962, NAA-NRAO, Tucson Operations, 36 Foot Telescope, Box 2; KIK interview with FDD, 14 September 2010, NAA-KIK, Oral Interviews. https://science.nrao.edu/about/publications/open-skies#section-10
4. J. Pawsey, Notes of Future Program at Green Bank, 17 July 1962, NAA-NRAO, Founding and Organization, Antenna Planning. https://science.nrao.edu/about/publications/open-skies#section-10
5. FDD to file, 7 February 1963, NAA-NRAO, Tucson Operations, 36 Foot Telescope, Box 2.
6. AUI-BOTXC, 15 May 1964.
7. KIK interview with FDD, op. cit.; AUI-BOTXC, 22 February 1962.
8. F. Callender (NRAO Business Manager) to F. Lowe [sic], 16 November 1963, NAA-NRAO, Tucson Operations, 36 Foot Telescope, Box 1.
9. As a general guideline the antenna surface accuracy should be at least as good as $\lambda/16$. At this level the antenna gain is reduced by ½ over a perfect surface.
10. AUI-BOTXC, 20 March 1964.
11. The unusually large f/D or focal length/diameter meant that the feed support legs would need to be twice as long as normal to reach the focal point. AUI-BOTXC, 15 October 1964.
12. P. Mezger, Principal Considerations of Radioastronomical Observations at Very High Frequencies, March 1964, NAA-NRAO, Tucson Operations, 36 Foot Telescope, Box 3. https://science.nrao.edu/about/publications/open-skies#section-10

13. DSH to F. Low, 9 March 1965, NAA-NRAO, Tucson Operations, 36 Foot Telescope, Box 2.

14. DSH to F. Low, 1 April 1965, NAA-NRAO, Tucson Operations, 36 Foot Telescope, Box 2.

15. P. Mezger, op.cit.

16. F. Low to eight prospective suppliers, 19 March 1964, NAA-NRAO, Tucson Operations, 36 Foot Telescope, Box 1.

17. NRAO Memorandum summarizing 23 January 1963 meeting between NRAO and Rohr. NAA-NRAO, Green Bank Operations, LSFT, Box 5. Bob Hall was then an antenna design engineer at Rohr. Earlier, while at Blaw Knox he had contributed to the design of the Green Bank 85 Foot antenna and then the 300 Foot transit dish. Later he joined NRAO as project manager for the construction of the GBT.

18. AUI-BOTXC, 15 October 1964.

19. AUI-BOTXC, 20 March 1964.

20. R. Hall to DSH, 26 October 1965, NAA-NRAO, Tucson Operations, 36 Foot Telescope, Box 2.

21. E. Conklin to DSH, HH, and WEH, 31 January 1971, NAA-NRAO, Tucson Operations, 36 Foot Telescope, Box 3.

22. FORTH: an Application-Oriented Language Programmer's Guide, NRAO Computer Division Internal Report No. 11, 1973. http://library.nrao.edu/public/memos/comp/CDIR_11.pdf

23. AUI-BOT, 19–20 October 1967.

24. AUI-BOTXC, 16 November 1967; DSH to T. Glennan, NAA-NRAO, Tucson Operations, 36 Foot Telescope, Box 5.

25. Bolometers are used to make an incoherent measurement of the energy falling on a cooled detector. Carefully designed filters define the bandwidth which is otherwise limited by the atmospheric window. The challenge is to design filters that are low loss within the desired frequency band, but have sufficiently high attenuation outside the window to eliminate responses due to fluctuations in tropospheric water vapor content.

26. M. Simon to WEH, 1 March 1971, NAA-NRAO, Tucson Operations, 36 Foot Telescope, Box 3.

27. K. Jefferts to WEH, 16 November 1972, NAA-NRAO, Tucson Operations, 36 Foot Telescope, Box 4.

28. DSH to M. Kundu, 26 February 1969, NAA-NRAO, Tucson Operations, 36 Foot Telescope, Box 4.

29. J. Payne to DEH, 7 April 1975; J. Payne to HH, 20 October 1979, NAA-NRAO, Tucson Operations, Site and Administration, Box 2.

30. A. Penzias to WEH, 27 February 1969, NAA-NRAO, Tucson Operations, 36 Foot Telescope, Box 4.

31. L. Snyder to DSH, 17 January 1978, NAA-NRAO, Tucson Operations, 36 Foot Telescope, Box 3.

32. KIK to DSH, 12 May 1978, NAA-NRAO, Tucson Operations, 36 Foot Telescope, Box 3.

33. E. Wetmore to E. Conklin, 21 July 1972, NAA-NRAO, Tucson Operations, 36 Foot Telescope, Box 3.

34. B. Turner, The NRAO 25-Meter Telescope: 2nd Status report, NAA-NRAO, Tucson Operations, 25 Meter Telescope, Box 2.

35. Phil Solomon via Marshall Cohen to KIK, 24 March 1981, NAA-KIK, VLBA History and Development.
36. W. Horne to HH, 11 February 1980, NAA-NRAO, Tucson Operations, 25 Meter Telescope, Box 3a.
37. Telegram from MAG to J. Jefferies, 15 November 1979, NAA-NRAO, Tucson Operations, 25 Meter Telescope, Box 3a.
38. P. Strittmatter to G. Huguenin, 7 September 1978, NAA-NRAO, Tucson Operations, 25 Meter, Box 3a. Huguenin was a member of the Astronomy Advisory Committee (as was one of the authors, KIK).
39. Draft report of the Subcommittee on Millimeter Wave Facilities of the Advisory Committee for Astronomical Sciences, NAA-NRAO, Tucson Operations, 25 Meter Telescope, Box 1.
40. L. Snyder to F. Johnson, 3 May 1982, NAA-NRAO, Tucson Operations, 25 Meter Telescope, Box 4.
41. MAG to MSR, 17 April 1980; MAG to M. Balister and JWF, 26 November 1980, NAA-NRAO, Tucson Operations, 36 Foot Telescope, Box 3.
42. JWF, Notes on Improving the 36-Foot Telescope, 24 October 1973, NAA-NRAO, Tucson Operations, 36 Foot Telescope, Box 3.
43. J. Marymor to WEH, 5 January 1980, NAA-NRAO, Directors' Office, NSF Correspondence.
44. NAA-NRAO, Tucson Operations, 36 Foot/12 Meter Telescope.
45. William (Bill) Irvine has written an undated personal account of the history of the FCRAO, NAA-KIK, Open Skies, Exploring the Millimeter Sky.
46. See "The LMT Book" written by W. Irvine for further details. http://www.lmtgtm.org/the-lmt-book/
47. Proposition Commune pour Ondes Observatoire sur Ondes Millimetriques, NAA-NRAO, Millimeter Array, Planning, Box 1. The early history of French radio astronomy leading to IRAM is given by Encrenaz et al. (2011).
48. E. Blum to DSH, 4 April 1975, Tucson Operations, 25 Meter Telescope, Box 2.
49. DSH to file, 21 May 1975, NAA-NRAO, Tucson Operations, 25 Meter Telescope, Box 2.
50. R. Wilson to PVB and others, 29 Oct 1982, NAA-NRAO, MMA, Planning, Box 1.
51. R. Wilson (Bell Labs), P. Solomon (Stony Brook), L. Snyder (Illinois), N. Scoville (Massachusetts), P. Vanden Bout (Texas), W. Welch (Berkeley), B. Ulich (MMT Observatory), F. Lovas (National Bureau of Standards), M. Kutner (Rensselaer), P. Palmer (Chicago), P. Goldsmith (Massachusetts), G. Knapp (Princeton), E. Churchwell (Wisconsin), A. Barrett (MIT), P. Thaddeus (Goddard Inst. for Space Studies), T. Phillips (Caltech), A. Stark (Bell Labs), and J. Bally (Bell Labs) to L.P. Bautz and F. S. Johnson, 29 October 1982, NAA-NRAO, Millimeter Array, Planning, Box 1.
52. Ibid.
53. L.P. Bautz to R. Wilson and Co-Signatories, 23 November 1982, NAA-NRAO, Millimeter Array, Planning, Box 1. Members of the Subcommittee were Alan Barrett, Chair (MIT), D. Downes, C. Lada, P. Palmer, L. Snyder, and W. Welch, with V. Pankonin as NSF Staff Liaison.
54. B. Turner to MSR, 17 February 1983, NAA-NRAO, Millimeter Array, Planning, Box 1.

55. Report of the Subcommittee on Millimeter- and Submillimeter-Wavelength Astronomy, NSF Astronomy Advisory Committee, April 1983, NAA-NRAO, Millimeter Array, Planning, Box 1. http://library.nrao.edu/public/memos/alma/memo009.pdf

56. F. Owen's memo, dated 10 September 1982, later became Millimeter Array Memo No. 1. NAA-NRAO, Millimeter Array, Planning, Box 1. http://library.nrao.edu/public/memos/alma/main/memo001.pdf

57. MSR to F. Owen, 3 March 1983, NAA-NRAO, Millimeter Array, Planning, Box 1.

58. F. Owen to MSR, 16 March 1983, NAA-NRAO, Millimeter Array, Planning, Box 1.

59. NAA-NRAO, Millimeter Array, Planning, Box 1.

60. Wootten and Schwab eds. 1988, *Science with a Millimeter Array*. http://library.nrao.edu/public/collection/02000000000303.pdf; Brown and Schwab eds. 1988, *MMA Design Study*. http://library.nrao.edu/public/collection/02000000000256.pdf

61. Report of the MMA Technical Advisory Committee, 13 April 1984, NAA-NRAO, Millimeter Array, Planning, Box 1.

62. M. Gordon, Millimeter-Wave Array Memo Series No. 25, 1 October 1984, NAA-NRAO, Millimeter Array, Planning, Box 1. http://library.nrao.edu/public/memos/alma/main/memo025.pdf

63. V. Pankonin to PVB, 4 November 1991, NAA-NRAO, Millimeter Array, Planning, Box 2.

64. R. Dickman to RLB and PVB, 30 March 1994, NAA-NRAO, Millimeter Array, Planning, Box 2.

65. M. Rieke (ACAST Chair), 6 January 1992, Resolution on the Millimeter-wave Array, NAA-NRAO, Millimeter Array, Planning, Box 2.

66. As with previous Decade Reviews, NRAO was well represented. K. Kellermann chaired the Radio Panel, which included NRAO staff members R. Fisher, M. Goss, J. Uson, as well as D. Heeschen as Vice Chair. Heeschen and R. Wilson (Bell Labs) were the two radio astronomers on the parent Survey Committee.

67. As Chair of the Radio Panel, KIK participated in most of the Parent Survey Committee meetings. There was little or no support among the Survey Committee members for the Southern Hemisphere 8 meter telescope. Both the Northern and Southern Hemisphere 8 meter telescopes were built as part of the US-UK Gemini project at a cost of approximately $184 million, half of which was born by the NSF. Construction was completed in 1999 and 2000 respectively, more or less simultaneously with construction funding for the GBT (Sect. 9.7).

68. MMA Design and Development Plan, September 1992, NAA-NRAO, Millimeter Array, Planning, Box 2.

69. Carbon fiber reinforced plastic (CFRP) has a very low thermal coefficient of expansion.

70. The NSF MRE budget was restricted to construction funding only.

71. Material presented to the NSB, September 1994, NAA-NRAO, Millimeter Array, Planning, Box 3.

72. MOU between NAOJ and NRAO for Cooperative Studies for the LSMA and the MMA, 12 June 1995; renewed March 1998 and 11 December 1998.

Following the formation of ALMA, the MOU was again renewed on 10 June 1999, and was signed by ESO, AUI, and NOAJ. NAA-RLB, ALMA.

73. R.L. Brown, The Atacama Array: A Possible LMSA-MMA Collaboration in Chile, MMA Management Document No. 3, December 1995. https://library.nrao.edu/public/memos/mma/MMA_MD_03.pdf

74. Proposal for Dutch Participation in a Large Millimeter Array, (undated), NAA-NRAO, Millimeter Array, Planning, Box 2.

75. N. Scoville, P. Vanden Bout, and W. Welch, Confidential memo on CARMA, the MMA, and potential Dutch participation, 31 January 1995. Also H. Butcher to R. Dickman, PVB, and RLB, 17 April 1996, NAA-NRAO, Millimeter Array, Planning, Box 2.

76. Confidential draft Prospectus for Foreign Participation in the Millimeter Array, 9 February 1995, NAA-NRAO, Millimeter Array, Planning, Box 2.

77. The European currency unit (Ecu) was the predecessor of the Euro and was worth about $1.35 at the time.

78. RLB to C. Madsen, 28 February 2012, NAA-RLB, ALMA.

79. MOU between the NSF, ESO, CNRS, MPG, NFRA, and PPARC, June 1999, NAA-NRAO, ALMA, Multi-Institutional Agreements.

80. Papers relating to the various MOUs, Coordinating and Liaison Committees may be found at NAA-NRAO, ALMA, Multi-Institutional Agreements.

81. Japanese participation in ALMA: A Proposal from NOAJ, 16 August 2002, NAA-NRAO, ALMA, Multi-National Agreements.

82. In 2014, the University of Arizona obtained the ESO ALMA prototype antenna and erected it on Kitt Peak to replace the old NRAO 12 Meter Telescope. The North American prototype was sent to Greenland to be used as part of the Event Horizon Telescope. (https://eventhorizontelescope.org/).

83. NRAO and ESO had previously agreed that all of the antennas would be of the same design and would be purchased from a single contractor, with consideration to secure "juste retour" vendors in the ESO Vertex bid. However, no formal agreement was ever established to achieve this goal.

84. The full story leading to the construction and operation of ALMA is a complex one beyond the scope of this book but is conveyed in a forthcoming book by Robert Dickman and Paul Vanden Bout.

BIBLIOGRAPHY

REFERENCES

Baars, J.W.M. 2013, *International Radio Telescope Projects, A Life Among their Designers, Builders, and Users* (Rheinbach: Baars)

Baars, J.W.M. et al. 1987, The IRAM 30-m Millimeter Radio Telescope on Pico Veleta, Spain, *A&A*, **175**, 319

Baars, J.W.M. et al. 1999, The Heinrich Hertz Telescope and the Submillimeter Telescope Observatory, *PASP*, **111**, 627

Bahcall, J.B. 1991, *The Decade of Discovery in Astronomy and Astrophysics* (Washington: National Academy Press)

Barger, A.J. et al. 1998, Submillimetre-Wavelength Detection of Dusty Star-Forming Galaxies at High Redshift, *Nature*, **394**, 248

Booth, R. et al. 1989, The Swedish-ESO Submillimetre Telescope (SEST), *A&A*, **216**, 315

Brown, R.L. 1990, *The Millimeter Array, Proposal to the National Science Foundation* (Washington: AUI)

Brown, R.L. and Schwab, R.R. eds. 1988, *Millimeter Array Design Concept – MMA Design Study, II* (Charlottesville: NRAO/AUI)

Chu, T.S. et al. 1978, The Crawford Hill 7 Meter Millimeter Wave Antenna, *BSTJ*, **57**, 1257

Dame, T.M. et al. 2001, The Milky Way in Molecular Clouds: A New Complete CO Survey, *ApJ*, **547**, 792

Downes, D. 1995, *Large Southern Array* (Grenoble: IRAM)

Encrenaz, P. et al. 2011, The Genesis of the Institute of Radio Astronomy at Milllimeter Wavelengths (IRAM), *JAHH*, **14**, 83

Epstein, E.E. 1977, The Aerospace Corporation, *BAAS*, **9**, 1

Field, G. 1982, *Astronomy and Astrophysics for the 1980's, Vol. 1* (Washington: National Academy of Sciences)

Field, G.B. and Hitchcock, J.L. 1966, The Radiation Temperature of Space at λ 2.6 mm and the Excitation of Interstellar CN, *ApJ*, **146**, 1

Findlay, J.W. and von Hoerner, S. 1972, *A 65-Meter Telescope for Millimeter Wavelengths* (Charlottesville: NRAO)

Gordon, M.A. 1984, A New Surface for an Old Scope, *S&T*, **67**, 326

Gordon, M.A. 2005, *Recollections of Tucson Operations: The Millimeter Wave Observatory of the National Radio Astronomy Observatory* (Dordrecht: Springer)

Greenstein, J.L. 1973, *Astronomy and Astrophysics for the 1970's, Report of the Astronomy Survey Committee, Vol. 2* (Washington: National Research Council)

Güsten, R. et al. 2006, The Atacama Pathfinder EXperiment (APEX) - A New Submillimeter Facility for Southern Skies, *A&A*, **454**, L13

Ho, P. et al. 2004, The Submillimeter Array, *ApJL*, **616**, L1

Holland, W.S. et al. 1999, SCUBA: A Common-User Submillimetre Camera Operating on the James Clerk Maxwell Telescope, *MNRAS*, **303**, 659

Holland, W.S. et al. 2013, SCUBA-2: The 10 000 Pixel Bolometer Camera on the James Clerk Maxwell Telescope, *MNRAS*, **430**, 2513

Ishiguro, M. et al. 1994, The Large Millimeter Array. In ASPC **59**, *Astronomy with Millimeter and Submillimeter Wave Interferometry*, ed. M. Ishiguro and J. Welch (San Francisco: ASP), 405

Keller, G. 1961, Report of the Advisory Panel on Radio Telescopes, *ApJ*, **134**, 927

Kellermann, K.I. ed. 1991, Report by the Radio Astronomy Panel of the Astronomy and Astrophysics Survey Committee. In *The Decade of Discovery in Astronomy and Astrophysics* (Washington: National Academy Press)

Kohno, K. 2005, The Atacama Submillimeter Telescope Experiment. In ASPC **344**, *The Cool Universe: Observing Cosmic Dawn*, ed. C. Lidman and D. Alloin (San Francisco: ASP), 242

Low, F.J. 1961, Low-Temperature Germanium Bolometer, *JOSA*, **51**, 1300

Low, F. and Davidson, A. 1965, Lunar Observations at a Wavelength of 1 Millimeter, *ApJ*, **142**, 1278

Mangum, J.G. et al. 2006, Evaluation of the ALMA Prototype Antennas, *PASP*, **118**, 1257

Moran, J.M. 2006, The Submillimeter Array. In ASPC **356**, *Revealing the Molecular Universe: One Antenna Is Never Enough*, ed. D.C. Backer et al. (San Francisco: ASP), 45

Moran, J.M. et al. 1984, *A Submillimeter-Wavelength Telescope Array: Scientific, Technical, and Strategic Issues: Report of the Submillimeter Telescope Committee of the Harvard-Smithsonian Center for Astrophysics* (Cambridge: Smithsonian Astrophysical Observatory)

Phillips, T.G. 2007, The Caltech Submillimeter Observatory, *IEEE/MTT-S Int. Microw. Symp.*, 1849

Plambeck, R.L. 2006, The Legacy of the BIMA Millimeter Array. In ASPC **356**, *Revealing the Molecular Universe: One Antenna Is Never Enough*, ed. D.C. Backer et al. (San Francisco: ASP), 3

Sargent, A. 1979, Molecular Clouds and Star formation. II - Star Formation in the Cepheus OB3 and Perseus OB2 Molecular Clouds, *ApJ*, **218**, 736

Sault, R.J. et al. 1995, A Retrospective View of MIRIAD. In ASPC **77**, *Astronomical Data Analysis Software and Systems IV*, ed. R.A. Shaw et al. (San Francisco: ASP), 433

Scoville, N. et al. 1994, The Owens Valley Millimeter Array. In ASPC **59**, *Astronomy with Millimeter and Submillimeter Wave Interferometry*, ed. M. Ishiguro and J. Welch (San Francisco: ASP), 10

Siringo, G. et al. 2009, The Large Apex Bolometer Camera LABOCA, *A&A*, **497**, 945

Stacey, J.M. and Epstein, E.E. 1964, Precision Radio-Astronomical Antenna for Millimeter Wavelength Observations, *AJ*, **69**, 558

Stark, A.A. et al. 1988, The Bell Laboratories CO Survey. In *Molecular Clouds in the Milky Way and External Galaxies*, Lecture Notes in Physics, **315**, ed. R. Dickman et al. (Heidelberg: Springer)

Tolbert, C.W. 1966, Observed Millimeter Wavelength Brightness Temperatures of Mars, Jupiter, and Saturn, *AJ*, **71**, 30

Tolbert, C.W. and Straiton, A.W. 1965, Investigation of Tau A and Sgr A Millimeter Wavelength Radiation, *AJ*, **70**, 177

Tolbert, C.W. et al. 1965, A 16-Foot Diameter Millimeter Wavelength Antenna System. Its Characteristics and Applications, *IEEE Trans. A&P*, **AP-13**, 225

Townes, C. 1957, Microwave and Radio-frequency Resonance Lines of Interest to Radio Astronomy. In *IAU Symposium No. 4: Radio Astronomy*, ed. H.S. van de Hulst (Cambridge: CUP), 92

Vanden Bout, P.A. et al. 2012, The University of Texas Millimeter Wave Observatory, *JAHH*, **15**, 232

Welch, J. et al. 1996, The Berkeley-Illinois-Maryland Association Millimeter Array, *PASP*, **108**, 98

Wilson, R.W. et al. 1970, Carbon Monoxide in the Orion Nebula, *ApJL*, **161**, L43

Wootten, A. and Schwab, F.R. eds. 1988, *Science with a Millimeter Array - MMA Design Studies, I* (Charlottesville: NRAO/AUI)

FURTHER READING

Giacconi, R. 2008, *Secrets of the Hoary Deep: A Personal History of Modern Astronomy* (Baltimore: The Johns Hopkins Press)

Leverington, D. 2017, *Observatories and Telescopes of Modern Times* (Cambridge: CUP)

CHAPTER 11

NRAO and Radio Astronomy in the Twenty-First Century

Following the inauspicious experience with the 140 Foot Telescope, NRAO apparently learned to manage big projects. The VLA and VLBA were built on schedule and on budget. But the Green Bank Telescope project was funded before the design was complete and was prematurely rushed into construction with unfortunate consequences to the cost and schedule. However, by the beginning of the twenty-first century NRAO was operating the most powerful radio telescopes in the world, the VLA, the VLBA, and the GBT, and had become the acknowledged leader in the evolution of radio astronomy from a technique to an astronomical-based science. As radio telescopes became more sophisticated and computer-aided, observations and reduction became more automated; radio astronomers evolved from experimenters to observers to data analysts. By the turn of the century, the traditional breed of radio astronomers was disappearing. NRAO users often no longer participated in the observing, and with the start of ALMA observations in 2011, often did not even participate in the planning of the observations or the reduction of data.

As the operation of the powerful new NRAO facilities demanded a greater and greater share of the National Science Foundation (NSF) astronomy budget, the university radio observatories were gradually closed, exacerbating the community's long standing love-hate relationship with NRAO. Many university researchers, unable to get sufficient NSF grants to observe with their students at NRAO, turned their attention elsewhere. The pressure for observing time gradually diminished, and the NSF began to discuss the partial divestment of the NRAO, once the NSF poster child.

11.1 New Discoveries and New Problems

Radio astronomy provided the first observations of the cosmos outside the traditional narrow optical window characteristic of all previous astronomical studies extending over many millennia. Today, astronomers study the infrared,

© The Author(s) 2020, corrected publication 2021
K. I. Kellermann et al., *Open Skies*, Historical & Cultural Astronomy,
https://doi.org/10.1007/978-3-030-32345-5_11

ultraviolet, X-ray, and gamma-ray portions of the electromagnetic spectrum as well as radio and optical observations, and are beginning to explore the non-electromagnetic Universe of gravity wave and neutrino astronomy. However, radio astronomy was the first of the new astronomies and captured most of the new discoveries. Karl Jansky and Grote Reber started it all in the 1930s, and after WWII, discoveries and recognitions followed, highlighted by Nobel Prizes awarded to eight different radio astronomers.[1]

These exciting new discoveries of the previously unrecognized nonthermal universe were largely outside the main stream of optical astronomy, with its traditional emphasis on solar and stellar astrophysics and other thermal phenomena. For many years, radio astronomy in the US developed separately from the astronomical community, primarily by people with backgrounds in radiophysics rather than astronomy. There was strong support from the federal government, first from the Department of Defense (DOD), in particular the Office of Naval Research (ONR) and the Air Force Office of Scientific Research (AFOSR), and later the NSF. Radio astronomers shared both an appreciation of instrumentation as well as a common feeling of not being fully accepted by the traditional astronomy community. Although competing with each other for limited funds, the real "enemy" was perceived to be the "optical astronomer." US radio astronomers became united in pursuing expensive new facilities, and they felt more at home within the URSI Commission J on Radio Astronomy rather than the American Astronomical Society (AAS) or the International Astronomical Union (IAU) (Sullivan 2009, p. 418).

The middle of the twentieth century saw the construction in the US of many DOD-funded university-based facilities, particularly the very successful Caltech Owens Valley Radio Observatory (Sect. 6.6). Later, the NSF funded the NRAO in Green Bank, the 36 Foot mm wave telescope, the VLA, the VLBA, and the GBT, and, starting in 1971, the Arecibo Observatory, but many other projects, some very promising, died somewhere along the complex funding route. These included the Owens Valley Array (Sect. 7.2), the Associates for Research in Astronomy 100 meter dish (Sect. 9.5), the CAMROC/NEROC 440 foot radome enclosed dish (Sect. 9.5), and the NRAO 25 meter millimeter-wave project (Sect. 10.3).

Due to limited NSF funds for facility operations, the big projects began to eat the ongoing smaller university radio astronomy programs. Those university programs were the breeding ground for the new generation of radio astronomers who designed, built, and operated the large new national facilities, and also provided the opportunity for developing the innovative new techniques and telescopes. The struggle between NRAO and the university-operated radio observatories continued throughout the history of NRAO, beginning in the mid-1950s during the lengthy debates about forming a national radio observatory. However, it was the big VLA and VLBA projects in the 1970s and 1980s that had the biggest impact on the university facilities.

As a result of the greatly improved sensitivity of radio telescopes (Fig. 11.1), radio astronomy was no longer constrained to study radio sources discovered

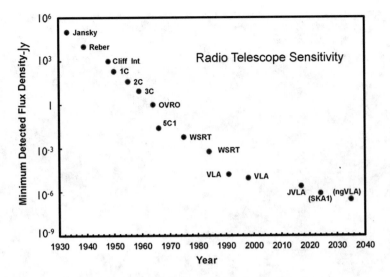

Fig. 11.1 Plot showing the weakest detected radio source as function of time since Jansky's 1933 radio detection of the Milky Way. The points refer in sequence to Jansky, Reber, the Australian Sea Interferometer, the Cambridge 1C, 2C, and 3C surveys, the Owens Valley Interferometer, the Cambridge 5C survey, the Westerbork Synthesis Radio Telescope, and the VLA. The last two points show the expected sensitivities of SKA-1 mid and the ngVLA respectively. In less than 100 years the sensitivity of radio telescopes has improved by a factor of about 100 million. Credit: NRAO/AUI/NSF

by radio surveys, but could study the radio emission from known cosmic objects. This was especially attractive to the large stellar astronomy community who could use observations of the thermal emission to determine the rate at which stars were losing mass. With the introduction of the VLA, astronomers, for the first time, could make images with resolutions comparable to those obtained from the biggest optical telescopes. VLA users studied solar system objects, stars, supernovae, the interstellar medium, galaxies, quasars, and cosmology. Radio observations became a tool for all astronomers, not just those trained in radio astronomy, and NRAO encouraged the growth of this broader user base. However, the VLBA, with its unique angular resolution, had no parallels at other wavelengths, and so had a limited user base, not only because it was perhaps more difficult to use, but also because the science was so different.

Early radio spectroscopy involved the laborious tuning of a single narrow-band instrument over a wide frequency range. With the subsequent enormous increases in sensitivity and the introduction of powerful multi-channel digital spectrometers, the impact to astronomy of the 140 Foot and later the 36 Foot and 12 Meter millimeter-wave telescopes was huge. They opened the new field of astrochemistry, exploring extreme conditions of temperature and pressure not easily duplicated in the laboratory, and the study of cold giant molecular

Fig. 11.2 JVLA radio image of the strong radio galaxy known as Hercules A superimposed on a Hubble Space Telescope optical image. The multi-frequency radio image made at 6 and 8 GHz (5 and 3.75 cm) shows the two-sided radio jet feeding a multiple bubble-like structure extending nearly one million light years from the supermassive black hole located at the center of the parent galaxy, suggesting a history of multiple outbursts. The lighter colored regions along the jet represent synchrotron emission from higher energy electrons and the darker red colors from lower energy electrons in the bubble rings. Credit: W. Cotton and R. Perley (NRAO/AUI/NSF), S. Baum (NASA) and C. O'Dea (RIT), and the Hubble Heritage Team (STScI/AURA)

clouds where new stars are formed. Using the VLA and eventually ALMA, the radio astronomy emphasis has slowly shifted from radio galaxies and quasars (Fig. 11.2) to the problems connected with the formation of stars, protostars, and planetary systems in dense regions obscured to light that can only be studied at radio wavelengths (Fig. 11.3).

11.2   RADIO ASTRONOMY AND OPTICAL ASTRONOMY

Since their founding in the 1950s, NRAO and its optical counterpart KPNO/NOAO[2] have developed in very different directions. From its beginning, NRAO concentrated on major facilities too costly for a single university, while KPNO/NOAO was founded to provide access to telescopes and clear skies access to astronomers without their own telescope or from eastern universities with cloudy skies.

Fig. 11.3 ALMA image of the dusty circumstellar disk surrounding the young Sun-like star HL Tau, located about 450 light-years from the Earth. This image made at 1 mm wavelength (300 GHz) covers an area about the size of our solar system and shows evidence for the early formation of a planetary system. Credit: ALMA (NRAO/ESO/NAOJ); C. Brogan, B. Saxton (NRAO/AUI/NSF)

The VLA, VLBA, GBT, and ALMA are all unique facilities, unrivaled anywhere else in the world. By contrast, the national optical astronomy facilities are less competitive. The two 8 meter (315 inch) Gemini telescopes are smaller than the Caltech/University of California 10 meter (394 inch) telescopes. The largest telescope at KPNO, the 4 meter (158 inch) Mayall Telescope, went into operation nearly two decades after the larger 5 m (200 inch) Palomar telescope. As was noted in the minutes of the November 1967 meeting of the AUI Executive Committee, "The Kitt Peak observatory has adopted the policy of acquiring additional small scale observing equipment whenever existing equipment is overloaded, but Dr. Heeschen does not consider this to be a function of NRAO".[3]

There are perhaps three reasons why NRAO and NOAO/KPNO developed along these different paths. For optical telescopes, the sensitivity depends only linearly on telescope diameter, whereas at radio wavelengths the sensitivity depends on the area of the aperture or the square of the diameter (Ekers 1978). Secondly, there are a very large number of stars in the sky with a wide range of astrophysical conditions (e.g., temperature, pressure, element abundance) that can be studied by modern optical spectroscopy using only moderate size telescopes, thus allowing significant research using only modest facilities. The third reason is perhaps more related to the historical and sociological development of radio and optical astronomy rather than to astrophysics. In the US, optical astronomy developed and thrived with support from philanthropists such as Charles Tyson Yerkes, William Johnson McDonald, and James Lick, as well as from the Carnegie and Rockefeller Foundations, and in a some cases state governments (e.g., Arizona, California). The philanthropists, of course, wanted their money to go to big noticeable telescopes. Yet only a privileged few astronomers at Yerkes, McDonald, Lick, Mount Wilson, and Palomar Observatories had access to these giant telescopes in the 60–200 inch class that were required to explore the exciting new developments in galaxy research and cosmology. Others had to be satisfied with using more modest facilities in generally less than ideal locations. Not surprisingly, by the middle of the twentieth century there was increasing tension between the "have" and "have not" optical astronomers, with the former being primarily at the large Mt. Wilson, Mt. Palomar, and Lick observatories which did not encourage competition from NOAO/KPNO.

Unlike optical astronomy which, as result of dramatic improvements in detector technology, has been able to effectively use telescopes as much as 100 years after their initial deployment, the US closed radio telescopes when they became less productive, sometimes when operating funds were no longer available, and in a few cases where acts-of-God led to the closing of otherwise competitive facilities. NRAO closed the 4-element Green Bank Interferometer, the Green Bank 140 Foot Telescope, and the NRAO 12 Meter mm dish[4] so that limited resources could be devoted to newer facilities, the VLA, GBT, and ALMA respectively. Many other, still productive US radio telescopes were closed due to the loss of operating funds.

11.3   NRAO AND THE US RADIO ASTRONOMY COMMUNITY

Since the first ideas were floated about establishing a national radio astronomy observatory, the "love-hate" relationship between the university-operated radio observatories and national radio astronomy facilities has been the powerful background against which US radio astronomy developed during the latter half of the twentieth century. For the most part, NRAO facilities were used and appreciated by the many radio astronomers located at universities aspiring to do research in the rapidly developing field of radio astronomy, but which had no facilities of their own. Indeed, some schools, such as the Universities of Maryland, Pennsylvania, and Iowa, and Virginia Tech, started new radio astronomy programs and hired new faculty based on the opportunity for their faculty and students to use NRAO telescopes. On the other hand, places such as Caltech, Stanford, Ohio State, and the University of Illinois with ambitious projects of their own, saw NRAO as a direct competitor for limited funds. It was again a case of "haves" and "have-nots." There were many more "have-nots" than "haves," but the "haves" had more clout in Washington.

This dichotomy came to a head with the NRAO/Caltech controversy, first over the VLA (Chap. 7) and then the VLBA (Chap. 8). In the end, even the "haves," sensing the growing competition from abroad, combined with their strong interest in having the best possible facilities for their research, supported funding for NRAO, even if it might be at the expense of their own facility. For example, as discussed in Sect. 7.4, MIT's Bernie Burke provided critical support in getting the VLA funded, even at the expense of the proposed MIT-Harvard CAMROC 440 foot dish. Similarly, as related in Sect. 8.7, Marshall Cohen and others at Caltech supported and helped design the VLBA, knowing that it would lead to the loss of funding for the Caltech/JPL VLBI correlator, then the leading instrument of its kind in the world.

The 1969 amendment to the 1970 Military Authorization Act, introduced by Washington Senator Mike Mansfield, limited the use of Defense Department funds to research that included "a direct and apparent relationship to a specific military function." The Mansfield Amendment had an immediate and long lasting impact on the future funding of US radio astronomy, and ultimately on all US astronomy. Although the amendment formally applied only to the 1970 DOD Authorization Bill, both military and civilian funding agencies were apparently unclear on the interpretation. Military laboratories were fearful of risking their military programs, so did not contest the amendment. The NSF was faced with absorbing new university-led activities previously supported by DOD, with an uncertain level of funding to support these new activities. The NSF struggled for years to keep a balance between operating the university radio astronomy observatories and NRAO. Nevertheless, with continued budget challenges resulting from the need to operate the powerful new instruments being constructed, NSF support of still-productive university radio observatories gradually declined, and with time so did the support for NRAO facilities. Some of the unique US radio telescopes that lost NSF funding included the University of Maryland Clark Lake decameter array, the Stanford 5 element array, the Owens Valley Interferometer and 130 foot antenna, the University of Arizona Radio Observatory, and the Caltech Sub-Millimeter Observatory. As discussed in Sect. 10.4, threatened with closure in 2007, the Berkley-Illinois-Maryland Millimeter Array at Hat Creek and the Caltech OVRO Millimeter Array combined to form CARMA, the Combined Array for Millimeter Astronomy, but it too was closed for lack of operating funds in 2015.

11.4   Conflict and Collaboration

David Munns (2013) has characterized the development of radio astronomy as the result of friendly collaborations by scientists united by a common culture, an appreciation of technical matters, and a need to deal with a perceived common "enemy," optical astronomers. While collaborations have been important in some areas such as VLBI (Kellermann and Cohen 1988; Kellermann and Moran 2001), the intense competition between Australian and UK radio astronomers and between Cambridge and Manchester radio astronomers in the 1950s and 1960s (Sect. 2.1) (Edge and Mulkay 1976) reflects an alternative interpretation, as does the continuing tension between NRAO and some US university-based radio observatories.

For example, the deferral by Doc Ewen and Ed Purcell to the Dutch and Australian H I teams to simultaneously publish their H I discovery is described by Munns (2013) and Sullivan (2009, p. 414) as evidence of the cooperative culture of radio astronomy. Only after their success did Ewen and Purcell offer to wait for a joint publication, knowing full well that they were the first and that they would be so recognized, as indeed they have been. It is perhaps reflective of the sense of competition and priority, that when Jan Oort wrote to Grote Reber in 1945 asking his advice on building a radio telescope, he only mentioned his interest in the distribution of interstellar gas in the Galaxy. Oort, of course, was already aware of the possibility of detecting the 21-cm line, and that was clearly much of his motivation.[5] But he apparently held that close to his chest.

As described in Chap. 7, VLBI, especially earth-to-space VLBI, has perhaps been the poster child for a technically complex, and successful scientific cooperation among individual scientists, institutions, and countries, including some of the earliest serious scientific collaborations between the US and the Soviet Union. Perhaps these collaborations were successful because they were not driven by a political desire to collaborate or build bridges between countries and political systems, but because the science required it. The first VLBI experiments came about following friendly, but nevertheless intense, competition between Canadian and US radio astronomers who can (and still do) argue about who was first, depending on one's definition of success and what constitutes VLBI.

As discussed by DeVorkin (2000), Harwit (2015), Sullivan (2009) and others, much of the early support for radio astronomy, particularly in the US, came from military interests. With the end of the Cold War and the reduced threat of a major war between super-powers, the political forces have moved in the direction of international collaboration in the construction of expensive new scientific research instruments. Radio astronomy, with its long history of international collaborations, has been well poised to exploit these opportunities. The European VLBI Network (Sect. 8.5) has been successfully driven by a combination of scientific goals, European unity, and competition with the US. ALMA (Sect. 10.7), with a price tag in excess of 1 billion US dollars, is the most recent example of a successful large global collaboration involving countries from four continents. The Square Kilometre Array (Sect. 11.7) is a very ambitious multi-billion Euro international program that has been challenged by the international aspect, differing views on the technical implementation, and national ambitions.

11.5   The National Radio Quiet Zone and Radio Frequency Spectrum Management[6]

Since its earliest years, radio astronomy has needed to deal with interference from both intended radio broadcast and communication transmissions as well as from unintended radio noise, generated, for example, by automobile ignitions, radars, or computers. As early as the 1930s, Karl Jansky reported radio interference from passing motor boats and from nearby medical diathermy machines. Grote Reber could only observe at night at 160 MHz, due to daytime interference from automobiles. Reber recognized the sensitivity limitations imposed by radar and the FM broadcast services, and was probably the first person to urge the creation of protected bands to study the astrophysical lines which he predicted might be observed in emission or in absorption. With the increasing use of the radio spectrum for a wide variety of communication and data transfer applications, the competition for frequency allocations has become intense. Radio frequency spectrum management has developed an extensive, often contentious, bureaucracy, with a rich collection of national and international agencies, committees, and acronyms.

Radio waves do not respect international borders, and since 1906 the allocation of the radio spectrum has been guided by the International Telecommunications Union (ITU), now a specialized agency of the United Nations. The 1927 Washington Conference established the International Radio Consultative Committee (CCIR) to coordinate technical studies and to develop international standards which are reviewed every three to five years at a World Radiocommunication Conference[7] (WRC). National administrations enable their own frequency allocations, generally, but not always, respecting the ITU/WRC guidelines.[8] Within the US, radio frequencies are assigned by the Federal Communications Commission (FCC) for non-federal government use and, for federal use, by the National Telecommunications and Information Administration (NTIA) based on input from the government Interdepartmental Radio Advisory Committee (IRAC).

In the early 1950s both URSI and the IAU discussed the need to protect some frequencies for radio astronomy. Although URSI passed a 1950 resolution asking that frequencies be reserved for radio astronomy, it was not initially adopted by the CCIR. In 1957, URSI set up the Sub-Commission on Radio Frequency Allocation for Radio Astronomy under the leadership of NRAO's John Findlay to prepare for the 1959 Geneva Administrative Radio Conference. With input from Findlay's committee, as well as from the IAU, the CCIR recommmended that (1) radio telescopes should be located at sites as free as possible from interference, (2) governments should protect frequencies used for radio astronomy in their countries, and (3) that there should be complete international protection from interference in the bands around the 327 MHz deuterium frequency, the 1420 MHz H I frequency, and the 1667 MHz hydroxyl (OH) frequency, as well as seven bands between 40 MHz (7.5 m) and 1 GHz (30 cm) reserved for continuum observations.

At the WRC meeting later that year in Geneva, radio astronomers made the case to reserve frequencies for passive radio astronomy. Jan Oort represented the IAU and Lloyd Berkner represented URSI as non-voting delegates. John Findlay was part of the official US delegation, which was dominated by commercial interests and so provided less than enthusiastic support for radio astronomy (Sullivan 1959). Following long and contentious negotiations, for the first time, largely due to the pressures from the Dutch delegation, the Radio Astronomy Service was recognized as a legitimate user of the radio spectrum. The 1959 WRC recommended that multiple bands covering about one percent of the radio spectrum be allocated for radio astronomy, with special protection in the 1420 to 1427 MHz hydrogen band (Findlay 1960, 1991). In 1992, the ITU also defined the level of harmful interference not be exceeded in the radio astronomy bands, but with the great improvements in the sensitivity of radio telescopes since that time, these approved levels were no longer adequate to protect radio astronomy from harmful interference.

After the 1959 WRC, following a recommendation from the US President's Science Advisory Committee, the National Academy of Sciences (NAS) held a series of meetings to begin a dialog between radio astronomers and the government. This resulted in the creation of the NAS Committee on Radio Frequencies (CORF) to represent the US radio astronomy community in reacting to FCC "Proposals of Proposed Rule Making." Starting with John Findlay and Hein Hvatum, CORF has nearly always included someone from NRAO to provide advice on proposed new threats to radio astronomy such as satellite broadcasting and communications, airport radars, and most recently the pervasive automobile collision control radars. In addition to CORF, the NSF Astronomy Division, like other government agencies and military branches, includes at least one person expert in spectrum management issues to protect from threats to each agencies' use of the spectrum. On the international front, the Inter-Union Committee on the Allocation of Frequencies (IUCAF) was set up in 1960 by the three international scientific unions, URSI, the IAU, and COSPAR,[9] to coordinate the requirements for the protection of radio astronomy, and to convey these priorities to the national and international bodies responsible for frequency allocations. Both CORF and IUCAF have paid particular attention to the protection of frequencies corresponding to especially interesting spectral lines.

In Sect. 3.4 we described the establishment of the West Virginia Radio Astronomy Zoning Act which was passed by the West Virginia legislature to protect NRAO from unintended radio noise. This was followed on 19 November 1958 by the FCC establishment of The National Radio Quiet Zone (NRQZ) to provide protection from radio transmissions that might cause harmful interference to NRAO and to the US Naval Station at Sugar Grove (Sect. 9.3). It is centered around Sugar Grove, West Virginia, and covers an area of approximately 13,000 square miles in Virginia and West Virginia. All applications for fixed radio transmissions located within the NRQZ are reviewed for potential interference to either the Green Bank or the Sugar Grove facilities before being licensed by either the FCC or NTIA. The combination of the regulatory protection afforded by the NRQZ, the low population density, and the geographic protection provided by the surrounding mountains have been crucial in keeping Green Bank radio astronomy observations relatively free from radio frequency interference (RFI).

However, with the increasing demands on the radio spectrum from commercial, government, personal, and military activities, RFI has become an even greater problem than ever for radio astronomy. Moreover, the proliferation of satellite-based transmissions, high-powered radio and TV broadcasts, and equally high-power global communications satellites, has left no place on earth immune from RFI. At the same time, radio astronomers are no longer content with observing around a few protected spectral line frequencies or within the narrow bands allocated for continuum observations. Spectroscopic observations are subject to redshifts due to the expansion of the Universe. The region just below 1.4 GHz is of particular interest for observing atomic hydrogen in distant galaxies, but this is an especially crowded part of the radio spectrum. Continuum observations are even more difficult. The JVLA, for example, now covers the entire spectrum between 1 and 50 GHz. A typical continuum observation may cover several GHz of bandwidth, but the protected radio astronomy bands are typically only a few tens of MHz wide and do not provide meaningful protection.

On the other hand, there are only a few radio observatories left in the US, and to a lesser extent in the world, and they do not each observe in all of the protected bands all of the time. Modern radio astronomy needs only local protection around each observatory, and only in those bands and for those times that they are being used. Fortunately, above a few hundred MHz, radio propagation is limited to about one hundred miles, so selective protection is possible. Except for a few bands containing important spectral lines such as the 1420–1427 MHz band or the 1.6 GHz OH band, the protected radio astronomy bands have become no more than bargaining chips. However, fearing that the powerful commercial interests will gobble up anything put on the table, CORF, IUCAF, and the radio astronomy community, have been reluctant to offer up these narrow protected bands in return for local, time-sensitive, protection over much larger bands. While it seems inevitable that the current paradigm for radio spectrum management will be replaced by a shared use of the spectrum policy, it remains to be seen who gets what share.[10]

11.6   THE TRANSITION TO "BIG SCIENCE"

Following the early work of Karl Jansky and Grote Reber, radio astronomy in the United States has been largely supported by federal or state funds with little contribution from universities, industry, or private supporters. There were few obstacles to pursuing new initiatives, and few formalities to inhibit creative ideas. As described in Sect. 8.1, the NRAO VLBI program was initially funded by the NRAO Director with no formal proposal or committee review, and has developed into a global industry and one of NRAO's major facilities (Sects. 8.8 and 8.9). The 36 Foot millimeter wave (Sect. 10.2) telescope was initially funded on the basis of a few paragraphs added to NRAO's 1964 budget request, although based on what was known at the time, there were only a few objects in the sky known to be strong enough to be detected with the proposed telescope. No one anticipated the explosion in millimeter wave spectroscopy that would result in an international billion dollar ALMA program (Sect. 10.7).

Since those early years of the 1930s through the 1950s and the golden years of discovery of the 1960s, the landscape has changed. Whether for building a new instrument, an individual research grant, or for observing time, proposals are expected to give the expected results in some detail, almost obviating the need for the proposed activity. Risky but high potential payoff projects are difficult to get through the success-oriented peer review system. Some researchers have learned to play the system by proposing projects that are already nearly done, and using the new resources to attack more speculative projects.[11] Proposals to the NSF, even for modest research grants, need to supply a detailed technical description, a budget, a discussion of the "Broader Impact" to society,[12] as well as the expected scientific results. Indeed, even before submitting a proposal to the NSF, one has to be familiar with the 181 page NSF "Proposal & Award Policies and Procedure Guide" as well as other instructional material.

Combined with the increasing cost and complexity of radio telescopes, the modern administration of science provides fewer opportunities for individual initiative or opportunity to make new discoveries. The early pioneers were able to develop a new idea, build their own equipment, observe, and interpret the results. There were few, if any previous papers that had to be read, understood, and cited. Single or two or three author papers were common. By contrast, modern radio astronomy programs often involve large teams of scientists, engineers, and software experts, sometime located in different countries. The four-level peer review filter of facility or instrument proposal, research grant proposal, telescope time preproposal, and journal referees often represents a significant challenge to new research ideas, new facilities, or publishing new ideas. At the same time, the responsibilities of reviewing proposals for grants, telescope time, and journal reviewing represent a significant burden on the community.

While it is tempting to look back on those early years with rose-colored glasses, it is sobering to recall that Karl Jansky (Sect. 1.1), George Southworth (Sect. 2.1), and John Bolton (Sect. 6.6), to name a few, were constrained from pursuing their radio astronomy interests by the more immediate needs of their employer. Grote Reber had to deal with an editor who delayed publication of his pioneering paper because the editor did not have the background to understand and appreciate it (Sect. 1.3). The personal sacrifices of people like Lovell, Bowen, Hachenberg, and Heeschen in taking on the challenges and risks of building the Jodrell Bank, Parkes, Effelsberg, and VLA telescopes respectively should not be minimized.

The early ideas of Menzel and Berkner that led to the formation of NRAO reflected a departure from the individualism of the pioneering radio astronomers such as Reber, Ryle, Lovell, Pawsey, Bolton, Wild, Mills, and Christiansen. The NRAO founders had envisioned a facility where university scientists would bring their own instrumentation to install and use on the large NRAO telescopes. To an extent this happened during the early years in Green Bank with the 300 Foot and then 140 Foot Telescopes as well as the 36 Foot millimeter antenna on Kitt Peak. All this changed with the VLA. As Dave Heeschen (1991) noted,

> Most radio astronomy in those days was done by radio physicists and engineers who invented, built, and then used their own instruments. New telescopes usually came about either because someone had a bright idea, developed it and then did whatever radio astronomy he could with it, or because someone asked a particular scientific question and then designed an instrument to answer it. These both proved very fruitful ways to proceed, especially, when the practitioners were as talented and clever as those of the 50's and 60's. The VLA did not come about this way, however. Our motives and goals were quite different. We set out from the beginning to build a flexible, general-purpose instrument for a broad scientific purpose, to be used by a lot of people other than, or in addition to, the designers. Many people considered that approach inappropriate, unimaginative, undesirable, overly expensive, and unneeded, and we had to cope with these criticisms for 10 or 15 years.

Obtaining government funding for big science projects remains challenging. At best there is a long series of hurdles that must be passed; at any point a project can be killed. At a 1963 Green Bank meeting to discuss options for building a large steerable radio telescope, Dave Heeschen listed the "filters" that any project must pass through as (a) the NSF, (b) the Office of Science and Technology, (c) the Bureau of the Budget, (d) the President's Science Advisory Committee, (e) Congress, and (f) the President. Of these, Heeschen considered the NSF, even with its multiple reviews "the easiest" and the Bureau of the Budget "the toughest."[13]

From the earliest years, NRAO pioneered the use of digital data and computer aided analysis. When John Bolton visited Green Bank in late 1964 he wrote back to Taffy Bowen in Australia,

> Green Bank is now very impressive as far as facilities are concerned, very depressing as far as surroundings and social life are concerned, has a large working staff who are expending a lot of energy but, considering the resources, is not overproductive. It is probably over-automated and over-digitized. If you make your observation by writing a set of instructions for a telescope operator to carry out, then write a set of instructions for a computer to extract some data from the results, then it is rather unlikely that you are going to find anything other than what you are looking for.[14]

Ironically, it was John Bolton who led the effort at Parkes to transition from analyzing chart recordings to the use of computer analysis of digitized telescope output. But his 1964 warning would be prophetic. At first, radio astronomers wrote their own programs to reduce the burden of manual data analysis. With time, the software became sufficiently sophisticated that it was written by others, often teams of people rather than a single individual, and increasingly the software teams included more people trained in computer science than in physics or astronomy. Observers were able to use these software packages to reduce their data, but were increasingly ignorant about how they worked. Indeed, with rare exceptions, no one person understood everything. NRAO was satisfying its expected role to provide easy-to-use facilities for students and scientists not trained in radio astronomy techniques. Multi-wavelength astronomy became the norm, and NRAO was expected to produce so-called "science-ready-data-products." Astronomers are no longer radio, optical, or X-ray astronomers. Instead, they are galactic or extragalactic astronomers, or they perhaps concentrate on planetary astronomy or exoplanets—and are equally familiar, or perhaps unfamiliar, as the case may be, with techniques at all wavelengths. The ambitious goal of providing science-ready-data-products for data of increasing complexity and sophistication remains a challenge.

Bolton's warning about computers inhibiting discoveries is debatable. Certainly many of the scientific areas of contemporary radio astronomy would not be possible without modern high-speed digital computers. This certainly includes essentially all interferometric arrays such as the VLA, ALMA, IRAM (NOEMA), the Westerbork telescope, Murchison Widefield Array (MWA), and the Australia Telescope Compact Array, and especially VLBI; but also most single-dish studies, including pulsar searching and timing, spectroscopy, and mapping of extended structures, would not be possible without modern digital data analysis. Further, there has been an explosion in the volume of data associated with many radio astronomy programs. Sophisticated computer clusters have replaced laptops, and multi-person software development teams have replaced individual programmers.

At the 1997 dedication of the Jansky Memorial monument at the former Bell Labs Holmdel site, Paul Vanden Bout pointedly summarized the situation.

> Large facilities require large teams to keep them going. We have moved very far from the days of Jansky and Reber where a lone individual made a big contribution..... [Users] treat these facilities as vending machines. They put in a proposal, decide whether they want their M&Ms with or without peanuts. They pull the right lever and out comes data. Preferably calibrated data. It's all too much for those of us who grew up thinking real radio astronomy was done with a chart recorder. So we sit around and lament this. On the other hand, I've noticed that the people who complain the loudest, are the very people who are lobbying for the big new telescopes. I guess that means that we recognize that it is inevitable that we should seize the future and get on with it. After all, if this is what it takes to make the field go, then we ought to do it. There will be surprises and discoveries.[15]

11.7   THE SQUARE KILOMETRE ARRAY (SKA)

Discussions about building a radio telescope with a very large collecting area go back at least to the early design of the Benelux Cross (Sect. 7.1) that had a collecting area of 600,000 square meters (Christiansen and Högbom 1961). Barney Oliver's Cyclops concept of one thousand 100 meter dishes (Oliver and Billingham 1973) had a geometric collecting area of about 10 million square meters. Later, Govind Swarup (1981, 1991) proposed the Giant Equatorial Radio Telescope (GERT) with more than 100,000 square meters of collecting area, while in the USSR, Yuri Parijskij (1992) discussed radio telescopes "with a collecting area of about 1 million square meters." A half a century ago, Grote Reber actually built an array in Tasmania with about a square kilometer of collecting area (Reber 1968). Reber's 2.1 MHz (144 m) array contained 192 half-wave east-west dipoles arranged in a 1075 meter (3526 foot) diameter circle. By properly phasing the elements, he could steer the 8 degree wide beam anywhere along the meridian. Reber's array had, and still has, the largest collecting area of any radio telescope ever built.[16]

First Years The start of the modern discussions directed toward building a next generation radio telescope, with an effective collecting area of one million square meters or one square kilometer, is generally considered to be Peter Wilkinson's (1991) paper on the "Hydrogen Array," which he discussed at a symposium held in Socorro, New Mexico, on the occasion of the tenth anniversary of the VLA dedication. Even earlier, Dutch radio astronomers, led by Robert Braun, were discussing a Square Kilometre Array Interferometer (SKAI) (Noordam 2013). In 1993 and 1994, URSI and the IAU set up the Large Telescope Working Group and the Future Large Scale Facilities Working Group, respectively, to coordinate discussion and planning among groups considering large scale radio and optical telescope projects. It was soon realized that the description of a "Square Kilometre Array Interferometer" was redundant and the term Square Kilometre Array or SKA was adopted. By agreement, the British spelling "kilometre" was adopted rather than the American spelling "kilometer."

The early strawman SKA concept consisted of a condensed nearly filled array for high sensitivity to low surface brightness complemented by three sparsely filled arms to improve the angular resolution. There was no shortage of ideas about how to implement the SKA, but most turned out to be unrealistic. The most straightforward approach considered was to build an array of parabolic dishes, much like the VLA or the Westerbork Synthesis Radio Telescope (WSRT), but with a much larger number of elements, perhaps taking advantage of Swarup's GMRT SMART concept (Sect. 6.6) to reduce costs. The Dutch proposal for a so-called "aperture array" of a very large number of simple antenna elements appeared very attractive. By properly combining the signals from the different elements, it would be possible to create multiple beams in the sky, and, moreover, to generate nulls in the response in the direction of interference. Australian radio astronomers suggested a field of Luneberg lenses[17] or parabolic cylinders. Canada proposed to build an array of 200 meter Large Adaptive Reflectors, each illuminated by feeds and receivers suspended from a tethered floating aerostat. China proposed to build some ten Arecibo-like spherical reflectors.

The challenge with all of these approaches was to keep the cost down to a level that could be realistically funded. An early cost estimate suggested that the aperture array could be built for $100–$200 million, but further development work indicated that the true cost of the SKA could be much higher. As the estimates of the construction costs continued to grow, there was agreement to revisit the specifications. Noting that the original requirement of a million square meters of collecting area was based on the 1991 assumption that the system temperature was likely to be about 50 K, the requirement was restated that the ratio of collecting area divided by system temperature (A/T) would be 20,000 m^2/deg. With anticipated system temperatures of 20 K, the required collecting area to meet the new specification was reduced to 400,000 square meters but the name "Square Kilometre Array" was nevertheless retained.

Although the SKA started out as the "Hydrogen Array," it was clear that with the planned sensitivity of the SKA it would impact almost every area of astronomy. Successive publications advertised the growing science case (Taylor and Braun 1999; Carilli and Rawlings 2004), culminating in the massive, two-volume, 1996-page book edited by Tyler Bourke (2015). Complementing the science publications, Hall (2005) and Smolders and Haarlem (1999) updated the status of the engineering design.

In his foreword to Peter Hall's book, Richard Schilizzi reported that

> The current schedule for the SKA foresees a decision on the SKA site in 2006, a decision on the design concept in 2009, construction of the first phase (international pathfinder) from 2010 to 2013, and construction of the full array from 2014 to 2020. The cost is estimated to be about 1000 M€.

International Organization Starting in 1999, an informal group of scientists from the US, the Netherlands, Australia, and Canada began to meet to develop plans for building the SKA. As there was no existing international structure to accommodate a multi-national project, one had to be created. At the IAU General Assembly in 2000 in Manchester, representatives from nine countries interested in the SKA signed a Memorandum of Understanding establishing the International SKA Steering Committee (ISSC) to provide broad oversight to the project. Ron Ekers (Fig. 11.4) from Australia, who had been the key person in getting the SKA program off the ground and in generating interest from the various national funding agencies, was elected as the first Chair of the ISSC. Four years later the ISSC created the International SKA Project Office (ISPO), with Richard Schilizzi (Netherlands) as ISPO Director, and a year later Peter Hall (Australia) became the Project Engineer. In 2008, a new Memorandum of Agreement created the SKA Science and Engineering

Fig. 11.4 Ron Ekers was as strong advocate for building the SKA as an international collaborative activity. He became the first Chair of the International SKA Steering Committee (ISSC). Credit: CSIRO Radiophysics photo archives

Committee (SSEC) and SKA Project Development Office (SPDO) replacing the ISSC and ISPO respectively, with Schilizzi continuing as SPDO Director.

The ISSC/SSEC met twice a year to review progress on the planning, costing, and siting of the SKA. Initially the Committee consisted of 18 members but was later expanded to 24 to accommodate the growing global interest in the SKA. One-third of the membership represented Europe, one-third the US, and one-third the rest of the world, reflecting the anticipated cost sharing. The ISSC/SSEC received input from international Science, Engineering, and Site Selection Advisory Committees, while the ISPO/SPDO was supported by a number of Working Groups and Task Forces within the ISPO.

The US SKA Consortium Recognizing the growing interest in the SKA, both internationally and within the United States, NRAO organized a meeting in Green Bank in 1998 to discuss how the US might participate in planning for the SKA. A few months later, a small group met at MIT and agreed to form the US SKA Consortium (USSKAC) to coordinate US participation in the SKA. Jackie Hewitt from MIT was elected as the Consortium Chair and Jill Tarter from the SETI Institute as the Vice Chair. Hewitt soon stepped down as Chair due to her other responsibilities, and was replaced by Tarter, then Yervant Terzian followed by James (Jim) Cordes, both from Cornell, and finally Patricia (Trish) Henning from the University of New Mexico.

The initial USSKAC membership consisted of Caltech, Cornell University, MIT/Haystack, NRL, the University of California, Berkeley, Ohio State University, and the SETI Institute. NRAO did not join originally due to some uncertainty about whether joining such an organization was allowed under the NRAO-NSF Cooperative agreement. Instead, Rick Fisher and one of the authors (Kellermann) were appointed as At-Large members of the USSKAC. In October 1999 AUI became a formal member of the USSKAC, and Fisher and Kellermann then continued as the AUI/NRAO representatives. Nevertheless, due to NSF concerns about NRAO's responsibilities to ALMA and the EVLA program, as well as community desire to keep the US SKA development within the university community, NRAO did not play a leadership role in the USSKAC activities, reflecting the continuing tensions between NRAO and the university groups.

The US SKA Technology Development Program (TDP), funded by the NSF and totaling about $13.5 million, supported the USSKAC between 2000 and 2011. The TDP was administered by Cornell with Jim Cordes as the PI. The USSKAC developed an antenna concept, investigated possible feeds, receivers, and backends, and studied the array configuration and RFI mitigation. In addition to the funded programs at Caltech, Cornell/NAIC, MIT/Haystack, the SETI institute, and the Universities of California and Illinois, in-kind support of technology development was provided by JPL, NRL, NRAO, and the University of New Mexico. For about a decade, the USSKAC met twice a year to review the progress with the TDP and to adopt a US position for the various issues being considered by the ISSC/SSEC. In addition to supporting technical developments within the US, the TDP also provided the funds to hold the semi-annual USSKAC meetings as well as to send six (then eight) representatives to the ISSC/SSEC meetings.

As one of its goals, the USSKAC developed a "strawman design" to build the SKA as an array of 4400 twelve meter diameter dishes known as a *Large-N Small-D* array. The antenna elements were configured to include 2,320 antennas within a 35 km diameter, and 160 stations, each containing nine antennas spread across North America, with baselines up to 3,500 km. Unlike the strawman proposals developed in other countries, the US concept included the full costs of instrumentation, site development, program management, software, etc., and was optimistically estimated to cost $1.41 billion. The US also included an operations plan based on VLA operations that included an operations budget of $61.5 million per year.[18]

Management and Funding From the beginning the SKA was planned as an international project, with the US, Europe, and the "rest-of-the-world" each contributing one-third of the costs. Owing to the anticipated NSF commitments to ALMA, Large Synoptic Survey Telescope (LSST), and Daniel K. Inoue Solar Telescope (DKIST), as well as to a large (24–30 m) optical telescope, it was understood by all that the NSF would not be able to contribute to SKA construction on the same time scale anticipated by other countries, but that a phased contribution seemed feasible, possibly with a US emphasis on the higher frequencies in a possible second phase of construction. Both NASA and the Department of Energy were approached, but showed little or no interest. The US Department of State expressed interest, but offered no money.

Financial support for SKA technical development came partly from the US TDP, as well as from other national sources, but the driving support came from the European Commission funding for the Square Kilometre Array Design Studies (SKADS, 2004 to 2009) led by Arnold Van Ardenne from the Netherlands and the Preparatory Phase for the SKA (PrepSKA, 2008 to 2012) program led by Phillip Diamond from the UK. PrepSKA was organized into seven Work Packages covering Management, SKA Design, Site Studies, Governance, Procurement, Funding, and Implementation. As the planning proceeded, there was a wide range of opinion on the SKA technology, the siting, and the organizational structure, leading to escalation in both the scope and cost of SKA.

Concerned that, due to other priorities, the national funding agencies could not meet the growing cost of the SKA on the ambitious time scale planned by the ISPO, the ISSC decided to split SKA construction into three phases. Phase 1, or SKA-1, would comprise only five to ten percent of the full SKA, and probably cost 15 to 20 percent of the full SKA, which was to be completed in Phase 2. It was clear that no single technology could cover the entire frequency range, so in 2005, the SKA was divided into SKA-Low, below a few hundred MHz, and SKA-Mid, up to a few GHz, using an aperture array and dish array respectively. A vaguely defined Phase 3 covering the higher frequencies, up to few tens of GHz, primarily financed by the US, was part of the longer term planning.

Reflecting the broad international interests and with the encouragement of ISSC members, in 2005 representatives of the various national funding agencies began to hold their own meetings, and in 2006 they formed the Informal Funding Agencies Group (IFAG). From the NSF, Wayne van Citters, Vernon Pankonin, and Jim Ulvestad participated at various times in IFAG meetings. Pankonin, in particular, was an active participant in many of the committees and working groups. In 2009, the funding agencies further formalized their role and created the Agencies SKA Group (ASG).

The ISPO Director reported to the ISSC/SSEC Chair, the Chair of the IFAG/FAWG, and the Chair of PrepSKA Board, each with their own goals and ambitions. This created an organizational challenge, not only for the ISPO Director, but for the broader project management issues as well. Management by committee, in this case multiple committees, was reminiscent of the committee management of the Green Bank 140 foot Telescope (Sect. 4.4), with the added challenge of international membership, each with their own agenda.

Open Skies The concept of Open Skies was widely accepted by the international radio astronomy community. Most of the world's major radio telescopes, including Westerbork and the Australia Telescope Compact Array, as well as the European VLBI Network and all of the NSF-funded radio observatories in the US, welcomed proposals from all scientists independent of their national affiliation. The ISSC naturally assumed that an Open Skies policy would carry over to the SKA, so was surprised that the national funding agencies expressed strong reservations. In 2007, the SSEC presented a resolution to the FAWG for an Open Skies concept for the SKA which read,

> Recognizing that open access to all qualified scientists, independent of institutional, national, or regional affiliation will give the best scientific returns, the ISSC believes that the allocation of SKA observing time or access to data obtained with the SKA should be based solely on merit, without regard to quotas, financial, or in-kind contributions to the construction or operation of the SKA.[19]

However, the IFAG responded that they could not sell this to their governments, who would ask what they were getting in return for spending tax money if their scientists did not have priority access to the SKA. While this same question is often asked of NRAO, as a matter of national policy, the Open Skies concept has remained in effect in the US for essentially all federally funded science projects.

Prototypes, Precursors, and Pathfinders Trying to demonstrate the merits of their proposed technology, a number of countries set out to build what were variously called SKA prototypes, precursors, or pathfinders. Others, seeing an opportunity to exploit the growing visibility of SKA, declared already planned or ongoing new facilities as SKA pathfinders.

The US *Large-N Small-D* concept grew out of the series of SETI Workshops held between 1997 and 1999 (Ekers et al. 2002) which led to the plan for the Allen Telescope Array (ATA) as an SKA Pathfinder for SETI as well as for radio astronomy. However, as a result of over-optimistic cost estimates, only 42 of the originally planned 350 six meter hydroformed dishes were built at the University of California's Hat Creek Observatory. China was able to build the Five-hundred-meter Aperture Spherical Telescope (FAST), which has the largest collecting area of any filled aperture radio telescope in the world[20] LOFAR[21] in the Netherlands and the MWA[22] in Western Australia, PAPER/HERA,[23] and the LWA[24] have all demonstrated the aperture-array technology at meter wavelengths. The Australia SKA Pathfinder (ASKAP)[25] is a novel array of 36 parabolic dishes, each containing a focal plane array. Perhaps the most impressive SKA Pathfinder is the South African MeerKat array of 64 antennas each 13.5 m in diameter which rivals the VLA in sensitivity.

Siting the SKA The competition for siting the SKA was intense and for nearly eight years drove much of the SKA discussion and planning. Australia, South

Africa, and Argentina each proposed sites with very low population densities and free of locally generated RFI. China proposed a site in the karst region in southern China in order to accommodate their proposed array of large spherical reflectors. The proposed US SKA Consortium *Large-N Small-D* concept was located in the US desert southwest and centered near the VLA, but with elements located throughout North America.

In 2004, the ISSC called for proposals to host the SKA. Due to a lack of resources, and perhaps lack of interest, the USSKAC, in spite of all its earlier work, did not respond with a formal proposal.[26] After careful review, the ISSC reduced the potential sites to Western Australia and the Karoo area of South Africa. The Chinese plan contained too few elements for satisfactory imaging and had high cloud cover, while the proposed Argentina site between two mountain ranges was too small to accommodate the planned configuration, and was located too close to the geomagnetic equator with corresponding ionospheric instabilities.

In 2011, both Australia and South Africa submitted extensively documented proposals to host the SKA and vigorously promoted the merits of their sites in the political as well as the scientific arena. Vernon Pankonin (NSF) chaired the SKA Siting Group (SSG) charged with overseeing the confrontational process. Following RFI tests at the two sites, extensive review by the SPDO, the SSEC, as well as by an external international SKA Site Advisory Committee (SSAC) chaired by James Moran from Harvard, the SKA was faced with a problem. Both Australian and South African radio astronomers had already invested huge resources in developing their sites, in getting their governments on board, and in preparing their proposals. Concerned that if the SKA were built in only one of these countries, the other country might drop out of the program, the new SKA Organization Board, after considering input from yet another advisory committee, the SKA Site Options Working Group (SSOWG), recommended a two-site solution. SKA-1 would include a mid-frequency array of dishes in South Africa, while Australia would host the low frequency aperture array as well as a survey telescope patterned after ASKAP.

Choosing the location of the SPDO and later the SKA Organization Headquarters was equally contentious. Cornell, Dwingeloo, and Manchester each put in a bid to host the SPDO. NRAO had submitted a Letter of Intent but withdrew, recognizing that the PrepSKA funding was not going to be spent in the US. In 2007, Manchester narrowly beat out Dwingeloo to host the SPDO, largely due to concerns that Dwingeloo was too far from an international airport. At this time, the SPDO optimistically anticipated that Phase I construction would begin by 2012. Manchester again narrowly defeated Dwingeloo and Bonn in 2011 to host the SKA Project Office during the pre-construction era, and UK funding enabled the construction of an elaborate £3.34 million headquarters building at Jodrell Bank. In a later competition to host the SKA permanent headquarters, the selection advisory panel[27] gave a close nod to Padua, Italy over Manchester, but the UK flexed its political muscle, and following a not too subtle threat to withdraw from the project, the SKA Global Headquarters ended up at Jodrell Bank (*Nature* editorial 2015).

The SKA Sans the US The first USSKAC TDP was funded following a favorable endorsement from the 2000 Decade Review, *Astronomy and Astrophysics in the New Millennium* (McKee and Taylor 2001), which suggested $22 million for SKA technology development among the top half of its recommendations for Moderate Initiatives, and specifically called attention to the significant nature of the international SKA collaboration. The 2005 report of the NSF Radio, Millimeter and Submillimeter Planning Group, chaired by Martha Haynes from Cornell, stated that "it is imperative for the future of meter to centimeter wave astronomy that the U.S. play a leadership role in the design and development of the SKA. To accomplish this, NSF must provide adequate support for the U.S. SKA technology development and demonstrator instrument programs".[28] In 2008, the AUI committee on Future Prospects for US Radio, Millimeter, and Submillimeter Astronomy, chaired by Richard McCray from the University of Colorado, gave further endorsement to "Develop the technologies for the era of Square Kilometer Array science, Develop, test, prototype, and implement the technologies required to achieve SKA-class science, [and] Review and assess the progress of the international SKA effort on a continuing basis."[29]

Encouraged by these earlier endorsements and the growing worldwide attention to the SKA, the USSKAC and the international SKA partners were optimistic that the SKA would get a green light from the 2010 Decade Review, *New Worlds New Horizons* (Blandford 2010). However, NRAO, the USSKAC, and the SPDO apparently sent mixed messages to the Blandford Committee. Knowing that the NSF was already committed to other large construction projects, NRAO led a proposal for a $40 million plan for technology development only for a North America Array (NAmA) that could lead up to construction of the high frequency component of the SKA with construction not starting until sometime after 2020. The proposed NAmA[30] had ten times the collecting area of the VLA with baselines comparable to those of the VLBA spread throughout the US. Meanwhile the SPDO and the SSEC were discussing a much earlier start of SKA construction and were not considering any sites in the US.

The National Academy's report (Blandford 2010) gave a strong endorsement to the scientific opportunities presented by the SKA as well as to the international nature of the project. But the Blandford Committee was concerned about the short time scale being discussed by the SKA project, and by the 1.7 billion Euro construction cost which they calculated to be underestimated by a factor of five or six.[31] As a result, the Committee's report did not include any recommendation for funding for either the SKA or the NAmA, which effectively eliminated any possibility of significant US participation in the SKA. The Committee did recognize that further development work was needed, and suggested that the SKA be revisited at a mid-term review.

Jacqueline Hewitt from MIT chaired a mid-term review in 2015, but the Committee only discussed the status of approved projects, and did not consider the SKA or any other new project.[32]

Earlier, the SKA had been better received by the European funding agencies, which placed the SKA and the planned Extremely Large ESO telescope on an equal basis, and, of course, both Australia and South Africa were committed to hosting their share of the SKA. While concerned about the absence of US expertise and funding, the international partners also recognized an opportunity to seize leadership in radio astronomy away from the US. In 2011, the funding agency representatives from Australia, Canada, Italy, New Zealand, the Netherlands, South Africa, and the UK[33] formed a new not-for-profit legal entity in the UK to be known as the "SKA Organization" with responsibility to oversee the "Pre-Construction" era SKA activities. The SSEC and the SPDO were dissolved effective 31 December 2011. The new SKAO Board that replaced the ASG agreed that SKA-1 should have a cost cap of 650 million Euros, a value not that different from what had been presented a few years earlier to the Blandford Committee for the full SKA project.

Each SKA Organization member country was expected to contribute 50,000 Euros to the activities of the SKAO, as well as to pledge in-kind technical development within their country using their own national resources. Although the USSKAC had no prospect of contributing money or significant in-kind resources, AUI President Ethan Schreier sent out some feelers to allow AUI to contribute to the SKAO and, on behalf of the USSKAC, to propose to the NSF for funding for the required technical development contribution. However, the AUI initiative was coolly received by both the SSEC and by the USSKAC. Reflecting the earlier agencies group concerns about Open Skies, the Founding Board adopted a no-pay no-play policy. Only scientists living and working in one of the SKA member countries would have access to the future SKA or to data from the SKA. US scientists were excluded from using the SKA, although it was noted that scientists from the SKAO Member countries traditionally had, and continued to have, free access to the NRAO VLA, VLBA, and the GBT.

John Womersley from the UK became the Chair of the new SKA Organization Founding Board, and in 2013, Phil Diamond became the new Director-General of the SKA Organization. In 2015, faced with continued project escalation and increasing cost, the SKAO was forced to a major re-baselining in order to keep to the original 650 million Euros (later inflated to 690 million) cost limit. Both SKA-1 Mid in South Africa and SKA-1 Low in Australia were descoped. Instead of starting fresh in South Africa, the SKAO decided instead to add 133 new SKA 15 m dishes to the 64 MeerKat 13.5 m dishes to form a 197 element array of dishes of mixed design. The planned survey telescope in Australia using phased array feeds based on the ASKAP Pathfinder was deferred, leaving Australia with only a descoped low frequency aperture array.

In March 2019, Ministers from Australia, China, Italy, the Netherlands, Portugal, South Africa, and the UK met in Rome to sign a treaty which established the Square Kilometre Array Observatory as an intergovernmental organization to build and operate the SKA. Guests from seven other countries, including NRAO Director Tony Beasley, were also present at the signing, but there was no official representation from the US. The agreement will enter into force after the governments of five signatories, including the hosts, Australia, South Africa, and the UK, have ratified the treaty.[34]

11.8   THE NEXT GENERATION VLA (NGVLA)

Following the disappointing rejection of Phase II of the EVLA proposal (Sect. 7.8) and the lack of endorsement of either the NAA or SKA by the Blandford Committee, the future of US centimeter radio astronomy appeared uncertain. The JVLA was, by any measure, the most powerful radio telescope in the world, but faced with the unclear future of the GBT and VLBA, as well as the growing prominence of the SKA, Tony Beasley, the new NRAO Director, was concerned about where US radio astronomy would be in 20 or 30 years. In order to delineate the science goals and help define US radio astronomy for the 2030s and beyond, Beasley organized a series of three national meetings between 2016 and 2018 on *U.S. Radio/Millimeter/Submillimeter Science Futures in the 2020s*, which was supported in part by the Kavli Foundation. The first Kavli meeting concentrated on defining the key science questions. Kavli II reviewed the various large and intermediate scale projects proposed to address the science questions posed in the first meeting, and Kavli III concentrated on defining the next generation VLA (ngVLA) which had emerged from the two earlier meetings.[35]

Starting in 2015, with funding from the NSF, NRAO, working with the US radio astronomy community, developed the science case (Murphy 2018), the performance specifications, and a reference design for the ngVLA, which will operate from 1.2 to 116 GHz (2.6 mm to 25 cm). With a total of 244 eighteen meter offset parabolic antenna elements spread throughout the US and 19 six meter dishes in a compact configuration, the ngVLA (Fig. 11.5) will have up to ten times better sensitivity than the JVLA and several times better sensitivity than SKA-1 Mid. In order to cover a wide range of surface brightness sensitivity and angular resolution, the antenna elements will be configured in four tiers and will have an angular resolution as good as 0.1 milliarcsec at the shortest wavelength.[36] When completed, the ngVLA will replace the current NRAO JVLA and VLBA.

Assuming both the ngVLA and SKA-1 are constructed, they will provide complementary science opportunities with locations in each hemisphere. SKA-1 Low in Australia will cover the frequency range from 50 to 350 MHz. SKA-1 Mid in South Africa will cover 0.35 to 15 GHz. The ngVLA will work from 1.2 GHz to 116 GHz with emphasis on the higher frequencies, and will fill in the gap between SKA-1 Mid and ALMA. In order to fully exploit these new opportunities, representatives of the SKA and ngVLA met in 2019 in Reykjavik, Iceland, to begin a discussion of a global alliance to allow the

Fig. 11.5 Artist's conception of the ngVLA. Credit: NRAO/AUI/NSF

exchange of observing time between the ngVLA and SKA-1, effectively continuing the Open Skies approach.

11.9   DIVESTMENT

As discussed in Sect. 8.10, following the NSF Senior and Portfolio Reviews, for two years the VLBA was operated separately from NRAO as the Long Baseline Observatory (LBO), but on 1 October 2018, the LBO was reintegrated into NRAO. As a result of the same NSF exercise, starting 1 October 2016, the Green Bank facilities became part of the new Green Bank Observatory (GBO), but was not later reintegrated into NRAO. Karen O'Neill, longtime NRAO Assistant Director for Green Bank Operations, became the Director of the GBO reporting to AUI Vice President Tony Beasley.

To help plan for any future GBO actions, the NSF initiated an Environmental Impact Study for five possible alternatives ranging from the preferred alternative of "Collaboration with Interested Parties for Continued Science and Education-focused Operations with Reduced NSF Funding" to the drastic "Demolition and Site Restoration."[37] Like the LBO, with the help of AUI the GBO was able to raise outside funds to help support its operations. GBO supporters included the *Breakthrough Listen* project (Sect. 5.7) and a contract with West Virginia University to support their GBT pulsar program, as well as various programs related to Space Situation Awareness. Although the non-NSF support for the GBO did not reach the NSF target goals, on 26 July 2019 the

NSF issued a Record of Decision that the GBO would continue science and education-focused operations but with reduced NSF funding.[38]

The NSF divestment of the GBO came at a price for radio astronomy. As a result of the decreased NSF funding, not only will astronomical observing with the GBT be reduced, but there will be less support for innovative new astronomical instrumentation such as PAFs and wideband spectrometers. Since the external funders are paying for GBT observing time, this means less time for Open Skies observing and, with time, it may get worse. The GBT is not only the largest fully steerable radio telescope in the world, it is the largest fully steerable microwave and millimeter wave antenna in the world, with a wide range of potential commercial, space, and defense related applications unrelated to radio astronomy. The pressure to use the GBT for other purposes, especially if defense related, combined with the NSF interest in constraining its related operating budget for astronomy, could lead to further loss of the GBT for astronomy.

NRAO started in Green Bank and for many years NRAO was synonymous with Green Bank. Hopefully, the visions of people like Lloyd Berkner, Richard Emberson, Dave Heeschen, and Alan Waterman will prevail, and Green Bank will to continue to provide world-leading radio astronomy facilities.

11.10   Lessons Learned

The near fatal consequences of the 140 Foot experience (Sect. 4.4) served as a good lesson for future NRAO projects. The VLA and VLBA were built only after years of design, development, and prototyping combined with broad community involvement. They were finished on budget and on schedule, more than met their design specifications, and with later enhancements have continued to provide unique capabilities for research. The 36 Foot millimeter telescope was finished within the budget allocation, but the completion was significantly delayed due to manufacturing challenges. Like the 140 Foot, the 36 Foot did not perform to the design specifications, but due to its excellent instrumentation provided by the Charlottesville CDL, along with competitive scheduling, it opened up a whole new field of millimeter-wave radio astronomy. By the time the GBT was built, NRAO apparently had forgotten the lessons of the 140 Foot Telescope, and the GBT construction was fraught with many of the same problems reflecting the same ambitious goals, over optimism about the complexity, cost, schedules, and perceived urgency that resulted in prematurely starting construction before the design was complete.

The GBT and VLA experiences were different. NRAO management ran a tight ship with the declared determination to finish the VLA on time and on budget. That they not only succeeded, but ended up with an instrument better in almost every way than specified in the proposal, is largely credited to Dave Heeschen's determination, but the tight control by Hein Hvatum, and some innovative contracting by Jack Lancaster were also crucial to the success of the VLA. The VLA also differed from the 36 Foot, 140 Foot and GBT projects, in that, aside from the commercial procurement of the relatively straight forward antenna elements, the VLA project was largely run "in-house," with NRAO as the prime contractor. If a small sub-contractor did not perform, they could be replaced. The VLBA, except for the record-playback system and the correlator, was a straightforward extension of existing instrumentation and practices. Many of same people involved in the VLA construction played the same role with the VLBA, which also was completed essentially on budget.

ALMA was a much more complex project. Interestingly, there was never any proposal for ALMA, nor was it ever considered by a decade review committee. Starting out in the ashes of the proposed 25 meter millimeter-wave telescope, ALMA began as the NRAO Millimeter-Array (MMA). As described in Sect. 10.7, following a series of trilateral negotiations with Japan and Europe, the MMA morphed into the more ambitious Atacama Large Millimeter/Submillimeter Array (ALMA). While ALMA was clearly a remarkable scientific success, the complexities of an international project, possibly coupled with unrealistic cost estimates, resulted in the 2004 so-called "re-baselining," with a reduction in the number of antenna elements, decreased initial instrumentation, delayed completion, and a large increase in the construction costs. Although the nature of the international ALMA agreement provides some long-term funding security, it is unlikely that the NRAO would again enter into an international project as an equal partner, but would likely be more favorably inclined toward participation either as the leading partner or as a minor partner.

NOTES

1. Martin Ryle and Antony Hewish in 1974, Arno Penzias and Robert Wilson in 1978, Russell Hulse and Joe Taylor in 1993, John Mather and George Smoot in 2006.
2. The Kitt Peak National Observatory (KPNO) was established in 1958, two years after NRAO. In 1982, KPNO was combined with the Cerro Tololo Inter-American Observatory and the National Solar Observatory to form the National Optical Astronomy Observatory (NOAO).
3. AUI-BOTXC, 16 November 1967.
4. Operation of the NRAO 12 Meter mm wavelength telescope on Kitt Peak by NRAO with NSF funding ceased in July 2000. However, with the support from the state of Arizona, the University of Arizona has continued to operate the telescope for astronomical observations.
5. J. Oort to GR, 30 August 1945, NAA-GR, Correspondence, General Correspondence I. https://science.nrao.edu/about/publications/open-skies#section-11
6. A detailed description of the events surrounding the 1959 Administrative Radio Conference is given by Findlay (1960). The NRQZ is described in https://science.nrao.edu/facilities/gbt/interference-protection/nrqz
7. Until 1993, it was called the World Administrative Radio Conference (WARC).
8. Military activities in all countries, including the US, are often the most egregious violators of ITU regulations.

9. The Committee on Space Research was established by the International Council of Scientific Unions (ICSU) in 1958 to promote international scientific research is space. During the Cold War period, COSPAR served as an important non-government conduit between the US and USSR on all issues of scientific space research.

10. The need for the shared use of the radio spectrum is discussed in the NAS report, *Spectrum Management for Science in the 21st Century*, 2010 (Washington: Nat. Acad. Press) https://doi.org/10.17226/12800. More details about radio spectrum management and the allocation of frequencies for radio astronomy are given in *The Handbook of Frequency Allocations and Spectrum Protection for Scientific Uses*, 2015 (Washington: Nat. Acad. Press). https://doi.org/10.17226/21774

11. In 2019, a Virginia Tech professor was found guilty of spending NSF grant money "for research Zhang knew had already been done in China. Zhang intended to use the grant funds for other … projects rather than for the projects for which the funds were requested." https://www.justice.gov/usao-wdva/pr/former-virginia-tech-professor-found-guilty-grant-fraud-false-statements-obstruction

12. Described in the NSF Proposal & Award Policies and Procedures Guide https://www.nsf.gov/publications/pub_summ.jsp?ods_key=nsf18001 as "full participation of women, persons with disabilities, and underrepresented minorities in science, technology, engineering, and mathematics (STEM); improved STEM education and educator development at any level; increased public scientific literacy and public engagement with science and technology; improved well-being of individuals in society; development of a diverse, globally competitive STEM workforce; increased partnerships between academia, industry, and others; improved national security; increased economic competitiveness of the US; and enhanced infrastructure for research and education."

13. 28–29 October 1963 meeting notes, NAA-NRAO, Founding and Organization, Antenna Planning, Box 2.

14. Bolton to Bowen, 1 December 1964. NAA-KIK Open Skies.

15. PVB at Jansky Monument Dedication, NAA-KGJ Additional Materials Related to Jansky.

16. Reber's array operated at relatively long wavelengths where a large effective collecting area can be built out of wire dipoles and is relatively cheap. It is much more complex and much more expensive to achieve the same collecting area at decimeter and centimeter wavelengths.

17. A Luneburg lens is made from a spherical dielectric with an index of refraction that varies radially from the center. It has the property to focus incoming radio emission at point that can then be detected with a horn and conventional radiometer. The idea for the SKA was to have a large number of feed-receiver arrangements all looking in different directions. However, the cost of building sufficiently large Luneburg lenses and the losses in the dielectric were both too great to seriously consider this technology for the SKA.

18. NAA-KIK, Square Kilometre Array, USSKAC.

19. SKA Report 84, Report of the SKA Operations Working Group, https://www.skatelescope.org/uploaded/37483_memo_84.pdf

20. The construction of the Five-hundred-meter Aperture Spherical Telescope (FAST) in southern China was completed in 2016. Patterned after the Arecibo

radio telescope, FAST has about twice the effective collecting area as Arecibo. Its 4,400 triangular panels form an active surface that can be adjusted to form a 300 m diameter parabolic surface pointing as much as 40 degrees from the zenith. https://arxiv.org/ftp/arxiv/papers/1105/1105.3794.pdf

21. The LOw Frequency ARray (LOFAR) began as a joint project of the US, the Netherlands, and Australia to construct a large aperture array to operate at meter wavelength below 250 MHz. An evaluation of potential sites and consideration of potential RFI suggested a preference for Australia, the US Southwest, or Europe in that order. However, Dutch radio astronomers were able to secure funding from the Dutch government to build LOFAR in the northeastern Netherlands as a multi-disciplinary sensor array to facilitate research in geophysics, computer sciences, and agriculture, as well as astronomy. International stations located throughout Europe have since been added.

22. The MWA is an international project including MIT/Haystack with partial funding from the NSF, and includes 256 tiles each containing 16 antenna elements. The MWA simultaneously images an area about 30 degrees in diameter at multiple frequencies in the range 70–230 MHz (1.3–4.3 m). http://www.mwatelescope.org/

23. The Precision Array to Probe the Epoch of Reionization (PAPER) was an aperture array SKA precursor first tested in Green Bank, further developed in Western Australian, and later built at the South African SKA site. PAPER was succeeded by the Hydrogen Epoch of Reionization Array (HERA). HERA, which abandoned the aperture array concept, consists of a large grid of 14 meter (46 foot) non-tracking dishes.

24. The University of New Mexico had been heavily involved in promoting the location of LOFAR in the US southwest. When the decision was made to build LOFAR in the Netherlands, the University of New Mexico proceeded to build the Long Wavelength Array (LWA) on the VLA site. http://www.phys.unm.edu/~lwa/index.html

25. Each 12 m diameter ASKAP antenna contains a 94-element dual polarization focal plane array that forms 36 independent beams on the sky. ASKAP operates in the frequency range 700 MHz to 1.8 GHz (16.7–43 cm) and is co-located with the MWA on the Australian SKA site in Western Australia. https://www.atnf.csiro.au/projects/askap/index.html

26. The plan to site the SKA in the US was considered inadequate by many due to the expected higher levels of RFI, but there was also concern among some US ISSC members that a US site would likely mean undesired control by NRAO. Because of changing staff, the University of New Mexico was not able to proceed with the planned study of site availability and the existing fiber network. At the time, NRAO was preoccupied with building both the EVLA and ALMA and could not spare the resources needed to develop a credible site proposal.

27. The headquarters selection panel included representatives from Australia, South Africa, the Netherlands, and ESO.

28. http://hosting.astro.cornell.edu/~haynes/rmspg/docs/rmspgreport.pdf

29. NAA-AUI.

30. https://www.nrao.edu/nio/naa/

31. Having seen huge cost escalation in projects recommended by previous decade surveys, especially for NASA space missions, the Blandford Committee was

more sensitive to cost estimates than was the case for previous Decade Reviews, and commissioned the Aerospace Corporation to provide independent cost estimates for projects being considered by the Committee. The Aerospace Corporation reported that the true cost of constructing the SKA was at least five times greater than what was presented to the Committee.

32. New Worlds, New Horizons: A Midterm Assessment. https://www.nap.edu/catalog/23560/new-worlds-new-horizons-a-midterm-assessment

33. Later, China and Sweden also joined the SKA Organization (SKAO), and India joined as only an Associate Member. Germany has not formally joined the SKAO but through the German Max Planck Society plans to contribute resources to MeerKat and SKA-1 Mid. The 2011 site proposal from Australia included long baseline sites in New Zealand, which were later dropped when the decision was made to locate SKA-1 Mid in South Africa. Faced with no clear role in the SKA and competing programs for limited resources, New Zealand decided in 2019 to withdraw from the SKAO.

34. The Anticipated SKA1 Science performance is given by Braun at https://astronomers.skatelescope.org/wp-content/uploads/2017/10/SKA-TEL-SKO-0000818-01_SKA1_Science_Perform.pdf

35. Kavli I was held in Chicago on 15–17 December 2016. https://science.nrao.edu/science/meetings/2015/2020futures/program, Kavli II was held in Baltimore, Maryland from 3–5 August 2017, http://www.cvent.com/events/u-s-radio-millimeter-submillimeter-science-futures-ii/custom-18-b7c37ec376c44055b80cfb2f5ef030b5.aspx, and Kavli III held in Berkley, California from 2–4 August 2017, http://www.cvent.com/events/u-s-radio-millimeter-submillimeter-science-futures-iii/custom-22-a7c6d735d0b141eca298518ce31cbaae.aspx

36. McKinnon, M. et al. ngVLA: The Next Generation Very Large Array, White Paper submitted to the Astro2020 Decade Review Committee. http://surveygizmoresponseuploads.s3.amazonaws.com/fileuploads/623127/5043187/137-fe4771da409a7413465c9bb1cb579ae7_McKinnonMarkM.pdf

37. NSF Environmental Impact Statement for the Green Bank Observatory, 22 February 2019. https://www.nsf.gov/mps/ast/env_impact_reviews/greenbank/eis/FEIS.pdf

38. https://www.nsf.gov/mps/ast/env_impact_reviews/greenbank/GBO_ROD_Final_72619.pdf

Bibliography

References

Blandford, R.D. ed. 2010, *New Worlds, New Horizons* (Washington: NAS Press)

Bourke, T. et al. eds. 2015, *Advancing Astrophysics with the Square Kilometre Array*, Proceedings of Science (Manchester: SKA Organization) www.skatelescope.org/books/

Carilli, C.C. and Rawlings, S. eds. 2004, *Science with the Square Kilometre Array* (Amsterdam: Elsevier)

Christiansen, W.N. and Högbom, J.A. 1961, The Cross Antenna of the Proposed Benelux Radio Telescope, *Nature*, **191**, 215

DeVorkin, D. 2000, Who Speaks for Astronomy? How Astronomers Responded to Government Funding after World War II, *Historical Studies in the Physical and Biological Sciences*, **31** (1), 55

Edge, D.O. and Mulkay, M.J. 1976, *Astronomy Transformed: the Emergence of Radio Astronomy in Britain* (New York: Wiley)

Ekers, R. 1978, The Convergence of Optical and Radio Techniques. In *ESO Conference Optical Telescopes of the Future*, ed. F. Pacini et al. (Geneva: ESO), 387

Ekers, R.D. et al. eds. 2002, *SETI 2020: A Roadmap for the Search for Extraterrestrial Intelligence* (Mountain View: SETI Press)

Findlay, J.W. 1960, Commission V On Radio Astronomy, Protecting Frequencies for Radio Astronomy, *URSI Information Bulletin*, **124**

Findlay, J.W. 1991, I.U.C.A.F. and Frequencies for Radio Astronomy. In ASPC **17**, *Light Pollution, Radio Interference, and Space Debris*, ed. D. L. Crawford (San Francisco: ASP), 194

Hall, P. ed. 2005, *The SKA; An Engineering Perspective* (Dordrecht: Springer)

Harwit, M. 2015, The Impacts of Military, Industrial, and Private Support on Modern Astronomy, paper 90.06 presented at 225th AAS meeting

Heeschen, D.S. 1991, Reminiscences of Early Days of the VLA. In ASPC **19**, *Radio Interferometry: Theory, Techniques, and Application*, ed. T.J. Cornwell and R.A. Perley (San Francisco: ASP), 150

Kellermann, K.I. and Cohen, M.H. 1988, The Origin and Evolution of the NRAO-Cornell VLBI System, *JRASC*, **82**, 24

Kellermann, K.I. and Moran, J. 2001, The Development of High-Resolution Imaging in Radio Astronomy, *ARAA*, **39**, 457

McKee, C.F. and Taylor, J.H. eds. 2001, *Astronomy and Astrophysics in the New Millennium* (Washington: NAS Press)

Munns, D.P.D. 2013, *A Single Sky: How an International Community Forged the Science of Radio Astronomy* (Cambridge: MIT Press)

Murphy, E. ed. 2018, *Science with the Next Generation Very Large Array*, ASP Monograph 7 (San Francisco: ASP)

Nature 2015, un-authored editorial, *Nature*, **519**, 129

Noordam, J. 2013, The Dawn of the SKAI: What Really Happened. In *Resolving the Sky – Radio Interferometry: Past, Present, and Future*, ed. M.A. Garrett and J.C. Greenwood (Manchester: SKA Organization), 68

Oliver, B.M. and Billingham, J. eds. 1973, *Project Cyclops: A Design Study for a System for Detecting Extraterrestrial Intelligent Life*, NASA CR114445 (originally published 1972, revised 1973, reprinted 1996 by the SETI League and the SETI Institute with additional material)

Parijiskij, Yu. N. 1992, Radio Astronomy of the Next Century, *Astronomy and Astrophysical Transactions*, **1**, 85

Reber, G. 1968, Cosmic Static at 144 Meters Wavelength, *J. Franklin Institute*, **285** (1), 1

Smolders, A.B. and van Haarlem, M.P. 1999, *Perspectives on Radio Astronomy: Technologies for Large Antenna Arrays* (Dwingeloo: ASTRON)

Sullivan, Walter 1959, *New York Times*, 20 September, 27

Sullivan, W.T. III 2009, *Cosmic Noise* (Cambridge: CUP)

Swarup, G. 1981, Proposal for an International Institute for Space Science and Electronics and for a Giant Equatorial Radio Telescope, *Bulletin of the Astronomical Society of India*, **9**, 269

Swarup, G. 1991, Giant Metrewave Radio Telescope (GMRT). In ASPC **19**, *Radio Interferometry, Theory, Techniques, and Applications*, ed. T.J. Cornwell and R.A. Perley (San Francisco: ASP), 376

Taylor, A.R. and Braun, R. eds. 1999, *Science with the Square Kilometre Array: A Next Generation World Radio Observatory* (Dwingeloo: ASTRON)

Wilkinson, P.N. 1991, The Hydrogen Array. In ASPC **19**, *Radio Interferometry, Theory, Techniques, and Applications*, ed. T.J. Cornwell and R.A. Perley (San Francisco: ASP), 428

Further Reading

Cruz-Pol, S. 2019, *RF Spectrum Management* (Mayagüez: Cruz-Pol)

DeBoer, D.R. 2013, Radio Frequencies: Policy and Management, *IEEE Trans. Geosciences and Remote Sensing*, **51**, 4918

Ekers, R.D. 2013, The History of the Square Kilometer Array (SKA) Born Global. In *Resolving the Sky – Radio Interferometry: Past, Present, and Future*, ed. M.A. Garrett and J.C. Greenwood (Manchester: SKA Organization), 68

Findlay, J.W. 1962, Protecting the Science of Radio Astronomy, *Science*, **137**, 829

Robinson, B. 1999, Frequency Allocation: The First Forty years, *ARAA*, **37**, 65

Sullivan, W.T. 2009, The History of Radio Telescopes, 1945-1990, *Experimental Astron*, **25**, 107

Correction to: Open Skies: The National Radio Astronomy Observatory and Its Impact on US Radio Astronomy

CORRECTION TO: K. I. KELLERMANN ET AL., OPEN SKIES,
HISTORICAL & CULTURAL ASTRONOMY,
HTTPS://DOI.ORG/10.1007/978-3-030-32345-5

A number of corrections were unfortunately missed during the proofing and correction process. The version supplied here has been updated.

FM page iv
In cover caption, changed from "Plains of San Agustin, 50 miles east of Socorro" to "Plains of San Agustin, 50 miles **west** of Socorro"

Page 39, line 14
Added diacriticals to François, Émile

Page 40, line 4 in the Japan section
Added **T.** Hatanaka

Page 42, paragraph 2, lines 7-8
Changed from: "On 12 February two German **destroyers**" to "On 12 February two German **warships**"

Page 47, paragraph 2, line 6
Changed "strong sources appear weaker than any given flux density level" to "strong sources appear weaker **at** any given flux density level"

Page 51, lines 8-9
Changed from: "Wurzburg antenna at Kootwijk **on the Dutch coast,** Muller and Oort." to "Wurzburg antenna at Kootwijk **in central Holland,** Muller and Oort."

Page 61, line 4
Changed from "inner 20 **meter** diameter" to "inner 20 **foot** diameter"

Page 63, Fig. 2.8 caption

The updated online version of the book can be found at
https://doi.org/10.1007/978-3-030-32345-5

© The Author(s) 2021
K. I. Kellermann et al., *Open Skies*, Historical & Cultural Astronomy,
https://doi.org/10.1007/978-3-030-32345-5_12

C1

Changed from: "Prof. Jan Oort and **Mrs. Oort** leaving the **1955** dedication of the Dwingeloo 25 meter radio telescope." to "Prof. Jan Oort and **Queen Juliana** leaving the **1956** dedication of the Dwingeloo 25 meter radio telescope."

Page 65, line 2

Changed "Haleakula" to "Haleakela"

Page 75 Leverington, D. reference

Changed "Observations and Telescopes of Modern Times" to "Observatories and Telescopes of Modern Times"

Page 81

Changed "Berkly" to "Berkeley"

Page 103

Removed extra "a" from "an urgent **a** need"

Page 104, middle paragraph, last line

Removed extra "a" from "as a major **a** program"

Page 130, line 5

Changed from "operated with ten miles" to "operated **within** ten miles"

Page 168, Fig. 4.5 caption

Changed from "Dedication of the 85 Foot Howard E. Tatel Telescope, 16 October **1968**."

to "Dedication of the 85 Foot Howard E. Tatel Telescope, 16 October **1958**."

Page 193, 6 lines from bottom

Added "the" in "…it is **the** largest equatorial"

Page 215, line 9

Changed from "debated the augments" to "debated the **arguments**"

Page 272, Fig. 6.5, last line of caption

Changed from "J. Findley" to "J. **Findlay**"

Page 275, 4 lines from bottom

Changed from "von Hoerner begin using" to "von Hoerner **began** using"

Page 282, last paragraph, line 4

Changed from "The WMAP satellite contained **20** NRAO HEMT amplifiers." to "The WMAP satellite contained **80** NRAO HEMT amplifiers."

Page 288-289

Changed from "Stephen Chu, a 1970 student, went on to win a Nobel Prize in Physics and later became Secretary of the Interior in Barack Obama's administration." to "Stephen Chu, a 1970 student, went on to win a Nobel Prize in Physics and later became Secretary of Energy in Barack Obama's administration."

Page 291, line 11

Changed from "known as VRO 46.26.01" to "known as **VRO 42.22.01**"

Page 291, line 3 of U Mich section

Added "to" in "used **to** locate Japanese ships"

Page 297, paragraph 2, line 10

Changed from "a large track of land" to "a large **tract** of land"

Page 306, 8 lines from end

Changed from "Instead it he sent it back with the terse note" to "Instead, he sent it back with the terse note"

Page 307, line 10

Deleted extra "that" from "large that **that** the atmosphere"

Page 340, line 6

Changed from "visiting from **Sweden** in 1967" to "visiting from **Holland** in 1967"

Page 396, line 10

Changed from "August 1967 URSI General Assembly held in Montreal." to "August 1967 URSI **Radio Science Meeting** held in Montreal"

Page 462, middle paragraph, 2 lines from end

Changed from "it would take a freak **1989** accident" to "it would take a freak **1988** accident."

Page 480, 2 lines from bottom

Changed from "von Hoerner commented that LSFT" to "von Hoerner commented that **LFST**"

Page 507, Fig. 9.12

Changed caption from "100 meter telescope concepts. Right: axially symmetrical design; Left: offset reflector 100 design." to "100-meter telescope design concepts. **Left: axially symmetrical design; Right: off-set reflector design.**"

Page 566, 7 lines from bottom

Changed from "the primary criteria" to "the primary **criterion**"

Page 571, fig. 10.9

Changed caption from ".... Fred Lo (NRAO Director), Bob Brown (NRAO), Catherine Cesarsky (ESO Director General) and Daniel Hofstadt (ESO)." to "... Fred Lo (NRAO Director), **Massimo Tarenghi (ESO),** Catherine Cesarsky..."

Page 601, bottom line

Added "of" in the line "Most **of** the world's major radio telescopes"

Page 640, index entry for Jansky, Curtis Moreau

Changed from "Jansky, Curtis Moreau" to "Jansky, **Cyril** Moreau"

Appendix A

Abbreviations and Acronyms Used in the Text

1hT	One hectare Radio Telescope
AAC	Astronomy Advisory Committee [NSF]
AAEO	Astronomy, Atmospheric, Earth and Ocean Sciences [NSF Division]
AAS	American Astronomical Society
ACAST	Advisory Committee for the NSF Division of Astronomical Sciences
ACC	ALMA Coordinating Committee
AEC	ALMA Executive Committee
AEG	Antenna Evaluation Group [ALMA]
AFOSR	Air Force Office of Scientific Research [US]
AGN	Active galactic nuclei
AHB	Alan H. Bridle
AIO	Arecibo Ionospheric Observatory
AIL	Airborne Instruments Laboratory
AIPS	Astronomical Image Processing System
ALMA	Atacama Large Millimeter Array, Atacama Large Millimeter/submillimeter Array
alt-az	Alt-azimuth
ANAS	Archives of the National Academy of Sciences
AOC	Array Operations Center [NRAO, see also DSOC]
AORG	Army Operational Research Group [UK]
APEX	Atacama Pathfinder Experimental Telescope
ARA	Associates for Research in Astronomy
ARC	ALMA Regional Center

© The Author(s) 2020
K. I. Kellermann et al., *Open Skies*, Historical & Cultural Astronomy,
https://doi.org/10.1007/978-3-030-32345-5

ARISE	Advanced Radio Interferometry between Space and Earth
ARO	Algonquin Radio Observatory
ARPA	Advanced Research Project Agency, later DARPA [US]
ASC	Astro Space Center [Russia]
ASG	Agencies SKA Group
ASIAA	Academica Sinica Institute for Astronomy and Astrophysics
ASICs	Application Specific Integrated Circuits
ASKAP	Australia SKA Prototype
AST	NSF Astronomy Division
ASTE	Atacama Submillimeter Telescope Experiment
ATA	Allen Telescope Array
ATM	Alan T. Moffet
AUI	Associated Universities, Inc.
AURA	Associated Universities for Research in Astronomy
AZ	Arizona
BFB	Bernard F. Burke
BGC	Barry G. Clark
BIMA	Berkeley-Illinois-Maryland Association
BOB	Bureau of the Budget [US]
BNL	Brookhaven National Laboratory
BOB	Bureau of the Budget
BSS	Beyond Southern Skies
CA	California
CAMROC	Cambridge Radio Observatory Committee
CARMA	Combined Array for Research in Millimeter-Wave Astronomy
CASA	Common Astronomical Software Application
CCIR	International Radio Consultative Committee
CDL	Central Development Laboratory [NRAO]
CfA	Harvard-Smithsonian Center for Astrophysics
CFRP	Carbon fiber reinforced plastic
CIW	Carnegie Institution of Washington
CLBA	Canadian Long Baseline Array
CMW	Campbell M. Wade
CNRS	Centre National de la Recherche Scientifique
COMSAT	Communications Satellite Corp
CONACyT	Consejo Nacional de Ciencia y Technologia
CORF	Committee on Radio Frequencies
COSPAR	Committee on Space Research
COSPUP	Committee on Science and Public Policy [NAS]
CRPL	Central Radio Propagation Laboratory [NBS]
CRS	Congressional Research Service
CRSI-TP	COMSAT-RSI Technical Products
CSIR	Council for Scientific and Industrial Research [Australia]

CSIRO	Commonwealth Scientific and Industrial Research Organization [Australia]
CSO	Caltech Submillimeter Observatory
CTS	Communications Technology Satellite [Canada]
DARPA	Defense Advanced Research Projects Agency [US]
DEW	Distant Early Warning
DKIST	Daniel K. Inoue Solar Telescope
DoD	Department of Defense [US]
DSH	David S. Heeschen
DSOC	Domenici Science Operations Center [see also AOC]
DTM	Department of Terrestrial Magnetism of the Carnegie Institution of Washington
ELT	Extremely Large Telescope
ENB	Ellen N. Bouton
EoR	Epic of Reionization
ESA	European Space Agency
ESO	European Southern Observatory
ESSCO	Electronic Space Systems Corporation
EVLA	Expanded Very Large Array
FAST	Five hundred meter Aperture Spherical Telescope
FCC	Federal Communications Commission
FCRAO	Five College Radio Astronomy Observatory
FET	Field Effect Transistor
FIRST	Fast Imaging Survey of the Twenty Centimeter Sky
GaAsFET	Gallium Arsenide Field Effect Transistor
GAS	George A. Seielstad
GBI	Green Bank Interferometer
GBO	Green Bank Observatory
GBT	Green Bank Telescope
GERT	Giant Equatorial Radio Telescope [India]
GJS	Gordon J. Stanley
GMRT	Giant Metrewave Radio Telescope [India]
GMT	Giant Magellan Telescope
GR	Grote Reber
GRT	Giant Radio Telescope [Australia]
GSMT	Giant Segmented Mirror Telescope
GVWG	Global VLBI Working Group
HALCA	Highly Advanced Laboratory for Communications and Astronomy
HEAO	High Energy Astronomy Observatory
HEMT	High Electron Mobility Transistor
HERA	Hydrogen Epoch of Reionization
HHT	Heinrich Hertz Telescope
HPBW	Half power beam width

HRMS	High Resolution Microwave Survey [previously called MOP]
HSA	High Sensitivity Array
IC	Integrated circuit
ICSU	International Council of Scientific Unions
IFAG	Informal Funding Agencies Group [SKA]
IKI	Institut Kosmicheskih Issledovanyi (Space Research Institute) [Russia]
INAOE	Instituto Nacional de Astrofísica, Óptica y Electrónica [Mexico]
IRAC	Interdepartmental Radio Advisory Committee
IRAM	Institut de Radioastronomie Millimétrique
ISAS	Institute of Space and Astronautical Science [Japan]
ISM	Interstellar medium
ISPO	International SKA Project Office
ISSC	International SKA Steering Committee
ITU	International Telecommunications Union
IUCAF	Inter-Union Committee on the Allocation of Frequencies
IVC	International Video Corporation
IVS	International VLBI Satellite
JAO	Joint ALMA Office
JCMT	James Clerk Maxwell Telescope
JVLA	Jansky Very Large Array
JIVE	Joint Institute for VLBI in Europe
JLG	Jesse L. Greenstein
JPL	Jet Propulsion Laboratory
JWF	John W. Findlay
KGJ	Karl G. Jansky
KIK	Kenneth I. Kellermann
L-SAT	[European] Large Satellite
LABOCA	Large Apex Bolometer Camera
LAD	Lee A. DuBridge
LADWP	Los Angeles Department of Water and Power
LIGO	Laser Interferometer Gravity-Wave Observatory
LMT	Large Millimeter Telescope
LNSD	Large Number Small Diameter
LOC	Library of Congress
LOFAR	LOw Frequency ARRay
LORAN	Long Range Navigation
LSA	Large Southern Array
LSMA	Large Millimeter and Submillimeter Array [Japan]
LSST	Large Synoptic Survey Telescope
LWA	Long Wavelength Array
MA	Massachusetts
MAG	Mark A. Gordon

MDC	MMA Development Consortium
MERLIN	Multi-element Radio Linked Interferometric Network
MIRIAD	Multichannel Image Resolution, Image Analysis, and Display
MIT	Massachusetts Institute of Technology
MK	Mark [MK I, MK II, MK III]
MMA	Millimeter Array
MMAOC	MMA Oversight Committee [NSF]
MoA	Memorandum of Agreement
MOP	Microwave Observing Project (later called HRMS)
MoU	Memorandum of Understanding
MPG	Max Planck Gesellschaft
MPIfR	Max-Planck-Institut für Radioastronomie
MPE	Mathematical, Physical and Engineering Research [NSF Directorate]
MPS	Mathematical and Physical Sciences [NSF Directorate]
MRE	Major Research Equipment [NSF]
MREFC	Major Research Equipment and Facilities Construction [NSF]
MSR	Morton S. Roberts
MUSA	Multiple Unit Steerable Array
MWA	Murchison Widefield Array
MWO	Millimeter Wave Observatory [Univ. Texas]
MWPO	Mount Wilson and Palomar Observatories
NAmA	North American Array
NAASC	North American ALMA Science Center
NAIC	National Astronomy and Ionospheric Center
NAPRA	North American Program in Radio Astronomy
NAS	National Academy of Sciences [US]
NASA	National Aeronautics and Space Administration
NASEM	National Academies of Science, Engineering, and Medicine [US]
NBS	National Bureau of Standards [US]
NEON	National Ecological Observatory Network
NEROC	North East Radio Observatory Corporation
NFRA	Netherlands Foundation for Research in Astronomy
NGST	Next Generation Solar Telescope
ngVLA	next generation Very Large Array
NIOC	Navy Information Operations Command [US]
NLSRT	New Large Steerable Radio Telescope
NM	New Mexico
NOAO	National Optical Astronomy Observatory
NOEMA	NOrthern Extended Millimeter Array [IRAM]
NOO	National Optical Observatory
NRAO	National Radio Astronomy Observatory

NRC	National Research Council [US]
NRL	Naval Research Laboratory
NRQZ	National Radio Quiet Zone
NSB	National Science Board [US]
NSF	National Science Foundation
NTIA	National Telecommunications and Information Administration
NTT	New Technology Telescope
NUG	Network Users Group
NV	Nevada
NVSS	NRAO VLA Sky Survey
OECD	Organization for Economic Co-operation and Development
OIR	Optical and Infrared
OMB	Office of Management and Budget [US]
ONR	Office of Naval Research [US]
ORINS	Oak Ridge Institute for Nuclear Studies
OSF	Operations Support Facility [ALMA]
OSO	Onsala Space Observatory
OSTP	Office of Science and Technology Policy [US]
OVA	Owens Valley Array
OVI	Owens Valley Interferometer
OVLBI	Orbiting Very Long Baseline Interferometry
OVRO	Owens Valley Radio Observatory
PAF	Phased Array Feed
PAPER	Precision Array to Probe the Epoch of Reionization
PPARC	Particle Physics and Astronomy Research Council [UK]
PrepSKA	Preparatory Phase for the SKA
PVB	Paul A. Vanden Bout
PSAC	President's Science Advisory Committee
QUASAT	Quasar Satellite
RFB	Robert F. Bacher
RFI	Radio frequency interference
RFP	Request for Proposals
RH	Robyn Harrison
RISC	RadioAstron International Science Council
RLB	Robert L. Brown
RRL	Radio recombination lines
RSI	Radiation Systems, Inc.
SAGMA	Scientific Advisory Group for Millimetre Astronomy
SAMURAI	Science of AGNs and Masers with Unprecedented Resolution in Astronomical Imaging
SAO	Smithsonian Astrophysical Observatory
SCUBA	Submillimetre Common-Use Bolometer Array
SDRA	Serendipitous Discoveries in Radio Astronomy

SDSS	Sloan Digital Sky Survey
SEB	Sierra E. Brandt
SEST	Swedish ESO Submillimetre Telescope
SERENDIP	Search for Extraterrestrial Radio Emissions from Nearby Developed Intelligent Populations
SETI	Search for Extraterrestrial Intelligence
SEU	Single event upsets
SIRTF	Space Infrared Telescope Facility
SIS	Superconductor-Insulator-Superconductor
SKA	Square Kilometre Array
SKADS	SKA Design Studies
SKAI	SKA Interferometer
SKAO	SKA Organization
SMA	Sub-Millimeter Array [Harvard-Smithsonian]
SMART	Stretched Mesh Attached to Rope Trusses
SMT	Submillimeter Telescope [MPIfR-UoA]
SPDO	SKA Project Development Office
SRI	Stanford Research Institute
SSAC	SKA Site Advisory Committee
SSB	Space Studies Board (of the NAS)
SSG	SKA Siting Group
SSOWG	SKA Site Options Working Group
SSTWG	SETI Science and Technology Working Group
STFP	Science and Technology Facilities Panel [UK]
SURA	Southeastern Universities Research Association
TAC	Time Allocation Committee
TDP	Technology Development Program [US SKA]
TDRSS	Tracking and Data Relay Satellite System
TMT	Thirty Meter Telescope
TSS	Targeted Search System
TRR	Theodore R. Riffe
TVA	Tennessee Valley Authority
TX	Texas
UK	United Kingdom
UNAM	Universidad Nacional Autonoma de México
UoA	University of Arizona
URSI	International Union of Radio Science
US	United States [of America]
USNO	United States Naval Observatory
USSKAC	US SKA Consortium
USSR	Union of Soviet Socialist Republics
VA	Virginia
VCR	Video Cassette Recorder
VISC	VSOP International Steering Committee
VLA	Very Large Array

VLBA	Very Long Baseline Array
VLBI	Very Long Baseline Interferometry
VLD	Very Large Dish
VLT	Very Large Telescope [ESO]
VSOP	VLBI Space Observatory Program
WIDAR	Wide-band Interferometric Digital Architecture
WRC	World Radiocommunication Conference
WSRT	Westerbork Synthesis Radio Telescope
WTS	Woodruff T. Sullivan III
WV	West Virginia
WVRAZA	West Virginia Radio Astronomy Zoning Act
WW II	World War II

CITATION ABBREVIATIONS FOR NRAO/AUI ARCHIVES MATERIALS

AUI-BOT	AUI Board of Trustees Minutes
AUI-BOTXC	AUI Board of Trustees Executive Committee Minutes
NAA	NRAO/AUI Archives
NAA-AHB	NRAO/AUI Archives, Papers of Alan H. Bridle
NRAO-AUI	NRAO/AUI Archives, Records of Associated Universities Inc.
NAA-BFB	NRAO/AUI Archives, Papers of Bernard F. Burke
NAA-DSH	NRAO/AUI Archives, Papers of David S. Heeschen
NAA-GR	NRAO/AUI Archives, Papers of Grote Reber
NAA-JDK	NRAO/AUI Archives, Papers of John D. Kraus
NAA-JWF	NRAO/AUI Archives, Papers of John W. Findlay
NAA-KIK	NRAO/AUI Archives, Papers of Kenneth I. Kellermann
NAA-MAG	NRAO/AUI Archives, Papers of Mark A. Gordon
NAA-MSR	NRAO/AUI Archives, Papers of Morton S. Roberts
NAA-NRAO	NRAO/AUI Archives, Records of the National Radio Astronomy Observatory
NAA-NRAO DO	NRAO/AUI Archives, Records of the National Radio Astronomy Observatory, Director's Office
NAA-PVB	NRAO/AUI Archives, Papers of Paul A. Vanden Bout
NAA-RLB	NRAO/AUI Archives, Papers of Robert L. Brown
NAA-WTS	NRAO/AUI Archives, Papers of Woodruff T. Sullivan III

CITATION ABBREVIATIONS FOR OTHER ARCHIVAL MATERIALS

CITA	California Institute of Technology Archives
CITA-ATM	Papers of Alan T. Moffet, California Institute of Technology Archives
CITA-GJS	Papers of Gordon J. Stanley, California Institute of Technology Archives
CITA-JLG	Papers of Jesse L. Greenstein, California Institute of Technology Archives
CITA-LAD	Papers of Lee A. DuBridge, California Institute of Technology Archives
CITA-RFB	Papers of Robert F. Bacher, California Institute of Technology Archives
DDE	Dwight D. Eisenhower Presidential Library
DTMA	Radio Astronomy Program Records, 1950–1976, Department of Terrestrial Magnetism, Carnegie Institution of Washington
HLA-IB	Papers of Ira Bowen, Huntington Library Archives
HUA	Harvard University Archives
LOC-ATW	Papers of Alan T. Waterman, Library of Congress
LOC-IIR	Papers of I.I. Rabi, Library of Congress
NAAustrl	National Archives of Australia
NAS-NRC-A	National Academy of Sciences, National Research Council, Archives
RCA	Research Corporation Archives

Appendix B

NRAO Timeline

5 May 1933	Karl G. Jansky's detection of radio waves announced in the *New York Times*.
Summer 1937	Grote Reber constructs first parabolic radio telescope in Wheaton, IL.
1940	Reber publishes *Cosmic Static* papers in ApJ and Proc. IRE.
18 July 1946	Charter from NY Board of Regents establishes AUI.
30 July 1946	Edward Reynolds becomes first AUI President.
19 November 1948	Frank D. Fackenthal becomes AUI President.
16 February 1951	Lloyd V. Berkner becomes AUI President.
April 1951	Alan T Waterman becomes first NSF Director.
26–27 December 1953	AAAS meeting in Boston includes review papers on radio astronomy.
4–7 January 1954	Washington Conference on Radio Astronomy discusses US radio astronomy.
3 May 1954	NSF establishes an Advisory Panel for Radio Astronomy.
20 May 1954	Berkner proposes asking NSF for money; Ad Hoc committee established.
8 February 1955	AUI receives $85K from NSF to begin work on radio astronomy facility.
13 December 1955	Steering Committee recommends Green Bank WV for observatory site.
26 July 1956	NSF plans to purchase land in Green Bank; appropriates initial $3.5 million.
9 August 1956	WV Radio Astronomy Zoning Act becomes first-ever protective legislation.
17 November 1956	NSF/AUI agreement establishes NRAO with AUI as the managing agency; Lloyd Berkner becomes Acting Director of the NRAO.
14 May 1957	Offices opened on Green Bank site.
17 October 1957	Dedication of Observatory in Green Bank.
Summer 1958	Construction begins on 85 Foot, Little Big Horn, first buildings.
14 August 1958	Ground breaking for 140 Foot Telescope.
16 October 1958	Dedication of the Howard E. Tatel 85 Foot Telescope.
19 November 1958	FCC establishes National Radio Quiet Zone (NRQZ).

(continued)

© The Author(s) 2020
K. I. Kellermann et al., *Open Skies*, Historical & Cultural Astronomy,
https://doi.org/10.1007/978-3-030-32345-5

13 February 1959	First observations with Tatel 85 Foot Telescope.
1 July 1959	Otto Struve becomes first NRAO Director.
11 April 1960	Project Ozma, first observations.
1 December 1960	Leland J. Haworth becomes AUI President.
1 April 1961	Edward Reynolds becomes AUI President.
21 April 1961	Isidor I. Rabi becomes AUI President.
27 April 1961	Ground breaking for 300 Foot Telescope.
1 December 1961	Otto Struve resigns as NRAO Director; David S. Heeschen becomes Acting NRAO Director.
17 December 1961	J.L. Pawsey appointed as NRAO Director, effective October 1, 1962.
March 1962	Pawsey visits Green Bank; first symptoms of illness appear.
1 October 1962	300 Foot Telescope operational with continuum receivers for 750 and 1400 MHz.
19 October 1962	Gerald F. Tape becomes AUI President.
19 October 1962	David S. Heeschen becomes NRAO Director.
30 November 1962	Pawsey dies after extended illness.
1962	Funding request for ~30 foot mm-wave telescope.
1963	First digital autocorrelator built by Sander Weinreb, used on Tatel Telescope.
July 1963	Leland J. Haworth becomes NSF Director
1 July 1963	Hein Hvatum becomes Head, Electronics Division; John W. Findlay becomes Deputy Director.
10 July 1963	Edward Reynolds becomes AUI President.
1964	*Ground-Based Astronomy: A Ten-Year Program issued.*
1 June 1964	First observations with 2-element interferometer.
1 December 1964	Theodore P. Wright becomes AUI President.
February 1965	140 Foot Telescope construction completed.
11 February 1965	NSF gives final approval to locate 36 Foot Telescope on Kitt Peak.
1 July 1965	Hein Hvatum becomes Asst. Director for Technical Services; Theodore R. Riffe becomes Asst. Director for Administration.
1 October 1965	T. Keith Glennan becomes AUI President.
13 October 1965	Dedication of 140 Foot Telescope in Green Bank.
20 October 1965	Sander Weinreb becomes Head, Electronics Division.
20 December 1965	NRAO takes occupancy of new building in Charlottesville.
Fall 1966	VLA Design Group begins work.
January 1967	2-volume VLA proposal submitted to NSF; 3rd volume added, January 1969.
Spring 1967	3-element interferometer completed in Green Bank.
8–9 May1967	First successful VLBI observations between Green Bank and Maryland Point.
1 October 1967	John W. Findlay becomes Asst. Director for Green Bank Operations.
January 1968	36 Foot Telescope begins operation on Kitt Peak.
January 1968	First international VLBI observations, between 140 Foot and Onsala, Sweden.
1 July 1968	Franklin A. Long becomes Acting AUI President.
1 July 1968	George Grove becomes Head, Tucson Operations.
1 May 1969	Gerald F. Tape becomes AUI President.
June 1969	Dicke panel recommendations include beginning VLA construction.
July 1969	William D. McElroy becomes NSF Director.
1 October 1969	Morton S. Roberts becomes Asst. Director for Green Bank Operations.
December 1969	Greenstein Committee advocates start of VLA in 1971.
1 January 1970	Edward (Ned) K. Conklin becomes Head, Tucson Operations.

(continued)

1 October 1970	David E. Hogg becomes Asst. Director for Green Bank Operations.
Spring 1971	Greenstein Committee final report names VLA as its highest priority.
February 1972	H. Guyford Stever becomes NSF Director.
August 1972	Congress approves VLA Project.
24 October 1972	Hein Hvatum becomes Associate Director for Technical Services.
25 October 1972	John (Jack) H. Lancaster becomes VLA Project Manager.
1 January 1973	VLA Construction Project begins.
1 October 1973	Mark A. Gordon becomes Asst. Director for Tucson Operations.
18 October 1973	E-Systems, Inc. awarded subcontract to design and fabricate VLA antennas.
1974	Introduction of cryogenic Schottky diode mixers for radio astronomy.
1 July 1974	William E. Howard III becomes Asst. Director for Green Bank Operations
31 March 1975	Service Building and Antenna Assembly Building completed at VLA site.
May 1975	VLA Project staff move from Charlottesville to VLA site and Socorro offices.
23 December 1975	Theodore R. Riffe becomes Associate Director for Administration.
18 February 1976	VLA antennas 1 and 2 obtain first fringes with 1.24 km baseline.
August 1976	Richard C. Atkinson becomes Acting NSF Director.
18 October 1976	Four-element array of VLA antennas obtains first fringes.
27–28 Nov. 1976	First real-time satellite-linked interferometry between Green Bank 140 Foot antenna and Algonquin Park 150 foot antenna.
1 January 1977	Kenneth I. Kellermann becomes Acting Asst. Director for Green Bank Operations.
May 1977	Richard C. Atkinson becomes NSF Director.
July 1977	*25 Meter Telescope for Millimeter Wavelengths* proposal submitted to NSF.
1 July 1977	Robert L. Brown becomes Asst. Director for Green Bank Operations.
7 August 1978	Campbell M. Wade becomes Acting Asst. Director for VLA Operations.
1 October 1978	Morton S. Roberts becomes NRAO Director.
1 December 1978	Campbell Wade becomes Asst. Director for VLA Operations.
March 1979	Creation of FITS (Flexible Image Transport System).
July 1979	Work begins on writing AIPS (Astronomical Image Processing Software).
1980	First use of cryogenic field-effect transistor (FET) amplifiers.
July 1980	Donald N. Langenberg becomes Acting NSF Director.
18 August 1980	J. Richard Fisher becomes Asst. Director for Green Bank Operations.
28 August 1980	Ronald D. Ekers becomes Asst. Director for VLA Operations.
10 October 1980	VLA dedication.
10 October 1980	Robert E. Hughes becomes AUI President.
November 1980	First public release of AIPS (Astronomical Image Processing Software).
December 1980	John B. Slaughter becomes NSF Director.
24 March 1981	First meeting of Working Group to upgrade 36 Foot Telescope.
14 September 1981	Martha P. Haynes becomes Asst. Director for Green Bank Operations.
1982	Field Committee recommends VLBA.
April 1982	NSF Astronomy Advisory Committee favors VLBA over 25 meter telescope.
May 1982	Proposal to construct and operate VLBA submitted to NSF.
15 July 1982	36 foot telescope closes for resurfacing and upgrade to 12 meter.
28–29 October 1982	First US workshop on synthesis array for mm-wave astronomy.
November 1982	Edward A. Knapp becomes NSF Director.
17 January 1984	First observations with 12 Meter Telescope.

(continued)

1 June 1984	Robert L. Brown becomes Asst. Director for Tucson Operations.
September 1984	Erich Bloch becomes NSF Director.
30 September 1984	Hein Hvatum becomes Acting NRAO Director.
1 October 1984	George A. Seielstad becomes Asst. Director for Green Bank Operations.
1 January 1985	Paul A. Vanden Bout becomes NRAO Director.
January 1985	Decision to build a combined VLA/VLBA operations center in Socorro.
7 February 1985	Robert L. Brown becomes Associate Director for Operations.
5 July 1985	David E. Hogg becomes Asst. Director for Tucson Operations.
February 1986	Construction on first VLBA antenna begins at Pie Town, NM.
1 November 1986	Darrel T. Emerson becomes Asst. Director for Tucson Operations.
1 September 1987	Michael Balister becomes Asst. Director for CDL.
1988	Introduction of niobium SIS mixers for radio astronomy.
February 1988	David S. Heeschen becomes Acting Asst. Director for VLA Operations.
April 1988	W. Miller Goss becomes Asst. Director for Socorro Operations.
15 November 1988	300 foot telescope collapses.
8 December 1988	Dedication of Array Operations Center, Socorro, NM.
July 1989	Green Bank Telescope (GBT) design funds received.
24 August 1989	Telemetry from Voyager Neptune flyby received by VLA.
July 1990	Millimeter Array proposal submitted to NSF.
September 1990	Frederick M. Bernthal becomes Acting NSF Director.
December 1990	GBT construction funds received.
March 1991	Walter E. Massey becomes NSF Director.
1 May 1991	Ground breaking for the GBT.
May 1991	Bahcall Committee recommends MMA as #2 ground-based astronomy project.
19 May 1992	David E. Hogg becomes Asst. Director for Green Bank Operations.
15 February 1993	F. Jay Lockman becomes Asst. Director for Green Bank Operations.
April 1993	Frederick M. Bernthal becomes Acting NSF Director.
April 1993	Construction finished on 10th and final VLBA antenna on Mauna Kea HI.
20 August 1993	VLBA dedication.
October 1993	Neal F. Lane becomes NSF Director.
November 1994	National Science Board (NSB) approves project development plan for MMA.
1996	Design of single-chip balanced and sideband-separating SIS mixers.
June 1996	NASA contracts with NRAO to design and build amplifiers WMAP.
8 July 1996	John Webber becomes Asst. Director for CDL.
18 March 1997	Lyle H. Schwartz becomes AUI President.
9 April 1998	Martha P. Haynes becomes Interim AUI President.
May 1998	National Science Board authorizes $26M for MMA design & development.
August 1998	Rita R. Colwell becomes NSF Director
1 January 1999	Philip R. Jewell becomes Asst. Director for Green Bank Operations.
June 1999	NSF/ESO sign MOU for joint design and development phase of ALMA.
1 July 1999	Riccardo Giacconi becomes AUI President.
7 February 2000	Robert L. Brown becomes NRAO Deputy Director.
May 2000	McKee/Taylor Committee supports ALMA and EVLA.
25 August 2000	Dedication of the GBT.
November 2000	EVLA Phase I Proposal submitted to NSF.
November 2001	U.S. Congress appropriates $12.5 million to initiate construction of ALMA.
15 November 2001	National Science Board approves construction of EVLA Phase I.

(continued)

15 December 2001	James S. Ulvestad becomes Asst. Director for New Mexico Operations.
1 June 2002	W. Miller Goss named Interim NRAO Director.
July 2002	End of NSF support for NRAO 12 Meter Telescope on Kitt Peak.
1 September 2002	Fred K.Y. Lo becomes NRAO Director.
May 2003	David E. Hogg becomes Interim Deputy Director.
6 November 2003	Groundbreaking for ALMA Operations Support Facility.
February 2004	Arden L. Bement, Jr. becomes Acting NSF Director.
1 September 2004	James J. Condon becomes Interim Deputy Director.
22 October 2004	Ethan J. Schreier becomes AUI President.
November 2004	Arden L. Bement, Jr. becomes NSF Director.
April 2005	Addition to NRAO's Edgemont Rd. building in Charlottesville completed.
20 June 2005	Philip R. Jewell becomes NRAO Deputy Director.
21 June 2005	Richard Prestage becomes Interim Asst. Director for Green Bank Operations.
1 February 2006	Richard Prestage becomes Asst. Director for Green Bank Operations.
1 October 2007	Robert L. Dickman becomes Asst. Director for New Mexico Operations.
10 May 2008	Karen O'Neill becomes Interim Asst. Director for Green Bank Operations.
1 October 2008	Karen O'Neill becomes Asst. Director for Green Bank Operations.
June 2010	Cora B. Marrett becomes Acting NSF Director.
October 2010	Subra Suresh becomes NSF Director.
1 January 2011	Richard Prestage becomes Asst. Director for CDL.
1 April 2011	Clair J. Chandler becomes Interim Asst. Director for New Mexico Operations.
1 August 2011	Dale A. Frail becomes Asst. Director for New Mexico Operations.
31 August 2011	Shing-Kuo Pan becomes Acting Asst. Director for CDL.
30 September 2011	ALMA begins Early Science observations.
31 March 2012	VLA is re-dedicated as the Karl G. Jansky Very Large Array (JVLA).
18 May 2012	Anthony Beasley becomes NRAO Director.
March 2013	Cora B. Marrett becomes Acting NSF Director.
13 March 2013	ALMA formally inaugurated.
8 February 2014	Robert L. Dickman becomes Interim Asst. Director for CDL.
March 2014	France A. Córdova becomes NSF Director.
7 October 2014	Robert L. Dickman becomes Asst. Director for CDL.
September 2015	Mark M. McKinnon becomes Interim Asst. Director for New Mexico Operations.
2 September 2016	Brent Carlson becomes Asst. Director for CDL.
1 October 2016	New 10-year NSF/AUI cooperative agreement for operation of VLA, NAASC, CDL; GBO and LBO each have separate NSF/AUI agreements.
31 October 2016	Bill Randolph becomes Interim Asst. Director for CDL.
December 2016	Mark M. McKinnon becomes Asst. Director for New Mexico Operations.
10 July 2017	Bert Hawkins becomes Asst. Director for CDL.
13 September 2017	ngVLA receives $11M in FY2018 NSF funding for design/development.
November 2017	Adam Cohen becomes AUI President.
23 October 2018	VLBA is reintegrated into NRAO.

Index

© The Author(s) 2020, corrected publication 2021
K. I. Kellermann et al., *Open Skies*, Historical & Cultural Astronomy,
https://doi.org/10.1007/978-3-030-32345-5

Printed in the United States
by Baker & Taylor Publisher Services